Supercritical Fluid Science and Technology

Volume 5

Hydrothermal and Supercritical Water Processes

Supercritical Fluid Science and Technology

Series Editor – Erdogan Kiran

Vol. 1 Supercritical Fluids and Organometallic Compounds: From Recovery of Trace Metals to Synthesis of Nanostructured Materials.
By Can Erkey

Vol. 2 High-Pressure Fluid Phase Equilibria: Phenomenology and Computation.
By Ulrich K. Deiters and Thomas Kraska

Vol. 3 Supercritical Fluid Science and Technology: Phase Equilibrium Engineering.
By Esteban Brignole and Selva Pereda

Vol. 4 Introduction to Supercritical Fluids: A Spreadsheet-based Approach.
By Richard Smith, Hiroshi Inomata, and Cor Peters

Supercritical Fluid Science and Technology

Volume 5

Hydrothermal and Supercritical Water Processes

Gerd Brunner
Hamburg University of Technology
Hamburg, Germany

Amsterdam • Boston • Heidelberg • London • New York • Oxford
ELSEVIER Paris • San Diego • San Francisco • Singapore • Sydney • Tokyo

Elsevier
Radarweg 29, PO Box 211, 1000 AE Amsterdam, The Netherlands
The Boulevard, Langford Lane, Kidlington, Oxford OX5 1GB, UK

British Library Cataloguing-in-Publication Data
A catalogue record for this book is available from the British Library

Library of Congress Cataloging-in-Publication Data
A catalog record for this book is available from the Library of Congress

ISBN: 978-0-444-59413-6
ISSN: 2212-0505

For information on all Elsevier publications visit our web site at www.store.elsevier.com

Printed and bound in Great Britain

14 15 16 17 18 10 9 8 7 6 5 4 3 2 1

Contents

Foreword

It is a pleasure to introduce the 5th volume in the Elsevier Book Series on Supercritical Fluid Science and Technology. This book on "Supercritical Water and Hydrothermal Processes" has been authored by Professor Gerd Brunner at the Hamburg University of Technology in Germany. Professor Brunner is one of the best known names in the field of supercritical fluids and their applications. He has vast experience and has been a leader in the field. Professor Brunner has also been a member of the Editorial Board of the Journal of Supercritical since 1988, and has been serving as the Regional Editor of the Journal for Europe since 2000.

Starting with the basics, the author presents comprehensive treatments of the application areas of importance. Chapter 1 provides a broad introduction to the essential features of supercritical water and highlights the various topics covered in the book. Just reading Chapter 1 alone, I am sure you will immediately appreciate the value of this book and share my view that this is an indispensable reference volume to all, in academia and industry, who are working with supercritical fluids or hydrothermal processes.

Chapter 2 is devoted to the thermodynamic, transport and electrical properties of water. Extensive data are presented on the volumetric properties, viscosity, thermal conductivity, diffusion, electrical conductivity, dielectric constant, refractive index and surface tension covering a wide range of temperatures and pressures. Numerous models and correlations are reviewed and discussed. Discussions on these properties and their variation with pressure are extended to mixtures of water in Chapter 3. Binary mixtures such as water + carbon dioxide and water + hydrocarbon, or ternary mixtures such as water + hydrocarbon + carbon dioxide or hydrogen are covered at length, and their phase behavior is discussed in terms of the Van Konynenburg and Scott classifications. Water–salt mixtures are also discussed.

Chapter 4 is devoted to thermal properties and heat transfer in water, which is discussed in different phase domains of water, including the gas, liquid, two-phase domain and the near critical region. Extensive documentation of the applicable models and heat transfer correlations are presented. Heat transfer in mixtures of water is also discussed with an emphasis on the greater importance of linking heat transfer with phase behavior as phase splitting is often encountered in systems involving mixtures.

Chapter 5 is devoted to water as a reaction medium. The chapter highlights the pressure-tunable nature of the properties of water and its use as variable reaction medium. A range of different reactions are covered with the majority of the discussion being devoted to hydrolysis reactions.

Chapter 6 provides a review of principles of extraction and mass transfer. The effect of factors such as pressure, temperature, solvent-solute ratio, solvent density, and particle size on extraction is discussed. Examples that involve extraction with hydrothermal or supercritical water are then presented. Among the specific applications discussed in detail are the extraction of organic compounds and heavy metals from soil, extraction of fats and proteins from bone material, and extraction of natural plant materials. Chapter 7 is devoted to reactive extractions with a focus on the extraction and liquefaction, or the gasification of coal, oil shale, and other fuel-related materials.

Chapter 8 is devoted to biomass conversions in supercritical or high-temperature water. The treatment is systematic, starting with detailed discussions of conversions of the monomeric constituent sugars (such as glucose, fructose and xylose) and the polymeric components that include starch, pectin, cellulose, hemicelluloses, and lignin. The chapter then covers in detail the conversion of lignocellulosic materials with a focus on biomass liquefaction and biomass gasification.

Chapter 9 presents depolymerization, or fragmentation of polymeric materials using high-temperature or supercritical water as the reaction medium. Among the specific applications discussed are the hydrolytic depolymerization of poly(ethylene terephthalate), nylon 6, polycarbonate, and poly(lactic acid); and the fragmentation of thermoplastics such as polyethylene and polystyrene, and of polymers encountered in fiber-reinforced plastic materials.

Chapter 10 provides a detailed treatment of oxidation reactions. Discussions are focused on hydrothermal flames and supercritical water oxidation for conversion or destruction of organic chemicals.

Chapter 11 describes the hydrothermal processing of inorganic materials for generation of fine particles, or for synthesis of metal oxides, or for control of crystal structure or morphology. Among the various metal oxides discussed is the formation of silica particles via crystallization from their water solutions.

Chapter 12 is devoted to corrosion phenomena that are of significance for process equipment in processes such as supercritical water oxidation or supercritical water based cooling systems for nuclear reactors. Corrosion principles are reviewed and corrosion behavior of stainless steel and various alloys is discussed. Protective approaches such as the use of transpiring wall reactors in supercritical water oxidation are also discussed.

The final chapter of the book describes various high-pressure laboratory systems for conducting experiments related to phase equilibria, reactions and extractions, and for equipment components that are involved in processes that involve hydrothermal or supercritical water.

This book is an authoritative account of all aspects of supercritical water and hydrothermal processes. It is rich with an extensive compilation of property data, correlations and examples of applications. I trust you will find this book to be a resource of high value in your research and teaching.

Erdogan Kiran
Series Editor
Blacksburg, VA
January 2014

Preface

This book, *Hydrothermal and Supercritical Water Processes*, presents an overview on the properties and the applications of water at elevated temperatures and pressures. It combines fundamentals with production process aspects.

Water is an extraordinary substance. It is necessary for life and is present everywhere. Water forms phenomena such as glaciers, the polar ice caps, rain, snow, fog, and contributes substantially to the greenhouse effect. Liquid water is available in underground reservoirs, in rivers, lakes, and most abundantly in the sea. Water is applied in numerous everyday activities and industrial processes.

At elevated temperatures (and pressures) the properties of water change dramatically due to the modifications of the molecular structure of bulk water that varies from a stable three-dimensional network, formed by hydrogen bonds at low and moderate temperatures, to an assembly of separated polar water molecules at high and supercritical temperatures. With varying pressure and temperature, water is transformed from a solvent for ionic species to a solvent for polar and non-polar substances.

This variability and an enhanced reactivity of water have led to many practical applications and to even more research activities, related to such areas as energy transfer, extraction of functional molecules, unique chemical reactions, biomass conversion and fuel materials processing, destruction of dangerous compounds and recycling of useful ones, the growth of monolithic crystals, and the preparation of metallic nanoparticles.

The result of the intensive research is an enormous number of scientific publications and practical applications at various stages. Most of the specialized areas of the application of water and the underlying fundamentals are covered in separate comprehensive reviews and handbooks. The original scientific papers and the compilations are both available in the literature. This book is intended as an introduction to the wide range of activities that are possible in aqueous mixtures. It is organized to allow understanding of the main features, outlines the main applications and provides access to further information. It is the intention to make enough data available that the book can be used as reference. Access to process applications is facilitated by pointing out the fundamental principle of engineering common to the very different special applications. The combination of fundamentals and detailed processing examples should lead to the awareness that these fundamentals basically determine all applications, whatever the nonconventional conditions of state or the detailed problems.

Hydrothermal and Supercritical Water Processes is a volume in a series published by Elsevier on "Supercritical Fluid Science and Technology". Therefore, there will be some overlap with other titles in this series. This book covers the most important areas of water at elevated temperatures from the point of view of the author. The different topics are presented as an overview based on the fundamentals. It is not a review and therefore does not cover all the literature, although the author has taken care to include all of the important aspects.

The book is written for undergraduate and graduate students of various disciplines, such as general chemistry, physical chemistry, geochemistry, applied physics, chemical engineering and related engineering disciplines, as well as for scientists and engineers interested in the topics presented here and being active in the field, and for all interested readers.

The production of this book needed support from many persons. I acknowledge all the help that I got in the last three years. First, I want to thank my wife Annemarie for her patience and encouragement. Then, I am very grateful to Prof. Erdogan Kiran, Virginia Tech, USA. As an editor of the book series, he asked me to write this book and followed its development with great enthusiasm and many useful comments. My thanks also go to the team at Elsevier: Derek Coleman, Susan Dennis, and Mohanapriyan Rajendran. They were very helpful and eventually produced this book from the accumulated pages. Last, but by far not least, I thank all my co-workers, who have contributed with their work. They laid the foundations for me to be able to write this book with their many PhD Dissertations, Master, Bachelor, and Diploma Theses. These are represented in the reference lists of the various chapters. The Institute of Thermal Separation Processes, directed by Prof. Irina Smirnova, of the Hamburg University of Technology provided valuable background support. I would also like to mention Ms. S. Meyer-Storckmann, who has been very helpful with providing literature.

It is my hope that this book will prove useful in extending the knowledge on and the application of hydrothermal and supercritical water.

Gerd Brunner
Hamburg
January 2014

Chapter 1

Introduction

Water is the most abundant and ubiquitous compound on planet Earth. By far, most of the amount of water occurs in the liquid state at temperatures of $T = 0–40\ °C$. The properties of water at these conditions are familiar. Nevertheless, a small part of water occurs at temperatures around $T = 100\ °C$ or higher, for example, in geysers. Around such hot-water sources, minerals precipitate. Furthermore, gem-like monocrystals, like rock-crystals, are most probably the product of hot-water processes. The steam engine, pioneer of industrialization, uses evaporation and condensation of water at high temperatures. Beyond these examples, high-temperature water is useful for many other applications. These applications are the main topic of this book, which is about the fundamentals and processes of hydrothermal water, at temperatures higher than $T = 100\ °C$, and water in the supercritical state, at temperatures $T > 374\ °C$ and pressures $P > 22$ MPa.

The properties of water vary from ambient conditions to critical conditions and above critical conditions over a remarkably wide range. From ambient to supercritical temperature, water changes its character from a solvent for ionic species to a solvent for nonionic species. Electrochemical properties vary substantially. For example, the dipole moment decreases from its high value at ambient conditions to a value common for normal solvents at supercritical conditions. But even in the critical region, water is still as polar as acetone. The pH value of liquid water decreases by three units with temperature increasing to about $T = 250\ °C$, thus providing many more H_3^+ ions for acid-catalyzed reactions. Just below the critical temperature, the ionic product of water changes tremendously, rendering near-critical and supercritical water a much less ionized compound than at ambient conditions. Reactivity of water increases in the neighborhood of the critical point with or without a catalyst. These, and other properties of hydrothermal and supercritical water, are presented in Chapter 2.

Most of the applications of water at high temperatures use the interaction of water with other compounds, as in chemical reactions, extraction processes, hydrothermal processes, and others. Properties of mixtures of water with other compounds must be known for these processes. Water at ambient conditions readily dissolves salts and is not miscible with nonpolar components such as hydrocarbons. At supercritical conditions, the solvent power of water for ionic

Supercritical Fluid Science and Technology, Vol. 5. http://dx.doi.org/10.1016/B978-0-444-59413-6.00001-7

species is dramatically reduced. On the other hand, supercritical water and hydrocarbons become completely miscible. This behavior of aqueous mixtures is discussed in Chapter 3 for major groups of compounds, such as gases, hydrocarbons, volatile and nonvolatile electrolytes, and metal oxides.

Experimental results are shown in Chapter 3 for phase equilibria with water. A systematic classification of equilibrium behavior is presented. Modeling of phase behavior and solubility is of great importance for processes. Even for aqueous mixtures, relatively simple equations of state are able to describe the phase behavior of aqueous systems for a wide range of conditions. Solubility of gases in water can be treated in many cases with Henry's law and some modifications. Electrolytes introduce the chemical reaction of ions into the description of phase behavior. For example, carbon dioxide is a weak electrolyte in aqueous solution, but can be of influence on reactions that depend on pH value. In addition to aqueous mixtures of inert gases and hydrocarbons with water, phase behavior and solubility of volatile electrolytes, nonvolatile electrolytes, metal oxides, and salts are treated in Chapter 3. Solubility of metal oxides in hydrothermal and supercritical water is the basis for hydrothermal processes leading to monocrystals and functional nanoparticles, treated in Chapter 11.

Water has a high heat capacity and an exceptional high enthalpy of evaporation. These properties are ideal for heat transport processes. Therefore, hydrothermal and supercritical water is vastly applied as a heat-transporting and heat-transferring medium, mostly in power plants, as discussed in Chapter 4.

In many processes involving water at hydrothermal and supercritical conditions, water is by far the main component. Due to the high-temperature level, recovery of energy from the process effluents is required for economic reasons. In most cases, heat transfer to or from pure water can be taken as a first approximation. Nevertheless, in many cases, mixtures of process compounds with water are processed and split into several phases during heat recovery, thus connecting phase behavior to heat transfer processes. Condensation of mixtures from high-temperature conversion processes shows that from the incoming gaseous mixture with decreasing temperature and pressure, a gaseous phase, a hydrocarbon rich liquid phase, and an aqueous liquid phase are formed. Such cases are shown in Chapter 4.

Chemical reactions in hydrothermal and supercritical water are carried out in a wide range of conditions. Properties of the reaction medium "water" can be adjusted without adding or removing compounds from the reaction mixture. Depending on conditions of state, water thus exerts totally different actions to other compounds. For reactions at hydrothermal conditions, pressure must be higher than the vapor pressure of water to keep the reaction in the liquid phase because gaseous water loses one of the most important properties for a reaction medium: solvent power.

Several properties of water change with increasing temperature. The ionic product increases up to a temperature of about $T = 250\ ^{\circ}C$ by three orders of magnitude, the dielectric constant drops from about 80 Debye to about 2 Debye, and density decreases. But up to relatively high temperatures, liquid water maintains essentially the familiar properties of a highly polar solvent. The dramatic change occurs in the neighborhood of the critical point, the so-called critical region. Within about $\Delta T = 20$ K, properties of water change in such a way that familiar properties of water are lost. For example, water loses its solvent power for ionic species and becomes a good solvent for non-polar components.

In the critical region, water has a density that is easily changed by pressure and temperature. While supercritical water largely loses its solvent power at low pressures, it remains a polar solvent. At higher pressures, a good solvent power and higher polarity can be reached for supercritical water. Thus, the near-critical and supercritical region is a working field for realizing different reaction conditions and for recovering products by changing conditions. Furthermore, water at near-critical and supercritical conditions is fully miscible with many organic and inorganic compounds, enabling reactions in a homogeneous medium. Therefore, supercritical water can be a reaction medium for reactions usually carried out in organic solvents.

Supercritical water is an excellent reaction medium for reactions requiring heterogeneous catalysts, since high diffusivity avoids mass-transfer limitation and high solvent power prevents coke formation and poisoning of the catalyst. The reaction rates of small free radicals are increased due to the high collision frequency. Reactions of high-molecular-mass free radicals, as they occur during pyrolysis, are slowed down by a cage effect caused by solvent molecules at high pressure. Reactions with many different compounds are described in Chapter 5.

Valuable compounds often occur in nature in materials that contain water, such as fresh plants. Extraction with hot water has been carried out for many centuries at ambient pressure. Liquid water at temperatures $T > 100$ °C and pressures $P > 0.1$ MPa extends these applications and makes available compounds that have not been accessible at low-pressure conditions. At near-critical and supercritical conditions, the effective extraction of contaminants from waste materials becomes feasible. The extraction process, characterized by solubility, kinetics, and mass transfer is treated in Chapter 6.

At lower temperatures, ionic and polar species will be extracted by water, thus creating an alternative to weak-polar supercritical solvents like CO_2. At higher temperatures, in particular, approaching the critical temperature, also nonpolar substances will be readily dissolved and extracted. In many cases, this is a physical process, where compounds are dissolved in a solvent, recovered with the solvent, and then separated from the solvent as product. But there are also many cases in which the solvent not only physically interacts with the substrate and the extract compounds but also reacts by breaking chemical bonds between substrate and extract compounds or by reacting with the compounds itself. In particular, with hydrothermal and supercritical water as a solvent, such reactions occur at elevated temperatures due to the properties and the increased reactivity of water. Useful products of this process can be the extract components, the residual substrate, or both. Examples treated in Chapter 6 are the cleaning of soil from unwanted compounds, the cleaning of bone materials, and the extraction of substances from plant materials.

Fuel materials as coal, oil shale, tar sands, and heavy bitumen can be processed with high-temperature water, as discussed in Chapter 7. Processing of fuel materials with water makes use of the advantageous properties of water, such as availability, environmental compatibility, solubility behavior, at ambient and high-temperature conditions. At low temperatures, hydrocarbons are nearly immiscible with water, while at high temperatures, the solubility increases until total miscibility occurs around the critical temperature. In parallel, the reactivity of hydrocarbons with water increases with temperature and hydrolytic and pyrolytic transformations are observed. Only simple hydrocarbons are relatively stable at near-critical and supercritical water conditions. With increasing temperature, other properties of hydrocarbons also change: viscosity is drastically reduced, surface tension decreases with the enhanced mutual solubility, and nonhydrocarbon components can react to compounds that are readily removed from the hydrocarbon mixture. The hydrocarbon mixture from the process is treated and upgraded to conventional hydrocarbon mixtures. Removal of solids, including fine particles, from effluents is possible in processes with near-critical and supercritical water.

Processing of biomass with water is an important topic since biomass is a renewable resource and usually contains a substantial amount of water. Therefore, processing of biomass with hydrothermal and supercritical water has an important role in the exploitation of this resource. It is treated in Chapter 8. Processing of biomass with high-temperature and supercritical water is carried out for different purposes: extraction and conversion for food purposes, derivation of useful compounds for chemical synthesis, acquiring compounds for fermentation, or transformation in order to use biomass for providing energy. In this respect, biomass is a renewable resource with low sulfur content and near-to-zero emission of CO_2 can be realized in the cycle of growing and conversion of biomass.

Biomass is a complex mixture of diverse compounds such as sugars, starch, cellulose, hemicellulose, lignin, and proteins. Investigation of the conversion of these compounds yields valuable information for processing natural materials that are mixtures of these compounds. Both aspects are discussed in Chapter 8.

The idea, to use as many compounds as possible from biomass, led recently to the vision of a biorefinery, in which the biomass feedstock is separated into all its valuable compounds. These compounds can be extracted, modified by hydrolysis or thermal pyrolysis, and then separated into gaseous, liquid, and solid fractions. Other ways to use biomass are liquefaction to a fuel-like liquid, or to gasify biomass to a gas that can be burnt or used as synthesis gas.

Another important topic of the future is the recycling of synthetic polymers. High-temperature and supercritical water are excellent reaction media for decomposition of synthetic polymers, since they do not introduce compounds that contaminate the products and water is easy to remove from the decomposition products. In Chapter 9, examples are discussed that illustrate the principles of depolymerization reactions with water.

Polymers from condensation polymerization processes, such as polyethylene terephthalate (PET), nylon, and polyurethane, are readily depolymerized to their monomers in supercritical water. Polymers from addition polymerization processes, such as phenol resin, epoxy resin, and polyethylene, can also be depolymerized in high-temperature and supercritical water. Composite polymer materials can be separated into useful fractions. Chlorinated polymers can be treated to remove halogens, beside the conversion of the polymers to useful degradation products. From biodegradable polymers, such as polylactic acid, the monomer can be recovered in high purity and yield.

Hydrothermal and supercritical water can be used for eliminating toxic and dangerous compounds by oxidation in an aqueous environment, see Chapter 10. Oxidation can be carried out as oxidation in a liquid aqueous environment in the temperature range from $T = 100$ to 374 °C, as "Wet Air Oxidation," or as oxidation in supercritical water, known as "Supercritical Water Oxidation," and as oxidation in so-called "Hydrothermal Flames." These oxidation reactions aim for the conversion of the organic compounds to totally oxidized end-products or such effluents that can be further treated with conventional wastewater cleaning methods. For some applications, for example, for conversion of methane to methanol, partial oxidation is of interest. So far, success is limited in supercritical water, as presented in Chapter 10.

Processing of inorganic substances with high-temperature liquid and supercritical water is generally named "Hydrothermal processing" (Chapter 11). Hydrothermal processes can be used in the synthesis of fine particles, single crystals, functional metal oxides, and more. For many of these processes, subcritical water conditions are sufficient for the reaction. Processing in near-critical and supercritical water is attractive because the density, viscosity, diffusivity, and dielectric constant of the reaction medium can be controlled through relatively small changes in temperature and pressure. Variation of the properties of the reaction medium makes it feasible to control particle size, crystal structure, and morphology. Reaction at supercritical water conditions is a combination of hydrolysis, condensation, and thermal decomposition that can be at least partly also be controlled by adjustment of the operating conditions.

The interaction of aqueous solutions with the materials of the experimental or production plants can lead to major problems due to corrosion. The aqueous solutions contain many reactive compounds, for example, chloride ions, under oxidative or reductive conditions, that can cause corrosion. Corrosion includes all reactions of components of an aqueous reaction mixture with the walls of the equipment of experimental or production facilities, like pipes, valves, fittings, and autoclaves, as discussed in Chapter 12.

Corrosion in hydrothermal and supercritical water processes depends on the properties of the reaction mixture, on the properties of the construction materials, and on the construction of the processing equipment. In general, no satisfactory solution to corrosion problems can be achieved by addressing only one aspect. On the contrary, only a comprehensive adjustment of all aspects influencing corrosion can lead to a process configuration that can be operated successfully and for a reasonable, extended time span.

Various types of corrosion occur, such as general corrosion, intergranular corrosion, pitting corrosion, and stress corrosion cracking, as discussed in Chapter 12. General corrosion occurs mainly in the near-critical and supercritical temperature range in a region, where the density is high. Corrosion rates below $T=320\ ^\circ C$ are very low. At temperatures $T>320\ ^\circ C$ and pressures higher than the critical pressure, corrosion rates increase considerably. At a pressure of $P=24$ MPa, high corrosion rates drop down to the level observed at lower temperatures just around the critical temperature of water, while at a pressure of $P=38$ MPa, corrosion rates remain high up to a temperature of about $T=450\ ^\circ C$, since at that temperature density reaches the limit for corrosion due to ionic processes.

Processes with hydrothermal and supercritical water need a reaction vessel that can withstand the operating conditions. Although very similar in principle, the pressure vessels are different with respect to size, purpose, and construction, as discussed in Chapter 13.

Although, there are many different processes, as discussed in Chapters 4–11, the basic processing steps for hydrothermal and supercritical water systems are similar. They comprise the following: the feed has to be introduced into the high-pressure vessel, operating conditions have to be established and maintained, the process has to be carried out that leads to the desired products, the reaction products have to be retrieved and separated, and the water as the working fluid has to be recycled or cleaned for discharge. Obviously, differently designed components are needed for laboratory scale and for demonstration or production scale, although the basic process remains the same. At all sizes, the function of the components must be fulfilled. While at laboratory scale, the main effort is concentrated on the reaction, at larger scale new problems arise and need solutions, such as handling of feed and products in large quantities, heat transfer, fluid dynamics, and treatment of solids.

Experimental processes typically are concerned mainly with the process step itself. Handling of educt and product components is of secondary importance, as long as conditions are maintained for operating the process appropriately. Production processes are obviously operated to deliver products, contrary to experimental processes, for which the determination of optimal process parameters is the main objective and produced sample products are used for analysis and for presentation. In the context of this book, production processes are only treated to characterize the differences to experimental processes. No systematic compilation of production processes with hydrothermal and supercritical water is presented due to limitation of space.

It seems, at a first glance, that production processes differ from experimental processes mainly by size. While in experimental processes, samples of the size of less than grams to maximal kilograms are produced, production processes may begin at a production rate of hundreds of kilograms for very high-priced materials and may reach a yearly production rate of hundreds of thousands of tons. This difference in size brings along a number of problems, mostly engineering problems. Other problems to be solved are the supply of the feed and the disposal of wastes as the major logistic problems, financing as an economical problem, and acquiring acceptance for the process and the product as a sociological problem.

From the point of view of technology, the main difference between an experimental process and a production process is complexity. This means that solving a major processing problem with hydrothermal and supercritical water does not mean that a production process is established. Many more steps, some of them of totally other aspects than hydrothermal processing, become important, and in many cases are decisive. One of the aspects that make production processes much more complex than experimental processes is the necessity to treat the process fluid, water, in order to recycle it into the process or to dispose it into the environment. Examples for processes with hydrothermal and supercritical water are presented in Chapter 13.

Chapter 2

Properties of Pure Water

In this chapter, the role of water in general, the main properties of water at standard and hydrothermal conditions, the structure of water, and phase behavior are addressed first. Then, individual properties of water are presented. The thermodynamic, transport, electrical properties are discussed in detail. For thermodynamic properties, the fundamental thermodynamic equations are briefly reviewed, followed by the individual properties like volumetric properties (and P,V,T-behavior), heat capacity, enthalpy, and entropy. Transport properties that are discussed include viscosity, thermal conductivity, and diffusivity. Electrical properties that are covered are ion product, electrical conductivity, and dielectric constant. This chapter is concluded with a brief discussion of refractive index and surface tension. Various tables and diagrams, useful for understanding and first conceptual designs are included throughout the chapter.

2.1 INTRODUCTION

Water is the most abundant and ubiquitous compound on planet Earth. It is essential for life and is a major constituent of all known life-forms. Water determines our lives in many aspects such as nurturing, cleaning, and influencing the weather. Water occurs on earth in three states of aggregation: as vapor, as liquid, and as solid, which are a consequence of our planet's orientation to our sun. Water in the form of water vapor is the most important compound for controlling temperatures on earth. The water content of our atmosphere is variable between about 0.5 mol% in cold dry air and 3–4 mol% in humid tropical air [1]. The water cycle, that is, the amount of water transported in the atmosphere, is about 2500 times larger than the carbon cycle. In its liquid form, water comprises more than 96% of the oceans which cover our earth to about 71% with a depth of about 4200 m for more than half of the oceans [2]. Liquid water in fresh water lakes and rivers is a minor contribution of about 0.1% to total liquid water on our planet. Water as solid ice in glaciers and ice-caps around the poles contain only about 1.7% of all water [3] but plays an important role for our climate. Hydrothermal water occurs in nature in thermal baths, in geysers, and in volcanic outlets below water level (black smokers).

Supercritical Fluid Science and Technology, Vol. 5. http://dx.doi.org/10.1016/B978-0-444-59413-6.00002-9

Water as the all-time and everywhere-available compound acts as a working fluid in many aspects, for example, from the use of its potential energy in ancient and modern machines to make mechanical and electrical energy available, as cooking fluid for preparing food, as solvent in the extraction of natural compounds, and not least of all, as processing fluid in power plants.

This chapter will concentrate on water at elevated temperatures and pressures, and on the applications of water at hydrothermal and supercritical conditions, as the title of the chapter states. The main area covered in this chapter is water and its applications at temperatures above $T > 100$ °C (373 K) up to the critical temperature of water ($T = 374$ °C, 647 K) and far beyond, and pressures above the saturation pressure of water up to around $P = 200$ MPa. Properties of water at ordinary or ambient conditions will be included for comparison.

The reader will get sufficient insight to understand or realize the effects that the properties of water and their dramatic changes exert on their environment and on other compounds. Data of properties of water will be provided to such an extent that the reader will be able to work with them without being forced to refer to other sources. This holds for understanding the effects and for using data for at least conceptual process design. It may be necessary to refer to fundamental treatises, to more extensive databases, or to other computational resources for detailed design purposes or for special effects. Therefore, the necessary and available sources will be addressed.

Pure water is a tasteless, odorless, colorless, clear liquid. It shimmers lightly blue in thick layers as does its solid form, that is, ice. Water in this chapter refers to the isotopic composition of water defined in the international standard called Vienna Standard Mean Ocean Water (VSMOW) [4]. Heavy water D_2O, that is, where hydrogen is 100% deuterium, is beyond the scope of this chapter. For more information, see IAPWS (International Association for the Properties of Water and Steam) releases at http://iapws.org/.

Properties of water are summarized in Table 2.1. A detailed description of the individual properties follows below. The values of properties in Table 2.1 are the recommended values issued by IAPWS. This organization has over several decades dedicated its work to the precise measurement, correlation, and distribution of best values of the properties of water. The IAPWS releases and guidelines are the most reliable data available for the properties of water and should be used in scientific and industrial work. They can be accessed at http://iapws.org/.

Properties of water vary over a remarkably wide range, from ambient conditions to critical conditions and above critical conditions. The values given in Table 2.2 emphasize this variation. The individual properties are discussed below in detail. Water from ambient to supercritical conditions changes its character from a solvent for ionic species to a solvent for nonionic species. Electrochemical properties, for example, the dipole moment, decrease from the high value at ambient conditions to a value at critical conditions that is

TABLE 2.1 Properties of Water (Hydrogen Oxide) (IAPWS, http://iapws.org/)

Molecular Formula		H_2O
Molar mass		18.015268 g/mol
Melting point at 1013 mbar		273.15 K (0 °C)
Boiling point at 1013 mbar		373.13 K (99.98 °C)
Triple point		273.16 K (0.01 °C)
		(611.657 ± 0.010) Pa
Density[a]	At 273.15 K (ice)	916.8 kg m^{-3}
	273.15 K (water)	999.818 kg m^{-3}
	277.15 K	999.97495 kg m^{-3}
	293.15 K	0.998.231 kg m^{-3}
Refractive index	0.589 μm, 273.15 K	
	0.1 MPa	1.33434
Critical pressure	P_c	22.064 MPa
Critical temperature	T_c	647.096 K
Critical density	ρ	322 kg m^{-3}
Enthalpy of fusion[a]		285.89 kJ mol^{-1}
Enthalpy of melting[a]	At 0 °C	6.010 kJ mol^{-1}
Enthalpy of boiling[a]	At 100 °C	40.651 kJ mol^{-1}
Dynamic viscosity (liq.)	298.15 K	889.7351×10^{-6} Pa s
Surface tension	298.15 K	71.98 mN m^{-1}
Electrical conductivity	298.15 K	0.055 μS cm^{-1}
Dielectric constant[b]		1.85498(9) Debyes
	or	$6.1875(3) \times 10^{-30}$ C m
Polarizability[c]		$1.494(7) \times 10^{-30}$ m^3
Specific gas constant *R*		0.46151805 kJ $kg^{-1}K^{-1}$

[a] *Römpp [5].*
[b] *Isolated H_2O molecule in ground vibrational state; experiments by Shostak, cited in IAPWS 2001.*
[c] *Mean total dipole polarizability in the static limit.*

common for normal solvents. Nevertheless, water in the critical region is as polar as acetone. Another example is the pH-value. It decreases by three units from ambient temperature to near-critical temperature, thus providing much more H_3^+-ions for acid-catalyzed reactions. Just below critical temperature,

TABLE 2.2 Properties of H_2O at Different Conditions of State

Property	Water at Normal Conditions	Near-Critical Water	Supercritical Water			Superheated Steam	
T (°C)	25	350	400	400	450	250	450
T (K)	298.15	623.15	673.15	673.15	723.15	523.15	723.15
P (MPa)	0.1	25	25	50	25	2	1.5
ρ (kg m^{-3})	997.45	625.45	166.54	577.79	108.98	8.9689	4.5624
λ (Wm^{-1} K^{-1} 10^{-3})	607.2	481.1	168.0	451.3	108.8	43.49	61.90
c_P (kJ kg^{-1} K^{-1})	4.179	6.994	13.27	6.789	5.043	2.555	2.174
η (Pa s 10^{-6})	890.45	72.81	29.18	68.02	28.96	17.85	26.51
ε (–)	78.5	14.865	3.8	12.16	1.745	1.03	1.03
pK_W (–)	14.0	11.551	16.556	11.557	18.135	11.2	

the ionic product changes tremendously and renders near-critical and supercritical water into a much less ionized compound than at ambient conditions. Reactivity of water increases in the neighborhood of the critical point, with or without a catalyst. These examples of specific properties of water are the basis for hydrolysis, hydrothermal transformations, oxidative processes, chemical reactions, and more that will be discussed in the following chapters [6].

2.2 STRUCTURE OF H_2O

Water with the general chemical formula H_2O has a central single oxygen atom covalently bonded to two hydrogen atoms. The angle of the two hydrogen atoms is about 104.4°. The distance between the centers of the oxygen atom and the hydrogen atoms is about 0.096 nm. These values are for an isolated molecule in the ground state and are approximate since the geometry is not fixed due to vibrations of the molecule [4]. Figure 2.1 shows a schematic diagram of a single water molecule.

The atoms in the water molecule differ in their electronegativity. Oxygen is strongly electronegative, 3.44 on the Pauling-scale, hydrogen less electronegative, 2.20. This difference polarizes the O—H bonding, forming electrical poles at different positions, resulting in a dipole of the water molecule. The dipole moment was determined as $\varepsilon = 1.85498(9)$ Debye or $6.1875(3) \times 10^{-30}$ C m [4].

The density distribution of the electrons results in an overlapping of the p-orbitals of the oxygen atom with the s-orbitals of the hydrogen atoms.

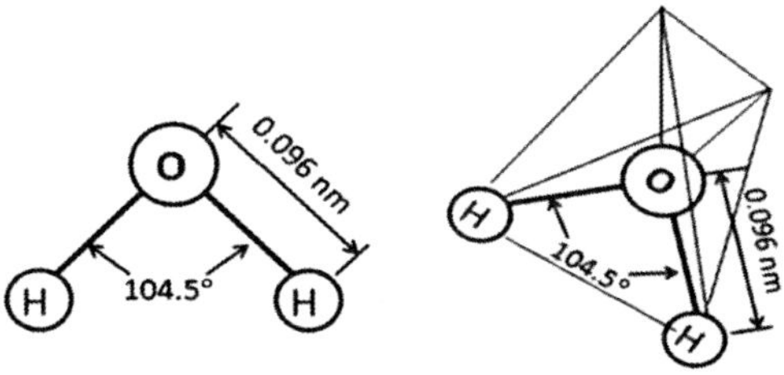

FIGURE 2.1 Molecular structure of a single water molecule (left) and tetrahedral structure (right).

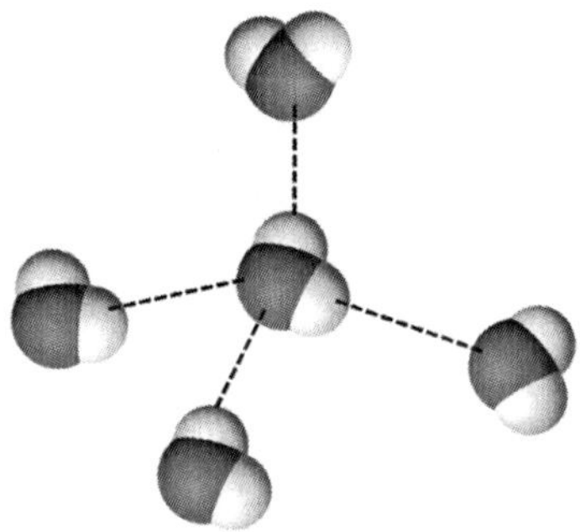

FIGURE 2.2 Model of hydrogen bonds between molecules of water. ----: hydrogen bonds. *Modified from Ref. [3].* (For color version of this figure, the reader is referred to the online version of this chapter.)

The result is a partial ionic character of the covalent bonding (Pauling, cited from Ref. [5]). The dipole polarizability of water, taken as the mean total dipole polarizability in the static limit, is $1.494(7) \times 10^{-30}\ m^3$. As a consequence of the dipole moment, and in combination with the dielectric properties, water dissolves polar substances, causes dissociation of dissolved compounds (salts, bases, acids), and forms complexes, hydrates, and hydrogen bonds [5]. Hydrogen bonds determine the structure of water as a highly ordered structure (similar to ice I) where hydrogen bonds hold together aggregates and single water molecules that are present at intermediate places in the lattice [5], see Figure 2.2.

By hydrogen bonding, each hydrogen atom is arranged between the oxygen atom to which it is covalently bonded and an oxygen atom on another molecule. This produces a highly directional bonding interaction with a strength near $20\ kJ\ mol^{-1}$, weaker than a chemical bond, but stronger than other intermolecular forces. Water molecules favor tetrahedral coordination with their neighbors. The tetrahedral coordination is nearly perfect in ice. In fluid water, groups of molecules prefer hydrogen-bonded configurations over a random orientation. At higher temperatures, the amount of hydrogen bonding decreases with temperature, favoring more random configurations. Still, some hydrogen bonding exists at high temperatures, up to more than 700 K [4]. More information on the microscopic structure of water is available at Seward and Driesner [7] and for supercritical water in the recently published book of Marcus [8].

Hydrogen bonds in water are responsible for values of properties uncommon to a substance comparable to H_2O, such as H_2S or NH_3. Water has a relatively high boiling point and a high critical point. Water has a high enthalpy of melting, far higher than other substances melting at low temperatures, a high enthalpy of vaporization with the effect that evaporation of water is energy intensive, and high values for surface tension, specific capacity of heat, absorption of sound, and viscosity [5].

Modern analytical facilities (neutron scattering, third-generation synchrotron light sources, and others) in combination with highly increased computing power permit investigations of the structure of aqueous media at the atomic level. Nevertheless, full understanding and description of hydrogen bonding in water is still not possible (Chialvo and Cummings [9,10]) and Soper *et al.* [11], cited from Ref. [7].

Liquid water expands with increasing temperature, accompanied by a decreasing viscosity and dielectric constant caused by configurational modifications and changes of the extent of hydrogen bonding. The properties of liquid to supercritical water are determined by changes in hydrogen bonding as a function of temperature and pressure [7].

At low temperatures and a density around $\rho = 1000\ \text{kg m}^{-3}$, the structure of water is characterized by short-ranged, tetrahedral ordering. This structure inherits configurational elements from ice [7]. Water molecules are present in the vertices of the tetrahedral positions. Water aggregates can comprise up to 100 molecules [12]. The formation of hydrogen bonds is exothermic; therefore, the equilibrium constant K_i decreases as the temperature increases [13], cited from Ref. [14].

The first important change in water structure upon increasing the temperature, as shown in computer simulations, occurs at a temperature of about $T = 150\ °C$ (423 K). Nearest neighbor water molecules that are not hydrogen bonded to the central water and sit in interstitial spaces of the tetrahedral framework exist at lower temperatures. They make up a major proportion of the nearest neighbors at high temperatures [7]. With an increase in temperature, a more random distribution of neighboring water molecules is expected from experimental results. From computer simulations, it can be concluded that at $T = 300\ °C$ still a significant number of these water molecules can be considered hydrogen bonded (e.g., Kalinichev and Bass [15], cited from Ref. [7]).

Near the critical point, the structure of water changes significantly. The network of hydrogen bonds is destroyed and water exists as separate clusters with a chain structure [14,16]. At the critical point, the kinetic energy of water molecules is only about 30% of the estimated hydrogen bond energy of 24 kJ mol^{-1} [17].

Supercritical water structure is still not really well known. While researchers agree to the drastic changes around the critical point of water, the degree of changes is still under discussion. It is agreed that in the supercritical region, molecules remain hydrogen bonded and may exist in clusters. Galkin and Lunin [14] report that computer simulation showed that the number of water molecules in the clusters strongly depends on the density of the supercritical fluid. With respect to the size of the remaining hydrogen-bonded clusters, different results have been published.

According to Tassaing *et al.* [18], only monomeric water is detected in SCW at $T=380\ °C$ (653 K) and low pressure (density) in the range of 2.5–5.0 MPa (10–50 kg m^{-3}). A progressive increase of the pressure (density) from 5.0 to 18.0 MPa (from 50 to 100 kg m^{-3}) shows the occurrence of dimers in the fluid. These findings are in agreement with previously reported results [19–22]:

- The density threshold from which dimers occur, at about 40–50 kg m^{-3}, is in qualitative agreement with other experimental [13] and theoretical work [20]. In addition, such a density threshold is also seen from NMR data [22] and molecular dynamics results [21] on rotational relaxation time.
- Upon increasing the density from 100 to 400 kg m^{-3}, the H-bonded species occurring in the fluid are mainly trimers.
- With the OH characteristic frequency of about 3410 cm^{-1}, the presence of larger clusters can be ruled out because their characteristic OH bond is situated at lower frequencies than those observed here. Therefore, supercritical water is essentially composed of monomers, dimers, and trimers at $T=380\ °C$ (653 K) and pressures (density) in the range of 20–25 MPa (100–400 kg m^{-3}). This conclusion is consistent with molecular dynamics simulations [12,23].

According to Seward and Driesner [7], supercritical water at $T=400\ °C$ (673 K) and 40 MPa has 29% of the hydrogen bonds that exist in ambient temperature, liquid water (Hoffmann and Conradi [24]). The degree of hydrogen bonding at $T=500\ °C$ (773 K) and 43.0 MPa decreases further to about 13% of the ambient value [7].

According to Galkin and Lunin [14] at $T=673$ K and a density of water of $\rho=850$ kg m^{-3}, up to 50% of the total amount of hydrogen bonds is broken [25]. In the low-density region ($T=673$ K, $\rho=167$ kg m^{-3}), water clusters comprise up to 10 molecules, but as the density increases ($T=772$ K, $\rho=528$ kg m^{-3}), the number of molecules in certain clusters can exceed 20. At a high density of supercritical water ($T=680$–771 K, $\rho=972$–1284 kg m^{-3}), all molecules are again interconnected by hydrogen bonds, but there are also a few clusters comprised of less than 15 molecules that temporarily are not involved in the common infinite network of H bonds ([16], cited from Ref. [14]).

Increased pressure at low temperatures tends to destroy the water structure, since at higher-density water molecules have to be placed in a nontetrahedral arrangement. The same pressure increase at high temperature promotes water structuring. At high temperatures, bulk density is usually much lower than $\rho = 1000\ \mathrm{kg\ m^{-3}}$. In this case, the higher kinetic energy of water molecules leads to a decreased structural order. With increasing pressure and increasing density, short-range order due to hydrogen bonding starts to increase again [7].

The addition of components such as cations, anions, and other components changes the structure of water. A discussion of the molecular effects is given by Seward and Driesner [7], and others, for example, Chialvo *et al.* [19].

Recently, an extended review on the structure of supercritical water, its bulk properties, and its molecular properties, and dynamics has been published by Marcus [8]. The vast literature on supercritical water is surveyed and major results presented. Marcus's book discusses the properties of water in more detail and from a more physical–chemical point of view than is possible within the scope of this chapter. A very thorough discussion is given also on the experimental methods to obtain properties of supercritical water and on the bulk and molecular properties of supercritical water.

2.3 PHASE DIAGRAM

The phase diagram of water is fundamental for understanding the behavior of water in its various states and for designing processes with water. Water can exist as solid, liquid, or gas. Water melts at ambient standard pressure of $P = 1013$ mbar at $T = 0\ °\mathrm{C}$ (273.15 K) and boils at $T = 99.8\ °\mathrm{C}$ (373.13 K). At higher temperatures, gaseous and liquid water coexist in equilibrium along the vapor-pressure curve, which ends at the critical point.

The P,T-behavior can best be shown in a phase diagram (Figure 2.3). The triple point, where solid, liquid, and gaseous water coexist in equilibrium, has been defined to occur at $T = 273.16$ K (0.01 °C) at $P = 611.657$ Pa [26]. The density of ice at $T = 273.15$ K (0 °C) is $\rho = 916.8\ \mathrm{kg\ cm^{-3}}$, that of liquid water at $T = 273.15$ K (0 °C) is $\rho = 999.818\ \mathrm{kg\ cm^{-3}}$, and at $T = 277.15$ K (4 °C) is $\rho = 999.97495\ \mathrm{kg\ cm^{-3}}$. The density of gaseous water is $\rho = 0.02\ \mathrm{kg\ m^{-3}}$ at $T = 273.15$ K (0 °C).

The phase transition between solid water (ice) and gaseous water below the triple point is called sublimation. The coexistence curve between solid and liquid water (melting curve) rapidly goes up to high pressures with slight temperature changes. The slope of the melting curve is negative. At higher pressures, ice melts at lower temperatures than at ambient pressure, a thermodynamic consequence of the lower density of ice compared to that of liquid water [4]. Ice exists in several forms. While ice I, the form stable at lower pressures, is of lower density than liquid water, other forms are of higher

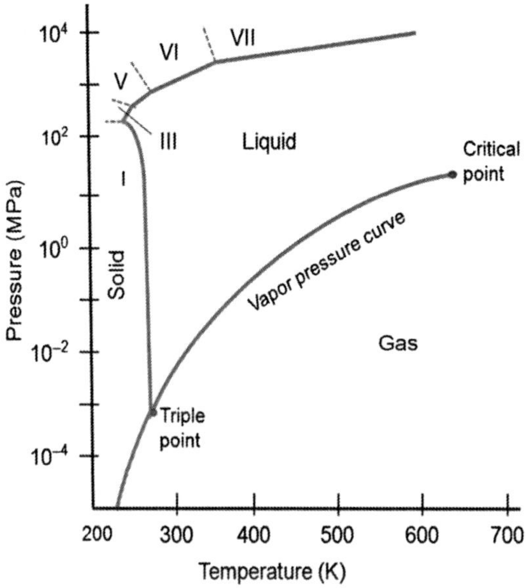

FIGURE 2.3 Phase diagram of water. *Modified from Ref. [4]*. (For color version of this figure, the reader is referred to the online version of this chapter.)

density than the coexisting liquid water. Consequently, the melting curve has a positive slope at high pressures. For more details, see [4]. Figure 2.3 shows a simplified phase diagram of water. It represents the regions where gaseous water (water vapor, steam), liquid water, and solid water (ice) exist. The solid lines represent conditions of state at which one state coexists with one other. Where solid lines meet, water exists in three states in an invariant equilibrium. At low temperatures (below the triple point), gaseous water and solid water coexist up to the triple point, where gaseous, liquid, and ice are in equilibrium. At high pressures, additional triple points exist due to different stable forms of ice. This region is largely beyond the scope of this chapter. More information can be obtained from Refs. [4,27].

2.3.1 Supercritical Region

Beyond the critical temperature and at elevated pressures, water (as other substances) changes its properties rapidly. In this chapter, this region is defined as "supercritical" for temperatures above the critical temperature T_c and pressures above the critical pressure P_c. Properties in this region deviate from normal behavior. In particular, near the critical point, water is less dense (about one-third) than liquid water at lower temperatures and becomes compressible. Density varies substantially with small changes in pressure and temperature. Dielectric constant (relative permittivity, i.e., the ratio of the permittivity of a substance to the permittivity of free space) is related to density and drops from a high value of about 80 at ambient temperature to about 1–10 (depending on pressure) at the critical temperature. At high pressures ($P > P_c$), regions of relative high, liquid-like density exist and enable water in the supercritical state to dissolve other substances, even such with a low volatility. The dissolved components can be precipitated by reducing the density (pressure) without phase change. More and detailed information on the P,V,T-behavior of water is presented in Section 2.4.2.

The supercritical region is readily defined by the critical values of temperature and pressure as lower limits. The region of strong variation of properties is not limited by these values. The region of enhanced changes of properties extends to much higher temperatures and pressures than the critical values and to lower temperatures than the critical temperature, well into the region where only a liquid exists. This liquid is compressible and often called "near-critical" liquid (or fluid). More details are presented below with the individual properties of water.

2.3.2 Meta-Stable States [4]

Phase diagrams show the equilibrium phases of water. Additionally, water can exist in metastable states beyond thermodynamic equilibrium for extended periods of time. Such states are (i) superheated liquid, which can be obtained if the liquid is not disturbed; (ii) supercooled liquid by careful cooling below the freezing temperature ([4], review: [28]); and (iii) supersaturated vapor or steam if cooled to a temperature where the equilibrium state is a liquid. In steam turbines, supersaturated steam may occur due to the very short time scales of the expansion [29].

2.4 THERMODYNAMIC PROPERTIES

Thermodynamic properties are volumetric properties and caloric properties. The main volumetric property is the relationship between volume V (or density $\rho = 1/V$), the pressure P, and the absolute temperature T. Caloric properties are determined by the addition or removal of thermal energy to a system, like internal energy U and specific heat capacities c_P and c_V. Very important are the combinations of volumetric and caloric properties in the thermodynamic functions or potentials.

2.4.1 Thermodynamic Functions

Thermodynamic functions, also called thermodynamic potentials, are internal energy (U), enthalpy (H), Helmholtz energy (A), and Gibbs energy (G). Changes in these functions in a closed system are derived from Gibbs's fundamental equations. The relations are given in Table 2.3.

TABLE 2.3 Thermodynamic Functions and Relations

Thermodynamic functions	Changes of thermodynamic functions
U = Internal energy	$dU = TdS - PdV$
$H = U + PV$ = Enthalpy	$dH = TdS + VdP$
$A = U - TS$ = Helmholtz energy	$dG = -SdT + VdP$
$G = U + PV - TS = H - TS$ = Gibbs energy	$dA = -SdT - PdV$
Useful relations	**Maxwell relations**
$\left(\frac{\partial U}{\partial S}\right)_V = T = \left(\frac{\partial H}{\partial S}\right)_P.$	$\left(\frac{\partial T}{\partial P}\right)_S = \left(\frac{\partial V}{\partial S}\right)_P.$
$\left(\frac{\partial H}{\partial P}\right)_S = V = \left(\frac{\partial G}{\partial P}\right)_T.$	$\left(\frac{\partial S}{\partial V}\right)_T = \left(\frac{\partial P}{\partial T}\right)_V.$
$\left(\frac{\partial U}{\partial V}\right)_S = -P = \left(\frac{\partial A}{\partial V}\right)_T$	$\left(\frac{\partial S}{\partial P}\right)_T = \left(\frac{\partial V}{\partial T}\right)_P.$
$\left(\frac{\partial A}{\partial T}\right)_V = -S = \left(\frac{\partial G}{\partial T}\right)_P$	$\left(\frac{\partial T}{\partial V}\right)_S = -\left(\frac{\partial P}{\partial S}\right)_V$
Heat capacities	**Gibbs–Helmholtz relation**
$c_V = \left(\frac{\partial U}{\partial T}\right)_V = T\left(\frac{\partial S}{\partial T}\right)_V$	$\Delta H = \Delta G - T\left(\frac{\partial(\Delta G)}{\partial T}\right)_P$
$c_P = \left(\frac{\partial H}{\partial T}\right)_P = T\left(\frac{\partial S}{\partial T}\right)_P$	**Influence of P or V on H or U**
$c_P - c_V = -T\left(\frac{\partial V}{\partial T}\right)_P^2\left(\frac{\partial P}{\partial V}\right)_T$	$\left(\frac{\partial H}{\partial P}\right)_T = V - T\left(\frac{\partial V}{\partial T}\right)_P$
$c_P - c_V = T\left(\frac{\partial V}{\partial T}\right)_P\left(\frac{\partial P}{\partial T}\right)_S$	$\left(\frac{\partial U}{\partial V}\right)_T = T\left(\frac{\partial P}{\partial T}\right)_V - P$
$\left(\frac{\partial c_V}{\partial V}\right)_T = T\left(\frac{\partial^2 P}{\partial T^2}\right)_V$	**Isothermal compressibility** $\kappa_T = -\frac{1}{V}\left(\frac{dV}{dP}\right)_T$
$\left(\frac{\partial c_P}{\partial P}\right)_T = -T\left(\frac{\partial^2 V}{\partial T^2}\right)_P$	**Thermal expansion coefficient** $\alpha = \frac{1}{V}\left(\frac{\partial V}{\partial T}\right)_P$

Enthalpy, internal energy, entropy, and others are important thermodynamic functions, which can be related to operating variables of processing equipment, for example, enthalpy to a temperature increase in a heat exchanger or increasing pressure in an autoclave. These properties can be derived from P,V,T-behavior [30] and are in more detail discussed in textbooks on thermodynamics [31]. The relations are listed in Table 2.4.

TABLE 2.4 Derivation of Thermodynamic Properties from *P,V,T*-Data [31]

$$U=\int_0^P\left[V-T\left(\frac{\partial V}{\partial T}\right)_{P,n}\right]dP-PV+\sum_i(n_ih_i^0),$$

$$H=\int_0^P\left[V-T\left(\frac{\partial V}{\partial T}\right)_{P,n}\right]dP+\sum_i(n_ih_i^0),$$

$$S=\int_0^P\left[\frac{nR}{P}-\left(\frac{\partial V}{\partial T}\right)_{P,n}\right]dP-R\sum_i[n_i\ln(y_iP)]+\sum_i(n_is_i^0),$$

$$A=\int_0^P\left[V-\frac{nRT}{P}\right]dP+RT\sum_i[n_i\ln(y_iP)]-PV+\sum_i[sn_i(h_i^0-Ts_i^0)]$$

$$G=\int_0^P\left[V-\frac{nRT}{P}\right]dP+RT\sum_i[n_i\ln(y_iP)]+\sum_i[n_i(h_i^0-Ts_i^0)]$$

$$\mu_i=\int_0^P\left[\bar{v}_i-\frac{RT}{P}\right]dP+RT\ln(y_iP)+h_i^0-Ts_i^0.$$

$\bar{v}_i=(\partial v/\partial n_i)_{T,p,n_j}$ = the partial molar volume of i, y_i = mole fraction of i, n = total number of moles, $\mu_i^0\ h_i^0-Ts_i^0$, n_i = number of moles i, s_i^0 = molar entropy of pure i as ideal gas at T and $p=0.1$ MPa, h_i^0 = molar enthalpy of pure i as ideal gas at T, R = gas constant; general.

Variations of these properties can be calculated in dependence of pressure, temperature, and composition of a mixture using equations of state (EOS). Accuracy of the results depends on the accuracy of the experimental data on which the calculation is based, and on the accuracy of the equation of state, representing the *P*,*V*,*T*-behavior.

2.4.2 *P, V, T*-Behavior

Knowledge on *P*,*V*,*T*-behavior is fundamental for understanding the applicability of hydrothermal and supercritical water. The equilibrium curves (sublimation, vapor–liquid) relate the applied conditions of state (*P*,*T*) to the stabile phases (liquid, gas, supercritical). In particular, the vapor-pressure curve determines the necessary pressure for liquid or gaseous conditions, a decisive condition of state for applications in the hydrothermal and supercritical region. Therefore, correlations are presented for the sublimation curve, the vapor-pressure curve, and the *P*,*V*,*T*-behavior in the saturated, liquid, gaseous, and supercritical regions. Precise empirical correlations, as well as molecular-thermodynamic-based correlations, will be presented. The latter are important for correlating mixture properties (see Chapter 3).

2.4.2.1 Sublimation Curve

The phase transition below the triple point is called sublimation. Experimental data of the sublimation pressure have been correlated by Wagner *et al.* [27] for the temperature range from $T = 190$ to 273.16 K.

$$\ln\left(\frac{P_{\text{subl}}}{P_{\text{t}}}\right) = a_1\left(1 - \Theta_{\text{t}}^{-1.5}\right) + a_2\left(1 - \Theta_{\text{t}}^{-1.25}\right) \quad (2.1)$$

where $\Theta_{\text{t}} = T/T_{\text{t}}$, $T_{\text{t}} = 273.16$ K, $P_{\text{t}} = 611.657$ Pa, $a_1 = -13.928169$, and $a_2 = 34.7078238$.

2.4.2.2 Vapor-Pressure Curve

Gaseous and liquid water coexist in equilibrium along the vapor-pressure curve. It separates the regions where liquid and gaseous water exists. Saturation pressures increase exponentially with increasing temperatures, see Figure 2.4. With increasing temperature, the density of the liquid phase decreases and that of the gaseous phase increases, see Figure 2.5. The vapor-pressure curve ends at the critical point, where gaseous and liquid water become identical. Accepted data for the critical constants of water are $P_{\text{c}} = 22.064$ MPa, $T_{\text{c}} = 373.946$ °C (647.096 K), $\rho_{\text{c}} = 322$ kg m^{-3} [26].

Vapor-pressure data for water have been determined experimentally to an accuracy of about 0.025%. The following equation, representing the data within this accuracy, was presented by Wagner and Pruß [33]:

$$\ln(P^{\text{s}}/P_{\text{c}}) = (T_{\text{c}}/T)\left(a_1\tau + a_2\tau^{1.5} + a_3\tau^3 + a_4\tau^{3.5} + a_5\tau^4 + a_6\tau^{7.5}\right) \quad (2.2)$$

where $\tau = 1 - T/T_{\text{c}}$, and the coefficients a_1–a_6 are listed in Table 2.5.

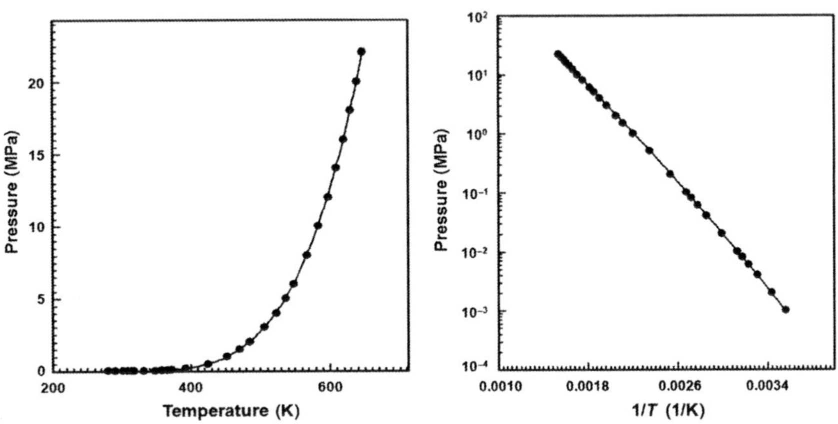

FIGURE 2.4 Vapor-pressure P^{sat} for H_2O as a function of temperature T (left) and as a function of $\log P$ versus $1/T$ (right). Points: Data IAPWS-95. Solid line calculated with the program PE [32] using the Peng–Robinson equation of state in the modification of Peneloux (explanation see text below). Parameters: $\omega = 0.326903$, $c/b = 0.281947$.

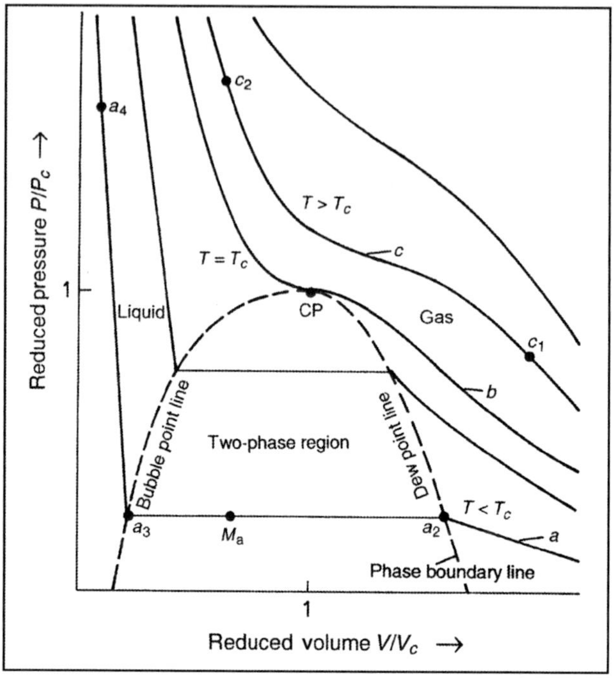

FIGURE 2.5 *P*,*V*,*T*-behavior of a fluid, for example, water. Solid curves: isotherms at selected temperatures, dashed curve: saturated liquid (left) and saturated vapor (right) as boundary line for the coexistence region of vapor and liquid water in equilibrium [30].

TABLE 2.5 Coefficients for Equation (2.2)

	a_i
1	−7.85951783
2	1.84408259
3	−11.7866497
4	22.6807411
5	−15.9618719
6	1.80122502

Beyond the critical point, the vapor-pressure curve is hypothetically extended as curve of P versus T along the critical isochore $\rho = \rho_c$, without a discontinuity in the slope $(\partial P/\partial T)_P$. The slope of the vapor-pressure as a function of T at the critical point is given by Equation (2.3) ([34], cited from Ref. [35]):

$$\left(\frac{\mathrm{d}P}{\mathrm{d}T}\right)_{\mathrm{sat}}^{\mathrm{c}}=\left(\frac{\partial P}{\partial T}\right)_{\rho}^{\mathrm{c}}=0.267\,\mathrm{MPa\,K^{-1}} \tag{2.3}$$

The vapor pressure is shown in Figure 2.4. Experimental data have been correlated with a simple equation of state (EOS) (see below). Those EOS represent vapor-pressure data well, can be adjusted to represent the critical point exactly, but are not able to represent gaseous and liquid phases alike. Accuracy in the liquid phase is much less than in the gaseous phase.

2.4.2.3 *Generalized* P,V,T-*Behavior*

A generalized P,V,T diagram in Figure 2.5 shows the dependence of volume on pressure for different temperatures. At subcritical temperatures ($T<T_c$, isotherm a), the volume of a gas decreases rapidly with increasing pressure, until the phase boundary line is reached (a_2). Crossing this line by decreasing the total volume causes the formation of liquid drops of specific volume a_3. As long as the total volume of the two-phase mixture lies in the two-phase region, a gas of volume a_2 and a liquid of volume a_3 coexist at constant pressure. Points within the two-phase region represent mixtures of gas and liquid, coexisting at constant pressure and temperature. The relative amounts of liquid and gas can be determined by the lever rule. At a_3, all the gas is condensed to liquid a_3. A further reduction of total volume results in a rapid increase in pressure ($a_3 \rightarrow a_4$), since compressibility of a liquid is very low compared to that of a gas.

At temperatures higher than critical ($T>T_c$, isotherm c), a gas (c_1) can be compressed to a liquid-like volume (c_2) without visible change of phases. The critical isotherm ($T=T_c$, isotherm b) has a horizontal tangent at the critical point. At the critical point, the liquid and the gas branch of the phase boundary line meet and phases become identical. Isotherms at $T>T_c$, but near T_c, are flat in the vicinity of the critical point. Compressibility in this region is high. Small changes in pressure or temperature cause large variations of (specific) volume or density [30].

2.4.3 Calculation of *P, V, T*-Behavior

Reliable and easily accessible P,V,T-data are essential for scientists and engineers. A graphic representation is useful for those getting acquainted with the properties of water and those who have to use these properties in process design. They need tools for a quick orientation and fast calculations, in particular, in conceptual design. Tables, like steam tables, are also useful for that purpose and provide a higher degree of accuracy additionally. For scientific work and design calculations in real processes, the best available correlation is just good enough. The IAPWS formulation for general and scientific and that for industrial use are the best ones for calculating properties of ordinary water as well as for heavy water. Precisely calculated values are necessary, for example, for steam cycles. IAPWS has put great effort in the formulation of equations that represent experimental data over a wide range of conditions [35]. IAPWS has developed two formulations for calculating the thermodynamic properties of water and steam: the IAPWS formulation 1995 for General and Scientific Use (IAPWS-95) [36–38] and the IAPWS formulation for Industrial Use (IAPWS-IF2007) [36,39,40]. The IAPWS-95 formulation is an equation for the specific Helmholtz energy, which is an optimized combination of polynomials and exponentials in terms of powers of the reduced temperature and density. It contains some nonanalytic terms in order to mimic the singular behavior of the thermodynamic properties in the vicinity of the critical point [35]. The formulation is not repeated here. Detailed formulations, tables, and computer programs are available from IAPWS and the authors (http://www.iapws.org/; [36,41,42]). Values calculated with IAPWS-95 form the basis for all property values in this chapter, if not stated otherwise. They serve as reference values since they represent the experimental data within experimental errors.

In the following, P,V,T-data and their correlation with EOS will be presented, from ideal-gas state to real behavior.

2.4.3.1 Ideal-Gas State

Water, as many other substances in gaseous state at low densities or high temperatures, behave as an ideal gas. The "ideal gas" is a model for the behavior of monoatomic spherical molecules exerting no forces, and the only interaction between molecules is a perfectly elastic collision. Nevertheless, this model is very valuable to evaluate gas-phase behavior at low densities and to serve as reference to evaluate real gas behavior. The volumetric behavior of an ideal gas is given by

$$P \cdot V = R \cdot T \quad \text{or} \quad \frac{P}{\rho R T} = 1 \tag{2.4}$$

with P is the pressure, V is the volume, ρ is the density, R is the gas constant, and T is the absolute temperature.

Real gases, like H_2O, are molecules rather than single atoms. While a monoatomic gas stores energy as translational energy only, a diatomic gas stores kinetic energy in translational, rotational, and vibrational forms. Therefore, energy-related properties of gases, behaving volumetric like the ideal gas, differ for each species. Molar-specific heats are proportional to the gas constant R. Enthalpy and internal energy of ideal gases depend on temperature only and are independent on pressure [43].

Energy-related properties (internal energy, enthalpy, etc.) can be estimated from calorimetric or acoustic measurements at low densities, or statistical mechanics and knowledge of the rotational and vibrational states of the molecule that can be obtained from spectroscopy. The statistical–mechanical calculations lead to a molar isobaric heat capacity c_P, which can be integrated to obtain other energy-related properties of the ideal gas [4].

$$\Delta U_{\text{ideal gas}} = \int_{T_1}^{T_2} c_V \mathrm{d}T \tag{2.5}$$

$$\Delta H_{\text{ideal gas}} = \int_{T_1}^{T_2} c_P \mathrm{d}T \tag{2.6}$$

For water, a representation covering temperatures from $T = 130$–2000 K is given by Cooper [44]:

$$\frac{c_P}{R} = b_0 + \sum_{i=1}^{5} \left(b_i \frac{x_i^2 \exp(-x_i)}{(1 - \exp(-x_i))^2} \right) \tag{2.7}$$

where $x_i = \beta_i / T$ and coefficients b_i and β_i are listed in Table 2.6.

2.4.3.2 *Real* P,V,T-*Behavior*

Figure 2.6 shows the density along isotherms and at saturation. At low temperatures, the liquid density changes only slightly with pressure due to the very low compressibility of liquid water. With increasing temperature, the liquid expands, affecting properties of water significantly.

Near the critical point, isobars and isotherms show a nearly vertical slope. In this region, a small change in temperature or pressure causes a large change in density. This is due to the transition from subcritical to supercritical conditions. At subcritical conditions, saturated liquid and gas have different densities at the same pressure and temperature. At the critical point, for the first time (with respect to temperature), a continuous isotherm exists, showing a vertical slope. Each property, defined by a derivative of this isotherm, as for example, the isothermal compressibility, $-(1/V)\ (\partial V/\partial P)_T$, diverges at the critical point [4]. Each property, thus defined, shows high values in the critical

TABLE 2.6 Coefficients for Equation (2.7)

i	b_i	β_i/K
0	4.00632	
1	0.012436	833
2	0.97315	2289
3	1.2795	5009
4	0.96956	5982
5	0.24873	17,800

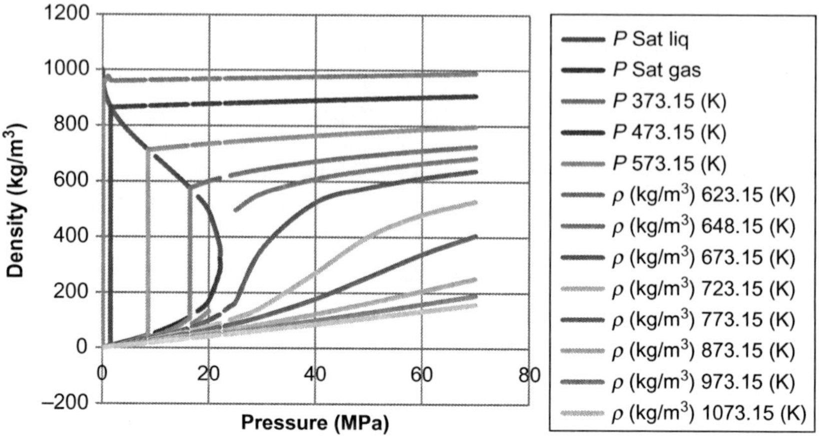

FIGURE 2.6 Density of water in dependence of pressure and temperature, shown as isotherms. *Data IAPWS-95 [36], cited from Ref. [45].* (For color version of this figure, the reader is referred to the online version of this chapter.)

region, above the critical temperature. Those high values reveal the increased change of density in the region adjacent to the critical point. They are typical for most properties in this region and are exploited in processes, for example, for the precipitation of dissolved products by changing the density or by exploiting these rapid changes in supercritical fluid processes in general. In that way, the critical enhancements and the critical region are well integrated in processing concepts. A detailed discussion of phenomena in the close vicinity of the critical point has been published by Anisimov *et al.* [35].

2.4.3.2.1 Calculation of Properties

Properties can be calculated by different methods. The most accurate representation of experimental data is possible with polynomial equations, containing empirical coefficients only. For many calculations in science and engineering, correlation methods are needed that contain information on molecular behavior, for example, to be able to calculate mixture properties. A common way is to define EOS that contain molecular information. In the following, first an empirical correlation is presented for properties of water at saturation and then P,V,T-values of water are correlated with EOS.

2.4.3.2.2 Properties at Saturation: Empirical Correlation

The density of the saturated phases for water is shown in Figure 2.7. Equations have been derived for saturation conditions and the most important thermodynamic properties (density, enthalpy, entropy) [4,33] that are much simpler than the IAPWS formulation and are still accurately representing experimental data.

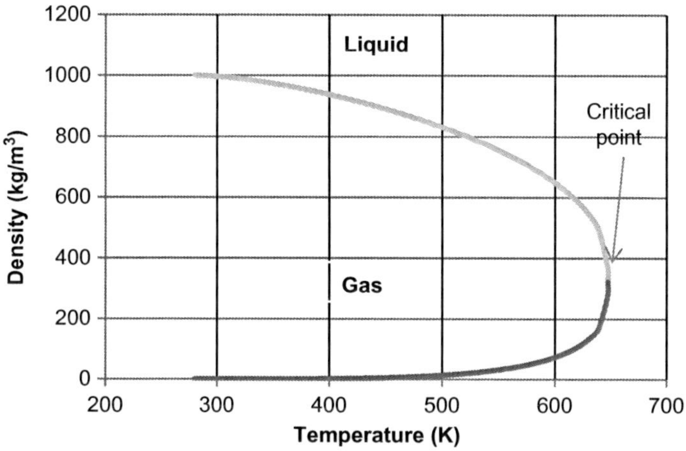

FIGURE 2.7 Density of liquid and gaseous water at saturation along the vapor-pressure curve. *Data IAPWS-95, taken from Ref. [45].* (For color version of this figure, the reader is referred to the online version of this chapter.)

TABLE 2.7 Coefficients for Equations (2.8) and (2.9)

i	b_i	c_i
1	1.99274064	−2.03150240
2	1.09965342	−2.68302940
3	−0.510839303	−5.38626492
4	−1.75493479	−17.2991605
5	−45.5170352	−44.7586581
6	-6.74694450×10^5	−63.9201063

The densities of the saturated liquid and vapor, ρ' and ρ'', on a mass basis are given by

$$\rho'/\rho_c = 1 + b_1\tau^{1/3} + b_2\tau^{2/3} + b_3\tau^{5/3} + b_4\tau^{16/3} + b_5\tau^{43/3} + b_6\tau^{110/3} \quad (2.8)$$

$$\ln(\rho''/\rho_c) = c_1\tau^{2/6} + c_2\tau^{4/6} + c_3\tau^{8/6} + c_4\tau^{18/6} + c_5\tau^{37/6} + c_6\tau^{71/6} \quad (2.9)$$

where $\tau = 1 - T/T_c$, T_c and P_c are the values recommended by IAPWS. The coefficients b_i and c_i are listed in Table 2.7. Values for saturation properties calculated by Equations (2.8) and (2.9) differ slightly from those from the IAPWS-95 formulation for thermodynamic properties. The differences are much smaller than the uncertainties in the formulations [4].

2.4.3.2.3 Correlation of *P,V,T*-Behavior Using EOS

Quantitative representation of the *P*,*V*,*T*-behavior of supercritical fluids can be achieved by EOS that take into consideration the properties of individual molecules. Cubic EOS and EOS derived from statistical thermodynamics are appropriate for most substances and their mixtures applied in scientific and engineering use, as long as these substances are nonpolar or moderately polar. For more strongly polar and associating compounds, like water, other relations must also be used, in particular for calculating liquid volume or density. Typically, cubic EOS represent gas-phase data well, but deviations from liquid-phase densities may be substantial (see below examples and comparison of some EOS). Excellent formulations, that is, empirical polynomial expressions, are available for water with the formulations of IAPWS, as mentioned above. These formulations deviate from experimental values within experimental errors only and are accepted for use as reference data. But those EOS contain many and physical nonsignificant parameters and, in general, cannot be used with mixtures. Therefore, their application is limited to special cases, for example, for calculations of process cycles for steam or refrigerants [30]. In the following, the Soave–Redlich–Kwong equation of state and the Peng–Robinson equation of state (PR EOS), two widely used EOS, will be discussed for correlating volumetric properties of water.

2.4.3.2.4 Analytical EOS

An analytical equation of state is an algebraic relationship between pressure, temperature, and molar volume. EOS should meet some conditions: The criterion of thermodynamic stability must be fulfilled at the critical point:

$$(\mathrm{d}P/\mathrm{d}V)_{T_c} = 0 \text{ and } \left(d^2P/\mathrm{d}V^2\right)_{T_c} = 0. \qquad (2.10)$$

Properties of a fluid should be represented by the equation of state from the liquid to the ideal-gas state. At very low pressure, the equation of state should approach the ideal-gas law. In 1873, Van der Waals [46] published his famous equation of state (Equation 2.11), based on the assumption of spherical molecules with attractive and repulsive forces depending on the relative distance of the molecules.

$$P = \frac{RT}{V-b} - \frac{a}{V^2}, \qquad (2.11)$$

with

$$V_c = 3b; \; P_c = \frac{a}{27b^2}; \; T_c = \frac{8a}{27Rb}. \qquad (2.12)$$

The equation consists of the repulsion term with parameter *b*, the covolume as parameter for representing repulsive forces, and the attraction term with parameter a representing attractive forces. Both parameters are related to the critical point, as stated in Equation (2.12), and are therefore dependent on the accuracy of these values. The Van der Waals equation of state can qualitatively account for most phenomena in P,V,T-behavior and phase equilibria (PE). Yet accuracy is not sufficient for practical purposes for most substances. Therefore, various modifications have been proposed, from which the PR EOS and the Redlich–Kwong–Soave equation of state, and their modifications, are used in this chapter. For a more comprehensive treatment of EOS, the reader is referred to textbooks [31,47].

The main disadvantage of early cubic EOS is a fixed temperature dependence of the attraction term. It is highly probable that different substances with different attractive forces exhibit a different temperature dependence of their attractive forces. For example, Van der Waals forces of nonpolar molecules depend differently on distance between the molecules than electrostatic forces of polar molecules. Various modifications, taking into consideration individual temperature dependence for a substance, have been proposed. One is the Soave–Redlich–Kwong equation of state [48] presented in Equations (2.13)–(2.17).

$$P=\frac{RT}{V-b}-\frac{a(T)}{V(V+b)}, \tag{2.13}$$

with

$$b=0.08664\frac{RT_c}{P_c}, \tag{2.14}$$

where $a(T)$ depends on temperature:

$$a(T)=a'(T_c)\alpha(T), \tag{2.15}$$

$$a'(T_c)=0.42748\frac{R^2T_c^2}{P_c}, \tag{2.16}$$

$$\alpha(T)=\left[1+\left(0.480+1.574\omega-0.176\omega^2\right)\left(1-(T/T_c)^{0.5}\right)\right]^2. \tag{2.17}$$

with

$$\omega=-\log P_r^{sat}-1, \quad \text{at } T_r=0.7,$$

$$P_r=\frac{P}{P_c}: \text{reduced pressure},$$

$$T_r=\frac{T}{T_c}: \text{reduced temperature}.$$

In this modification, component parameters are also linked to the critical point, yet information on vapor pressure, given by the acentric factor ω, introduces some information on the interaction forces at about 1/10 of the critical pressure. The Soave–Redlich–Kwong equation of state was the first modified equation of state where the parameter a was made temperature dependent in such a way that the vapor pressure curve could be reproduced well. The equation of state requires three input parameters per pure compound: T_c, P_c, and ω [32]. The Soave–Redlich–Kwong equation of state was further modified by several researchers. Peneloux *et al.* [49] proposed a volume translation that does not affect the calculation of vapor pressures. The Soave–Redlich–Kwong equation of state, in the modification by Peneloux is given by:

$$P = \frac{RT}{\tilde{v} - \tilde{b}} - \frac{a(T)}{\tilde{v}\left(\tilde{v} + \tilde{b}\right)}, \tag{2.18}$$

with

$$\tilde{v} = v + c \text{ and } \tilde{b} = b + c, \tag{2.19}$$

with c as volume translation parameter optimized to represent critical data. Parameters of pure components are given by:

$$a(T) = a_c \alpha(T), \text{ with } a_c = 0.42747 \frac{R^2 T_c^2}{P_c} \tag{2.20}$$

$$\tilde{b} = 0.08664 \frac{RT_c}{P_c} \tag{2.21}$$

$$\alpha(T_r) = \left(1 + m\left(1 - \sqrt{T_r}\right)\right)^2; \; T_r = T/T_c \tag{2.22}$$

$$m = 0.48 + 1.574\omega - 0.16\omega^2 \tag{2.23}$$

The PR EOS [30], Equation (2.24), is the most successful alternative.

$$P = \frac{RT}{V - b} - \frac{a(T)}{V(V + b) + b(V - b)}. \tag{2.24}$$

$$b = 0.07780 \frac{RT_c}{P_c}; \tag{2.25}$$

$$a(T) = a(T_c)\alpha(T_R, \omega); \tag{2.26}$$

$$a(T_c) = 0.45724 \frac{R^2 T_c^2}{P_c}; \tag{2.27}$$

$$\alpha = \left[1 + \beta\left(1 - T_r^{0.5}\right)\right]^2, \tag{2.28}$$

with

$$\beta = 0.37464 + 1.54226\omega - 0.26992\omega^2. \tag{2.29}$$

The PR EOS has the same characteristics with respect to parameter determination as the Soave modification of the Redlich–Kwong equation of state.

Parameters are linked to the critical point. Some information on interacting forces and shape of the molecules is introduced by the acentric factor [30]. It is known to give slightly better values of liquid densities than the Soave–Redlich–Kwong equation of state. The PR EOS requires three inputs per compound: T_c, P_c, and the acentric factor ω [32]. The equation was modified by Peneloux *et al.* [49] who suggested a volume translation which does not affect the calculation of vapor pressures. The PR EOS, in the modification of Peneloux *et al.*, is given by:

$$P = \frac{RT}{\tilde{v} - \tilde{b}} - \frac{a(T)}{\tilde{v}^2 + 2\tilde{b}\tilde{v} - \tilde{b}^2}, \tag{2.30}$$

with

$$\tilde{v} = v + c \text{ and } \tilde{b} = b + c,$$

The empirical parameter c is volume translation, optimized to represent critical data. Pure-component parameters are obtained from:

$$a(T_c) = 0.45724 \frac{R^2 T_c^2}{P_c}, \text{ with } a(T) = a(T_c)\,\alpha(T_r, \omega) \tag{2.31}$$

$$\tilde{b} = 0.0778 \frac{RT_c}{P_c} \tag{2.32}$$

2.4.3.2.5 Correlation of *P,V,T*-Behavior of Water with EOS

P,V,T-behavior of water, as for other pure components, can be calculated by EOS. Calculation programs are available as commercial programs or can be individually established. A program for calculating vapor pressure, *P,V,T*-data, and PE in binary and ternary mixtures was established by the group of the author. It is called "PE" and can be downloaded for free from the Website of the Institute [32]. PE offers about 40 EOS with up to seven different mixing rules. The program further offers subroutines to determine pure-component equation of state parameters based on pure-component vapor pressures and densities at saturation conditions. Graphical output routines allow a visual comparison of experimental and calculated data of the vapor pressure curve and *P,V,T*-data. As a useful feature, it is possible to enforce pure-component parameters that reproduce the critical temperature and pressure of the compound during pure-component parameter optimization.

The following diagrams show correlations of density of water dependent of temperature and pressure. In Figures 2.8 and 2.9, results obtained with the original EOS, and in Figure 2.10, results obtained with the PR EOS modified by Peneloux *et al.* [49] (PR-Peneloux VT, with VT for "volume translation").

With the modified EOS, a very good representation of volumetric data in the gaseous, near-critical, and supercritical region can be obtained. Values for

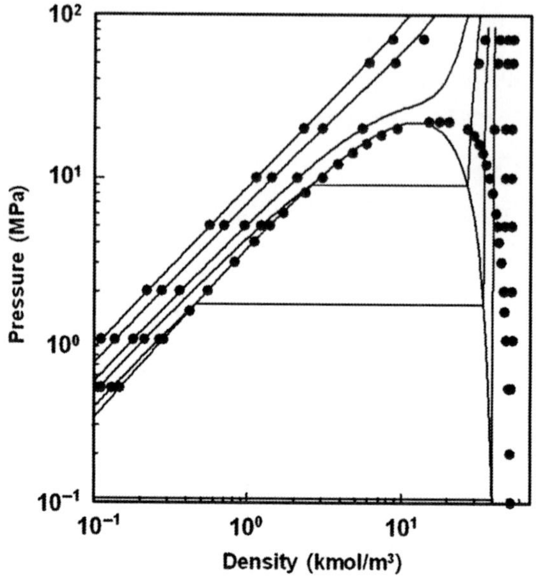

FIGURE 2.8 *P,V,T*-behavior of H_2O, EOS Soave–Redlich–Kwong. Points: IAPWS-95; parameter: $T_c = 647.096$ (K), $P_c = 22.064$ (MPa), $\omega = 0.315178$ (–); calculated with PE 2000 [32]. Deviations: P_{sat}, 4.7%; V_{liq}, 42.2%; V_{gas}, 11.5%.

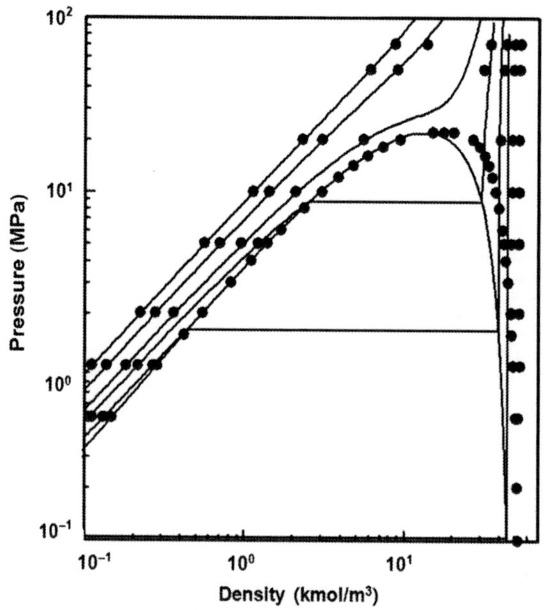

FIGURE 2.9 *P,V,T*-behavior of H_2O, EOS Peng–Robinson. Points: IAPWS-95; parameter: $T_c = 647.096$ (K), $P_c = 22.064$ (MPa), $\omega = 0.327118$ (–); calculated with PE 2000 [32]. Deviations: P_{sat}, 2.6%; V_{liq}, 27.0%; V_{gas}, 8.6%.

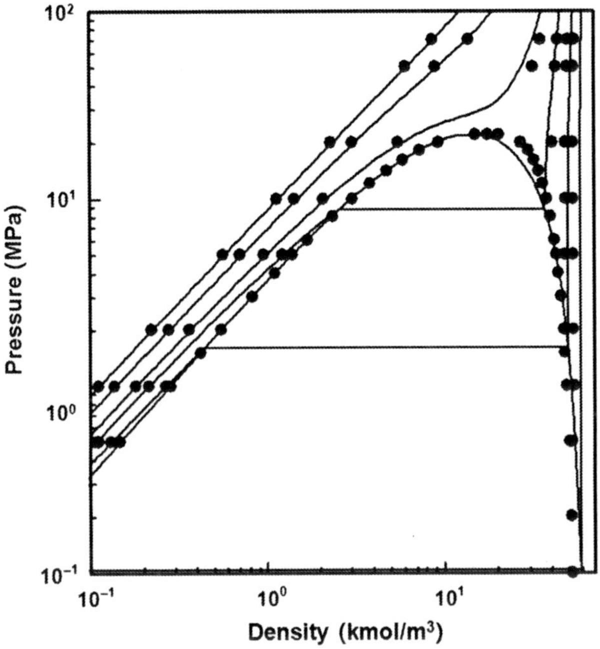

FIGURE 2.10 *P,V,T*-behavior of H_2O, EOS PR-VT Peneloux [49]. Points: IAPWS-95; parameter: $T_c = 647.096$ (K), $P_c = 22.064$ (MPa), Calculated with PE 2000 [32]. Optimized parameters for $\Delta T_c = 0$; $\Delta P_c = 0$; $\omega = 0.326903$; $c/b = 0.281947$; deviations: P_{sat}, 2.6%; V_{liq}, 14.2%; V_{gas}, 6.1%.

the liquid or at high densities in the supercritical region are not as good, but sufficient accurate for most applications. The original EOS behave similar in the gaseous region but perform worse in the liquid and high-density region. The modified EOS are sufficiently accurate for calculating properties and PE of mixtures, as is demonstrated in Section 2.3 on mixtures of substances with water. In this context, EOS based on statistical thermodynamics, such as SAFT, will also be discussed.

2.4.3.3 Calculating Thermodynamic Properties of H_2O in the Critical Region

An accurate equation of state is needed for reliable calculations of the thermodynamic properties in the critical region, because density and thermodynamic properties change rapidly with pressure and temperature [4]. The critical region may be defined to a density range from $\rho = 100$ to $600\ \mathrm{kg\ m^{-3}}$ at a pressure range from $P = 20$ to 30 MPa, temperatures from $T = 650$ to 720 K at a density of $\rho = 100\ \mathrm{kg\ m^{-3}}$ and $T = 625$ to 650 K at a density of $\rho = 600\ \mathrm{kg\ m^{-3}}$ [35].

EOS have been developed that reproduce the theoretically predicted power laws for the singular asymptotic critical thermodynamic behavior and incorporate crossover to analytic behavior away from the critical point [34]. Nevertheless, the use of IAPWS-95 is recommended when other EOS are not sufficient. IAPWS-95 is much simpler than the formulations reproducing very near-critical regions, and it represents the thermodynamic properties over a wide range, inside and outside the critical region. Furthermore, IAPWS-95 represents thermodynamic property data in the critical region within experimental accuracy except for some c_V-data in the near-vicinity of the critical point [35].

2.4.4 Joule–Thomson Effect of Water

In many cases, pressure changes are part of the process. From experience it is known (since Joule and Thomson carried out their experiments in the nineteenth century [50]) that temperature of most gases decreases when pressure is lowered in a flowing gas by throttling. This effect is called the Joule–Thomson effect and is easily experienced while experimenting with carbon dioxide [30]. The Joule–Thomson coefficient η_{JT} is defined as the variation of temperature with pressure at constant enthalpy:

$$\eta_{JT} = \left(\frac{\partial T}{\partial P}\right)_H . \tag{2.33}$$

It can be derived from volumetric properties:

$$\eta_{JT} = \frac{T(\partial V/\partial T)_P - V}{c_P} . \tag{2.34}$$

The Joule–Thomson coefficient of an ideal gas is zero. In real gases, the Joule–Thomson coefficient is different from zero and depends on pressure and temperature. For $\eta_{JT} > 0$, temperature decreases, and for $\eta_{JT} < 0$, temperature increases during an expansion. The temperature at which the sign of the Joule–Thomson coefficient changes is the inversion- or Boyle temperature, which itself depends on pressure. The inversion temperature of most gases is above ambient temperature, but for hydrogen, the inversion temperature is about $T = -80$ °C. Values for the Joule–Thomson coefficient are listed in Table 2.8 for H_2O. Although the Joule–Thomson coefficient is not a thermodynamic property for itself, the effect is important for practical purposes and values for the Joule–Thomson coefficient help to realize the changes occurring during processes. From the table, it can be seen that during expansion of liquid water the temperature increases. In the gaseous and near-critical state, temperature of water during expansion drops.

2.4.5 Thermodynamic Properties of Water at Saturation

Correlations for enthalpies and entropies of water at saturation are presented in the following equations [4,33]. The enthalpies of the saturated liquid and

TABLE 2.8 Joule–Thomson Coefficient η_{JT} for H_2O

Liquid Water			Gaseous Water (steam)		
T (K)	P (bar)	η_{JT} (K/bar)	T (K)	P (bar)	η_{JT} (K/bar)
293.15	1–250	−0.0224	379.15	1.2	5.8223
318.15	1–500	−0.0216	493.65	21.6	2.0705
338.15	100–500	−0.0200	514.15	31.8	1.1438
357.15	1–400	−0.0186	523.25	1.023	1.6443
			528.15	41.4	1.5458
			536.25	46.1	1.4599
Critical and supercritical region			548.15	51	1.3291
638.75	220.75	0.1150	563.15	71.6	1.181
646.55	220.40	0.2632	599.65	11.9	0.88
647.75	232.15	0.1891	604.15	126.7	0.848
649.45	251.90	0.1451	610.25	136.7	0.801
650.95	401.9	0.0526	615.95	146.7	0.764
			623.65	161.7	0.69
			628.65	171.6	0.6708
			646.65	212.0	0.5365
			646.85	3.9	0.833
			649.64	220.8	0.507
			654.95	1.2	0.8023

Data from Franz and Grigull [51].

vapor, H' and H'', and the saturated liquid and vapor entropies, S' and S'', are given on a mass basis by:

$$H' = \alpha + \frac{1}{\rho'}\frac{dP^s}{dT} \tag{2.35}$$

$$H'' = \alpha + \frac{T}{\rho''} + \frac{dP_s}{dT} \tag{2.36}$$

$$S' = \phi + \frac{1}{\rho'}\frac{dP^s}{dT} \tag{2.37}$$

$$S'' = \phi + \frac{1}{\rho''} + \frac{1}{\rho''}\frac{dP^s}{dT}, \tag{2.38}$$

TABLE 2.9 Coefficients for Equations (2.39) and (2.40) [33]

i	d_i
1	-5.65134998×10^{-8}
2	2690.66631
3	127.287297
4	−135.003439
5	0.981825814

where dP^s/dT, the derivative of the vapor-pressure curve, is obtained by differentiating Equation (2.2). In order to compute enthalpies and entropies at saturation, the auxiliary quantities α and Φ are defined by

$$\alpha/\alpha_0 = d_a + d_1\Theta^{-19} + d_2\Theta + d_3\Theta^{4.5} + d_4\Theta^5 + d_5\Theta^{54.5} \quad (2.39)$$

$$\phi/\phi_0 = d_\phi + \frac{19}{20}d_1\Theta^{-20} + d_2 \ln\Theta + \frac{9}{7}d_3\Theta^{3.5} + \frac{5}{4}d_4\Theta^4 + \frac{109}{107}d_5\Theta^{53.5} \quad (2.40)$$

where $\theta = T/T_c$, $a_0 = 1000\ \mathrm{J\,kg^{-1}}$, $\Phi_0 = \alpha_0/T_c$, $d_\alpha = -1135.905627715$, $d_\Phi = 2319.5246$, and d_1–d_5 are listed in Table 2.9. The values of d_α and d_Φ are chosen to satisfy the convention that the internal energy and entropy are assigned a value of zero for the saturated liquid at the triple point.

2.4.6 Heat Capacity (Specific Heat) of Water

Heat capacity or specific heat is a measure of the heat energy needed to produce an increase in temperature. The amount of heat required depends on the conditions at which the heating occurs. Heat capacity or specific heat of liquid water is substantially higher than that of most other liquid substances. When water is heated, the energy absorbed causes the hydrogen bonds to break. As the energy for these processes is not available to increase the kinetic energy, considerable heat is needed to raise the temperature. Values of the specific heat are needed extensively for planning and calculating heat transfer processes (see Section 2.4 on heat transfer). Therefore, beside the definition and the connection to volumetric behavior, in Figure 2.11 the dependence of specific heat of water on pressure and temperature and in Table 2.10 numerical values for quick calculations are presented.

In typical gases, heat capacity c_P decreases with temperature and increases with pressure. In supercritical fluids, near to the critical point, this dependency is ambivalent. Depending on the conditions, c_P can rise or decrease with pressure, in that way behaving as a liquid or a gas. Because hydrothermal and supercritical water is applied, while undergoing major changes of pressure

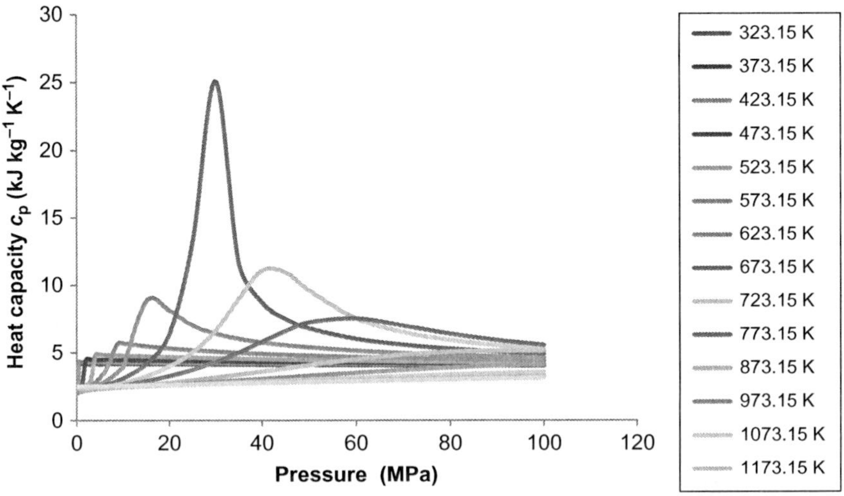

FIGURE 2.11 Heat capacity of water. *Data IAPWS-95; cited from Ref. [52].* (For color version of this figure, the reader is referred to the online version of this chapter.)

and temperature, behavior of c_P seems to be better related to density. On the other hand, since P and T are adjustable parameters in a process, dependence on P and T must be familiar [30].

Most commonly used is the specific heat at constant pressure: c_P

$$c_P = \left(\frac{\partial H}{\partial T}\right)_P = T\left(\frac{\partial S}{\partial T}\right)_P \tag{2.41}$$

and the specific heat at constant volume c_V:

$$c_V = \left(\frac{\partial U}{\partial T}\right)_V = T\left(\frac{\partial S}{\partial T}\right)_V \tag{2.42}$$

The difference between the specific heat capacity at constant pressure c_P and the specific heat capacity at constant volume c_V can be derived from volumetric properties (Table 2.3) and is related to the compressibility by

$$c_P - c_V = \frac{T}{\rho}\left(\frac{\partial P}{\partial T}\right)_\rho^2 \kappa_T \tag{2.43}$$

An overview of the behavior of c_P of water in the critical region is shown in Figure 2.12. Values for c_P are reduced with the heat capacity of the liquid at boiling conditions at 0.1 MPa for the liquid state, and values for the gaseous state are reduced with the heat capacity of the ideal gas. The heat capacity at constant pressure of liquid water increases near the boiling temperature. The increase is more pronounced, the nearer the boiling temperature is to the critical temperature. Heat capacity of the gas exhibits the greatest deviations from

TABLE 2.10 Heat Capacity at Constant Pressure, c_P, for Water, in kJ kg^{-1} K^{-1} (IAPWS-95, cited from Ref. [52])

P (MPa)	323.15 K	373.15 K	423.15 K	473.15 K	523.15 K	573.15 K	623.15 K	673.15 K	723.15 K	773.15 K	873.15 K	973.15 K	1073.15 K	1173.15 K	1273.15 K
0.1	4.181	2.042	1.98	1.975	1.99	2.013	2.04	2.07	2.102	2.135	2.203	2.273	2.343	2.412	2.478
0.5	4.181	4.216	4.312	2.138	2.081	2.069	2.078	2.097	2.121	2.149	2.212	2.279	2.348	2.415	2.481
1	4.179	4.215	4.31	2.4	2.213	2.147	2.128	2.132	2.147	2.168	2.224	2.287	2.353	2.419	2.484
2	4.177	4.213	4.307	4.486	2.555	2.328	2.239	2.205	2.199	2.208	2.247	2.303	2.364	2.427	2.49
3	4.115	4.21	4.304	4.481	3.047	2.55	2.364	2.286	2.255	2.248	2.272	2.318	2.375	2.435	2.496
4	4.173	4.208	4.301	4.475	4.856	2.828	2.507	2.373	2.314	2.291	2.297	2.335	2.386	2.444	2.502
5	4.11	4.206	4.298	4.469	4.843	3.181	2.672	2.468	2.376	2.335	2.322	2.351	2.398	2.452	2.508
6	4.168	4.204	4.294	4.464	4.83	3.642	2.861	2.572	2.442	2.381	2.348	2.368	2.409	2.46	2.514
7	4.166	4.201	4.291	4.458	4.817	4.272	3.082	2.686	2.512	2.428	2.375	2.385	2.421	2.469	2.521
8	4.164	4.199	4.288	4.453	4.804	5.196	3.342	2.811	2.586	2.478	2.402	2.402	2.433	2.477	2.527
9	4.162	4.197	4.285	4.448	4.792	5.724	3.652	2.948	2.665	2.53	2.43	2.419	2.444	2.486	2.534
10	4.16	4.195	4.282	4.442	4.78	5.676	4.028	3.1	2.749	2.584	2.458	2.437	2.456	2.494	2.54
15	4.149	4.184	4.268	4.417	4.724	5.47	8.838	4.177	3.26	2.891	2.61	2.529	2.518	2.538	2.573
20	4.139	4.174	4.254	4.394	4.674	5.311	8.138	6.371	3.982	3.269	2.778	2.627	2.583	2.584	2.606
25	4.129	4.163	4.24	4.371	4.629	5.182	6.994	13.27	5.043	3.737	2.961	2.729	2.649	2.63	2.64
30	4.119	4.154	4.227	4.35	4.588	5.075	6.401	25.08	6.655	4.312	3.16	2.836	2.716	2.677	2.674
35	4.109	4.144	4.215	4.331	4.551	4.984	6.022	11.67	8.982	5.006	3.373	2.945	2.785	2.724	2.709

40	4.1	4.135	4.202	4.312	4.516	4.906	5.752	8.717	11.08	5.799	3.597	3.057	2.854	2.772	2.744
45	4.091	4.125	4.191	4.294	4.484	4.837	5.547	7.483	10.94	6.601	3.829	3.17	2.923	2.819	2.778
50	4.083	4.117	4.18	4.277	4.454	4.775	5.384	6.789	9.595	7.239	4.062	3.283	2.992	2.866	2.812
60	4.066	4.1	4.158	4.245	4.4	4.67	5.138	6.011	7.547	7.534	4.5	3.503	3.126	2.957	2.879
70	4.05	4.083	4.138	4.216	4.352	4.584	4.959	5.571	6.505	6.992	4.846	3.705	3.253	3.045	2.943
80	4.035	4.068	4.119	4.189	4.309	4.51	4.82	5.28	5.912	6.383	5.057	3.88	3.369	3.126	3.004
90	4.021	4.053	4.101	4.164	4.271	4.447	4.708	5.071	5.529	5.908	5.138	4.022	3.472	3.201	3.061
100	4.008	4.039	4.084	4.141	4.236	4.391	4.615	4.911	5.261	5.557	5.122	4.129	3.561	3.269	3.113

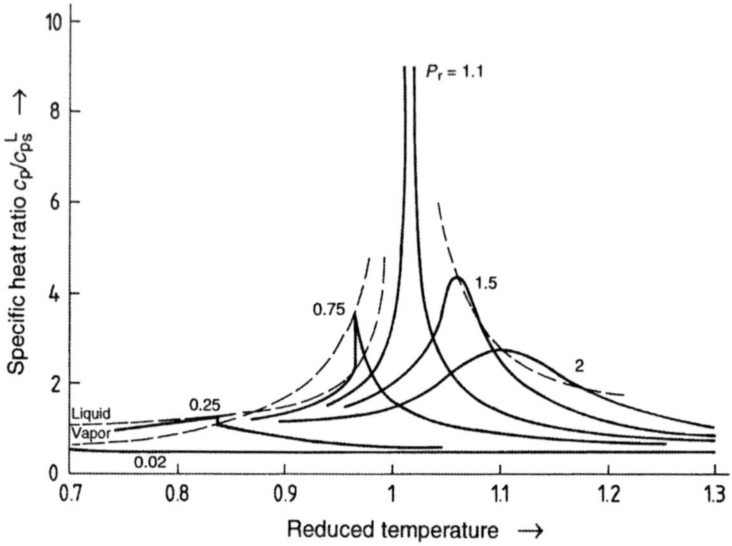

FIGURE 2.12 Specific heat of water at constant pressure in the critical region [30]. Reduced parameters are explained in the text.

ideal gas near the critical point. But these deviations decrease rapidly with increasing temperature. For pressures lower than the critical, the ideal gas heat capacity is approached with increasing temperature. Between heat capacities of saturated gas and boiling liquid, a gap occurs, which widens when the critical pressure is approached. At pressures higher than the critical ($P_r > 1$), c_P runs through a maximum, which becomes flatter and is shifted to higher temperatures, with increasing pressures. At the critical point, the heat capacity is infinite [30].

2.4.7 Enthalpy

Enthalpy is the sum of the internal energy of the system plus the product of the pressure of the gas in the system and its volume: $H = U + PV$. Therefore, enthalpy of a gas decreases with pressure at constant temperature. Enthalpy of the liquid phase at constant temperature remains constant for moderate pressures. Only at high pressures of more than 100 MPa does the influence of pressure on thermodynamic properties of most liquids become substantial.

For a process at constant pressure, the change of enthalpy is

$$\Delta H_P = \int_{T_1}^{T_2} c_P \, \mathrm{d}T \tag{2.44}$$

The reference point for enthalpy is $T = 0\,°\mathrm{C}$, where the value of the enthalpy is set to zero. Figure 2.13 shows a plot for the enthalpy of water. Values for the enthalpy can be found in the steam tables below or in more

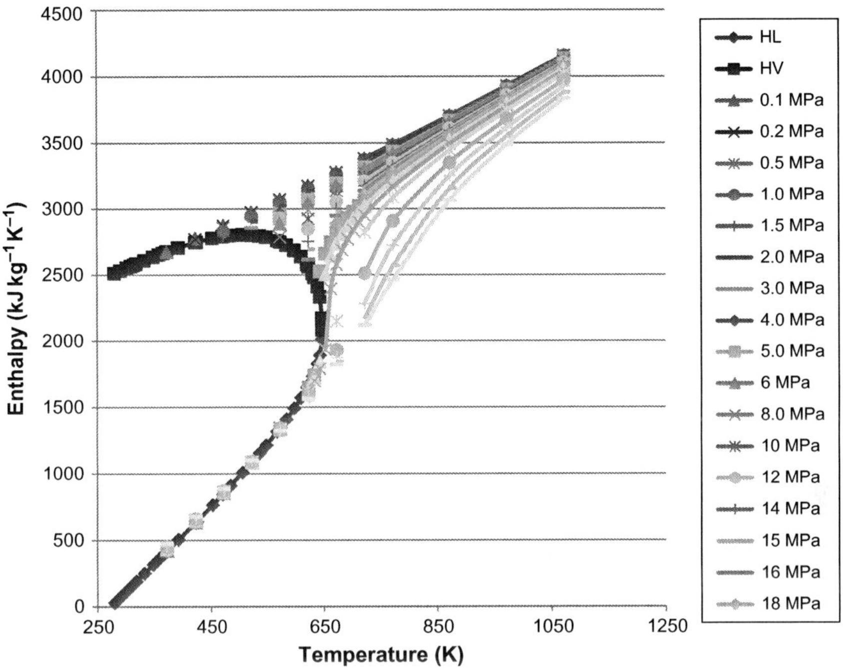

FIGURE 2.13 Enthalpy of water. *Data IAPWS-95, cited from Ref. [52].* (For color version of this figure, the reader is referred to the online version of this chapter.)

detailed steam tables in the literature. Enthalpy increases with pressure and temperature. The difference between the enthalpies of the saturated vapor and liquid, the enthalpy of vaporization (ΔH_V) is about 2500 kJ kg^{-1} at the triple point, decreases with temperature and steeply declines near the critical point. The enthalpy of vaporization of water is high. Therefore, phase changes in processes require high amounts of energy.

2.4.8 Entropy

Entropy is a thermodynamic property related to the ability to convert one form of energy to others. In particular, important is entropy for planning of processes in which high amounts of heat are created or used in combination with kinetic energy processes. There is a vast literature on the general importance of entropy. A discussion is beyond the scope of this chapter. Nevertheless, in the context of hydrothermal and supercritical water and its applications, the definition of entropy as given in Equation (2.45) is essential:

$$dS \equiv \frac{dQ_{\text{rev}}}{T} \tag{2.45}$$

Entropy changes are calculated with the reference point of zero at $T = 0\,°C$:

$$\Delta S_{1\to 2} = \int_1^2 \frac{dQ_{rev}}{T} \tag{2.46}$$

In Figure 2.14, the entropy of water is shown in relation to its dependence on temperature and pressure. The left part mainly is dominated by the values for the entropy of the saturated gaseous state (upper line) and saturated liquid state (lower line). Entropy increases with temperature (at constant pressure). Increasing pressure leads to a higher degree of order in the molecular arrangement. Therefore, entropy decreases with increasing pressure.

Changes of conditions of state of water as a working fluid are common in many applications. Due to the high amount of energy involved, it is useful to get a quick overview on the energy involved with the changes of state. This can be carried out in a temperature–entropy-diagram (*T*,*S*-diagram). Entropy changes show irreversible losses. Thermodynamic work needed for a reversible cyclic process can be determined by the area that is included by the process cycle in a *T*,*S*-diagram. In Figure 2.15, a schematic *T*,*S*-diagram is plotted and explained. A *T*,*S*-diagram can be downloaded from Wikipedia: *T*,*S*-diagrams for water.

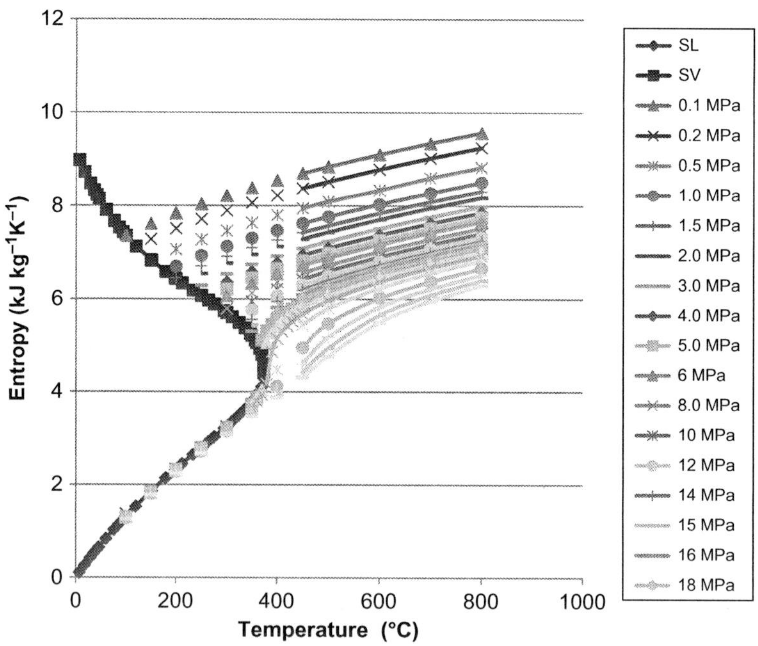

FIGURE 2.14 Entropy of water. *Data from IAPWS-95, cited from Ref. [52].* (For color version of this figure, the reader is referred to the online version of this chapter.)

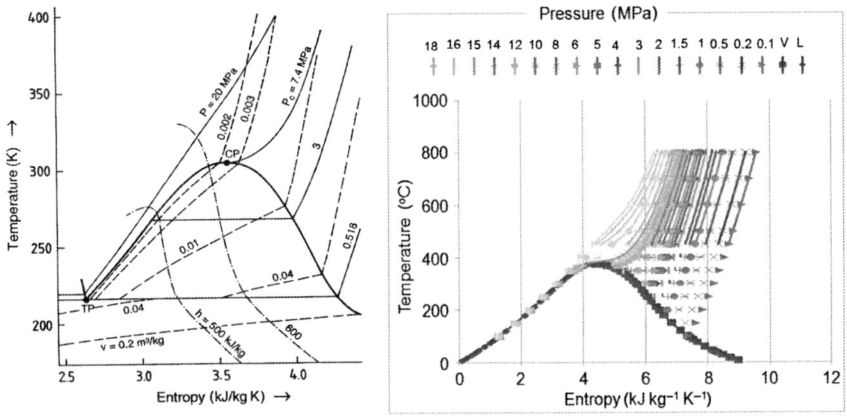

FIGURE 2.15 Left: *T,S*-diagram for CO_2 [30]. Right: *T,S*-diagram for H_2O (derived from Figure 2.14). (For color version of this figure, the reader is referred to the online version of this chapter.)

2.4.9 Steam Tables

Thermodynamic properties of water can be calculated with high accuracy using software prepared by IAPWS, or the authors who have developed the IAPWS formulations. For scientific research and industrial process design, the IAPWS formulation provides the necessary tool [36,42]. For practical applications, tables of thermodynamic properties, called "steam tables," can be used. They are useful for quick conceptual design purposes and understanding thermodynamic processes. In this chapter, steam tables for the hydrothermal, the critical, and the supercritical region are presented. They are derived from steam tables, established by Harvey [45] using the IAPWS-95 formulation. They can be downloaded from the internet: NISTIR 5078 [45] (Tables 2.11 and 2.12).

2.5 TRANSPORT PROPERTIES

2.5.1 Viscosity

The viscosity of a fluid is a measure of its ability to transfer momentum across the direction of flow in a nonequilibrium situation. Viscosity is defined by a fluid that flows laminar or free of turbulences with a velocity gradient vertical to the direction of flow. This gradient in velocity induces a shear stress. By relating shear stress to an area and dividing it by the velocity gradient, the viscosity of the fluid is obtained, as given by Equation (2.47).

$$\eta = \frac{\tau}{du/dy}, \tag{2.47}$$

with τ is the shear stress per unit of area, u is the flow velocity, and y is the coordinate vertical to direction of flow.

TABLE 2.11 Properties of Water at Saturation [52]

		Density		Enthalpy		Entropy	
P (MPa)	T (K)	ρ_L (kg/m^3)	ρ_V (kg/m^3)	H_L (kJ/kg)	H_V (kJ/kg)	S_L (kJ/(kg K))	S_V (kJ/(kg K))
0.001	280.12	999.86	0.00774	29.3	2513.7	0.10591	8.9749
0.002	290.645	998.64	0.01493	73.43	2532.9	0.26056	8.7226
0.004	302.11	995.92	0.02875	121.39	2553.7	0.42239	8.4734
0.006	309.309	993.59	0.04214	151.48	2566.6	0.52082	8.329
0.008	314.659	991.54	0.05525	173.84	2576.2	0.59249	8.2273
0.01	318.956	989.83	0.06817	191.81	2583.9	0.6492	8.1488
0.02	333.235	983.13	0.13075	251.42	2608.9	0.83202	7.9072
0.04	349.025	974.3	0.25044	317.62	2636.1	1.0261	7.669
0.06	359.076	967.99	0.36607	359.91	2652.9	1.1454	7.5311
0.08	366.636	962.93	0.47914	391.71	2665.2	1.233	7.4339
0.1	372.756	958.63	0.59034	417.5	2674.2	1.3028	7.3588
0.2	393.36	942.94	1.1291	504.7	2706.2	1.5302	7.1269
0.5	424.981	915.29	2.668	640.09	2748.1	1.8604	6.8207
1	453.028	887.13	5.145	762.52	2777.1	2.1381	6.585
1.5	471.437	866.65	7.5924	844.56	2791	2.3143	6.443

2	485.527	849.8	10.042	908.5	2798.3	2.4468	6.339
3	507.003	821.9	15.001	1008.3	2803.2	2.6455	6.1856
4	523.504	798.37	20.09	1087.5	2800.8	2.7968	6.0696
5	537.091	777.37	25.351	1154.6	2794.2	2.921	5.9737
6	548.735	758	30.818	1213.9	2784.6	3.0278	5.8901
8	568.158	722.2	42.507	1317.3	2758.7	3.2081	5.745
10	584.147	688.42	55.463	1408.1	2725.5	3.3606	5.616
12	597.825	655.18	70.106	1491.5	2685.4	3.4967	5.4939
14	609.816	621.22	87.069	1571	2637.9	3.6232	5.3727
16	620.505	584.99	107.42	1649.7	2580.8	3.7457	5.2463
18	630.142	543.54	133.3	1732.1	2509.8	3.8718	5.1061
20	638.899	490.19	170.5	1827.2	2412.3	4.0156	4.9314
22	646.855	369.77	274.16	2011.3	2173.1	4.2945	4.5446
22.064	647.096	322	322	2084.3	2084.3	4.407	4.407

TABLE 2.12 Properties of Liquid Compressed Water and Superheated Steam [52]

	0.1 MPa (T_s=99.606 °C)			0.2 MPa (T_s=120.210 °C)			0.5 MPa (T_s=151.831 °C)		
T (K)	ρ (kg/m^3)	H (kJ/kg)	S (kJ/(kg K))	ρ (kg/m^3)	H (kJ/kg)	S (kJ/(kg K))	ρ (kg/m^3)	H (kJ/kg)	S (kJ/(kg K))
368.15	0.58967	398.1	1.2504						
373.15	0.58967	2675.8	7.361	958.4	419.24	1.3701	958.54	419.47	1.3069
423.14	0.51636	2776.6	7.6148	1.0418	2769.1	7.281	917.02	632.19	1.8418
473.15	0.46031	2875.5	7.8356	0.92551	2870.7	7.5081	2.3528	2855.8	7.061
523.15	0.4156	2974.5	8.0346	0.8341	2971.2	7.71	2.1078	2961	7.2724
573.15	0.37895	3074.5	8.2172	0.75975	3072.1	7.8941	1.9135	3064.6	7.4614
623.15	0.34832	3175.8	8.3866	0.69785	3173.9	8.0644	1.7539	3168.1	7.6346
673.15	0.3223	3278.6	8.5452	0.64544	3277	8.2236	1.6199	3272.3	7.7955
723.15	0.29992	3382.8	8.6946	0.60044	3381.6	8.3734	1.5056	3377.7	7.9465
773.15	0.28046	3488.7	8.8361	0.56135	3487.7	8.5152	1.4066	3484.5	8.0892
873.15	0.24827	3705.6	9.0998	0.49677	3704.8	8.7792	1.2436	3702.5	8.3343
973.15	0.22272	3929.4	9.3424	0.44557	3928.8	9.022	1.1149	3927	8.5977
1073.15	0.20194	4160.2	9.5681	0.40396	4159.8	9.2479	1.0104	4158.4	8.824

	1.0 MPa ($T_s = 179.878\ °C$)			1.5 MPa ($T_s = 198.287\ °C$)			2.0 MPa ($T_s = 212.377\ °C$)		
T (K)	ρ (kg/m^3)	H (kJ/kg)	S (kJ/(kg K))	ρ (kg/m^3)	H (kJ/kg)	S (kJ/(kg K))	ρ (kg/m^3)	H (kJ/kg)	S (kJ/(kg K))
373.15	958.77	419.84	1.3065	959	420.22	1.3061	959.24	420.59	1.3057
423.14	917.31	632.5	1.8412	917.59	632.81	1.8407	917.87	633.12	1.8401
473.15	4.8539	2828.3	6.6955	7.5498	2796	6.4536	865	852.45	2.3298
523.15	4.2965	2943.1	6.9265	6.5785	2923.9	6.7111	8.9689	2903.2	6.5475
573.15	3.8762	3051.6	7.1246	5.8925	3038.2	6.9198	7.9677	3024.2	6.7684
623.15	3.5398	3158.2	7.3029	5.3594	3148	7.1036	7.215	3137.7	6.9583
673.15	3.2615	3264.5	7.4669	4.9256	3256.5	7.271	6.6131	3248.3	7.1292
723.15	3.0262	3371.3	7.62	4.5624	3364.8	7.4262	6.1146	3358.2	7.2866
773.15	2.824	3479.1	7.7641	4.2524	3473.7	7.5718	5.6921	3468.2	7.4337
873.15	2.4931	3698.6	8.031	3.7484	3694.7	7.8405	5.0097	3690.2	7.7043
973.15	2.233	3924.1	8.2755	3.3544	3921.1	8.086	4.479	3918.2	7.9509
1073.15	2.0227	4156.1	8.5024	3.0368	4153.8	8.3135	4.0528	4151.5	8.179

	3.0 MPa ($T_s = 233.853\ °C$)			4.0 MPa ($T_s = 250.354\ °C$)			5.0 MPa ($T_s = 263.941\ °C$)		
T (K)	ρ (kg/m^3)	H (kJ/kg)	S (kJ/(kg K))	ρ (kg/m^3)	H (kJ/kg)	S (kJ/(kg K))	ρ (kg/m^3)	H (kJ/kg)	S (kJ/(kg K))
373.15	959.71	421.34	1.305	960.17	422.1	1.3042	960.63	422.85	1.3034
423.14	918.44	633.74	1.839	919	634.36	1.8379	919.56	634.98	1.8368

Continued

TABLE 2.12 Properties of Liquid Compressed Water and Superheated Steam [52]—Cont'd

	3.0 MPa (T_s=233.853 °C)			4.0 MPa (T_s=250.354 °C)			5.0 MPa (T_s=263.941 °C)		
T (K)	ρ (kg/m³)	H (kJ/kg)	S (kJ/(kg K))	ρ (kg/m³)	H (kJ/kg)	S (kJ/(kg K))	ρ (kg/m³)	H (kJ/kg)	S (kJ/(kg K))
473.15	865.7	852.86	2.3282	866.51	853.27	2.3267	867.26	853.68	2.3251
523.15	14.159	2856.5	6.2893	798.92	1085.8	2.7935	800.09	1085.7	2.791
573.15	12.318	2994.3	6.5412	16.987	2961.7	6.3639	22.053	2925.7	6.211
623.15	11.043	3116.1	6.7449	15.044	3093.3	6.5843	19.242	3069.3	6.4516
673.15	10.062	3231.7	6.9234	13.618	3214.5	6.7714	17.29	3196.7	6.6483
723.15	9.269	3344.8	7.0856	12.493	3331.2	6.9386	15.792	3317.2	6.821
773.15	8.6062	3457.2	7.2359	11.568	3466	7.0922	14.581	3434.7	6.9781
873.15	7.5503	3682.8	7.5103	10.115	3674.9	7.3705	12.706	3666.8	7.2605
973.15	6.7383	3912.2	7.759	9.0109	3906.3	7.6214	11.297	3900.3	7.5136
1073.15	6.0903	4146.9	7.9885	8.1352	4142.3	7.8523	10.188	4137.7	7.7458

	6.0 MPa (T_s=275.585 °C)			8.0 MPa (T_s=295.008 °C)			10 MPa (T_s=310.997 °C)		
T (K)	ρ (kg/m³)	H (kJ/kg)	S (kJ/(kg K))	ρ (kg/m³)	H (kJ/kg)	S (kJ/(kg K))	ρ (kg/m³)	H (kJ/kg)	S (kJ/(kg K))
373.15	961.1	423.6	1.3026	962.02	425.11	1.3011	962.93	426.62	1.2996
423.14	920.11	635.61	1.8357	921.22	636.86	1.8335	922.32	638.11	1.8313
473.15	868	854.09	2.3235	869.48	854.94	2.3205	870.94	855.8	2.3174

523.15	801.23	1085.7	2.7886	803.49	1085.7	2.7839	895.7	1085.8	2.7792
573.15	27.632	2885.5	6.0703	41.188	2786.5	5.7937	715.29	1343.3	3.2488
623.15	23.668	3043.9	6.3357	33.361	2988.1	6.1321	44.564	2924	5.9459
673.15	21.088	3178.2	6.5432	29.117	3139.4	6.3658	37.827	3097.4	6.2141
723.15	19.17	3302.9	6.7219	26.182	3273.2	6.5579	33.578	3242.3	6.4219
773.15	17.646	3423.1	6.8826	23.942	3399.5	6.7266	30.478	3375.1	6.5995
873.15	15.322	3658.7	7.1693	20.643	3642.4	7.0221	26.057	3625.8	6.9045
973.15	13.597	3894.3	7.4246	18.239	3882.2	7.2821	22.937	3870	7.1693
1073.15	12.248	4133.1	7.6582	16.39	4123.8	7.5184	20.564	4114.5	7.4085

	12 MPa (T_s = 324.675 °C)			14 MPa (T_s = 336.666 °C)			15 MPa (T_s = 342.155 °C)		
T *(K)*	ρ *(kg/m³)*	**H** *(kJ/kg)*	**S** *(kJ/(kg K))*	ρ *(kg/m³)*	**H** *(kJ/kg)*	**S** *(kJ/(kg K))*	ρ *(kg/m³)*	**H** *(kJ/kg)*	**S** *(kJ/(kg K))*
373.15	963.84	428.12	1.298	964.75	429.63	1.2965	965.2	430.39	1.2958
423.14	923.41	639.37	1.8292	924.49	640.63	1.8271	925.03	641.27	1.826
473.15	872.37	856.67	2.3144	873.8	857.55	2.3114	874.5	857.99	2.31
523.15	807.86	1085.9	2.7747	809.98	1086	2.7702	811.03	1086.1	2.768
573.15	799.55	1341.2	3.2401	723.6	1339.2	3.2319	725.55	1338.3	3.2279
623.15	58.068	2848.1	5.7609	75.577	2753.1	5.559	87.1	2693.1	5.4437

Continued

TABLE 2.12 Properties of Liquid Compressed Water and Superheated Steam [52]—Cont'd

	12 MPa (T_s=324.675 °C)			14 MPa (T_s=336.666 °C)			15 MPa (T_s=342.155 °C)		
T (K)	ρ (kg/m³)	H (kJ/kg)	S (kJ/(kg K))	ρ (kg/m³)	H (kJ/kg)	S (kJ/(kg K))	ρ (kg/m³)	H (kJ/kg)	S (kJ/(kg K))
673.15	47.38	3052	6.0764	58.003	3002.3	5.9459	63.812	2975.7	5.8819
723.15	41.41	3209.8	6.3028	49.743	3175.7	6.1946	54.121	3157.9	6.1434
773.15	37.275	3350	6.4903	44.357	3324.1	6.3932	48.014	3310.8	6.348
873.15	31.594	3608.9	6.8054	37.252	3591.8	6.7191	40.127	3583.1	6.6796
973.15	27.694	3857.7	7.0753	32.509	3845.3	6.9941	34.939	3839.1	6.9572
1073.15	24.768	4105.1	7.3173	29.003	4095.8	7.2391	31.132	4091.1	7.2037

	16 MPa (T_s=347.355 °C)			18 MPa (T_s=356.992 °C)		
T (K)	ρ (kg/m³)	H (kJ/kg)	S (kJ/(kg K))	ρ (kg/m³)	H (kJ/kg)	S (kJ/(kg K))
373.15	965.65	431.14	1.295	966.55	432.66	1.2935
423.14	925.57	641.9	1.825	926.63	643.17	1.8229
473.15	875.2	858.44	2.3085	876.59	859.35	2.3056
523.15	812.06	1086.2	2.7685	814.09	1086.4	2.7615
573.15	727.46	1337.4	3.224	731.16	1335.8	3.2164
623.15	102.4	2617	5.3045	587.2	1658.7	3.7547
673.15	70.021	2947.6	5.8179	83.924	2886.4	5.6883

723.15	58.654	3139.7	6.0937	68.239	3101.8	5.9975
773.15	51.752	3297.3	6.3046	59.488	3269.7	6.2223
873.15	43.034	3574.4	6.6421	48.945	3556.8	6.572
973.15	37.385	3832.9	6.9224	42.321	3820.4	6.8579
1073.15	33.269	4086.3	7.1703	37.566	4076.9	7.1089

	20 MPa (T_s = 365.749 °C)			22 MPa (T_s = 373.705 °C)			25 MPa		
T (K)	ρ (kg/m³)	H (kJ/kg)	S (kJ/(kg K))	ρ (kg/m³)	H (kJ/kg)	S (kJ/(kg K))	ρ (kg/m³)	H (kJ/kg)	S (kJ/(kg K))
373.15	967.44	434.17	1.292	968.32	435.68	1.2906	969.65	437.95	1.2883
423.14	927.69	644.45	1.8208	928.74	645.73	1.8187	930.3	647.66	1.8156
473.15	877.97	860.27	2.3027	879.32	861.2	2.2999	881.33	862.61	2.2956
523.15	816.09	1086.7	2.7573	818.05	1086.9	2.7532	820.92	1087.4	2.7471
573.15	734.71	1334.4	3.2091	738.13	1333	3.2021	743.02	1331.3	3.1919
623.15	600.64	1646	3.729	611.67	1635.9	3.7075	625.45	1623.9	3.6804
633.15	548.01	1740.1	3.8787	568.14	1719.4	3.8404	589.31	1698.6	3.7993
643.15	144.44	2526.5	5.1097	492.96	1842.5	4.0332	540.46	1789.8	3.9423
653.15	121.07	2659.4	5.3149	163.31	2504.5	5.0555	450.82	1935.7	4.1671
663.15	108.81	2747.2	5.4483	135.52	2643.9	5.2675	215.18	2395.7	4.86
673.15	100.5	2816.9	5.5525	121.13	2735.8	5.4051	166.54	2578.6	5.14

Continued

TABLE 2.12 Properties of Liquid Compressed Water and Superheated Steam [52]—Cont'd

	20 MPa (T_s=365.749 °C)			22 MPa (T_s=373.705 °C)			25 MPa		
T (K)	ρ (kg/m³)	H (kJ/kg)	S (kJ/(kg K))	ρ (kg/m³)	H (kJ/kg)	S (kJ/(kg K))	ρ (kg/m³)	H (kJ/kg)	S (kJ/(kg K))
683.15	94.255	2876.2	5.64	111.48	2808.4	5.5122	145.28	2687.1	5.3
693.15	89.278	2928.7	5.7163	104.28	2870	5.6018	131.94	2769.4	5.4197
703.15	85.158	2976.4	5.7847	98.582	2924.5	5.6798	122.36	2837.8	5.5176
713.15	81.652	3020.4	5.8469	93.886	2973.7	5.7494	114.96	2897.3	5.6016
723.15	78.609	3061.7	5.9043	89.907	3019.2	5.8127	108.98	2950.6	5.6759
773.15	67.598	3241.2	6.1446	76.116	3211.8	6.0705	89.744	3165.9	5.9642
873.15	54.991	3539	6.5075	61.175	3521	6.4473	70.72	3493.5	6.3637
973.15	47.318	3807.8	6.799	52.378	3795.1	6.7447	60.084	3776	6.6702
1073.15	41.895	4067.5	7.0531	46.253	4058	7.002	52.848	4043.8	6.9322

	30 MPa			40 MPa			50 MPa		
T (K)	ρ (kg/m³)	H (kJ/kg)	S (kJ/(kg K))	ρ (kg/m³)	H (kJ/kg)	S (kJ/(kg K))	ρ (kg/m³)	H (kJ/kg)	S (kJ/(kg K))
373.15	971.82	441.74	1.2847	976.1	449.33	1.2775	980.27	456.94	1.2705
423.14	932.86	650.88	1.8106	937.86	657.42	1.8008	942.7	664.02	1.7912
473.15	884.62	865.02	2.2888	890.94	870	2.2755	896.97	875.19	2.2628
523.15	825.56	1088.4	2.7373	834.29	1090.7	2.7187	842.41	1093.5	2.7013
573.15	750.66	1328.9	3.176	764.36	1325.6	3.1473	776.48	1324	3.1218
623.15	643.95	1608.8	3.6436	671.86	1588.8	3.5871	693.25	1576.1	3.5431

673.15	357.43	2152.8	4.4757	523.34	1931.4	4.1145	577.79	1874.4	4.0029
723.15	148.43	2821	5.4421	270.89	2511.8	4.9448	402.04	2284.7	4.5896
773.15	115.07	3084.7	5.7956	177.84	2906.5	5.4744	257.07	2722.6	5.1762
873.15	87.377	3446.7	6.2373	123.62	3350.4	6.017	163.72	3252.5	5.8245
973.15	73.242	3743.9	6.5598	100.71	3679.1	6.374	129.59	3614.6	6.2178
1073.15	63.99	4020	6.83	86.799	3972.6	6.6612	110.22	3925.8	6.5225

	60 MPa			70 MPa		
T *(K)*	ρ *(kg/m³)*	**H** *(kJ/kg)*	**S** *(kJ/(kg K))*	ρ *(kg/m³)*	**H** *(kJ/kg)*	**S** *(kJ/(kg K))*
373.15	984.34	464.56	1.2637	988.32	472.19	1.2569
423.14	947.38	670.69	1.782	951.93	677.43	1.773
473.15	902.74	880.55	2.2057	908.29	886.07	2.239
523.15	850.03	1096.8	2.685	857.21	1100.5	2.6696
573.15	787.39	1323.5	3.0986	797.36	1323.9	3.0773
623.15	710.88	1567.5	3.5065	726.03	1561.6	3.4748
673.15	612.42	1843.2	3.9317	638.41	1822.9	3.8779
723.15	479.51	2180.2	4.414	528.42	2123.7	4.3084
773.15	338.73	2570.3	4.9356	405.97	2466.1	4.766
873.15	206.91	3156.8	5.6527	251.58	3067.4	5.5002
973.15	159.62	3551.3	6.0814	190.39	3490.3	5.9599
1073.15	134.12	3880	6.4033	158.31	3835.7	6.2981

The viscosity defined by Equation (2.47) is called the dynamic viscosity with dimensions of Pa s or kg m^{-1} s^{-1}. The ratio of the dynamic viscosity to the mass density is called kinematic viscosity, defined as η/ρ, with units of m^2 s^{-1}.

Fluids are discerned by their behavior of viscosity in dependence on velocity gradient and shear stress. Viscosity of the so-called Newtonian fluids is independent on shear stress and velocity gradient and depends only on conditions of state (P,V,T). Water, in general, and most pure liquids, simple mixtures, and gases are Newtonian fluids. Non-Newtonian fluids, on the other hand, are characterized by a viscosity depending on either shear stress or shear rate [30].

Viscosity of water depends weakly on temperature and strongly on density. Influence of pressure on the viscosity of gases is strong only in certain ranges of temperature and pressure. In general, influence of pressure is low at high reduced temperatures ($T_r = T/T_c$) and at low reduced pressures ($P_r = P/P_c$).

Influence of pressure can be derived from a generalized presentation of viscosity, as shown in Figure 2.16 [30]. Viscosity at supercritical conditions decreases with temperature at constant pressure to a minimum and then increases with temperature. The viscosity minimum is shifted to higher

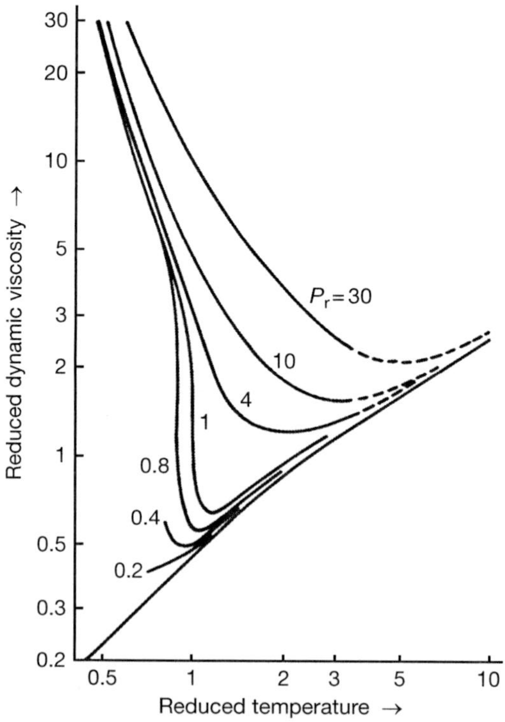

FIGURE 2.16 Generalized viscosity behavior. *After Ref. [53].*

temperatures with increasing pressure. At temperatures below the minimum, the supercritical fluid behaves, with respect to viscosity, like a liquid: Viscosity decreases with temperature. At temperatures above the minimum, the supercritical fluid behaves, with respect to viscosity, like a gas: Viscosity increases with temperature. Near the critical point, viscosity of the supercritical fluid decreases with temperature.

According to Childs and Hanley [54], a criterion, presented in Equation (2.48), can be formulated for dense gas regions, where the influence of pressure on viscosity must be taken into account. Viscosity for gases at pressures lower than that given by Equation (2.48) can be determined by normal correlations without considering pressure.

$$P/P_c > 0.183(T/T_c) - 1. \tag{2.48}$$

The influence of pressure on the value of viscosity is less than 1%. Otherwise, a correction for pressure is considered necessary. For the viscosity of water, a correlation has been developed by IAPWS, covering low- and high-density regions (see below).

Viscosities of liquids and gases are quite different. Viscosities of liquids are far higher than those of gases but decrease rapidly with increasing temperature. Viscosity of gases under ambient conditions is about $\eta \approx 10^{-5}$ Pa s. Dynamic viscosities of liquids cover a broad range. The viscosity of liquid *n*-heptane at 20 °C is: $\eta = 4 \times 10^{-4}$ Pa s, for *n*-hexadecane 35×10^{-4} Pa s, for water 10×10^{-4} Pa s, for methylene chloride 1.8×10^{-4} Pa s (all at 20 °C).

Viscosity of gases at low pressure is due to transfer of impulse energy by individual impacts between molecules, moving at random between layers of different velocity in the flowing medium. In liquids, an impulse transfer by a similar mechanism is also effective but is superimposed by the intermolecular forces of the densely packed liquid molecules. The density of liquids is sufficiently high that the average distance of two molecules is in the range of the force fields of the molecules. Based on molecular properties, so far no correlation could be developed for the calculation of the viscosity of liquids. IAPWS released a correlation for the viscosity of water, based on experimental data, applying empirical equations (Table 2.13 and Figure 2.17).

2.5.1.1 Correlations for the Viscosity of Water

Correlations for estimating viscosity are available for most substances. For water, the best correlation is the formulation developed by IAPWS, the IAPWS-2008 [55]. Water is an important calibration fluid for viscometry (the measurement of viscosity), and the International Organization for Standards (ISO) provides a standard value for the viscosity of water at $T = 20$ °C and $P = 101.325$ kPa (standard atmospheric pressure). The calibration value is 1.0016 mPa s with an uncertainty of 0.17% (ISO, 1998). This value agrees

TABLE 2.13 Dynamic Viscosity of Water

	Dynamic Viscosity Water (10^{-6} kg m^{-1} s^{-1})														
P (MPa)	*323.15 K*	*373.15 K*	*423.15 K*	*473.15 K*	*523.15 K*	*573.15 K*	*623.15 K*	*673.15 K*	*723.15 K*	*773.15 K*	*873.15 K*	*973.15 K*	*1073.15 K*	*1173.15 K*	*1273.15 K*
0.1	547.1	12.27	14.18	16.18	18.22	20.29	22.37	24.45	26.52	28.57	32.61	36.55	40.37	44.08	47.66
0.5	547.1	282	182.5	16.05	18.14	20.24	22.34	24.44	26.52	28.58	32.63	36.57	40.39	44.09	47.68
1	547.2	282.1	182.7	15.89	18.04	20.18	22.31	24.42	26.51	28.58	32.64	36.59	40.41	44.11	47.69
2	547.4	282.4	182.9	134.5	17.85	20.07	22.25	24.4	26.51	28.6	32.68	36.63	40.46	44.15	47.73
3	547.6	282.7	183.2	134.7	17.67	19.97	22.2	24.38	26.52	28.62	32.72	36.68	40.5	44.2	47.77
4	547.8	282.9	183.4	135	106.2	19.88	22.16	24.37	26.53	28.65	32.76	36.73	40.55	44.24	47.8
5	547.9	283.2	183.7	135.2	106.5	19.8	22.12	24.37	26.55	28.68	32.81	36.78	40.6	44.29	47.84
6	548.1	283.5	183.9	135.5	106.7	19.73	22.1	24.37	26.57	28.71	32.86	36.83	40.65	44.33	47.89
7	548.3	283.7	184.2	135.7	107	19.68	22.09	24.39	26.6	28.75	32.91	36.89	40.71	44.38	47.93
8	548.5	284	184.4	136	107.3	19.65	22.09	24.41	26.64	28.8	32.96	36.94	40.76	44.44	47.97
9	548.7	284.3	184.7	136.2	107.6	86.12	22.11	24.44	26.68	28.85	33.02	37.01	40.82	44.49	48.02
10	548.9	284.5	184.9	136.5	107.8	86.52	22.15	24.48	26.73	28.91	33.09	37.07	40.88	44.54	48.07
15	549.8	285.9	186.2	137.7	109.2	88.4	22.94	24.93	27.13	29.29	33.46	37.42	41.21	44.84	48.33
20	550.8	287.2	187.4	138.9	110.5	90.12	69.33	26.03	27.81	29.85	33.92	37.84	41.58	45.17	48.62
25	551.8	288.5	188.6	140	111.7	91.72	72.81	29.18	28.96	30.64	34.49	38.32	42	45.54	48.95
30	552.9	289.9	189.9	141.2	112.9	93.22	75.51	44	30.85	31.73	35.17	38.86	42.47	45.95	49.3

35	554	291.2	191.1	142.3	114.1	94.64	77.79	55.79	34.03	33.19	35.97	39.47	42.97	46.38	49.68
40	555.1	292.5	192.3	143.5	115.3	95.99	79.8	61.31	39.02	35.12	36.9	40.13	43.52	46.85	50.08
45	556.2	293.9	193.4	144.6	116.4	97.29	81.6	65.07	45.06	37.56	37.95	40.86	44.1	47.34	50.51
50	557.4	295.2	194.6	145.7	117.5	98.53	83.27	68.02	50.5	40.49	39.14	41.65	44.72	47.86	50.95
60	559.9	297.9	197	147.8	119.6	100.9	86.26	72.65	58.45	47.03	41.85	43.38	46.04	48.95	51.89
70	562.4	300.5	199.3	149.9	121.7	103.1	88.93	76.33	63.98	53.13	44.93	45.3	47.47	50.12	52.88
80	565.1	303.2	201.5	152	123.7	105.2	91.37	79.46	68.26	58.22	48.2	47.34	48.98	51.34	53.92
90	567.9	305.8	203.8	154	125.6	107.2	93.62	82.22	71.8	62.46	51.49	49.54	50.54	52.59	54.97
100	570.8	308.4	206	156	127.5	109.1	95.73	84.71	74.86	66.07	54.67	51.61	52.12	53.86	56.04

Data IAPWS, cited from WA [52].

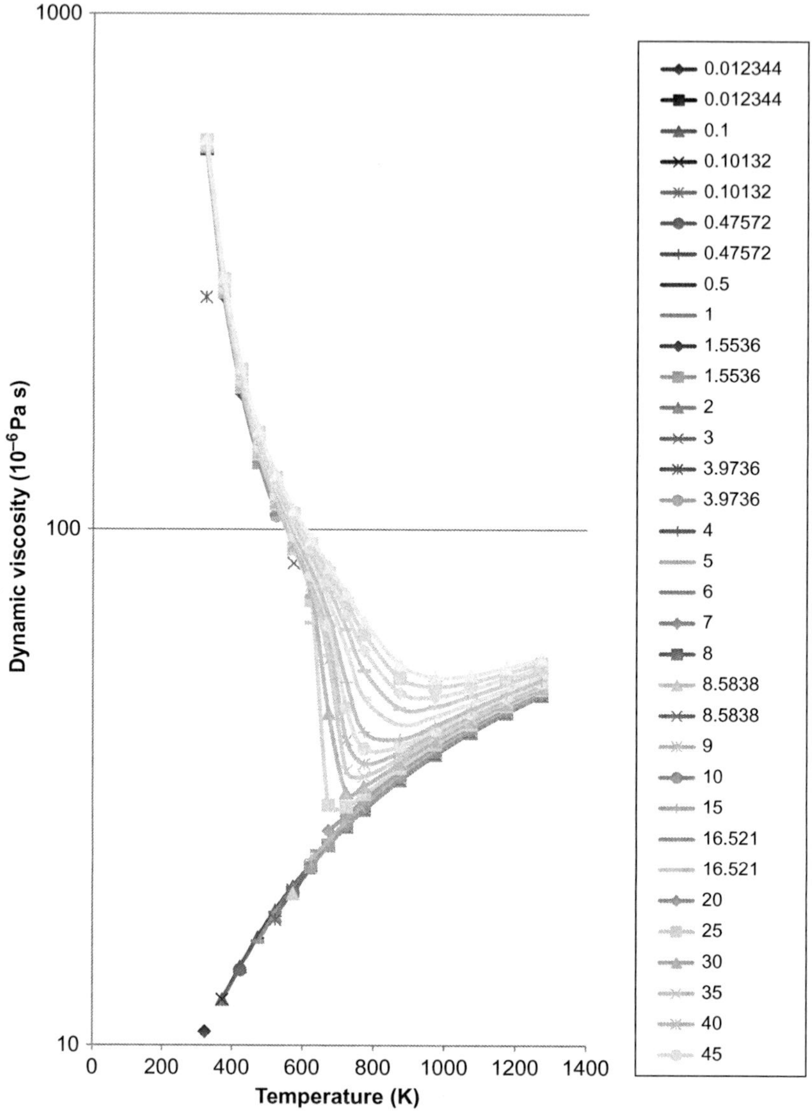

FIGURE 2.17 Viscosity of water at various pressures [MPa]. *Data IAPWS-2008, cited from Ref. [52].* (For color version of this figure, the reader is referred to the online version of this chapter.)

with the result calculated from the IAPWS viscosity formulation. The uncertainty in the IAPWS formulation is specified by tables giving the uncertainty of the viscosity for a grid of temperature–pressure state points as determined from the available data. In general, these uncertainties range from about 1% to 3% in the broad region covered by the formulation [4].

2.5.1.2 Recommended Correlating Equation for the Viscosity of Water

The viscosity is represented by Equation (2.49):

$$\eta_r = \eta_{r,0}(T_r) \times \eta_{r,1}(T_r, \rho_r) \times \eta_{r,2}(T_r, \rho_r) \tag{2.49}$$

The first factor, $\eta_{r,0}$, of the product represents the viscosity in the dilute-gas limit and is given by

$$\eta_{r,0}(T_r) = \frac{100\sqrt{T_r}}{\sum_{i=0}^{3} \frac{H_i}{T_r^i}} \tag{2.50}$$

with coefficients H_i given in Table 2.14. The second factor, $\eta_{r,1}$, represents the contribution to viscosity due to finite density:

$$\eta_{r,1}(T_r, \rho_r) = \exp\left[\rho_r \sum_{i=0}^{5} \left(\frac{1}{T_r} - 1\right)^i \sum_{j=0}^{6} \left[H_{ij}(\rho_r - 1)^j\right]\right] \tag{2.51}$$

with coefficients H_{ij} given in Table 2.15. The third factor, $\eta_{r,2}$, represents the critical enhancement of the viscosity.

Values used in the correlation are:

- T is the absolute temperature (International Temperature Scale of 1990), ρ is the density, P is the pressure, and η is the viscosity.
- Reference constants: Critical constants according to the IAPWS release [55]; viscosity: $\eta^0 = 1.00 \times 10^{-6}$ Pa s.
- Dimensionless variables: $T_r = T/T_c$, $\rho_r = \rho/\rho_c$, $P_r = P/P_c$, $\eta_r = \eta/\eta^0$.
- Range of validity: $0 < P < P_t$ and $273.16\ \text{K} \leq T \leq 1173.15\ \text{K}$, $P_t \leq P \leq 300$ MPa, and $T_m(P) \leq T \leq 1173.15$ K.

T_m is the pressure-dependent melting temperature and P_t is the triple-point pressure. IAPWS-95 should be used to determine the densities used as input to Equation (2.49).

TABLE 2.14 Coefficients H_i for $\eta_{r,0}(T_r)$

i	H_i
0	1.67752
1	2.20462
2	0.6366564
3	−0.241605

TABLE 2.15 Coefficients H_{ij} for $\eta_{r,1}(T_r,\rho_r)$

i	j	H_{ij}
0	0	5.20094×10^{-1}
1	0	8.50895×10^{-2}
2	0	-1.08374
3	0	-2.89555×10^{-1}
0	1	2.22531×10^{-1}
1	1	9.99115×10^{-1}
2	1	1.88797
3	1	1.26613
5	1	1.20573×10^{-1}
0	2	-2.81378×10^{-1}
1	2	-9.06851×10^{-1}
2	2	-7.72479×10^{-1}
3	2	-4.89837×10^{-1}
4	2	-2.57040×10^{-1}
0	3	1.61913×10^{-1}
1	3	2.57399×10^{-1}
0	4	-3.25372×10^{-2}
3	4	6.98452×10^{-2}
4	5	8.72102×10^{-3}
3	6	-4.35673×10^{-3}
5	6	-5.93264×10^{-4}

The critical enhancement is only significant in a very small region of density and temperature around the critical point. Although exactly at the critical point the viscosity is infinite, the enhancement term contributes an amount greater than 2% of the full viscosity only within the following boundaries: $645.91\ \mathrm{K} < T < 650.77\ \mathrm{K}$, $245.8\ \mathrm{kg\ m^{-3}} < \rho < 405.3\ \mathrm{kg\ m^{-3}}$. Because the critical enhancement is insignificant except in this small region around the critical point, complexity and computing time may be reduced by omitting the critical enhancement for applications outside this region. This can be done by setting $\mu_2 = 1$ (IAPWS-2008) [55].

2.5.2 Thermal Conductivity

The thermal conductivity of a fluid is a measure of its ability to transfer heat via a temperature gradient. The quantity of energy flowing by conduction is proportional to the temperature gradient (Fourier-law of conduction).

$$J_{\mathrm{Q}} = -\lambda \frac{\mathrm{d}T}{\mathrm{d}z} \tag{2.52}$$

where J_{Q} is the heat flux (energy per unit time and unit area), λ is the thermal conductivity, and $\mathrm{d}T/\mathrm{d}z$ is the temperature gradient. The thermal conductivity is considered independent of the magnitude of the temperature gradient but depends on temperature and density. The units of thermal conductivity are Watt per meter per degree Kelvin ($\mathrm{Wm^{-1}\,K^{-1}}$) [4].

Thermal conductivity of gases is connected with the molecular transport of energy and is therefore also connected to viscosity and heat capacity. With increasing pressure, thermal conductivity of gases increases at constant temperature. The influence of pressure is low up to moderate pressures. An overview is presented in Figure 2.19. Beyond normal variation with pressure and temperature, values for thermal conductivity show great variations with pressure and temperature in the critical region. Thermal conductivity at the critical point is infinite. In general, at supercritical conditions thermal conductivity of water increases with pressure at constant temperatures. Variations are great in the critical region. With increasing temperature, thermal conductivity of water decreases, runs through a minimum, and then increases with temperature. The minimum is shifted with increasing pressure to higher temperatures (Figure 2.19).

While thermal conductivity of most organic liquids is in the range of $\lambda = 0.1\text{–}0.2\ \mathrm{W\,m^{-1}\,K^{-1}}$ at temperatures below boiling temperature, water and other strongly polar molecules have a higher thermal conductivity by a factor of 2–3. Thermal conductivity of liquids is not strongly influenced by pressure. In general, thermal conductivity of liquids increases up to 100 MPa by a factor of 2–3 [30].

Thermal conductivity for water is presented in the following in diagrams and data tables. In Figure 2.18 and Table 2.16, the thermal conductivity of water at saturation conditions is presented. Source of the data is the release on thermal conductivity published by the IAPWS in 2008 (IAPWS-2008) [56].

The figures show the general behavior of the thermal conductivity of water. Thermal conductivity of gaseous water increases moderately with temperature along the vapor and liquid saturation boundaries. In the critical region, the increase becomes steeper with a critical divergence to infinity at the critical point. Thermal conductivity of the liquid is unusual. While most fluids show a decrease in thermal conductivity with increase of temperature along the saturated liquid (well below the critical temperature), water exhibits a maximum value near $T = 400$ K. Below this temperature, the isobaric

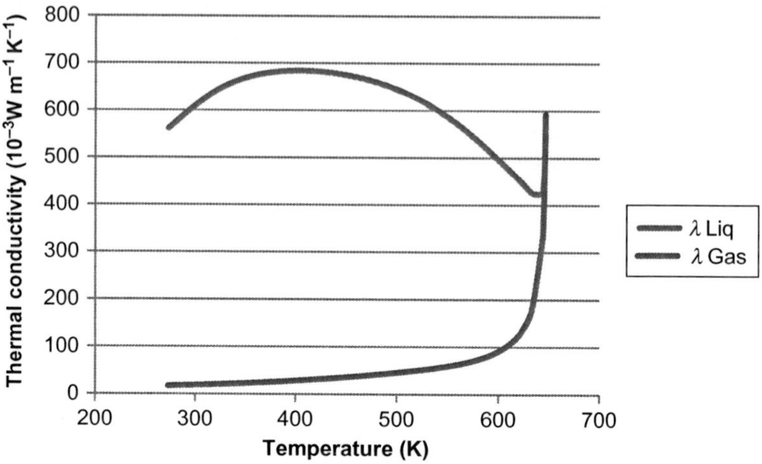

FIGURE 2.18 Thermal conductivity of water at saturation conditions. *Data: IAPWS-2008 [56].* (For color version of this figure, the reader is referred to the online version of this chapter.)

thermal conductivity versus temperature curves have a positive slope. This behavior may be related to the temperature dependence of the hydrogen-bonding phenomenon. Thermal conductivity increases steadily with pressure at constant temperature (Figure 2.19). The variation in the extended critical region is remarkable and must be considered closely for applications.

2.5.2.1 *Correlation of Thermal Conductivity of Water*

Commonly used correlations are not really suitable for calculating the thermal conductivity of water. Under the guidance of IAPWS, standard formulations for the thermal conductivity of water as a function of temperature and density have been developed and published (IAPWS, 1998) [56]. Two sets of equations for the thermal conductivity have been proposed: a computationally simpler formulation for industrial use, and the more complicated and computer-time-consuming equations for general and scientific use. Since this chapter is oriented to applications, only the formulation for industrial use is presented. It is sufficient. Accuracy is generally estimated as 1.5–4% with higher deviations near the critical point [56]. Thermal conductivity is represented by the following equation:

$$\lambda_r = \lambda_{r,0}(T_r) + \lambda_{r,1}(\rho_r) + \lambda_{r,2}(T_r, \rho_r) \tag{2.53}$$

The formulation is presented in reduced variables: $T_r = T/T_c$; $\rho_r = \rho/\rho^0$; $\lambda_r = \lambda/\lambda^0$, with $T_c = 647.26$ K, $\rho^0 = 317.7$ kg m^{-3}; $\lambda^0 = 1$ W m^{-1} K^{-1}. The reference values differ from the commonly applied critical values for calculating reduced variables, with T as absolute temperature according to the International Temperature Scale of 1990, ρ density, and λ thermal conductivity.

TABLE 2.16 Thermal Conductivity of Water at Saturation

T (K)	P (MPa)	λ_{Liq}	λ_{Gas}	T (K)	P (MPa)	λ_{Liq}	λ_{Gas}
273.16	0.0006177	561	17.07	483.15	1.908	657	42.09
283.15	0.001228	580	17.62	493.15	2.32	649.7	44.17
293.15	0.002339	598.4	18.23	503.15	2.797	641.3	46.38
303.15	0.004247	615.4	18.89	513.15	3.347	631.8	48.73
313.15	0.007385	630.6	19.6	523.15	3.976	621.2	51.27
323.15	0.01235	643.6	20.36	533.15	4.692	609.2	54.04
333.15	0.01995	654.3	21.19	543.15	5.503	595.9	57.12
343.15	0.0312	663.1	22.07	553.15	6.417	581.1	60.62
353.15	0.04741	670	23.01	563.15	7.442	565	64.72
363.15	0.07018	675.3	24.02	573.15	8.588	547.4	69.67
373.15	0.1014	679.1	25.1	583.15	9.865	528.8	75.86
383.15	0.1434	681.7	26.24	593.15	11.284	509.2	83.94
393.15	0.1987	683.2	27.47	603.15	12.858	489.1	95
403.15	0.2703	683.7	28.76	613.15	14.601	468.6	110
413.15	0.3615	683.3	30.14	623.15	16.529	447.4	134.1
423.15	0.4762	682	31.6	633.15	18.666	425.8	181.8
433.15	0.6182	680	33.13	643.15	21.044	425.1	324.5
443.15	0.7922	677	34.75	644.15	21.297	438.5	368.1
453.15	1.0028	673.3	36.45	645.15	21.554	467.5	439.1
463.15	1.255	668.8	38.24	646.15	21.814	548.2	592.2
473.15	1.555	663.3	40.11				

Source: IAPWS-2008 [56].

The range of validity is:

$273.15\ \text{K} \le T \le 773.15\ \text{K},\ P \le 100\ \text{MPa};$
$773.15\ \text{K} < T \le 923.15\ \text{K},\ P \le 70\ \text{MPa};$
$923.15\ \text{K} < T \le 1073.15\ \text{K},\ P \le 40\ \text{MPa}.$

The term $\lambda_{r,0}(T_r)$ represents the thermal conductivity of steam in the ideal-gas limit and has the form

$$\lambda_{r,0}(T_r) = \sqrt{T_r}\sum_{k=0}^{3} a_k T_r^k, \tag{2.54}$$

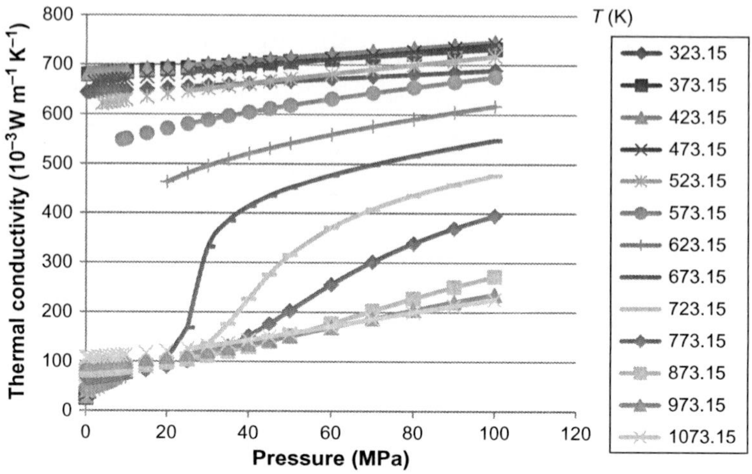

FIGURE 2.19 Thermal conductivity of water. *Data source: IAPWS-2008 [56]*. (For color version of this figure, the reader is referred to the online version of this chapter.)

with the coefficients a_k: $a_0=0.0102811$; $a_1=0.0299621$; $a_2=0.0156146$; $a_3=-0.00422464$.

The function $\lambda_{r,1}(\rho_r)$ is defined by

$$\lambda_{r,1}(\rho_r)=b_0+b_1\rho_r+b_2\exp\left[B_1(\rho_r+B_2)^2\right], \tag{2.55}$$

with coefficients b_i and B_i: $b_0=-0.397070$; $b_1=0.400302$; $b_2=1.06$; $B_1=-0.171587$; $B_2=2.392190$.

The function $\lambda_{r,2}(T_r,\rho_r)$ is defined by

$$\lambda_{r,2}(T_r,\rho_r)=\left(\frac{d_1}{T_r^{10}}+d_2\right)\rho_r^{9/5}\exp\left[C_1\left(1-\rho_r^{14/5}\right)\right] + d_3 S\rho_r^{Q}\exp\left[\left(\frac{Q}{1+Q}\right)\left(1-\rho_r^{1+Q}\right)\right]+d_4\exp\left(C_2T_r^{3/2}+\frac{C_3}{\rho_r^5}\right), \tag{2.56}$$

with Q and S are given by:

$$\Delta T_r=|T_r-1|+C_4, \tag{2.57}$$

with

$$Q=2+\frac{C_5}{\Delta T_r^{3/5}} \tag{2.58}$$

and

$$S=\begin{cases}\dfrac{1}{\Delta T_{\mathrm{r}}} & \text{for } T_{\mathrm{r}}\geq 1\\ \dfrac{C_6}{\Delta T_{\mathrm{r}}^{3/5}} & \text{for } T_{\mathrm{r}}<1\end{cases} \tag{2.59}$$

with coefficients d_i and C_i: $d_1=0.0701309$; $d_2=0.0118520$; $d_3=0.00169937$; $d_4=-1.02$; $C_1=0.642857$; $C_2=-4.11717$; $C_3=-6.17937$; $C_4=0.00308976$; $C_5=0.0822994$; $C_6=10.0932$.

Thermal conductivity of water shows large increases in a relatively broad region around the critical point, as can be seen from the isotherms in Figure 2.19. These increases are represented in the correlations for the thermal conductivity, as well as in the numerical database (see Table 2.17). Thus, it is considered relevant for practical calculations. For accurate calculations of the thermal conductivity in the vicinity of the critical point, the interpolating equation for general and scientific use should be adopted as published in the IAPWS release (IAPWS-2008) [56].

2.5.3 Diffusion

Diffusion is transport of matter without convection or mechanically induced mixing. If for our purposes we restrict diffusion to being induced only by concentration differences, then the diffusion coefficient is the proportionality factor between the diffusion potential (concentration difference) and diffusive flow (mass flow). Practical interest is mainly on the diffusion of solutes in water. While viscosity and thermal conductivity of gases and liquids differ by one to two orders of magnitude, diffusion coefficients of gases and liquids differ by four orders of magnitude:

$$\frac{\lambda_{\mathrm{L}}}{\lambda_{\mathrm{G}}}=10\text{–}100;\quad \frac{\eta_{\mathrm{L}}}{\eta_{\mathrm{G}}}=10\text{–}100;\quad \frac{D_{\mathrm{L}}}{D_{\mathrm{G}}}=10^{-4}. \tag{2.60}$$

Up to moderate pressures, the diffusion coefficient of gases is reversely proportional to pressure or density. At higher pressures, mostly self-diffusion coefficients have been determined by experiment. They have been correlated by Dawson *et al.* [30,57]:

$$\frac{D\cdot\rho}{(D\cdot\rho)^0}=1+0.053432\rho_{\mathrm{r}}-0.030182\rho_{\mathrm{r}}^2-0.029725\rho_{\mathrm{r}}^3, \tag{2.61}$$

$$0.8<T_{\mathrm{r}}<1.9;\ \ 0.3<P_{\mathrm{r}}<7.4, \tag{2.62}$$

with D = self-diffusion coefficient at T and ρ, ρ = density; $(D\cdot\rho)^0=D\rho$ at T, but low pressure; and ρ_{r} = reduced density.

At higher densities, D decreases markedly. In this range, the presented correlation is only a rough approximation.

TABLE 2.17 Thermal Conductivity of Water (10^{-3} Wm^{-1} K^{-1})

	T (K)												
P (MPa)	323.15	373.15	423.15	473.15	523.15	573.15	623.15	673.15	723.15	773.15	873.15	973.15	1073.15
0.1	643.6	25.08	28.85	33.28	38.17	43.42	48.97	54.76	60.77	66.98	79.9	93.38	107.3
0.5	643.8	679.3	682.1	34.93	39.18	44.09	49.45	55.14	61.09	67.26	80.14	93.6	107.5
1	644	679.6	682.4	37.21	40.52	44.96	50.07	55.61	61.49	67.61	80.44	93.88	107.8
2	644.7	680.2	683.1	663.8	43.49	46.81	51.37	56.62	62.31	68.33	81.07	94.45	108.3
3	644.9	680.7	683.8	664.7	47	48.87	52.78	57.69	63.19	69.1	81.73	95.06	108.8
4	645.4	681.2	684.5	665.5	621.4	51.19	54.31	58.83	64.12	69.9	82.41	95.69	109.4
5	645.8	681.8	685.1	666.3	622.5	53.85	55.99	60.06	65.1	70.74	83.13	96.35	109.9
6	646.3	682.3	685.8	667.3	624	57	57.85	61.38	66.14	71.62	83.87	97.02	110.5
7	646.8	682.9	686.5	668.1	625.2	60.83	59.92	62.81	67.24	72.55	84.64	97.72	111.2
8	647.2	683.4	687.1	669	626.5	65.77	62.27	64.35	68.42	73.52	85.44	98.44	111.8
9	647.7	684	687.8	669.8	627.7	548.6	64.96	66.04	69.66	74.54	86.27	99.19	112.5
10	648.2	684.5	688.4	670.6	628.8	550.7	68.09	67.88	70.99	75.61	87.14	99.98	113.2
15	650.5	687.2	691.8	674.9	634.8	561.4	100.9	80.66	79.16	81.83	91.92	104.2	116.9
20	652.8	689.9	695.1	679	640.5	571.3	463.2	105.4	90	89.84	97.55	109.1	121.2
25	655.2	692.6	698.4	683.2	646.1	580.3	481.1	168	108.8	100.2	104.1	114.6	126

30	657.5	695.3	701.7	687.2	651.6	588.8	496	331.4	136	113.6	111.7	120.8	131.3
35	659.8	698	704.9	691.3	656.8	596.8	509	384.3	176.1	130.6	120.3	127.5	137
40	662.1	700.7	708.2	695.3	662	604.3	520.7	413.9	226.6	151.5	129.9	134.8	143.2
45	664.4	703.3	711.4	699.2	667	611.5	531.5	434.8	275.1	176	140.5	142.6	149.7
50	666.7	706	714.6	703.1	671.9	618.3	541.5	451.3	314.9	202.7	152	150.9	156.6
60	671.3	711.2	721	710.9	681.5	631.2	559.6	477.4	370.5	255.5	176.9	168.5	170.7
70	675.8	716.4	727.2	718.5	690.7	643.2	575.9	498.5	408	301.3	202.7	186.7	185.2
80	680.3	721.6	733.5	726	699.7	654.4	590.7	516.8	435.7	338.8	227.6	204.3	199.5
90	684.7	726.7	739.7	733.4	708.5	665.1	604.3	533.2	457.7	369.5	250.8	220.9	213
100	689	731.7	745.8	740.8	717.1	675.3	617	548.2	476.2	394.9	272.1	236	225.4

Source: IAPWS-2008 [56].

Diffusion in dense fluids can be connected with viscosity by the Stokes–Einstein equation. According to this equation, the self-diffusion coefficient for molecules with diameter σ in a medium of viscosity η is obtained from:

$$D = \frac{kT}{C_{SE}\pi\sigma\eta} \tag{2.63}$$

where k is the Boltzmann factor and C_{SE} is the Stokes–Einstein constant. Experimental self-diffusion coefficients indicate that the Stokes–Einstein relation is applicable if σ is not considered as constant (see below).

The self-diffusion coefficient of pure water near atmospheric pressure at $T = 25\ °C$ is approximately $2.3 \times 10^{-9}\ m^2\ s^{-1}$ [58]. Experimental data have been published by Krynicki *et al.* [58], Holz *et al.* [59], Easteal (1984) [60], Easteal *et al.* [61], Weingärtner [62], and Lamb *et al.* [63]. Boero *et al.* [64] determined the self-diffusion coefficient of supercritical water using a first-principles molecular dynamics technique. Their results at $T = 673$ K for the self-diffusion coefficient of water are: $D_w = (46.2 \pm 0.6) \times 10^{-9}\ m^2\ s^{-1}$ at a density of $\rho = 730\ kg\ m^{-3}$ and $D_w = (103.5 \pm 2.1) \times 10^{-9}\ m^2\ s^{-1}$ at a density of $\rho = 320\ kg\ m^{-3}$ agree with the values reported by Lamb *et al.* [63], as discussed by Corti *et al.* [65].

Experimental data are shown in Figure 2.20 for self-diffusion coefficients of water below the critical temperature, in Figure 2.21 for self-diffusion coefficients for subcritical and supercritical temperatures, in Table 2.18, corresponding to Figure 2.20, for self-diffusion coefficients of water below the critical temperature, and in Table 2.19 for self-diffusion coefficients of water above the critical temperature.

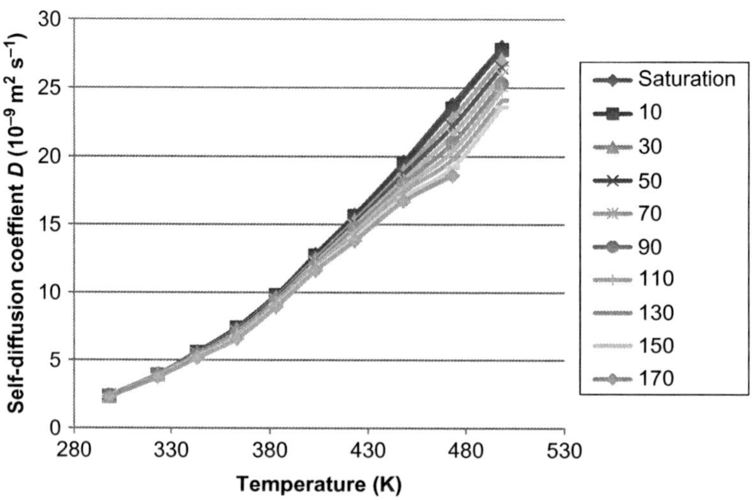

FIGURE 2.20 Self-diffusion coefficient of water below the critical temperature at different pressures [MPa]. *Data from Krynicki* et al. *[58], see Table 2.18.* (For color version of this figure, the reader is referred to the online version of this chapter.)

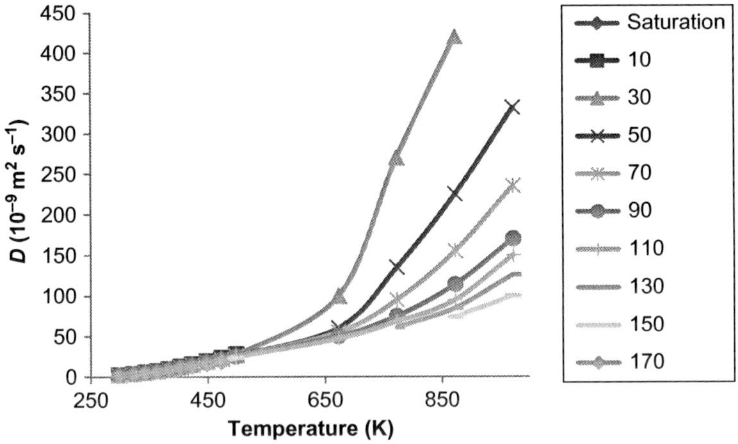

FIGURE 2.21 Self-diffusion coefficient of water at subcritical and supercritical temperatures at different pressures in MPa. Data above T_c have been interpolated from values published by Lamb *et al.* [63]. *Data from Krynicki* et al. *[58] (below* T_c*) and from Lamb* et al. *[63] (above* T_c*), see Tables 2.18 and 2.19.* (For color version of this figure, the reader is referred to the online version of this chapter.)

2.5.3.1 Experimental Data for the Self-Diffusion Coefficient of Water

The self-diffusion coefficient of water increases with temperature and decreases with pressure. The increase with temperature is monotonous in the subcritical and the supercritical region with a smooth transition between these regions. At supercritical conditions, the self-diffusion coefficient of water goes up to high values, thus favoring transport processes in this region.

2.5.3.2 Correlations for the Self-Diffusion Coefficient of Water

Self-diffusion coefficients of water can be correlated in the temperature range between $T=0$ and 100 °C applying the following equation (Equation 2.64) with an accuracy better than 1% [59]:

$$D = D_0[(T/T_s) - 1]^{\gamma} \tag{2.64}$$

with $D_0 = 1.635 \times 10^{-8} \pm 2.242 \times 10^{-11}$ ($m^2\,s^{-1}$), $T_s = 215.05 \pm 1.20$(K), $\gamma = 2.063 \pm 0.051$.

From ambient temperature to about $T = 500$ K, Krynicki *et al.* [58] found that at constant pressure the term $D\eta/T$ of Equation (2.63) is constant within the limits of the experimental error:

$$\frac{D\eta}{T} = \text{const.} = (6.9 \pm 0.4)\,10^{-15}\,\text{N}\,\text{K}^{-1}\,(\text{at}\,P = \text{const.}). \tag{2.65}$$

Marcus [66] calculated for supercritical water, under the assumption of $C_{SE} = \text{const.} = 4$ (slipping limit), the dependence of the hydrodynamic

TABLE 2.18 Self-Diffusion Coefficient of Water Below the Critical Temperature

	T (K)			Self-Diffusion Coefficient of H_2O (10^{-9} m^2 s^{-1})						
P (MPa)	298.15	323.15	343.15	363.15	383.15	403.15	423.15	448.15	473.15	498.15
Saturation	2.3	3.89	5.61	7.42	9.81	12.8	15.7	19.6	23.8	28
10	2.31	3.9	5.6	7.4	9.78	12.7	15.6	19.4	23.5	27.8
30	2.34	3.92	5.55	7.28	9.66	12.5	15.3	19	22.8	27.1
50	2.37	3.95	5.5	7.2	9.57	12.3	15.1	18.7	22.2	26.5
70	2.39	3.95	5.44	7.09	9.44	12.2	14.9	18.3	21.6	25.9
90	2.39	3.94	5.4	6.97	9.38	12	14.7	18	21	25.3
110	2.38	3.93	5.31	6.89	9.21	11.9	14.4	17.7	20.4	24.8
130	2.36	3.89	5.28	6.79	9.13	11.8	14.2	17.4	19.8	24.1
150	2.33	3.8	5.21	6.67	9.01	11.7	14	17	19.2	23.6
170	2.3	3.74	5.15	6.59	8.92	11.6	13.8	16.7	18.6	

Data from Krynicki *et al.* [58].

TABLE 2.19 Self-Diffusion Coefficient of Water at Supercritical Temperatures

	T (K)			Self-Diffusion Coefficient H_2O (10^{-9} m^2 s^{-1})			
P (MPa)	673.15		773.15		873.15		973.15
19.9	286	25.6	326	33.6	361	39.7	471
22.1	243	31.4	251	40.3	282	48	346
24.1	213	35.9	213	46.7	242	56.8	285
26.5	161	40.3	189	58.4	194	73.1	226
28	120	49.1	141	69.6	156	89.3	172
29.1	112	55.1	125	80.9	125	106	155
31.4	89.7	69.1	95.6	107.4	97	145.5	108
37.8	70	91	74.5	145.9	76.4		
56.5	57.1	131.3	61.7				
105.6	47.4						

Data from Lamb *et al.* [63].

diameter of the water molecule σ on pressure and temperature in the supercritical region from $T=673$ to 973 K:

$$\sigma = \frac{2k_B T}{\pi C_{SE} D\eta} = 427 - 0.367\,T + (-0.125 + 0.00116\,T)P, \qquad (2.66)$$

with σ in pm, T in K, P in MPa.

The hydrodynamic diameter σ calculated by Marcus using Equation (2.66) (under the assumption of $C_{SE}=\text{const.}=4$) was in fair agreement with the diameter calculated with Equation (2.63) from experimental data. A detailed discussion on the self-diffusion of water in connection with the molecular structure and hydrogen bonding of water has been reported by Marcus [66].

2.6 ELECTRICAL PROPERTIES

2.6.1 Electrical Conductivity of Water

Electrical conductivity Λ is defined as the inverse of the specific resistance r: $\Lambda = 1/r$ in units $(\Omega\ \text{cm})^{-1}$ or S cm^{-1} with $S = 1/\Omega$. Pure water is able to conduct electricity to a very slight extent due to its self-ionization. The conductivity of water is commonly used as a measure of its purity. Perfect purity corresponds to a conductivity of approximately $\Lambda = 0.055\ \mu\mathrm{S\ cm^{-1}}$ for the liquid at $T = 25\ °\mathrm{C}$ [4]. The electrical conductivity of pure water as a function of temperature and pressure can be obtained from the known values of K_w, the ion product of water (see below). Table 2.20 summarizes the predictions up to $T = 673$ K and $P = 1000$ MPa in the high-density region. In Figure 2.22, values for the electrical conductivity are presented graphically.

2.6.1.1 Correlation of Electrolytic Conductivity of Water

IAPWS has released a guideline on electrolytic conductivity in 1990 [67]. The dimensionless electrolytic conductivity (κ_r) of water is calculated from:

$$\kappa_r = 10^{-3} \Lambda_{0,r} K_{w,r}^{1/2} \rho_r \tag{2.67}$$

with dimensionless variables: $\kappa_r = \kappa/\kappa^{ref}$, $\Lambda_{0,r} = \Lambda_0/\Lambda^{ref}$, $K_{w,r} = K_w/K_w^{ref}$, $\rho_r = \rho/\rho^{ref}$, κ is the electrolytic conductivity (specific conductance) of water, K_w is the ion product of water, ρ is the density of water, and Λ_0 is the molar conductivity of H^+ plus OH^- ions in water.

Reference values: $\kappa^{ref} = 10^2\,\mathrm{S\,m^{-1}}$, $\Lambda^{ref} = 10^{-4}\,\mathrm{m^2\,S\,mol^{-1}}$, $K_w^{ref} = 1\ (\mathrm{mol\ kg^{-1}})^2$, $\rho_{ref} = 10^3\ \mathrm{kg\ m^{-3}}$.

The ion product of water (K_w) at a given density and temperature can be obtained from the Release on the Ionization Constant of H_2O [68], see below. The dimensionless molar conductivity ($\Lambda_{0,r}$) is calculated from:

$$\Lambda_{0,r} = \left(\rho_{h,r-\rho_r}\right) \Lambda_{\infty,r} / \rho_{h,r} \tag{2.68}$$

with

TABLE 2.20 Specific Conductivity ($\mu\mathrm{S\ cm^{-1}}$) of Sub- and Supercritical Water [65]

P (MPa)	298 K	373 K	473 K	573 K	673 K
Saturation	0.055	0.765	2.99	2.41	
50	0.0686	0.942	4.08	4.87	1.17
100	0.0836	1.13	5.22	7.8	4.91
200	0.117	1.53	7.65	14.1	14.3
400	0.194	2.45	13.1	28.9	39.2
600	0.291	3.51	19.5	46.5	71.3
800	0.416	4.67	26.7	66.9	110
1000		5.92	34.8	90.2	155

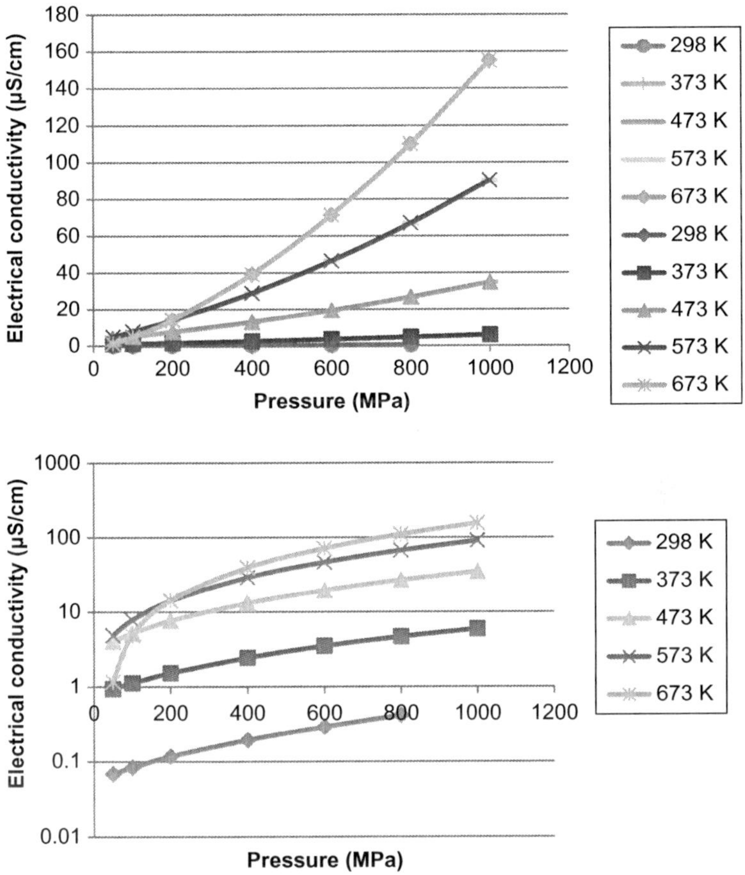

FIGURE 2.22 Electrolytic conductivity of water. *Data from Corti* et al. *[65]*. (For color version of this figure, the reader is referred to the online version of this chapter.)

$$\Lambda_{\infty,\mathrm{r}} = A_0 - \left[A_1^{-1} + \sum_{i=0}^{3} A_{i+2} t_\mathrm{r}^{i+1}\right]^{-1} \tag{2.69}$$

$$\rho_{\mathrm{h,r}} = B_0 - \left[B_1^{-1} + \sum_{i=0}^{2} B_{i+2} t_\mathrm{r}^{i+1}\right]^{-1} \tag{2.70}$$

with dimensionless variable: $t_\mathrm{r} = (T - T^0)/t^0$, t = Celsius temperature of water, $t = T - T^0$, with T the absolute temperature. The superscripts "0" denote the following reference values: $t^0 = 1$ °C, $T^0 = 273.15$ K. The dimensionless variables are listed in Table 2.21. The range of application is $0 < t$ (°C) < 800, $600 < \rho$ (kg m^{-3}) < 1200, with a maximum pressure of $P = 1000$ MPa. The

TABLE 2.21 Coefficients for Equations (2.69) and (2.70)

A_0	1850.	B_0	16.0
A_1	1410.	B_1	11.6
A_2	2.16417×10^{-6}	B_2	3.26×10^{-4}
A_3	1.81609×10^{-7}	B_3	-2.30×10^{-6}
A_4	-1.75297×10^{-9}	B_4	1.10×10^{-8}
A_5	7.20708×10^{-12}		

uncertainty is estimated not to exceed 5% for liquid water temperatures up to $T=374$ °C; or 15% for temperatures from $T=374$–800 °C and $P=1000$ MPa.

2.6.2 Dielectric Constant

The dielectric constant ε is a property of an electrical insulating material equal to the ratio of the capacitance C of a capacitor filled with the given material to the capacitance of an identical capacitor in a vacuum, C_0, without the dielectric material. By definition, the dielectric constant is a number without dimensions. The static dielectric constant also describes the ability of a solvent to be polarized by an electric field. A higher dielectric constant is associated with a greater ability to dissolve electrolytes or polar compounds. Liquid water has a higher dielectric constant than most liquids because of its polarity and the association of molecules due to hydrogen bonds [4]. The value of the dielectric constant at room temperature ($T=25$ °C, 298.15 K) is 1.00059 for air, 2.25 for paraffin, and 78.2 for water. The dielectric constant increases for most substances with temperature. For liquid water, the static dielectric constant decreases rapidly at higher temperatures, corresponding to the lowering of the density and the extent of hydrogen bonding. At the critical point, a value of about 5 is approached [4]. The static dielectric constant increases with pressure at constant temperature (Figure 2.23). The static dielectric constant is also frequency dependent. More information on the dependence of ε on frequency is given in Ellison *et al.* [70].

Values for the static dielectric constant ε of water are presented in Figure 2.23 for saturated liquid and vapor, in Figure 2.24 in dependence of pressures up to $P=500$ MPa and temperatures up to $T=823.15$ K, and in Figure 2.25 for selected values of pressure, depicting better the behavior in the critical region. Experimental values have been evaluated by Uematsu and Franck [69] and Fernández *et al.* [71]. The values tabulated in Table 2.20 for saturated liquid water and vapor and in Table 2.21 for pressurized water and

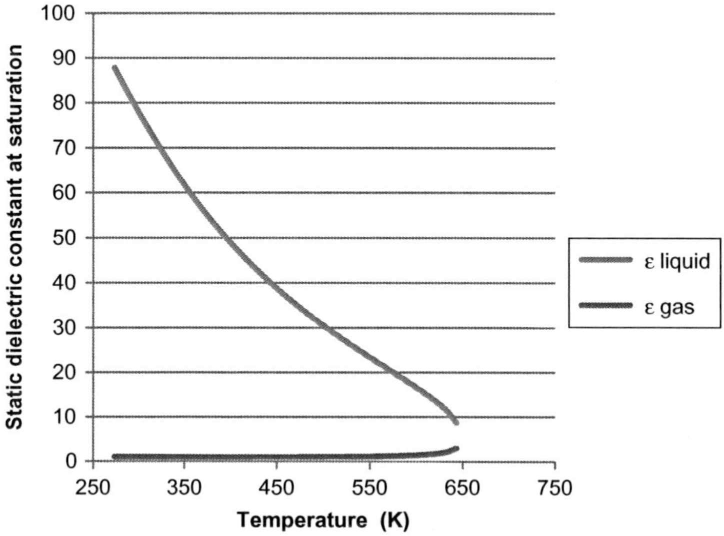

FIGURE 2.23 Static dielectric constant for saturated liquid and gaseous water. *Data from Uematsu and Franck [69].* (For color version of this figure, the reader is referred to the online version of this chapter.)

steam have been taken from Uematsu and Franck [69]. A correlation has been released by the IAPWS for the static dielectric constant of water representing the best available standard and is presented below [72] (Tables 2.22 and 2.23).

2.6.2.1 Correlationof the Static Dielectric Constant of Water

The equation for ε is derived from the following expressions (IAPWS-1997) [72]:

$$\frac{\varepsilon-1}{\varepsilon+2}=A\frac{\varepsilon}{(2\varepsilon+1)(\varepsilon+2)}+B, \tag{2.71}$$

where A and B are given by

$$A=\frac{N_A\mu^2}{\varepsilon_0 k}\frac{\rho g}{T},\quad B=\frac{N_A\alpha}{3\varepsilon_0}\rho \tag{2.72}$$

The physically correct root of Equation (2.71) for the dielectric constant is

$$\varepsilon=\frac{1+A+5B+\sqrt{9+2A+18B+A^2+10AB+9B^2}}{4-4B} \tag{2.73}$$

with g is the Harris–Alder g-factor; k is the Boltzmann constant; M_w is the molar mass of water; N_A is the Avogadro number; P is the pressure; T is the absolute temperature, ITS-90; T_c is the critical temperature; α is the mean molecular polarizability of the isolated water molecule; ε is the dielectric constant (relative

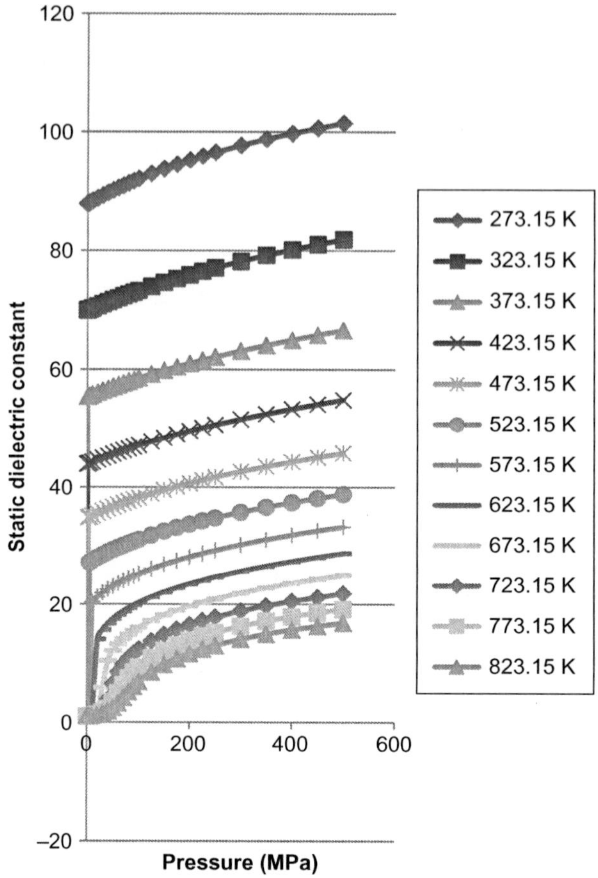

FIGURE 2.24 Static dielectric constant of water and steam. *Data from Uematsu and Franck [69]*. (For color version of this figure, the reader is referred to the online version of this chapter.)

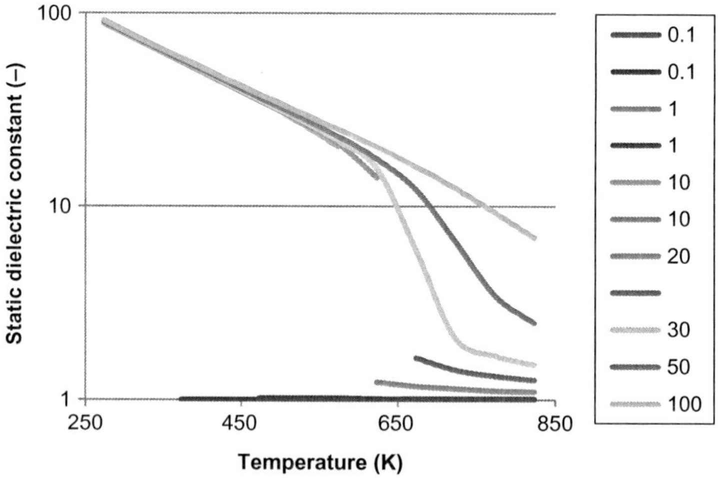

FIGURE 2.25 Static dielectric constant of water and steam. Selected isobars. *Data from Uematsu and Franck [69]*. (For color version of this figure, the reader is referred to the online version of this chapter.)

TABLE 2.22 Static Dielectric Constant for Saturated Liquid and Gaseous Water

T (K)	ε_{liquid}	ε_{gas}	*T* (K)	ε_{liquid}	ε_{gas}
273.15	87.81	1	463.15	36.44	1.03
283.15	83.99	1	473.15	34.74	1.04
293.15	80.27	1	483.15	33.11	1.05
303.15	76.67	1	493.15	31.53	1.06
313.15	73.22	1	503.15	30.01	1.07
323.15	69.9	1	513.15	28.53	1.09
333.15	66.73	1	523.15	27.08	1.11
343.15	63.7	1	533.15	25.68	1.13
353.15	60.81	1	543.15	24.3	1.15
363.15	58.05	1	553.15	22.94	1.18
373.15	55.41	1	563.15	21.6	1.22
383.15	52.9	1.01	573.15	20.26	1.27
393.15	50.5	1.01	583.15	18.92	1.33
403.15	48.22	1.01	593.15	17.56	1.4
413.15	46.03	1.01	603.15	16.17	1.5
423.15	43.94	1.01	613.15	14.72	1.64
433.15	41.95	1.02	623.15	13.16	1.85
443.15	40.03	1.02	633.15	11.36	2.19
453.15	38.2	1.03	643.15	8.7	3

Data from Uematsu and Franck [69].

permittivity); ε_0 is the permittivity of free space; μ is the dipole moment of the isolated water molecule; ρ is the density; and ρ_c is the critical density.

Constants used in the dielectric constant correlation:

Parameter	Value
Permittivity of free space, ε_0	$[4 \times 10^{-7}\ \pi(299{,}792{,}458)^2]^{-1}\ C^2\ J^{-1}\ m^{-1}$
Mean molecular polarizability, α	$1.636 \times 10^{-40}\ C^2\ J^{-1}\ m^2$
Molecular dipole moment, μ	6.138×10^{-30} C m
Boltzmann constant, k	$1.380658 \times 10^{-23}\ J\ K^{-1}$
Avogadro number, N_A	$6.0221367 \times 10^{23}\ mol^{-1}$
Molar mass of water, M_w	$0.018015268\ kg\ mol^{-1}$

TABLE 2.23 Static Dielectric Constant of Water and Steam

	273.15 K	323.15 K	373.15 K	423.15 K	473.15 K	523.15 K	573.15 K	623.15 K	673.15 K	723.15 K	773.15 K	823.15 K
P (MPa)	273.15	323.15	373.15	423.15	473.15	523.15	573.15	623.15	673.15	723.15	773.15	823.15
0.1	87.81	69.91	1	1	1	1	1	1	1	1	1	1
0.5	87.83	69.92	55.43	43.95	1.01	1.01	1.01	1.01	1.01	1	1	1
1	87.86	69.94	55.44	43.96	1.02	1.02	1.02	1.01	1.01	1.01	1.01	1.01
2.5	87.93	69.99	55.5	44.02	34.79	1.06	1.04	1.04	1.03	1.03	1.02	1.02
5	88.05	70.09	55.59	44.12	34.9	27.15	1.11	1.08	1.07	1.06	1.05	1.04
10	88.28	70.27	55.76	44.3	35.11	27.43	20.39	1.23	1.17	1.14	1.11	1.1
20	88.75	70.63	56.11	44.66	35.52	27.95	21.24	14.07	1.64	1.42	1.32	1.26
30	89.2	70.98	56.44	45.01	35.91	28.43	21.95	15.66	5.91	2.07	1.68	1.51
40	89.64	71.32	56.77	45.34	36.28	28.87	22.56	16.72	10.46	3.84	2.34	1.9
50	90.07	71.66	57.08	45.67	36.63	29.28	23.1	17.55	12.16	6.57	3.45	2.48
60	90.49	71.98	57.39	45.98	36.96	29.67	23.58	18.24	13.28	8.53	4.9	3.26
70	90.9	72.3	57.69	46.28	37.28	30.03	24.02	18.84	14.16	9.87	6.31	4.2
80	91.29	72.62	57.98	46.57	37.59	30.37	24.43	19.37	14.88	10.88	7.5	5.16
90	91.67	72.92	58.27	46.86	37.89	30.7	24.81	19.85	15.5	11.7	8.47	6.06
100	92.04	73.22	58.55	47.14	38.17	31.01	25.17	20.29	16.05	12.39	9.29	6.88
125	92.89	73.93	59.19	47.78	38.86	31.78	26.03	21.26	17.21	13.77	10.88	8.53

150	93.71	74.62	59.82	48.4	39.5	32.46	26.77	22.09	18.16	14.85	12.07	9.8
175	94.48	75.27	60.42	48.98	40.1	33.09	27.45	22.83	18.98	15.74	13.04	10.81
200	95.2	75.89	61	49.54	40.66	33.67	28.07	23.49	19.69	16.51	13.86	11.65
225	95.87	76.5	61.55	50.08	41.2	34.22	28.64	24.09	20.33	17.19	14.56	12.38
250	96.51	77.08	62.08	50.59	41.7	34.74	29.17	24.65	20.91	17.8	15.19	13.01
300	97.69	78.17	63.1	51.55	42.65	35.69	30.15	25.65	21.94	18.85	16.25	14.07
350	98.75	79.19	64.05	52.45	43.52	36.56	31.02	26.53	22.83	19.74	17.14	14.93
400	99.72	80.13	64.94	53.3	44.33	37.36	31.81	27.32	23.62	20.52	17.89	15.66
450	100.6	81.02	65.78	54.1	45.1	38.09	32.54	28.04	24.32	21.2	18.55	16.28
500	101.42	81.84	66.57	54.85	45.82	38.78	33.21	28.7	24.96	21.82	19.14	16.83

Data from Uematsu and Franck [69].

TABLE 2.24 Coefficients N_h and Exponents i_h and j_h of Equation (2.76) for the g-factor

h	N_h	i_h	j_h
1	0.978224486826	1	0.25
2	−0.957771379375	1	1
3	0.237511794148	1	2.5
4	0.714692244396	2	1.5
5	−0.298217036956	3	1.5
6	−0.108863472196	3	2.5
7	$0.949327488264\times 10^{-1}$	4	2
8	$-0.980469816509\times 10^{-2}$	5	2
9	$0.165167634970\times 10^{-4}$	6	5
10	$0.937359795772\times 10^{-4}$	7	0.5
11	$-0.123179218720\times 10^{-9}$	10	10
12	$0.196096504426\times 10^{-2}$		

The Harris and Alder g-factor is given by

$$\frac{\epsilon-1}{\varepsilon+2}=\frac{N_{\mathrm{A}}\rho}{3}\left(\frac{\alpha}{\varepsilon_0}+\frac{g\mu^2}{3kT\varepsilon_0}\frac{9\varepsilon}{(2\varepsilon+1)(\varepsilon+2)}\right) \tag{2.74}$$

Values of the Harris and Alder g-factor g can be determined from values of ε with the following equation:

$$g=\left(2+\frac{1}{\varepsilon}\right)\frac{kT}{3\mu^2}\left[\frac{3\varepsilon_0}{N_{\mathrm{A}}\rho}(\varepsilon-1)-\alpha(\epsilon+2)\right]. \tag{2.75}$$

The representation of g is

$$g=1+\sum_{h=1}^{11}N_h\left(\frac{\rho}{\rho_{\mathrm{c}}}\right)^{i_h}\left(\frac{T_{\mathrm{c}}}{T}\right)^{j_h}+N_{12}\frac{\rho}{\rho_{\mathrm{c}}}\left(\frac{T}{228K}-1\right)^{-1.2} \tag{2.76}$$

The values of N_h, i_h, and j_h are given in Table 2.24.

2.7 MISCELLANEOUS PROPERTIES OF WATER

2.7.1 Refractive Index

The refractive index of a substance is a measure of the speed of light in that substance. It is expressed as a ratio of the speed of light in vacuum relative

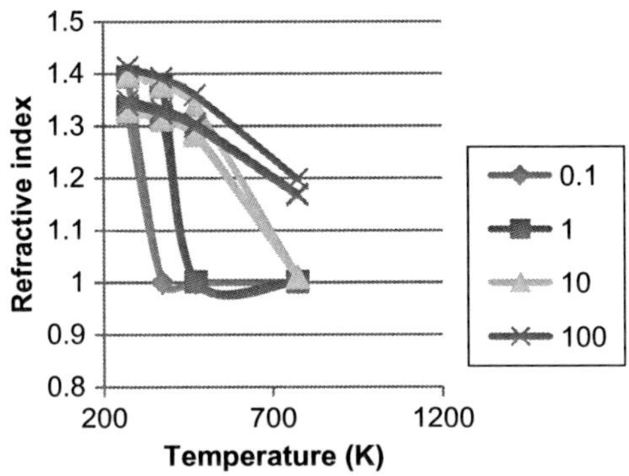

FIGURE 2.26 Refractive index of water at different pressures in MPa. The different lines at constant pressure show the influence of the wavelength (see table below). Lines only connect data. Values below 1 have no physical significance. *Data IAPWS [75]*. (For color version of this figure, the reader is referred to the online version of this chapter.)

to that in the considered medium: n = velocity of light in a vacuum/velocity of light in medium. The refractive index depends on frequency. Visible and near-visible frequencies have been studied extensively and are covered here due to their importance. Propagation, reflection, and absorption of light in water pose problems. Solution of the problems requires knowledge of the refractive index. Furthermore, water is an excellent reference for relative refractive index measurements in other liquids, and the refractive index itself is used as a local probe of properties, such as the density, that cannot readily be measured *in situ* [73,74].

The refractive index of water in the visible region is not qualitatively different from other fluids, with values in the dense liquid near 1.3. IAPWS has released a standard formulation [75]. In Figure 2.26, the refractive index of water is shown and some selected values are listed in Table 2.25. More data are available from Schiebner *et al.* [74] or can be calculated by Equation (2.75).

2.7.1.1 Correlation for the Refractive Index of Water

The refractive index is represented by the following equation:

$$\frac{n^2-1}{n^2+2}\frac{1}{\rho_r} = a_0 + a_1\rho_r + a_2T_r + a_3\lambda_r^2T_r + \frac{a_4}{\lambda_r^2} + \frac{a_5}{\lambda_r^2-\lambda_{r,UV}^2} + \frac{a_6}{\lambda_r^2-\lambda_{r,IR}^2} + a_7\rho_r^2 \tag{2.77}$$

with n is the refractive index with respect to vacuum; P_{sat} is the saturation pressure; T is the absolute temperature, ITS-90; t is the temperature in °C; λ is the wavelength of light; and ρ is the mass density.

Reference constants: $T^{ref} = 273.15$ K, $\rho^{ref} = 1000$ kg m^{-3}, $\lambda^{ref} = 0.589$ μm.
Dimensionless variables: $T_r = T/T^{ref}$, $\rho_r = \rho/\rho^{ref}$, $\lambda_r = \lambda/\lambda^{ref}$.

TABLE 2.25 Refractive Index of Water

	T (K)	*P* (MPa)			
Wavelength (μm)		0.1	1	10	100
0.2265	273.15	1.394527	1.394711	1.396526	1.412733
	373.15	1.0002168	1.375622	1.377286	1.391983
	473.15	1.0001683	1.0017754	1.338299	1.35933
	773.15	1.0001008	1.0010155	1.0109906	1.198312
0.589	273.15	1.334344	1.334494	1.335969	1.349101
	373.15	1.0001876	1.318725	1.320084	1.332057
	473.15	1.0001456	1.0015359	1.287891	1.305191
	773.15	1.0000871	1.0008773	1.0094939	1.170231
1.01398	273.15	1.326135	1.326279	1.32771	1.340435
	373.15	1.0001837	1.311257	1.312577	1.324202
	473.15	1.0001427	1.0015052	1.281529	1.298369
	773.15	1.0000856	1.0008619	1.0093267	1.167119

Data IAPWS [75].

TABLE 2.26 Coefficients for the Formulation, Equation (2.77)

$a_0 = 0.244257733$	$a_4 = 1.58920570 \times 10^{-3}$
$a_1 = 9.74634476 \times 10^{-3}$	$a_5 = 2.45934259 \times 10^{-3}$
$a_2 = -3.73234996 \times 10^{-3}$	$a_6 = 0.900704920$
$a_3 = 2.68678472 \times 10^{-4}$	$a_7 = -1.66626219 \times 10^{-2}$
$\lambda_{r,UV} = 0.2292020$	$\lambda_{r,IR} = 5.432937$

The coefficients a_0–a_7 and the constants λ_{UV}, λ_{IR} are given in Table 2.26.

Equation (2.77) for the refractive index of water may be applied in the following range: Temperature: $-12 \leq t$ (°C) ≤ 500, density $0 \leq \rho$ (kg m^{-3}) ≤ 1060, wavelength $0.2 \leq \lambda$ (μm) ≤ 1.1. Calculated values are also in good agreement with recent experimental results [76] in liquid water at wavelengths up to 1.9 μm. At wavelengths further into the UV and IR, there are absorption features that complicate the refractive index behavior [4].

2.7.2 Surface Tension

Surface tension (surface free energy) σ (N m^{-1} or J m^{-2}) of water at the gas–liquid interface is due to the attraction between the molecules being directed away from the surface as surface molecules are more attracted to the molecules within the liquid than they are to molecules of the gas at the surface. The vapor–liquid surface tension of water is higher than that of most other liquids: For water at $T=300$ K, surface tension is about $\sigma=71.7$ mN m^{-1}, compared to $\sigma=20$–40 mN m^{-1} for typical organic liquids [4].

The data for the surface tension of water have been correlated in an IAPWS release (IAPWS, 1994) [77]. The correlation is based on data evaluated by Vargaftik [78]. Results of the correlation are shown in Figure 2.27 and Table 2.27. The value of the surface tension for water drops rapidly ($\sigma=0.155$ mN m^{-1} K^{-1} at $T=25$ °C) with increasing temperature from the high value at ambient conditions to zero at the critical point. There are two inflection points, at $T=277.15$ and 535.15 K [79].

2.7.2.1 Correlation of the Surface Tension of Water

The IAPWS recommended the following equation for the correlation of values of surface tension of water [77]:

$$\sigma = B\tau^{\mu}(1+b\tau) \tag{2.78}$$

where σ is the surface tension, $\tau=1-T/T_c$, T is the absolute temperature, $T_c=647.096$ K, $B=235.8$ mN/m, $B=-0.625$, and $\mu=1.256$.

This equation is valid between the triple point ($T=0.01$ °C) and reference temperature, T_c. The uncertainty of the calculated values of the surface tension of water is on the order of 0.5% at low temperatures, increasing to 1% and beyond as the temperature approaches T_c. Equation (2.78) also describes well the surface tension of a water–air interface at low and moderate temperatures and pressures [4].

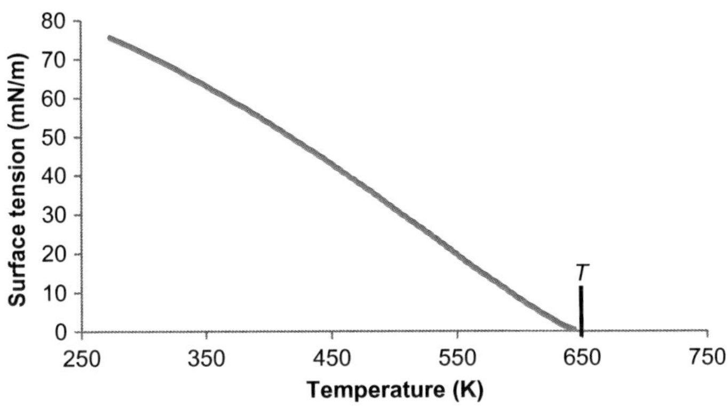

FIGURE 2.27 Surface tension of water. *Data from Ref. [77].* (For color version of this figure, the reader is referred to the online version of this chapter.)

TABLE 2.27 Surface Tension of Water

T (K)	SFT (mN/m)	T (K)	SFT (mN/m)
273.55	75.59	483.64	35.22
283.16	74.16	503.38	30.43
292.78	72.75	524.74	25.63
303.59	71.07	543.67	21.14
313.62	69.66	562.11	16.62
323.48	67.9	582.8	12.13
343.27	64.4	592.68	9.88
362.3	60.86	602.33	7.73
383.31	56.99	612.33	5.72
403.89	52.76	620.23	4.18
424.69	48.35	632.57	1.92
443.68	44.39	642.41	0.49
463.74	39.74	643.54	0.4
473.49	37.51		

Data from Ref. [77].

2.7.3 Ionic Dissociation Constant of Water

Water dissociates to some extent into positive and negative ions. The reaction can be written as

$$2H_2O \Leftrightarrow H_3O^+ + OH^- \tag{2.79}$$

This equilibrium is important for electrochemical reactions in hydrothermal and supercritical water as is discussed in the following sections. Since the proton hydration is complete, this reaction may be used as a representative model for both liquid and vapor phases. The equilibrium constant for the ionization of water K_w can be defined as

$$K_w = \frac{a_{H_3O^+} a_{OH^-}}{a_{H_2O}^2}, \tag{2.80}$$

with a_i as the dimensionless activity of the species i. Since it can be assumed that the sum of the chemical potentials of H^+ and H_3O^+ is formally the same as the twofold of the chemical potential of H_2O at any temperature and density (pressure) so that K_w is numerically equal to

$$K_w = \frac{a_{H^+} a_{OH^-}}{a_{H_2O}}, \qquad (2.81)$$

Equation (2.81) is most commonly used in the literature [80]. The activity of H_2O can be taken as unity. Ion concentrations are commonly measured in molality units. Then, the equilibrium constant K_w of the reaction is very close to 10^{-14} for liquid water at ambient temperature, or the familiar value of 7 for the pH of neutral water [4]. The equilibrium concentrations of the ions vary with pressure and temperature. In general, at higher temperatures dissociation increases (favoring proton-based catalytic reactions), while lowering the density strongly reduces the dissociation. K_w of the saturated liquid increases with temperature about three orders of magnitude, then decreases rapidly when approaching the critical point, parallel to the decrease of the density. The dissociation in low-density steam is very small.

Available experimental data have been analyzed by Marshall and Franck [81], Pitzer [82], and Bandura and Lvov [83]. The IAPWS has released a formulation for the correlation of the ionic constant of water which is the standard for calculating the ionization constant of water [80]. The formulation is presented in the following after the graphic and tabulated representation of data for the ionization constant of water, derived from that release, see Figures 2.28 and 2.29, and Table 2.28.

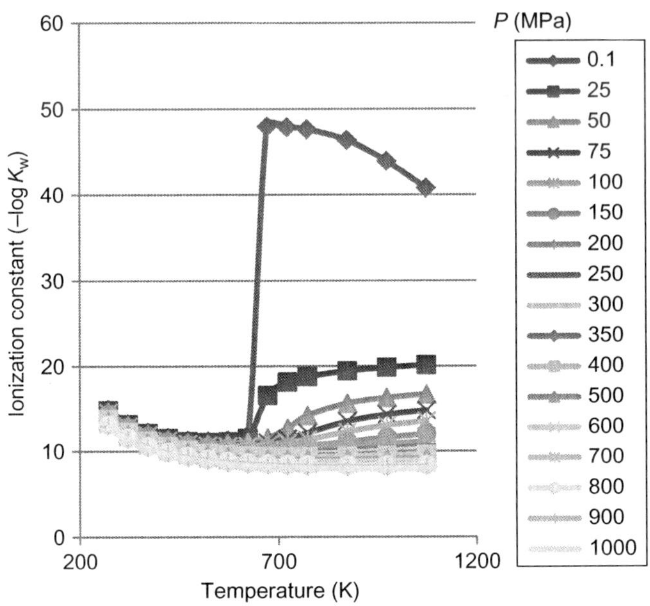

FIGURE 2.28 Negative logarithm (base 10) of the ionization constant of water K_w. Data IAPWS release on ionic constant [80]. (For color version of this figure, the reader is referred to the online version of this chapter.)

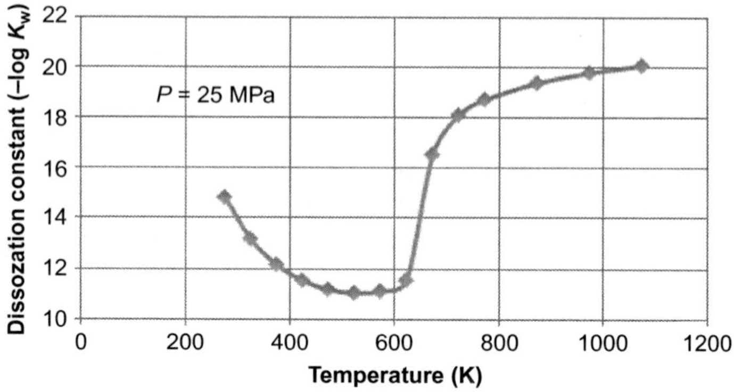

FIGURE 2.29 Negative logarithm (base 10) of the ionization constant of water K_w for a pressure of $P=25$ MPa. Data IAPWS release on ionic constant [80]. (For color version of this figure, the reader is referred to the online version of this chapter.)

Figure 2.28 shows the ionization constant as a function of pressure and temperature in isobars. K_w increases (the negative logarithm of K_w or the pH-value decreases) with temperature up to the critical region for all pressures. At low and moderate pressures, K_w decreases, very rapidly in the low pressure region (compare the 0.1 MPa-isobar) and more moderately but substantially in the denser near-critical and supercritical region (compare the 25 MPa-isobar in Figure 2.28). This behavior has tremendous effects on the solubility of substances as will be discussed in context with water as solvent in Section 2.3. In Table 2.28, the calculated values from the IAPWS release are reproduced in order to make readily available precise values of the ionization constant of water.

2.7.3.1 *IAPWS-Correlation for the Ionization Constant of Water*

The equation for the ionization constant of water, K_w, provided in the IAPWS release is a semiempirical equation for K_w as a function of density and temperature. The equation represents values of $pK_w \equiv -\log_{10}(K_w)$ within experimental uncertainties for a temperature range from $T=25$–$800\,°C$ and densities from $\rho=0$–1.25 g cm^{-3}.

The ionization constant of water, K_w, as a function of temperature and density can be represented by the following equation:

$$pK_w = -2n\left[\log_{10}(1+Q) - \frac{Q}{Q+1}\rho\left(\beta_0+\beta_1 T^{-1}+\beta_2\rho\right)\right] + pK_w^G + 2\log_{10}\frac{m^0 M_w}{G}, \quad (2.82)$$

$$Q = \frac{\rho}{\rho^0}\exp\left(\alpha_0+\alpha_1 T^{-1}+\alpha_2 T^{-2}\rho^{2/3}\right), \quad (2.83)$$

TABLE 2.28 Negative Logarithm (Base 10) of the Ionization Constant of Water K_w

MPa	273.15	323.15	373.15	423.15	473.15	523.15	573.15	623.15	673.15	723.15	773.15	873.15	973.15	1073.15
0.1	14.946	13.264	12.252	11.641	11.31	11.205	11.339	11.92	47.961	47.873	47.638	46.384	43.925	40.785
25	14.848	13.181	12.165	11.543	11.189	11.05	11.125	11.551	16.566	18.135	18.758	19.425	19.829	20.113
50	14.754	13.102	12.084	11.45	11.076	10.898	10.893	11.076	11.557	12.71	14.195	15.621	16.279	16.693
75	14.665	13.026	12.006	11.364	10.974	10.769	10.715	10.802	11.045	11.491	12.162	13.507	14.301	14.791
100	14.58	12.953	11.933	11.283	10.88	10.655	10.568	10.6	10.744	11.005	11.381	12.296	13.04	13.544
150	14.422	12.815	11.795	11.135	10.713	10.458	10.327	10.295	10.345	10.464	10.642	11.117	11.613	12.032
200	14.278	12.687	11.668	11	10.564	10.289	10.131	10.062	10.063	10.119	10.22	10.513	10.853	11.171
250	14.145	12.567	11.549	10.876	10.43	10.14	9.963	9.869	9.839	9.859	9.917	10.112	10.36	10.609
300	14.021	12.453	11.437	10.76	10.306	10.005	9.814	9.702	9.651	9.646	9.677	9.81	9.998	10.199
350	13.906	12.346	11.331	10.651	10.191	9.881	9.679	9.554	9.487	9.465	9.476	9.567	9.712	9.877
400	13.797	12.243	11.23	10.548	10.083	9.766	9.555	9.42	9.341	9.305	9.302	9.361	9.475	9.613
500	13.595	12.052	11.042	10.356	9.884	9.557	9.332	9.182	9.086	9.031	9.007	9.024	9.094	9.191
600	13.411	11.875	10.868	10.181	9.703	9.369	9.135	8.974	8.866	8.798	8.761	8.749	8.79	8.861
700	13.24	11.71	10.705	10.018	9.537	9.197	8.956	8.787	8.67	8.593	8.546	8.514	8.536	8.587
800	13.08	11.556	10.553	9.865	9.381	9.037	8.791	8.616	8.493	8.409	8.354	8.308	8.314	8.352
900	12.93	11.41	10.41	9.721	9.236	8.888	8.638	8.458	8.33	8.24	8.18	8.122	8.117	8.144
1000	12.788	11.272	10.273	9.585	9.098	8.748	8.495	8.311	8.178	8.084	8.019	7.952	7.939	7.957

Data source: IAPWS release on ionic constant [80].

$$K_w = \frac{a_{H_3O^+} a_{OH^-}}{a_{H_2O}^2}, \tag{2.84}$$

where M_w is the molar mass of water and α_0, α_1, α_2, β_0, β_1, β_2 are empirical parameters, given in Table 2.29. The last term in Equation (2.82) converts the ionization constant of water from the ideal-gas standard state (used for K_w^G) to the molal standard state (used for K_w). The ionization constant equation includes the equilibrium constant of the ionization reaction in the ideal-gas state K_w^G. The ideal-gas ionization constant of water is approximated by a temperature function (Table 2.30):

$$pK_w^G = \gamma_0 + \gamma_1 T^{-1} + \gamma_2 T^{-2} + \gamma_3 T^{-3} \tag{2.85}$$

where γ_0, γ_1, γ_2, γ_3 are empirical coefficients given in Table 2.29; ρ is the mass density, g cm^{-3}; ρ^0 is the normalizing mass density = 1 g cm^{-3}; $G = 1000$ g kg^{-1}; K_w is the ionization constant of water at $m^0 = 1$ mol kg^{-1} (molal standard state), dimensionless; K_w^G = ionization constant of water at

TABLE 2.29 Empirical Coefficients of Equation (2.85)

Coefficient	Value	Units
γ_0	6.141500×10^{-1}	–
γ_1	4.825133×10^{4}	K
γ_2	-6.770793×10^{4}	K^2
γ_3	1.010210×10^{7}	K^3

TABLE 2.30 Empirical Coefficients of Equations (2.82) and (2.83)

Coefficient	Value	Units
n	6	–
α_0	−0.864671	–
α_1	8659.19	K
α_2	−22786.2	(g cm^{-3})$^{-2/3}$ K^2
β_0	0.642044	(g cm^{-3})$^{-1}$
β_1	−56.8534	(g cm^{-3})$^{-1}$ K
β_2	−0.375754	(g cm^{-3})$^{-2}$

$\rho = 0\ g\ cm^{-3}$ (ideal-gas standard state), dimensionless; M_w is the molar mass of water = 18.015268 g mol^{-1}; m^0 is the standard molality = 1 mol kg^{-1}; n is the ion coordination number, dimensionless, n fixed at 6; P_s is the vapor–liquid saturation pressure; t is the Celsius temperature, °C; and T is the absolute temperature, K.

REFERENCES

[1] D. Möller, J. Feichtner, H. Hermann, Wasser—Das umtriebige Elexier, in: R. Zellner (Ed.), Chemie über den Wolken, Wiley-VCH, Weinheim, Germany, 2011, p. 128.

[2] Wikipedia, http://en.wikipedia.org/wiki/Ocean, 2011 (accessed 5.12.11).

[3] Wikipedia, http://en.wikipedia.org/wiki/Water, 2011 (accessed 5.12.11).

[4] A.H. Harvey, D.G. Friend, Physical properties of water: aqueous systems at elevated temperatures and pressures, in: D.A. Palmer, R. Fernandez-Prini, A.H. Harvey (Eds.), Physical Chemistry in Water, Steam and Hydrothermal Solutions, Elsevier, Amsterdam, The Netherlands, 2004, pp. 1–27, Chapter 1.

[5] J. Falbe, M. Regitz (Eds.), in: Wasser: Römpp Chemie Lexikon, vol. 6, Georg Thieme, Stuttgart, 1995.

[6] G. Brunner, Near critical and supercritical water. Part I. Hydrolytic and hydrothermal processes, J. Supercrit. Fluids 47 (2009) 373–381.

[7] T.M. Seward, T. Driesner, Hydrothermal solution structure: experiments and computer simulations, in: D.A. Palmer, R. Fernândez-Prini, A.H. Harvey (Eds.), Aqueous Systems at Elevated Temperatures and Pressures: Physical Chemistry in Water, Steam and Hydrothermal Solutions, Elsevier, Amsterdam, The Netherlands, 2004, pp. 149–182, Chapter 5.

[8] Y. Marcus, Supercritical Water. A Green Solvent: Properties and Uses, Wiley, Hoboken, NJ, USA, 2012.

[9] A.A. Chialvo, P.T. Cummings, Microstructure of ambient and supercritical water. Direct comparison between simulation and neutron scattering experiments, J. Phys. Chem. 100 (1996) 1309–1316.

[10] A.A. Chialvo, P.T. Cummings, Molecular-based modeling of water and aqueous solutions at supercritical conditions, Adv. Chem. Phys. 109 (1999) 115–205.

[11] A.K. Soper, F. Bruni, M.A. Ricci, Site-site pair correlation functions of water from 25 to 400 °C. Revised analysis of new and old diffraction data, J. Chem. Phys. 106 (1997) 247–254.

[12] T.I. Mizan, P.E. Savage, R.M. Ziff, Fugacity coefficients for free radicals in dense fluids: HO_2 in supercritical water, in: K.W. Hutchenson, N.R. Foster (Eds.), Innovations in Supercritical Fluids: Science and Technology, vol. 608, American Chemical Society, Washington, DC, 1995, p. 47.

[13] identical with [18]G.V. Bondarenko, Y.E. Gorbaty, An infrared study of water vapour in the temperature range 573–723 K. Dimerization enthalpy and absorption intensities for monomer and dimer, Mol. Phys. 74 (1991) 639–647.

[14] A.A. Galkin, V.V. Lunin, Subcritical and supercritical water: a universal medium for chemical reactions, Russ. Chem. Rev. 74 (2005) 21–35.

[15] A.G. Kalinichev, J.D. Bass, Hydrogen bonding in supercritical water. 2. Computer simulations, J. Phys. Chem. A 101 (1997) 9720–9727.

[16] A.G. Kalinichev, S.V. Churakov, Size and topology of molecular clusters in supercritical water: a molecular dynamics simulation, Chem. Phys. Lett. 302 (1999) 411–417.

[17] G.E. Walrafen, M.R. Fisher, M.S. Hokmabadi, W.H. Yang, Temperature dependence of the low- and high-frequency Raman scattering from liquid water, J. Chem. Phys. 85 (1986) 6970–6982.

[18] T. Tassaing, Y. Danten, M. Besnard, Supercritical water: local order and molecular dynamics, Pure Appl. Chem. 76 (1) (2004) 133–139.

[19] (a) A.A. Chialvo, P.T. Cummings, J.M. Simonson, R.E. Mesmer, Solvation in high-temperature electrolyte solutions. I. Hydration shell behavior from molecular simulation, J. Chem. Phys. 110 (1999) 1064–1074. (b) A.A. Chialvo, P.T. Cummings, J.M. Simonson, R.E. Mesmer, Solvation in high-temperature electrolyte solutions. II. Some formal results, J. Chem. Phys. 110 (1999) 1075–1086.

[20] B. Guillot, Y. Guissani, Simulation of the far infrared spectrum of liquid water and steam along the coexistence curve, in: G.C. Tabisz, M.N. Neuman (Eds.), Collision and Interaction-Induced Spectroscopy, vol. 452, Kluwer Academic, Dordrecht, 1995, p. 129.

[21] N. Yoshii, H. Yoshie, S. Miura, S. Okazaki, A molecular dynamics study of sub- and supercritical water using a polarizable potential model, J. Chem. Phys. 109 (1998) 4873–4884.

[22] N. Matubayasi, N. Nakao, M. Nakahara, Structural study of supercritical water. III. Rotational dynamics, J. Chem. Phys. 114 (2001) 4107–4115.

[23] J. Marti, Analysis of the hydrogen bonding and vibrational spectra of supercritical model water by molecular dynamics simulations, J. Chem. Phys. 110 (1999) 6876.

[24] M.M. Hoffman, M.S. Conradi, Are there hydrogen bonds in supercritical water? J. Am. Chem. Soc. 119 (1997) 3811–3817.

[25] C.H. Uffindell, A.I. Kolesnikov, J.-C. Li, J. Mayers, Lattice dynamic calculated using a range of water potentials, Phys. B 444 (2000) 276–278.

[26] IAPWS, IAPWS guideline on the use of fundamental physical constants and basic constants of water, in: The International Association for the Properties of Water and Steam, Gaithersburg, Maryland, USA, September, 2001. http://www.iapws.org.

[27] W. Wagner, A. Saul, A. Pruß, International equations for the pressure along the melting curve and the sublimation curve of ordinary water substance, J. Phys. Chem. Ref. Data 23 (1994) 515–527.

[28] H.E. Stanley, S.V. Buldyrev, M. Canpolat, O. Mishima, M.R. Sadr-Lahyijany, R. Scala, F.W. Starr, The puzzling behaviour of water at very low temperature, Phys. Chem. Chem. Phys. 2 (2000) 1551–1558.

[29] F. Maršík, J. Hrubý, P. Demo, Z. Kožíšek, V. Petr, M. Kolovratník, Binary homogeneous nucleation in selected aqueous vapor mixtures, in: D.A. Palmer, R. Fernandez-Prini, A.H. Harvey (Eds.), Aqueous Systems at Elevated Temperatures and Pressures: Physical Chemistry in Water, Steam and Hydrothermal Solutions, Elsevier, Amsterdam, The Netherlands, 2004, pp. 205–242, Chapter 7.

[30] G. Brunner, Gas Extraction, Steinkopff, Darmstadt, Springer, New York, 1994.

[31] S.I. Sandler, Chemical and Engineering Thermodynamics, second ed., Wiley, New York, 1989.

[32] O. Pfohl, S. Petkov, G. Brunner, PE 2000. A Powerful Tool to Correlate Phase Equilibria, Herbert Utz, München, Germany, 2000. Program available from, www.tu-harburg.de/V8/gruppe-prof-smirnova/veroeffentlichungen.html.

[33] W. Wagner, A. Pruß, International equations for the saturation properties of ordinary water substance—revised according to the International Temperature Scale of 1990, Addendum to J. Phys. Chem. Ref. Data 16 (1987) 893, J. Phys. Chem. Ref. Data 22 (1993) 783–787.

[34] A. Kostrowicka Wyczalkowska, Kh.S. Abdulkadirova, M.A. Anisimov, J.V. Sengers, Thermodynamic properties of H_2O and D_2O in the critical region, J. Chem. Phys. 113 (2000) 4985–5002.

[35] M.A. Anisimov, J.V. Sengers, J.M.H. Levelt-Sengers, Near-critical behavior of aqueous systems, in: D.A. Palmer, R. Fernandez-Prini, A.H. Harvey (Eds.), Aqueous Systems at Elevated Temperatures and Pressures: Physical Chemistry in Water, Steam and Hydrothermal Solutions, Elsevier, Amsterdam, The Netherlands, 2004, pp. 30–71, Chapter 2.
[36] International Association for the Properties of Water and Steam, Release on the IAPWS Formulation 1995 for the Thermodynamic Properties of Ordinary Water Substance for General and Scientific Use, 1995, available from: http://www.iapws.org.
[37] A.H. Harvey, A.P. Peskin, S.A. Klein, NIST/ASME Steam Properties, NIST Standard Reference Database 10, Version 2.2, National Institute of Standards and Technology, Gaithersburg, MD, 2000.
[38] W. Wagner, A. Pruß, The IAPWS Formulation 1995 for the Thermodynamic Properties of Ordinary Water Substance for General and Scientific Use, J. Phys. Chem. Ref. Data 31 (2002) 387–535.
[39] W. Wagner, J.R. Cooper, A. Dittmann, J. Kijima, H.-J. Kretzschmar, A. Kruse, R. Mareš, K. Oguchi, H. Sato, I. Stöcker, O. Šifner, Y. Takaishi, I. Tanishita, J. Trübenbach, Th. Willkommen, The IAPWS industrial formulation 1997 for the thermodynamic properties of water and steam, ASME J. Eng. Gas Turbines Power 122 (2000) 150–182.
[40] W.T. Parry, J.C. Bellows, J.S. Gallagher, A.H. Harvey, ASME International Steam Tables for Industrial Use, ASME Press, New York, USA, 2000.
[41] Roles of Various IAPWS Documents Concerning the Thermodynamic Properties of Ordinary Water Substance, Advisory Note No. 2, The International Association for the Properties of Water and Steam, Kyoto, 2004.
[42] Revised Release on the IAPWS Industrial Formulation 1997 for the Thermodynamic Properties of Water and Steam (The revision only relates to the extension of region 5 to 50 MPa), The International Association for the Properties of Water and Steam, Lucerne, 2007.
[43] M.V. Sussman, Elementary General Thermodynamics, Robert E. Krieger Publishing Co., Malabar, Florida, USA, 1989, pp. 42–43.
[44] J.R. Cooper, Representation of the ideal-gas thermodynamic properties of water, Int. J. Thermophys. 3 (1982) 35–43.
[45] A.H. Harvey, Thermodynamic Properties of Water: Tabulation from the IAPWS Formulation 1995 for the Thermodynamic Properties of Ordinary Water Substance for General and Scientific Use; downloaded from NISTIR 5078.
[46] J.D. Van der Waals, Over de continuiteit van den gas- en vloeistoftoestand, Ph.D. Dissertation, University of Leiden, The Netherlands, 1873.
[47] J.M. Prausnitz, R.N. Lichtenthaler, E. Gomes de Azevedo, Molecular Thermodynamics of Fluid-Phase Equilibria, third ed., Prentice Hall, Inc., Upper Saddle River, NJ, USA, 1999.
[48] G.S. Soave, Equilibrium constants from a modified Redlich-Kwong equation of state, Chem. Eng. Sci. 33 (1972) 225–229.
[49] Á. Peneloux, E. Rauzy, R. Fréze, A consistent correction for Redlich-Kwong-Soave volumes, Fluid Phase Equilib. 8 (1982) 7–23.
[50] W. Thomson, J.P. Joule, On the thermal effects of fluids in motion, Philos. Trans. R. Soc. Lond. 143 (1853) 357–365, (1854), 321–364.
[51] G. Franz, U. Grigull, Messungen des Joule-Thomson-Koeffizienten von Wasser und Wasserdampf (Measurement of the Joule-Thomson-Coefficient for Water and Steam), Wärme Stoffübertragung 5 (1972) 181–190.
[52] VDI Wärmeatlas, 8. ed., 1997, Stoffwerte von Wasser, DB 1–DB 14.
[53] O.A. Uyehara, K.M. Watson, A universal viscosity correlation, Natl. Petrol. News 36 (1944) R-714.

[54] G.E. Childs, H.J.M. Hanley, Applicability of dilute gas port property tables to real gases, Cryogenics 8 (1968) 94–97.

[55] The International Association for the Properties of Water and Steam, Release on the IAPWS Formulation 2008 for the Viscosity of Ordinary Water Substance, Berlin, 2008, available from: http://www.iapws.org.

[56] The International Association for the Properties of Water and Steam: Revised Release on the IAPS Formulation 1985 for the Thermal Conductivity of Ordinary Water Substance, Berlin, 2008, available from: http://www.iapws.org.

[57] R.F. Dawson, F. Khoury, R. Kobayashi, Self-diffusion measurements in methane by pulsed nuclear magnetic resonance, AIChE J. 16 (1970) 725–729.

[58] K. Krynicki, C.D. Green, D.W. Sawyer, Pressure and temperature dependence of self-diffusion in water, Faraday Discuss. Chem. Soc. 66 (1978) 199–208.

[59] M. Holz, S.R. Heil, A. Sacco, Temperature-dependent self-diffusion coefficients of water and six selected molecular liquids for calibration in accurate ^{1}H NMR PFG measurements, Phys. Chem. Chem. Phys. 2 (2000) 4740–4742.

[60] A.J. Easteal, A general empirical relationship between tracer or self-diffusion coefficients of liquids and pressure, AIChE J. 30 (1984) 641–642.

[61] A.J. Easteal, W.W. Price, L.A. Woolf, Diaphragm cell for high-temperature diffusion measurements. Tracer diffusion coefficients for water to 363 K, J. Chem. Soc. Faraday Trans. 1 85 (1989) 1091–1097.

[62] H. Weingärtner, Self-diffusion in liquid water—a reassessment, Z. Phys. Chem. Neue Fol. 132 (1982) 129–149.

[63] W.J. Lamb, G.A. Hoffmann, J. Jonas, Self diffusion in compressed supercritical water, J. Chem. Phys. 74 (1981) 6875–6880.

[64] M. Boero, K. Terakura, T. Ikeshoji, C.C. Liew, M. Parinello, Water at supercritical conditions: a first principles study, J. Chem. Phys. 115 (2001) 2219–2277.

[65] H.R. Corti, L.N. Trevani, A. Anderko, Transport properties in high temperature and pressure ionic solutions, in: D.A. Palmer, R. Fernandez-Prini, A.H. Harvey (Eds.), Aqueous Systems at Elevated Temperatures and Pressures: Physical Chemistry in Water, Steam and Hydrothermal Solutions, Elsevier, Amsterdam, The Netherlands, 2004, pp. 322–375, Chapter 10.

[66] Y. Marcus, On transport properties of hot liquid and supercritical water and their relationship to the hydrogen bonding, Fluid Phase Equilib. 164 (1999) 131–142.

[67] The International Association for the Properties of Water and Steam: Electrolytic Conductivity (Specific Conductance) of Liquid and Dense Supercritical Water from 0 °C to 800 °C and Pressures up to 1000 MPa; 1990, available from: http://www.iapws.org.

[68] The International Association for the Properties of Water and Steam: Release on the Ionization Constant of H_2O, 2007, available from: http://www.iapws.org.

[69] M. Uematsu, E.U. Franck, Static dielectric constant of water and steam, J. Phys. Chem. Ref. Data 9 (1980) 1291–1306.

[70] W.J. Ellison, K. Lamkaouchi, J.-M. Moreau, Water: a dielectric reference, J. Mol. Liq. 68 (1996) 171–279.

[71] D.P. Fernández, A.R.H. Goodwin, E.W. Lemmon, J.M.H. Levelt-Sengers, R.C. Williams, A formulation for the static permittivity of water and steam at temperatures from 238K to 873K at pressures up to 1200MPa, including derivatives and Debye-Huckel coefficients, J. Phys. Chem. Ref. Data 26 (1997) 1125–1169.

[72] The International Association for the Properties of Water and Steam (IAPWS): Release on the Static Dielectric Constant of Ordinary Water Substance for Temperatures from 238 K to 873 K and Pressures up to 1000 MPa (IAPWS-1997), available from: http://www.iapws.org.

[73] P. Schiebener, J. Straub, J.M.H. Levelt Sengers, J.S. Gallagher, Refractive index of water and steam as function of wavelength, temperature and density, J. Phys. Chem. Ref. Data 19 (1990) 677–717.

[74] P. Schiebener, J. Straub, J.M.H. Levelt Sengers, J.S. Gallagher, Erratum: refractive index of water and steam as function of wavelength, temperature and density, J. Phys. Chem. Ref. Data 19 (1990) 1617.

[75] The International Association for the Properties of Water and Steam: Release on the Refractive Index of Ordinary Water Substances a Function of Wavelength, Temperature and Pressure, Erlangen, 1997.

[76] J.E. Bertie, Z. Lan, Infrared intensities of liquids XX: the intensity of the OH stretching band of liquid water revisited, and the best current values of the optical constants of $H_2O(l)$ at 25 °C between 15,000 and 1 cm^{-1}, Appl. Spectrosc. 50 (1996) 1047–1057.

[77] IAPWS Release on Surface Tension of Ordinary Water Substance, September 1994, available from: http://www.iapws.org.

[78] N.B. Vargaftik, B.N. Volkov, L.D. Volyak, International tables of the surface tension of water, J. Phys. Chem. Ref. Data 12 (1983) 817–820.

[79] Wikipedia, http://en.wikipedia.org/wiki/Surface_tension.

[80] The International Association for the Properties of Water and Steam: Release on the Ionization Constant of H_2O, Lucerne, 2007, available from: http://www.iapws.org.

[81] W.L. Marshall, E.U. Franck, Ion product of water substance, 0–1000C, J. Phys. Chem. Ref. Data 10 (1981) 295–304.

[82] K.S. Pitzer, Self-ionization of water at high-temperature and the thermodynamic properties of the ions, J. Phys. Chem. 86 (1982) 4704–4708.

[83] A.V. Bandura, S.N. Lvov, The ionization constant of water over a wide range of temperatures and densities, J. Phys. Chem. Ref. Data 35 (2006) 15–30.

Chapter 3

Properties of Mixtures with Water

3.1 INTRODUCTION

This chapter covers the properties of mixtures of various compounds with water. Major groups of such compounds are gases; nonelectrolytes, for example, hydrocarbons; volatile and nonvolatile electrolytes; and metal oxides. Thermodynamic properties are discussed in general and mixture behavior is shown for excess volumes. The influence of salts on dissociation of water is discussed. Transport properties such as electric conductivity, diffusion, viscosity, and thermal conductivity of aqueous mixtures are important for designing hydrothermal processes and are reviewed. The main emphasis of this chapter is on phase behavior of aqueous mixtures. Fundamentals of phenomenological behavior are explained for binary and ternary systems. Calculation of phase equilibria with equations of state (EOS) is a helpful tool for obtaining an overview on phase behavior of specific systems. EOS are shown to be applicable for aqueous systems for a wide range of conditions and for binary and for multicomponent systems. Solubility of volatile nonelectrolytes (gases) in water is an often needed parameter in process calculations. It can be treated in many cases with Henry's law and some modifications. Phase behavior of nonelectrolytes near the critical point of water is a requirement for an overview on process conditions and is discussed for nonelectrolytes and electrolytes. Aqueous systems with alkanes and aromatic compounds in binary and ternary systems are of major importance for hydrothermal processes and are treated in some detail. Phase equilibrium with electrolytes introduces the chemical reaction of ions into the description of phase behavior. Physical and chemical equilibrium of electrolytes in water is covered for volatile electrolytes, like carbon dioxide as a weak electrolyte, and for nonvolatile electrolytes, like metal oxides and salts.

Supercritical Fluid Science and Technology, Vol. 5. http://dx.doi.org/10.1016/B978-0-444-59413-6.00003-0

3.2 PROPERTIES OF MIXTURES

3.2.1 Thermodynamics of Mixtures

Properties of mixtures very often differ substantially from pure component properties that are constituents of the mixture. A common example is the dissolution of supercritical gases in liquids, where these gases exist in liquid state in the mixture, while they cannot exist as liquid at supercritical conditions as pure components.

A naive assumption is that a thermodynamic mixture property can be determined from pure component properties (at same conditions of state) with the sum of pure component properties weighted with their fractions in the mixture. This expectation is in general not true [1]. In most cases, mixture properties differ from the pure component property of each component.

For an intensive state variable d, the mixing property Δd_{mix} is the difference between the real mixture property d (at given values of P, T, and x_i) and the arithmetic mean value of the real pure component property d_i at the same values of P and T, as given in Equation (3.1),

$$\Delta d_{\text{mix}} = d - \sum (d_i x_i) \tag{3.1}$$

or for the extensive state variable D,

$$\Delta D_{\text{mix}} = n\Delta d_{\text{mix}} = nd - \sum (n_i d_i) \tag{3.2}$$

The mixture property d can be calculated from the arithmetic mean value by defining a special mixture variable the partial molar variable $\overline{d}_i$:

$$d \equiv \sum (x_i \overline{d}_i) \tag{3.3}$$

$$D = nd = \sum (n_i \overline{d}_i) \tag{3.4}$$

$$\overline{d}_i = \left(\frac{\partial (nd)}{\partial n_i} \right)_{T,P,n_j} \tag{3.5}$$

with n as the number of moles in the mixture and n_i as the number of moles of component i in the mixture.

Mixing properties and partial molar variables are connected by Equation (3.6):

$$\Delta d_{\text{mix}} = \sum \left[x_i \left(\overline{d}_i - d_i \right) \right] \tag{3.6}$$

The partial molar variables are not equal to pure component values of the same variable. Mixing properties are zero ($\Delta d_{\text{mix}} = 0$) only in cases in which all partial molar variables are equal to the real values of the pure component properties.

As stated in Chapter 2, water in the gaseous state and at low densities or high temperatures behaves as an ideal gas (Equation 2.4). The real behavior of gaseous mixtures can be related to an ideal reference state given by the ideal gas mixture (IGM) by Equations (3.7)–(3.9):

$$(PV)^{\mathrm{IGM}} = (n_1 + n_2 + \cdots + n_N)RT = nRT \tag{3.7}$$

$$U^{\mathrm{IGM}}(T,n) = \sum_{i=1}^{N} n_i u_i^{\mathrm{IG}}(T) \tag{3.8}$$

$$\bar{s}_i^{\mathrm{IGM}}(T,P,x_i) - s_i^{\mathrm{IG}}(T,P) = -R\ln\frac{P_i}{P} = -R\ln x_i \tag{3.9}$$

The mixture of ideal gases is characterized by the lack of interaction forces between the molecules of the components. Therefore, for the volumetric and energetic variables of the mixture, the mixing property is equal to zero. For the entropy, the Gibbs energy, and the Helmholtz energy, the mixing properties are not zero, as given in Table 3.1.

The behavior of real liquids is far from the behavior of an IGM. For real liquids, it is more convenient to use the ideal liquid mixture (ILM) as a reference. In the ideal liquid mixture, the partial molar properties are equal to the pure component properties (Equations 3.10 and 3.11):

$$\bar{h}_i^{\mathrm{IM}}(T,P,x_i) = h_i(T,P) \cdot h_i(T,P) = \text{enthalpy of the pure component} \tag{3.10}$$

TABLE 3.1 Partial Molar Properties and Mixing Properties for Ideal Gas Mixtures (IGMs)

Partial Molar Properties	Mixing Properties
$\bar{v}_i^{\mathrm{IGM}}(T,P,x_i) = v_i^{\mathrm{IG}}(T,P)$	$\Delta v_{\mathrm{mix}}^{\mathrm{IGM}} = 0$
$\bar{u}_i^{\mathrm{IGM}}(T,x_i) = u_i^{\mathrm{IG}}(T)$	$\Delta u_{\mathrm{mix}}^{\mathrm{IGM}} = 0$
$\bar{h}_i^{\mathrm{IGM}}(T,x_i) = h_i^{\mathrm{IG}}(T)$	$\Delta h_{\mathrm{mix}}^{\mathrm{IGM}} = 0$
$\bar{s}_i^{\mathrm{IGM}}(T,P,x_i) = s_i^{\mathrm{IG}}(T,P) - R\ln x_i$	$\Delta s_{\mathrm{mix}}^{\mathrm{IGM}} = -R\sum_{i=1}^{N} x_i \ln x_i$
$\bar{g}_i^{\mathrm{IGM}}(T,P,x_i) = g_i^{\mathrm{IG}}(T,P) + RT\ln x_i$	$\Delta g_{\mathrm{mix}}^{\mathrm{IGM}} = RT\sum_{i=1}^{N} x_i \ln x_i$
$\bar{a}_i^{\mathrm{IGM}}(T,P,x_i) = a_i^{\mathrm{IG}}(T,P) + RT\ln x_i$	$\Delta a_{\mathrm{mix}}^{\mathrm{IGM}} = RT\sum_{i=1}^{N} x_i \ln x_i$

Where *v*, volume; *u*, internal energy; *h*, enthalpy; *s*, entropy; *g*, Gibbs energy; *a*, Helmholtz energy; IGM, ideal gas mixture; IG, ideal gas.

TABLE 3.2 Relations for Partial Molar Quantities, Mixing Properties, and Excess Properties for Ideal Mixtures (IMs) and Real Mixtures

Ideal Mixture	Real Mixture
$\overline{v}_i^{\mathrm{IM}} = v_i$ $\Delta v_{\mathrm{mix}}^{\mathrm{IM}} = 0$	$\overline{v}_i^{\mathrm{E}} = \overline{v}_i - v_i$ $v^{\mathrm{E}} = \Delta v_{\mathrm{mix}}$
$\overline{u}_i^{\mathrm{IM}} = u_i$ $\Delta u_{\mathrm{mix}}^{\mathrm{IM}} = 0$	$\overline{u}_i^{\mathrm{E}} = \overline{u}_i - u_i$ $u^{\mathrm{E}} = \Delta u_{\mathrm{mix}}$
$\overline{h}_i^{\mathrm{IM}} = h_i$ $\Delta h_{\mathrm{mix}}^{\mathrm{IM}} = 0$	$\overline{h}_i^{\mathrm{E}} = \overline{h}_i - h_i$ $h^{\mathrm{E}} = \Delta h_{\mathrm{mix}}$
$\overline{s}_i^{\mathrm{IM}} = s_i - R\ln x_i$ $\Delta s_{\mathrm{mix}}^{\mathrm{IM}} = -R\sum x_i \ln x_i$	$\overline{s}_i^{\mathrm{E}} = \overline{s}_i - s_i + R\ln x_i$ $s^{\mathrm{E}} = \Delta s_{\mathrm{mix}} + R\sum x_i \ln x_i$
$\overline{g}_i^{\mathrm{IM}} = g_i + RT\ln x_i$ $\Delta g_{\mathrm{mix}}^{\mathrm{IM}} = RT\sum x_i \ln x_i$	$\overline{g}_i^{\mathrm{E}} = \overline{g}_i - g_i - RT\ln x_i$ $g^{\mathrm{E}} = \Delta g_{\mathrm{mix}} - RT\sum x_i \ln x_i$
$\overline{a}_i^{\mathrm{IM}} = a_i + RT\ln x_i$ $\Delta a_{\mathrm{mix}}^{\mathrm{IM}} = RT\sum x_i \ln x_i$	$\overline{a}_i^{\mathrm{E}} = \overline{a}_i - a_i - RT\ln x_i$ $a^{\mathrm{E}} = \Delta a_{\mathrm{mix}} - RT\sum x_i \ln x_i$

$$\overline{v}_i^{\mathrm{IM}}(T,P,x_i) = v_i(T,P)\cdot v_i(T,P) = \text{molar volume of the pure component} \quad (3.11)$$

The difference between properties of a real mixture and an ideal mixture (IM) is called excess property (Equation 3.12):

$$d^{\mathrm{E}} = \Delta d_{\mathrm{mix}} - \Delta d_{\mathrm{mix}}^{\mathrm{IM}} = \Delta d_{\mathrm{mix}} - \Delta d_{\mathrm{mix}}^{\mathrm{IGM}} \quad (3.12)$$

$\Delta d_{\mathrm{mix}}^{\mathrm{IM}} = \Delta d_{\mathrm{mix}}^{\mathrm{IGM}}$ if the mixture is an IGM, for example, $g^{\mathrm{E}} = \Delta g_{\mathrm{mix}} - \Delta g_{\mathrm{mix}}^{\mathrm{IM}}$ (Table 3.2).

3.2.2 Properties of Mixtures with Water

3.2.2.1 P,V,T-*Behavior and Excess Volume*

Molar volumes have been measured as functions of pressure, temperature, and composition in the homogeneous range for mixtures of water with argon [2,3], xenon [4,5], hydrogen [6], carbon dioxide, methane [7], and benzene [8]. In Figures 3.1 and 3.2, the excess volumes of water–hydrogen mixtures, water–methane mixtures, and water–carbon dioxide mixtures are shown. At moderate pressures, the excess volumes are large with a maximum lying at about 70 mol % of water for the water–hydrogen system. With increasing pressure, the excess

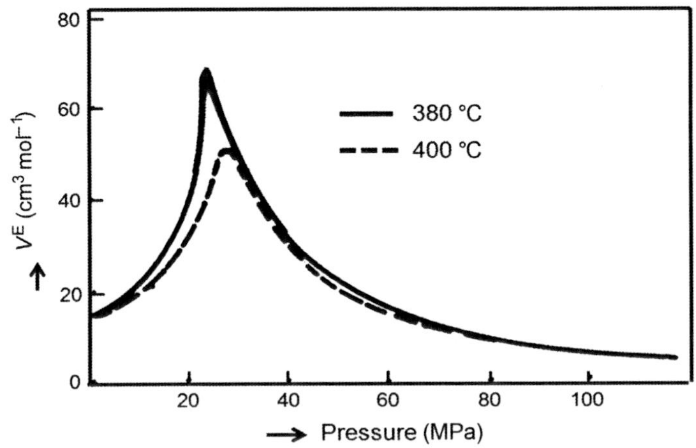

FIGURE 3.1 Molar excess volume of water–methane [7]. *Modified from Ref. [4].*

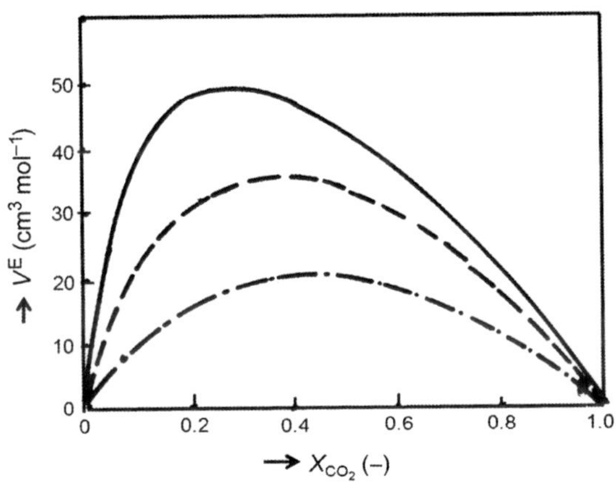

FIGURE 3.2 Molar excess volumes for CO_2–H_2O; $T = 400\,°C$; P: ——— 30 MPa, – – – 25 MPa, —•— 20 MPa. *Modified from Ref. [9].*

volumes first become greater (compare Figure 3.1 for water–methane) and then reach a maximum and become smaller with increasing pressure [7,10].

3.2.2.2 Ionic Dissociation

The equilibrium constants for the ionization reaction of a large number of electrolytes were determined from the concentration dependence of the molar conductivity. Figure 3.3 shows the negative logarithm for the ionization constant for NaCl in aqueous solutions as a function of pressure for constant temperatures as an example. Ionization constants of solutes generally decrease if the temperature is raised at constant density and increase with increasing density at constant temperature. A linear relationship exists between the

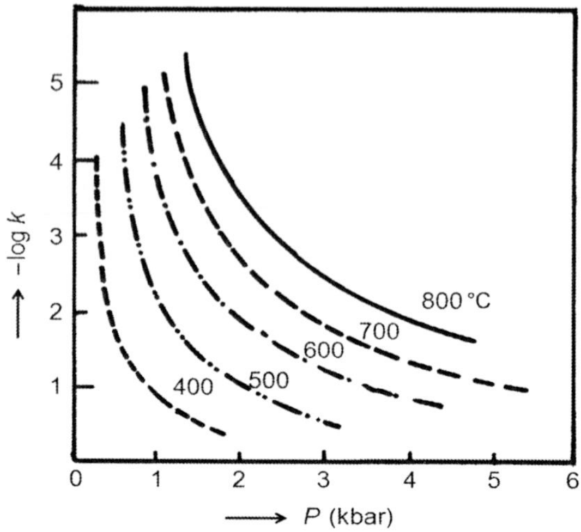

FIGURE 3.3 Equilibrium constant for the reaction $NaCl \leftrightarrow Na^+ + Cl^-$. *Modified from Ref. [10].*

logarithms of the ionization constant and of the density of water over a broad range of density at a given temperature. The slope is only slightly temperature-dependent. Thus, the linear relationship may be used for inter- and extrapolation purposes [10].

The ion product of water (K_w) may be influenced by adding electrolytes. The addition of up to 3 mol kg^{-1} NaCl or KCl to pure water at 25 °C not only has a noticeable effect on the proton activity but also leaves the ion product of water essentially unchanged. But at an increased temperature of $T=300$ °C and saturation pressure conditions, log K_w is enhanced by 1.7 units in a 3 M solution relative to pure water. Because of both the strong increase of the ion product of water with rising temperature and density and the decrease of ionization constants of solute with increasing temperature, hydrolysis reaction becomes more important in supercritical aqueous solution. The equilibrium constant for the hydrolysis reaction $Cl^- + H_2O \rightarrow HCl + OH^-$, for example, is by about three orders of magnitude larger at $T=500$ °C and $P=200$ MPa ($\rho=700$ kg m^{-3}) than at normal conditions [10].

3.2.2.3 Transport Properties

The presence of ionic solutes generates new diffusion coefficients and also modifies to some extent the transport properties of water in the solution. This section 3.2.2.3, deals mainly with those transport properties that are a direct consequence of the presence of ionic solutes, that is, electric conductivity and solute diffusion. A detailed survey of transport properties of aqueous solutions at elevated pressure and temperature was published by Corti *et al.* [11]. The following section is mainly based on this review.

3.2.2.3.1 Electric Conductivity

The molar conductance of a dilute aqueous solution is almost zero at high temperatures and low density. With increased density up to about $\rho = 200\ \text{kg m}^{-3}$, the conductivity rises slowly, then more rapidly, passes through a minimum at about $\rho = 600\ \text{kg m}^{-3}$, and finally decreases at still higher density. Such density dependence is generally found and is caused by the ionization behavior of electrolytes in water. In very dilute water, neutral molecules are stable; at intermediate densities, molecules and ions are present in comparable amounts; and at the maximum of the conductivity, ions are the prevailing species in the ionization equilibrium. High mobility of the ions at this condition increases the molar conductance by one order of magnitude compared to that at ordinary conditions. With further increase of density, the mobility of the ions is lowered and the molar conductance decreases parallel to the reciprocal viscosity [10].

The electric conductivity of pure water can be obtained from the known values of K_w, the ion product in molal scale [12], and the limiting conductivities of the hydrogen ion $\lambda^0_{H^+}$ and the hydroxide ion $\lambda^0_{OH^-}$ (Equation 3.13):

$$\Lambda^0_w = c_{H^+}\lambda^0_{H^+} + c_{OH^-}\lambda^0_{OH^-} = K_w^{1/2}\rho\left(\lambda^0_{H^+} + \lambda^0_{OH^-}\right) \tag{3.13}$$

with ρ as density.

The presence of ions increases the electric conductivity. The contribution of the solvent itself, of the H^+ and the OH^- ions, to the total conductivity can be estimated from Equation (3.13).

The values for the ion limiting conductivity and the self-diffusion coefficient can be derived from the simple hydrodynamic Stokes model where the viscous friction ζ_v on a spherical object of radius r moving through a continuum solvent of viscosity η_0 is given by

$$\zeta_v = A\pi r\eta_0 \tag{3.14}$$

A is a constant that depends on the boundary conditions (four for slip and six for stick conditions, respectively). By using this expression for friction, it is possible to obtain expressions for the limiting ionic conductivity:

$$\lambda^0_i = \frac{z^2 eF}{A\pi r\eta_0} \tag{3.15}$$

which is known as the Nernst–Einstein equation.

According to Equation (3.15), a plot of the product $\lambda^0_i\eta_0$ versus r^{-1} should yield a straight line for all ions bearing the same charge z, in a given solvent such as water. Experimental data show that for the smaller ions, the product is lower than that predicted by the Nernst–Einstein equation when crystallographic radii are used for the ions.

For the limiting transport coefficients of ions in water, there is no sufficient model available. An empirical temperature-dependent equation was proposed [13] to predict the ion limiting conductances at temperatures up to 473 K:

$$\log_{10}(\lambda^0\eta_0) = A + B/T \tag{3.16}$$

Values of the parameters A and B are tabulated for several ions [13]. This model is suitable for predicting the temperature dependence of the limiting conductivity along the saturation line. Since no density dependence is included, it cannot be used for calculating conductivity away from the saturation line.

In most practical applications, complex species such as metal halide or metal hydroxide complexes play an important role. Very little experimental information is available about the limiting conductivities of complexes. An estimation of the limiting ionic conductivities of ions formed by the association of ions of unsymmetrical electrolytes, such as $CaCl^+$ or $NaSO_4^-$, can be obtained from Equation (3.15) by assuming that the volume of the complex ions is equal to the sum of the volume of the n constituent simple ions [14]:

$$\frac{z_{\text{complex}}}{\lambda^0_{\text{complex}}} = \left[\sum_{i=1}^{n}\left(\frac{z_i}{\lambda_i^0}\right)^3\right]^{1/3} \tag{3.17}$$

This equation is considered suitable for predicting ionic conductivities of complex ions at room temperature [11].

The influence of the concentration (c) on the conductance of an electrolyte in very dilute solutions is represented by Equation (3.18):

$$\Lambda = \Lambda^0 - S \cdot c^{1/2} \tag{3.18}$$

where S is the limiting-law slope, which in the case of symmetrical electrolytes can be expressed as $S = \alpha\Lambda^0 + \beta$, where α and β are given by

$$\alpha = \frac{82.046 \times 10^4 z^2}{(\epsilon T)^{3/2}} \tag{3.19}$$

$$\beta = \frac{8.2487 z}{\eta(\epsilon T)^{1/2}} \tag{3.20}$$

with the units of S being S cm^2 mol$^{-3/2}$ dm$^{3/2}$ (the molar concentration is usually expressed as mol dm^{-3}) and the water viscosity, η, expressed in Pa s [11].

The concentration dependence of the molar conductivity in aqueous solutions at temperatures near or above the critical point of water is dominated by the association constant. For a detailed discussion of the molar conductivity in aqueous solutions, the reader is referred to the article of Corti *et al.* [11].

Electric Conductivity in Multicomponent Mixtures To calculate the electric conductivity of multicomponent mixtures, it is necessary to use a mixing rule that utilizes the conductivities, either experimentally obtained or calculated, of binary subsystems containing one cation and one anion. Such mixing rules can be written in terms of molar fractions or solute fractions with equivalent ionic strength and the specific conductivity of constituent binary subsystems:

$$\kappa(K) = a_1\kappa_1(K) + a_2\kappa_2(K) \tag{3.21}$$

where a_1 and a_2 are the fractions of binary subsystems 1 and 2, respectively, and the specific conductivities of the binary subsystems (K_1 and K_2) are evaluated at constant concentration (K), which can be either constant total molarity, constant equivalent concentration, or constant ionic strength.

A general mixing rule for multicomponent systems has been proposed [14] wherein the contributions of constituent binary cation–anion pairs are averaged:

$$\kappa = c_{eq}\sum_{M=1}^{N_c}\sum_{X=1}^{N_A} f_M f_X\left[\lambda_{M(X)}(I) + \lambda_{X(M)}(I)\right] = c_{eq}\sum_{M=1}^{N_c}\sum_{X=1}^{N_A} f_M f_X \Lambda^0_{MX}(I) \tag{3.22}$$

where c_{eq} is the total equivalent concentration; f_M and f_X are the equivalent fractions of the cation and anion, respectively; $\lambda_{M(X)}$ is the conductivity of cation M in the presence of anion X; $\lambda_{X(M)X(M)}$ is the conductivity of anion X in the presence of cation M; and N_c and N_A are the total numbers of cations and anions, respectively.

The equivalent fractions are defined as

$$f_i = \frac{|z_i|c_i}{c_{eq}} \tag{3.23}$$

The equivalent concentration c_{eq} is given by

$$c_{eq} = \sum_{M}^{N_c} c_M|z_M| = \sum_{X}^{N_A} c_X|z_X| \tag{3.24}$$

The conductivities $\lambda_{M(X)M(X)}$ and $\lambda_{X(M)X(M)}$ are defined at constant molar ionic strength I. For this purpose, these quantities are calculated at the concentrations of the ions in a binary pair MX given by

$$c_M = \frac{2I}{|z_M|(|z_M| + |z_X|)};\ c_X = \frac{2I}{|z_X|(|z_M| + |z_X|)} \tag{3.25}$$

Equation (3.25) has been derived to satisfy the condition of a constant ionic strength. This mixing rule gives accurate predictions for the electric conductivity of mixed systems, as illustrated in Figure 3.4 for the system NaCl–$MgCl_2$–H_2O. In this example, the conductivities in the binary subsystems Na^+–Cl^- and Mg^{2+}–Cl^- were calculated with effective ionic radii [14] and the conductivities of the mixed system were obtained using Equations (3.22)–(3.25). Experimental data to test the mixing rule at high temperatures over

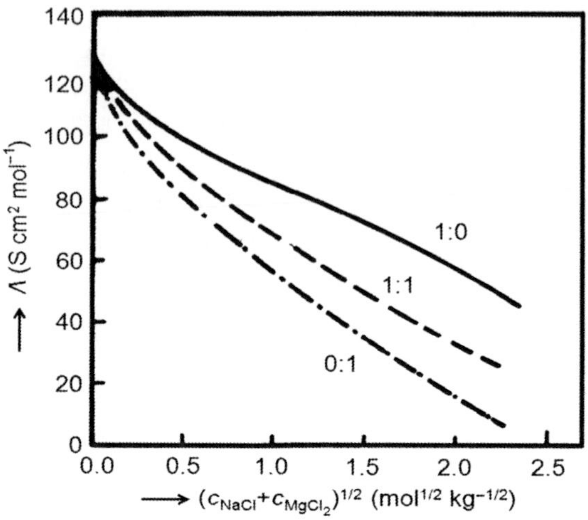

FIGURE 3.4 Electric conductivity in an aqueous mixture: NaCl–$MgCl_2$–H_2O at 298.15 K. NaCl:$MgCl_2$: ——— 1:0, – – –: 1:1, —•—: 0:1. *Modified from Ref. [11].*

substantial concentration ranges are not available. Nevertheless, the mixing rule should be equally applicable at high and low temperatures [15].

3.2.2.3.2 Diffusion

Using the Stokes formulation for friction (Equation 3.14), the limiting diffusion coefficient can be derived as

$$D_i^0 = \frac{kT}{A\pi r\eta_0} \tag{3.26}$$

which is known as the Stokes–Einstein equation.

For diffusion, it is necessary to distinguish between self-diffusion and mutual diffusion. In mutual diffusion, due to the requirement of electric neutrality, positive and negative ions move along the concentration gradient at the same speed. Therefore, in a binary solution, there is only one mutual diffusion coefficient. In self-diffusion, the electric neutrality constraint does not apply. Self-diffusion coefficients are defined separately for all species in the solution. Thus, mutual diffusion coefficients are usually measured for the whole electrolyte, whereas self-diffusion coefficients are obtained for individual species [11].

Self-Diffusion The concentration dependence of self-diffusion has been studied with the methods of statistical mechanics. According to the results, the relaxation effect causes the deviation of the self -diffusion coefficient from its value at infinite dilution

$$D_i = D_i^0\left(1+\frac{\delta k_i}{k_i}\right) \tag{3.27}$$

where $\frac{\delta k_i}{k_i}$ is the relaxation term and is given by

$$\frac{\delta k_i}{k_i} = -\frac{\kappa_D z_i^2 e^2}{3\varepsilon kT}\left(1-\sqrt{d}\right) \tag{3.28}$$

where κ_D is the inverse Debye length (Equation 3.29), z_i is the charge, and ε is the dielectric constant.

$$\kappa_D = \left(8\pi e^2 N_A/(\varepsilon kT)\right)^{1/2} I^{1/2} \tag{3.29}$$

In the simple case of a tracer species 1 in an electrolyte containing ions 2 and 3, the function takes the form

$$d = \frac{|z_1|}{|z_1|+|z_2|}\left(\frac{|z_2|\lambda_2^0}{|z_1|\lambda_2^0+|z_2|\lambda_1^0} + \frac{|z_3|\lambda_3^0}{|z_1|\lambda_3^0+|z_3|\lambda_1^0}\right) \tag{3.30}$$

where λ_i^0 denotes the limiting conductivity of ion i.

This model is valid only for dilute solutions. Anderko and Lencka [16] proposed a model for self-diffusion that is applicable to concentrated aqueous electrolyte solutions and for ionic and nonionic species. In systems with substantial ionic concentration, long-range interionic forces and short-range forces can be combined with all other interparticle forces on the same basis and can be treated as short-range forces. Solution properties can be calculated by methods similar to those for nonelectrolytes [11]. Therefore, Equation (3.27) can be modified as follows:

$$D_i = D_i^0\left(\frac{D_i^{HS}}{D_i^0}\right)\left(1+\frac{\delta k_i}{k_i}\right) \tag{3.31}$$

where the first term in parentheses denotes the hard-sphere contribution. Figure 3.7 shows the behavior of diffusing species in the system $LiCl + H_2O + O_2$ at $T = 298$ K to illustrate the general behavior of self-diffusion as a function of concentration. As shown by the dotted lines, the model can predict the composition dependence up to ca. 1 M using crystallographic radii (Figure 3.5) [11].

Mutual Diffusion For the volume-fixed diffusion coefficient in electrolyte or nonelectrolyte solutions, a general expression has been proposed for a binary system. The mutual diffusion coefficient, D_V, is given by

$$D_V = \frac{\eta_B^0}{\eta}\left(x_B D_{AB}^0 + x_A D_{BB}^0\right)\frac{\partial \ln a_A}{\partial \ln x_A} \tag{3.32}$$

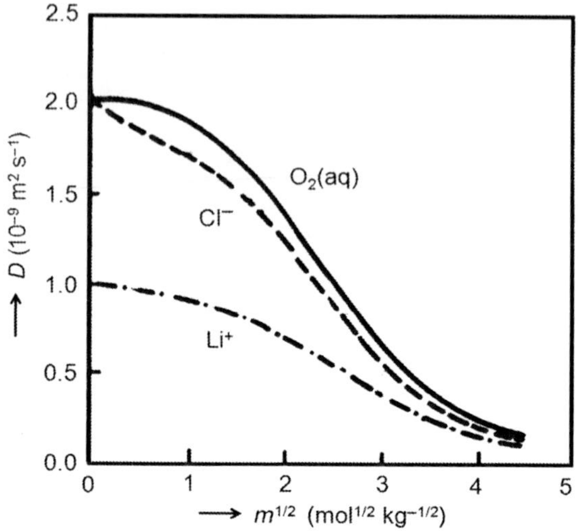

FIGURE 3.5 Self-diffusion coefficients for Li+, Cl^-, and O_2 in the system $LiCl + H_2O + O_2$ at 298.15 K. *Modified from Ref. [11].*

where x denotes the mole fraction, η is the viscosity, D^0_{AB} is the tracer diffusion coefficient of A at infinite dilution in B, and D^0_{BB} is the self-diffusion coefficient of B in B. Because of symmetry, the same value of D_V can be obtained by switching the subscripts A and B.

The effect of concentration on both diffusion and viscosity is analogous. Therefore, data or correlations for viscosity can be used to predict the concentration dependence of mutual diffusion. It is generally observed over wide concentration ranges that the largest effects on mutual diffusivity are due to the thermodynamic term $\partial \ln a_A / \partial \ln x_A$ and the viscosity correction. In dilute binary solutions, the concentration dependence of mutual diffusion is primarily due to the electrophoretic effect. In concentrated solutions, the electrophoretic effect becomes numerically small in comparison to the thermodynamic term.

A predictive correlation was proposed that is valid up to fairly high concentrations. It is assumed that the diffusing entity is a hydrated solute with the hydration number h:

$$D_{MX} = \left(D^0_{MX} + \Delta_1 + \Delta_2\right)\left(1 + m\frac{\mathrm{d}\ln\gamma}{\mathrm{d}m}\right)(1 - M_w hm)\left[1 + M_w m\left(\frac{vD^0_w}{D^0_{MX}} - h\right)\right]\frac{\eta_0}{\eta} \tag{3.33}$$

where Δ_1 and Δ_2 are the electrophoretic corrections, the second term in parentheses is the thermodynamic term expressed in terms of molality, D^0_w is the self-diffusion coefficient of pure water, η_0 is the viscosity of pure water, η is the viscosity of the solution, and v is the number of ions that result from the dissociation of the solute. The electrophoretic terms are given, for a binary solution, by

$$\Delta_n = kTA_n \frac{\left(z_1^n t_2^0 + z_2^n t_1^0\right)^2}{a^n |z_1 z_2|} \tag{3.34}$$

where t_i^0 are transport coefficients at infinite dilution, which can be obtained from ionic limiting conductivities, and the coefficients A_n are the functions of the dielectric constant and viscosity of the solvent. Equation (3.33) has been shown to have good predictive capabilities for concentrations up to several mol (kg H_2O)$^{-1}$. Typical deviations are within 1% for concentrations up to 1 molal and 2–3% up to several molal [11]. Figure 3.6 shows the D_{MX}/D_{MX}^0 ratios calculated with Equation (3.33) using the approach of Lindsay [17].

Diffusion in Multicomponent Systems

Self-Diffusion In the case of self-diffusion, the Stefan–Maxwell formalism of diffusion has been used to derive a mixing rule that makes it possible to predict self-diffusivities of both ionic and neutral solution species in multicomponent solutions as long as they can be obtained for the constituent binary systems (i.e., systems containing one salt or one molecular solute in water) [11]. To derive this mixing rule, it was assumed that a multicomponent mixture contains N_c cations, N_A anions, and N_N neutral solutes. It is assumed that the amounts of the cation and anion in the hypothetical single-solute solution should be proportional to the concentrations of the cation and the anion in the multicomponent solution. Furthermore, the hypothetical single-solute

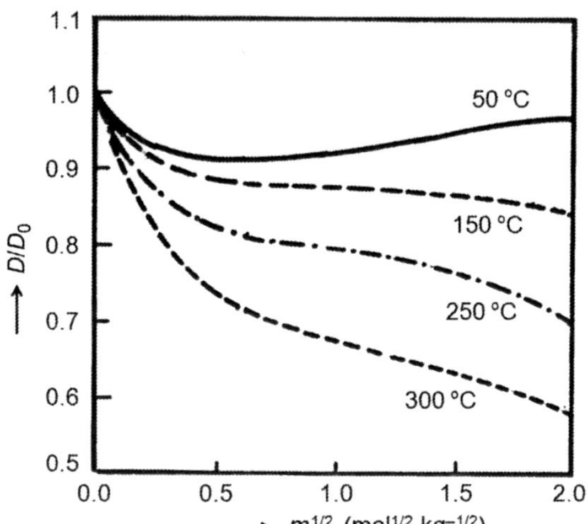

FIGURE 3.6 Concentration dependence of mutual diffusion coefficients of NaCl in H_2O. *Modified from Ref. [11].*

solution should be electrically neutral. Expressions that satisfy these conditions are given by Anderko and Lencka [16].

Mutual Diffusion It should be noted that mutual diffusion in multicomponent solutions is not a simple extension of binary diffusion. When more than one salt is present in the same solution, the restriction that anions and cations must diffuse with the same speed is lifted. This has important implications. On the phenomenological level, there may be a substantial modification in the main diffusion coefficients D_{ii} and large cross coefficients D_{ij}. On the molecular level, electrostatic interactions manifest themselves in both the electrophoretic and relaxation effects, whereas mutual diffusion in a binary solution is affected only by the electrophoretic effect.

The prediction of mutual diffusivities requires the simultaneous availability of diffusion coefficients, electric conductivity, and transference numbers for constituent binary subsystems at the temperature of interest. Such data are available for a limited number of systems, usually only at room or slightly elevated temperatures, which limits the usefulness at high temperatures. For details, the reader is referred to the discussion by Corti *et al.* [11].

3.2.2.3.3 Viscosity of Electrolyte Solutions

Proposed models for the viscosity of water–electrolyte mixtures extend only to small values of electrolyte concentrations. The viscosity of dilute solutions is not much different from that of the pure solvent. Therefore, the practical usefulness of these equations is limited. An extension to a higher but still low concentration was proposed by Jones and Dole [11,18]. For multicomponent solutions, the Jones–Dole equation can be written as

$$\eta_r = 1 + AI^{1/2} + \sum_i c_i B_i \tag{3.35}$$

where η_r is the reduced viscosity, that is, the viscosity of the solution related to the viscosity of the solvent at the same conditions. B_i are the Jones–Dole coefficients for each individual ion. This equation is valid for concentrations up to 0.1 molal, for some systems up to higher concentrations. The B_i coefficients are characteristic for each ion and are additive for electrolytes. The ionic coefficients can be determined from those for individual solutes with the often used convention $B_{K^+} = B_{Cl^-}$. For concentrations up to several molal, the Jones–Dole equation was extended by adding a quadratic term [19]:

$$\eta_r = 1 + Ac^{1/2} + Bc + Dc^2 \tag{3.36}$$

Based on this concept, a general expression for multicomponent systems valid up to saturation for most aqueous systems was proposed [20]:

$$\eta_r = 1 + AI^2 + \sum_i c_i B_i + \sum_i \sum_j (f_i f_j D_{ij} I^2) \tag{3.37}$$

where f_i and f_j are fractions of species i and j and D_{ij} is the interaction parameter between i and j. The fractions f_i are defined as modified molar fractions, that is,

$$f_i = \frac{c_i/l_i}{\sum_k (c_k/l_i)} \tag{3.38}$$

where the factor l_i is the greater of $|z_i|$ or 1. For concentrated solutions, the parameter D_{ij} in Equation (3.37) depends on the ionic strength:

$$D_{ij} = d_1 + d_2 I + d_3 \exp\left[0.08\left(I/\text{mol kg}^{-1}\right)^{3/2}\right] \tag{3.39}$$

where d_1, d_2, and d_3 are empirical parameters. The parameters d_2 and d_3 are required only for systems with a substantial ionic strength (usually above 5 molal). They have a weak temperature dependence, which can be expressed by a simple exponential function:

$$d_i = d_{i,0} \exp[d_{i,1}(T - 273.15\text{K})] \tag{3.40}$$

Equations (3.37)–(3.40) are suitable for modeling viscosity up to the saturation limit in wide temperature ranges, as shown in Figure 3.7 for NaCl–H_2O.

Viscosity in Multicomponent Systems Unlike electric conductivity and diffusivity, viscosity can be calculated from a model that is directly applicable to multicomponent solutions. For dilute solutions, the Jones–Dole equation (Equation 3.35) can be rigorously written for multicomponent systems because of the additivity of the B coefficient for individual ions. Also, the semiempirical species–species interaction contribution that extends the Jones–Dole equation to concentrated solutions (Equation 3.37) is given in a

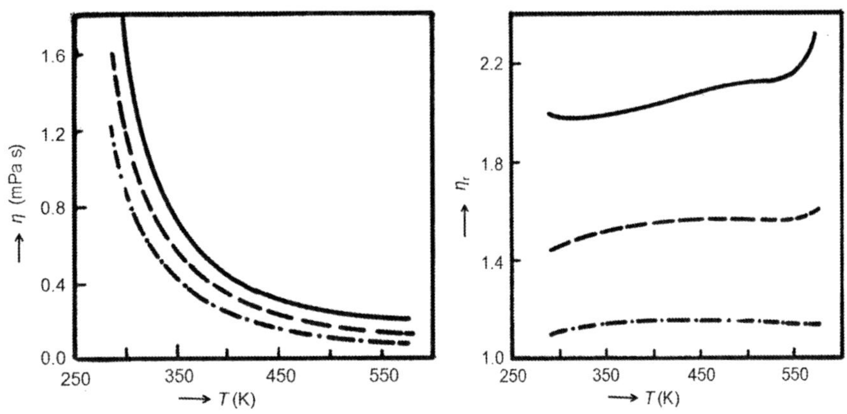

FIGURE 3.7 Viscosity of NaCl–H_2O solutions; molarity: ——— 6.038 m, – – – 3.5161 m, —•— 1.0661 m. *Modified from Ref. [11].*

multicomponent form. Lencka *et al.* [20] verified the performance of this equation for selected systems containing multiple salts and obtained good agreement with experimental data.

3.2.2.3.4 Thermal Conductivity of Electrolyte Solutions

Thermal conductivity of dilute and moderately concentrated electrolyte solutions is described by a simple linear equation in the molar concentration proposed by Riedel [21] at room temperature:

$$\lambda = \lambda_0 + \sum_i \alpha_i c_i \tag{3.41}$$

where λ_0 is the thermal conductivity of pure water and α_i is the contribution of ion i.

McLaughlin extended this equation to $T = 373$ K by assuming that the thermal conductivity of the electrolyte solutions has the same temperature dependence as pure water. For calculating the temperature dependence of the thermal conductivity, McLaughlin proposed the equation for the thermal conductivity (in W m^{-1} K^{-1}) [22]:

$$\lambda(T,m) = \frac{1.1622\,\lambda_0(T)}{\lambda_0(T_0)}\left[0.515 - \alpha_s \frac{1000\rho(T,m)m}{1000 + M_s m}\right] \tag{3.42}$$

where λ is the thermal conductivity (in W m^{-1} K^{-1}), ρ is the density of the solution, M_s is the molar mass of the salt, α_s is the sum of the α-coefficients of the anion and cation, λ_0 is the thermal conductivity of pure water, and $T_0 = 293$ K.

Most of the data on the thermal conductivity of electrolyte solutions at high temperature and pressure were reported during the last decade by Abdulagatov and Magomedov [23] up to $T = 473$ K and $P = 100$ MPa. The thermal conductivities of the salt solutions decrease with increasing concentration, except for NaF, Na_2CO_3, Na_2SO_4, Na_3PO_4, and some alkaline hydroxides (Li, Na, and K). The coefficients a_i are tabulated for several ions and they are negative for most of the ions, except for OH^-, F^-, SO_4^{2-}, PO_4^{3-}, and CrO_4^{2-} [23]. Nagasawa *et al.* [24] analyzed Equation (3.42) for the case of NaCl solutions in the range of $T = 273$–353 K and concentrations up to 5 mol kg^{-1} and concluded that the disagreement between experimental and calculated values is 2% at most [11].

For electrolyte solutions up to $P = 100$ MPa, the thermal conductivity at constant temperature and concentration increases almost linearly with pressure [23,24] with a slope quite similar to that observed for pure water. For the pressure dependence of the thermal conductivity of electrolyte solutions, a correlation was proposed [25]:

$$\lambda(P,m) = \lambda(P_0,m)\frac{\lambda_0(P)}{\lambda_0(P_0)} \tag{3.43}$$

where λ_0 is the thermal conductivity of pure water and $P_0=0.1$ MPa. It was found that this simple correlation could reproduce experimental values for several electrolytes within 2% [23]. The temperature dependence of the thermal conductivity at constant pressure and concentration shows a maximum for the thermal conductivity at a temperature between $T=400$ and 420 K that is almost independent on pressure and shows a similar behavior as pure water [11].

3.3 PHASE BEHAVIOR OF MIXTURES WITH WATER

3.3.1 Thermodynamics of Phase Equilibrium

With respect to phase behavior (phase equilibrium), it is of interest to know the thermodynamic relations between the coexisting phases of a heterogeneous system and between equilibrium pressures and temperatures, the composition, and the volume, the enthalpy, and the entropy of the various phases. In the following, the basic equations are repeated [26].

Equilibrium is established when there are no further changes of the variables with time. An isolated system is in equilibrium if the following equations are satisfied:

$$(\delta S)_U \leq 0; \quad (\delta U)_S \geq 0; \quad (\delta G)_{T,P} \geq 0 \tag{3.44}$$

with δ indicating any change, S for the entropy, U for the internal energy, and G for the Gibbs energy. The variations of the concentration (in number of moles) are represented by the chemical potential μ (Equation 3.45):

$$\mu_i = (\partial U/\partial n_i)_{S,V,n} = (\partial A/\partial n_i)_{T,V,n} = (\partial H/\partial n_i)_{S,P,n} = (\partial G/\partial n_i)_{T,P,n} \tag{3.45}$$

Then, the equation for equilibrium in a system of m components and π phases is

$$\begin{aligned} T^1 &= T^2 = \cdots = T^\pi \\ P^1 &= P^2 = \cdots = P^\pi \\ \mu_1^1 &= \mu_1^2 = \cdots = \mu_1^\pi \\ &\vdots \\ \mu_m^1 &= \mu_m^2 = \cdots = \mu_m^\pi \end{aligned} \tag{3.46}$$

Equation (3.46) states that equilibrium is established if the values for the temperature, the pressure, and the chemical potentials are equal in all the different phases. A relation between the degrees of freedom F, the number of components m, and the number of phases π can be derived, commonly known as the phase rule, for nonreacting systems [26]:

$$F = m + 2 - \pi \tag{3.47}$$

In a binary two-phase system, one concentration in one phase in addition to temperature or pressure can be specified. At given values of pressure and temperature, the compositions of the two phases are fixed. In a ternary two-phase system in addition to temperature and pressure, one concentration in one phase has to be specified in order to fix the problem for a unique solution.

The chemical potential is not easily understood as a physical quantity like pressure and temperature. Lewis therefore defined a function f, which he called fugacity. In the rigorous equation $(\partial\mu_i/\partial P)_T = V_i$ for V_i, the volume of an ideal gas $V_i = (RT)/P$ is incorporated and the equation is integrated at constant temperature. The resulting equation is the relation between the chemical potential of an ideal gas and pressure:

$$\mu_i - \mu_i^0 = RT\ln\left(P/P^0\right) \tag{3.48}$$

with μ_i the chemical potential of component i, μ_i^0 the chmical potential of component i at a reference state, P the system pressure, P^0 a reference pressure (same pressure as for the reference chemical potential), and T the temperature.

For real gases and substances, the real behavior is by definition represented by the *fugacity f*, replacing pressure P as variable:

$$\mu_i - \mu_i^0 = RT\ln\left(f_i/f_i^0\right) \tag{3.49}$$

With fugacity, the equilibrium condition for the chemical potential can be transformed. For a gaseous (v) and a liquid (l) phase, the relation between chemical potential and fugacity is

$$\mu_{i,\mathrm{v}} - \mu_{i,\mathrm{v}}^0 = RT\ln\left(f_{i,\mathrm{v}}/f_{i,\mathrm{v}}^0\right) \quad \text{and} \quad \mu_{i,\mathrm{l}} - \mu_{i,\mathrm{l}}^0 = RT\ln\left(f_{i,\mathrm{l}}/f_{i,\mathrm{l}}^0\right) \tag{3.50}$$

Substituting these equations into the equilibrium equation, we obtain

$$\mu_{i,\mathrm{v}}^0 + RT\ln\left(f_{i,\mathrm{v}}/f_{i,\mathrm{v}}^0\right) = \mu_{i,\mathrm{l}}^0 + RT\ln\left(f_{i,\mathrm{l}}/f_{i,\mathrm{l}}^0\right) \tag{3.51}$$

If the standard states are equal, the equilibrium relation in terms of fugacity is readily obtained (Equation 3.52). But even if the standard states are not equal, they can be transformed by the relation between fugacity and chemical potential:

$$f_{i,\mathrm{v}} = f_{i,\mathrm{l}} \tag{3.52}$$

In international chemical engineering literature, fugacity is used, while in more fundamental approaches, chemical potential is preferred. If we define the fugacity coefficient φ_i,

$$\varphi_i = f_i/(y_i P) \tag{3.53}$$

equilibrium conditions for a gas–liquid system can be written as

$$\varphi_i^{\mathrm{v}} y_i = \varphi_i^{\mathrm{l}} x_i \tag{3.54}$$

The fugacity coefficient can be calculated from volumetric properties:

$$RT\ln\varphi_i = \int_0^P [\overline{v}_i - (RT)/P]\,\mathrm{d}P \tag{3.55}$$

where $\overline{v}_i = (\partial V/\partial n_i)_{T,V,n_j}$ is the partial molar volume of component i.

Since most EOS, representing volume V, are explicit in P, a transformation for independent variables V and T may be more useful:

$$RT\ln\varphi_i = \int_V^{\infty} \left[(\partial P/\partial n_i)_{T,V,n_j} - (RT)/V\right]\mathrm{d}V - RT\ln z \tag{3.56}$$

where $z=(PV)/(RT)$ is the compressibility factor. Besides the different independent variables, there is furthermore the difference between these two equations: $\overline{v}_i$ is a partial molar quantity while $\mathrm{d}P/\mathrm{d}n$ is not.

The equations for the equilibrium phases can be combined and the fugacity coefficient eliminated. For a liquid and a gaseous phase in equilibrium, we obtain

$$RT\ln(x_i/y_i) = \int_{P^+}^{P} [\overline{v}_i(y_i) - \overline{v}_i(x_i)]\,\mathrm{d}P \tag{3.57}$$

where $\overline{v}_i(y_i)$ and $\overline{v}_i(x_i)$ are the partial molar volumes of component i in the vapor and the liquid phase, respectively, and P^+ is a reference pressure, low enough that ideal behavior can be assumed.

Now, the equations for phase equilibrium can be solved, provided a function $V=f(T,P,n_i)$, an EOS, is available that allows to calculate integrals and derivatives with sufficient accuracy.

Different types of mixtures with water occur and are of interest for process applications. Gases dissolve in liquid water. They form physical solutions or react to some extent with water. Important examples are oxygen or nitrogen for nonreacting species, carbon dioxide for a slightly reactive species, and ammonia for a highly reactive species. Phase equilibrium (physical equilibrium) will be treated from low to supercritical temperatures, covering the range that occurs in processes. Chemical equilibrium will be addressed separately for reactive species.

Mixtures of water and nonelectrolytes behave very differently at low, high, or supercritical temperatures. Interaction of volatile and nonvolatile components with water will be treated. Electrolytes are dissolved readily by water at subcritical temperatures but are only slightly soluble at supercritical temperatures up to moderate pressures. Solubility of electrolytes will be discussed since it plays an important role in supercritical water oxidation and hydrothermal processes. The influence of salts on phase equilibria will also be covered in some detail [27]. As far as possible, some statements on structure of aqueous solutions will be presented. Status of dissolved gases, change of mixture properties with temperature, in particular near-critical and supercritical temperatures, and structure of electrolyte solutions at lower and supercritical temperatures will be discussed. Critical curves will also be discussed. They represent the limits for the coexistence of more than one phase. The number of coexisting phases is determined by the phase rule. For a nonreacting system, the phase rule (Equation 3.47) determines the number of degrees of freedom F for a mixture of m components and π coexisting phases. According to Equation (3.47), a binary system may exist in up to four phases at equilibrium. An overview on binary phase equilibria will be discussed first.

3.3.2 Phase Equilibrium in Binary Aqueous Systems

Within this section, an overview on binary aqueous phase behavior is given. Water can be the compound with the higher critical temperature, as in CO_2–H_2O, or the compound with the lower critical temperature, as in H_2O–NaCl. Furthermore, within a series of homologue compounds, the role of the water as component can change, as is discussed for water–hydrocarbon systems in Section 3.4.3. In the following, phenomenology of phase behavior will be explained: first, phase equilibria with only fluid phases and, then, equilibria with fluid and solid phases.

3.3.2.1 The Phenomenology of Binary Gas–Liquid and Gas–Liquid–Liquid Systems

Using EOS, different types of phase behavior have been calculated by systematically varying the parameters. van Konynenburg and Scott [28] suggested five types that can be defined by characteristic values formulated with the parameters a and b of the van der Waals EOS. This simple equation is able to represent the various types of phase behavior, although agreement between experimental and calculated results is not good. Rowlinson and Swinton [29] reviewed this work and added a sixth class, which can only be predicted by angle-dependent potential functions. The six types of phase behavior of binary systems are shown in Figure 3.8 and are briefly discussed later in the text. Only P,T-projections are presented, since the other projections can be derived. The following discussion is taken from Brunner "Gas Extraction" [26]:

> *Type I*: continuous, uninterrupted gas–liquid critical curve. Azeotropic behavior is possible.

If the components of a binary system are not too different in molecular weight and chemical structure, they are miscible in the liquid phase, and the vapor–liquid coexistence region is limited by a continuous critical curve, leading from the critical point of one component to the critical point of the other component. Such a system is classified as type I. Its P,T-diagram is shown in Figure 3.8a. The phase equilibrium shown is characteristic for binary systems of a supercritical component and a compound of medium volatility.

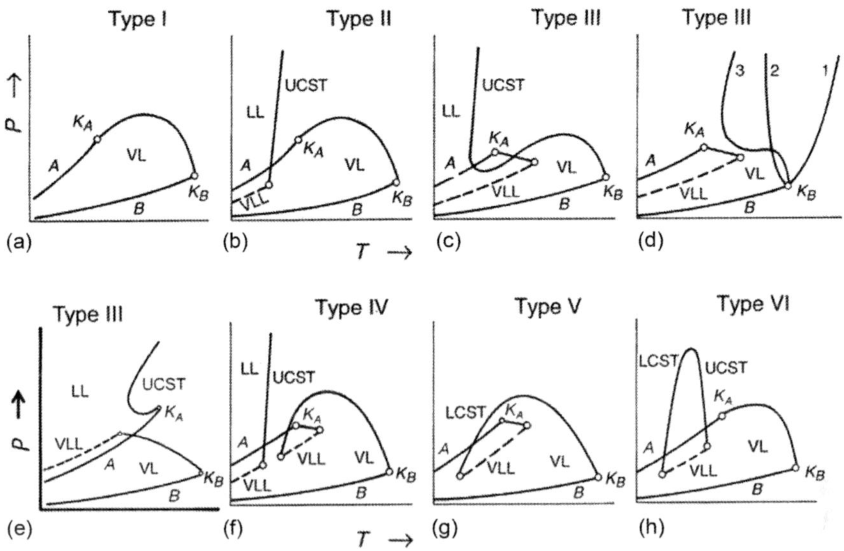

FIGURE 3.8 The six types of binary phase behavior of gas–liquid systems. *P*,*T*-projections of phase boundary lines (*P*,*T*-diagrams) for different classes of phase behavior of binary systems. VLL, vapor–liquid–liquid equilibrium; VL, vapor–liquid equilibrium; LL, liquid–liquid equilibrium; UCST, upper critical solution temperature; LCST, lower critical solution temperature; *A*, component *A* (with the lower critical temperature); *B*, component *B* (with the higher critical temperature); K_A, critical point of component *A*; K_B, critical point of component *B*. See text for examples. A seventh class, reported in Ref. [27], is not included in the overview (see discussion). *Modified from Ref. [26].*

At the limiting values for the composition ($x=0$ and $x=1$), vapor-pressure curves of components *A* and *B* separate regions of a gas from that of a liquid. The vapor-pressure curves end at the critical point of each component (K_A and K_B). The critical curve connects the critical points and limits the two-phase region, which extends between the vapor-pressure curves and below the critical line.

Example: H_2O–NH_3

Type II: continuous, uninterrupted gas–liquid critical curve. Azeotropic behavior is possible. Three-phase line liquid–liquid–gas ends below the critical temperature of the more volatile component. The liquid–liquid critical curve is nearly vertical.

As differences in the physical and chemical properties of components become greater, immiscibility in the liquid phase may occur. The effect of increasing differences in the properties of compounds can be demonstrated for a series of homologous compounds. For paraffins with ethane as the more volatile compound in each binary, liquid immiscibility comes up with components of more than 19 carbon atoms. In many cases, miscibility gaps extend to the critical region. McGlashan and Schneider [30] systematically investigated and reviewed this phase behavior and the following discussion is partly based on their work.

In systems with an uninterrupted gas–liquid critical curve, at lower temperatures, immiscibility in the liquid phase can emerge. This phase behavior is classified as type II (Figure 3.8b). The upper critical solution temperature (UCST) may increase with pressure or decrease with pressure. At temperatures below the UCST, liquid–liquid equilibria exist, which extend to the three-phase line vapor–liquid–liquid equilibrium (VLL). At pressures below the VLL line, a gaseous and a liquid phase coexist at equilibrium. At temperatures higher than the UCST and pressures higher than the vapor pressure of *A* or higher than pressures of the critical curve, there is complete miscibility between components *A* and *B*.

Examples

Water–biphenyl [31]
Water–naphthalene [32]
Water–tetrahydronaphthalene [33]

Nonaqueous systems:
n-Octane–carbon dioxide

Type III: systems with a divided gas–liquid critical curve. Both legs start at the critical point of one of the components. One leg ends at the endpoint of the three-phase line. The other leg extends to higher temperatures and pressures, in some cases passing through a temperature minimum and a pressure minimum. Type III systems are subdivided according to Figure 3.8c–e. They are discussed in more detail in Sections 3.4.2 and 3.4.3.

With decreasing tendency of mutual miscibility, the region of immiscibility extends to higher temperatures, as does the critical curve (see Figure 3.8c). The critical curve is divided into two branches. This phase behavior is classified as type III. The gas–liquid critical curve extends from the gas–liquid critical point of one component and runs to the UCST. The second leg extends from the critical point of the other component. In some systems, there is a pressure minimum of the critical curve at relatively low temperatures (see Figure 3.8c). If the mutual solubility of both the components is slightly higher than for a type III system shown in Figure 3.8c, then the critical curve may extend to the liquid–liquid line and divide it into two parts, as shown in Figure 3.11f. This phase behavior is classified as type IV [29] (see succeeding text). The gas–liquid critical curve can tend to be at higher temperatures until in curve 1 (Figure 3.8d) and the total critical curve extends to temperatures beyond the critical temperature of the less volatile compound. At pressures above this curve and temperatures above this critical temperature, the so-called gas–gas equilibria exist. Examples are d_1, helium–ammonia and helium–xenon; d_2, ammonia–nitrogen and water–carbon dioxide; and d_3, *n*-hexadecane–carbon dioxide.

Considering the behavior of one solvent (e.g., H_2O) with a series of homologue compounds, the phase behavior can change. This was shown by Brunner in his systematic investigation of phase behavior of different compounds with water, for example, for the *n*-alkane–water systems [36] (see Figures 3.8e and 3.28). For *n*-alkanes with less than 26 carbon atoms (<C26), one leg of the discontinuous liquid–gas critical curve connects the liquid–gas UCEP (upper critical endpoint) to the hydrocarbon critical point. The other leg extends from the critical point of water to higher pressures after passing through a *P* and *T* minimum. For *n*-alkanes with 28 and more carbon atoms ($\geq$C28), the liquid–gas critical curve beginning at the hydrocarbon critical point extends to higher pressures after passing through a *P* and *T* minimum. The other leg of the critical curve extends from the critical point of water to the liquid–gas UCEP. Such a phase behavior can be expected for high-molecular-weight hydrocarbons with critical temperatures substantially higher than that of water. The reason is that the density of the organic liquid remains relatively constant while the density of the aqueous liquid decreases significantly near the critical temperature of water [33].

Further examples

Water–*trans*-decalin
Water–benzene
Water–methyl-substituted benzenes (toluene, *o*-xylene, *p*-xylene, and 1,3,5-trimethylbenzene) [33]

Type IV: systems with three critical curves according to Figure 3.8f. The third critical curve begins at the critical point of the component with the higher critical temperature and ends at the critical endpoint of the three-phase line.

Example: methane–*n*-hexene

If no liquid–liquid immiscibility at low temperatures occurs before crystallization, then phase behavior is as shown in Figure 3.8g, which is classified as type V. The lower part of the critical curve between component *B* and the VLL line has the character of a lower critical solution temperature (LCST). The character of the critical curve changes gradually with increasing temperature to a vapor–liquid equilibrium (VL) critical curve, which is also true for some of the critical curves of type III and the critical curve of type IV phase behavior.

Type V: systems with two critical curves as type IV, but without liquid–liquid critical curve.

Examples

Ethane–ethanol
Methane–*n*-hexane

At low temperatures, liquid–liquid immiscibility is possible in a type I system. Such a phase diagram is classified as type VI. Type VI systems are only found in systems where one or both components exhibit self-association due to hydrogen bonding, and in the mixture, there is additional strong intercomponent hydrogen bonding [29]. Such mixtures are important with the application of hydrothermal and supercritical water.

Type VI: systems with liquid immiscibility over a certain range of temperature and critical curves like that shown in Figure 3.8h.

Examples

D_2O–2-methylpyridine
Water–*n*-butanol
Water–2-butanol

A further discussion of aqueous systems in the critical region is presented in Sections 3.4.2 and 3.4.3. Between all the types shown in Figure 3.8, there are intermediate cases of two- or three-phase region and can be used to derive estimates of the phase behavior of multicomponent systems.

3.3.2.2 Binary Systems with Solid Phases

A solid phase often occurs in binary mixtures of components of very different volatility, in addition to fluid phases. Such equilibria exist in systems where the melting point temperature of the low-volatile component is higher than the critical temperature of water. Phase behavior of such a binary mixture is presented in Figure 3.9 for the case where the solid phase does not extend to the critical curve and in Figure 3.10 for the case where the solid phase extends beyond the critical curve and divides it into two sections. To simplify understanding, it is assumed that there is no immiscibility in the liquid phases. Furthermore, the discussion is restricted to near-critical and higher temperatures for component *A*, water, excluding solid phases of *A*.

Figure 3.9 presents the three-dimensional plot and the *P*,*T*-projection of the phase behavior of a binary system, consisting of a volatile component (component *A*, water) and a nonvolatile component (component *B*). The melting point temperature of the nonvolatile component is higher than the critical temperature of the volatile component. Three cuts at constant temperature illustrate phase equilibrium in different regions. Curve $(V+L)_A$ is the vapor-pressure curve of component *A* and ends at its critical point K_A and at lower temperatures at the triple point A_t of component *A* (not shown). The three-phase line $V+L+B$, the vapor-pressure curve of saturated solutions of *B*, runs

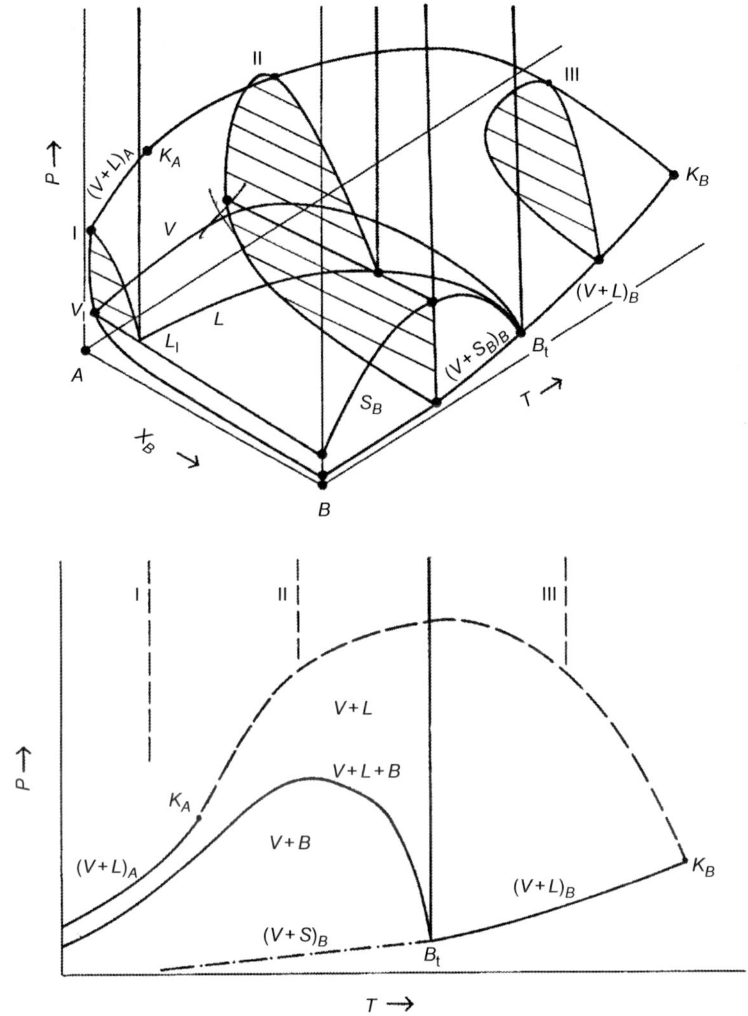

FIGURE 3.9 Phase equilibrium in a binary system. The melting point temperature of the low-volatile component is higher than the critical temperature of the solvent (water). Three-dimensional plot (upper part) and P,T-diagram (lower part). *Reproduced from Ref. [26].*

to a maximum pressure and then declines to the vapor pressure of B at its triple point B_t. From this point, the vapor-pressure lines $(V+L)_B$ and $(V+S)_B$ proceed. The vapor-pressure curve of liquid component B ends at the critical point K_B. The critical points of both pure components are connected by the critical curve K_AK_B, which limits the two-phase area for VL.

Lines V_lB_t and L_lB_t represent the gas–liquid equilibrium of the three-phase equilibrium $V+L+B$. This line proceeds to the triple point of component B. In the region marked $V+B$, a gaseous phase coexists with solid B. If the

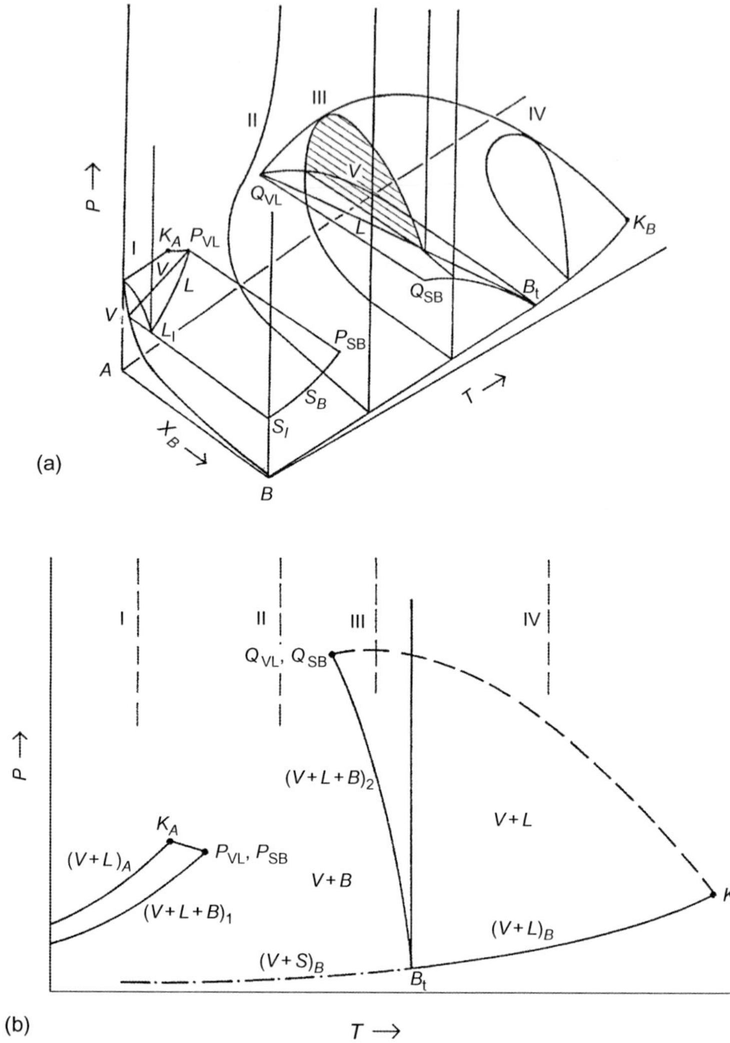

FIGURE 3.10 Phase equilibrium in a binary system. The melting point temperature of the low-volatile component is higher than the critical temperature of the solvent. The solubility curve divides the critical line into two parts. Three-dimensional plot (part a) and *P*,*T*-diagram (part b). *Reproduced from Ref. [26].*

pressure is raised at constant temperature in this region to values at which solid *B* melts, the gas–liquid equilibrium region marked $V+L$ is reached. An example is the system H_2O–NaCl, although only part of the whole phase diagram can be investigated due to the high temperatures.

Figure 3.10 presents the three-dimensional plot and the *P*,*T*-projection of the phase behavior of a binary system, in which the solid region extends beyond the critical curve. Four cuts at constant temperature illustrate phase equilibrium in different regions. Line $L_l P_{VL}$ represents the solubility of B in liquid A, which decreases in the region of the critical temperature of A. Line $V_l P_{VL}$ represents the solubility of B in gaseous A, which increases in the same region until both lines meet at the critical endpoint P_{VL}. In the P,T-projection, the critical endpoint is represented by a single point, while in the three-dimensional plot, it can be seen that there are two endpoints of different composition, namely, P_{VL}, representing the endpoint of VL, and P_{SB}, representing the endpoint for solid B. Lines $V_l P_{VL}$, $L_l P_{VL}$, and $S_l P_{SB}$ fall on a single line $(V+L+B)_1$ in the P,T-projection.

The melting point of component B is lowered by component A, as line $(V+L+B)_2$ shows, starting from the triple point B_t of pure B. The pressure necessary to keep the volatile component in solution rapidly increases, as can be seen from the right-hand side of part b of Figure 3.10. The coexisting gaseous phase increases in density. More and more of component A dissolves in the coexisting liquid phase, until the compositions become equal at the critical endpoint Q. At the critical endpoint, a single fluid phase Q_{VL} is in equilibrium with solid B at Q_{SB}. Lines of the coexisting phases fall on a single line $(V+L+B)_2$ in the P,T-projection, as was discussed in the preceding text for the lower temperature part of the diagram. An example for this behavior is the system H_2O–SiO_2 [37].

In this type of systems, equilibrium between a gaseous and a solid phase is found not only at pressures below the three-phase line $V+L+B$ but also at temperatures between P and Q. In this region, no liquid phase exists. Only a gaseous and a solid phase coexist at equilibrium (see Figure 3.10, cut II). Pressure can be enhanced to very high values in order to achieve high solubilities of the low-volatile component in the supercritical solvent. In the pressure region, to where an extension of the critical lines would lead, a relatively large increase in solubility of B in A with pressure can be achieved. This is due to the transformation of phase equilibrium corresponding to cuts I and III to that of cut II. At temperatures higher than the melting point of B, a gaseous and a liquid phase coexist. The region between P and Q can be used for extraction of solid compounds from solid substrates, insoluble in the solvent, and the region at higher temperatures than the melting temperature, for gas–liquid separations.

More on water–salt systems is presented in this chapter in Sections 3.5–3.7. The formation of solid phases of component A, the solvent (water), has been omitted in this discussion, since this region is not of importance in the context of this book. The interested reader is referred to review articles, for example, the one by Koningsveld and Diepen [38].

Understanding of phenomenological phase behavior is fundamental for selecting operating conditions for processes. Such processes nearly always involve more than two components. Phase behavior of such multicomponent systems can often be explained by the binary systems that are the basis for multicomponent systems. In order to understand the effect of adding an additional component to a binary system, in the following section, phase behavior of ternary systems is discussed for some examples.

3.3.3 Phase Equilibria of Ternary Systems

In the context of this book, processes are of interest where the phase behavior of aqueous systems is exploited. For the dissolution and precipitation of one pure component, binary systems, as discussed earlier in the text, provide sufficient information on equilibrium. Even a mixture of more than two components of similar volatility may be treated with sufficient accuracy as a pseudo-binary system. But in many cases, phase behavior of three different components must be considered, for example, for the separation of mixtures into compounds, the modification of the properties of a solvent, or the common case that a mixture of compounds is in contact with a supercritical solvent. Therefore, in the following, an example of phase behavior of three-component systems will be discussed shortly. More on ternary systems is contained in the following paragraph, in context with the calculation of phase equilibria.

In Figure 3.11, the phase behavior of the ternary system CO_2–n-$C_{16}H_{34}$–H_2O is shown for a constant temperature of $T = 473.15$ K (200 °C) and for a pressure of $P = 20$ MPa. The pressure dependence and phase equilibrium at

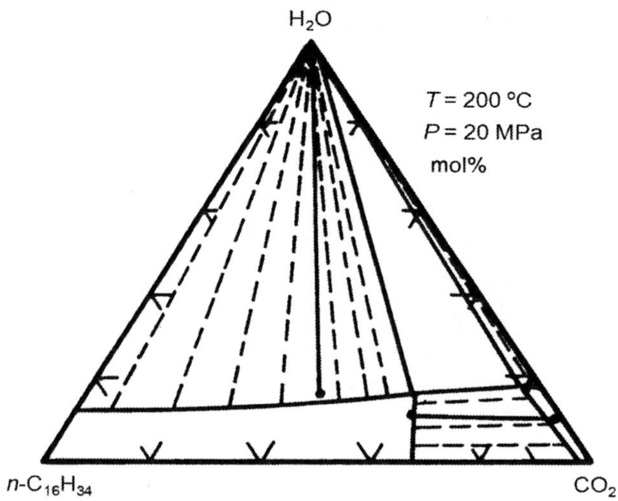

FIGURE 3.11 Phase behavior of the ternary system CO_2–n-$C_{16}H_{34}$–H_2O at $T = 200$ °C and $P = 20$ MPa, mol%. Experimental results (points at the end of full lines connecting equilibrium phases) and calculation results using the Peng–Robinson equation of state. The calculation procedure is discussed. Parameters: $k_{12} = 0.2584$; $k_{13} = 0.0173$; $k_{23} = 0.1020$. The phase behavior is explained in more detail in Figure 3.12. *Modified from Ref. [35].*

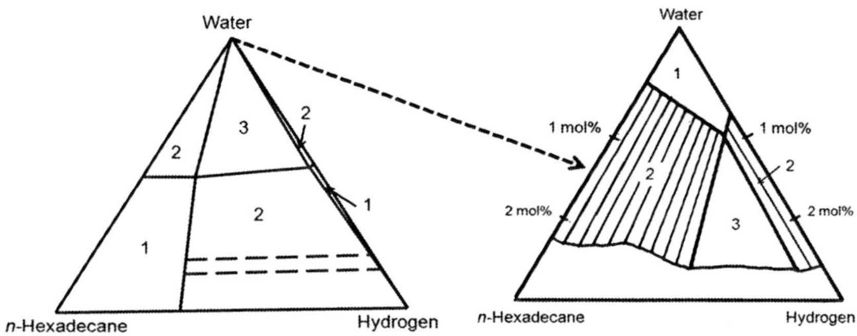

FIGURE 3.12 Phase equilibrium in the ternary system H_2–H_2O–$C_{16}H_{34}$. $T = 300\ °C$ and $P = 20$ MPa. The number of coexisting phases is indicated in the diagram. On the right side, the upper part is enlarged [34].

a temperature of $T = 573.15$ K (300 °C) are shown in Figure 3.41. Such phase equilibrium may be important for hydrothermal processes where hydrocarbons are transformed and the products have to be recovered at low pressures. It is interesting to note that the phase behavior at pressures below $P = 25$ MPa shows three two-phase regions (the third one is a small area on the right side of the triangle) and a three-phase area, as shown on a larger scale in Figure 3.12 for the ternary system H_2–H_2O–$C_{16}H_{34}$.

A good part of the phase diagram can be constructed with the information given by the binary systems. But the extent of three-phase equilibria cannot be concluded from binary phase equilibria only. Here, besides the experimental information on ternary systems, the modeling of phase equilibria with EOS becomes important.

3.3.4 Calculation of *P,V,T*-Behavior and Phase Behavior of Mixtures with Water

Empirical correlations with a great number of coefficients of no physical meaning are inappropriate for modeling *P,V,T*-behavior and phase behavior of mixtures. Only for very few substances, the coefficients of the empirical EOS have been determined and practically no rules for combining the pure component parameters to mixture parameters are known. Therefore, in this section, EOS will be discussed for the application to mixtures. The EOS have already been presented in Chapter 2 and are now extended to describe mixtures. The application will be restricted to the Soave–Redlich–Kwong EOS and the Peng–Robinson EOS and a volume-translation modification of these EOS. Examples will be presented using the free available program for phase equilibrium calculations PE [39] that was already introduced in Chapter 2. Further information on this topic can be obtained from [40–44].

3.3.4.1 EOS for Mixtures

For the purposes of calculating phase equilibria in mixtures at elevated temperatures and pressures, EOS are by far the most useful equations up to now. Therefore, within the scope of this book, only EOS are considered. Suitable EOS are presented in Chapter 2.

For a liquid and a gaseous phase in equilibrium, we obtain

$$RT\ln(x_i/y_i) = \int_{P^+}^{P} [\bar{v}_i(y_i) - \bar{v}_i(x_i)]\,\mathrm{d}P \tag{3.57}$$

For the calculation of phase equilibria, it is now assumed that an EOS is able to represent the properties of all the components of the mixture and the properties of the mixture as a pseudo-component (one-fluid theory) [26]. The properties of the mixture are calculated by applying rules by which the properties of the mixture are related to the pure component properties, the so-called mixing rules. Then, the parameters of the EOS are mixture parameters. For the Soave–Redlich–Kwong EOS,

$$P = \frac{RT}{V - b_m} - \frac{a_m(T)}{V(V + b_m)} \tag{3.58}$$

and for the Peng–Robinson EOS,

$$P = \frac{RT}{V - b_m} - \frac{a_m(T)}{V(V + b_m) + b_m(V - b_m)} \tag{3.59}$$

where $a_m(T)$ and b_m are the mixture parameters of the EOS. The modified EOS are defined for mixtures in an analogous way.

The relation between the mixture parameters and the pure component parameters, that is, the dependence of mixture parameters on composition, called the mixing rule, has to be defined in such a way that using as simple a relation as possible, mixture properties are correctly represented. Most common are quadratic mixing rules that are for a binary mixture:

$$a_m(T) = x_1^2 a_{11}(T) + 2x_1x_2a_{12}(T) + x_2^2 a_{22}(T) \tag{3.60}$$

$$b_m = x_1^2 b_{11} + 2x_1x_2b_{12} + x_2^2 b_{22} \tag{3.61}$$

where x stands for the concentration in one phase; x_i, a_{ii}, and b_{ii} refer to pure component values; and b_{12} and $a_{12}(T)$ are binary parameters. The mixing rules for the other phases are analog. The binary parameters are mostly written in the form

$$a_{12}(T) = [a_{11}(T)\cdot a_{22}(T)]^{0.5}(1 - k_{12}) \tag{3.62}$$

$$b_{12} = 0.5(b_{11} + b_{22})(1 - l_{12}) \tag{3.63}$$

The absolute values $|k_{12}|$ and $|l_{12}|$ are small compared to 1. They are fitted to binary experimental data, usually concentrations of phases in heterogeneous equilibrium. In many cases, especially in systems with nonpolar or weak polar compounds, one binary parameter in the attractive term (k_{12}) is sufficient. Two binary parameters in the mixing rules (k_{12} and l_{12}) may lead to an indefinite number of paired values of these parameters, which represent the data with about the same deviations [26].

Binary systems of nonpolar and weakly polar compounds, and surprisingly also water, are correlated well by the presented EOS, over a wide range of temperatures and pressures, as can be seen in the examples. In the critical region of the binary systems, there are greater deviations from experimental values.

The importance of binary systems is their relative simplicity and the deep insight into principal phase behavior that can be extracted from binary systems. For the representation of multicomponent system, it is common practice to use binary interactions only and to calculate multicomponent properties with binary interaction coefficients [26]. The mixing rules for a multicomponent system with m components are then given by Equations (3.64)–(3.66), with the parameter α for the temperature dependence of the attractive parameter a_{ii} and if only one binary interaction parameter k_{ij} is employed:

$$a_m(T) = \sum_i^m \sum_j^m \left(x_i x_j a_{ij}^+\right) \tag{3.64}$$

$$a_{ij}^+ = \left(a_{ii} a_{jj}\right)^{0.5} \left(1 - k_{ij}\right) \quad \text{for } i \neq j \tag{3.65}$$

$$a_{ii} = a_i T^{\alpha} \tag{3.66}$$

$$b_m = \sum_{i=1}^m (x_i b_i) \tag{3.67}$$

For practical correlation purposes, this procedure sometimes must be modified. For ternary systems with immiscibility in two binary systems, the phase equilibrium calculation will yield sufficiently accurate results using only interaction coefficients determined from binary systems. But for ternary systems with immiscibility in only one binary system, the results often are not adequate. Therefore, the binary interaction coefficients have to be fitted to ternary data. In general, for design purposes, pseudo-binary interaction coefficients should be calculated from multicomponent data, if experimental data are available [26].

The mixing rules, presented earlier in the text, are simple but proved to be inadequate for asymmetrical systems and polar components. The introduction of composition-dependent mixing rules is one way to overcome these problems. The most general one is to introduce the excess free enthalpy (G_E) into the mixing rule. For the excess free enthalpy, any model, especially any activity coefficient model, can be employed. Wong and Sandler [45] proposed to use the excess free energy (A_E) instead of the free excess enthalpy, because the introduction of G_E violates the quadratic composition dependence of the second virial coefficient. Using the excess free energy to introduce the composition dependence of the interactions, Wong and Sandler formulated mixing rules that satisfy the second virial composition dependence [46].

The Wong–Sandler mixing rule defines mixture parameters a and b by the following equations:

$$b-\frac{a}{RT}=\sum_i\sum_j\left(x_ix_j\left(b-\frac{a}{RT}\right)_{ij}\right) \tag{3.68}$$

For the binary interaction coefficient, the following term was proposed:

$$\left(b-\frac{a}{RT}\right)_{ij}=(b_i+b_j)/2-\sqrt{a_ia_j}(1-k_{ij})/(RT) \tag{3.69}$$

Mathias *et al.* [47] proposed a new mixing rule (Equation 3.70; called Mathias–Klotz–Prausnitz mixing rule, MKP) that is particularly useful for modeling systems containing similar components. Using this mixing rule yields the same calculation results if the mole fraction x_i of a component is arbitrarily split into two mole fractions x_i' and x_i'' with $x_i'+x_i''=x$. Many mixing rules in this case produce different results. The effect is called Kistenmacher syndrome [48] and is important for the calculation of phase equilibria of similar compounds. If this effect is avoided, real systems with similar components, for example, isomers, can be correlated [44,47].

Combining the Wong–Sandler mixing rule with the MKP mixing rule, the following equations are obtained:

$$a=\sum_{i=1}^{N}\sum_{j=1}^{N}\left[x_ix_j\sqrt{a_ia_j}\left(1-k_{ij}^{\mathrm{MKP}}\right)\right]+\sum_{i=1}^{N}x_i\left(\sum_{j=1}^{N}x_j\left(\sqrt{a_ia_j}\lambda_{ji}^{\mathrm{MKP}}\right)^{1/3}\right)^3 \tag{3.70}$$

with

$$k_{ji}^{\mathrm{MKP}}=k_{ij}^{\mathrm{MKP}} \text{ and } \lambda_{ji}^{\mathrm{MKP}}=\lambda_{ij}^{\mathrm{MKP}}$$

$$b_{ij}=\frac{b_i+b_j}{2}(1-l_{ij}) \text{ with } l_{ji}=l_{ij} \tag{3.71}$$

with k, l, and λ as interaction coefficients that are fitted to experimental data:

$$b=\sum_{i=1}^{N}\sum_{j=1}^{N}\left[x_ix_j\frac{b_i+b_j}{2}(1-l_{ij})\right]+\frac{1}{RT}\sum_{i=1}^{N}x_i\left(\sum_{j=1}^{N}x_j\left(\sqrt{a_ia_j}\lambda_{ji}^{\mathrm{MKP}}\right)^{1/3}\right)^3 \tag{3.72}$$

This mixing rule is also implemented in the phase calculation program PE [39].

3.3.4.2 Examples for Phase Equilibrium Calculations

In the following figures, examples for calculation of phase equilibria with EOS are presented. In the diagrams, the EOS and the interaction parameters are indicated. Interaction parameters were determined by fitting to experimental phase equilibrium data. For ternary systems consisting of carbon dioxide–water–cresol isomers, the binary systems are calculated (Tables 3.3 and 3.4; Figures 3.13 and 3.14).

Experimental data for the binary systems can be excellently correlated. The interaction coefficients from these calculations were used without further fitting parameters to correlate ternary equilibria. For the system phenol–water, which is totally miscible at 373.15 K, the interaction parameters were determined by minimizing the deviations of the liquid phase of the ternary system with fixed interaction parameters for the other binary systems.

TABLE 3.3 Interaction Parameters for the Peng–Robinson Equation of State for the Correlation of Binary Systems Consisting of Phenol, Toluene and Cresol-Isomers with Carbon Dioxide

	k_{ij} (–)	λ_{ij} (–)	Δx (%)	Δy (%)
CO_2 + *o*-cresol	0.1211	0.1325	6.8	4.5
CO_2 + *m*-cresol	0.1378	0.1057	2.1	3.9
CO_2 + *p*-cresol	0.0998	0.1000	4.6	7.0
CO_2 + phenol	0.0906	0.0822	2.6	3.1
CO_2 + toluene	0.0798	–	0.3	7.9

Δx and Δy are the deviations of calculated values from experimental values [44].

TABLE 3.4 Interaction Parameters for the Peng–Robinson Equation of State for the Correlation of Binary Systems Consisting of Carbon Dioxide and Cresol-Isomers with Water

	k_{ij} (–)	λ_{ij} (–)	Δx^{L1} (%)	Δx^{L2} (%)
H_2O + *o*-cresol	−0.0520	0.0569	0.6	1.1
H_2O + *m*-cresol	−0.0721	0.0812	1.1	0.8
H_2O + *p*-cresol	−0.0903	−0.0435	1.3	0.1
CO_2 + H_2O	0.0774	−0.2151	3.4	3.7

Δx and Δy are the deviations of calculated values from experimental values [44].

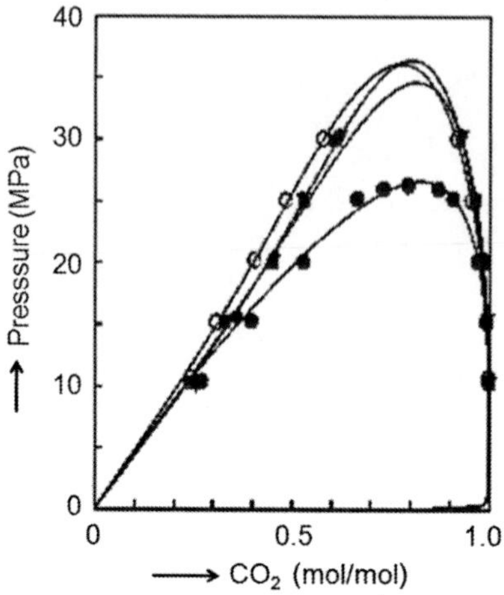

FIGURE 3.13 Binary *P,x*-diagrams for phenol and cresol isomers with carbon dioxide at $T = 373.15$ K. Experimental data: ○, phenol; ●, *o*-cresol; ■, *m*-cresol; ▼, *p*-cresol. Calculation with the Peng–Robinson equation of state; interaction parameters, see Table 3.3 [44].

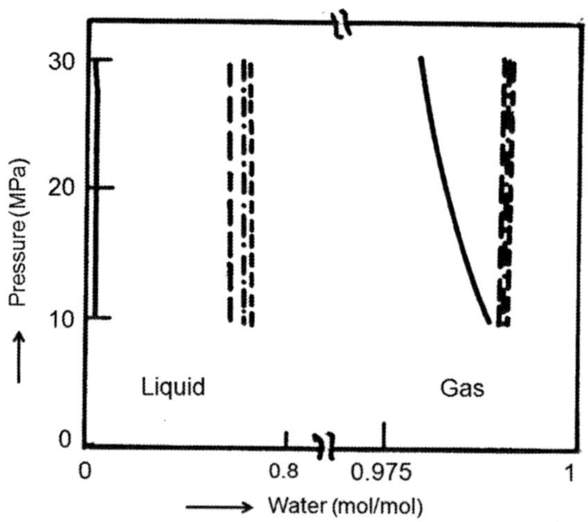

FIGURE 3.14 Binary *P,x*-diagrams for carbon dioxide and cresol isomers with water at $T = 373.15$ K (enlarged scale). ———, CO_2; ———, *o*-cresol; —•—, *m*-cresol; – – –, *p*-cresol. Calculation with the Peng–Robinson equation of state; interaction parameters, see Table 3.4 [44].

Calculated values for ternary systems with toluene, *o*-cresol, *m*-cresol, or *p*-cresol are in good coincidence with experimental data. These values are results from predictive calculations since no ternary data were used for determining the parameters. Relative deviations are listed in Table 3.5. The mean deviation is around 13%. This value seems high, but the result is due to small values for the mole fractions. Absolute deviations in the water-rich phases are

TABLE 3.5 Ternary Systems Consisting of Benzene Derivates–Water–Carbon Dioxide at $T = 373.15$ K

	Mean Deviation (%)
o-Cresol	10.6
m-Cresol	7.1
p-Cresol	15.8
Toluene	17.2
Phenol	(10.7)[1]
	−0.0067/0.0937

Mean deviations in ternary systems for the individual components for calculated values, using binary interaction parameters only, from experimental values [44].

[1] $k/\lambda_{water,phenol}$.

<0.5 mol% and, in the carbon dioxide-rich phases, <0.8 mol%. The results of the calculations are shown in Figure 3.15.

The ternary systems water–benzene derivatives–carbon dioxide can be predicted with high accuracy. Three-phase equilibria could be confirmed experimentally, with the exception of the one in the system water–phenol–carbon dioxide. Even in this case, the errors are minimal since the composition of the gas phases is nearly identical. Since the chemical potential must be equal at equilibrium, the same holds for the chemical potential for the liquid phase [44].

A second example for a ternary is shown for the system water–*n*-hexadecane–hydrogen. Such equilibria may occur in energy-related processes. In Figure 3.16, the ternary equilibrium is shown for a pressure $P = 20$ MPa and temperatures $T = 200$–350 °C. The phase boundary lines and the tie lines have been calculated with a modified Redlich–Kwong EOS using an early version of the program PE [34,49]. The variation of the phase boundary lines and the two-phase and three-phase areas with temperature can be clearly seen, which is helpful in designing equipment for heat transfer. It is interesting to note that the different cubic EOS yield about the same results. Therefore, the choice of a special EOS of this type is not critical.

A third example for ternary systems is the system water–benzene–hydrogen as shown in Figure 3.17. Pressure and temperature dependence is shown in Figure 3.37.

In conclusion, it can be stated that cubic EOS (van der Waals-type EOS) can be used for correlating phase equilibria with hydrothermal and subcritical water. They can be used as a predictive tool for conceptual design processes. The combination of the examples shown in the preceding text opens the view on the complexity of phase equilibria occurring during transformations of

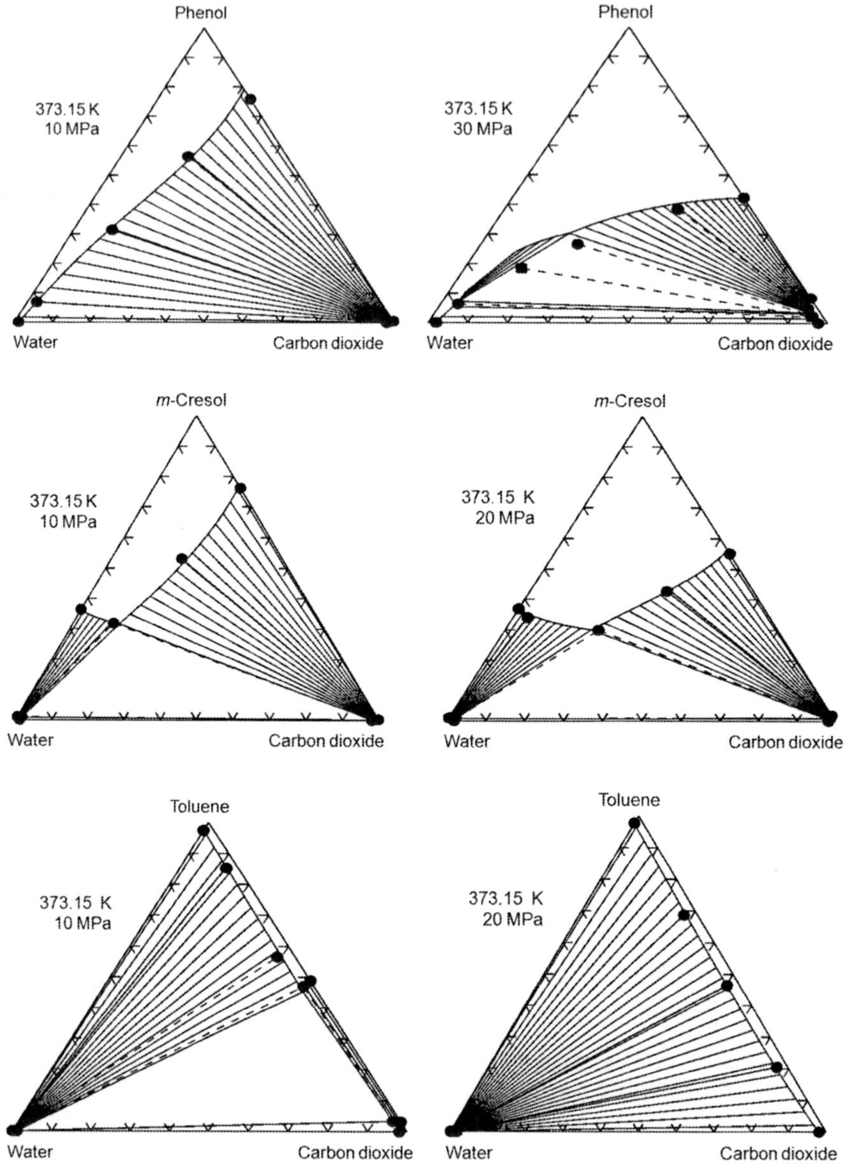

FIGURE 3.15 Ternary systems water–benzene derivatives–carbon dioxide at $T = 373.15$ K and $P = 10$–30 MPa, concentration (mol/mol). Points are experimental values. The phase boundary lines and the tie lines have been predicted with the Peng–Robinson EOS using the program PE [39]. The interaction parameters for phenol–water were determined differently (see text). The predicted three-phase equilibrium in the system water–phenol–carbon dioxide could not be confirmed by subsequent experiments [44].

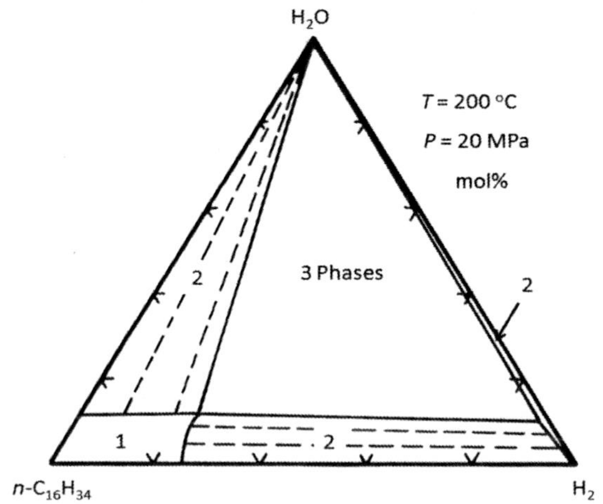

FIGURE 3.16 Phase equilibrium in the system water–n-hexadecane–hydrogen at $P=20$ MPa and $T=473.15$ (200 °C). Pressure and temperature dependence is shown in Figure 3.39. *Modified from Ref. [34].*

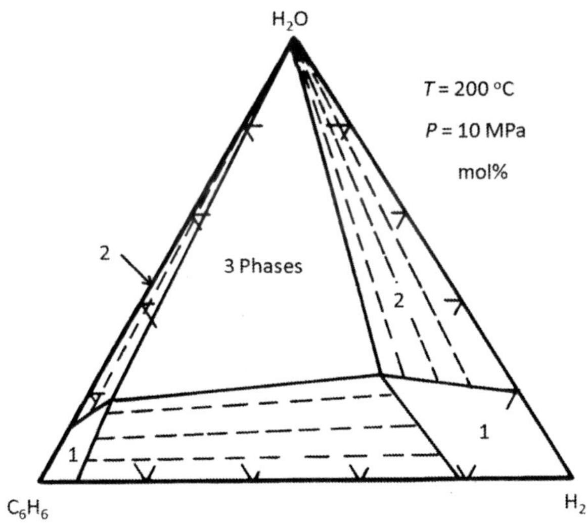

FIGURE 3.17 Phase equilibrium in the ternary system water–benzene–hydrogen. $T=473.15$ K (200 °C); $P=10$ MPa, calculated with a modified Redlich–Kwong equation of state. Pressure and temperature dependence is shown in Figure 3.37. *Modified from Refs. [34,49].*

hydrocarbons or carbon-containing material with water. For those cases, the calculation of multicomponent phase equilibria with EOS is extremely helpful. More examples of such equilibria will be presented further down in this chapter.

3.4 WATER AND NONELECTROLYTES

In the following, solubility of nonelectrolyte gases in water using mainly Henry's law is treated. In a short overview, phase behavior of gases with water and the critical region are addressed. Phase behavior of water and organic nonelectrolyte compounds, mostly hydrocarbons, is discussed for binary and some ternary systems.

3.4.1 Solubility of Nonelectrolyte Gases in Water

In this section, the solubility of nonelectrolyte gases in water is discussed. The solubility of nitrogen in water serves as an example. Critical curves, phase behavior at the critical point, and general phase behavior are reviewed. Thermodynamics of aqueous mixtures of gases, quantitative calculation of the solubility with Henry's law, and the distribution coefficient K_D are presented using the International Association for the Properties of Water and Steam (IAPWS) formulation.

3.4.1.1 Thermodynamics (Henry's Law) and Solubility of Gases

The solubility of gases in water and their distribution between coexisting liquid and gaseous phases are fundamental properties for processes with hydrothermal and supercritical water. In many cases, the volatile nonelectrolyte is the main component of the gaseous phase and water is the main component of the coexisting liquid phase. The two fluid phases are very different in density and in composition. This situation of dilute equilibrium phases occurs if the partial pressure of water is much lower than that of the volatile nonelectrolyte and the aqueous liquid phase is a dilute solution [51]. Nevertheless, the underlying thermodynamic equations are that for equilibrium:

$$f_{i,\mathrm{v}} = f_{i,\mathrm{l}} \tag{3.52}$$

The fugacity in the gaseous phase can be written as

$$f_{i,\mathrm{v}} = \varphi_i y_i P \tag{3.73}$$

with φ_i as the fugacity coefficient of component i, y_i as the mole fraction of component i in the gaseous phase, and P as the total pressure.

Using the concept of activity coefficient for the liquid phase, the fugacity in the liquid phase $f_{i,\mathrm{l}}$ can be written as

$$f_{i,\mathrm{l}} = \gamma_i x_i f_i^0 \tag{3.74}$$

with γ_i as the activity coefficient of component i, x_i as the mole fraction of component i in the liquid phase, and f_i^0 as the fugacity at a standard state.

At conditions where the gaseous phase can be considered as ideal, no interaction forces are active and the fugacity coefficient $\varphi_i = 1$. With the definition for the partial pressure $P_i = y_i P$, the fugacity for the gaseous phase is equal to P_i. In the liquid phase, for the simplest case, the activity coefficient γ_i can be considered to be equal to one and the standard state for the fugacity to be the saturation pressure P_i^{SAT}. With these simplifications, the equilibrium between the gaseous phase and the liquid phase can be written as

$$P_i = x_i P_i^{\mathrm{SAT}} \tag{3.75}$$

This is called the ideal solubility. The concentration in the liquid phase, the solubility of component i, depends only on the partial pressure of component i in the gaseous phase and the saturation pressure, hence on the temperature. The solvent is not of influence. Calculating the ideal solubility yields results mostly in the right order of magnitude, but of course not accurate ones. Therefore, it can only be used for very rough estimations.

Early in the nineteenth century, it was already found that for a specified constant temperature, the ratio of the partial pressure, $P_2 = y_2 P$, of a volatile nonelectrolyte (gas) to its solubility in the liquid solution x_2 is nearly constant:

$$k_{\mathrm{H}} = \frac{P_2}{x_2} \approx \mathrm{const} \tag{3.76}$$

This relation is now well known as Henry's law with the constant k_{H} named as Henry's constant [52]. A rigorous definition of Henry's constant is [40]

$$k_{\mathrm{H,solvent}} = \lim_{x_i \to 0} \frac{f_{i,l}}{x_i} \quad \text{(at constant temperature and pressure)} \tag{3.77}$$

For conditions where Equation (3.77) is applicable, the simple law of Henry becomes a very useful predictive tool [51]. Henry's constant $k_{\mathrm{H,solvent}} = k_{\mathrm{H,s}}$ depends on temperature and at high pressures also on pressure. For a discussion of pressure and temperature dependence, Henry's law can be written more generally as

$$f_i = \varphi_i y_i P = k_{\mathrm{H,s}} x_i \tag{3.78}$$

The pressure dependence of the fugacity in the liquid phase is [40]

$$\left(\frac{\partial \ln k_{\mathrm{H,s}}}{\partial P}\right) = \frac{\overline{v}_i^{\infty}}{RT} \tag{3.79}$$

with $\overline{v}_i^{\infty}$ as the partial molar volume of component i at infinite dilution.

Equation (3.79) can be integrated with the assumption that the fugacity of i at constant T and P is proportional to x_i [40]:

$$\ln \frac{f_i}{x_i} = \ln k_{\mathrm{H},i,\mathrm{s}}^{P^{\mathrm{ref}}} + \frac{\int_{P^{\mathrm{ref}}}^{P} \overline{v}_i^{\infty} \, \mathrm{d}p}{RT} \tag{3.80}$$

with P^{ref} as a reference pressure, which can be set as equal to the saturation pressure $P^{\mathrm{ref}} = P_1^{\mathrm{SAT}}$. At temperatures well below the critical temperature of the solvent, the partial molar volume $\overline{v}_i^{\infty}$ can be considered as independent of pressure. With subscript 1 referring to the solvent and subscript 2 referring to the solute, Equation (3.80) can be written [40] as

$$\ln\frac{f_2}{x_2} = \ln k_{\mathrm{H},2.1}^{P_1^{\mathrm{SAT}}} + \frac{\overline{v}_i^{\infty}\left(P - P_1^{\mathrm{SAT}}\right)}{RT} \tag{3.81}$$

This equation is known as the Krichevskii–Kasarnovsky equation [53]. It can represent solubility data of slightly soluble gases up to high pressures, as can be seen from the example of the solubility of nitrogen in water, shown in Figure 3.21. The application of this equation is limited to small values of the solubility and temperatures well below the critical temperature of the solvent (Figure 3.18).

The influence of concentration can be taken into account by introducing a concentration-dependent activity coefficient using the simple Margules equation:

$$\ln\gamma_1 = \frac{A}{RT}x_2^2 \text{ and } \gamma_2 = \frac{A}{RT}\left(x_1^2 - 1\right), \text{ with } \gamma_2 \rightarrow 1 \text{ for } x_2 \rightarrow 0 \tag{3.82}$$

Then, Equation (3.81) is transformed to

$$\ln\frac{f_2}{x_2} = \ln k_{\mathrm{H},2.1}^{P_1^{\mathrm{SAT}}} + \frac{A}{RT}\left(x_1^2 - 1\right) + \frac{\overline{v}_i^{\infty}\left(P - P_1^{\mathrm{SAT}}\right)}{RT} \tag{3.83}$$

Equation (3.83) is known as the Krichevsky–Ilinskaya equation [55]. It is applicable to the solubility of light gases for an extended range of concentration [40].

At higher temperatures, approaching the critical temperature of water, it becomes more useful to use the equilibrium distribution of the volatile non-electrolyte, K_D, between the two fluid phases (see succeeding text). As a more

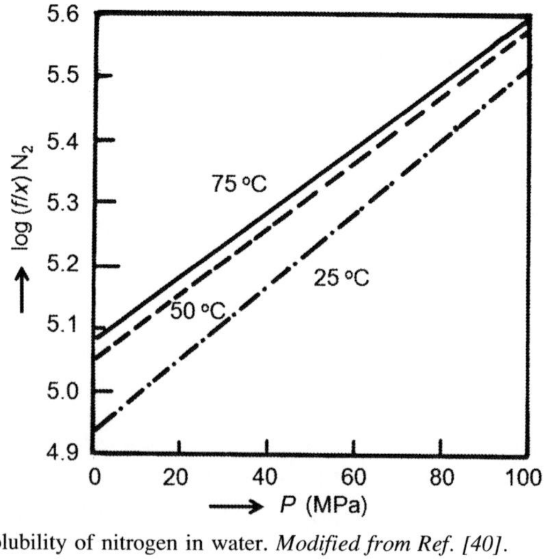

FIGURE 3.18 Solubility of nitrogen in water. *Modified from Ref. [40].*

comprehensive alternative, calculation of the two-phase equilibrium with EOS, as discussed in the text earlier, is applicable.

It is now of interest to look at real solubility data. In Figure 3.19, the solubility of nitrogen in water is shown in dependence of pressure at constant temperatures. The solubility isotherms are not exact straight lines, as they should be in case Equation (3.77) would be strictly valid. Nevertheless, a straight line is a good approximation for the pressure dependence of solubility over a wide pressure range, much more so, if the pressure range of interest is small. The temperature dependence of solubility for nitrogen in water is shown in Figure 3.20.

For nitrogen in water, as for aqueous solutions of volatile nonelectrolytes (gases) in general, beginning at ambient temperature, gas solubility decreases, passes through a minimum, and then increases as the temperature approaches

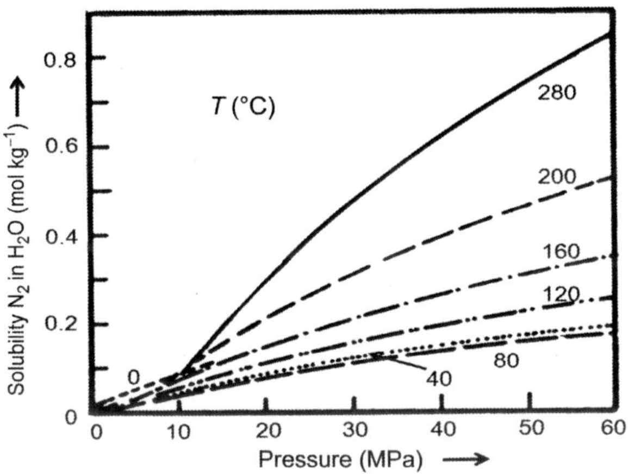

FIGURE 3.19 Solubility of nitrogen in H_2O. See Table 3.6. *Data from Ref. [56].*

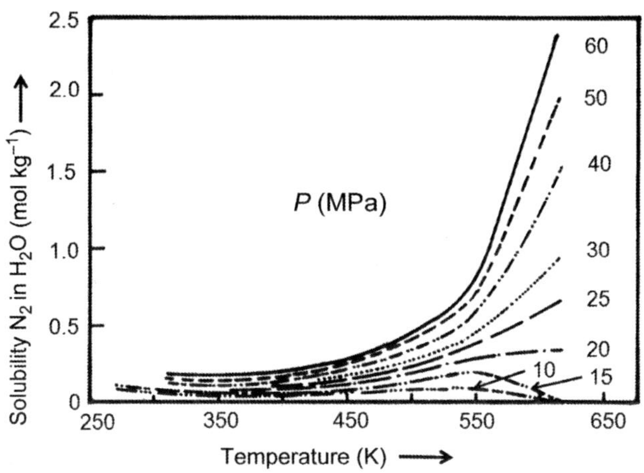

FIGURE 3.20 Solubility of nitrogen in H_2O in dependence on temperature. See Table 3.6. *Data from Ref. [56].*

TABLE 3.6 Calculated Nitrogen Solubility in Water (mol kg^{-1}) [56]

P	*T* (K)								
(MPa)	*273.15*	*313.15*	*353.15*	*393.15*	*433.15*	*473.15*	*513.15*	*553.15*	*613.15*
0.1	0.001	0.0005	0.0002	0	0	0	0	0	0
0.5	0.005	0.0027	0.0021	0.0015	0	0	0	0	0
1	0.01	0.0054	0.0043	0.004	0.0025	0	0	0	0
2.5	0.0242	0.0133	0.011	0.0112	0.012	0.0089	0	0	0
5	0.0462	0.0258	0.0216	0.0229	0.0274	0.0317	0.0243	0	0
10	0.0842	0.0485	0.0412	0.0448	0.0562	0.0743	0.0934	0.0861	0
15	0.1161	0.0686	0.0592	0.065	0.0827	0.1134	0.1566	0.1974	0.0265
20		0.0867	0.0756	0.0835	0.1071	0.1493	0.2144	0.2994	0.3441
25		0.1031	0.0908	0.1008	0.1297	0.1823	0.2672	0.3926	0.6471
30		0.1184	0.105	0.1169	0.1507	0.2127	0.3155	0.4779	0.936
35		0.1327	0.1183	0.132	0.1702	0.2408	0.3597	0.5558	1.212
40		0.1463	0.131	0.1462	0.1885	0.2667	0.4001	0.627	1.4744
45		0.1593	0.1432	0.1598	0.2057	0.2907	0.4371	0.692	1.7253
50		0.1719	0.1549	0.1727	0.2218	0.3129	0.4711	0.7513	1.9648
55		0.1826	0.1695	0.194	0.253	0.3337	0.5022	0.8055	2.1937
60		0.1963	0.1774	0.1971	0.2516	0.353	0.5308	0.8549	2.4125

the critical temperature of water, as shown for nitrogen in water in Figure 3.20 and in Table 3.6. The increase of the isothermal solubility with pressure around the critical pressure is mainly due to the dramatic density change with variation of pressure.

At sufficiently high temperature, compounds of very different polarity can form homogeneous mixtures. At these conditions, water is miscible with nonpolar gases and liquid hydrocarbons. Besides the mutual solubility of water and volatile nonelectrolytes, phase behavior is of interest, in particular the limit of coexisting phases. This information can be shown in *P*,*T*-projections of the critical curves (see this chapter, Section 3.4.2; Figure 3.24).

3.4.1.2 Henry's Constant and Solubility and the Equilibrium Distribution Coefficient Using the Krichevsky Parameter

Henry's constant is finite on the path along the vapor-pressure curve of the solvent. At the critical point, a finite limit is reached [58,59]. The initial slope of the critical line starting at the critical point characterizes the phase behavior near the critical point. The value of this slope is named the Krichevsky parameter and can be determined by the following equation [54,60,61]:

$$K_{\mathrm{Kr}} = \left(\frac{\partial P}{\partial x}\right)^{\mathrm{c}}_{v,T} = \left(\frac{\mathrm{d}T}{\mathrm{d}x}\right)^{\mathrm{c}}_{\mathrm{CRL}} \left[\left(\frac{\mathrm{d}P}{\mathrm{d}T}\right)^{\mathrm{c}}_{\mathrm{CRL}} - \left(\frac{\mathrm{d}P}{\mathrm{d}T}\right)^{\mathrm{c}}_{\sigma}\right] \tag{3.84}$$

with CRL for critical line and $(\mathrm{d}P/\mathrm{d}T)^{\mathrm{c}}_{\sigma}$ as the limiting slope of the saturation pressure at the critical point of the solvent. Using the Krichevsky parameter, Henry's constant and the distribution equilibrium constant can be written as

$$RT\ln\left(\frac{k_{\mathrm{H}}}{f_1}\right) \cong C + K_{\mathrm{Kr}}(\rho_1 - \rho_{\mathrm{c}})/\rho_{\mathrm{c}}^2 \tag{3.85}$$

with C as a solute-dependent constant, f_1 as the fugacity of the pure-solvent liquid phase, ρ_1 as the liquid-phase density of the solvent, and ρ_{c} as the critical density of the solvent.

The infinite-dilution distribution equilibrium constant K_{D} is related to the Krichevskii parameter by [58,62]

$$RT\ln K_{\mathrm{D}} = 2K_{\mathrm{Kr}}(\rho_l - \rho_{\mathrm{c}})/\rho_{\mathrm{c}}^2 \tag{3.86}$$

According to Equations (3.85) and (3.86), $T\ln(k_{\mathrm{H}}/f_1)$ and $T\ln K_{\mathrm{D}}$ are linear in the solvent density with a slope given by ($1\times$ or $2\times$) $K_{\mathrm{Kr}}/R\rho_{\mathrm{c}}^2$. The linearity is verified by experiments as shown in Figure 3.21 [59]. The experimental slope of $T\ln(k_{\mathrm{H}}/f_1)$ differs from the expected value of $K_{\mathrm{Kr}}/R\rho_{\mathrm{c}}^2 = 0.68$ K L mol^{-1}, since the asymptotic conditions are not satisfied [54].

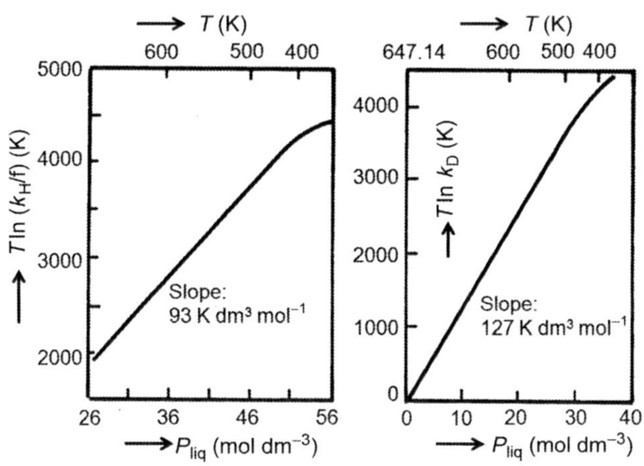

FIGURE 3.21 Linear dependence of Henry's constant and the infinite-dilution distribution coefficient K_{D} of nitrogen in water on the liquid density of water over large ranges of temperature. The asymptotic slopes should be, respectively, equal to and twice the value of $K_{\mathrm{Kr}}/R\rho_{\mathrm{c}}^2$, approximately 68 K dm^3 mol^{-1}. *Modified from Ref. [54].*

3.4.1.3 Henry's Constant and Vapor–Liquid Distribution Constant for Gases in H_2O (IAPWS Formulation)

For dilute aqueous solutions, Henry's law can be applied, as already stated in the preceding text. Approaching the critical temperature of water, the vapor–liquid distribution constant is more helpful. The IAPWS has published a guideline on both parameters [63], which is cited in the following:

Henry's constant k_H is defined by

$$k_H = \lim_{x_2 \to 0} \frac{f_2}{x_2} \tag{3.87}$$

with f_2 as the liquid-phase fugacity and x_2 as the mole fraction of the solute.

The vapor–liquid distribution constant K_D is defined by

$$K_D = \lim_{x_2 \to 0} \frac{y_2}{x_2} \tag{3.88}$$

with y_2 as the vapor-phase solute mole fraction in equilibrium with the liquid.

Henry's constant k_H is given as a function of temperature by

$$\ln\left(k_H / P_1^{SAT}\right) = \frac{A}{T_r} + \frac{B\tau^{0.355}}{T_r} + CT_r^{-0.41} \exp\tau \tag{3.89}$$

with $\tau = 1 - T_r, T_r = T/T_{c1}$, T_{c1} is the critical temperature of the solvent as recommended by IAPWS, 647.096 K for H_2O, and P_1^{SAT} is the vapor pressure of the solvent at the temperature of interest. P_1^{SAT} is calculated from the correlation of Wagner and Pruss [64] for H_2O:

$$\ln\left(P_1^{SAT} / P_{c1}\right) = T_r^{-1} \sum_{i=1}^{n} \left(a_i \tau^{b_i}\right) \tag{3.90}$$

with P_{c1} for the critical pressure of the solvent (22.064 MPa for H_2O). Values of a_i and b_i are listed in Table 3.7.

TABLE 3.7 Coefficients for Equation (3.90) for H_2O

i	a_i	b_i
1	−7.85951783	1
2	1.84408259	1.5
3	−11.7866497	3
4	22.6807411	3.5
5	−15.9618719	4
6	1.80122502	7.5

TABLE 3.8 Parameters for Equation (3.89)

Solute	A	B	C	T_{min} (K)	T_{max} (K)
He	−3.52839	7.12983	4.47770	273.21	553.18
Ne	−3.18301	5.31448	5.43774	273.20	543.36
Ar	−8.40954	4.29587	10.52779	273.19	568.36
Kr	−8.97358	3.61508	11.29963	273.19	525.56
Xe	−14.21635	4.00041	15.60999	273.22	574.85
H_2	−4.73284	6.08954	6.06066	273.15	636.09
N_2	−9.67578	4.72162	11.70585	278.12	636.46
O_2	−9.44833	4.43822	11.42005	274.15	616.52
CO	−10.52862	5.13259	12.01421	278.15	588.67
CO_2	−8.55445	4.01195	9.52345	274.19	642.66
H_2S	−4.51499	5.23538	4.42126	273.15	533.09
CH_4	−10.44708	4.66491	12.12986	275.46	633.11
C_2H_6	−19.67563	4.51222	20.62567	275.44	473.46
SF_6	−16.56118	2.15289	20.35440	283.14	505.55

Correlation of Henry's constants for aqueous solutions [63].

Values of the coefficients A, B, and C in Equation (3.89) for each system considered are listed in Table 3.8, along with the minimum and maximum temperatures of the data to which the correlations were fitted.

The vapor–liquid distribution constant K_D is given as a function of temperature by

$$\ln K_D = qF + \frac{E}{T/K} f(\tau) + \left(F + G\tau^{2/3} + H\tau\right)\exp\left(\frac{273.15 - (T/K)}{100}\right) \tag{3.91}$$

where $(\tau) = (\rho_1^{SAT}(l)/\rho_{c1}) - 1$, $q = -0.023767$, $\rho_1^{SAT}(l)$ is the liquid density along the vapor–liquid saturation boundary, ρ_{c1} is the critical density of the solvent, and $f(\tau)$ is taken from Wagner and Pruss [64]:

$$f(\tau) = \sum_{i=1}^{n} \left(c_i \tau^{d_i}\right) \tag{3.92}$$

Values of c_i and d_i are listed in Table 3.9. For Equation (3.91), the function $f(\tau) = (\rho_1^{SAT}(l)/\rho_{c1}) - 1$ is needed, not the value of ρ_{c1} itself. The values of the coefficients E, F, G, and H in Equation (3.91) for each system considered are listed in Table 3.10. The minimum and maximum temperatures for

TABLE 3.9 Coefficients for Equation (3.92) for H_2O [63]

i	c_i	d_i
1	1.99274064	1/3
2	1.09965342	2/3
3	−0.510839303	5/3
4	−1.75493479	16/3
5	−45.5170352	43/3
6	-6.7469445×10^5	110/3

TABLE 3.10 Parameters for Correlation of Vapor–Liquid Distribution Constants with Equation (3.91)

Solute	E	F	G	H
He	2267.4082	−2.9616	−3.2604	7.8819
Ne	2507.3022	−38.6955	110.3992	−71.9096
Ar	2310.5463	−46.7034	160.4066	−118.3043
Kr	2276.9722	−61.1494	214.0117	−159.0407
Xe	2022.8375	16.7913	−61.2401	41.9236
H_2	2286.4159	11.3397	−70.7279	63.0631
N_2	2388.8777	−14.9593	42.0179	−29.4396
O_2	2305.0674	−11.3240	25.3224	−15.6449
CO	2346.2291	−57.6317	204.5324	−152.6377
CO_2	1672.9376	28.1751	−112.4619	85.3807
H_2S	1319.1205	14.1571	−46.8361	33.2266
CH_4	2215.6977	−0.1089	−6.6240	4.6789
C_2H_6	2143.8121	6.8859	−12.6084	0
SF_6	2871.7265	−66.7556	229.7191	−172.7400

Solvent is H_2O unless otherwise stated [63].

these fits are identical to those listed in Table 3.8. Root-mean-square deviations for both $\ln k_H$ and $\ln K_D$ are in most cases below 5%. For details, see Ref. [63]. In Table 3.11, the values for $\ln k_H$ for the solubility of various gases in water are listed [63], and in Figure 3.22, $\ln k_H$ is plotted for some of these

TABLE 3.11 Calculated Values of $\ln(k_H/1\ \text{GPa})$ for Solutes at Selected Temperatures [63] (in H_2O Unless Otherwise Noted)

Solute	300 K	400 K	500 K	600 K
He	2.6576	2.1660	1.1973	*−0.1993*
Ne	2.5134	2.3512	1.5952	*0.4659*
Ar	1.4061	1.8079	1.1536	*0.0423*
Kr	0.8210	1.4902	0.9798	*0.0006*
Xe	0.2792	1.1430	0.5033	*−0.7081*
H_2	1.9702	1.8464	1.0513	−0.1848
N_2	2.1716	2.3509	1.4842	0.1647
O_2	1.5024	1.8832	1.1630	−0.0276
CO	1.7652	1.9939	1.1250	*−0.2382*
CO_2	−1.7508	−0.5450	−0.6524	−1.3489
H_2S	−2.8784	−1.7083	−1.6074	*−2.1319*
CH_4	1.4034	1.7946	1.0342	−0.2209
C_2H_6	1.1418	1.8495	*0.8274*	*−0.8141*
SF_6	3.1445	3.6919	2.6749	*1.2402*

Italics denote extrapolation beyond the range of fitted data.

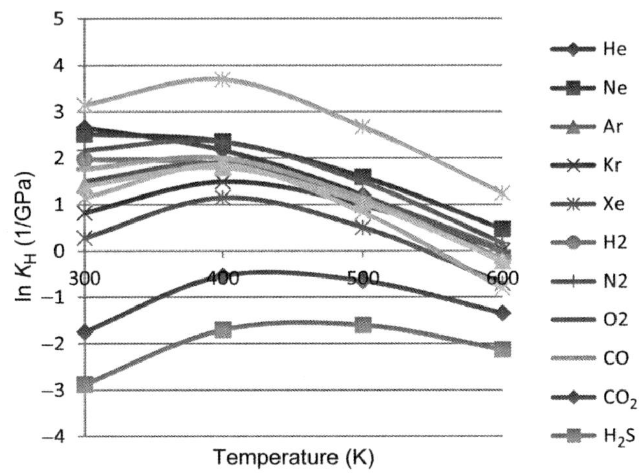

FIGURE 3.22 Values for Henry's constant ($\ln k_H$) for various gases [63]. (For color version of this figure, the reader is referred to the online version of this chapter.)

TABLE 3.12 Calculated Values of ln K_D for Solutes at Selected Temperatures (in H_2O Unless Otherwise Noted)

Solute	300 K	400 K	500 K	600 K
He	15.2250	10.4364	6.9971	*3.8019*
Ne	15.0743	10.6379	7.4116	*4.2308*
Ar	13.9823	10.0558	6.9869	*3.9861*
Kr	13.3968	9.7362	6.8371	*3.9654*
Xe	12.8462	9.4268	6.3639	*3.3793*
H_2	14.5286	10.1484	6.8948	3.7438
N_2	14.7334	10.6221	7.2923	4.0333
O_2	14.0716	10.1676	6.9979	3.8707
CO	14.3276	10.2573	7.1218	*4.0880*
CO_2	10.8043	7.7705	5.2123	2.7293
H_2S	9.6846	6.5840	4.2781	*2.2200*
CH_4	13.9659	10.0819	6.8559	3.7238
C_2H_6	13.7063	10.1510	*6.8453*	*3.6493*
SF_6	15.7067	11.9887	8.5550	*4.9599*

Italics denote extrapolation beyond the range of fitted data [63].

gases. In Table 3.12, values for ln K_D for the distribution of various gases between liquid and gaseous phase in aqueous solutions are listed [63], and in Figure 3.23, ln K_D is plotted for some of these gases.

The solubility of a gas in water is affected by the presence of other solutes. The most important are electrolytes causing the solubility to decrease (salting-out) or increase (salting-in). Furthermore, some dissolving gases ionize in water, slightly like CO_2 or fully like HCl. Then, additional species determine the equilibrium distribution between phases and in particular solubility in the liquid phase. These aspects and the influence of mixed solvents are discussed in the sections on CO_2 and NH_3 in water (Section 3.5) and on salts in water (Section 3.7).

3.4.2 Phase Behavior and Critical Curves

Binary aqueous systems containing volatile nonelectrolytes usually have interrupted critical curves (corresponding to type III systems in the classification of von Konynenburg and Scott). The upper branch begins at the critical point of pure water and either has a temperature minimum, sometimes also a pressure minimum, or proceeds monotonously to higher temperature and

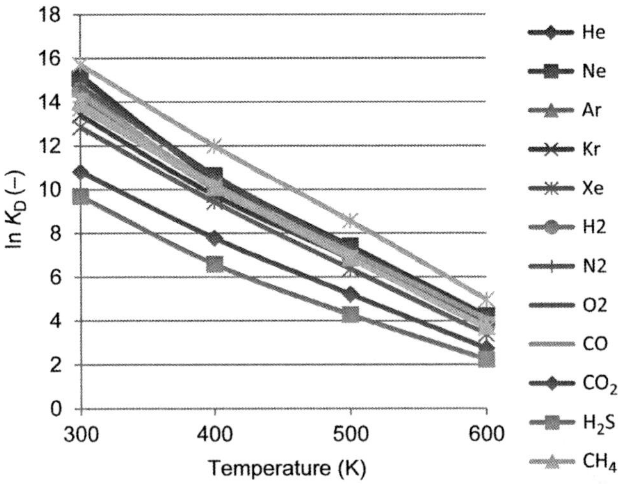

FIGURE 3.23 Values for K_D for various gases [63]. (For color version of this figure, the reader is referred to the online version of this chapter.)

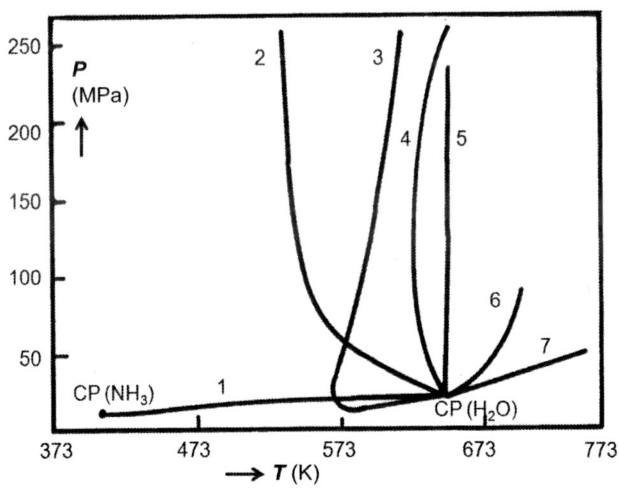

FIGURE 3.24 Critical curves for binary systems: (1) NH_3–H_2O, (2) CO_2–H_2O, (3) C_6H_6–H_2O, (4) CH_4–H_2O, (5) H_2–H_2O, (6) HCl–H_2O, (7) NaCl–H_2O. CP(H_2O) and CP(NH_3) are the critical points of water and ammonia. Two phases coexist below or to the left. *Modified from Ref. [57].*

pressures [4] (see Figure 3.24). An example is the system water–hydrogen [6]. At temperatures below the critical curve, two phases coexist, one rich in water and one rich in hydrogen. The two-phase area extends to very high pressures. Up to 200 MPa, the mutual solubility decreases with pressure. The critical curve rises almost vertically to high pressures in the P,T-projection of Figure 3.24.

Critical curves of some binary aqueous systems are shown in Figure 3.24. Two phases coexist below or to the left of the lines. For aqueous systems of volatile nonelectrolytes, the critical curves run to very high pressures, indicating that pressure is of much less influence on the system than temperature. The critical pressures are much lower for aqueous electrolyte systems, like NH_3.

The two-phase area extends in most cases to the left side of the P,T-projections (lower temperatures) and occurs at pressures below the critical curve. In systems with a critical curve exhibiting a minimum in temperature, from that minimum upward, the two-phase area lies above the critical curve (with respect to pressure).

In case the critical curves run to temperatures higher than the critical temperature of water, no liquid phase can exist. All mixtures consist of two supercritical fluids, and the water-rich mixture and the mixture rich in the volatile nonelectrolyte can be addressed as a dense gas. In the two-phase area, above the critical temperature of water, only these dense gases can exist. Therefore, this phenomenon sometimes is called "gas–gas immiscibility." For critical curves that directly run to higher temperatures, the gas–gas immiscibility begins at the critical point of water. This case is called gas–gas immiscibility of the first kind.

In systems with a critical curve that has a minimum in temperature, gas–liquid equilibrium occurs in the two-phase area below the critical temperature of water. If the critical curves cross the critical temperature of water at pressures above the temperature minimum, again a gas–gas immiscibility occurs in the two-phase area, now called gas–gas immiscibility of the second kind.

Similar to hydrogen, oxygen and nitrogen behave with water. For water–oxygen, the critical curve has a slight temperature minimum around 640.15 K and 75 MPa [65]. For the system water–nitrogen, the minimum of the critical curve occurs at almost the same pressure and temperature [65]. Henry's constant at $T = 298.15$ K, and hence the solubility for water–nitrogen, is twice as high as for water–oxygen. Near the critical point of water, however, two Henry's constants are almost equal [4].

The critical curves separate the single-phase region of binary systems from the two-phase region. At temperatures above the critical temperature of water ($T > 673$ K), hydrocarbons, hydrogen, and oxygen can be mixed with supercritical water in any proportion. In particular, adding oxygen to mixtures of water and methane, combustion can be initiated and electrolysis of supercritical water can proceed without the formation of the gas phase on electrodes [57].

Besides the type of fluid phase and the boundaries of vapor–liquid and vapor–solid equilibrium, the course of the critical lines originating at the critical point of water is of importance. When a small amount of volatile nonelectrolyte is added to water, a critical line develops, since the critical parameters of the solution depend on the concentration. The initial slopes of the critical line are related by the following equation:

$$K_{\mathrm{Kr}} \equiv \left(\frac{\partial P}{\partial x}\right)^{\mathrm{c}}_{v,T} = \left(\frac{\mathrm{d}T}{\mathrm{d}x}\right)^{\mathrm{c}}_{\mathrm{CRL}} \left[\left(\frac{\mathrm{d}P}{\mathrm{d}T}\right)^{\mathrm{c}}_{\mathrm{CRL}} - \left(\frac{\mathrm{d}P}{\mathrm{d}T}\right)^{\mathrm{c}}_{\mathrm{SAT}}\right] \quad (3.93)$$

The limiting slope of the saturation pressure at the solvent critical point for water is given by Equation (3.93a):

$$\left(\frac{\mathrm{d}P}{\mathrm{d}T}\right)^{\mathrm{c}}_{\mathrm{SAT}} = \left(\frac{\partial P}{\partial T}\right)^{\mathrm{c}}_{\rho} = 0.267\,\mathrm{MPa\,K^{-1}} \quad (3.93a)$$

The qualitative critical behavior of the thermodynamic properties of dilute solutions is dominated by K_{Kr} [58]. The values of the Krichevskii parameter for a number of aqueous solutions are presented in Table 3.13.

The general behavior of critical lines for some classes of binary aqueous systems near the critical point of water can be seen in Figure 3.24. To the left from the critical isochor of water, the values of the Krichevskii parameter are positive, while to the right of the critical isochor of water, the values of the Krichevskii parameter are negative. In aqueous solutions of nitrogen, oxygen, and noble gases, the critical lines run to high pressure with a slightly negative initial slope. Solutions of carbon dioxide and alkanes have critical lines that pass through a temperature minimum before turning to higher temperatures and pressures. All these binary systems are of type III with positive values of the Krichevskii parameter. For aqueous solutions of nonvolatile salts that strongly interact with water, the critical line runs to high pressures and temperatures for small amounts of salt added. Aqueous alkali halides have a vapor–liquid critical line that is not interrupted (type I systems). In the type I system water–ammonia, the critical line runs to lower pressures and temperatures [58].

TABLE 3.13 Krichevskii Parameter for Some Aqueous Solutions

Solute	$K_{\mathrm{Kr}} \times 10^{-3}$ (MPa)	Solute	$K_{\mathrm{Kr}} \times 10^{-3}$ (MPa)
Helium	0.17	Ethane	0.16
Argon	0.17	Propane	0.17
Hydrogen	0.17	*n*-Hexane	0.17
Nitrogen	0.18	Ethene	0.15
Oxygen	0.17	Benzene	0.097
Carbon dioxide	0.12	Toluene	0.10
Ammonia	0.041	Silica	−0.19
Methane	0.16	Sodium chloride	−0.61

Selected values from Ref. [58].

3.4.3 Water and Organic Nonelectrolyte Compounds (Water and Hydrocarbons)

Water and nonpolar organic compounds mix poorly at low temperatures. But solubility of nonpolar compounds in water changes drastically at high temperatures. Approaching the critical temperature, most organic compounds are highly soluble in water, if not totally miscible. This will be discussed in detail, since it is a major feature of near-critical and supercritical water in processes. In competition to enhanced solubility, organic compounds at high temperatures become instable and may chemically react due to thermal cracking and hydrolysis. In this section, physical equilibrium is discussed only, while chemical changes are treated separately. The variation of mutual solubility with water is shown for illustration of the temperature dependence in Figure 3.25 for some hydrocarbons.

The increase in mutual solubility of water and hydrocarbons is remarkable. Water dissolves at high temperatures (around $T = 550$ K) in lubricating oil to about 20 wt% and solubility in water raises for the branched hydrocarbons through a temperature increase of 200 K, by three orders of magnitude. Since the phase behavior of water and hydrocarbons is of greatest importance for the topics of this book, a detailed description of phase equilibrium follows.

3.4.3.1 Phase Behavior of Organic Nonelectrolyte Compounds with Water

Phase behavior of organic nonelectrolyte compounds with water usually falls within the types II and III according to the definition of Scott and van Konynenburg. This phase behavior for binary systems of water and an organic nonelectrolyte compound is described with reference to Figures 3.26 and 3.27.

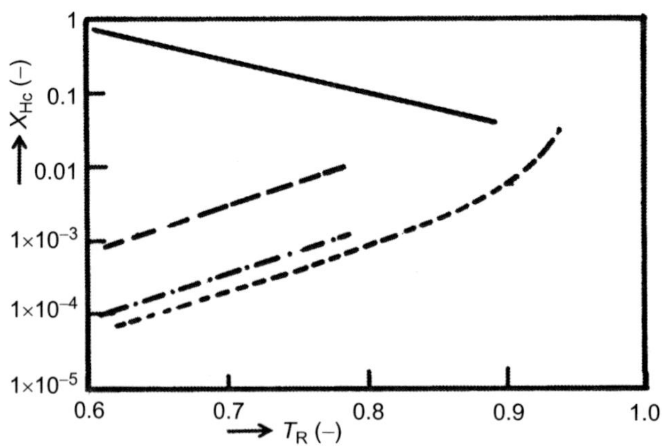

FIGURE 3.25 Temperature dependence of the solubility of water–hydrocarbon systems, $P \approx 25$ MPa. ———, H_2O in lubricating oil; — — —, methylbenzene in H_2O; —•—, dimethylcyclohexane in H_2O; – – –, real weathered contamination (from diesel fraction) in H_2O [66].

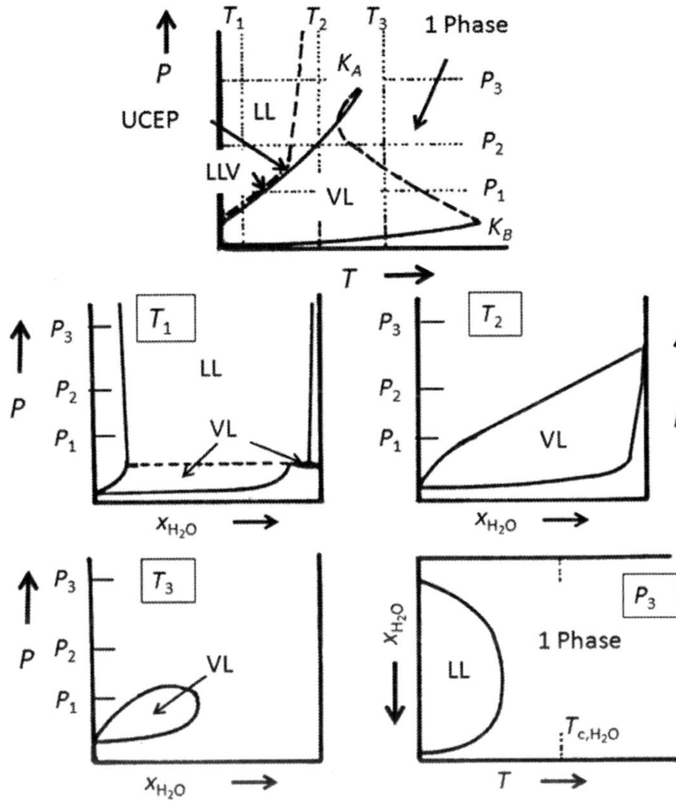

FIGURE 3.26 Phase behavior of water–hydrocarbon systems of type II (see text).

3.4.3.1.1 Phase Behavior According to Type II

Type II systems are characterized by an uninterrupted critical curve gas–liquid and a liquid–liquid immiscibility region that ends below the critical temperature of the more volatile component. For the example of Figure 3.26, water is the more volatile component. The vapor-pressure curves (———) of water and the nonelectrolyte end at their critical points. In a type II system, the critical curve (— — —) of the gas–liquid two-phase area leads from the critical point of water (K_A) to the critical point of the other component (K_B). The critical curve liquid–liquid begins at the critical endpoint (UCEP) of the three-phase line (—●—) and extends nearly vertical to higher pressures.

The mutual solubility of the compounds can be shown in cuts of the P,T-diagram at constant temperature (P,x-cuts) or at constant pressure (T,x-cuts). For water–hydrocarbon systems of type II, those cuts show the following characteristic features (compare Figure 3.26).

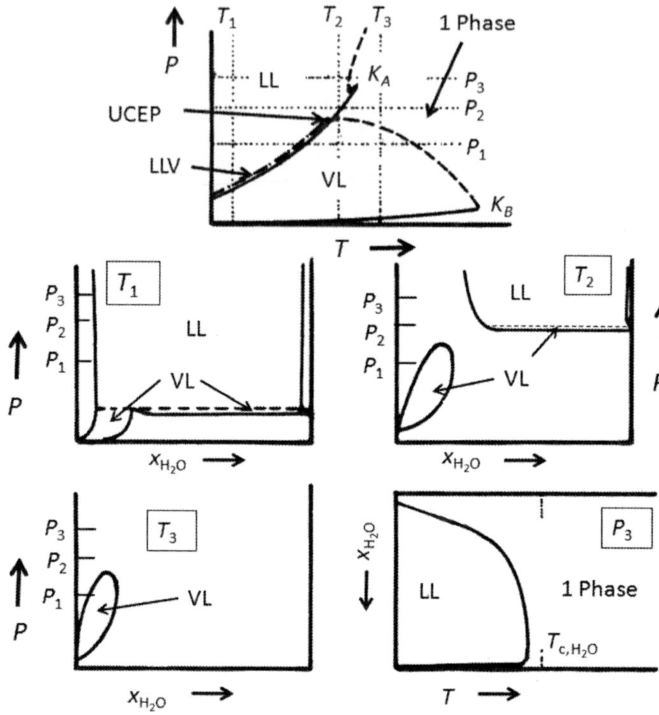

FIGURE 3.27 Phase behavior of water–hydrocarbon systems of type III (see text).

Mutual solubility increases with increasing temperature. The liquid–liquid immiscibility gap becomes smaller until both liquid phases become identical at the UCEP of the three-phase line. At temperatures above the UCEP, but below the gas–liquid critical line (e.g., T_2), gas–liquid equilibria exist that extend over the entire range of concentrations.

At temperatures below the UCEP and at low pressures, gas–liquid equilibria start from the pure components vapor pressures and end in an extended liquid–liquid immiscibility gap (T_1). At the three-phase line, a gas phase and two liquid phases coexist. In the example shown in Figure 3.26, the three-phase line lies above the vapor-pressure curve of H_2O (the more volatile component), a behavior typical for water–hydrocarbon mixtures.

At temperatures above the critical temperature of water, gas–liquid equilibria exist. The miscibility gap gets smaller with increasing temperature until it vanishes at the critical point of the hydrocarbon.

The examples for a type II behavior are water–biphenyl [31] and water–naphthalene [32]. More examples are presented at the detailed discussion of the behavior of water and hydrocarbons in the critical region.

3.4.3.1.2 Phase Behavior According to Type III

Type III systems are characterized by an interrupted critical curve gas–liquid. One leg begins at the critical point of the more volatile component (K_A, H_2O), while the other leg starts at the critical point of the less volatile component (K_B, hydrocarbon). Both critical lines do not meet. Depending on the system, either of the critical lines starting from critical points of the pure components can end on the three-phase line liquid–liquid–gas (compare Figure 3.27 and the more detailed discussion in Section 3.4.3.2). For the example shown in Figure 3.27, the critical line starting from the critical point of the hydrocarbon component ends on the three-phase line. Such behavior is typical for many water–hydrocarbon systems.

P,*x*-cuts and *T*,*x*-cuts of type III systems can be derived analog to the behavior of type II systems from the *P*,*T*-phase diagram.

Examples for such phase behavior are *n*-alkanes–H_2O [36], benzene [31] and benzene derivatives–H_2O [32,67], and saturated ring systems of hydrocarbons like cyclohexane [31] and decahydronaphthalene (decalin) [67].

3.4.3.2 Phase Behavior of Hydrocarbon–Water Systems in the Critical Region

3.4.3.2.1 Water and Alkanes

Phase equilibria of binary system of *n*-alkanes and water at temperatures near the critical temperature of water have been determined by various authors [36]: methane–water [68,69], ethane–water [70], propane–water [71,72], butane–water [70,73], pentane–water [74], hexane–water [75,76], and heptane + water [31,74]. All these show gas–gas equilibria of the second type.

In the following, for some *n*-alkane–water systems, phase behavior in the range of critical temperature and critical endpoints is presented. Most of the data have been provided by the group of Erwin Brunner. In the tables and figures, the following abbreviations are used: *P*, pressure; *T*, temperature; llg, liquid–liquid–gas three-phase point or line; lg, liquid–gas critical point; UCEP, upper critical endpoint; cp1, critical point of water; and cp2, critical point of the alkane (Table 3.14).

For *n*-alkane–water mixtures, an interrupted critical curve of type III has been found. A liquid–liquid–gas three-phase line occurs and terminates at a liquid–gas UCEP. For methane–water, the liquid–liquid–gas three-phase line and the liquid–gas UCEP are not existent since there is no common range of temperatures where both substances are liquid. For *n*-alkanes, the phase behavior is shown in Figure 3.28 illustrating the variation with alkane chain length.

The behavior in the critical region for *n*-alkane–water systems has been systematically studied by Brunner. The following discussion is based on his fundamental investigation [36].

TABLE 3.14 Measured Values for the llg Three-Phase Line and the Two Branches of the Interrupted Critical Curve for *n*-Hexane, *n*-Decane, *n*-Hexadecane, and *n*-Triacontane–Water

Hexane–H_2O			*n*-Decane–H_2O			*n*-Hexadecane–H_2O			*n*-Triacontane–H_2O		
T *(K)*	P *(MPa)*		T *(K)*	P *(MPa)*		T *(K)*	P *(MPa)*		T *(K)*	P *(MPa)*	
419	1.125	llg	465.8	1.473	lg	456.3	1.1	llg	502.5	2. 76	llg
436.1	1.645		494	2.686		486.9	2.08		532.1	4.62	
456.3	2.49		523.3	4.63		517.9	3.72		562	7.36	
476.8	3.75		543.2	6.495		548.4	6.14		590.2	11.02	
485.6	4.363		562.8	9		578.4	9.65		609.9	14.32	
493.5	5.02		567.6	9.655	UCEP	589.6	11.34		629.9	18.3	
496.4	5.282	UCEP	569.4	9.6	lg	593.3	11.95		635.3	19.46	
499	4.98	lg	574.8	9.33		596.8	12.56		635.9	19.63	x
501.6	4.525		580.2	8.88		599.9	13.12		636.7	19.8	
504.4	3.995		585.2	8.3		603.2	13.77		637.9	20.09	
506.4	3.46		590.6	7.58		605.6	14.27		638.5	20.27	
507.9	3.03	cp2	596.1	6.79		608.3	14.87	UCEP	639.8	20.63	
647.14	22.064	cp1	600.9	5.97		623.6	15.3	lg	641.1	21.03	
642.2	22.52	lg	605.9	4.945		632	15.235		641.6	21.35	UCEP
640	22.76		610.4	3.923		641.6	14.88		641.7	21.3	lg

636.7	23.2		613.6	3.224		651.8	14.065		642.1	21.37	
633.7	23.82		617.65	2.103	cp2	660.6	12.97		642.6	21.43	
631.4	24.59		647.14	22.064	cp1	670.4	11.47		643.1	21.5	
629.9	25.35		644.5	22	lg	679.9	9.88		643.9	21.63	
627.9	27.78		643.2	21.95		689.1	7.99		644.8	21.73	
627.6	31.5		641	21.9		722	1.444	cp2	645.6	21.87	
629.8	36.6		633.3	22.44		647.14	22.064	cp1	646.3	21.96	
632.2	40.8		632	23.74		646.2	21.98	lg	646.7	22.03	
634.3	43.9		634.8	28.28		645.3	21.88		647.14	22.064	cp1
636.2	47.3		638.7	32.18		644.5	21.78		857	0.83	cp2
			642.9	36.1		643.9	21.7		670.8	24.9	lg
			649.1	42.07		642.6	21.56		661.4	24.27	
						641	21.42		655.6	23.95	
						639.8	21.3		653	24.13	
						638.2	21.24		654.5	25.25	
						637.3	21.26		657.8	26.85	
						636.3	21.6		662.1	28.75	
						636.6	21.95		665.3	30.2	

Continued

TABLE 3.14 Measured Values for the llg Three-Phase Line and the Two Branches of the Interrupted Critical Curve for *n*-Hexane, *n*-Decane, *n*-Hexadecane, and *n*-Triacontane–Water—Cont'd

Hexane–H_2O		*n*-Decane–H_2O		*n*-Hexadecane–H_2O		*n*-Triacontane–H_2O	
T *(K)*	P *(MPa)*	T *(K)*	P *(MPa)*	T *(K)*	P *(MPa)*	T *(K)*	P *(MPa)*
				637.7	23.05		
				638.8	23.85		
				640.8	25.08		
				643.3	26.55		
				646.3	28.5		
				650.7	31.1		

Data from Ref. [36].

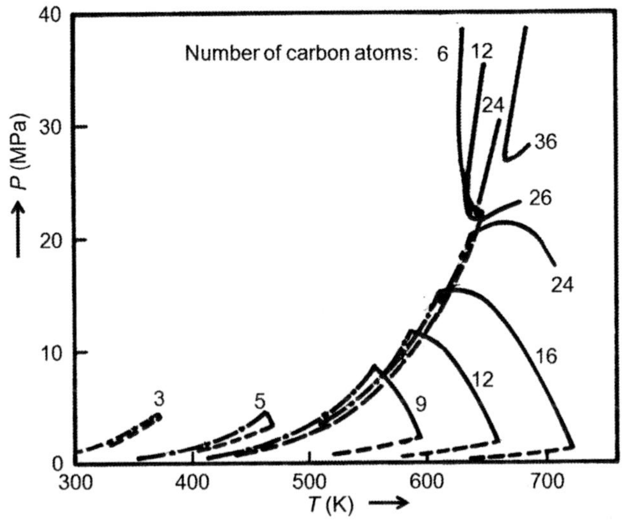

FIGURE 3.28 Phase diagram for selected *n*-alkanes. ———, vapor pressure of H_2O and of *n*-alkane; —•—, three-phase line liquid–liquid–gas; ———, critical line of the two-phase area. *Modified from Ref. [36].*

Characteristic for *n*-alkane–water systems is that the liquid–liquid–gas three-phase pressure is higher than the vapor pressures of the individual components, resulting in heterogeneous azeotropy.

The critical curves and the liquid–liquid–gas three-phase lines change with the carbon number of the *n*-alkane. The alkanes' critical temperatures cover a wide range, CH_4, $T_c = 190.6$ K, and $C_{36}H_{74}$, $T_c = 894$ K, with the consequence that short-chain alkanes up to C_5 have a higher volatility than water, *n*-hexane has about the same volatility as water, and alkanes with more than six carbon atoms have a volatility lower than that of water.

All *n*-alkane–water mixtures have an interrupted critical curve with a liquid–gas UCEP, where the gas phase and the lower dense liquid phase become identical. For C_1 (CH_4) to C_{26} ($C_{26}H_{54}$), the gas phase becomes identical with the *n*-alkane-rich liquid phase at the liquid–gas UCEP, since the water-rich liquid phase always has a higher density than the alkane-rich liquid phase. The critical curve begins at the liquid–gas UCEP and ends at the critical point of the *n*-alkane.

The other part of the interrupted critical curve for *n*-alkane–water mixtures up to C_{26} begins at the critical point of water. The critical curve runs through a temperature minimum, for C_7–C_{26} also through a pressure minimum, and then rises steeply to high pressures with increasing temperature.

No liquid–liquid UCEP occurs in the *n*-alkane–water mixtures due to the very wide liquid–liquid miscibility gap. In mixtures of aromatic compounds and water, an infinite miscibility occurs, for example, for naphthalene–water [32], diphenyl–water [31], and 1,2,3,4-tetrahydronaphthalene–water [77].

When the liquid–gas UCEP reaches temperatures near the critical temperature of water, the density of the water-rich liquid phase l_1 decreases faster with rising temperatures than that of the alkane-rich liquid phase l_2. The water-rich phase l_1 has the lower density before the UCEP is reached. A reversal of the liquid phases occurs (barotropic effect). Therefore, the water-rich liquid phase l_1 becomes identical with the gas phase at the l_1g-UCEP and the l_1g-critical curve that begins at the l_1g-UCEP ends at the critical point of water cp1. The barotropic effect was found for *n*-alkane–water mixtures with alkanes $>C_{28}$. For more information, the reader is referred to the original literature [36].

3.4.3.2.2 Water and Aromatic Hydrocarbons

For mixtures of water and aromatic compounds, a similar study to the behavior of *n*-alkane–water systems provides information on higher-molecular-weight hydrocarbons, mostly aromatic character, from benzene to pyrene [33].

Benzene Derivatives: Water Water and aromatic hydrocarbons with a single ring like benzene, methylbenzene (toluene), 1,2-dimethylbenzene (*o*-xylene), 1,4-dimethylbenzene (*p*-xylene), and 1,3,5-trimethylbenzene have a type III phase behavior. The critical curve is interrupted and a liquid–liquid–gas three-phase line terminates at a liquid–gas UCEP, where the organic liquid phase and the gas phase become identical. At the critical point of water, one leg of the critical curve starts and runs through a pressure and a temperature minimum and then rises steeply to high pressures. From the liquid–gas UCEP, the other leg of the liquid–gas critical curve runs to the critical point of the aromatic hydrocarbon.

The liquid–liquid critical curves are shifted to higher temperatures with increasing alkyl content and decreasing aromaticity. The gradients of the liquid–gas critical curves are similar. The location of each curve is determined by the critical temperature and pressure of the hydrocarbon [33]. *P*,*T*-plots for binary systems of these compounds with water are shown in Figure 3.29.

In the following, experimental values for the critical curves and the three-phase lines for some systems are listed in tables and visualized in diagrams (Table 3.15).

3.4.3.2.3 Water and Hydrocarbons with Multiple Aromatic Rings

Water and aromatic hydrocarbons with multiple rings like 4,2-naphthalene, 1-methylnaphthalene, acenaphthene, fluorene, anthracene, phenanthrene, pyrene, diphenyl, diphenylmethane, 1,2-diphenylethane, and 1,4-diphenylbenzene show a type II phase behavior [33] with a continuous liquid–gas critical curve and a liquid–liquid critical curve that originates at the liquid–liquid UCEP and rises steeply to higher pressures.

The liquid–gas critical curve starts from the critical point of water and runs through a minimum of temperature and pressure. The liquid–gas critical

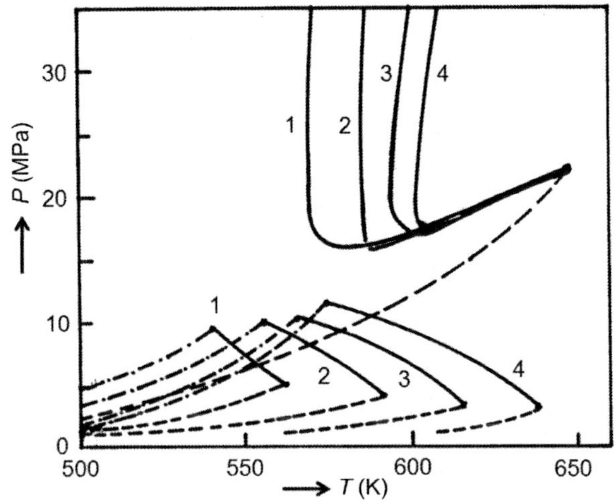

FIGURE 3.29 *P,T*-projections for some binary mixtures of methyl-substituted benzene derivatives and water. (1) Benzene, (2) toluene, (3) *o*-xylene, and (4) *p*-xylene. *Modified from Ref. [33].*

curve ends at the critical point of the aromatic hydrocarbon after passing through a point of inflection and a pressure maximum. Due to thermal instability, this end of the critical curve may not be reached experimentally. The liquid–liquid critical curves are shifted to higher temperatures with the number of aromatic rings. All multiple aromatic ring compounds are completely miscible in compressed liquid water due to the interactions between the π-bonded electrons of aromatic compounds and water [33].

In Figure 3.30, *P,T*-projections for some binary systems of multiple-ring aromatic hydrocarbons and water are shown, and in Table 3.16, the experimental data for anthracene–water and phenol–water are listed.

Type II phase behavior was also found for binary systems of water and phenol, benzophenone, diphenyl ether, 1-chloronaphthalene, and benzyl [33].

The transition from type II phase behavior to type III phase behavior has been demonstrated on two-ring compounds [33]. While the fully aromatic naphthalene shows type II behavior and the semiaromatic tetralin (1,2,3,4-tetrahydronaphthalene) still shows type II behavior, the nonaromatic decalin (transdecahydronaphthalene) shows type III behavior. For decalin, the liquid–liquid critical curve is shifted to such high temperatures that intersection with the liquid–gas critical curve occurs and type III phase behavior follows [33,67,78].

For indene, a barotropic effect occurs on the liquid–liquid–gas three-phase line at 568 K. Above 568 K, the indene-rich liquid phase is the denser phase. At the liquid–gas UCEP, the water-rich liquid phase becomes identical with the gas phase. Therefore, the liquid–gas critical curve that extends from the liquid–gas UCEP ends at the critical point of water. The other leg of the liquid–gas critical curve would extend from the critical temperature of the

TABLE 3.15 Benzene, Toluene, *o*-Xylene, and *p*-Xylene–Water

Benzene–Water			Toluene–Water			*o*-Xylene–Water			*p*-Xylene–Water		
T *(K)*	P *(MPa)*		T *(K)*	P *(MPa)*		T *(K)*	P *(MPa)*		T *(K)*	P *(MPa)*	
438.9	1.545	llg	459.4	1.785	llg	434.1	0.812	llg	420.1	0.576	llg
468.5	2.805		479.5	2.686		459	1.442		450.3	1.207	
488.6	4.05		494.4	3.568		474.1	1.989		480.2	2.314	
509	5.73		509.6	4.59		497.3	3.152		509.9	4.09	
521.7	6.925		519.6	5.57		517.6	4.57		526.7	5.48	
530.7	8.065		530.5	6.665		532.2	5.855		609.3	5.2	
536.6	8.79		541.1	7.93		547.4	7.495		616.19	3.511	cp2
541.6	9.45	1g-UCEP	546.9	8.675		565	9.87		647.1	22.064	cp1
540.8	9.145	1g	551.9	9.365		570.2	10.64		643.9	21.7	1l+1g
541.9	8.62		552.6	9.475		575.1	11.41	lg-UCEP	638.3	21.08	
546.6	7.94		554.4	9.73		578.5	10.83	lg			
550.3	7.31		555.7	9.915		584	10.53				
554.2	6.59		556.1	9.98	1g-UCEP	589.5	10.15				
557.9	5.89		555.1	9.825	1g	595.3	9.57				
560.8	5.257		554.7	9.725		601.4	8.89				

562.12	4.898	cp2	556.7	9.55		606.7	8.2	
647.1	22.064	cp1	557.7	9.43		610.4	7.64	
641.1	21.57	ll+lg	559.1	9.28		614.5	6.97	
632.7	20.9		561.5	9.06		618.2	6.3	
620.9	19.86		563.7	8.83		621.2	5.72	
607	18.52		566.8	8.49		624.1	5.085	
592.2	16.89		570.5	8.02		627.2	4.415	
582.9	16.03		573.7	7.55		629.9	3.963	
573.2	16.7		577.1	7.03		630.29	3.374	cp2
570.1	19.09		580.6	6.44		647.1	22.064	cp1
569.1	24.3		583.6	5.89		645.5	21.85	ll+1g
570.1	31.3		586.6	5.325		641.9	21.44	
572.3	38.4		589.4	4.75		638	21	
573.9	44.2		591.8	4.106	cp2	634.7	20.57	
576.3	53.7		647.1	21.064	cp1	630.5	20.07	
			645.1	21.85	ll+lg	626.5	19.55	
			641.1	21.5		622.1	18.94	
			639.2	21.25		616	18.12	
			634	20.7		609.6	17.2	

Continued

TABLE 3.15 Benzene, Toluene, *o*-Xylene, and *p*-Xylene–Water—Cont'd

Benzene–Water		Toluene–Water		*o*-Xylene–Water		*p*-Xylene–Water	
T (K)	P (MPa)	T (K)	P (MPa)	T (K)	P (MPa)	T (K)	P (MPa)
		628.8	20.15	605.8	16.6		
		620.3	19.15	599.7	15.78		
		612.9	18.42	595.1	15.39		
		593.6	16.3	591.4	16.25		
		586.9	16.48	590.8	18.84		
		585.8	30.3	591.7	22.4		
		589.5	39.9	593.7	26.6		
				595.7	30.85		
				598.6	35.8		
				601	40.7		
				603.9	46.3		
				606	50.4		

Values for the critical curves and the three-phase line [33].

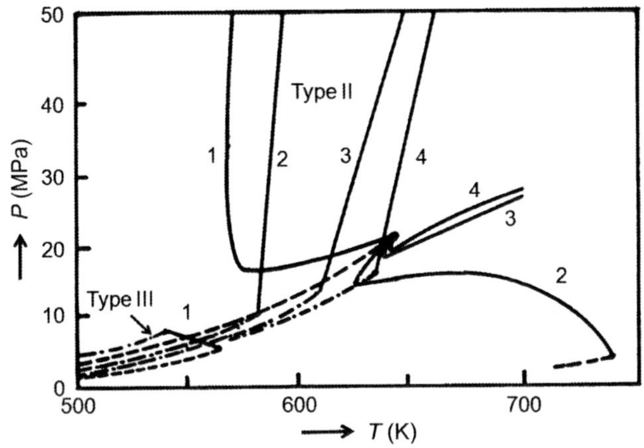

FIGURE 3.30 *P*,*T*-projections for some binary systems of multiple-ring aromatic hydrocarbons and water: (1) benzene, (2) naphthalene, (3) anthracene, and (4) pyrene. *Modified from Ref. [33].*

hydrocarbon but could be proven by experiment. An analog transition within a binary family occurs also with the alkane–water systems [36].

In Figure 3.31, the *P*,*T*-projections of the phase behavior of water with naphthalene, tetralin, decalin, and indene are shown.

In conclusion of this discussion, the type of phase behavior can be derived solely from the UCEP. If the mixture has a liquid–liquid UCEP, then type II phase behavior occurs (in rare cases type IV). If the mixture has a liquid–gas UCEP, then an interrupted liquid–gas critical curve exists and type III phase behavior occurs. If the organic-rich liquid phase becomes critical at the liquid–gas UCEP, the liquid–gas critical curve runs to the critical point of the hydrocarbon. If the aqueous liquid phase becomes critical at the UCEP, then the liquid–gas critical curve runs to the critical point of water [33].

3.4.3.3 Solubility Behavior

Besides information on general phase behavior, the solubility of compounds in the various phases is of importance for processes. Such information will be discussed for characteristic systems. Most data are available for hydrocarbons with less than 10 carbon atoms.

Solubility of five hydrocarbon–water systems (benzene, *n*-heptane, *n*-pentane, 2-methylpentane, and toluene) is known up to the critical solution temperature and 80 MPa [74]. A sharp maximum for the solubility occurs near the critical solution temperature at about 30 MPa. At higher pressures, solubility decreases, and a second two-phase region appears at high pressures. Solubility of heavy hydrocarbons in near-critical water was determined by visual observation [79]. Mutual solubilities of benzene, cyclohexane, and *n*-hexane and water at temperatures up to $T = 428$ K were measured and correlated with Henry's law

TABLE 3.16 Anthracene–Water and Phenol–Water

Anthracene–Water			Phenol–Water		
T (K)	P (MPa)		T (K)	P (MPa)	
521.6	3.938	llg	694.2	5.9	cp2
551.4	6.34		693.6	5.95	lg
569.5	8.27		690	6.72	
579.7	9.53		685.3	7.96	
590.3	10.98		677.3	9.8	
596.7	11.96		667.4	11.78	
601.5	12.68		657.9	13.25	
608.1	13.65	ll-UCEP	648.1	14.72	
			641.5	15.56	
755.4	27.15	lg	637.3	16.22	
685.6	25.4		634.8	16.76	
670.7	23.99		633.7	17.91	
664.1	23.14		636.2	19.08	
655.9	22.09		639.7	20.19	
648	20.95		641.7	20.79	
643.6	20.34		644.5	21.56	
640.8	19.96		645.7	21.84	
638.5	19.89		647.1	22.064	cp1
641.7	20.88				
643.9	21.45				
647.1	22.064	cp1			
869.3	2.87	cp2			

Values for the critical curves and the three-phase line [33].

for the solubility of hydrocarbons in water and a modified Redlich–Kwong EOS for the solubility of water in hydrocarbons [80].

For some systems of major importance, as water with benzene, toluene, 3-methylpentane, *n*-decane, and *n*-hexadecane, more detailed information is presented in the following.

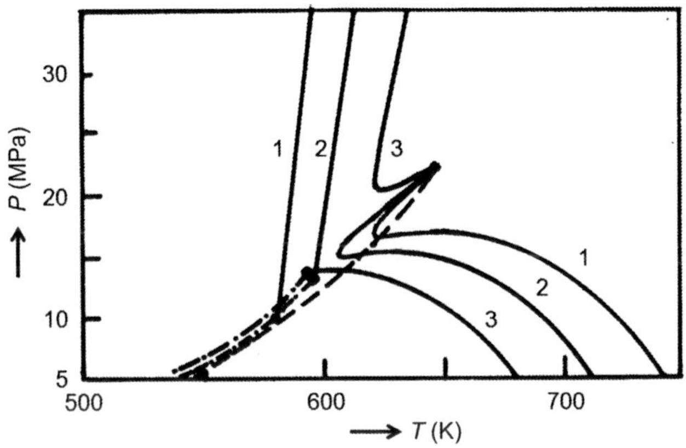

FIGURE 3.31 *P*,*T*-projections of the phase behavior of water with (1) naphthalene, (2) tetralin, and (3) decalin. *Modified from Ref. [33].*

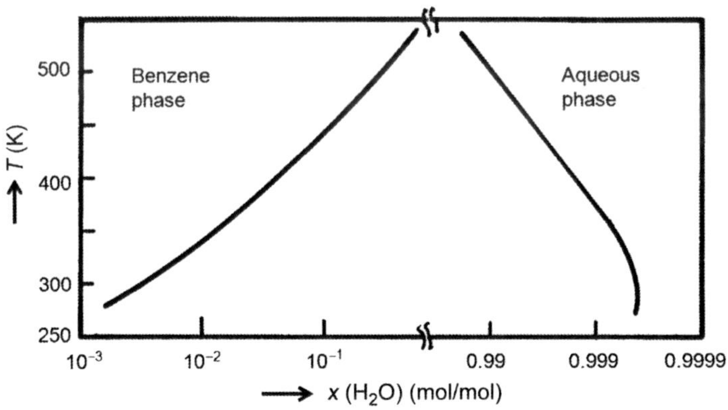

FIGURE 3.32 Mutual solubility of water and benzene. *Data from Neeely [82].*

3.4.3.3.1 Benzene–Water

The phase behavior and mutual solubility of the system benzene–water were investigated by several groups, that is, Refs. [33,81]. A detailed study of solubility in the benzene–water system is reported by Neely [82], where the available literature has been critically surveyed. Mutual solubility of benzene–water is shown in Figure 3.32.

3.4.3.3.2 Toluene–Water

The phase behavior and mutual solubility of the system toluene–water were reported with a discussion of the literature by Neely [82]. Solubility in both phases is presented in Figure 3.33.

3.4.3.3.3 Water–3-Methylpentane

Data for water–3-methylpentane are presented in Figure 3.34 from the study by Neely [82].

3.4.3.3.4 *n*-Decane–Water

For the system *n*-decane–water, Wang and Chao [83] have published some data, presented in Figure 3.35.

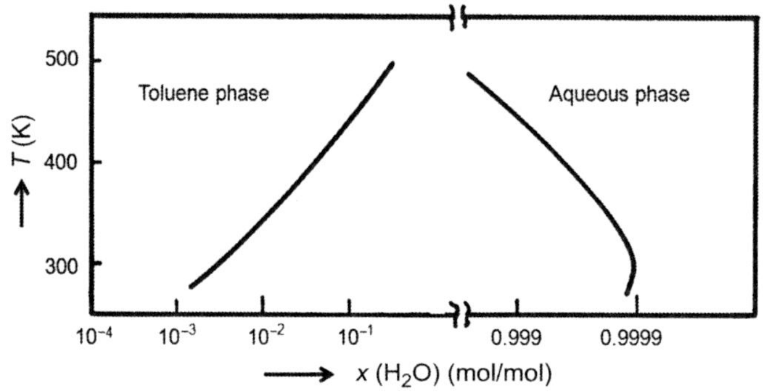

FIGURE 3.33 Mutual solubility of toluene–water. *Data from Neely [82].*

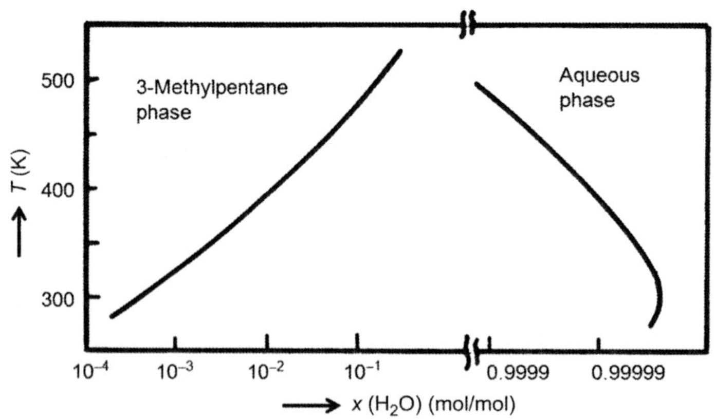

FIGURE 3.34 Mutual solubility of water and 3-methylpentane. *Data from Neely [82].*

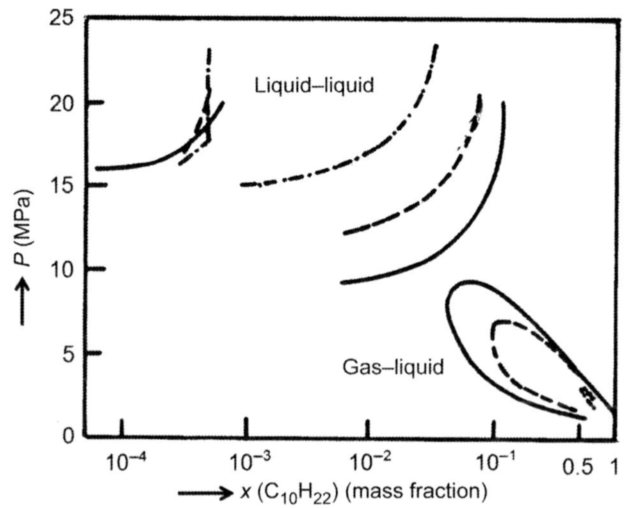

FIGURE 3.35 Phase equilibria in the system *n*-decane–water. ———, $T = 573.15$ K; – – –, $T = 593.15$ K; —•—, $T = 613.15$ K. *Data from Ref. [83].*

3.4.3.3.5 *n*-Hexadecane–Water

For the system *n*-hexadecane–water, only very few data are available. Mutual solubility at $P = 20$ MPa and $T = 573.15$ K in the aqueous phase is H_2O 99.99 mol%, *n*-C16 <0.01 mol%; in the organic phase, H_2O 46.32 mol%, *n*-C16 53.68 mol% [84]. Dohrn [49] carried out measurements in the system hydrogen–water–*n*-hexadecane. At a low content of hydrogen, the solubility at $P = 20$ MPa and $T = 573.15$ K for the aqueous phase was H_2O 99.5 mol%, *n*-C16 0.5 mol%, and in the organic phase, H_2O 48.4 mol%, *n*-C16 51.4 mol %, with about 0.4 mol% hydrogen in the organic phase. For $P = 10$ MPa and $T = 573.15$ K, the solubility in the organic phase was H_2 1.32 mol%, H_2O 53.52 mol%, *n*-C16 45.16 mol%, and in the aqueous phase, H_2 0.13 mol%, H_2O 99.45 mol%, *n*-C16 0.42 mol%.

3.4.3.4 Ternary Systems

Information on binary systems often provides sufficient information on equilibrium of ternary systems. Similar hydrocarbons and water may be treated as binary systems, if interest is mainly in the phase behavior. Even a mixture of more than two components of similar volatility may be treated with sufficient accuracy as a pseudo-binary system. But in many cases, phase behavior of three different components must be considered, for example, for the separation of mixtures into compounds or for compounds that differ substantially. Then, the system cannot be treated as quasi-binary.

In cases, where properties of the components differ more or less, all possible types of phase behavior, as discussed in the preceding text, may occur in each binary system, including transition types. An interesting example is the transition from a type II equilibrium into a type III equilibrium that has been observed in a ternary system of water with varying concentrations of tetralin and decalin [33].

For ternary systems, in addition to the characteristics of binary equilibria, phenomena occurring in the three-component area have to be considered. Discussions on the phase behavior of ternary systems, including the critical region of the system, have been worked out, that is, by Sadus [41]. In the following, phase behavior of some aqueous three-component systems will be treated.

3.4.3.4.1 Water–Benzene–Hydrogen and Water–Toluene–Hydrogen

The system water–benzene–hydrogen was investigated in a temperature range of $T = 473$–573 K and pressures from $P = 10$ to 30 MPa [49,50]. The phase behavior is shown in Figures 3.36 and 3.37. Experimental data are listed in Table 3.17.

In Figure 3.36, ternary phase equilibrium of H_2O–C_6H_6–H_2 is shown for $T = 473$ K and $P = 10$ MPa. Besides the one- and two-phase areas, the system exhibits a ternary phase region with a water-rich, a benzene-rich, and a hydrogen-rich phase. The one-phase region at high water concentrations is very small.

At $T = 573$ K, benzene and hydrogen are fully miscible. Benzene and water are fully miscible at $T = 573$ K and pressures from $P = 15$ to 40 MPa. At $P = 10$ MPa and $T = 573$ K, the system has only a small miscibility gap at high water concentrations. With increasing pressure, the water–hydrogen miscibility gap increases [6]. As a consequence, the miscibility of the system

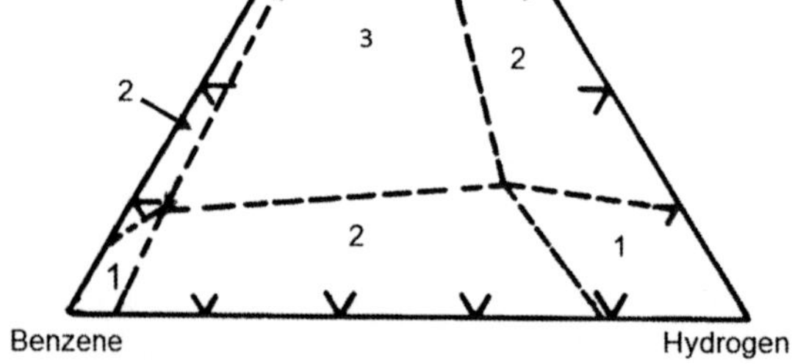

FIGURE 3.36 Phase behavior of H_2O–C_6H_6–H_2 at $T = 473$ K and at $P = 10$ MPa, mole fractions. Numbers indicate the number of coexisting phases [50].

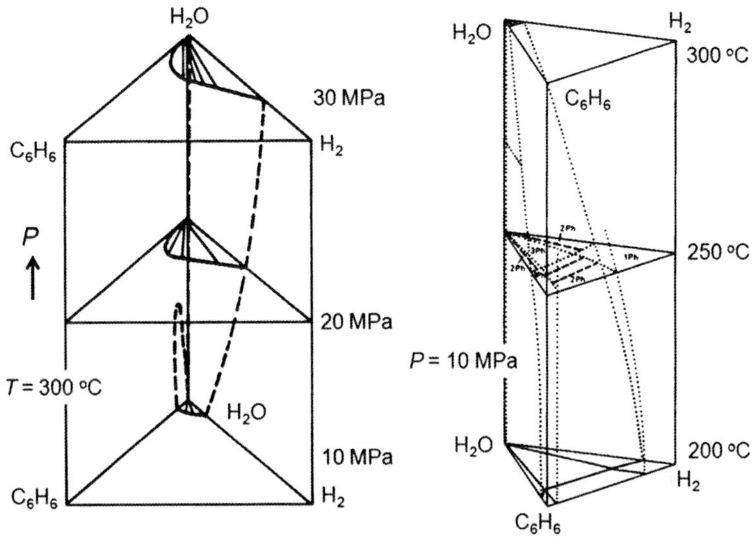

FIGURE 3.37 Phase behavior of H_2O–C_6H_6–H_2 at $T=573$ K (left) [49] and at $P=10$ MPa (right), mole fractions [50].

decreases from $P=20$ to 30 MPa. At about $P=40$ MPa, the miscibility for benzene–water ends, and the ternary system has again two binary miscibility gaps (not shown in Figure 3.37), see the phase behavior of benzene–water (Figure 3.37 (left)). The right-hand side of Figure 3.37 shows the temperature dependence of the phase behavior of H_2O–C_6H_6–H_2. The ternary diagram for $T=523$ K is derived from the binary systems without ternary experimental data as an example for the deduction of ternary data from binary equilibrium data. At least qualitative information can be derived for the ternary system [49,50].

3.4.3.4.2 The System Water–Toluene–Hydrogen

The system water–toluene–hydrogen has been measured at $P=10$ MPa in a temperature range from $T=373$ to 573 K [50]. In Figure 3.38, ternary phase equilibrium is shown for $T=503$ K at $P=10$ MPa (left). The temperature dependence of the phase equilibrium is shown in Figure 3.38 (right). Experimental values are listed in Table 3.18.

The ternary system water–toluene–hydrogen exhibits three-phase regions. The miscibility gap at $T=373$ K and $P=10$ MPa is very wide. Water dissolves more in the toluene-rich phase than toluene dissolves in the water-rich phase. The water-rich one-phase area is smaller than 1 mol%. At $T=573$ K, $P=10$ MPa, the three-phase area is no longer existent. The solubility of toluene in water has increased substantially and the solubility of water in the toluene-rich phase.

TABLE 3.17 Phase Equilibrium Data of H_2 (1)–H_2O (2)–C_6H_6 (3) (mol/mol) [49,50]

Gaseous Phase			Liquid Phase		
y_1	y_2	y_3	x_1	x_2	x_3
		$T=573.15$ K	$P=10$ MPa		
0.1413	0.8587	–	0.0015	0.9985	–
0.0844	0.8664	0.0492	0.0012	0.9963	0.0025
0.0624	0.8746	0.0630	0.0010	0.9936	0.0054
–	0.9085	0.0915	–	0.9982	0.0018
		$T=573.15$ K	$P=20$ MPa		
0.4710	0.5290	–	0.0040	0.9960	–
0.3480	0.5641	0.0879	0.0084	0.9801	0.0115
0.1950	0.6252	0.1798	0.0085	0.9761	0.0154
0.1576	0.6382	0.2042	0.0080	0.9580	0.0340
0.0955	0.6598	0.2447	0.0088	0.9471	0.0441
		$T=573.15$ K	$P=30$ MPa		
0.6010	0.3990	–	0.0160	0.9840	–
0.3215	0.5100	0.1685	0.0192	0.9594	0.0214
0.2460	0.5426	0.2114	0.0208	0.9439	0.0343
0.1926	0.5675	0.2399	0.0194	0.9431	0.0375

H_2 (1)–H_2O (2)–C_6H_6 (3)

T = *473.15 K*, P = *10 MPa*								
Gaseous phase			*Organic phase*			*Aqueous phase*		
y_1	y_2	y_3	x_1	x_2	x_3	x_1	x_2	x_3
0.7689	–	0.2311	0.0682	–	0.9318	–	–	–
0.5244	0.2399	0.2357	0.0480	0.1850	0.7670	0.0060	0.9820	0.0120

3.4.3.4.3 The System Water–*n*-Hexadecane–Hydrogen

The system water–*n*-hexadecane–hydrogen has been measured at pressures up to $P=30$ MPa in a temperature range from $T=373$ to 623 K [49]. In Figure 3.39, the pressure dependence of the ternary phase equilibrium is shown for $T=573$ K at 10 MPa $<P<$ 30 MPa (left) and the temperature

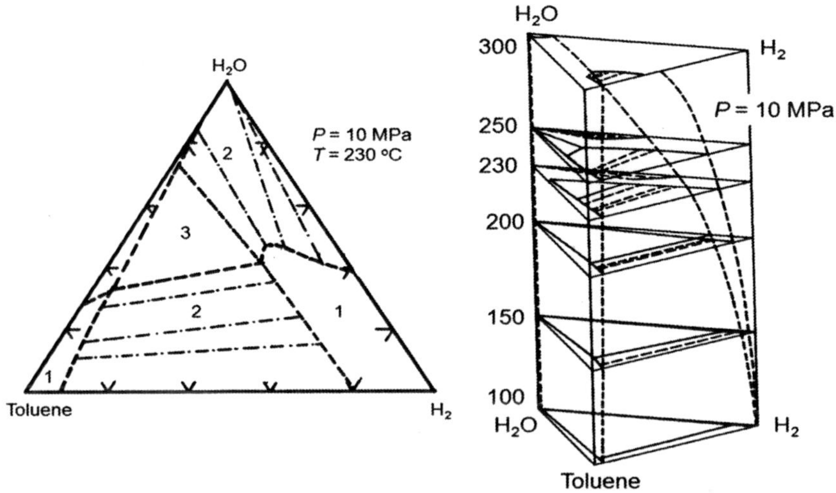

FIGURE 3.38 Phase equilibrium in the system water–toluene–hydrogen at $T = 503$ K, $P = 10$ MPa (left) and at 10 MPa, 373 K < T < 573 K (right), mole fractions. Numbers indicate the number of coexisting phases [50].

dependence at $P = 20$ MPa for 473 K < T < 623 K. Experimental values are listed in Table 3.19.

The system water–*n*-hexadecane–hydrogen consists of three equilibrium phases over a wide range of pressure and temperature. At $P = 20$ MPa and $T = 573$ K, the following phases exist:

- Three one-phase areas: one at high concentrations of hexadecane, one with high concentrations of hydrogen, and one with very high concentrations of water (>99 mol%). The hexadecane-rich phase dissolves up to 43 mol% water and up to 34 mol% hydrogen. In the hydrogen-rich phase, up to 53 mol% water but only 2.3 mol% hexadecane is dissolved in equilibrium. The water-rich phase is very small. It is enlarged in Figure 3.40 [34].
- Three two-phase areas: a gas–liquid equilibrium of the hydrogen-rich phase and the hexadecane-rich phase, a gas–liquid equilibrium of the hydrogen-rich-phase and the water-rich phase, and a liquid–liquid equilibrium of the water-rich and the hexadecane-rich phase.
- One three-phase area in between the two-phase areas.

The three-phase area often extends over most of the concentration ranges of the components. The mutual solubility increases with temperature and at high temperatures above about $T = 600$ K only two-phase regions exist.

3.4.3.4.4 The System Water–*n*-Hexadecane–Carbon Dioxide

The system water–*n*-hexadecane–carbon dioxide has been measured at pressures up to $P = 30$ MPa in a temperature range from $T = 473$ to 573 K [35].

TABLE 3.18 Phase Equilibrium in the System Toluene (1)–Water (2)–Hydrogen (3) [50]

Gaseous Phase			Organic Phase			Aqueous Phase		
y_1	y_2	y_3	x_1	x_2	x_3	x_1	x_2	x_3
				(mol/mol)				
				$T=373.15$ K, $P=10$ MPa				
–	0.0270	0.9730	–	–	–	–	0.9986	0.0014
0.0013	–	0.9987	0.9511	–	0.0489	–	–	–
0.0088	0.1180	0.8792	0.8807	0.0762	0.0431	0.0012	0.9973	0.0015
				$T=423.15$ K, $P=10$ MPa				
0.0221	0.0750	0.9029	0.8917	0.0523	0.0560	–	–	–
0.0421	0.1278	0.8301	0.7940	0.1549	0.0511	–	–	–
0.0095	0.0803	0.9102	–	–	–	0.0032	0.9950	0.0018
0.0278	0.0829	0.8893	–	–	–	0.0030	0.9951	0.0019
0.0327	0.1059	0.8614	0.7900	0.1533	0.0567	0.0007	0.9973	0.0020
				$T=473.15$ K, $P=10$ MPa				
0.1102	0.1926	0.6972	–	–	–	0.0020	0.9959	0.0021
0.1134	0.1823	0.7043	–	–	–	0.0120	0.9859	0.0021
0.1199	0.1749	0.7052	–	–	–	0.0020	0.9960	0.0020
0.1047	0.1071	0.7882	0.8501	0.0853	0.0646	–	–	–
0.1076	0.1258	0.7666	0.8412	0.0979	0.0609	–	–	–
0.1038	0.1819	0.7143	0.7715	0.1707	0.0578	0.0040	0.9940	0.0020
				$T=503.15$ K, $P=10$ MPa				
0.0631	0.4102	0.5267	–	–	–	0.0019	0.9958	0.0023
0.1349	0.4561	0.4090	–	–	–	0.0033	0.9943	0.0024
0.1916	–	0.8084	0.9099	–	0.0901	–	–	–
0.2094	0.1533	0.6373	0.8199	0.1090	0.0711	–	–	–
0.2119	0.2413	0.5468	0.7742	0.1615	0.0643	–	–	–
0.2165	0.3535	0.4300	0.6716	0.2761	0.0524	–	–	–
0.2099	0.4063	0.3838	0.6218	0.3303	0.0480	0.2543	0.7269	0.0188
				$T=523.15$ K, $P=10$ MPa				
0.0709	0.5642	0.3649	–	–	–	0.0045	0.9933	0.0022
0.1147	0.6353	0.2500	–	–	–	0.0454	0.9925	0.0021

TABLE 3.18 Phase Equilibrium in the System Toluene (1)–Water (2)–Hydrogen (3) [50]—Cont'd

Gaseous Phase			Organic Phase			Aqueous Phase		
y_1	y_2	y_3	x_1	x_2	x_3	x_1	x_2	x_3
				(mol/mol)				
0.0611	0.5378	0.4011	–	–	–	0.0021	0.9957	0.0022
0.0937	0.5931	0.3132	–	–	–	0.0124	0.9863	0.0013
0.2944	–	0.7056	0.9014	–	0.0986	–	–	–
0.2809	0.1953	0.5238	0.8142	0.1041	0.0817	–	–	–
0.2978	0.2704	0.4318	0.7686	0.1617	0.0697	–	–	–
0.3071	0.5199	0.1730	0.5499	0.4154	0.0347	0.0528	0.9430	0.0042
				$T=573.15$ K, $P=10$ MPa				
0.6509	–	0.3491	0.8779	–	0.1221	–	–	–
0.6136	–	0.3864	0.8700	–	0.1300	–	–	–
0.6596	–	0.3404	0.8705	–	0.1295	–	–	–
0.6736	0.1935	0.1329	0.7489	0.1629	0.0882	–	–	–
0.6324	0.0923	0.2753	0.8258	0.0656	0.1086	–	–	–
0.6489	0.0903	0.2608	0.8193	0.0691	0.1116	–	–	–
0.6345	0.0910	0.2745	0.8208	0.0709	0.1083	–	–	–
0.6666	0.1272	0.2062	0.7981	0.1031	0.0988	–	–	–
0.6327	0.1892	0.1781	0.7576	0.1508	0.0916	–	–	–
0.6366	0.1585	0.2049	0.7699	0.1331	0.0970	–	–	–

In Figure 3.41, the pressure dependence of the ternary phase equilibrium is shown for $T=473$ and 573 K at 10 MPa $<P<$ 30 MPa. The phase boundary lines were calculated with the Peng–Robinson EOS [35]. Experimental values are listed in Table 3.20.

3.4.3.4.5 Water–*n*-Hexadecane–Benzene, Water–*n*-Hexadecane–Toluene, and Water–*n*-Hexadecane–*n*-Hexane

Liquid–liquid equilibria are rarely measured at elevated pressure and temperature. For the previously discussed ternary systems, such data have been

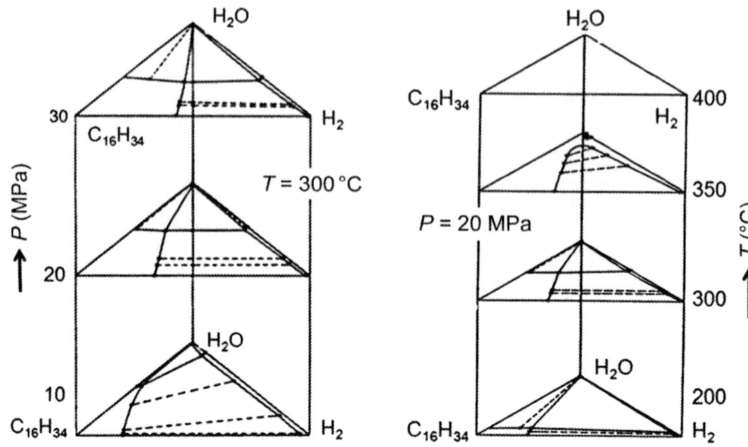

FIGURE 3.39 Pressure dependence of the ternary phase equilibrium for $T = 300\,°C$ at $10\,MPa < P < 30\,MPa$ (left), temperature dependence at 20 MPa for $200\,°C < T < 350\,°C$ K (right), mole fractions [49].

determined at $P = 20\,MPa$ and $T = 573\,K$ in the systems of water and *n*-hexadecane with benzene, toluene, and *n*-hexane [84]. The experimental data have been correlated with the Peng–Robinson EOS and a modified Redlich–Kwong EOS called HPW (Hederer–Peter–Wenzel) EOS. Experimental results are shown in Figure 3.42 and tabulated in Table 3.21. The parameters for correlation of the phase equilibria are listed in Table 3.22.

3.5 WATER AND ELECTROLYTES

Water and electrolyte systems are of great importance not only at ambient temperatures but also for hydrothermal systems. Phase behavior of water–electrolyte systems differs from that of water–nonelectrolyte systems in the way that in the aqueous phase, dissociation reactions to ionic species occur. This chemical equilibrium has to be considered in addition to the physical equilibrium. Phase behavior depends strongly on temperature since at high temperature, the electrolyte nature of water changes.

The general framework for the description of electrolytes is treated. For equilibria of water and volatile electrolytes, the equilibrium of CO_2–H_2O will be discussed in some detail: first, physical vapor–liquid phase equilibrium and, then, the chemical equilibrium in the liquid phase. CO_2–H_2O equilibrium is the representative example for a volatile weak electrolyte only slightly soluble in H_2O under most conditions. NH_3–H_2O equilibria will be covered as an example for weak volatile electrolytes that are fully miscible with H_2O. Both specific equilibria are of major importance for applications. Other volatile electrolytes, like H_2S, and various organic compounds acting as electrolytes in aqueous systems are presented, including the influence of salts on the

TABLE 3.19 System Hydrogen (1) Water (2)–*n*-Hexadecane (3) (mol/mol) [49]

Gaseous Phase			Organic Phase			Aqueous Phase		
y_1	y_2	y_3	x_1	x_2	x_3	x_1	x_2	x_3
				(mol/mol)				
			$T=573.15$ K, $P=10$ MPa					
0.9770	–	0.0230	0.2000	–	0.8000	–	–	–
0.9052	0.0739	0.0209	0.1746	0.0526	0.7728	–	–	–
0.3863	0.5889	0.0248	0.0847	0.2980	0.6174	–	–	–
0.1088	0.8562	0.0349	0.0132	0.5352	0.4516	0.0013	0.9945	0.0042
0.1058	0.8669	0.0272	–	–	–	0.0012	0.9980	0.0008
0.1413	0.8586	–	–	–	–	0.0015	0.9984	–
			$T=573.15$ K, $P=20$ MPa					
0.9850	–	0.0015	0.3400	–	0.6600	–	–	–
0.8500	0.1320	0.0018	0.3033	0.1211	0.5756	–	–	–
0.7925	0.1894	0.0181	0.2747	0.1760	0.5493	–	–	–
0.4684	0.5086	0.0230	0.1577	0.4753	0.3670	0.0087	0.9898	0.0015
0.4584	0.5132	0.0184	–	–	–	0.0075	0.9920	0.0005
0.4710	0.5290	–	–	–	–	0.0040	0.9960	–
–	–	–	0.0036	0.4823	0.5141	0.0006	0.9947	0.0047
			$T=573.15$ K, $P=30$ MPa					
0.9850	–	0.0150	0.4300	–	0.5700	–	–	–
0.8605	0.1261	0.0134	0.3774	0.1393	0.4833	–	–	–
0.8510	0.1374	0.0116	0.3803	0.1459	0.4738	–	–	–

Continued

TABLE 3.19 System Hydrogen (1) Water (2)–*n*-Hexadecane (3) (mol/mol) [49]—Cont'd

Gaseous Phase			Organic Phase			Aqueous Phase		
y_1	y_2	y_3	x_1	x_2	x_3	x_1	x_2	x_3
				(mol/mol)				
0.5910	0.3930	0.0160	0.2867	0.3623	0.3510	0.0156	0.9839	0.0005
0.6010	0.3990	–	–	–	–	0.0160	0.9840	–
–	–	–	0.1238	0.3891	0.4871	0.0075	0.9915	0.0010
			$T = 473.15$ K, $P = 20$ MPa					
0.9980	–	0.0020	0.2520	–	0.7480	–	–	–
0.9227	0.0749	0.0024	0.2205	0.0908	0.6887	–	–	–
0.8936	0.1046	0.0018	0.2245	0.1203	0.6552	0.0052	0.9943	0.0005
0.9280	0.0720	–	–	–	–	0.0020	0.9980	–
–	–	–	0.2241	0.1109	0.6650	0.0036	0.9955	0.0009
–	–	–	0.1769	0.1165	0.7066	0.0055	0.9930	0.0015
			$T = 623.15$ K, $P = 20$ MPa					
0.9770	–	0.0230	0.3684	–	0.6316	–	–	–
0.6394	0.3235	0.0371	0.2662	0.2664	0.4874	–	–	–
0.4985	0.4477	0.0538	0.2426	0.3083	0.4491	–	–	–
0.3134	0.6283	0.0583	0.1749	0.4819	0.3432	–	–	–
0.1769	0.7463	0.0768	0.1222	0.6042	0.2736	–	–	–
0.0720	0.9280	–	–	–	–	0.0200	0.9980	–
0.0492	0.9306	0.0202	–	–	–	0.0540	0.9423	0.0037

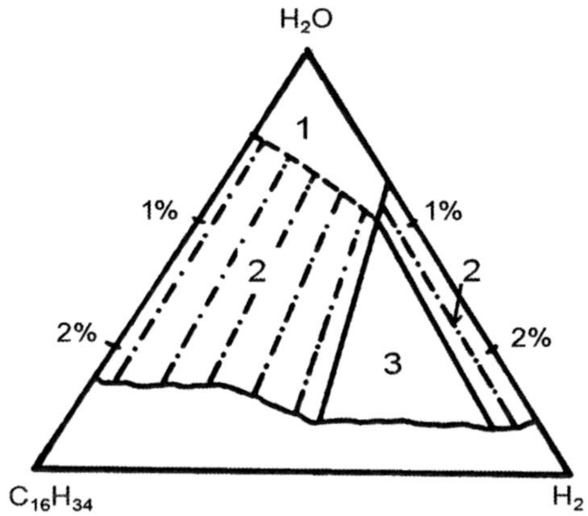

FIGURE 3.40 System water–*n*-hexadecane–hydrogen. Water-rich phase at $P=20$ MPa and $T=300$ °C, mole fractions [34].

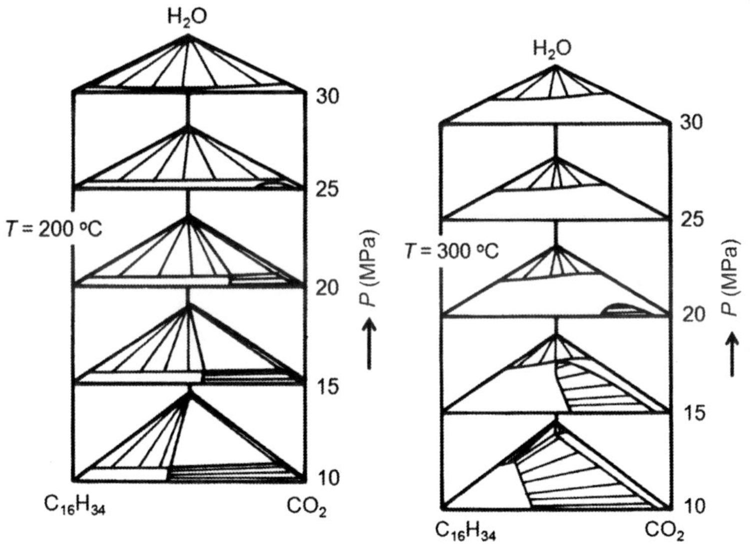

FIGURE 3.41 Phase equilibrium in the system water–*n*-hexadecane–carbon dioxide. $T=200$ °C (left), $T=300$ °C (right), mole fractions [35].

solubility of volatile electrolytes in water. Ternary systems are covered shortly. The section is concluded with a discussion of phase equilibria of water and nonvolatile electrolytes (salts).

3.5.1 Equilibria with Dissociating Compounds

Compounds dissolved at equilibrium in water can be present in the gaseous and in the liquid phase. In the liquid phase, they can dissociate and

TABLE 3.20 Phase Equilibrium in the System Water (1)–*n*-Hexadecane (2)–Carbon Dioxide (3) [35]

Gaseous Phase			Organic Phase			Aqueous Phase		
y_1	y_2	y_3	x_1	x_2	x_3	x_1	x_2	x_3
				(mol/mol)				
				$T = 473.15$ K, $P = 20.1$ MPa				
0.000	0.011	0.989	0.000	0.307	0.693	0.000	0.000	0.000
0.020	0.012	0.968	0.026	0.299	0.675	0.000	0.000	0.000
0.101	0.011	0.888	0.108	0.271	0.621	0.000	0.000	0.000
0.180	0.000	0.820	0.000	0.000	0.000	0.970	0.000	0.030
0.000	0.000	0.000	0.162	0.412	0.426	0.981	0.009	0.010
0.176	0.021	0.803	0.172	0.255	0.573	0.979	0.008	0.013
				$T = 573.15$ K, $P = 10.1$ MPa				
0.000	0.038	0.962	0.000	0.572	0.428	–	–	–
0.191	0.031	0.778	0.115	0.535	0.350	–	–	–
0.346	0.058	0.596	0.195	0.530	0.275	–	–	–
0.414	0.049	0.538	0.257	0.501	0.242	–	–	–
0.483	0.051	0.466	0.316	0.493	0.191	–	–	–
0.628	0.080	0.292	0.396	0.484	0.120	–	–	–
0.921	0.000	0.079	0.996	0.000	0.004	–	–	–
				$T = 573.15$ K, $P = 20.1$ MPa				
–	0.062	0.938	–	0.326	0.674	–	–	–
0.074	0.083	0.843	0.025	0.291	0.684	–	–	–
0.610	–	0.390	–	–	–	0.971	–	0.029
0.571	0.136	0.293	–	–	–	0.981	–	0.019
0.582	0.158	0.260	–	–	–	0.983	–	0.017
0.539	0.375	0.086	–	–	–	0.991	–	0.009
				$T = 573.15$ K, $P = 30.1$ MPa				
0.546	–	0.454	–	–	–	0.946	–	0.054
0.506	0.133	0.361	–	–	–	0.959	0.003	0.038
0.491	0.187	0.322	–	–	–	0.960	0.001	0.039
0.433	0.428	0.139	–	–	–	0.978	0.008	0.014

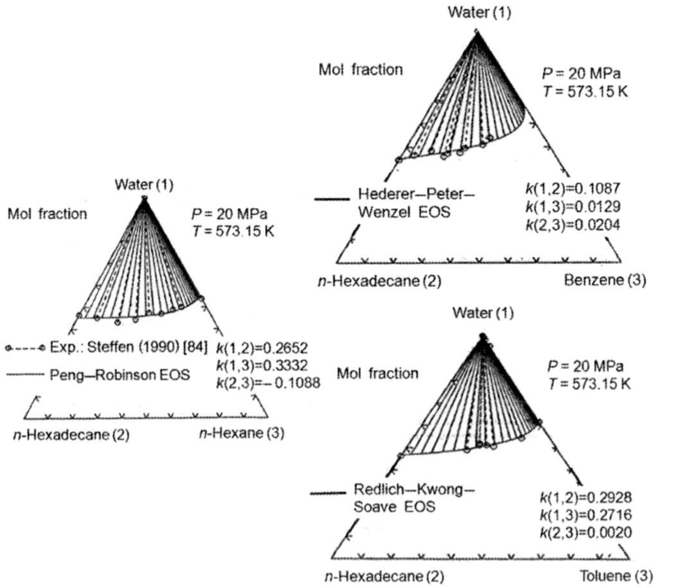

FIGURE 3.42 Phase equilibrium in systems of water and n-hexadecane with benzene, toluene, and *n*-hexane at 20 MPa and 573 K [84].

form ionic species. For a first discussion, it may be assumed that these ionic species are not present in the gaseous phase, but only in the liquid phase. If temperature and pressure are high and the gaseous phase approaches the properties of the liquid phase, this simplifying assumption must be dropped.

Thermodynamic equilibrium of an electrolyte in water must be described as combined physical and chemical equilibrium, for example, for the weak and volatile electrolyte CO_2:

$$\begin{gathered}\text{Gas-phase } CO_2 \\ \downarrow \\ \text{Liquid-phase } CO_2 + H_2CO_3\,(\text{aqueous}) \leftrightarrow H^+ + HCO_3^- \leftrightarrow 2H^+ + CO_3^{2-}\end{gathered} \tag{3.94}$$

where gas-phase CO_2 to liquid-phase CO_2 represents the physical equilibrium.

The physical part is analogous to the solubility of nondissociating compounds and the chemical part consists of one or more chemical reactions in series or in parallel leading to new ionic species. For example, a binary solution, such as sodium chloride in water, contains water, a cation, and an anion, three species but only two components since the concentrations of cation and anion are not two independent variables. The concentration of one ion fixes the other because the material balance and electroneutrality must be satisfied. For strong electrolytes (salts), ionization is complete, and for weak electrolytes (e.g., carbon dioxide), ionization is only partial. For a thermodynamic

TABLE 3.21 Phase Equilibrium in Systems of Water–*n*-Hexadecane, Benzene (3), Toluene, and *n*-Hexane at $P = 20$ MPa and $T = 573$ K (mol/mol) [84]

Lower Phase			Upper Phase		
x_1	x_2	x_3	y_1	y_2	y_3
Water (1)–n-hexadecane (2)–benzene (3)					
0.9999	–	<0.0001	0.4632	–	0.5368
0.9991	0.0008	0.0001	0.4692	0.0453	0.4755
0.9982	0.0017	0.0001	0.4625	0.0939	0.4436
0.9970	0.0028	0.0002	0.4560	0.1492	0.3948
0.9937	0.0061	0.0002	0.4793	0.1960	0.3247
0.9926	0.0073	0.0001	0.4938	0.2102	0.2960
0.9903	0.0097	<0.0001	0.5116	0.2552	0.2332
0.9856	0.0142	0.0002	0.5380	0.2727	0.1893
0.9850	0.0148	0.0002	0.5508	0.2879	0.1613
Water (1)–n-hexadecane (2)–toluene (3)					
0.9999	–	0.0001	0.4632	–	0.5368
0.9999	0.0001	0.0001	0.4908	0.2053	0.3039
0.9952	0.0046	0.0002	0.5120	0.2351	0.2529
0.9945	0.0053	0.0002	0.5204	0.2317	0.2479
0.9925	0.0073	0.0002	0.5129	0.2595	0.2277
0.9924	0.0075	0.0001	0.5071	0.2853	0.2077
0.9839	0.0159	0.0002	0.5437	0.3618	0.0945
0.9540	0.0460	–	0.6162	0.3838	–
Water (1)–n-hexane (2)–n-hexadecane (3)					
0.9999	–	0.0001	0.4632	–	0.5368
0.9997	0.0002	0.0001	0.4498	0.00786	0.4716
0.9984	0.0004	0.0012	0.4299	0.1705	0.3996
0.9989	0.0009	0.0002	0.4422	0.2392	0.3186
0.9987	0.0012	0.0001	0.4717	0.2813	0.2470
0.9984	0.0014	0.0002	0.4690	0.3319	0.1991
0.9952	0.0044	0.0004	0.4882	0.3704	0.1414
0.9935	0.0058	0.0007	0.4991	0.3978	0.1031
0.9962	0.0038	–	0.5380	0.4620	–

TABLE 3.22 Water–*n*-Hexadecane–Benzene, Water–*n*-Hexadecane–Toluene, and Water–*n*-Hexadecane–*n*-Hexane

	Water	Benzene	Toluene	*n*-Hexane	*n*-Hexadecane
T_c (K)	647.3	562.1	591.7	507.4	717.0
P_c (MPa)	22.09	4.89	4.11	2.97	1.42
ω (–)	0.334	0.212	0.257	0.296	0.742
Hederer–Peter–Wenzel EOS parameter					
a (bar^2 mol^{-2} K$^{-\alpha}$)	114.6	531.3	671.5	967.0	41,261.3
b (cm^3 mol^{-1})	16.07	75.32	91.99	108.51	274.92
α (–)	−0.47909	−0.52954	−0.51570	−0.58769	−0.92199
Modified Peng–Robinson EOS parameter					
a_c (bar^2 mol^{-2})	6.306	19.350	25.643	25.793	97.631
b (cm^3 mol^{-1})	19.55	74.07	90.57	106.36	268.97
x_0 (–)	0.93742	0.58209	0.55997	0.60939	0.86464

Pure-component parameters for correlation of the experimental data [84].

description, chemical equilibria and in addition material balances and the constraint of electroneutrality must be considered [40].

Electroneutrality requires in a solution of an electrolyte that the number of moles of the individual ionic species cannot be varied independently. When dissolved in a high-dielectric-constant solvent like water, an electrically neutral electrolyte $M_{v^+}X_{v^-}$ is dissociated into v^+ positive ions (cations), each with a charge z^+, and v^- negative ions (anions) of charge z^-. Charges are given in normalized units where $z^+=1$ for a proton. Electrolytic dissociation is represented by [40]

$$M_{v^+}X_{v^-} \leftrightarrow v^+M^{z^+} + v^-X^{z^-} \tag{3.95}$$

Electroneutrality requires

$$v^+z^+ + v^-z^- = 0 \tag{3.96}$$

Chemical equilibrium is represented by

$$\mu_{M_{v+}X_{v^-}} = v^+\mu_{M^{z+}} + v^-\mu_{X^{z-}} \tag{3.97}$$

The chemical potential $\mu_{M_{v+}X_{v+}}$ for the electrolyte is

$$\mu_{M_{v+}X_{v^-}} = \mu^*_{M_{v+}X_-} + RT\ln\left(m_+^{v^+}m_-^{v^-}\right) + RT\ln\left(\gamma_+^{v^+}\gamma_-^{v^-}\right) \tag{3.98}$$

$$\mu_{M_{\nu+}X_{-}} = \mu^*_{M_{\nu+}X_{-}} + RT\ln\left(a_+^{\nu^+} a_-^{\nu^-}\right), \tag{3.99}$$

with $\mu^*_{M_{\nu+}X_{\nu-}}$ for the chemical potential at standard state, m for molality as measure for concentration (m = moles of solute per kilogram of solvent), γ for the activity coefficient, and a for the activity.

The ionic molality $m_\pm$, the mean activity coefficient $\gamma_\pm$, and the mean ionic activity $a_\pm$ are defined by [40]

$$m_\pm = \left(m_+^{\nu^+} m_-^{\nu^-}\right)^{1/\nu} \tag{3.100}$$

$$\gamma_\pm = \left(\gamma_+^{\nu^+} \gamma_-^{\nu^-}\right)^{1/\nu} \tag{3.101}$$

$$a_\pm = \left[(a_+)^{\nu^+} (a_-)^{\nu^-}\right]^{1/\nu} = m_\pm \gamma_\pm^{(m)}. \tag{3.102}$$

By definition, $\gamma_\pm^{(m)}$ is unity at zero molality for all electrolytes. A more detailed discussion, in particular, on the standard state can be found in the textbook by Prausnitz *et al.* [40].

In addition, the Brønsted definition of acids and bases may be useful [85]. Dissolved compounds (A and B) are considered capable of releasing hydrogen ions or hydroxide ions:

$$\mathrm{H}A(\mathrm{aq}) \leftrightarrow \mathrm{H}^+(\mathrm{aq}) + A^-(\mathrm{aq}) \tag{3.103}$$

and

$$B(\mathrm{aq}) + \mathrm{H_2O(l)} \leftrightarrow B\mathrm{H}^+(\mathrm{aq}) + \mathrm{OH}^-(\mathrm{aq}) \tag{3.104}$$

The ionization product A^-(aq) is the conjugate base of HA(aq), because it behaves as a base in reacting with water to form HA(aq):

$$A^-(\mathrm{aq}) + \mathrm{H_2O(l)} \leftrightarrow \mathrm{H}A(\mathrm{aq}) + \mathrm{OH}^-(\mathrm{aq}) \tag{3.105}$$

Analog BH$^+$(aq) is the conjugate acid of B(aq). The equilibrium quotient of the acid ionization reaction, reaction (3.103), is defined as

$$Q_{1\mathrm{a}} = \frac{m_{A^-} m_{\mathrm{H}^+}}{m_{\mathrm{H}A}} \tag{3.106}$$

where m_{A^-}, m_{H^+}, and $m_{\mathrm{H}A}$ are molalities of the species A^-(aq), H^+(aq), and HA(aq), respectively. The equilibrium quotient $Q_{1\mathrm{a}}$ is a function of temperature, pressure, and ionic strength. The equilibrium constant of reaction (3.103) refers to infinitely dilute solutions in the hypothetical 1 mol kg^{-1} standard state, and thus, it is not a function of ionic strength:

$$K_{1\mathrm{a}} = \frac{a_{A^-} a_{\mathrm{H}^+}}{a_{\mathrm{H}A}} \tag{3.107}$$

where a_{A^-}, $a_{\mathrm{H}A}$, and a_{H^+} are activities in molality units. The corresponding expressions may be written for the ionization of bases, defined as Q_b and K_b.

The ratio between molalities and activities of solutes is defined as the activity coefficient, $\gamma = a/m$, so that

$$\log K_{1a} = \log Q_{1a} + \log\left(\frac{\gamma_{A^-}\gamma_{H^+}}{\gamma_{HA}}\right) \quad (3.108)$$

where γ_{A^-}, γ_{H^+}, and γ_{HA} are activity coefficients of A^-(aq), H^+(aq), and HA(aq). Since the last term in Equation (3.108) vanishes as ionic strength approaches zero, $\log K_{1a}$ is usually obtained from experimentally determined $\log Q_{1a}$ values by extrapolation to infinite dilution.

3.5.1.1 *Solubility of Electrolytes in Steam*

Solubility of electrolytes in steam is of great importance in hydrothermal processes and in power plants. The partitioning of an electrolyte CA between liquid water and the dense gaseous phase of water can be divided into the two steps of ion association (K_a) in the liquid and the distribution (partitioning) (K_v) of the neutral ion pair between the liquid and the gaseous phase [86] (see also Equation 3.94):

$$C^+(aq) + A^-(aq) \leftrightarrow CA(aq) \quad (3.109)$$

$$CA(aq) \leftrightarrow CA(v) \quad (3.110)$$

Combining both equilibrium constants yields

$$K_d = K_a K_v$$

For a number of electrolytes, empirical correlations have been derived, based on carefully evaluated experimental results [86]. They are listed for some important electrolytes in the following:

Hydrochloric acid:

$$\log K_d(\mathrm{HCl}) = -13.4944 - \frac{934.466}{T} - 11.0029 \log \rho_w + 5.4847 \log T \quad (3.111)$$

where ρ_w is the density of pure water (g cm^{-3}).

Near the critical temperature of water, a more theoretical-based expression may be preferred that relates the slope of the pressure to the mole fraction in the liquid phase [87]:

$$T \ln K_d(\mathrm{HCl}) = \frac{2B}{R}\left(\rho_{H_2O} - \rho_{(H_2O)_c}\right) + C \quad (3.112)$$

where $\rho_{H_2O}, \rho_{(H_2O)_c}$ are the densities in mol dm^{-3} of liquid water at T and the critical density of water. Equation (3.112) can be applied for the distribution of nonionic species also and is recommended for high-temperature data extrapolation to the critical temperature of water [86].

Sodium hydroxide:

$$\log K_{\mathrm{w}} K_{\mathrm{d}}(\mathrm{NaOH}) = 4.65 - 11{,}001/T \tag{3.113}$$

Ammonia and ammonium chloride:

$$\log K_{\mathrm{d}}(\mathrm{NH_3}) = -0.056 - \frac{458.71}{T} + 1.3267 \log \rho_{\mathrm{w}} \tag{3.114}$$

$$\log K_{\mathrm{d}}(\mathrm{NH_4Cl}) = -1.875 - \frac{1248.65}{T} + 23.401 \log \rho_{\mathrm{w}} \tag{3.115}$$

In Figure 3.43, partitioning constants for 1:1 electrolytes are plotted. H_2SO_4 (1:1) refers to the partitioning constant for sulfuric acid according to the equilibrium

$$\mathrm{H_2SO_4(aq)} \leftrightarrow \mathrm{HSO_4^-(aq)} + \mathrm{H^+(aq)} \tag{3.116}$$

3.5.2 Water and Volatile Electrolytes

Volatile electrolytes in aqueous solution exist in ionic and molecular, nondissociated form. In the gaseous phase, for most cases, only the molecular form must be considered. Calculation of vapor–liquid equilibria requires the solution of phase equilibrium equations for the molecular species, chemical equilibrium equations for the liquid phase, and material balances [40], as discussed in Section 3.5.1.

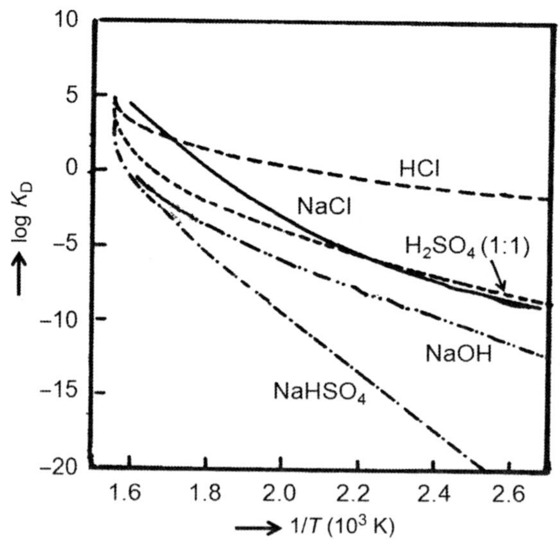

FIGURE 3.43 Partitioning constants for some electrolytes between the liquid aqueous phase and steam. *Modified from Ref. [86].*

At low concentrations, phase equilibrium between the gaseous phase and the liquid phase is determined by Henry's constant, k_H, and the molality of the molecular solute (see Section 3.4.1). This molality is strongly influenced by the chemical dissociation equilibrium, characterized by the dissociation constant, K. Phase equilibrium is determined by the molecular concentration of the electrolyte in the liquid phase, since ions are not volatile. For water as the solvent and a volatile electrolyte, VL is described by $f_w^V = f_w^L$ for water and $f_i^V = f_i^L$ for the volatile electrolyte, with f for fugacity. The equilibrium for water can be expressed as

$$\varphi_w^V y_w P = a_w P_w^s \varphi_w^s \int_{P_w^s}^{P} \frac{\overline{v}_w - v_w^s}{RT} dP \tag{3.117}$$

where a_w is the activity of water, P_w^s is the saturation pressure of pure water, φ_w^s is the fugacity coefficient of pure water at saturation, $\overline{v}_w$ is the partial molar volume of water, and v_w^s is the molar volume of water at saturation, all at system temperature.

For phase equilibrium, only the molecular electrolyte (subscript M) is relevant. Therefore, the equation for equilibrium is

$$\varphi_{MX,M}^V y_{MX,M} P = \gamma_{MX,M}^{(m)} m_{MX,M} k_{H,MX,M} \tag{3.118}$$

In Equation (3.118), $y_{MX,M} P$ is the partial pressure of the molecular weak electrolyte and $\varphi_{MX,M}^V$ is the vapor-phase fugacity coefficient; $m_{MX,M}$ is the molality of the electrolyte in molecular form; $\gamma_{MX,M}^{(m)}$ is the activity coefficient; and $k_{HMX,M}$ is Henry's constant for the molecular solute. Henry's constant k_H depends on temperature. When pressure is high, there is also an effect of pressure on k_H according to the rigorous thermodynamic equation $\frac{\partial \ln k_{H,i}}{\partial p} = \overline{v}_i^\infty / RT$, where i stands for solute and $\overline{v}_i^\infty$ is the partial molar volume of molecular i in the solvent at infinite dilution [40].

For temperatures high enough that Henry's law cannot be applied, the common equations for VL have to be used, for example,

$$\varphi_{MX,M}^V y_{MX,M} = \varphi_{MX,M}^l x_{MX,M} \tag{3.119}$$

$$RT \ln \varphi_{MX,M} = \int_0^P \left(\overline{v}_{MX,M} - \frac{RT}{P} \right) dP \tag{3.120}$$

with $\overline{v}_{MX,M}$ as the partial molar volume of the molecular form of the electrolyte.

The overall electrolyte concentration m_{MX} of the weak electrolyte in the liquid phase is derived from a mass balance on the molecular form $m_{MX,M}$ and the chemically reacted forms m_{M^-}:

$$m_{MX} = m_{MX,M} + m_{M^-} \tag{3.121}$$

For CO_2 dissolved in water,

$$m_{CO_2} = m_{CO_2,M} + m_{H_2CO_3} + m_{HCO_3^-} + m_{CO_3^{2-}} \tag{3.122}$$

The electroneutrality condition in the liquid phase combines the concentration of cations to those of anions

$$\sum_i (z_i m_i) = 0 \tag{3.123}$$

Chemical equilibrium connects the molecular concentration of the nondissociated electrolyte and the ionic concentrations of the weak electrolyte; for example, for the reaction,

$$K = \frac{a_+^{\nu+} a_-^{\nu-}}{a_{MX}} = \frac{m_+^{\nu+} m_-^{\nu-}}{m_{MX,M}} \frac{\left(\gamma_\pm^{(m)}\right)^\nu}{\gamma_{MX,M}^{(m)}} \tag{3.124}$$

where $\gamma_\pm^{(m)}$, is the mean ionic activity coefficient. For dilute solutions, all activity coefficients can be set to unity.

At higher temperatures, near the critical temperature of the solvent, it is advantageous to treat the solute and solvent on the same basis and correlate phase equilibria with an EOS as discussed in the preceding text. For further information, the reader is referred to Fernández-Prini *et al.* [51], where the modeling approaches are discussed.

In general, including electrolytes in an EOS involves adding electrostatic terms to the EOS for the uncharged system. Binary salt–water parameters are adjusted to data for single salts in water. Salt–gas parameters may be fitted to Setchenov constants to describe salting-out or salting-in effects. Such work on high-pressure gas solubility [88–92] has achieved qualitative and in some cases quantitative success for modeling vapor–liquid equilibria of gases with aqueous electrolytes up to temperatures of about $T = 500$ K and high pressures. Another approach involves an electrolyte model based on excess Gibbs energy into an EOS [93–95]. At high temperatures ($T > 573$ K), EOS approaches that neglect the dissociation of the electrolyte can be successful [96].

3.5.2.1 *Water and CO_2*

The system water–carbon dioxide is considered in the following, including ambient temperature because of the general importance of the CO_2–H_2O equilibrium. Physical phase equilibrium is treated first. Then, the influence of dissociation on equilibria of water with carbon dioxide is reviewed up to critical and supercritical temperatures, and the influence of salts (ionic strength) is discussed.

3.5.2.1.1 Phase Behavior of CO_2–H_2O (Physical Equilibrium)

Carbon dioxide is only slightly dissolving and dissociating in water. Since CO_2–H_2O equilibrium is of major importance for environmental and production processes, it will be discussed here in some detail, including temperatures below 373 K.

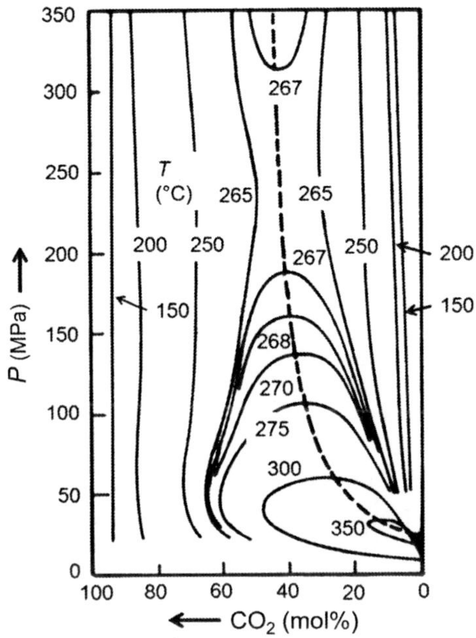

FIGURE 3.44 *P,x*-diagram of the system CO_2–H_2O. ———, critical curve. *Modified from Ref. [97].*

In Figure 3.44, the CO_2–H_2O equilibrium is shown for temperatures below the critical temperature of H_2O.

In the system CO_2–H_2O, the barotropic effect occurs at $P=80$ MPa and $T=323$ K, at $P=125$ MPa and $T=373$ K, and at $P=200$ MPa and $T=523$ K. At these conditions, the CO_2-rich phase is the one with greater density and therefore is the lower phase in a vessel.

The equilibrium between CO_2 and H_2O is of type III according to the scheme of van Konynenburg and Scott [28] with a broken critical curve (Table 3.23 and Figure 3.45). The lower critical curve starts at the critical point of pure CO_2. The upper critical curve starts at the critical point of H_2O and proceeds to lower temperatures and higher pressures passing through a minimum at $T=539.15$ K and $P=24.5$ MPa. At temperatures below the minimum, the critical curve runs to higher temperatures and much higher pressures. From the critical temperature of H_2O to the temperature minimum of the critical curve, vapor–liquid equilibria exist at pressures below the critical curve. At pressures and temperatures above the minimum of the critical line, also vapor–liquid equilibria exist, without the phases getting similar, as long as the critical curve is not crossing the critical temperature of H_2O (compare Figure 3.44). In this region, the mutual solubility of CO_2 and H_2O is first increasing and then again decreasing, widening the miscibility gap. The solubility of carbon dioxide in water is listed in Table 3.24 using data from various sources [97,98,100–103].

TABLE 3.23 Upper Critical Curve of the System CO_2–H_2O

T (K)	P (MPa)	x_{CO_2} (mol/mol)
647.3	22.13	0
643.15	23.5	0.025
633.15	27.9	0.075
623.15	33.3	0.125
613.15	37.3	0.155
603.15	42.7	0.190
593.15	48.5	0.220
583.15	53.9	0.242
573.15	60.8	0.267
563.15	71.6	0.295
553.15	91.7	0.330
548.15	108.0	0.347
543.15	137.0	0.367
541.15	162.0	0.385
539.15	245.0	0.415
541.15	≈360	0.435
UCEP		
304.65	≈7.4	≈0.9965
304.19	7.38	1.000

Data from Tödheide and Franck [97].

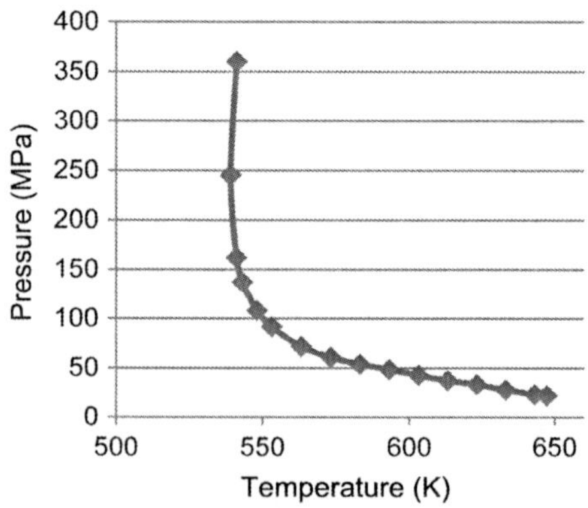

FIGURE 3.45 Upper critical curve of the system CO_2–H_2O. Minimum: 539.15 K, 245.0 MPa, $x_{CO_2} = 41.5$mol%. *Data from Tödheide and Franck [97].* (For color version of this figure, the reader is referred to the online version of this chapter.)

TABLE 3.24 Solubility of Carbon Dioxide in Water

P (MPa)	x_{CO_2} (mol/mol)			P (MPa)	x_{CO_2} (mol/mol)	
T (K)=273.15				T (K)=423.15[a]		
2.53	0.0247			10.0	0.0135	
				20.0	0.0215	
T (K)=285.15				30.0	0.0260	
5.07	0.0288			40.0	0.290	
10.14	0.0298			50.0	0.032	
15.21	0.0311			60.0	0.0345	
20.27	0.0322			70.0	0.0370	
30.41	0.0332			80.0	0.0390	
				90.0	0.0405	
T (K)=298.15				100.0	0.0420	
5.07	0.0220			150.0	0.0480	
10.14	0.0257	0.0327[b]				
40.55	0.3127			T (K)=473.15[a]		
50.69	0.03217			10.0	0.013	
				20.0	0.026	0.024[c]
T (K)=323.15				30.0	0.034	
5.07	0.0140	0.0138[d]	0.0137[e]	40.0	0.041	
10.14	0.0207	0.0203[d]	0.0205[e]	50.0	0.047	0.045[c]
15.21	0.0224	0.0219[d]		60.0	0.052	
20.27	0.0236	0.0230[d]	0.023[c]	70.0	0.056	
30.41	0.0254	0.0247[d]		80.0	0.058	
40.55	0.0269	0.0262[d]	[c]	90.0	0.061	
50.0			0.028[c]	100.0	0.063	0.061[c]
60.82	0.0297			150.0	0.072	0.073[c]
70.96	0.0310	0.0301[d]				
100.0			0.035[c]	T (K)=523.15[a]		
150.0			0.038[c]	10.0	0.012	

Continued

TABLE 3.24 Solubility of Carbon Dioxide in Water—Cont'd

P (MPa)	x_{CO_2} (mol/mol)			*P* (MPa)	x_{CO_2} (mol/mol)	
T (K) = 373.15				20.0	0.027	0.028[c]
5.07	0.00824	0.0082[d]		30.0	0.042	
10.14	0.0143	0.0138[d]		40.0	0.055	
15.21	0.0184	0.0175[d]		50.0	0.064	0.070[c]
20.27	0.0208	0.0199[d]	0.020[c]	60.0	0.074	
30.41	0.0239	0.0230[d]		70.0	0.084	
50.0			0.028[c]	80.0	0.094	
40.55	0.0262	0.0256[d]		90.0	0.104	
70.96	0.0312	0.0322[d]		100.0	0.115	0.120[c]
100.0			0.036[c]	150.0	0.144	0.149[c]
150.0			0.041[c]			
T (K) = 543.15[a]				*T* (K) = 598.15[a]		
20.0	0.027	0.026[c]		15.0	0.010	
30.0	0.046			20.0	0.026	
40.0	0.067			25.0	0.043	
50.0	0.087	0.087[c]		30.0	0.061	
60.0	0.106			35.0	0.082	
70.0	0.125			40.0	0.112	
80.0	0.145			P_c = 43.5	x_c = 0.180	
90.0	0.166					
100.0	0.190	0.190[c]				
P_c = 123.0	x_c = 0.288	P_c = 137.0 x_c = 0.367[c]				
				T (K) = 623.15[a]		
T (K) = 573.15[a]				20.0	0.015	0.008[c]
10.0	0.004			25.0	0.040	
15.0	0.016			30.0	0.064	0.051[c]
20.0	0.029			P_c = 32.5	x_c = 0.105	P_c = 33.3 x_c = 0.125[c]

TABLE 3.24 Solubility of Carbon Dioxide in Water—Cont'd

P (MPa)	x_{CO_2} (mol/mol)	*P* (MPa)	x_{CO_2} (mol/mol)
25.0	0.041		
30.0	0.054		
35.0	0.068		
40.0	0.084		
45.0	0.104		
50.0	0.127		
55.0	0.170		
$P_c=57.5.0$	$x_c=0.236$		

[a]*Data from Takenouchi and Kennedy [98].*
[b]*Data from King* et al. *[99].*
[c]*Data from Tödheide and Franck [97]. Values at 20.27 MPa were reported in Ref. [97] for 20 MPa.*
[d]*Data from Wiebe [100] (Table 2) given in (mol/mol). Data include Poynting correction. Data in the first column are taken from Wiebe [100] (Table 1) given there in cm³ CO_2(STP)/g H_2O, transformed to mol/mol with* $M_{H_2O}=18.02$, $M_{CO_2}=44.01$, and $\rho_{H_2O}(\mathrm{STP})=1.977\,\mathrm{g/cm^3}$.
[e]*Data from Bamberger* et al. *[101], at slightly different pressures, 5.06 and 10.09 MPa.*
Data from various authors (see legend).

At the low-temperature side, a critical curve runs from the critical point of CO_2 to the critical endpoint of the three-phase equilibrium gas–liquid–liquid. Below the three-phase line, liquid–liquid equilibria exist in the H_2O-rich concentration region and vapor–liquid equilibria exist in the CO_2-rich concentration region, while above the three-phase line equilibria are of the kind liquid–liquid (Figure 3.8d).

Calculation of solubility and distribution of the compounds between phases is often necessary in dilute systems. The IAPWS guideline on Henry's constant is a reliable database for Henry's constant and for the equilibrium distribution coefficient for CO_2 in H_2O [63]. Both parameters are shown in Figures 3.46 and 3.47. Discrete values have been presented in Tables 3.11 and 3.12.

3.5.2.1.2 Carbon Dioxide Equilibria in Aqueous Solution (Chemical Equilibria)

In this part, the chemical part of aqueous solutions of CO_2 in H_2O is discussed. CO_2 is dissolved in water in molecular form, reacts to some part to carbonic acid H_2CO_3, and dissociates in two steps into one hydrogen carbonate ion and one carbonate ion and the respective hydrogen ions (or hydronium ions H_3O^+, simplified to H^+) (Equations 3.94 and 3.125–3.128). The solution

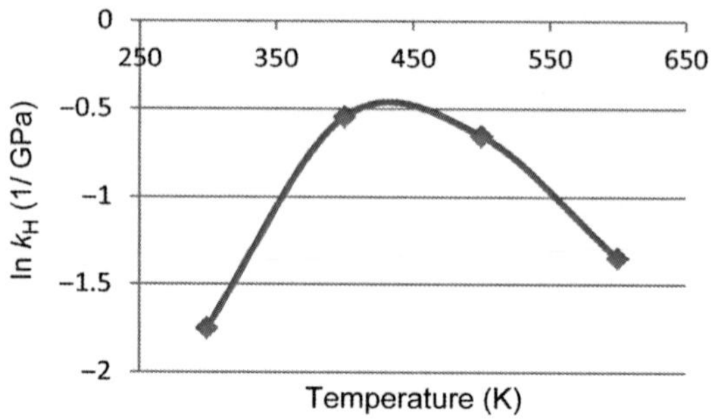

FIGURE 3.46 Henry's constant ln k_H for CO_2 in H_2O. *Data from the IAPWS guideline [63].* (For color version of this figure, the reader is referred to the online version of this chapter.)

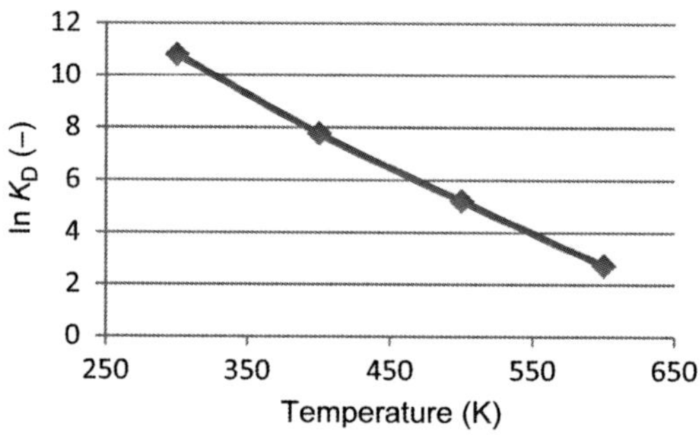

FIGURE 3.47 Equilibrium distribution coefficient for CO_2 in H_2O. *Data from the IAPWS guideline [63].* (For color version of this figure, the reader is referred to the online version of this chapter.)

of CO_2 in H_2O can be described with Henry's law as long as the temperature is not too close to the critical temperature. Values for Henry's constant are shown in Figure 3.46 and can be derived from the IAPWS formulation [63] or from Table 3.11. At temperatures above the temperature limit for the application of Henry's law, equilibrium must be calculated by incorporating both the liquid and the gaseous phase as discussed in the text earlier. Equilibrium distribution coefficients at infinite dilution are shown in Figure 3.47 and can be calculated in detail from the IAPWS formulation [63] or taken from Table 3.12.

A solution of CO_2 results in the following reactions:

$$CO_2^*(aq) + H_2O(l) \leftrightarrow H_2CO_3(aq) \tag{3.125}$$

with $CO_2^*(aq)$ as the total amount of CO_2. The reaction equilibrium of Equation (3.125) is heavy on the left side. The concentration of the hydrated molecular form of carbonic acid decreases with increasing temperature or ionic strength and increases slightly at high pressure ($P > 100$ MPa). At ambient temperature, the ratio of carbonic acid H_2CO_3 to CO_2 is less than 1% [85].

Therefore, it is reasonable to combine both concentration values to one $CO_2^*(aq)$:

$$m_{CO_2^*} = m_{CO_2} + m_{H_2CO_3} \tag{3.126}$$

Aqueous carbon dioxide ionizes to form hydrogen carbonate (bicarbonate) and carbonate ions according to the reactions:

$$CO_2^*(aq) + H_2O(l) \leftrightarrow H^+(aq) + HCO_3^-(aq) \tag{3.127}$$

$$HCO_3^-(aq) + OH^-(aq) \leftrightarrow CO_3^{2-}(aq) + H_2O(l) \tag{3.128}$$

The first ionization constant K_1 of $CO_2^*(aq)$ and the second ionization constant K_2 of $HCO_3^-(aq)$ are listed in Table 3.25 for temperatures up to $T=573$ K and ionic strength in NaCl solutions up to $I=5$. Data for the first ionization constant have been measured by Patterson *et al.* [104] and by Read [105]. The second ionization constant has been measured by Patterson *et al.* [106]. The solution chemistry of aqueous carbon dioxide has been reviewed by Palmer and van Eldik [107].

It is of interest to determine the carbonate ions in dependence of pH value ($pH = -\log H^+$). In an aqueous solution of CO_2 containing in addition only simple acids like HCl or bases like NaOH for the variation of the pH value, the concentration of the ions can be calculated in dependence of the pH value from Equations (3.129)–(3.132) [108]:

for $HCO_3{}^-$,

$$\log HCO_3{}^- = \log K_1 + \log CO_2 + pH \tag{3.129}$$

$$\log HCO_3{}^- = \log K_1 + \log(1/k_H) + \log P_{CO_2} + pH \tag{3.130}$$

for $CO_3{}^{2-}$,

$$\log CO_3{}^{2-} = \log K_1 + \log K_2 + \log CO_2 + 2pH \tag{3.131}$$

$$\log CO_3{}^{2-} = \log K_1 + \log K_2 + \log(1/k_H) + \log P_{CO_2} + 2pH \tag{3.132}$$

The concentrations of the species of the CO_2–H_2O equilibrium are plotted in Figure 3.48 in dependence of pH value. Although the diagram is valid only for 25 °C, far from hydrothermal temperatures, it is very illustrative. The hydrogen ion H^+, important for acid-catalyzed reactions, dominates at low pH values up to about pH 5.65. The carbonate ions exist only in certain pH ranges. $HCO_3{}^-$ ions dominate equilibrium ions in a pH range from 6 to about 9, while $CO_3{}^{2-}$ ions only exist at pH values higher than 7, in the basic range. It is also interesting to note that the concentration of molecular CO_2 in the liquid phase is about three orders of magnitude higher than that of molecular carbonic acid (H_2CO_3).

TABLE 3.25 Equilibrium Constants for the Dissociation of CO_2 in H_2O and in H_2O–NaCl Solutions, I (mol kg^{-1}), $pK_1 = -\log K_1$, $pK_2 = -\log K_2$ [85]

T (K)	pK_1 ($I=0$)	pQ_1 ($I=0.1$)	pQ_1 ($I=1.0$)	pQ_1 ($I=5.0$)
273.15	6.569	6.348	6.165	6.239
298.15	6.349	6.119	5.938	6.061
323.15	6.279	6.037	5.849	5.982
348.15	6.305	6.048	5.842	5.954
373.15	6.397	6.122	5.889	5.951
398.15	6.539	6.242	5.974	5.960
423.15	6.721	6.399	6.088	5.975
448.15	6.938	6.588	6.227	5.993
473.15	7.189	6.804	6.384	6.008
498.15	7.470	7.045	6.555	6.016
523.15	7.783	7.306	6.731	6.005
548.15	8.498	7.582	6.902	5.961
573.15	8.125	7.864	7.052	5.864

T (K)	pK_2 ($I=0$)	pQ_2 ($I=0.1$)	pQ_2 ($I=1.0$)	pQ_2 ($I=5.0$)
273.15	10.627	10.181	9.831	10.052
298.15	10.337	9.869	9.480	9.614
323.15	10.180	9.687	9.257	9.308
348.15	10.117	9.593	9.118	9.088
373.15	10.120	9.559	9.035	8.924
398.15	10.171	9.568	8.989	8.796
423.15	10.255	9.606	8.966	8.689
448.15	10.365	9.661	8.954	8.590
473.15	10.491	9.725	8.942	8.485
498.15	10.630	9.788	8.914	8.355
523.15	10.777	9.838	8.852	8.176

The equilibrium concentration of hydrogen ions, and hence the pH value, can be influenced by variation of the CO_2 pressure and of temperature since the concentration of dissolved CO_2 depends on these properties. Furthermore, the equilibrium constant is temperature- and pressure-dependent. Variation

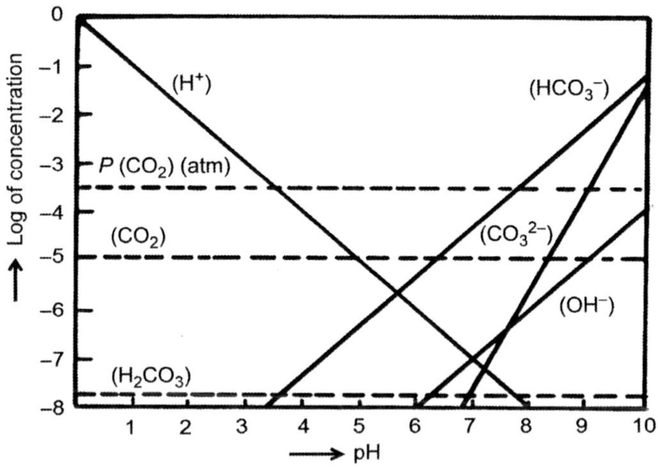

FIGURE 3.48 Concentration of species of the CO_2–H_2O equilibrium for a partial pressure of CO_2 of $P_{CO_2} = 10^{-3.5}$ atm and $T = 298$ K. pH is varied by the addition of a strong base or acids. *Modified from Ref. [108].*

TABLE 3.26 First Dissociation Constant K_1 of the Equilibrium $[H^+][HCO_3^-] = K_1[CO_2]$ $(10^{-7}$ mol kg$^{-1})$ [105]

K_1 $(10^{-7}$ mol kg$^{-1})$					
T (K)	298.15	372.85	423.15	473.45	523.30
***P* (MPa)**					
0.1	4.32	3.815	1.73	0.561	0.164
20	5.38	4.68	2.21	0.756	0.2415
40	6.65	5.705	2.8	1.007	0.3555

due to temperature changes can be calculated by the equation of van't Hoff. Variation due to pressure is expressed as changes of the activity coefficients. Up to a pressure of $P = 100$ MPa, the influence of pressure on the activity coefficients can be neglected, as Ellis [109] found in his investigation of the first dissociation constant up to pressures of $P = 250$ MPa. Table 3.26 lists equilibrium dissociation constants in dependence of pressure and temperature and Figure 3.49 shows these dissociation constants to temperatures of about $T = 600$ K [108]. The dissociation constant K_1 increases with increasing pressure at constant temperature and decreases with temperature (at constant pressure). An increase in temperature shifts the reaction equilibrium to the left side and an increase in pressure causes a shift to the right side.

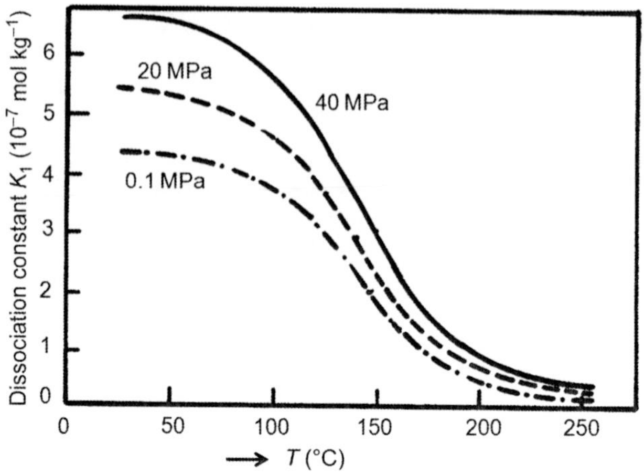

FIGURE 3.49 Dissociation constant K_1 for the equilibrium of CO_2–H_2O [105,110].

The amount of hydrogen ions, H^+, can be determined from Equations (3.133) to (3.138):

$$H^+ = OH^- + HCO_3{}^{2-} + 2CO_3{}^{2-} \tag{3.133}$$

$$K_w = H^+ \cdot OH^- \tag{3.134}$$

$$K_1 = (H^+ \cdot HCO_3{}^-)/CO_{2,aq} \tag{3.135}$$

$$H^+ = \frac{K_w}{H^+} + \frac{K_1\, CO_{2,aq}}{H^+} \tag{3.136}$$

$$H^+ = \sqrt{K_w + K_1 CO_{2,aq}} \tag{3.137}$$

$$[H^+] = \sqrt{K_w + K_2 [CO_{2,aq}]} \tag{3.138}$$

For the calculation of the amount of hydrogen ions or pH value ($pH = -\log H^+$), an input on the amount of dissolved molecular CO_2 is needed. Such solubility data have been determined by several authors [98,102]. Using the first dissociation constant, pH values for the liquid phase of the CO_2–H_2O equilibrium have been calculated and are shown in Figure 3.50.

The pH value of the liquid phase of the CO_2–H_2O equilibrium is dependent on pressure and temperature. The amount of hydrogen ions (H^+) decreases with increasing temperature at constant pressure, that is, the pH value increases and the acidity decreases. For the influence of pressure on the pH value, differing experimental data on solubility are available. According to the data of Wiebe and Gaddy [102] and Toews *et al.* [111], at temperatures up to $T = 100$ °C, acidity increases or pH value decreases with pressure, but experimental values of both authors do not coincide within a reasonable error. Contrary to that result, calculated pH values from the experimental

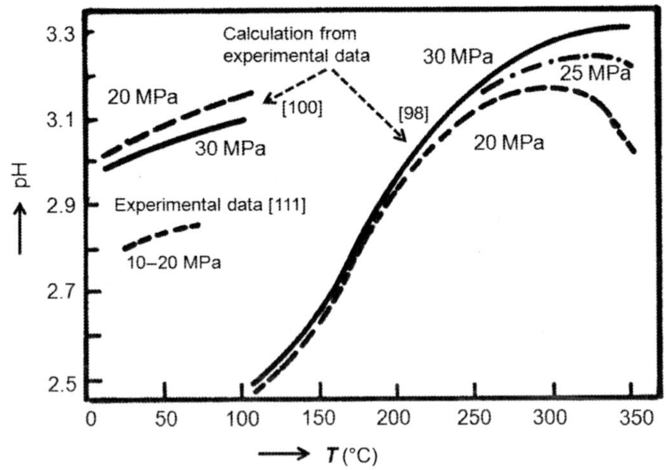

FIGURE 3.50 pH value for CO_2–H_2O, calculated with the first dissociation constant. *Modified from Ref. [110].*

solubility data of Takenouchi and Kennedy [98] show a decrease of acidity or an increase in pH value with increasing pressure. At temperatures above $T=473.15$ K, this effect is substantial. Furthermore, at a temperature of about $T=373.15$ K, the only region where data from all three investigations can be compared, the pH values differ substantially. Variation is nearly one pH unit or one order of magnitude for the concentration of the hydrogen ions, H^+, that are responsible for chemical effects. Investigations of other authors showed results that confirm the range of pH values determined from the data of Takenouchi und Kennedy. Meyssami *et al.* [112] measured pH values of 3.2–3.6 at $P=1$–35 MPa and $T=32$–42 °C, and Schaef and McGrail [113] reported pH values from 3.1 to 3.3 at $P=5.5$ MPa and $T=27$–70 °C. According to that data, a pH value of pH 3.2 ± 0.1 can be assumed [114] in the temperature range of $T=523.15$ K (250 °C) to $T=593.15$ K (320 °C) and pressures up to $P=25$ MPa.

The liquid aqueous phase is not necessarily saturated with CO_2 according to the equilibrium values. In Figure 3.51, the pH value of the liquid phase of CO_2–H_2O is shown as a function of carbon dioxide concentration for $T=250$ °C and $P=25$ MPa. The pH decreases by two orders of magnitude up to the saturation level as compared to pure water at the same conditions. Due to the logarithmic character of the underlying equation, the pH initially drops sharply and then moderately when CO_2 concentration is increased [115]. Addition of carbon dioxide beyond 50–75% saturation does not significantly contribute to a further decrease of pH value, but may be of importance for catalyzed reactions since pH (or H^+) is on a logarithmic scale and reactions may be catalyzed proportional to the amount of H^+.

In conclusion, adding CO_2 to an aqueous solution enhances the availability of hydrogen ions and as such catalyzes hydrolysis reactions. Concentration of

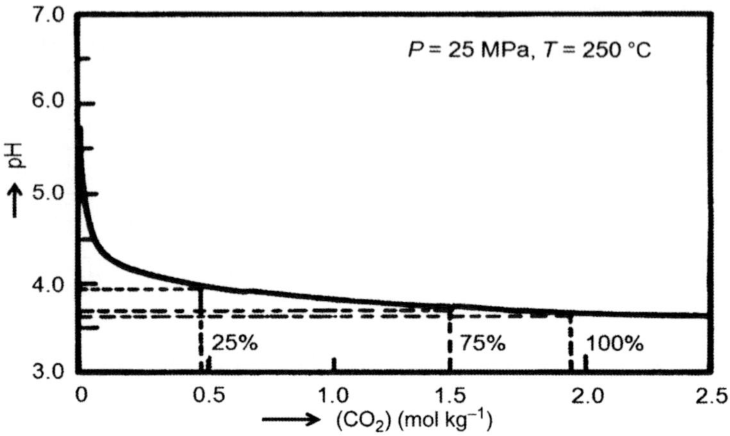

FIGURE 3.51 pH of water as a function of CO_2 concentration, related to the saturation concentration [115].

hydrogen ions in an aqueous system can be changed easily by increasing or decreasing CO_2 partial pressure.

3.5.2.1.3 Influence of Salts (Ionic Strength)

Aqueous solutions often contain ions of other components that are of influence on the equilibria and the pH value. In Table 3.27, the influence of KCl on the dissociation of water is illustrated. In this case, the addition of KCl to pure H_2O has only little effect on the ionization of H_2O. Another example for low temperatures is shown in Figure 3.52 for the influence of $MgCl_2$ ions on the pH value. In this case, the influence is much greater. The addition of $MgCl_2$ lowers the pH value about one to two pH units.

Ionic strength also influences solubility and hence Henry's constant. In Figure 3.53, the influence of NaCl on the solubility of CO_2 in H_2O is shown. With increasing ionic strength, the solubility of CO_2 decreases but follows the general behavior of the solubility of CO_2 in pure H_2O with respect to temperature.

3.5.2.2 Hydrochloric Acid–Water

Hydrochloric acid in water (HCl(aq)) is a strong electrolyte but becomes a weak acid at elevated temperatures and pressures. The association of hydrochloric acid becomes important under subcritical conditions at temperatures above about $T = 523.15$ K. Values of $-\log K_1$ (pK_1) for the association reaction ionization of HCl(aq) are presented in Table 3.28 [85]:

$$H^+(aq) + Cl^-(aq) \leftrightarrow HCl(aq) \tag{3.139}$$

The density model (Equation 3.111 or 3.141) can be used to describe the factors that affect the association of HCl(aq) in the sub- and supercritical regions (see Table 3.28) [120,121].

TABLE 3.27 Dissociation of Water at Saturation Pressure in KCl Media, I (mol kg^{-1}) [116]

	$\log_{10} K_w$				
T (K)	I=0	I=0.1	I=0.5	I=1.0	I=3.0
273.15	14.941	14.740	14.684	14.725	15.004
298.15	13.993	13.781	13.717	13.753	14.005
323.15	13.272	13.047	12.973	13.003	13.226
348.15	12.709	12.468	12.382	12.404	12.596
373.15	12.264	12.004	11.903	11.916	12.074
398.15	11.914	11.631	11.512	11.514	11.634
423.15	11.642		11.193	11.181	11.259
448.15	11.441		10.943	10.907	
473.15	11.302		10.727	10.680	
498.15	11.222	10.804	10.563	10.490	
523.15	11.196	10.723	10.429	10.323	
548.15	11.224	10.676	10.310	10.160	9.892
573.15	11.301	10.653	10.187	9.975	9.572

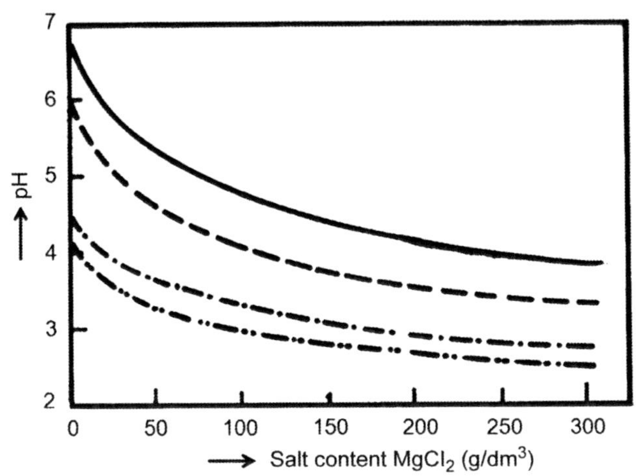

FIGURE 3.52 Influence of concentration of $MgCl_2$ on the pH value of CO_2–H_2O equilibria: ——, $P_{CO_2}=3\times10^{-6}$ MPa, $T=50$ °C; – – –, $T=20$ °C; —•—, $P_{CO_2}=2$ MPa, $T=50$ °C; —••—••—, $T=20$ °C. *Modified from Ref. [117].*

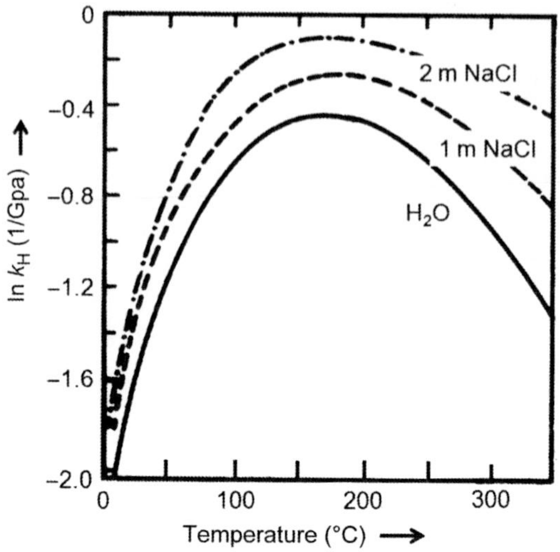

FIGURE 3.53 Henry's constant for CO_2–H_2O–NaCl systems. *Modified from Ref. [108]. Based on data from Ref. [118].*

TABLE 3.28 Molal Ion Association Constants of Aqueous HCl Solutions [119]

T (K)	*P* (MPa)	ρ_w (g cm^{-3})	pK_1
473.15	10.02	0.871	0.43
573.15	10.17	0.7159	0.97
623.15	27.80	0.6365	1.68
643.15	28.89	0.5724	2.29
653.15	31.15	0.5441	3.07
663.15	30.46	0.4772	3.42
673.15	30.76	0.3876	4.15
683.15	30.56	0.27	5.12

3.5.2.3 Ammonia–Water

The system ammonia–water is important for refrigeration cycles, power plant cycles, and biomass pretreatment due to its high solubility, volatility, buffer properties, and the ability to form complexes with metal ions. It is used in the treatment of boiler water and is present in boiler water as a breakdown product of hydrazine (which is added as a reducing agent) [85].

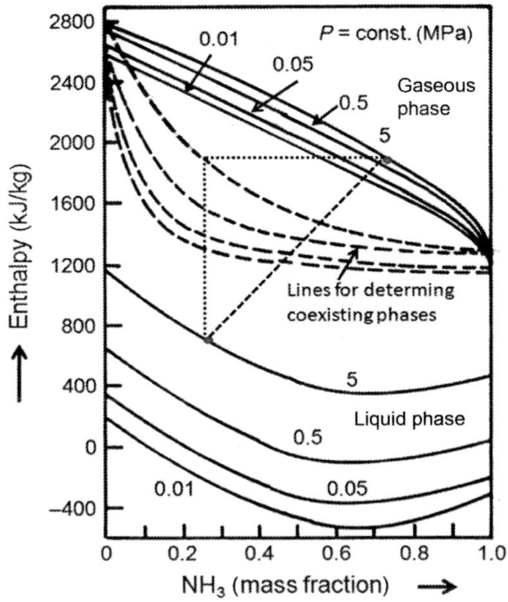

FIGURE 3.54 Enthalpy–concentration diagram for ammonia–water. *Modified from Ref. [122].*

Phase equilibrium of ammonia–water is of type I with a continuous critical curve from the critical point of ammonia ($T_c = 405.40$ K, $\rho_c = 225$ kg m^{-3}) to the critical point of water ($T_c = 647.096$ K, $\rho_c = 322$ kg m^{-3}) as shown in Figure 3.24. VL is shown in Figure 3.54 in an enthalpy-concentration diagram for different pressures. The upper lines represent the dew-point lines and the lower lines represent the bubble-point lines. The intermediate lines are construction lines to connect gaseous and liquid phases in equilibrium by drawing a horizontal line from a dew-point line until it crosses the construction line and then drawing a vertical line from that point until it crosses the bubble-point line. The endpoints at the dew-point line and at the bubble-point line connect the equilibrium phases. For the application of the enthalpy-concentration diagram to processes, the reader is referred to the standard textbooks on *Engineering Thermodynamics or on Separation Processes.*

Thermodynamic properties of the system ammonia–water can be calculated with a formulation published by IAPWS [123] based on the publication by Tillner-Roth and Friend [124], accessible from the IAPWS Web site at http://www.iapws.org. The formulation is not repeated here.

The ionization of ammonia bas been studied by a number of authors (see Ref. [85]). The values measured for the neutralization reaction by Hitch and Mesmer [125] are summarized in Table 3.29 together with the influence of ionic strength if KCl is added [85]:

TABLE 3.29 Equilibrium Constants for the Ionization of Ammonia in Aqueous KCl (I in mol kg^{-1}) at Saturation Pressure [85,125]

T (K)	log K_1 (I=0)	log Q_1 (I=0.1)	log Q_1 (I=1.0)	log Q_1 (I=3.0)
273.15	−4.864	−4.648	−4.460	−4.401
298.15	−4.752	−4.526	−4.336	−4.291
323.15	−4.732	−4.495	−4.296	−4.253
348.15	−4.772	−4.521	−4.306	−4.256
373.15	−4.856	−4.587	−4.351	−4.284
398.15	−4.976	−4.685	−4.422	−4.330
423.15	−5.128	−4.814	4.516	−4.393
448.15	−5.311	−4.969	−4.633	−4.470
473.15	−5.525	−5.151	−4.768	−4.559
498.15	−5.770	−5.357	−4.918	−4.654
523.15	−6.047	−5.584	−5.077	−4.747
548.15	−6.355	−5.827	−5.233	−4.825
573.15	−6.694	−6.079	−5.371	−4.870

$$NH_3(aq) + H_2O(l) \leftrightarrow NH_4^+(aq) + OH^-(aq) \quad (3.140)$$

Values for the ionization constant log K_1 with temperature and pressure can be also obtained from the density model (Equation (3.141); Table 3.31) up to 1073.15 K [85].

3.5.3 Other Inorganic and Organic Compounds

Ionization reactions of some inorganic and organic electrolytes are listed in Table 3.30. Sulfuric acid, H_2SO_4(aq), is considered to be a strong acid, which ionizes according to the reaction given in Table 3.30. First, dissociation constant of H_2SO_4(aq) has been reported by Xiang *et al.* [126] from T=350 to 400 °C. The ionization of hydrogen sulfate has been studied up to T=250 °C by Dickson *et al.* [127]. The values for log $_{10}Q_{2a}$ at ionic strengths in NaCl(aq) up to 5.0 mol kg^{-1} are tabulated in Table 3.31. The hydrogen sulfate ion is a moderately strong acid, occurring in power plants as an impurity. Sodium salts of phosphoric acid are used as pH-controlling agents in boiler water for power plants. Boric acid is used as a pH control agent in the primary

TABLE 3.30 Ionization Reactions of Inorganic and Organic Electrolytes [85]

$H_2SO_4(aq) \leftrightarrow HSO_4^-(aq) + H^+(aq)H_2SO_4(aq) \leftrightarrow HSO_4^-(aq) + H^+(aq)$
$HSO_4^-(aq) \leftrightarrow H^+(aq) + SO_4^{2-}(aq)$
$H_3PO_4(aq) + OH^-(aq) \leftrightarrow H_2PO_4^-(aq) + H_2O(l)$
$H_2PO_4^-(aq) + OH^-(aq) \leftrightarrow HPO_4^{2-}(aq) + H_2O(l)$
$H_2S(aq) \leftrightarrow H^+(aq) + HS^-(aq)$
$HNO_3(aq) \leftrightarrow H^+(aq) + NO_3^-(aq)$
$B(OH)_3(aq) + OH^- \leftrightarrow B(OH)_4^-(aq)$
$Si(OH)_4(aq) + OH^-(aq) \leftrightarrow SiO(OH)_3^-(aq) + H_2O(l)$
$Li^+(aq) + OH^-(aq) \leftrightarrow LiOH(aq)$
$Na^+(aq) + OH^-(aq) \leftrightarrow NaOH(aq)$
$K^+(aq) + OH^-(aq) \leftrightarrow KOH(aq)$
$CH_3COOH(aq) \leftrightarrow CH_3COO^-(aq) + H^+(aq)$
$HCOOH(aq) \leftrightarrow H^+(aq) + HCOO^-(aq)$

coolant of pressurized water reactors due to its high neutron cross section. Ionization of silicic acid is an important reaction in geochemical reactions. It is responsible for the enhanced solubility of quartz and silicate minerals. From the alkali metal hydroxides, LiOH is used as pH control agent in primary circuits in nuclear reactors, and NaOH and KOH are components of fluids formed at supercritical condition in geologic processes. Acetic and formic acids are formed in hydrothermal processes. They are thermal stable and highly corrosive [85].

Ionization constants for inorganic acids and bases and organic acids can be represented by fitting the experimental data to an empirical model:

$$\log K = q_1 + \frac{q_2}{T} + q_3 \ln T + q_4 T + q_5 T^2 + \frac{q_6}{T^2} + (q_7 + q_8 T + q_9/T) \log \rho_w \quad (3.141)$$

with T as the temperature in K and q_n as fitting parameters for each acid, listed in Table 3.31. For further information on specific electrolytes and source for experimental data, the reader is referred to the article of Tremaine *et al.* [85].

TABLE 3.31 Parameter for Ionization Constants According to Equation (3.141) [85]

Acids	$q_1 \times 10^{-2}$	$q_2 \times 10^{-4}$	$q_3 \times 10^{-1}$	$q_4 \times 10^{2}$	$q_5 \times 10^{4}$	$q_6 \times 10^{-5}$	$q_7 \times 10^{-1}$	$q_8 \times 10^{2}$	$q_9 \times 10^{-3}$	T (°C)
CO_2(aq)	−5.22461	2.96882	8.18401	−8.96488	0	−20.4679	0	0	0	0–300
H_3PO_4	−2.53198	1.76558	3.94277	−3.25405	0	−8.10134	0	0	0	0–300
$H_2PO_4^-$	−2.46045	1.71569	3.77345	−3.22082	0	−8.97579	0	0	0	0–300
HSO_4^-	5.627097	−1.32737	−10.2515	24.77538	−1.11703	0	0	0	0	0–250
H_2S	7.82439	−2.05657	−14.2742	36.1261	−1.6722	0	0	0	0	25–350
HNO_3	−5.41813	3.73852	7.5445	0	0	0	3.1264	0	0	250–319
$B(OH)_3$	−0.36260	0.364518	0.505527	0	0	0	1.64914	−2.3917	0	5–380
$Si(OH)_4$	−0.18401	0.234669	0.257979	0	0	0	0	0	0	0–300
HCl	0.0195	−0.13033	0	0	0	0	−0.9876	0	0	200–410
HCl	0.02638	−0.15188	0	0	0	0	−0.9078	0	0	200–600
LiOH	0.02094	−0.07936	0	0	0	0	−0.8192	0	1.3852	50–600
NaOH	0.02068	−0.06566	0	0	0	0	−1.2432	0	5.8769	100–600
KOH	0.02302	−0.09547	0	0	0	0	−0.9862	0	3.9935	100–600
Formic	3.879698	−1135.86	0	−0.01283	0	0	0	0	0	25–200
Acetic	−350.02	18.325	55.327	−0.06351	0	−1,110,500	0	0	0	0–300
Benzoic	−615.105	34,570.79	96.26395	−0.09761	0	−2,170,870	0	0	0	0–250

3.6 WATER AND METAL OXIDES

In this section, phase equilibria of water with nonvolatile electrolytes (salts) and nonelectrolytes at elevated temperature and pressure will be reviewed. Solubility of metal oxides in hydrothermal solutions will be shortly addressed.

Interaction of metal oxide and hydroxide minerals with hydrothermal solutions in the liquid state is of interest for industrial processes. Phase behavior and solubility in such systems are complicated due to the special properties of metal oxides and similar compounds (like SiO_2). In this section, only the condensed state of the aqueous systems is considered. Solubility in dense gaseous and supercritical solutions is treated in Section 3.7.

Surface chemistry affects the rates and mechanisms of metal oxide dissolution and precipitation. Both aspects are comprehensively reviewed by Wesolowski *et al.* [128].

Metals exposed to aqueous solutions generally form an oxide surface film, with a few exceptions (Au, Pt, and Cu under mild reducing conditions). Such a film may passivate the surface toward further oxidation.

Metal oxides formed in the presence of water are often hydrated regardless of whether the hydrous phase is thermodynamically stable relative to the pure oxide or a less hydrated phase. The aluminum system may serve as an example [128]:

$$Al_2O_3(s) + H_2O(l) \leftrightarrow 2AlOOH(s) \tag{3.142}$$

$$Al_2O_3(s) + 3H_2O(l) \leftrightarrow 2Al(OH)_3(s) \tag{3.143}$$

$$AlOOH(s) + H_2O(l) \leftrightarrow Al(OH)_3(s) \tag{3.144}$$

Metastable phases may be of importance in such systems. Dehydrated phases can maintain metastable equilibrium with aqueous solutions at lower temperatures and higher pressures. On the other hand, more hydrated phases may be metastable at increased temperatures and lower pressures. A hydrated phase can precipitate from a supersaturated solution or be formed by oxidation of a metal surface in contact with water, even if the hydrated phase is metastable relative to a less hydrated or anhydrous phase, due to kinetic barriers to the formation of the more stable phase. Metastable dehydrated phases often form a poorly crystalline, hydrated surface layer when exposed to hydrothermal solutions, affecting solubility. An example is corundum, α-Al_2O_3, which rapidly forms a series of AlOOH and $Al(OH)_3$ surface layers when exposed to low-pressure water vapor at ambient temperature [128].

Many metal oxides and hydroxides occur as two or more polymorphs of identical composition but different crystalline structure. Quartz (α-SiO_2) is the most stable form of silica under hydrothermal conditions but will transform to other morphologies with increasing temperature and pressure. Transition between these polymorphs may be neither rapid nor reversible. Hysteresis is common. Aqueous metal oxide systems, as all closed systems, eventually approach a state of stable equilibrium with the least possible energy configuration at given conditions. Chemical potential is then equal in all phases and modifications. Equilibrium thermodynamics also applies to systems containing metastable phases. These metastable components can control reactions. Under hydrothermal conditions, metastable states may approach stable states in a timescale similar to an industrial process but the metastable may remain even at conditions where only the thermodynamically stable state should exist. Under these aspects, hydrothermal processes involving aqueous systems of metal oxides (and other solids) must be carefully analyzed with respect to the actually occurring species [128].

A representative solubility reaction for the dissolution of anhydrous, single-metal oxides in an aqueous solution is

$$\mathrm{AlOOH(boehmite)} + 3\mathrm{H}^+(\mathrm{aq}) \leftrightarrow \mathrm{Al}^{3+}(\mathrm{aq}) + 2\mathrm{H_2O(l)} \tag{3.145}$$

or in general form

$$\frac{1}{x}\mathrm{Me}_x\mathrm{O}_a(\mathrm{OH})_{xz-2a}(\mathrm{cr}) + (z)\mathrm{H}^+(\mathrm{aq}) \leftrightarrow \mathrm{Me}^{z+}(\mathrm{aq}) + \left(z-\frac{a}{x}\right)\mathrm{H_2O(l)} \tag{3.146}$$

The equilibrium constant, defining the solubility product, is

$$K_{s0} = \frac{\mathrm{Me}^{z+}(\mathrm{aq})\cdot \mathrm{H_2O}^{(z-a/x)}}{\left[\mathrm{Me}_x\mathrm{O}_a(\mathrm{OH})_{xz-2a}(\mathrm{cr})\right]^{1/x}\cdot[\mathrm{H}^+(\mathrm{aq})]^z} \tag{3.147}$$

In dilute solutions, the solubility product can be expressed as

$$K_{s0} = \frac{\mathrm{Me}^{z+}(\mathrm{aq})}{[\mathrm{H}^+(\mathrm{aq})]^z} \quad \text{or rearranged:} \log[\mathrm{Me}^{z+}(\mathrm{aq})] = \log K_{s0} - z\mathrm{pH} \tag{3.148}$$

The concentration of total dissolved aluminum in equilibrium with γ-AlOOH (boehmite) in dependence of pH value is shown in Figure 3.55. At $T = 373.15$ K, the solubility curves approach straight lines with slopes of −3. At higher temperature ($T = 573.15$ K) and higher pH value, the solubility

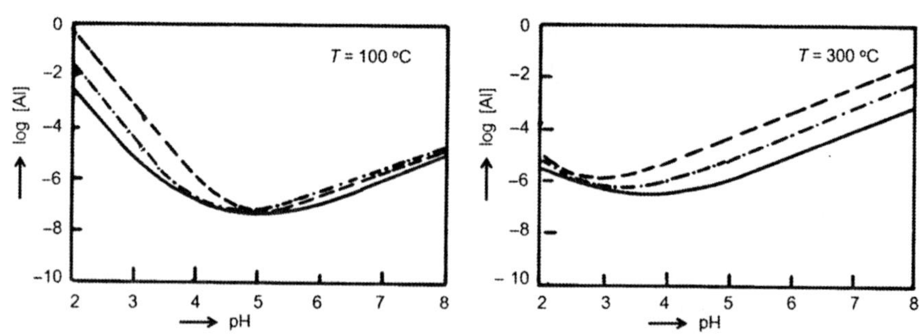

FIGURE 3.55 Solubility of total aluminum in equilibrium with boehmite as a function of pH_m. Total solubility: —— in water, —•— in 0.3 mol kg^{-1} NaCl, – – – in 5.0 mol kg^{-1} NaCl. *Modified from Ref. [128].*

deviates from this relationship, mainly due to hydrolysis of Al^{3+}. Furthermore, the solvation structure is changed [128].

In consequence of the changing solvation structure, the coordinated water molecules undergo stepwise acid dissociation [128]:

$$Me^{z+}(aq) + yH_2O(l) \leftrightarrow Me(OH)_y^{z-y}(aq) + yH^+(l) \tag{3.149}$$

$$K_{hy} = \frac{Me(OH)_y^{z-y} \cdot (H^+)^y}{a_w^y \cdot Me^{z+}} \tag{3.150}$$

In addition to reaction (3.145), γ-AlOOH (boehmite) undergoes stepwise hydrolysis [128]:

$$AlOOH(s) + 2H^+(aq) \leftrightarrow Al(OH)^{2+}(aq) + H_2O(l) \tag{3.151}$$

$$AlOOH(s) + H^+(aq) \leftrightarrow Al(OH)_2{}^+(aq) \tag{3.152}$$

$$AlOOH(s) + H_2O(l) \leftrightarrow Al(OH)_3{}^0(aq) \tag{3.153}$$

$$AlOOH(s) + 2H_2O(l) \leftrightarrow Al(OH)_4{}^- + H^+(aq) \tag{3.154}$$

with the equilibrium constants

$$K_{s0} = \frac{a_w^2 \cdot Al^{3+}}{(H^+)^3} \tag{3.155}$$

$$K_{s1} = \frac{a_w \cdot Al(OH)^{2+}}{(H^+)^2} \tag{3.156}$$

$$K_{s2} = \frac{Al(OH)_2{}^+}{H^+} \tag{3.157}$$

$$K_{s3} = \frac{Al(OH)_3{}^0}{a_w} \tag{3.158}$$

$$K_{s4} = \frac{Al(OH)_4{}^- \cdot H^+}{a_w^2} \tag{3.159}$$

where subscripts sy correspond to the number of deprotonated water molecules (y) in the solvation sphere of the dissolved metal cation. These subscripts are linked with metal ion hydrolysis products (x,y) for the species $Me_x(OH)_y^{xz-y}$. From Equations (3.150), (3.155)–(3.158), it follows that

$$K_{hy} = \frac{K_{sy}}{K_{s0}} \tag{3.160}$$

The stepwise hydrolysis terminates with the aluminate ion $Al(OH)_4{}^-$, which contains four hydroxides in tetrahedral configuration around the central metal ion with no undissociated water molecules in the nearest- neighbor shell [128]. The concentration of dissolved single-metal, single-valence state oxide is closely approximated by the equilibrium constant of the isocoulombic, pH-independent reaction:

$$(1/x)\mathrm{Me}_x\mathrm{O}_y(\mathrm{OH})_{xz-2y}(\mathrm{cr}) + (y/x)\mathrm{H_2O(l)} \leftrightarrow \mathrm{Me(OH)}_z^0(\mathrm{aq}) \qquad (3.161)$$

For isocoulombic reactions of this type, the equation

$$\log K_{sz} = \frac{\mathrm{Me(OH)}_z^0}{a_\mathrm{w}^{y/x}} \approx \log \mathrm{Me(OH)}_z^0 = \log Q_{sz} \qquad (3.162)$$

is expected to apply over a wide range of solution ionic strength. This quantity is plotted for a number of metal oxides in Figure 3.56 as a function of temperature and pressures near P_{sat}.

Figure 3.56 illustrates why zirconium (i.e., zircaloy), titanium, and alloys high in Ni, Cr, and Fe (stainless steel, Hastelloy, and Monel) are the preferred construction materials for corrosion-resistant hydrothermal equipment, since their passivating oxide surface films are extremely insoluble at near-neutral pH [128].

Isocoulombic reactions are ionization reactions written in such a way that the charges of the reactants and products are the same and they are symmetric with respect to ionic change. Thus, many of the effects of ionic strength, temperature, and pressure can be minimized [128]. For example, the ionization of an acid may be written as

$$\mathrm{H}A(\mathrm{aq}) + \mathrm{OH}^-(\mathrm{aq}) \leftrightarrow \mathrm{H_2O(l)} + A^-(\mathrm{aq}) \qquad (3.163)$$

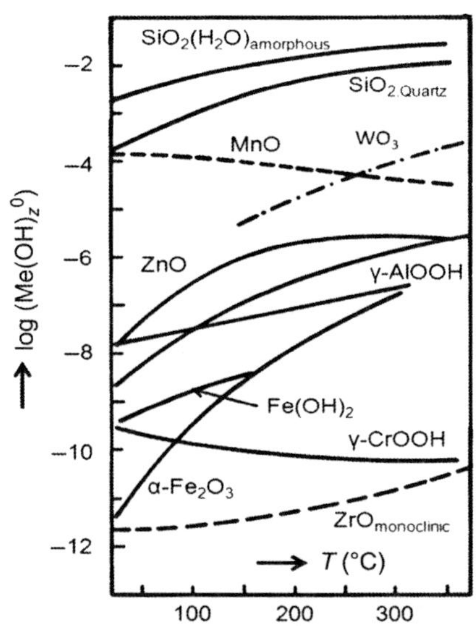

FIGURE 3.56 Temperature dependence of log $\mathrm{Me(OH)}_z^0 \approx \log K_{sz}$ for some reactions. *Modified from Ref. [128].*

For a solubility reaction or a metal ion hydrolysis reaction, it can often be assumed that the heat capacity change is zero and constant over the temperature range of interest, and pressure effects near P_{sat} may be ignored. Then, the Gibbs energy of formation ΔG_T^0 can be calculated from thermodynamic properties known at a single temperature and pressure, usually $T=298.15$ K and $P=0.1$ MPa:

$$\Delta G_T^0 = \Delta H_{298.15}^0 - T\Delta S_{298.15}^0 \tag{3.164}$$

The equilibrium constant then is a simple function of temperature:

$$\log K = a + b/T \tag{3.165}$$

Many isocoulombic aqueous reactions have heat capacity changes near zero and are also often found to be insensitive to pressure changes at temperatures up to $T=523$ K. They are closely approximated by Equation (3.165). Plots of log K_{s0} in dependence of $1/T$ are almost linear over a very wide range of temperatures. The temperature dependence can be described quite accurately by assuming a constant mean value of Δc_p^0 between conditions of T and P and reference conditions.

A discussion of quantitative empirical and semiempirical approaches to model solubility of inorganic compounds in hydrothermal and supercritical water is given by Leusbrock *et al.* [129]. The approach is based on phase equilibrium between the solid salt and the aqueous phase. The detailed discussion is beyond the scope of this book, but in Chapter 11, such a semiempirical correlation is used for modeling the solubility of metal oxides in water.

3.7 WATER AND NONVOLATILE ELECTROLYTES (SALTS)

Phase equilibria of systems of water and nonvolatile electrolytes (salts) are characterized by crystalline hydrates up to moderate temperatures up to about $T=473.15$ K. At higher temperatures, the crystalline hydrates no longer exist and liquid–liquid immiscibility, gas–liquid equilibria, and critical phenomena prevail. Water–salt systems at elevated temperatures and water–organic, organic, and other systems with components of different volatility are characterized by similar topological types of phase diagrams. Information on heterogeneous equilibria, as presented in Section 3.4, can also be applied to water–salt mixtures.

Beyond phase diagrams, solubility is important for hydrothermal processes. For example, the solubility of SiO_2 (quartz, α-SiO_2) in water is shown in Figure 3.57. SiO_2 is dissolved by water to some extent. This solubility changes tremendously in the neighborhood of the critical temperature. At relatively low pressures, with increasing temperature, the solubility drops to practically zero. At relatively moderate and high pressures, the solubility is constant and even increasing.

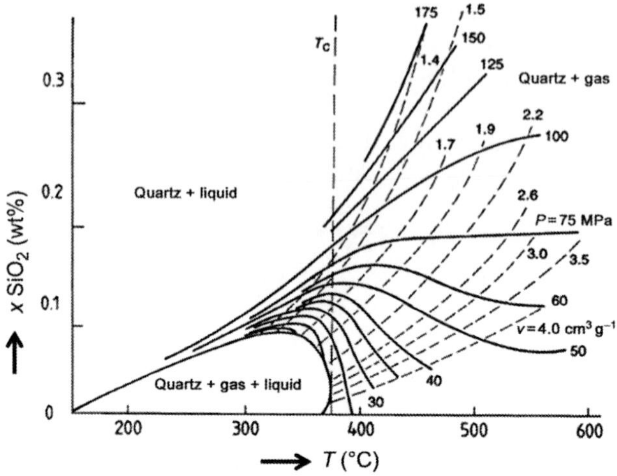

FIGURE 3.57 Solubility of SiO_2 in water. *Data from Ref. [37].*

3.7.1 Binary Systems

For binary systems with only fluid phases, the classification described in Sections 3.3 and 3.4 can be applied. If a solid phase occurs at temperatures and pressures of interest for hydrothermal applications or up to the critical region of the more volatile component, then the explanations given in Section 3.3.2.2 and in the following discussion can be used to interpret the phase behavior.

The commonly accepted classification of six classes has been derived using EOS with varying parameters. The original study by van Konynenburg and Scott [28] applied the van der Waals EOS. Valyashko [27] cites a number of other studies, carried out also with different EOS. With a Lennard–Jones-type EOS, a class VII was reported that is a combination of classes V and VI without new fundamental features. It is obvious that miscibility and immiscibility control the phenomena of phase behavior. Therefore, for the purpose of this book, it is considered sufficient to include the two types of phase behavior, where the solid phase extends beyond the critical temperature of the more volatile component (Section 3.3.2.2), that is, such with a continuous critical curve and such with an interrupted critical curve. Modifications of solid phases will not be considered. Between the different classes of phase behavior, there are transit regions, as has been demonstrated by systematic variation of the properties [30] (compare Section 3.4 on water–hydrocarbon systems). For a more detailed discussion of phase equilibria of water with salts, the reader is referred to the work of Valyashko [27].

Water–salt systems with a continuous critical curve are [27], for example, NH_3–H_2O, H_2O–NaCl, H_2O–KOH, H_2O–K_2CO_3 [130,131], H_2O–$CaCl_2$ [132,133], H_2O–CsCl [134,135], H_2O–KCl [132,135], H_2O–KH_2PO_4

[136,137], H_2O–KNO_3 [138,139], H_2O–LiCl [135,140], H_2O–$MgCl_2$ [132], H_2O–NaCl [141–143], H_2O–Na_2WO_4 [142,144], and H_2O–UO_2SO_4 [145,146].

Systems with an interrupted critical curve are [27], for example, H_2O–Al_2O_3 [147,148], H_2O–$BaSO_4$ [149,150], H_2O–$CaCO_3$ [151,152], H_2O–Fe_3O_4 [153,154], H_2O–K_2SO_4 [155], H_2O–Na_2SO_4 [156], H_2O–PbO [157], H_2O–SiO_2 [158].

A more extensive list on investigated water–salt systems has been compiled by Valyashko [27].

The solubility of NaCl in steam has been studied extensively. A summarizing discussion has been published by Palmer *et al.* [86]. Equation (3.166) [159] represents the experimental data reasonably well:

$$\log c_{\mathrm{NaCl}} = 8.637 - \frac{2100.7}{T} + 3.66 \log \rho_{\mathrm{w}} \tag{3.166}$$

The solubility of compounds in the aqueous phase for systems with an interrupted critical curve is about two orders of magnitude lower than those of with a continuous critical curve. Solubility is reduced drastically with increasing temperature. Examples of solubility curves are given in Figure 3.58 [27].

From solubility data, the type of system (continuous or interrupted critical curve) can be derived, if measurements are reported up to the melting temperature of the salt, at least up to $T = 400$ °C [27]. If the concentration of a salt in the liquid phase increases with increasing temperature, at about $T = 647$ K, and diverges from the composition of the equilibrium vapor solution, the system has a continuous critical curve (see the example H_2O–NaCl shown in Figure 3.58). That is the case in Figure 3.58 for Na_2SeO_4 and the compounds with solubility curves on its right side. All these binary systems have a continuous critical curve. If, on the other hand, the concentration of a salt in the liquid phase decreases with increasing temperature, at about $T = 647$ K, and the

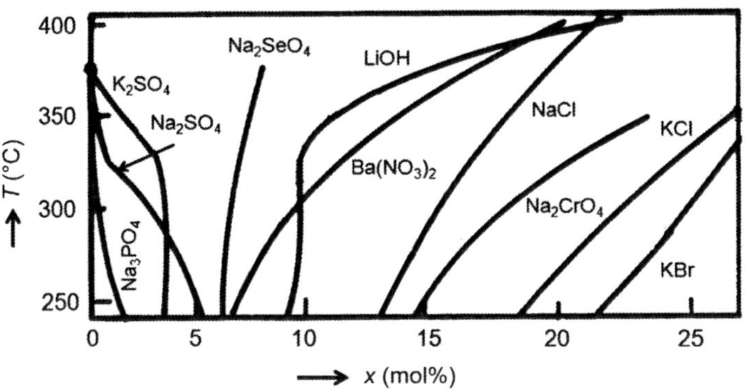

FIGURE 3.58 Composition of the aqueous saturated solutions of the indicated solids. *Modified from Ref. [27].*

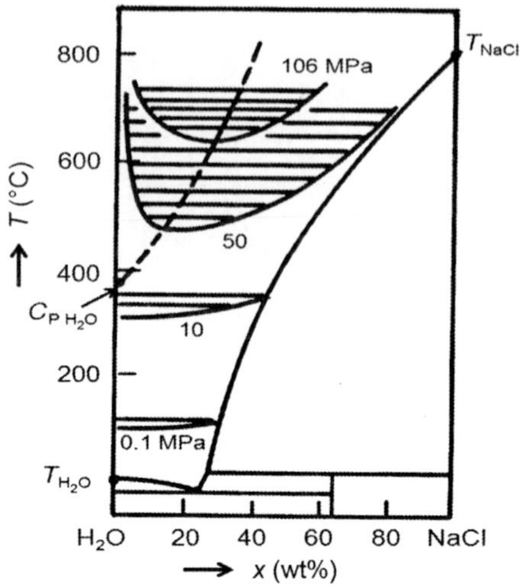

FIGURE 3.59 *T,x*-projection of the H_2O–NaCl phase diagram. ———, critical curve; ——, solubility curves and isobaric cross sections of two-phase equilibria. *Modified from Ref. [27].*

compositions of liquid and vapor converge, then the system has an interrupted critical curve [27].

The solubility of NaCl in H_2O (Figure 3.59) increases with temperature. The isobaric cross sections of the liquid–gas equilibrium extend to pure water for temperatures below the critical temperature of water (T_{c,H_2O}) and detach from the pure water axis for temperatures above T_{c,H_2O}.

The phase diagram of the system Na_2SO_4–H_2O (Figure 3.60) has two three-phase regions terminated by the critical endpoints *P* and *Q*. They are separated by the region of gas (supercritical)–solid equilibrium, indicated by the solubility isobars at $P=20$–120 MPa. At the second critical endpoint *Q*, the two liquid phases become identical in equilibrium with a solid phase. The critical curve terminates liquid–liquid equilibria, and the high-temperature three-phase region is a liquid–liquid–solid equilibrium that extends from the critical endpoint *Q* to the triple point of pure Na_2SO_4 ($T_{Na_2SO_4}$) with a continuous transformation of the equilibrium with two liquid phases (liquid–liquid–solid) into a liquid–gas equilibrium (gas–liquid–solid) with increasing temperature and decreasing pressure [27].

Three-phase equilibrium curves (gas–liquid–solid) with a maximum vapor pressure of the saturated solution are shown in Figure 3.61 for various systems with a continuous critical curve. In these systems, over a wide range of parameters, gas–liquid equilibria and gas–liquid critical points occur. Binary liquid–gas critical curves for various water–salt systems are shown in Figure 3.62.

For the critical line of H_2O–NaCl, the IAPWS has issued a guideline that serves as reference. Values are listed in Table 3.32. For calculations, the guideline can be downloaded from IAPWS [160,161].

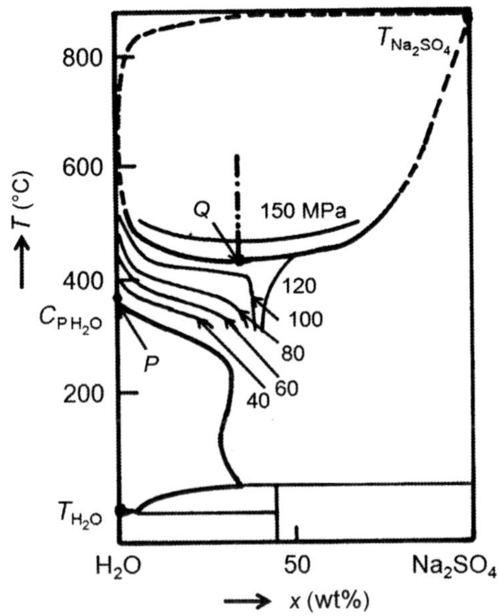

FIGURE 3.60 T,x-projection of the H_2O–Na_2SO_4 phase diagram. —•—••—, critical curve; ——, solubility curves and isobaric cross sections of two-phase equilibria. *Modified from Ref. [27].*

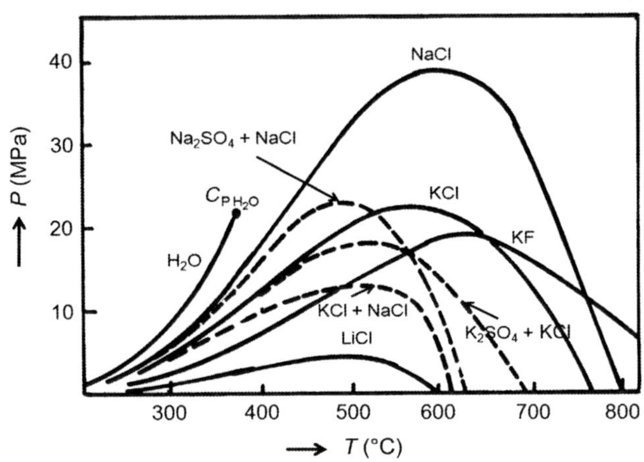

FIGURE 3.61 Three-phase equilibrium liquid–gas–solid (——) in binary systems with a continuous critical curve, and four-phase equilibria (eutonic curves) liquid–gas–solid 1–solid 2 (- - -) in ternary systems. *Modified from Ref. [27].*

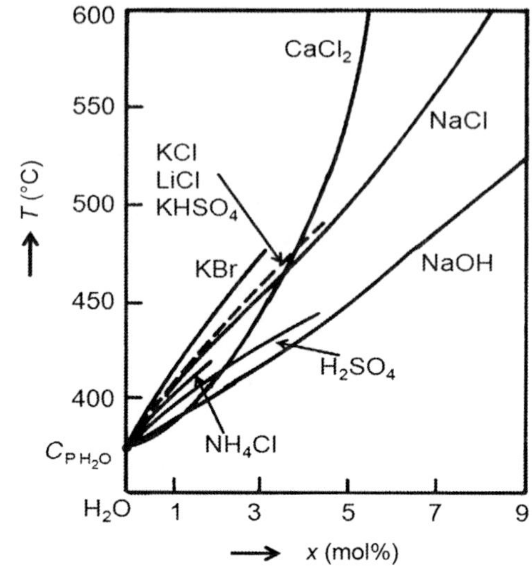

FIGURE 3.62 Binary liquid–gas critical curves for various water–salt systems. *Modified from Ref. [27].*

3.7.2 Ternary Systems and the Influence of Salts on Phase Equilibrium

The influence of salts on phase behavior of volatile and nonvolatile components is of major interest for hydrothermal processes. Salts may be in the feed mixture that is to be processed or may be added intentionally for achieving specific effects, for example, as catalysts. In the following, it will be shown how salts influence the general phase behavior.

For ternary systems, there is no topological scheme systemizing the phase behavior as for binary systems. The simple reason is the abundance of different types possible. The major features of phase behavior in ternary systems are mainly determined by the types of phase diagrams of the constituting binary subsystems. These types can be used for anticipating the main features of ternary systems and for a classification of ternary phase behavior. Therefore, the ternary systems will be classified according to the phase behavior of the binary subsystems with either a continuous critical curve or an interrupted critical curve. Ternary systems may have up to three of either binary type.

3.7.2.1 Ternary Systems with Two Binary Water–Salt Subsystems with Continuous Critical Curves

Phase behavior in ternary water–salt systems where both binary water–salt and anhydrous subsystems show a continuous critical curve and have been investigated to some extent the following: H_2O–NaCl–KCl [140,162–164], H_2O–KCl–KBr [165], H_2O–KCNS–KCl (KBr and KI) [166],

TABLE 3.32 Critical Curve of the Binary System H_2O–NaCl [160]

y (wt%)	T_c (K)	P_c (MPa)	ρ_c (kg cm^{-3})
0	647.096	22.0640	322.000
0.2	652.676	23.2243	334.646
0.4	654.946	23.7153	346.047
0.6	657.490	24.2780	356.447
0.8	659.567	24.7466	366.026
1	661.612	25.2160	374.916
1.2	663.629	25.6864	383.219
1.4	665.619	26.1574	391.016
1.6	667.583	26.6289	398.371
1.8	669.521	27.1004	405.336
2	671.435	27.5717	411.955
4	689.402	32.2408	464.982
6	705.804	36.7989	503.707
8	721.450	41.3101	534.482
10	737.152	45.9160	560.062
12	753.719	50.7860	581.901
14	771.963	56.0834	600.860
16	792.692	61.9375	617.492
18	816.717	68.4216	632.167
20	844.849	75.5545	645.148
22	877.898	83.3683	656.623
24	916.673	92.1443	666.733
26	961.986	103.007	675.582
28	1014.65	119.224	683.251
30	1075.46	148.807	689.801

H_2O–NH_4NO_3–$Pb(NO_3)_2$ [167], H_2O–NaCl–NaOH [168], H_2O–LiOH–NaOH, H_2O–LiOH–KOH, and H_2O–LiOH–RbOH [169]. The general feature of these systems is the appearance of eutonic equilibria, where two solid phases of components coexist with liquid and vapor solutions. Figure 3.63

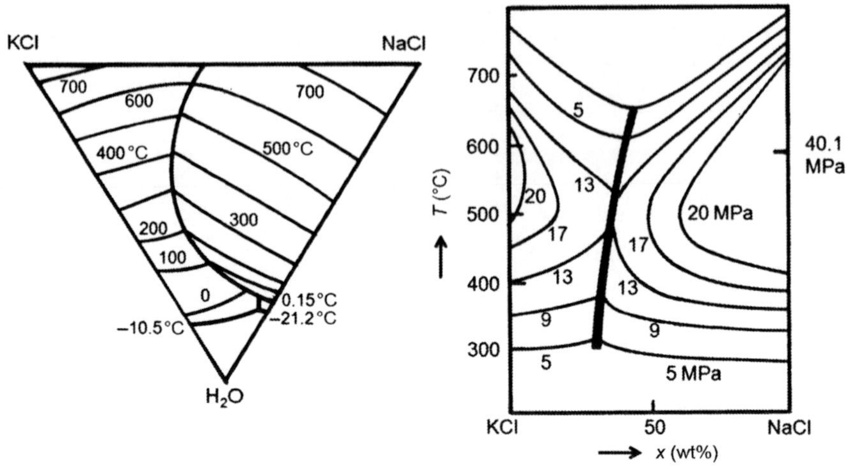

FIGURE 3.63 Solubility isotherms liquid–gas–solid in the system KCl–NaCl–H_2O (all binary systems with continuous critical curve), left, and T–x* projection of isobaric cross sections, right. $W^*_{KCl} = 100\,W_{KCl}/(W_{KCl} + W_{NaCl})$, where W_{KCl} and W_{NaCl} are the weight amounts of KCl and NaCl in aqueous solution. —, composition of eutonic solutions saturated with two solid phases at vapor pressure; ——, isothermal composition of solid-saturated liquid solutions at vapor pressure (left) and composition of solid-saturated liquid solution at constant vapor pressure (right). *Modified from Ref. [27].*

shows the projection of three-phase solubility surfaces in the system NaCl–KCl–H_2O. Vapor-pressure behavior in the ternary water–salt system is shown in Figure 3.63 as a set of isobaric cross sections of the saturated solution surfaces [27].

3.7.2.2 *Ternary Systems with Binary Water–Salt Subsystems of Continuous and Interrupted Critical Curves*

The general feature of ternary systems with binary water–salt systems of different type is the change of the sign of the temperature dependence of solubility from negative to positive in ternary solutions. As a result, an intersection of solubility isotherms occurs (Figure 3.64). Eutonic solutions are characterized by a positive temperature dependence of solubility and a pressure maximum on the eutonic curves (see the eutonic curves for the systems KCl–K_2SO_4–H_2O and NaCl–Na_2SO_4–H_2O in Figure 3.64). The negative temperature dependence of solubility is typical of dilute hydrothermal solutions with an interrupted critical curve. Temperature dependence of solubility becomes positive when the total solution concentration increases up to 2–15 mol% of salt [170,171].

Available experimental data for various properties of hydrothermal electrolyte solutions show the change of property behavior from "water-like" to "melt-like" [170].

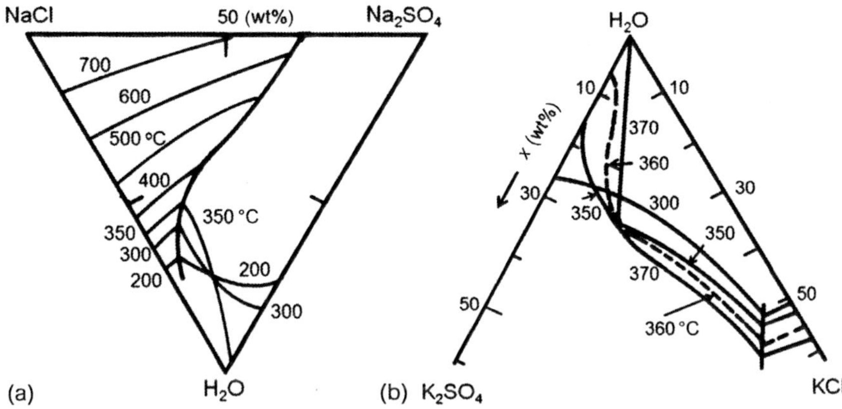

FIGURE 3.64 Solubility isotherms at vapor pressure in the systems $NaCl–Na_2SO_4–H_2O$ (a) and $KCl–K_2SO_4–H_2O$ (b). ——, composition of solid-saturated liquid solutions at vapor pressure and composition of liquid solutions saturated with two solid phases at vapor pressure. *Modified from Ref. [27].*

3.7.2.3 Ternary Systems with Two Binary Water–Salt Subsystems with Interrupted Critical Curves

Two types of ternary phase behavior were established in the experimental studies of ternary systems with two binary water–salt subsystems with interrupted critical curves. For most aqueous systems with high melting point oxides and aluminosilicates, such as $NaAlSi_3O_8–KAlSi_3O_8–H_2O$, $SiO_2–NaAlSi_3O_8–H_2O$, $SiO_2–KAlSi_3O_8–H_2O$, $SiO_2–CaAl_2Si_2O_8–H_2O$, and $CaO–SiO_2–H_2O$ [172,173], the phase behavior of ternary mixtures is very similar to the binary ones.

Supercritical fluid equilibria, observed for ternary mixtures at any ratio of salt components, show that a three-phase immiscibility region is retained in metastable conditions and in binary subsystems. In these ternary systems, the eutonic curves consist of two separated low- and high-temperature branches.

Another type of phase behavior takes place when the metastable immiscibility region becomes stable in ternary solutions. The joint solubility of the two salt components increases drastically. The eutonic curve is separated from the solubility curves in the binary water–salt subsystems by the immiscibility region. Temperature dependence of solubility in eutonic solutions becomes positive, and critical phenomena are absent. Such behavior was found in the ternary system $H_2O–SiO_2–Na_2Si_2O_5$ [174] and in the system $H_2O–K_2SO_4–KLiSO_4$ [175].

For a more detailed discussion of ternary systems of water–salt systems of the previously discussed type, the reader is referred to the work of Valyashko [27].

3.7.2.4 Ternary Aqueous Salt Systems with Volatile Components

In many cases, mutual solubility decreases in the presence of salts (salting-out effect). In ternary systems, the solubility curves can be derived qualitatively

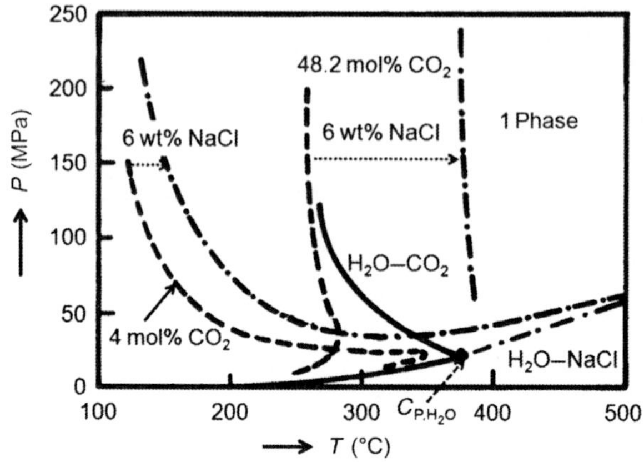

FIGURE 3.65 Influence of NaCl on the phase equilibrium of H_2O–CO_2. *Modified from Ref. [176].*

from the binary systems but the equilibrium is shifted to higher temperatures and pressures. An example is the system CO_2–H_2O–NaCl shown in Figure 3.65. The influence of the salt is more pronounced for higher gas concentrations. The addition of 6 wt% NaCl to a mixture of H_2O with 4 mol% CO_2 shifts the two-phase area 50 K to higher temperatures, while for a solution of 48.2mol% CO_2 in H_2O, the phase boundary line is shifted 100 K to higher temperatures [177,178].

The addition of a salt to a binary hydrocarbon–water system is of more influence on the phase equilibrium than adding a gas to a binary water–salt mixture. 1 wt% NaCl added to a mixture of 30 wt% C_2H_6 in H_2O shifts the phase boundary the same temperature difference as adding 10% of C_2H_6 [179]. As has been mentioned already, salts with an interrupted critical curve are less soluble in the gaseous phase than those with a continuous critical curve. The combination of salts with an interrupted and a continuous critical curve, for example, NaCl and Na_2SO_4 in water, leads to a ternary system with a continuous three-phase curve and to a reduction of the miscibility gap [130,176].

For the system water–methane–sodium chloride that was investigated up to $T = 773$ K and $P = 250$ MPa [180], similar shifts to higher temperatures of the boundary curves have been observed by an addition of only 0.53 and 2.61 mol% of sodium chloride (relative to water) to methane solutions of 17 and 51 mol% methane, respectively.

The ternary phase equilibrium of the system H_2O–CH_4–NaCl is shown in Figure 3.66 in triangular cuts for two temperatures and in dependence of temperature for $P = 200$ MPa in a triangular prism. Most parts of the diagrams are extrapolations from the few measurements available, but they make clear the influence of salt concentration and temperature on the phase behavior [4,180].

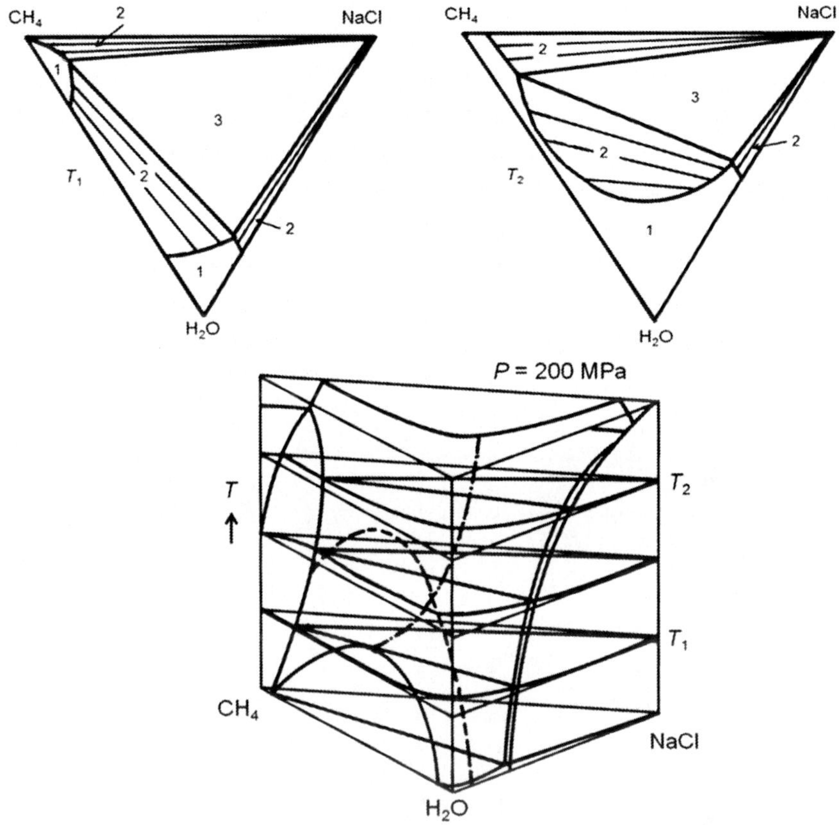

FIGURE 3.66 The ternary system H_2O–CH_4–NaCl. Numbers indicate the number of coexisting phases. *Modified from Ref. [180].*

In this chapter, an overview on the behavior of water and many substances has been presented. In the context of a textbook concerned with the applications of hydrothermal and supercritical water, this overview can only be a starting point for detailed information on specific mixtures. More detailed information on most topics of this chapter is covered in more detail in the excellent book by Palmer *et al.* [181].

REFERENCES

[1] S.I. Sandler, Chemical and Engineering Thermodynamics, second ed., Wiley, New York, 1989, p. 252.

[2] H. Lentz, E.U. Franck, Water-argon system at high temperatures and pressures, Ber. Bunsenges. Phys. Chem. 73 (1969) 28–35.

[3] H.J. Greenwood, The system $NaAlSi_2O_6$-H_2O-argon: total pressure and water pressure in metamorphism, J. Geophys. Res. 66 (1961) 3923–3946.

[4] E.U. Franck, Fluids at high pressures and temperatures, Pure Appl. Chem. 59 (1987) 25–34.

[5] E.U. Franck, H. Lentz, H. Welsch, The system water–xenon at high pressures and temperatures, Z. Phys. Chem. Neue Folge 93 (1974) 95–108.

[6] T.M. Seward, E.U. Franck, The system hydrogen–water up to 440 °C and 2500 bar pressure, Ber. Bunsenges. Phys. Chem. 85 (1981) 2–7.

[7] V.M. Shmonov, R.J. Sadus, E.U. Frank, High-pressure phase equilibria and supercritical pVT-data of the binary water–methane mixture to 723 K and 200 MPa, J. Phys. Chem. 97 (1993) 5054–5059.

[8] A.G. Faramarzi, Dissertation, University of Karlsruhe, Karlsruhe, 1982.

[9] J.G. Blencoe, J.C. Seitz, L.M. Anovitz, The CO_2-H_2O system. II. Calculated thermodynamic mixing properties for 400 °C, 0–400 MPa, Geochim. Cosmochim. Acta 63 (1999) 2393–2408.

[10] K. Tödheide, Hydrothermal solutions, Ber. Bunsenges. Phys. Chem. 86 (1982) 1005–1016.

[11] H.R. Corti, L.N. Trevani, A. Anderko, Transport properties in high temperature and pressure ionic solutions, in: D.A. Palmer, R. Fernández-Prini, A.H. Harvey (Eds.), Aqueous Systems at Elevated Temperatures and Pressures: Physical Chemistry in Water, Steam and Hydrothermal Solutions, Elsevier, Amsterdam, The Netherlands, 2004, pp. 321–375.

[12] IAPWS, The International Association for the Properties of Water and Steam: Release on the Ionization Constant of H_2O, Available from: http://www.iapws.org, 2007.

[13] B.S. Smolyakov, G.A. Veselova, Limiting equivalent conductance of Li^+, Na^+, K^+, Rb^+, Cs^+, Elektrokhimiya 11 (1975) 700–704.

[14] A. Anderko, M.M. Lencka, Computation of electrical conductivity of multicomponent aqueous systems in wide concentration and temperature ranges, Ind. Eng. Chem. Res. 36 (1997) 1932–1943.

[15] A.C. Sharygin, I. Mokbel, C. Xiao, R.H. Wood, Test of equation for the electrical conductance of electrolyte mixtures. Measurements of association of NaCl(aq) and Na_2SO_4(aq) at high temperatures, J. Phys. Chem. B 105 (2001) 229–237.

[16] A. Anderko, M.M. Lencka, Computation of electrical conductivity of multicomponent aqueous systems in wide concentration and temperature ranges, Ind. Eng. Chem. Res. 37 (1998) 2878–2888.

[17] W.T. Lindsay, Estimation of diffusion coefficients for electrolytes in hot water, Topical Report for Research Project S146-1, EPRI, Palo Alto, CA, 1980, cited from [11].

[18] G. Jones, M. Dole, Viscosity of aqueous solutions of strong electrolytes with special reference to barium chloride, J. Am. Chem. Soc. 51 (1929) 2950–2964.

[19] M. Kaminsky, Experimentelle Untersuchungen über die Konzentrations- und Temperaturabhängigkeit der Zähigkeit wäßriger Lösungen starker Elektrolyte, Z. Phys. Chem. Neue Folge 12 (1957) 206–231.

[20] M.M. Lencka, A. Anderko, S.J. Sanders, R.D. Young, Modeling viscosity of multicomponent electrolyte solutions, Int. J. Thermophys. 19 (1998) 367–378.

[21] L. Riedel, The thermal conductivity of aqueous solutions and strong electrolytes, Chem. Ing. Tech. 23 (1951) 59–64.

[22] E. McLaughlin, The thermal conductivity of liquids and dense gases, Chem. Rev. 64 (1964) 389–428.

[23] I.M. Abdulagatov, U.B. Magomedov, Thermal conductivity of aqueous KI and KBr solutions at high temperatures and high pressures, J. Solution Chem. 30 (2001) 223–235.

[24] Y. Nagasawa, H. Okada, J. Suzuki, A. Nagashima, Absolute measurements of the thermal conductivity of aqueous NaCl solutions at pressures up to 40 MPa, Ber. Bunsenges. Phys. Chem. 87 (1983) 859–866.

[25] R.M. DiGuilio, A.S. Teja, The thermal conductivity of lithium bromide–water salt solutions, Ind. Eng. Chem. Res. 31 (1992) 1081–1085.
[26] G. Brunner, Gas Extraction. An Introduction to Fundamentals of Supercritical Fluids and the Application to Separation Processes, Steinkopff, Springer, Darmstadt, New York, 1994.
[27] V.M. Valyashko, Phase equilibria of water–salt systems at high temperatures and pressures, in: D.A. Palmer, R. Fernández-Prini, A.H. Harvey (Eds.), Aqueous Systems at Elevated Temperatures and Pressures: Physical Chemistry in Water, Steam and Hydrothermal Solutions, Elsevier, Amsterdam, The Netherlands, 2004, pp. 597–641 (Chapter 15).
[28] P.H. van Konynenburg, R.L. Scott, Critical lines and phase equilibria in binary van der Waals mixtures, Philos. Trans. R. Soc. Lond. A 298 (1980) 495–540.
[29] J.S. Rowlinson, F.L. Swinton, Liquids and liquid mixtures, third ed., Butterworths, London, 1982.
[30] M.L. McGlashan, G.M. Schneider, High-pressure phase diagrams and critical properties of fluid mixtures, Chem. Thermodyn. 2 (1978) 105–146.
[31] K. Bröllos, K. Peter, G.M. Schneider, Fluide Mischsysteme unter hohem Druck, Phasengleichgewichte und kritische Erscheinungen in den binären Systemen cyclohexan-H_2O, *n*-heptan-H_2O, biphenyl-H_2O und benzol-D_2O bis 420 °C und 3000 bar, Ber. Bunsenges. Phys. Chem. 74 (1970) 682–686.
[32] Z. Alwani, G.M. Schneider, Phasengleichgewichte, kritische Erscheinungen und PVT-Daten in binären Mischungen von Wasser mit aromatischen Kohlenwasserstoffen bis 420 °C und 2200 bar, Ber. Bunsenges. Phys. Chem. 73 (1969) 294–301.
[33] E. Brunner, M.C. Thies, G.M. Schneider, Fluid mixtures at high pressures: phase behavior and critical phenomena for binary mixtures of water with aromatic hydrocarbons, J. Supercrit. Fluids 39 (2006) 160–173.
[34] R. Dohrn, Phasengleichgewichte in Mehrkomponentensystemen aus Wasserstoff, Wasser und Kohlenwasserstoffen, Ph.D. Dissertation, Hamburg University of Technology, Hamburg, Germany, 1986; Fortschrittsberichte VDI, 3/123, VDI-Verlag, Düsseldorf, 1986.
[35] J. Teich, Phasengleichgewichte bei erhöhten Drücken und Temperaturen in ternären Systemen aus Wasserstoff, Kohlendioxid, Wasser und Kohlenwasserstoffen, Ph.D. Dissertation, Hamburg University of Technology, Hamburg, Germany, 1992.
[36] E. Brunner, Fluid mixtures at high pressures IX. Phase separation and critical phenomena in 23 (*n*-alkane + water) mixtures, J. Chem. Thermodyn. 22 (1990) 335–353.
[37] G.C. Kennedy, A portion of the system silica-water, Econ. Geol. 45 (1950) 629–653.
[38] R. Koningsveld, G.A.M. Diepen, Supercritical phase equilibria involving solids, Fluid Phase Equilibr. 10 (1983) 159–172.
[39] O. Pfohl, S. Petkov, G. Brunner, PE, A Powerful Tool to Correlate Phase Equilibria, Herbert Utz, München, Program available from: www.tu-harburg.de/V8/gruppe-prof-smirnova/veroeffentlichungen.html, 2000.
[40] J.M. Prausnitz, R.N. Lichtenthaler, E. Gomes de Azevedo, Molecular Thermodynamics of Fluid-Phase Equilibria, third ed., Prentice Hall, Upper Saddle River, NJ, 1999.
[41] R.J. Sadus, High pressure phase behaviour of multicomponent fluid mixtures, Elsevier, Amsterdam, The Netherlands, 1992.
[42] S. Sandler, Equations of state for phase equilibrium computations, in: E. Kiran, J.M. H. Levelt-Sengers (Eds.), Supercritical Fluids—Fundamentals for Application, NATO ASI, Kluwer, Dordrecht, NL, 1994, pp. 147–175.
[43] R. Dohrn, Berechnung von Phasengleichgewichten mit Hilfe von Zustandsgleichungen, Habilitationsschrift, Hamburg University of Technology, Hamburg, Germany, 1994.

[44] O. Pfohl, Messung und Berechnung von Phasengleichgewichten mit nahe- und überkritischem Kohlendioxid sowie assoziierenden Komponenten im Hochdruckbereich, Ph.D. Dissertation, Hamburg University of Technology, Hamburg, Germany, 1998.

[45] D.S.H. Wong, S. Sandler, A theoretically correct mixing rule for cubic equations of state, AIChE J. 38 (1992) 671–680.

[46] H. Orbey, Mixing rules for the estimation of vapor–liquid equilibrium of highly non-ideal mixtures using cubic equations of state, in: E. Kiran, J.M.H. Levelt-Sengers (Eds.), Supercritical Fluids—Fundamentals for Application, NATO ASI, Kluwer, Dordrecht, NL, 1994, pp. 177–188.

[47] P.M. Mathias, H.C. Klotz, J.M. Prausnitz, Equation of state mixing rules for multicomponent mixtures: the problem of invariance, Fluid Phase Equilibr. 67 (1991) 31–44.

[48] M.L. Michelsen, H. Kistenmacher, On composition-dependent interaction coefficients, Fluid Phase Equilibr. 58 (1990) 229–230.

[49] R. Dohrn, G. Brunner, Phase equilibria in ternary and quaternary systems of hydrogen, water and hydrocarbons at elevated temperatures and pressures, Fluid Phase Equilibr. 29 (1986) 535–544.

[50] E. Liebold, Zur Abscheidung von Kohlenwasserstoffen aus wasserdampfhaltigen Gasgemischen bei höheren Temperaturen und Drücken, Ph.D. Dissertation, Hamburg University of Technology, Hamburg, Germany, 1988.

[51] R. Fernández-Prini, J.L. Alvarez, A.H. Harvey, Aqueous solubility of volatile nonelectrolytes, in: D.A. Palmer, R. Fernández-Prini, A.H. Harvey (Eds.), Aqueous Systems at Elevated Temperatures and Pressures: Physical Chemistry in Water, Steam and Hydrothermal Solutions, Elsevier, Amsterdam, The Netherlands, 2004, pp. 73–98 (Chapter 3).

[52] W. Henry, Experiments on the quantity of gases absorbed by water, at different temperatures, and under different pressures, Philos. Trans. R. Soc. Lond. 93 (1803) 29–274.

[53] I.R. Krichevsky, J.S. Kasarnovsky, Thermodynamical calculations of solubilities of nitrogen and hydrogen in water at high pressures, J. Am. Chem. Soc. 57 (1935) 2168.

[54] J.M.H. Levelt Sengers, Solubility near the solvent's critical point, J. Supercrit. Fluids 4 (1991) 215–222.

[55] I.R. Krichevsky, A.A. Ilinskaya, Partial molal volumes of gases dissolved in liquids (the thermodynamics of dilute solutions of nonelectrolytes), Zh. Fiz. Khim. 19 (1945) 621.

[56] R. Sun, W. Hu, Z. Duan, Prediction of nitrogen solubility in pure water and aqueous NaCl solutions up to high temperature, pressure, and ionic strength, J. Solution Chem. 30 (2001) 561–573.

[57] A.A. Galkin, V.V. Lunin, Subcritical and supercritical water: a universal medium for chemical reactions, Russ. Chem. Rev. 74 (2005) 21–35.

[58] M.A. Anisimov, J.V. Sengers, J.M.H. Levelt Sengers, Near-critical behavior of aqueous systems, in: D.A. Palmer, R. Fernández-Prini, A.H. Harvey (Eds.), Aqueous Systems at Elevated Temperatures and Pressures: Physical Chemistry in Water, Steam and Hydrothermal Solutions, Elsevier, Amsterdam, The Netherlands, 2004, pp. 29–71 (Chapter 2).

[59] M.L. Japas, J.M.H. Levelt Sengers, Gas solubility and Henry's law near the solvent's critical point, AIChE J. 35 (1989) 705–713.

[60] I.R. Krichevskii, Thermodynamics of critical phenomena in infinitely dilute binary solutions, Russ. J. Phys. Chem. 41 (1967) 1332–1338.

[61] J.M.H. Levelt Sengers, Critical behavior of fluid: concepts and applications, in: E. Kiran, J. M.H. Levelt-Sengers (Eds.), Supercritical Fluids—Fundamentals for Application, NATO ASI, Kluwer, Dordrecht, NL, 1994, pp. 3–38.

[62] A.H. Harvey, R. Crovetto, J.M.H. Levelt Sengers, Limiting vs. apparent critical behavior of Henry's constants and K factors, AIChE J. 36 (1990) 1901–1904.

[63] International Association for the Properties of Water and Steam, IAPWS, Guideline on the Henry's Constant and Vapor-Liquid Distribution Constant for Gases in H_2O and D_2O at High Temperatures, Kyoto, Japan, Available from: http://www.iapws.org, 2004.
[64] W. Wagner, A. Pruss, International equations for the saturation properties of ordinary water substance. Revised according to the international temperature scale of 1990, J. Phys. Chem. Ref. Data 22 (1993) 783–787.
[65] M.L. Japas, E.U. Franck, High pressure phase equilibria and PVT-data of the water–nitrogen system to 573 K and 250 MPa, Ber. Bunsenges. Phys. Chem. 89 (1985) 793–800.
[66] G. Brunner, B. Misch, A. Firus, K. Nowak, Cleaning of soil with supercritical water and supercritical carbon dioxide, in: R. Stegmann, G. Brunner, W. Calmano, G. Matz (Eds.), Treatment of Soil—Fundamentals, Analysis, Applications, Springer, Berlin, 2001, pp. 491–517.
[67] R. Jockers, G.M. Schneider, Fluid mixtures at high pressures. Fluid phase equilibria in the systems fluorobenzene +water, 1,4-difluorobenzene + water, and 1,2,3,4-tetrahydronaphthalene + decahydronaphthalene(trans) + water up to 360 MPa, Ber. Bunsenges. Phys. Chem. 82 (1978) 576–582.
[68] H. Welsch, Ph.D. Thesis, University of Karlsruhe, Karlsruhe, Germany, 1973.
[69] R.G. Sultanov, V.G. Skripka, A.Y. Namiot, Gasov. Prom. 16 (1971) 6, ibid. 17 (1972) 6.
[70] A. Danneil, K. Tödheide, E.U. Franck, Verdampfungsgleichgewichte und kritische. Kurven in den Systemen Äthan/Wasser und *n*-Butan/Wasser bei hohen Drücken, Chem. Ing. Tech. 39 (1967) 816–822.
[71] M.R. Sanchez, R. Coll, An. Quim. Int. Ed. 74 (1978) 1329.
[72] Th.W. de Loos, A.J.M. Wijen, G.A.M. Diepen, Phase equilibria and critical phenomena in fluid (propane + water) at high pressures and temperatures, J. Chem. Thermodyn. 12 (1980) 193–204.
[73] D.S. Tsiklis, V.Ya. Maslennikova, The proceedings of the USSR academy of sciences (Russian: Доклады Академии наук СССР), Dokl. Akad. Nauk SSSR, DAN SSSR 157 (1964) 417.
[74] J.F. Connolly, Solubility of hydrocarbons in water near the critical solution temperatures, J. Chem. Eng. Data 11 (1966) 13–16.
[75] E. Brunner, Fluid mixtures at high pressures I. Phase separation and critical phenomena of 10 binary mixtures of (a gas + methanol), J. Chem. Thermodyn. 17 (1985) 671–679.
[76] E. Brunner, W. Hültenschmidt, G. Schlichthärle, Fluid mixtures at high pressures IV. Isothermal phase equilibria in binary mixtures consisting of (methanol + hydrogen or nitrogen or methane or carbon monoxide or carbon dioxide), J. Chem. Thermodyn. 19 (1987) 273.
[77] R. Jockers, R. Paas, G.M. Schneider, Fluid mixtures at high pressures. Liquid-liquid phase equilibria and critical phenomena in the systems 1,2,3,4-tetrahydronaphthalene + water, decahydronaphthalene(cis) + water, decahydronaphthalene(trans) + water, and methane +trifluoromethane up to 2600 bar, Ber. Bunsenges. Phys. Chem. 81 (1977) 1093–1096.
[78] S.P. Christensen, M.E. Paulaitis, Phase equilibria for tetralin–water and 1-methylnaph-thalene–water mixtures at elevated temperatures and pressures, Fluid Phase Equilibr. 71 (1992) 63–83.
[79] N.D. Sanders, Visual observation of the solubility of heavy hydrocarbons in near-critical water, Ind. Eng. Chem. Fund. 25 (1986) 169–171.
[80] C. Tsonopoulos, G.M. Wilson, High-temperature mutual solubilities of hydrocarbons and water. Part I: benzene, cyclohexane and *n*-hexane, AIChE J. 29 (1983) 990–999.
[81] Ch.J. Rebert, W.B. Kay, The phase behavior and solubility relations of the benzene–water system, AIChE J. 5 (1959) 285–289.

[82] B.J. Neely, Aqueous Hydrocarbon Systems: Experimental Measurements and Quantitative Structure-Property Relationship Modeling, Ph.D. Dissertation, Oklahoma State University, 2007.
[83] Q. Wang, K.-C. Chao, Vapor-liquid and liquid–liquid equilibrium and critical states of water + *n*-decane mixtures, Fluid Phase Equilibr. 59 (1990) 207–215.
[84] G. Brunner, A. Steffen, R. Dohrn, High pressure liquid–liquid equilibria in ternary systems containing water, benzene, toluene, *n*-hexane and *n*-hexadecane, Fluid Phase Equilibr. 82 (1993) 165–172.
[85] P. Tremaine, K. Zhang, P. Bénézeth, C. Xiao, Ionization equilibria of acids and bases under hydrothermal conditions, in: D.A. Palmer, R. Fernández-Prini, A.H. Harvey (Eds.), Aqueous Systems at Elevated Temperatures and Pressures: Physical Chemistry in Water, Steam and Hydrothermal Solutions, Elsevier, Amsterdam, The Netherlands, 2004, pp. 441–492 (Chapter 13).
[86] D.A. Palmer, J.M. Sinonson, J.P. Jensen, Partitioning of electrolytes to steam and their solubilities in steam, in: D.A. Palmer, R. Fernández-Prini, A.H. Harvey (Eds.), Aqueous Systems at Elevated Temperatures and Pressures: Physical Chemistry in Water, Steam and Hydrothermal Solutions, Elsevier, Amsterdam, The Netherlands, 2004, pp. 409–439 (Chapter 12).
[87] J. Alvarez, H.R. Corti, R. Fernández-Prini, M.L. Japas, Distribution of solutes between coexisting steam and water, Geochim. Cosmochim. Acta 58 (1994) 2789–2798.
[88] G. Jin, M.D. Donohue, An equation of state for electrolyte-solutions. 1. Aqueous systems containing strong electrolytes, Ind. Eng. Chem. Res. 27 (1988) 1073–1084.
[89] A.H. Harvey, J.M. Prausnitz, Thermodynamics of high-pressure aqueous systems containing gases and salts, AIChE J. 35 (1989) 635–644.
[90] K. Aasberg-Petersen, E. Stenby, A. Fredenslund, Prediction of high-pressure gas solubilities in aqueous mixtures of electrolytes, Ind. Eng. Chem. Res. 30 (1991) 2180–2185.
[91] Y.-X. Zuo, T.-M. Guo, Extension of Patel–Teja equation of state to the prediction of the solubility of natural gas in formation water, Chem. Eng. Sci. 46 (1991) 3251–3258.
[92] Y. Wu, J.M. Prausnitz, Phase equilibria for systems containing hydrocarbons, water, and salt: an extended Peng–Robinson equation of state, Ind. Eng. Chem. Res. 37 (1998) 1634–1643.
[93] E. Zhao, B.C.-Y. Lu, Representation of electrolyte solution properties by means of the Peng–Robinson–Stryjek–Vera equation of state, Ind. Eng. Chem. Res. 37 (1998) 1616–1624.
[94] I. Vanderbeken, S. Ye, B. Bouyssière, H. Carrier, P. Xans, Ability of the MHV2 mixing rule to describe the effect of salt on gas solubility in brines at high temperature and high pressure, High Temp.-High Press. 31 (1999) 653–663.
[95] J. Li, M. Topphoff, K. Fischer, J. Gmehling, PSRK group contribution equation of state: revision and extension III, Ind. Eng. Chem. Res. 40 (2001) 3703–3710.
[96] J.J. Kosinski, A. Anderko, Equation of state for high-temperature aqueous electrolyte and nonelectrolyte systems, Fluid Phase Equilibr. 183 (2001) 75–86.
[97] K. Tödheide, E.U. Franck, Das Zweiphasengebiet und die kritische Kurve im System Kohlendioxid—Wasser bis zu Drucken von 3500 bar, Z. Phys. Chem. Neue Folge 37 (1963) 387–401.
[98] S. Takenouchi, G. Kennedy, The binary system H_2O-CO_2 at high temperatures and pressures, Am. J. Sci. 262 (1964) 1055–1074.
[99] M.B. King, A. Mubarak, J.D. Kim, T.R. Bott, The mutual solubilities of water with supercritical and liquid carbon dioxide, J. Supercrit. Fluids 5 (1992) 296–302.

[100] R. Wiebe, The binary system carbon dioxide–water under pressure, Chem. Rev. 29 (1941) 475–481.

[101] A. Bamberger, G. Sieder, G. Maurer, High-pressure (vapor+liquid) equilibrium in binary mixtures of (carbon dioxide+water or acetic acid) at temperatures from 313 to 353 K, J. Supercrit. Fluids 17 (2000) 97–110.

[102] R. Wiebe, V. Gaddy, The solubility of carbon dioxide in water at various temperatures from 12 °C to 40 °C and at pressures to 500 atmospheres. Critical phenomena, J. Am. Chem. Soc. 62 (1940) 815–817.

[103] C.H. Kwon, C.-H. Lee, J.W. Kang, Calculation of phase equilibrium for water+carbon dioxide system using nonrandom lattice fluid equation of state, Korean J. Chem. Eng. 27 (2010) 278–283.

[104] C.S. Patterson, G.H. Slocum, R.H. Busey, R.E. Mesmer, Carbonate equilibria in hydrothermal systems: first ionization of carbonic acid in NaCl media to 300 °C, Geochim. Cosmochim. Acta 46 (1982) 1653–1663.

[105] A.J. Read, The first ionization constant of carbonic acid from 25 to 250 °C and to 2000 bar, J. Solution Chem. 4 (1975) 53–70.

[106] C.S. Patterson, R.H. Busey, R.E. Mesmer, Second ionization of carbonic acid in NaCl media to 250 °C, J. Solution Chem. 13 (1984) 647–661.

[107] D.A. Palmer, R. van Eldik, The chemistry of metal carbonate and carbon dioxide complexes, Chem. Rev. 83 (1983) 651–731.

[108] J.N. Butler, Carbon Dioxide Equilibria and Their Applications, Addison-Wesley Publishing Company, Reading, MA, 1982.

[109] A.J. Ellis, The effect of pressure on the first dissociation constant of "carbonic acid" J. Chem. Soc. (1959) 3689–3699.

[110] B. Misch, Reinigung mischkontaminierten Bodenmaterials und kontinuierliche Extrak-tion von Feststoffen mit überkritischen Fluiden, Ph.D. Dissertation, Hamburg University of Technology, Hamburg, Germany, Available from the author at brunner@tuhh.de, 2001.

[111] K. Toews, R. Shroll, C.M. Wai, M.G. Smart, pH-defining equilibrium between water and supercritical CO_2. Influence on SFE of organics and metal chelates, Anal. Chem. 67 (1995) 4040–4043.

[112] B. Meyssami, M.O. Balaban, A.A. Teixeira, Prediction of pH in model systems pressurized with carbon dioxide, Biotechnol. Prog. 8 (1992) 149–154.

[113] H.T. Schaef, B.P. McGrail, Direct measurements of pH and dissolved CO_2 in H_2O-CO_2 brine mixtures to supercritical conditions, in: Proceedings of the 7th International Conference on Greenhouse Gas Control Technologies, Vancouver, Canada, 2004.

[114] T. Rogalinski, Hydrolyse von Proteinen in Wasser und Wasser-Kohlendioxid-Systemen unter erhöhten Drücken und Temperaturen, Ph.D. Dissertation, Hamburg University of Technology, Hamburg, Available from the author at brunner@tuhh.de, 2006.

[115] T. Albrecht, Conversion of Lignocellulosic Materials and Model Compounds in Sub- and Supercritical Water, Ph.D. Dissertation, Hamburg University of Technology, Hamburg, Germany, available from the author at brunner@tuhh.de, 2006.

[116] F.H. Sweeton, R.E. Mesmer, C.F. Baes Jr., Acidity measure—study using ion exchange and isotachophoresis. Polyhedron ments at elevated temperature. VII. Dissociation of water, J. Solution Chem. 3 (1974) 191–214.

[117] G. Härtel, S. Peter, W. Tunn, Über die Acidität wäßriger Lösungen der Chloride des Natriums, Calciums und Magnesiums in Anwesenheit von Kohlendioxid bei hohen Drücken, Erdöl Erdgas Kohle 96 (1980) 92.

[118] (a) A.J. Ellis, R.M. Golding, The solubility of carbon dioxide above 100 °C in water and in sodium chloride solutions, Am. J. Sci. 261 (1963) 47–60; (b) A.J. Ellis, W.A.J. Mahon, Chemistry and Geothermal Systems, Academic Press Inc., New York, 1977.

[119] P.C. Ho, D.A. Palmer, M.S. Gruszkiewicz, Conductivity measurements of dilute aqueous HCl solutions to high temperatures and pressures using a flow-through cell, J. Phys. Chem. B 105 (2001) 1260–1266.

[120] R.E. Mesmer, W.L. Marshall, D.A. Palmer, J.M. Simonson, H.F. Holmes, Thermodynamics of aqueous association and ionization reactions at high temperatures and pressures, J. Solution Chem. 17 (1988) 699–718.

[121] R.E. Mesmer, D.A. Palmer, J.M. Simonson, in: K.S. Pitzer (Ed.), Activity Coefficients in Aqueous Solutions, second ed., CRC Press, Boca Raton, FL, 1991 (Chapter 8).

[122] S. Kherris, M. Makhlouf, D. Zebbar, O. Sebbane, Contribution Study of the Thermodynamics Properties of the Ammonia–Water Mixtures, Available from: www.doiserbia.nb.rs/img/doi/0354-9836/2011%20OnLine-First/0354.

[123] International Association for the Properties of Water and Steam, Guideline on the IAPWS Formulation 2001 for the Thermodynamic Properties of Ammonia-Water Mixtures, IAPWS, Available from: http://www.iapws.org, 2001.

[124] R. Tillner-Roth, D.G. Friend, A Helmholtz free energy formulation of the thermodynamic properties of the mixture {water+ammonia}, J. Phys. Chem. Ref. Data 27 (1998) 63.

[125] B.F. Hitch, R.E. Mesmer, The ionization of aqueous ammonia to 300 in KC1 media, J. Solution Chem. 5 (1976) 667–680.

[126] T. Xiang, K.P. Johnston, W. Wofford, E.F. Gloyna, Spectroscopic measurements of pH in aqueous sulfuric acid and ammonia from sub- to supercritical conditions, Ind. Eng. Chem. Res. 35 (1996) 4788–4795.

[127] A.G. Dickson, D.J. Wesolowski, D.A. Palmer, R.E. Mesmer, Dissociation constant of bisulfate ion in aqueous sodium chloride solutions to 250 °C, J. Phys. Chem. 94 (1990) 7978–7985.

[128] D.J. Wesolowski, S.E. Ziemniak, L.M. Anovitz, M.L. Machesky, P. Bénézeth, D.A. Palmer, Solubility and surface adsorption characteristics of metal oxides, in: D.A. Palmer, R. Fernández-Prini, A.H. Harvey (Eds.), Aqueous Systems at Elevated Temperatures and Pressures: Physical Chemistry in Water, Steam and Hydrothermal Solutions, Elsevier, Amsterdam, The Netherlands, 2004, pp. 493–595 (Chapter 14).

[129] I. Leusbrock, S.J. Metz, G. Rexwinkel, G.F. Versteeg, Quantitative approaches for the description of solubilities of inorganic compounds in near-critical and supercritical water, J. Supercrit. Fluids 47 (2008) 117–127.

[130] H. Baierlein, Zur Löslichkeit von Salzen in überkritischem Wasserdampf, Ph.D. Dissertation, University of Erlangen-Nürnberg, Erlangen, Germany, 1983.

[131] D.O.H. Duah, Das Phasenverhalten von H_2O-Salzmischungen bei hohen Drücken und Temperaturen, Ph.D. Dissertation, University of Erlangen-Nürnberg, Erlangen, Germany, 1988.

[132] S.I. Tkachenko, K.I. Shmulovich, Dokl. Akad. Nauk (Russ.) 326 (1992) 1055–1059.

[133] J.L. Bischoff, R.J. Rosenbauer, R.O. Fournier, The generation of HCl in the system $CaCl_2$-H_2O: vapor–liquid relations from 380–500 °C, Geochim. Cosmochim. Acta 60 (1996) 7–16.

[134] M.A. Urusova, V.M. Valyashko, S.V. Petrenko, D.A. Bichikov, Zh. Neorg. Khim. 39 (1994) 2068–2078.

[135] M. Dubois, A. Weisbrod, A. Shtuka, Experimental determination of the two-phase (liquid and vapour) region in water–alkali chloride binary systems at 500 and 600 °C using synthetic fluid inclusions, Chem. Geol. 115 (1994) 227–238.

[136] W.T. Wofford, P.C. Dell'Orco, E.F. Gloyna, Solubility of potassium hydroxide and potassium phosphate in supercritical water, J. Chem. Eng. Data 40 (1995) 968–973.

[137] W.L. Marshall, C.E. Hall, R.E. Mesmer, The system dipotassium hydrogen phosphate–water at high temperatures (100–400 °C); liquid-liquid immiscibility and concentrated solutions, J. Inorg. Nucl. Chem. 43 (1981) 449–455.

[138] A. Benrath, F. Gjedebo, B. Schiffers, Über die Löslichkeit von Salzen und Salzgemi-schen in Wasser bei Temperaturen oberhalb von 100 °C, Z. Anorg. Allg. Chem. 231 (1937) 285–297.

[139] P. Dell'Orco, H. Eaton, T. Reynold, S. Buelow, The solubility of 1:1 nitrate electrolytes in supercritical water, J. Supercrit. Fluids 8 (1995) 217–227.

[140] M.I. Ravich, F.E. Borovaya, Izvest. Fiziko-Khim. Anal. 20 (1950) 165–183.

[141] S. Sourirajan, G.C. Kennedy, The system H_2O–NaCl at elevated temperatures and pressures, Am. J. Sci. 260 (1962) 115–141.

[142] K.G. Kravchuk, K. Tödhelde, Z. Phys. Chem. 193 (1996) 139–150.

[143] W.D. Gunter, I.-M. Chou, S. Girsperger, Phase relations in the system NaCl-KCl-H_2O. Part II: differential thermal analysis of the halite liquids in the NaCl-H_2O binary above 450 °C, Geochim. Cosmochim. Acta 47 (1983) 863–873.

[144] M.A. Urusova, V.M. Valyashko, Zh. Neorg. Khim. 21 (1976) 2805–2810.

[145] W.L. Marshall, J.S. Gill, Aqueous systems at high temperature—XI: effect of pressure on liquid–liquid immiscibility in the system UO_2SO_4-H_2O and liquid-supercritical fluid equilibria in the system UO_2SO_4-H_2SO_4-H_2O; 290–430 °C, 75–330 bars, J. Inorg. Nucl. Chem. 25 (1963) 1033–1040.

[146] W.L. Marshall, J.S. Gill, Effect of pressure on liquid–liquid immiscibility of high temperature aqueous solution mixtures of uranyl sulfate and sulfuric acid, 280–450 °C, 75–1800 bars, J. Inorg. Nucl. Chem. 36 (1974) 2303–2312.

[147] J.V. Walter, Experimental determination and interpretation of the solubility of corundum in H2O between 350 and 600 °C from 0.5 to 2.2 kbar, Geochim. Cosmochim. Acta 61 (1997) 4955–4964.

[148] K.H. Becker, L. Cemic, K.E. Langer, Solubility of corundum in supercritical water, Geochim. Cosmochim. Acta 47 (1983) 1573–1578.

[149] B. Strübel, Neues Jahr. Mineral. Monat. 7/8 (1967) 223–333.

[150] C.W. Blount, Barite solubilities and thermodynamic quantities up to 300 °C and 1400 bars, Am. Mineral. 62 (1977) 942–957.

[151] S.D. Malinin, Trudy 6 Sovesch. po Eksper. i Tekhnich. Mineral. i Petrogr (Russ.), Izd. AN SSSR, Moscow, 1962103–107.

[152] H. Schloemer, Hydrothermale Untersuchungen über das System CaO–MgO–CO_2–H_2O, Neues Jahr. Mineral. Monat. (1952) 129–135.

[153] I.-M. Chou, H.P. Eugster, Solubility of magnetite in supercritical chloride solutions, Am. J. Sci. 277 (1977) 1296–1314.

[154] S.E. Ziemniak, M.E. Jones, K.E.S. Combs, Magnetite solubility and phase stability in alkaline media at elevated temperatures, J. Solution Chem. 24 (1995) 837–877.

[155] M.I. Ravich, F.E. Borovaya, Zh. Neorg. Khim. 13 (1968) 1418–1425.

[156] M.I. Ravich, F.E. Borovaya, Zh. Neorg. Khim. 9 (1964) 952–974.

[157] K. Sue, Y. Hakuta, R.L. Smith, T. Adschiri, K. Arai, Solubility of lead(II) oxide and copper(II) oxide in subcritical and supercritical water, J. Chem. Eng. Data 44 (1999) 1422–1426.

[158] G. Kennedy, G. Wasserburg, M. Heard, R.C. Newton, The upper three-phase region in the system SiO_2–H_2O, Am. J. Sci. 260 (1962) 500–521.

[159] M.A. Styrikovich, Steam solutions, Vestn. Akad. Nauk SSSR 12 (1969) 70–77.
[160] The International Association for the Properties of Water and Steam (IAPWS), Guide-Line on the Critical Locus of Aqueous Solutions of Sodium Chloride, Prague, 2000.
[161] A.A. Povodyrev, M.A. Anisimov, J.V. Sengers, W.L. Marshall, J.M.H. Levelt Sengers, Critical locus of aqueous solutions of sodium chloride, Int. J. Thermophys. 20 (1999) 1529–1545.
[162] I.-M. Chou, S.M. Sterner, K.S. Pitzer, Phase relations in the system NaCl-KCl-H_2O: IV. Differential thermal analysis of the sylvite liquidus in the KCl-H_2O binary, the liquidus in the NaCl-KCl-H_2O ternary, and the solidus in the NaCl-KCl binary to 2 kb pressure, and a summary of experimental data for thermodynamic-PTX analysis of solid-liquid equilibria at elevated P-T conditions, Geochim. Cosmochim. Acta 56 (1992) 2281–2293.
[163] R.O. Fournier, J.M. Thompson, Composition of steam in the system NaCl-KCl-H_2O-quartz at 600 °C, Geochim. Cosmochim. Acta 57 (1993) 4365–4375.
[164] S.M. Sterner, D.L. Hall, R.J. Bodnar, Synthetic fluid inclusions. V. Solubility relations in the system NaCl-KCl-H_2O under vapor-saturated conditions, Geochim. Cosmochim. Acta 52 (1988) 989–1005.
[165] M.I. Ravich, F.E. Borovaya, Fazovyye ravnovesiya v troynykh, Izvest. Fiziko-Khim. Anal. 19 (1949) 69–81 (translated in Fluid Inclusions).
[166] P. Dingemans, Recl. Trav. Chim. Pays Bas 58 (1939) 559–581.
[167] P. Dingemans, Recl. Trav. Chim. Pays Bas 64 (1945) 194–204.
[168] M.I. Ravich, F.E. Borovaya, E.I. Luk'yanova, V.M. Elenevskaya, Izvest. Fiziko-Khim. Anal. 24 (1954) 280–298.
[169] L.S. Itkina, Lithium, Rubidium and Cesium Hydroxides (Russ.), Nauka, Moscow, 1973.
[170] V.M. Valyashko, Phase Equilibria and Properties of Hydrothermal Systems (Russ.), Nauka, Moscow, 1990.
[171] V.M. Valyashko, Phase behavior in binary and ternary water–salt systems at high temperatures and pressures, Pure Appl. Chem. 62 (1990) 2129–2138.
[172] A.L. Boettcher, P.J. Wyllie, The system CaO-SiO_2-CO_2-H_2O: III. A second critical end-point on the melting curve at 32.5 kbars and 515 °C, Geochim. Cosmochim. Acta 33 (1969) 611–632.
[173] D.B. Stewart, Schweiz. Mineral. Petrogr. Mitt. 47 (1967) 35–61.
[174] V.M. Valyashko, K.G. Kravchuk, Dokl. Akad. Nauk SSSR 242 (1978) 1104–1107.
[175] M.I. Ravich, V.M. Valyashko, Zh. Neorg. Khim. 14 (1969) 1650–1654.
[176] A. Firus, G. Brunner, Decontamination of soil and oxidation of the extracts by/in supercritical water: the underlying phase equilibria of binary and multicomponent systems, in: M. Perrut, G. Brunner (Eds.), Proceedings 3rd International Symposium on Supercritical Fluids, Strasbourg, vol. 3, 1994, pp. 195–200, ISBN: 2-905267-23-8.
[177] M. Gehrig, Phasengleichgewichte und pVT-Daten ternärer Mischungen aus Wasser, Kohlendioxid und Natriumchlorid bis 3 kbar und 550 °C, Dissertation, University of Karlsruhe, Karlsruhe, Germany, 1980.
[178] Ch. Schmidt, Experimental Study of the PVTX Properties in Part of the Ternary System H_2O-NaCl-CO_2, Virginia Polytechnic Institute and State University, Blacksburg, VA, 1997.
[179] T. Michelberger, Phasengleichgewichte fluider, ternärer, salzhaltiger Systeme von 300 K bis 800 K und bis zu hohen Drücken, Ph.D. Dissertation, University Karlsruhe, Karlsruhe, 1989.

[180] T. Krader, E.U. Franck, Phase equilibria and critical phenomena of the ternary system water–methane–sodium chloride to 500 °C and 2500 bar, Physica, 139–140 (1986) 66–69.

[181] D.A. Palmer, R. Fernández-Prini, A.H. Harvey (Eds.), Aqueous Systems at Elevated Temperatures and Pressures: Physical Chemistry in Water, Steam and Hydrothermal Solutions, Elsevier, Amsterdam, 2004.

Chapter 4

Heat Transfer

4.1 INTRODUCTION

Water has an outstanding role as working fluid for heat transfer processes due to its application in electric power generation. Near-critical and supercritical conditions have been applied for some time in conventional power generation processes. Recently, the supercritical region has become an area of increased research and intended application for nuclear reactor power generation processes. In other processes involving water at hydrothermal and supercritical conditions, water is the prevailing component and, due to the elevated temperature level, recovery of energy from process fluids becomes adamant for economic reasons. In this context, heat transfer to pure water can often be taken as first approximation. Yet, in many cases, mixtures of process compounds with water must be considered, often connecting phase behavior to heat transfer processes.

The following chapter presents a short overview on heat transfer to pure water in various important regions like liquid, gas, near-critical, boiling and condensation, and supercritical conditions. Emphasis is on heat transfer as it occurs in hydrothermal and supercritical water processes. This means convective heat transfer in tubes, mostly at turbulent conditions and heat transfer during flow through beds of solids. The influence of other compounds in an aqueous mixture on heat transfer is discussed, in particular, heat transfer and phase behavior for aqueous systems containing water, a hydrocarbon compound, and a gas.

4.2 HEAT TRANSFER TO PURE WATER

4.2.1 Overview on Heat Transfer Regions

The discussion of heat transfer is divided into four parts: gas and part of the compressible liquid (Region I), liquid (Region II), the two-phase region (Region III), and the near-critical region (Region IV) [1]. The regions are illustrated in Figure 4.1.

For regions I–III, methods for calculating heat transfer can be found in numerous textbooks. Only equations for the most important cases will be presented in this book. The near-critical region (Region IV) will be covered in

Supercritical Fluid Science and Technology, Vol. 5. http://dx.doi.org/10.1016/B978-0-444-59413-6.00004-2

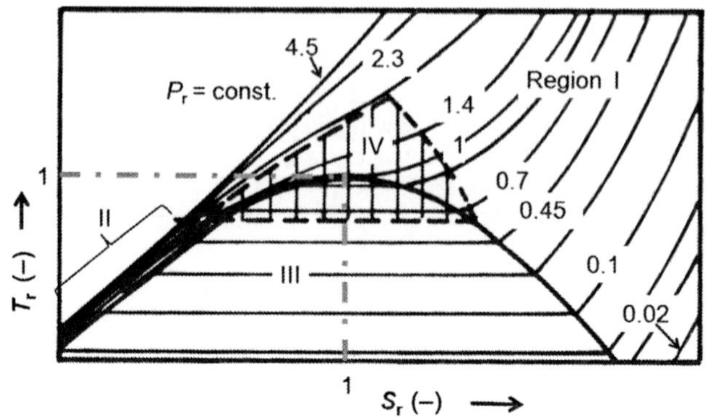

FIGURE 4.1 Regions for heat transfer in a pure fluid. Region I: Gas and a small part of the liquid region. Region II: Noncompressible liquid. Region III: Two-phase region. Region IV: Near-critical region. *Modified from Ref. [1]*. (For color version of this figure, the reader is referred to the online version of this chapter.)

more detail. Some characteristics of this region are: Regions I–III lie adjacent to the critical region. It is not possible to set precise boundaries for the near-critical heat transfer since the transitions are not abrupt. The extent of the influence of the critical point on heat transfer must be evaluated (compare Chapter 2.4.3, last section, for a definition of the critical region). Furthermore, the approach of the fluid toward the critical point is of influence and must also be considered. Heat transfer in the regions adjacent to the critical region can be correlated by conventional methods.

4.2.2 Heat Transfer to Gas and Liquid (Regions I and II)

4.2.2.1 Convective Heat Transfer in Tubes, Laminar Flow

Flow in tubes is laminar below a Reynolds number of $Re = 2300$ ($Re = u\,d_i/v$, u is the flow velocity, d_i is the inner diameter, and v is the kinematic viscosity). Above this value, the flow is considered to be turbulent, but full turbulent flow is prevalent only at $Re > 10^4$. In the transition region, flow patterns influence the regime.

The heat transfer coefficient α for a certain length of the tube is defined as

$$q = \alpha \Delta T_{\ln}, \tag{4.1}$$

with q as the heat flux, α as the heat transfer coefficient, and $\Delta T_{\ln}$ as the logarithmic temperature difference, defined by Equation (4.2).

$$\Delta T_{\ln} = \frac{(T_W - T_E) - (T_W - T_A)}{\ln(T_W - T_E/T_W - T_A)}, \tag{4.2}$$

with T_E is the temperature at the inlet, T_A is the temperature at the outlet of the pipe of the flowing medium, and T_W is the wall temperature.

The heat transfer coefficient for homogeneous mixtures is also defined in Equation (4.1), where the conditions of state and the properties of the mixture have to be inserted.

The following equations on convective heat transfer in tubes are recommended in Ref. [2]:

For a laminar flowing medium at *constant wall temperature* of the tube at fully developed laminar flow, the local heat transfer, expressed as Nusselt number, ($Nu = \alpha d_i/\lambda$, λ is the thermal conductivity) at the length x is:

$$Nu_{x,T_W} = 3.66, \tag{4.3}$$

for small values of $Re\,Pr(d_i/x)$ with $Pr = \nu/a$, ν is the kinematic viscosity, a is the temperature conductivity, and T_W is the wall temperature, and

$$Nu_{x,T_W} = 1.077\left(Re\,Pr\frac{d_i}{x}\right)^{1/3} \tag{4.4}$$

for large values of $Re\,Pr(d_i/x)$.

For the average heat transfer up to the length l of the tube of diameter d_i, the following equations are proposed:

$$Nu_{m,T_W} = 3.66, \tag{4.5}$$

for small values of $Re\,Pr(d_i/x)$, and

$$Nu_{m,T_W} = 1.615\left(Re\,Pr\frac{d_i}{x}\right)^{1/3} \tag{4.6}$$

for large values of $Re\,Pr(d_i/x)$.

For the cases of developing thermal and hydrodynamic flow, the reader is referred to textbooks.

For a laminar flowing medium with *constant specific heat flux q* and fully developed thermal and hydrodynamic laminar flow, the local heat transfer at the length x can be determined from

$$Nu_{x,q} = 4.364, \tag{4.7}$$

for small values of $Re\,Pr(d_i/x)$, and

$$Nu_{x,q} = 1.302\left(Re\,Pr\frac{d_i}{x}\right)^{1/3}, \tag{4.8}$$

for large values of $Re\,Pr(d_i/x)$.

For the average heat transfer up to the length l of the tube, the following equations are valid:

$$Nu_{m,q} = 4.364, \tag{4.9}$$

for small values of $RePr(d_i/x)$, and

$$Nu_{m,q} = 1.953\left(RePr\frac{d_i}{x}\right)^{1/3} \tag{4.10}$$

for large values of $RePr(d_i/x)$.

Heat transfer is different for cooling or heating if properties of the flowing medium are dependent on temperature. For gases, the heat transfer coefficient varies less than 10% in the temperature range of $0.5 < T/T_W < 2.0$. For liquids, the influence is more pronounced and can be taken into account by

$$Nu = Nu_{m,T}\left(\frac{Pr}{Pr_w}\right)^{0.11}, \tag{4.11}$$

4.2.2.2 Convective Heat Transfer in Tubes, Turbulent Flow

For fully developed turbulent flow, the equation given by Gnielinski [3] is recommended:

$$Nu_{m,T} = \frac{(\xi/8)RePr}{1 + 12.7\sqrt{\xi/8}\,(Pr^{2/3} - 1)}\left[1 + \left(\frac{d_i}{l}\right)^{2/3}\right] \tag{4.12}$$

with

$$\xi = (1.8\log Re - 1.5)^{-2} \tag{4.13}$$

Equation (4.12) is applicable for $10^4 \le Re \le 10^6$, $0.6 \le Pr \le 1000, d_i/l \le 1$.

In the transition range, an interpolating equation between laminar and turbulent flow can be applied

$$Nu_m = (1-\gamma)Nu_{m,2300} + \gamma Nu_{m,T,10^4} \tag{4.14}$$

with

$$\gamma = \frac{Re - 2300}{10^4 - 2300} \text{ and } 0 \le \gamma \le 1 \tag{4.15}$$

$Nu_{m,2300}$ is the Nusselt number resulting for the given conditions (constant wall temperature or constant heat flux) at $Re = 2300$ from Equation (4.6) or (4.10). $Nu_{m,T,10^4}$ is the Nusselt number from Equation (4.12) for $Re = 10^4$. Properties of the fluid have to be taken at the average temperature between inlet and outlet of the tube. The influence of temperature-dependent properties can be taken into account by Equation (4.11), in the range $0.1 < Pr/Pr_W < 10$.

Beside Equation (4.12), the well-known Dittus–Boelter equation [4] is widely used:

$$Nu_b = 0.023Re_b^{0.8}Pr_b^{0.4} \tag{4.16}$$

In the near-critical region (Region IV, Figure 4.1), standard techniques of correlating data break down. The ordinary Dittus–Boelter equation (Equation 4.16) cannot be used to correlate near-critical forced convection phenomena. Correlations for free convection are not consistent with pool boiling data, standard boiling equations exhibit discontinuities, and oscillations are commonplace [1] (see Section 4.3).

4.2.3 Boiling of Pure Water

Thermodynamic equilibrium of gas–liquid mixtures is assumed for boiling in the two-phase region at stationary flow in tubes. The boiling temperature T corresponds to the pressure P of the mixture. The mixture is characterized by its gaseous part y, determined by the total enthalpy $h=(h_m-h_\mathrm{l})/\Delta h_\mathrm{v}$, with h_m as the total (specific) enthalpy, h_l as the enthalpy of the liquid at saturation, Δh_v as the enthalpy of vaporization.

Boiling can occur in the flowing mixture in two different regimes: convective boiling and bubble boiling. During convective boiling, the heat transfer coefficient α does not depend on the heat flux q or the temperature difference between bulk (T_s) and wall (T_w): $\Delta T=T_\mathrm{w}-T_\mathrm{s}$. In the bubble-boiling regime, the heat transfer coefficient depends on the heat flux. This can be used for discerning the two regimes. Temperature, heat transfer coefficient, and heat flux are averaged values for the circumference of the tube. Values for the averaged heat transfer coefficient depend on the flow regime in the tube. Therefore, the flow regime must be considered for the calculation of heat transfer coefficients. The various possible flow regimes for horizontal and vertical tubes are illustrated in Figure 4.2 [5].

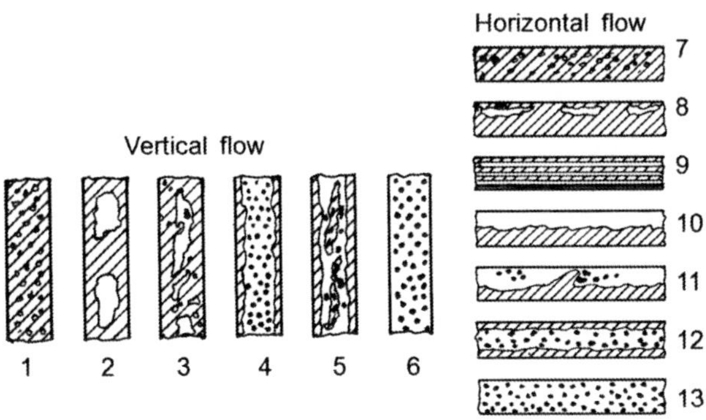

FIGURE 4.2 Flow regimes in vertical tubes (left) and horizontal tubes (right). 1–3: bubbles, 4–5: ring flow and drops, 6: drops only; 7, 8: bubbles, 9–11: layered flow, 12, 13: ring flow and drops. *Modified from Ref. [5].*

4.2.3.1 *Forced Convective Boiling of Pure Fluids*

Convective boiling or still boiling occurs in case of wall temperatures not high enough to initiate bubbles. During convective boiling, the liquid evaporates at the surface of the liquid phase.

4.2.3.1.1 Forced Convective Boiling in Vertical Tubes

During convective boiling, the heated wall of the tube is wetted for all flow regimes. The heat transfer coefficient $\alpha_{z,c}$ can be calculated with Equation (4.17) [5]:

$$\frac{\alpha_{z,c}}{\alpha_{l,0}} = \left\{ (1-h)^{0.01} \left[(1-h)^{1.5} + 1.9h^{0.6} \left(\frac{\rho'}{\rho''} \right)^{0.35} \right]^{-2.2} + h^{0.01} \left[\frac{\alpha_{g,0}}{\alpha_{l,0}} \left(1 + 8(1-h)^{0.7} \right) \left(\frac{\rho'}{\rho''} \right)^{0.67} \right]^{-2} \right\}^{-0.5} \tag{4.17}$$

$\alpha_{g,0}$ and $\alpha_{l,0}$ are the local heat transfer coefficients for the gaseous and the liquid fluid, respectively, calculated from the equations for convective heat transfer, considering hydrodynamic and thermal conditions. Properties of the fluid are those at saturation.

Equation (4.17) is valid in the range $3.5 \le \rho'/\rho'' \le 5000, 10^{-3} \le P/P_c \le 0.8$, with P_c as the critical pressure.

4.2.3.1.2 Forced Convective Boiling in Horizontal Tubes

Due to gravity, the liquid can flow in a layer at the bottom of the tube, especially for flow with ripples and layers. Wall temperatures may be higher at the top than at the bottom of the tube, in particular, for constant heat flow. The heat transfer coefficient, averaged over the circumference of the tube, is therefore dependent on the degree of wetting of the wall and on the thermal conductivity of the wall. The local heat transfer coefficient can be calculated from Equation (4.18) [5]:

$$\frac{\alpha_{z,c}}{\alpha_{l,0}} = \left\{ (1-h)^{0.01} \left[(1-h)^{1.5} + 1.2h^{0.4} \left(\frac{\rho'}{\rho''} \right)^{0.37} \right]^{-2.2} + h^{0.01} \left[\frac{\alpha_{g,0}}{\alpha_{l,0}} \left(1 + 8(1-h)^{0.7} \right) \left(\frac{\rho'}{\rho''} \right)^{0.67} \right]^{-2} \right\}^{-0.5} \tag{4.18}$$

$\alpha_{g,0}$ and $\alpha_{l,0}$ are the local heat transfer coefficients for the gaseous and the liquid fluid, respectively, calculated from the equations for convective heat transfer, considering hydrodynamic and thermal conditions. Properties of the fluid are those at saturation.

Equation (4.18) is valid for $3.5 \leq \rho'/\rho'' \leq 1500, 5 \times 10^{-3} \leq P/P_c \leq 0.8$.

4.2.3.2 *Bubble Boiling*

4.2.3.2.1 Bubble Boiling in Vertical Tubes

As long as critical boiling conditions are not approached, the wall is always wetted. The heat transfer coefficient can be calculated from Equation (4.19) [5]:

$$\frac{\alpha_{z,b}}{\alpha_0} = C_l \left(\frac{q}{q_0}\right)^n F(P/P_c) F(d) F(W) F(\dot{m}, y) \tag{4.19}$$

C_l represents the properties of the fluid, $F(W)$ represents the properties of the heat transferring wall, $F(d)$ represents tube dimensions, $F(P/P_c)$ represents the influence of the boiling pressure, $F(\dot{m},y)$ represents the influence of the flow regime, and $(q/q_0)^n$ represents the influence of heat flow. The reference value α_0 is the heat transfer coefficient of pool boiling.

$$n = 0.8 - 0.1 \times 10^{(0.76P/P_c)} \text{ for water and inorganic fluids.} \tag{4.20}$$

$$F(P/P_c) = 2.816(P/P_c)^{0.45} + \left(3.4 + \frac{1.7}{1 - (P/P_c)^7}\right)(P/P_c)^{3.7} \tag{4.21}$$

$$F(\dot{m}, y) = 1 \text{ for bubble boiling,} \tag{4.22}$$

$$F(d) = (d_0/d)^{0.4} \text{ with } d_0 = 10^{-2} \text{ m.} \tag{4.23}$$

$$F(W) = (R_a/R_{a0})^{0.133} \tag{4.24}$$

Considering all the influences, the heat transfer coefficient for bubble boiling in vertical tubes can be calculated from

$$\frac{\alpha_{z,b}}{\alpha_0} = C_l \left(\frac{q}{q_0}\right)^n \left[2.816(P/P_c)^{0.45} + \left(3.4 + \frac{1.7}{1 - (P/P_c)^7}\right)(P/P_c)^{3.7}\right] \left(\frac{d_0}{d}\right)^{0.4} \left(\frac{R_a}{R_{a0}}\right)^{0.133} \tag{4.25}$$

Values with subscript 0 are reference values: $d_0 = 1 \times 10^{-2}$ m, $R_{a0} = 1 \times 10^{-6}$ m. α_0 is the reference value for $P/P_c = 0.1$ and $R_{a0} = 1 \times 10^{-6}$ m. For water, $C_l = 0.72$, $\alpha_0 = 6400\ \text{W/(m}^2\text{K)}$.

Equation (4.25) can be applied for $0.001 \leq P/P_c \leq 0.985$, $1 \leq d(\text{mm}) \leq 32$, $0.05 \leq R_a(\mu\text{m}) \leq 5$.

For a medium flowing from upside down, the heat transfer coefficient is smaller

$$\alpha_{z,dw} = 0.75\alpha_{z,b}. \tag{4.26}$$

Alternatively, the following equation can be applied, if data are missing [6]:

$$Nu = 0.1\left[\frac{q_0 d_0}{\lambda_l T_s}\right]^{0.674}\left[\frac{\rho_g}{\rho_l}\right]^{0.156}\left[\frac{\Delta h_v d_0^2}{a_l^2}\right]^{0.371}\left[\frac{a_l^2 \rho_l}{\rho d_0}\right]^{0.350}\left[\frac{\eta_l c_{P,l}}{\lambda_l}\right]^{-0.16} \tag{4.27}$$

with subscript l, boiling liquid; g, saturated gas; λ, heat conductivity; ρ, density; a, temperature conductivity; η, dynamic viscosity; Δh_v, specific enthalpy of evaporation; d_0 diameter of bubbles departing from the heat transferring surface:

$$d_0 = 0.0149\beta\left(\frac{2\sigma}{g\left(\rho_l - \rho_g\right)}\right)^{0.5} \tag{4.28}$$

with β, wetting angle; for water 35°.

4.2.3.2.2 Bubble Boiling in Horizontal Tubes

Critical boiling is reached much earlier in horizontal tubes than in vertical tubes. The average heat transfer coefficient is dependent on mass flow $\dot{m}$ and gaseous fraction y [5]:

$$F(\dot{m}, y) = \left(\frac{\dot{m}}{\dot{m}_0}\right)^{0.25}\left[1 - \left(\frac{P}{P_c}\right)^{0.1}\left(\frac{q}{q_{c,b}}\right)^{0.3} y\right] \tag{4.29}$$

with

$$q_{c,b} = 2.79 q_{c,0.1}\left(\frac{P}{P_c}\right)^{0.4}\left(1 - \left(\frac{P}{P_c}\right)\right) \tag{4.30}$$

$$q_{c,0.1} = 0.13\Delta h_{v,0} \rho''^{0.5}_0 \left(\sigma_0 g\left(\rho'_0 - \rho''_0\right)\right)^{0.25}, \tag{4.31}$$

$\dot{m}_0 = 100$ kg/(m^2 s).

Properties are to be determined at $P/P_c = 0.1$.

4.2.3.3 Heat Transfer Crisis

Boiling liquids can transfer large amounts of heat at small temperature differences between wall and bulk liquid. The maximum heat flux is limited by changes in the heat transfer regimes. The so-called crisis in heat transfer through boiling is characterized in case of a fixed heat flux (e.g., by electrical heating) by an abrupt increase of the wall temperature if the heat flux is further increased, and in case of constant wall temperature, by an abrupt decrease of the heat flux, if the wall temperature is increased further [7].

The deterioration of the heat transfer is due to the fact that heat transfer is no longer from wall to liquid but from wall to gas, either by the formation of a gaseous layer between wall and bulk liquid (departure from nucleate boiling) or by drying out of the wall (dry-out).

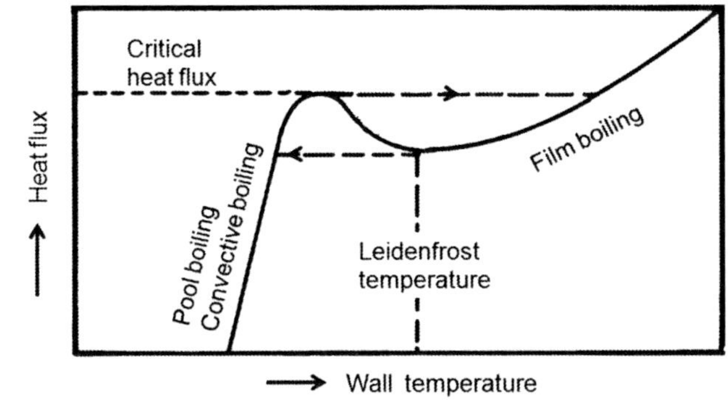

FIGURE 4.3 Heat transfer regimes for two-phase boiling. *Modified from Ref. [5].*

The Leidenfrost temperature marks the wall temperature limiting the unstable boiling range. At higher temperatures, the formation of a gas layer is stable and film boiling occurs with heat transfer through a stable gas layer. At lower temperatures, the gas layer is unstable. Figure 4.3 shows a schematic drawing of the heat transfer regimes for two-phase boiling.

In the following, the case of water flowing in vertical tubes is considered. It is of importance to be able to decide whether the heat transfer crisis is caused by departure from nucleate boiling or by dry-out. For both cases, equations have been proposed, based on extensive experimental data [5].

4.2.3.3.1 Departure from Nucleate Boiling

For the critical heat flow, Equation (4.32) can be applied [7]:

$$q_c = 10^3\left[10.3 - 17.5\frac{P}{P_c} + 8\left(\frac{P}{P_c}\right)^2\right]\left(\frac{8\times10^{-3}}{d}\right)^{0.5} \left(\frac{\dot{m}}{1000}\right)^{0.68(P/P_c)-1.2y-0.3} \cdot e^{-1.5y} \tag{4.32}$$

in the range 29 bar $\leq P \leq$ 196 bar, 500 kg/(m^{2}s) $\leq \dot{m} \leq$ kg/(m^{2}s), 0 $K \leq \Delta T \leq$ 75K, 4×10^{-3}m $\leq d \leq 25\times10^{-3}$m.

For the critical gaseous part

$$y_c = \frac{\ln\left(\frac{\dot{m}}{1000}\right)\left(0.68\frac{P}{P_c} - 0.3\right) - \ln q_c + \ln C}{1.2\ \ln\frac{\dot{m}}{1000} + 1.5} \tag{4.33}$$

with

$$C = 10^3\left[10.3 - 17.5\left(\frac{P}{P_c}\right) + 8\left(\frac{P}{P_c}\right)^2\right]\left(\frac{8\times10^{-3}}{d}\right)^{0.5} \tag{4.34}$$

For dry-out, the following equations are recommended [7].

$$q_c = 1.8447 \times 10^8 y^{-8} \dot{m}^{-2.664} (d \times 1000)^{-0.56} e^{0.1372P} \quad \text{for } 4.9 \leq P(\text{bar}) \leq 29.4, \tag{4.35}$$

$$q_c = 2.0048 \times 10^{10} y^{-8} \dot{m}^{-2.664} (d \times 1000)^{-0.56} e^{-0.0204P} \quad \text{for } 29.4 \leq P(\text{bar}) \leq 98, \tag{4.36}$$

$$q_c = 1.1853 \times 10^{12} y^{-8} \dot{m}^{-2.664} (d \times 1000)^{-0.56} e^{-0.0636P} \quad \text{for } 98 \leq P(\text{bar}) \leq 196. \tag{4.37}$$

Range of validity: $200 \leq \dot{m}\,(\text{kg/m}^2) \leq 5000$, $4 \times 10^{-3} \leq d(\text{m}) \leq 32 \times 10^{-3}$.
For the critical vapor content,

$$y_c = 10.795 q^{-0.125} \dot{m}^{-0.333} (d \times 1000)^{-0.07} e^{-0.01715P} \quad \text{for } 4.9 \leq P(\text{bar}) \leq 29.4, \tag{4.38}$$

$$y_c = 19.398 q^{-0.125} \dot{m}^{-0.333} (d \times 1000)^{-0.07} e^{-0.00255P} \quad \text{for } 29.4 \leq P(\text{bar}) \leq 98, \tag{4.39}$$

$$y_c = 32.302 q^{-0.125} \dot{m}^{-0.333} (d \times 1000)^{-0.07} e^{-0.00795P} \quad \text{for } 98 \leq P(\text{bar}) \leq 196. \tag{4.40}$$

For the validity of these equations, a stable hydrodynamic state must exist. Pulsating flow can bring along a much earlier beginning of the heat transfer crisis. To establish the limit for the heat transfer crisis, both equations have to be solved. The critical value for the heat transfer crisis is the lower value of both solutions.

4.2.4 Boiling of Mixtures

During boiling of pure fluids in tubes, the boiling temperature decreases due to pressure drop. In mixtures, the boiling temperature can rise due to preferred evaporation of the lower boiling components.

In mixtures, the two boiling regimes (pool boiling and nucleate boiling) exist also. For convective heat transfer, the same equations as for pure fluids can be applied, if the properties of the mixture are used for the calculation. Mixture properties are then determined according to well-known procedures from Chemical Engineering Thermodynamics, as well as the composition of the mixture. Special attention has to be paid for high viscous mixtures due to an additional heat transfer resistance in the liquid phase. For more information, the reader is referred to the literature, for example [5].

4.3 HEAT TRANSFER IN THE CRITICAL AND SUPERCRITICAL REGION (REGION IV)

4.3.1 Near-Critical Heat Transfer

Early experiments of Schmidt *et al.* [8] found free and natural convection in the vicinity of the critical point to exhibit a sharp increase in heat transfer coefficient. Other researchers found a decrease in heat transfer coefficients. Further experimental investigation brought results that explained this behavior: Shitsman [9] and Yamagata *et al.* [10] showed that heat transfer coefficients exhibit a minimum near the critical enthalpy. They also reported pressure oscillations in the same region. A maximum in heat transfer coefficient was found in experiments, in which the temperature difference between the wall and the bulk fluid was small. Hsu [11] suggested an explanation: For small temperature differences, the situation is similar to nucleate boiling with very good heat transfer. For large temperature differences, the situation is similar to film boiling with poor heat transfer, thus the minimum [1].

It is now well established that during heat transfer in the near-critical and supercritical region, heat transfer coefficients can be enhanced or reduced. Enhancement of heat transfer is used in several applications for chemical processing and conversion of energy. Inadequate layout of the boiler tubes can lead to deteriorated heat transfer (DHT) with unallowable high temperatures of the tube wall [10,12].

4.3.1.1 *Free and Natural Convection*

For free and natural convection, away from the critical point, McAdams' basic equation [13] can be applied

$$Nu_{\mathrm{f}} = C(Gr\,Pr)^{n} \tag{4.41}$$

with $C=0.53$, $n=¼$ for $10^3<Gr\,Pr<10^9$.

Closer to the critical point, modifications of the basic form are employed, for example, as proposed by Larson *et al.* [14]:

$$Nu_x = C Gr_{X=0.5}^{1/3} Pr_{X=0.5}^{0.247} \left(\frac{T_\infty}{T_{\mathrm{w}} - T_\infty}\right)^{0.137}, \tag{4.42}$$

in the range $1.01<P/P_{\mathrm{c}}<1.03, 0.98<T/T_{\mathrm{c}}<1.04$, with X as dimensional length.

Heat transfer coefficients show an enhancement in free convection pools and natural convection loops near the critical point. The enhancement occurs over a limited temperature range, and peaks are somewhat away from the critical point. Boiling-like behavior in near-critical natural convection heat transfer is not accompanied by the marked increase in the heat-transfer coefficient associated with nucleate boiling. Data of both the pools and natural convection loops have been successfully correlated by variable property modifications to the standard equations for the given configuration.

Although pool experiments have produced support for a pseudo-two-phase modeling of a near-critical fluid, single-phase correlations may be successful, for example, Equations (4.43) and (4.44) [15]. Oscillations are observed in natural and forced convection systems [1].

$$Nu = 16Re^2 Pr Gr^{-1}\left(\frac{l_T}{L_T}\right)\left(\frac{d}{\xi}\right)\text{(laminar)} \tag{4.43}$$

$$Nu = 0.079Re^{11/4} Pr Gr^{-1}\left(\frac{l_T}{L_T}\right)\left(\frac{d}{\xi}\right)\text{(turbulent)} \tag{4.44}$$

ξ is the dimensionless distance δ/b_H, L_T is one half of total loop length, l_T is the test section length, δ is the boundary layer thickness, b_H is the channel half width.

4.3.1.2 Forced Convection

For forced convection, modifications to the Dittus–Boelter equation have been proposed. An equation directly designed for use near the critical point is the one by Miropolskii and Shitsman [16]:

$$Nu_b = 0.023Re_b^{0.8} Pr_{min}^{0.4} \tag{4.45}$$

with

$$Pr_{min} = \begin{cases} Pr_b & \text{for } Pr_b < Pr_w \\ Pr_w & \text{for } Pr_w < Pr_b \end{cases} \tag{4.46}$$

This modification of the Dittus–Boelter equation represents data with remarkable results. In Figure 4.4, the heat transfer coefficient is shown for

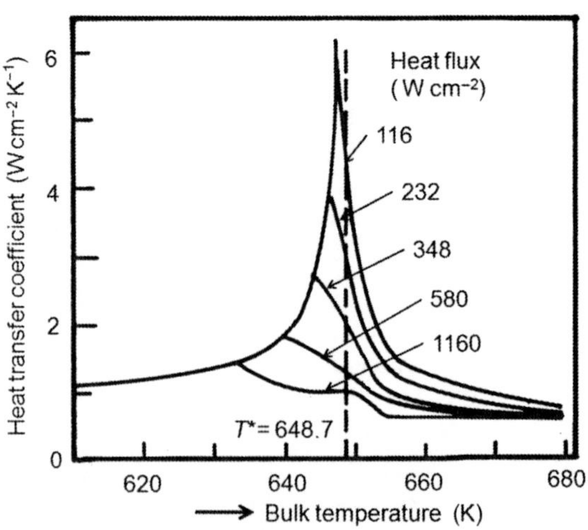

FIGURE 4.4 Calculated heat coefficient of water in the near-critical region according to Equation (4.45) [16]. $P/P_c = 1.052$, $\dot{m} = 100$ g/(cm^2 s), $d = 30$ mm. *Modified from Ref. [1].*

near-critical pressure and near-critical temperature. A sharp peak appears near the temperature $T^* = 648.7$ K. The peak decreases with increased heat flow until it disappears at about an order of magnitude higher than heat flow. There is evidence of a strong similarity to boiling, and there seems to be little discontinuity in going from high-pressure boiling to the supercritical region [1].

In bended tubes, an increase of about 2:1 in the heat transfer coefficient on the concave side can be expected as a result of curvature for gases and fluid systems at high T_w/T_b (T_w, wall temperature; T_b, bulk temperature). Devices which drive the high-density core to the wall or break up the low-density boundary layer, such as twisted tapes and spiral grooves, enhance near-critical heat transfer significantly [1].

4.3.1.2.1 Oscillations

Oscillations in near-critical flow occur due to the large density changes of a near-critical fluid. The critical frequency for the thermal-acoustic and system oscillations can be predicted by utilizing concepts of mechanics. Onset of thermal-acoustic and system oscillations may be predicted with a fair degree of reliability by conventional stability criteria [1].

4.3.2 Pseudocritical Properties

The unusual property behavior in the near-critical region leads to a definition of a so-called pseudocritical point. The pseudocritical temperature is defined as the temperature where the specific heat c_P reaches a maximum for a given supercritical pressure. Table 4.1 lists parameters for the pseudocritical point [17].

In order to use pseudoboiling models, certain two-phase quantities in addition to the vapor pressure curve must be defined. Typical for this situation is the latent heat of vaporization. Figure 4.5 shows a way of defining a pseudo-heat of vaporization from the enthalpy curves. Other pseudo-two-phase quantities can be approximated in the same way [1].

4.3.3 Heat Transfer to Supercritical Water

In this section, heat transfer to supercritical water is treated. Supercritical in this context is taken literally, that is, temperature and pressure are above the critical values. Nevertheless, phenomena occurring in the near-critical region below the critical parameters will be addressed also. Although the heat transfer problems can be loosely classified into the four regimes, as illustrated in Figure 4.1, it is not sufficient just to know what regime is of interest. The process path within these regimes must also be known [17]. The following statements are mainly derived from Pioro and Duffey [17].

TABLE 4.1 Pseudocritical Points of Water [17]

Pressure (MPa)	Pseudocritical Temperature (K)	Peak Value of Specific Heat (kJ/(kg K))
23	650.65	284.3
24	654.65	121.9
25	658.05	76.4
26	661.65	55.7
27	665.15	43.9
28	668.55	36.3
29	671.85	30.9
30	675.05	27.0
31	678.15	24.1
32	681.25	21.7
33	684.15	19.9
34	687.05	18.4
35	689.85	17.2

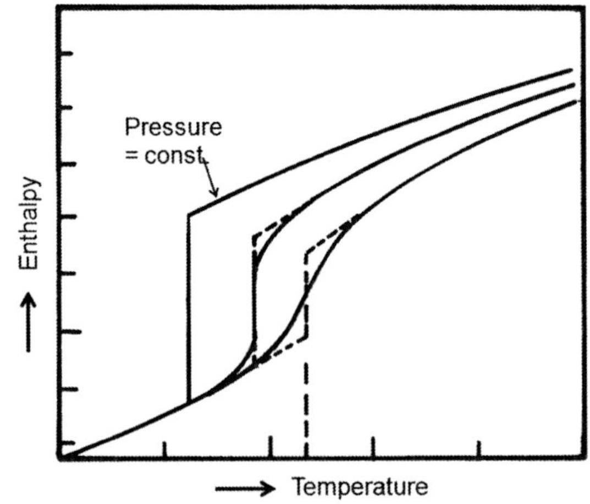

FIGURE 4.5 Pseudo-two-phase properties (schematic). Extrapolation procedure for equivalent latent enthalpy of vaporization. *Modified from Ref. [1].*

4.3.3.1 Experimental Heat Transfer to Water at Supercritical Pressures

4.3.3.1.1 Heat Transfer in Vertical Circular Tubes and Coils

In Figure 4.6, profiles of temperature and heat transfer coefficients are shown along the heated length of a vertical tube [18] for two different values of heat flux at a supercritical pressure of 24 MPa and near-critical, but supercritical temperatures. At the lower heat flux, DHT is followed downstream in the tube by a (small) region of improved heat transfer, with the consequence of a lower wall temperature (Figure 4.6, left). At higher heat flux, DHT is less pronounced and not followed by a marked region of improved heat transfer (Figure 4.6, right). Heat transfer coefficient calculated by the Dittus–Boelter correlation reveals the fundamental deviation of this equation in the critical region.

Pioro and Duffey [17] present an overview on the experimental results of investigations on heat transfer to water in vertical tubes at supercritical pressures. Their results are summarized in the following:

At supercritical pressure, the heat transfer coefficient for temperatures within the so-called pseudocritical region has a peak that decreases with pressure, compare Figure 4.7.

Increased heat transfer near the so-called pseudocritical point leads to a maximum value of the heat transfer coefficient [10]. The maximum decreases with increasing heat flux and pressure, see Figure 4.8. These results are confirmed by Pis'mennyy *et al.* 2005, in a detailed investigation [21], cited from [17].

Other investigations on up-flow in vertical tubes found DHT [22,23]. One type of DHT occurred in the entrance region ($L/D \leq 40$–60) and is due to the

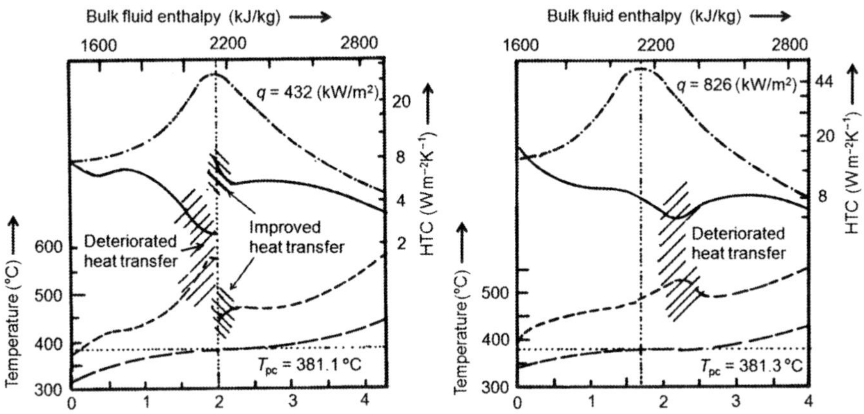

FIGURE 4.6 Temperature and heat transfer profiles along the heated length of a vertical tube, $d = 10$ mm. —•— Dittus–Boelter correlation, ——— Heat transfer coefficient, – – – inside wall temperature, — — — bulk fluid temperature. Left: $P_{in} = 24.0$ MPa, $\dot{m} = 503$ kg/(m^2 s). Right: $P_{in} = 24.0$ MPa, $\dot{m} = 1000$ kg/(m^2 s). *[19] cited from [18]. Modified from Ref. [17].*

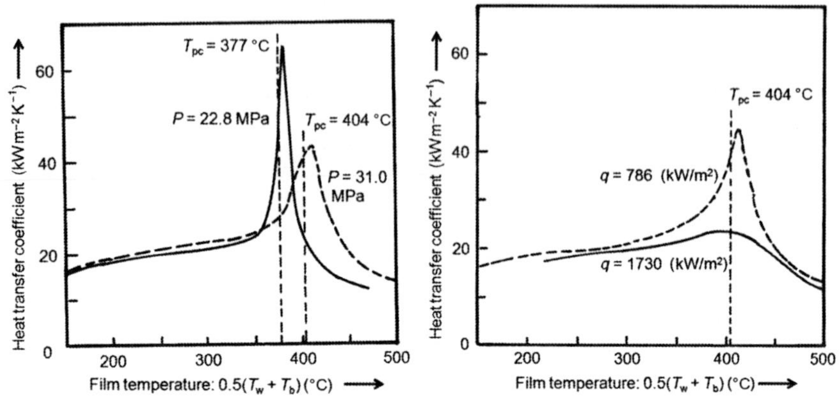

FIGURE 4.7 Left: Effect of pressure on heat transfer coefficient, ——— $P=22.8$ MPa, — — — $P=31.0$ MPa. Right: Effect of heat flux on heat transfer coefficient, — — — $q=786\ \mathrm{kW/m^2}$, ——— $q=1730\ \mathrm{kW/m^2}$. *[20], Modified from Ref. [17].*

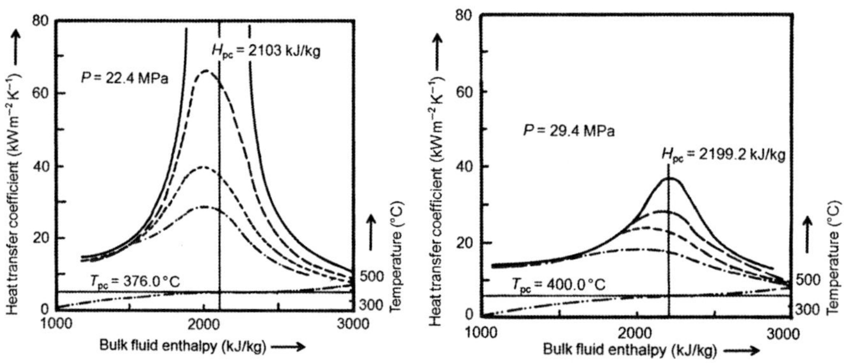

FIGURE 4.8 Heat transfer coefficient for water flowing upward in vertical tubes, —••— bulk fluid temperature, ——— $q=233\ \mathrm{kW/m^2}$, — — — $q=465\ \mathrm{kW/m^2}$, •••• $q=698\ \mathrm{kW/m^2}$, —•— $q=930\ \mathrm{kW/m^2}$, $d=10$ mm. Left: $P=22.6$ MPa, $\dot{m}=1177–1189\ \mathrm{kg\ m^{-2}\ s^{-1}}$, Right: $P=29.4$ MPa, $\dot{m}=1114–1126\ \mathrm{kg\ m^{-2}\ s^{-1}}$. *Modified from Ref. [17].*

flow structure in this region. This type of DHT is found mainly at low mass flows and high heat flows. It disappears at high mass flows.

The other type of DHT occurs at any position in the tube in a certain enthalpy range at high heat flows when the wall temperature exceeds the so-called pseudocritical temperature. For DHT, the ratio of heat flow and mass flow is used as indicator, with a limiting value of $q/\dot{m} > 0.4\,\mathrm{kJ/kg}$ [22] or $q/\dot{m} > 0.49\,\mathrm{kJ/kg}$ [24], $q/\dot{m} > 0.95–1.05\,\mathrm{kJ/kg}$ for stable flow and $q/\dot{m} > 0.68–0.9\,\mathrm{kJ/kg}$ for pulsating flow [25] with q (kW/m^2) and $\dot{m}$ (kg/(m^{2}s)).

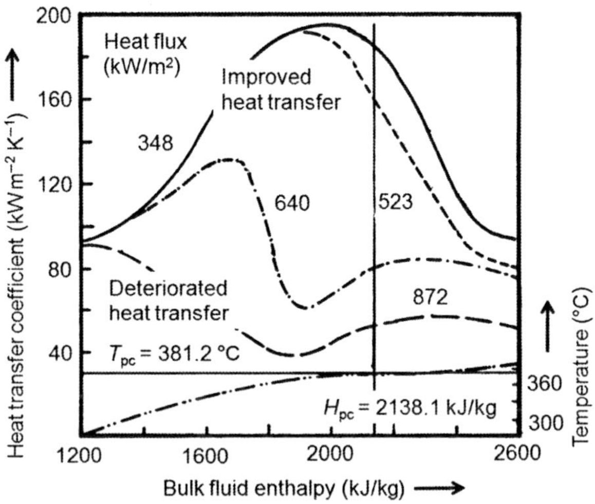

FIGURE 4.9 Variation of heat transfer coefficient of water flowing in tubes. —••— Bulk fluid temperature. *Modified from Ref. [17].*

For down-flow, no DHT was found in the entrance region and wall temperature variations were smaller than that for up-flow [26].

In Figure 4.9, experimental results are shown with regions for improved and DHT [24].

At supercritical pressures (and temperatures), a phenomenon similar to boiling was found [27] and attributed to the large differences in fluid density: high fluid density below the so-called pseudocritical point and low fluid density above. In this region, low-density fluid prevents the contact of high-density fluid to the heating wall, causing DHT and an effect similar to film boiling. It is argued that a drastic decrease in density occurs with a rapid expansion near the wall. This causes a component of flow velocity normal to the wall, resulting in the above-mentioned density layers. In addition, unpredictable heat transfer was observed when the so-called pseudocritical temperature was between the temperature of the heated wall and the bulk of the fluid. Parts of the low-density fluid are transported into the cooler bulk of the high-density fluid. There, the low-density fluid collapses, generating pressure pulses that can coincide with the frequency of the fluid column in the tube and cause high amplitude pressure fluctuations [17,28,29]. A discussion of the fundamentals of this phenomenon was presented by Carles, who named it the "piston effect" [30].

4.3.3.1.2 Heat Transfer in Horizontal Tubes

Heat transfer in horizontal tubes is affected by gravity, leading to temperature differences between the lower and the upper parts of the tube. The size of the differences depends, beside on the conditions of state, on enthalpy, mass flow, and heat flow. The temperature difference is directly proportional to heat flow, while it decreases with increasing mass flow. DHT also occurs in horizontal tubes, although being smoother compared to that in vertical tubes with a higher temperature increase on the upper part of the tube than on the lower part [17].

4.3.3.2 Prediction of Heat Transfer to Water at Supercritical Pressures

4.3.3.2.1 Forced Convection

For the calculation of heat transfer coefficients, empirical correlations based on experimental data are used. The Dittus–Boelter equation [4], Equation (4.16), was recommended for forced convection at subcritical pressures.

$$Nu_b = 0.0243 Re_b^{0.8} Pr_b^{0.4} \tag{4.16}$$

The equation was used also at supercritical pressure with good agreement to experimental data [31]. Near the critical point, Equation (4.16) may give wrong results since the dependence on the fast varying properties cannot be represented (see Figure 4.6). For water and carbon dioxide, the following equation has been developed [17]:

$$Nu_x = C Re_x^{0.77} Pr_w^{0.55} \tag{4.47}$$

with $C = 0.0266$ for water, $C = 0.0375$ for carbon dioxide. *Nu*- and *Re* numbers are calculated at boiling temperature T_b for $(T_{pc} - T_b)/(T_w - T_b) < 0$, pseudocritical temperature for $0 \leq (T_{pc} - T_b)/(T_w - T_b) \leq 1.0$, wall temperature, for $(T_{pc} - T_b)/(T_w - T_b) > 1.0$.

Equation (4.47) is based on experimental data up to pressures of $P = 34.5$ MPa for water but does not account for the peak in thermal conductivity. Several other equations have been proposed. They were evaluated by Jackson [32]. It was found that the correlations proposed by Krasnoshchekov *et al.* [33] and the modified version by Jackson [32] were the most accurate ones within a deviation of ±25%.

The equation proposed by Krasnoshchekov *et al.* 1967 [33] is:

$$Nu = Nu_0 \left(\frac{\rho_w}{\rho_b}\right)^{0.3} \left(\frac{\bar{c}_p}{c_{pb}}\right)^n \tag{4.48}$$

with

$$Nu_0 = \frac{\xi/8 Re_b \overline{Pr}}{12.7\sqrt{\xi/8}\left(\overline{Pr}^{2/3} - 1\right) + 1.07} \tag{4.49}$$

$$\xi = \frac{1}{(1.82 \log Re_b - 1.64)^2} \tag{4.50}$$

The exponent n in Equation (4.48) is

$$n=0.4 \text{ for } \frac{T_w}{T_{pc}} \le 1 \text{ or } \frac{T_b}{T_{pc}} \ge 1.2; \tag{4.51}$$

$$n=n_1=0.22+0.18\frac{T_w}{T_{pc}} \text{ for } 1 \le \frac{T_w}{T_{pc}} \le 2.5 \tag{4.52}$$

$$n=n_1+(5n_1-2)\left(1-\frac{T_b}{T_{pc}}\right) \text{ for } 1 \le \frac{T_b}{T_{pc}} \le 1.2 \tag{4.53}$$

The range of application is $8\times10^4 < Re_b < 5\times10^5$, $0.85 < \overline{Pr}_b < 65$, $0.90 < \rho_w/\rho_b < 1.0$, $0.02 < \overline{c}_p/c_{pb} < 4.0$, $0.9 < T_w/T_{pc} < 2.5$, $4.6\times10^4 < q < 2.6\times10^6$, with q in W/m^2, $x/D \ge 15$.

The equation proposed by Jackson [32] is:

$$Nu=0.0183Re_b^{0.82}Pr_b^{0.5}\left(\frac{\rho_w}{\rho_b}\right)^{0.3}\left(\frac{\overline{c}_p}{c_{p,b}}\right)^n \tag{4.54}$$

with

$$n=0.4 \text{ for } T_b < T_w < T_{pc} \text{ and for } 1.2\cdot T_{pc} < T_b < T_w, \tag{4.55}$$

$$n=0.4+0.2\left(\frac{T_w}{T_{pc}}-1\right) \text{ for } T_b < T_{pc} < T_w, \tag{4.56}$$

$$n=0.4+0.2\left(\frac{T_w}{T_{pc}}-1\right)\left[1-5\left(\frac{T_b}{T_{pc}}-1\right)\right] \text{ for } T_{pc} < T_b < 1.2\cdot T_{pc} \text{ and } T_b < T_w. \tag{4.57}$$

with $\overline{c}_p$ as the averaged specific heat between T_w and T_b, and subscripts pc, pseudocritical; b, bulk; w, wall.

Recently, Zahlan *et al.* [36] presented an analysis of heat transfer to supercritical water in bare vertical tubes and compared equations of Dittus–Boelter [4], Bishop *et al.* [34], Swenson *et al.* [20] with their new correlation.

The equation proposed by Bishop *et al.* [34] is

$$Nu_b=0.0069Re_b^{0.9}\overline{Pr}_b^{0.66}\left(\frac{\rho_w}{\rho_b}\right)^{0.43}\left(1+2.4\frac{D}{L}\right) \tag{4.58}$$

in the range $22.8 < P$ (MPa) < 27.6, $555.15 < T_b$ (K) < 800.15, $651 < \dot{m}$ (kg/(m^{2}s)), $0.31 < q$ (MW/m^2) < 3.46.

The equation proposed by Swenson *et al.* [20] is

$$Nu_w=0.00459Re_w^{0.923}\overline{Pr}_w^{0.613}\left(\frac{\rho_w}{\rho_b}\right)^{0.231} \tag{4.59}$$

in the range $22.8 < P$ (MPa) < 41.4, $348.15 < T_b$ (K) < 849.15, $366.15 < T_w$ (K) < 922.15, $542 < \dot{m}$ (kg/(m^{2}s)) < 2150, correlating experimental data within $\pm$ 15%.

The equation proposed by Mokry *et al.* [35] is:

$$Nu = 0.0061 Re_b^{0.94} \overline{Pr}_b^{0.684} \left(\frac{\rho_w}{\rho_b}\right)^{0.564} \tag{4.60}$$

with $\overline{Pr}_b$ as the averaged Prandtl number:

$$\overline{Pr}_b = \frac{(H_w - H_b)\mu_b}{k_b(T_w - T_b)} \tag{4.61}$$

μ_b is the dynamic viscosity, Pa·s, k_b is the thermal conductivity, W/(mK), subscript w refers to wall, subscript b refers to bulk.

In Figure 4.10, the equation proposed by Mokry *et al.* [35] is used for the correlation of temperature and heat transfer coefficient profiles for different heat fluxes and mass flows. The correlation shows a good fit for the experimental data within a wide range of flow conditions with an uncertainty of about $\pm$25% for calculated heat transfer coefficients and about $\pm$15% for calculated wall temperature [35].

For calculating the minimum heat flux at which the DHT regime appears, the following equation was proposed by Mokry *et al.* [35]:

$$q_{dht} = 58.97 + 0.745G\,(\text{kW/m}^2) \tag{4.62}$$

Zahlan *et al.* [36] have evaluated various correlations and compared the calculated values from Equation (4.60) (Mokry *et al.* [35]) for the heat transfer

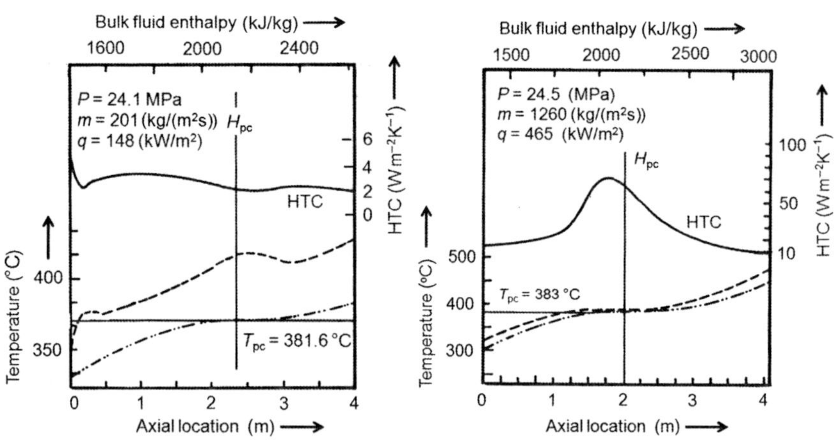

FIGURE 4.10 Temperature profiles and heat transfer coefficients along a circular tube ($L = 4$ m) at different flow conditions. ——— Heat transfer coefficient, – – – inside wall temperature, —••— bulk fluid temperature. Left: $P_{in} = 24.1$ MPa, $\dot{m} = 201$ kg m^{-2} s^{-1}, $q_{av} = 148$ kW m^{-2}, $d = 10$ mm. Right: $P_{in} = 24.5$ MPa, $\dot{m} = 1260$ kg m^{-2} s^{-1}, $q_{av} = 233$ kW m^{-2}, $d = 7.5$ mm. *Modified from Ref. [17].*

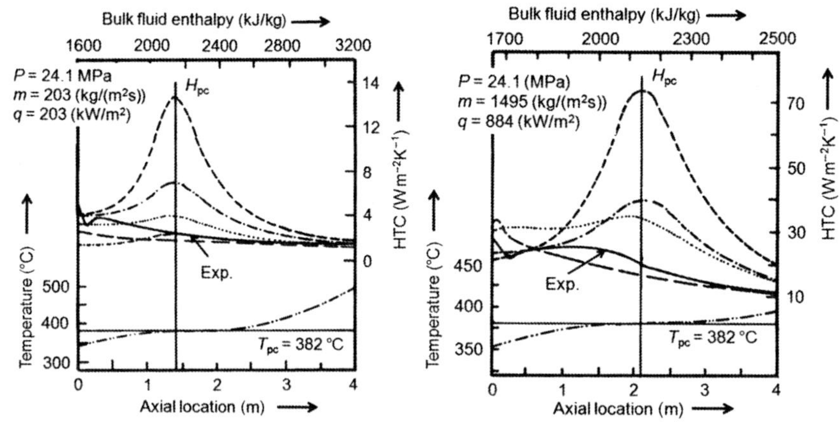

FIGURE 4.11 Comparison of correlations to values calculated with Equation (4.60). Left: $P_{in} = 24.1$ MPa, $\dot{m} = 203$ kg m^{-2} s^{-1}, $q_{av} = 203$ kW m^{-2}. Right: $P_{in} = 24.1$ MPa, $\dot{m} = 1405$ kg m^{-2} s^{-1}, $q_{av} = 884$ kW m^{-2}. Tube diameter $d = 10$ mm, ——— Experimental heat transfer coefficient, — — — Dittus–Boelter correlation (Equation 4.16), —•— Jackson correlation (Equation 4.54), •••• Bishop *et al.* correlation (Equation 4.58), ———— Swenson *et al.* correlation (Equation 4.59), – – – Mokry correlation; identical with experimental values right of H_{pc} (Equation 4.60), —••— bulk fluid temperature. Mokry *et al.* [35]. *Modified from Ref. [17].*

coefficient. This comparison, shown in Figure 4.11, with the other equations reveals that the Dittus–Boelter correlation significantly overestimates experimental heat transfer coefficients in the near-critical range [35]. Correlations proposed by Bishop *et al.* [34] and Jackson [32] deviate also substantially from the experimental data in the near-critical range. The equation proposed by Swenson *et al.* [20] represents experimental data better at low mass flow ($\dot{m} = 500$ kg/m^{2}s) than the three aforementioned equations but deviates from experimental data within the entrance region and at higher mass flows [35].

For a more detailed analysis of heat transfer in the near-critical and supercritical region, the reader is referred to the cited literature.

4.4 CONDENSATION

4.4.1 Condensation of Pure Fluids

A gas near thermodynamic saturation conditions is liquefied, if it is cooled. This condensation is connected with combined heat and mass transfer, since in addition to the heat removed by cooling, the condensing gas must be provided from the bulk of the gas phase. If the condensed liquid completely wets the heat transfer surface, a liquid film forms and condensation takes place at the phase boundary liquid–gas. This film condensation is the case for most technical situations [37].

Condensation is characterized by the following steps:

- The condensing gas must be transported to the surface of the liquid.
- Condensation occurs at the surface of the liquid.
- Enthalpy of condensation has to be transported to the cooled heat transfer surface.

The local heat transfer coefficient is defined by

$$\alpha_{F,x} = \frac{q_x}{T_I - T_{w,x}}, \tag{4.63}$$

with q_x, local heat flow vertical to the heat transfer surface, T_I:, temperature at the surface liquid–gas, and $T_{w,x}$,: temperature at the heat transfer surface.

The dimensionless Nusselt number is defined by

$$Nu_{F,x} = \frac{\alpha_{F,x} l}{\lambda_F} \tag{4.64}$$

with l as characteristic length

$$l = \sqrt[3]{\frac{\nu_F^2}{g}} \tag{4.65}$$

The energy balance is

$$k_x(T_s - T_{CM})dA = \Delta h_v d\dot{m}_F. \tag{4.66}$$

With T_s as the boiling temperature and T_{CM} as the temperature of the cooling medium.

For a heat transfer coefficient not too strongly dependent on axial length x, this equation can be directly integrated and the heat transfer area A can then be calculated with an average temperature difference.

$$A = \frac{\Delta h_v \dot{m}_{F,x=L}}{k(T_s - T_{CM})_{lm}} \tag{4.67}$$

with the overall heat transfer coefficient k

$$\frac{1}{k} = \frac{1}{\alpha_F} + \frac{s}{\lambda_W} + \frac{1}{\alpha_{CM}} \tag{4.68}$$

and the average temperature difference $(T_s - T_{CM})_{lm}$

$$(T_s - T_{CM})_{lm} = \frac{(T_s - T_{CM,in}) - (T_s - T_{CM,out})}{\ln(T_s - T_{CM,in})/(T_s - T_{CM,out})} \tag{4.69}$$

For cases where the heat transfer coefficient is strongly dependent on x, local values have to be used [37].

For the condensation on vertical surfaces and a gas velocity that is not influencing the condensate film, the following equations are reported [37] for the average heat transfer coefficient along the cooling surface.

Laminar flow (in the condensate film)

$$Nu_{\mathrm{F,lam}} = 0.925\left(\frac{1-\rho_{\mathrm{G}}/\rho_{\mathrm{F}}}{Re_{\mathrm{F,l}}}\right)^{1/3}, \tag{4.70}$$

with ρ_G and ρ_F as the density of the gas and the liquid.

Turbulent flow:

$$Nu_{\mathrm{F,turb}} = \frac{0.020 Re_{\mathrm{F,l}}^{7/24} Pr_{\mathrm{F}}^{1/3}}{1 + 20.52 Re_{\mathrm{F,l}}^{-3/8} Pr_{\mathrm{F}}^{-1/6}} \tag{4.71}$$

Laminar and turbulent flow in the condensate film can be combined

$$Nu_{\mathrm{F}} = \sqrt[1.2]{(f_{\mathrm{r}} Nu_{\mathrm{F,lam}})^{1.2} + (Nu_{\mathrm{F,turb}})^{1.2}}\, f_{\eta} \tag{4.72}$$

with

$$f_{\mathrm{r}} = \begin{cases} 1 \text{ for } Re_{\mathrm{F}} < 1 \\ Re_{\mathrm{F}}^{0.04} \text{ for } Re_{\mathrm{F}} \geq 1 \end{cases} \tag{4.73}$$

$$f_{\eta} = \left(\frac{\eta_{\mathrm{F,S}}}{\eta_{\mathrm{F,W}}}\right)^{1/4} \tag{4.74}$$

Equations (4.71)–(4.74) can be applied in the range

$$Re_{\mathrm{F}} < 10^4, 0.5 < Pr_{\mathrm{F}} < 500, 0.2 < \eta_{\mathrm{F,s}}/\eta_{\mathrm{F,W}} < 5.$$

For higher gas flow velocities, interaction between gaseous and liquid phase influences the heat transfer process. The influence is expressed as shear stress τ_G and is different for cocurrent and counter-current flow of the phases.

For cocurrent flow, the liquid film gets thinner and its velocity is enhanced [37]:

$$Nu_{\mathrm{F},x}^{*} = Nu_{\mathrm{F},x}\left(1 + \tau_{\mathrm{G}}^{*}\right)^{1/3} \tag{4.75}$$

with

$$\tau_{\mathrm{G}}^{*} = \frac{\tau_{\mathrm{G}}}{\rho_{\mathrm{F}} g \delta_{\mathrm{F}}^{*}}, \tau_{\mathrm{G}} = \frac{\xi_r}{8}\rho_{\mathrm{G}} u_{\mathrm{G}}^2 \tag{4.76}$$

where τ_G, shear stress at the condensate film surface; δ_F^*, thickness of the condensate film; ρ_G, density of the gaseous phase; u_G^2, average flow velocity in the gaseous phase.

For counter-current flow, shear stress at the surface of the liquid condensate film slows down the velocity of the down-flowing liquid film and the film gets thicker. Equation (4.77) can be applied for the laminar and turbulent flow.

$$Nu_{\mathrm{F},x}^{*} = Nu_{\mathrm{F},x}\left(1 - \tau_{\mathrm{G}}^{*}\right)^{1/3}. \tag{4.77}$$

The maximum condensate quantity is limited by hydrodynamic conditions. At fixed gas flow velocities and high amounts of condensate, entrainment and eventually flooding can occur. During condensation, a film forms at the heat transferring surface. Film thickness of the smooth laminar film can be described by Nusselt's equation [38] with acceptable accuracy [39,40].

$$\Psi_{\mathrm{L}} = \frac{6}{Re_{\mathrm{L}}} \tag{4.78}$$

with ψ_{G} as the dimensionless gas resistance factor.

$$\psi_{\mathrm{G}} = \frac{\Delta P d_{\mathrm{h}}}{\rho_{\mathrm{G}}(u_{\mathrm{G}} - u_{\mathrm{L}})^2 L}, \tag{4.79}$$

The shear stress in the liquid surface is proportional to the pressure drop. Relating the pressure drop to the impact pressure of the gas flow gives the dimensionless gas resistance factor ψ_{G} (Equation 4.79) with $u_{\mathrm{G}} - u_{\mathrm{L}}$, difference between velocity of the gas and liquid phase (slip velocity); d_{h}, hydraulic diameter; L, total length.

The influence of the gas flow on the flow regime can be taken into account by using the modified *Re* number [39,40].

$$Re_{\mathrm{L}}^{*} = Re_{\mathrm{L}}(1 + \psi_{\mathrm{G}})^{1/3}. \tag{4.80}$$

For the limiting gas flow (flooding), a simple semiempirical correlation, originally proposed by Wallis [41], was found to be applicable after adjusting the coefficients with experimental data [40].

$$j_{\mathrm{G}}^{*} = \frac{1.1457}{\left(1 + 0.4222\sqrt{\Phi}\right)^2}, \tag{4.81}$$

with

$$j_{\mathrm{G}}^{*} = \frac{u_{\mathrm{G}}}{\varepsilon}\sqrt{\frac{\rho_{\mathrm{G}}}{g d_{\mathrm{H}}(\rho_{\mathrm{L}} - \rho_{\mathrm{L}})}} \tag{4.82}$$

$$\Phi = \frac{u_{\mathrm{L}}}{u_{\mathrm{G}}}\left(\frac{\rho_{\mathrm{L}}}{\rho_{\mathrm{G}}}\right)^{0.5} \tag{4.83}$$

j_{G}^{*} is a modified Froude number relating the impact pressure to the difference between liquid head and buoyancy, ϵ is the void volume, equal to 1 for falling film columns, smaller for a column with mass transfer packing.

4.4.2 Condensation of Mixtures

For condensation of a pure gas, the temperature at the surface where the condensation takes place must be lower than that of the gas. The heat flow from the bulk of the gas to the condensing surface is only determined by the transport of heat in the condensate film. A convective flow vertical to the flow of the gas is directed from the bulk of the gas phase to the condensate surface where the condensation occurs. The driving force for the process is the temperature difference between the saturation temperature of the gas and the heat transferring wall.

Condensation of mixtures is somewhat more complicated than condensation of pure components, since thermodynamic behavior of the mixtures determines the phase transition from gas to liquid. Components of a gaseous mixture may have a volatility of the same order of magnitude. Then, all components are liquefied in a comparable quantity. For components that differ largely in their volatility, like, for example, water and carbon dioxide, water will be liquefied preferably and the condensation of carbon dioxide may be negligible for many applications. But in principle, the much more volatile component (carbon dioxide) is also transferred into the liquid state, only at a much lower quantity. Furthermore, the condensate of a gaseous mixture may be a homogeneous mixture or may be split into more than one phase. From this simple analysis, it can be concluded that condensation in mixtures is determined by heat transfer, mass transfer due to the transport of condensing components from the bulk of the gaseous phase to the liquid condensate, and phase equilibrium between the gas and the liquid. Phase equilibrium, at specified conditions of state, determines the composition and the quantity of liquid and gas.

In a condensation process of a mixture, the composition of the gaseous mixture changes along the surface of the condensate. Components of lower volatility condense preferably with a resulting concentration profile vertical to the condensate surface. This profile adds to the heat transfer resistance and may be rate-determining.

In a pure component gas, condensation can only occur if the temperature at the surface of the condensate and the heat transferring wall is lower than the saturation temperature. This is no longer true for gaseous mixtures. In such mixtures, depending on the phase equilibrium, condensation can occur if the temperature difference between wall and gaseous mixture is positive (heating) or negative (cooling).

The condensation of mixtures can be calculated using the correlations for the heat transfer coefficient derived for pure component condensation, as well for the heat transfer from the liquid film at the wall, as for the heat transfer coefficient in the bulk of the gaseous phase, provided the specific conditions for composition and mass flow for the mixture are considered. Even for two liquid phases in the condensate, assuming a homogeneous phase with the appropriate mixture composition, applying correlations derived for pure component condensation has lead to satisfying results [42,43].

Hydrothermal processes for transformation of hydrocarbons, biomass, organic waste, and others often lead to a gaseous mixture from which the products are recovered by condensation. This process can be carried out by heat transfer, for example, cooling to ambient temperature, or pressure reduction.

Beside the recovery of the products, it is important to recover the energy involved in the high-temperature hydrothermal processes. Therefore, heat transfer processes are of great importance in such processes, often determining the economy of the process.

4.4.2.1 Condensation in Binary Mixtures

4.4.2.1.1 Condensation by Temperature Changes

Condensation by changing the temperature is the most important mode for recovering products and energy. Condensation can take place by either decreasing the temperature or increasing the temperature, which is one of the major differences to condensation of a pure gas. This will first be illustrated on a binary system and later on a ternary system [44].

In Figure 4.12, a temperature-composition diagram (T,x diagram) of a binary mixture is shown. The left diagram shows phase equilibrium for three pressures. At P_1, both components are subcritical, at P_2, the component with the higher volatility is supercritical, the other component still subcritical, at P_3, both components are supercritical. The two-phase area extends no longer to the concentration limits, meaning that two phases (as necessary for condensation) exist only in a limited range of concentrations.

In the right diagram, condensation processes are illustrated. Condensation, caused by temperature reduction, is shown in the process from A to B, where a liquid phase of composition C and a gaseous phase of composition D are formed.

An example for the reverse case, a condensation by increase of temperature is illustrated in the temperature change from E to F with I as condensate and J as the resulting gaseous mixture. This process is also called retrograde

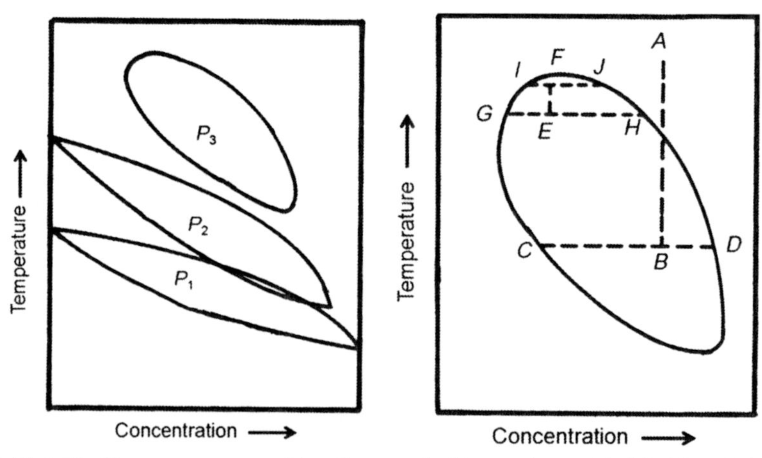

FIGURE 4.12 Temperature-composition diagram of a binary mixture. *Modified from Ref. [44].*

condensation. The temperature increase results in a liquid phase richer in the higher volatile component and a gas phase leaner in the higher volatile component. Beside the change of the composition, the quantity of the phases changes also. For the total composition remaining constant, the ratio of the phases changes with $FI/FJ<EG/EH$. Total composition is now nearer to the composition of the liquid (condensate) phase and the amount of condensate is increased.

Retrograde condensation by temperature increase is a phenomenon near the critical region of mixtures and may occur during processes. For the recovery of products, condensation by reduction of temperature is the normal case.

The condensation process as shown in Figure 4.12 (right, A to $B \rightarrow C+D$) is an isobaric process like in a vessel with totally mixed gas phase. If the gaseous mixture flows in a tube, gas and condensate flow relative to one another and the heat and mass transfer leads to concentration changes according to phase equilibrium in flow direction, as illustrated for cocurrent flow in Figure 4.13.

In a vertical condenser, gas and condensate flow downward in the cocurrent mode. Due to the lower temperature at the outside of the condenser wall, the gas is cooled and condensate is formed. At first, the condensate is rich in the low volatile component. With progressing condensation through lowering the temperature, the amount of condensate increases and the concentration of the component of higher volatility increases also, following the phase boundary line. The condensation ends when all the gaseous phase is liquefied at point E (Figure 4.13).

It is assumed that at the condensate surface (the phase boundary) gas and liquid are in equilibrium. For the condensation process and the necessary heat and mass transfer, the gas phase must have a slightly higher temperature than that at the interface, as well as the temperature of the liquid must be slightly lower than the equilibrium temperature. For calculating heat and mass

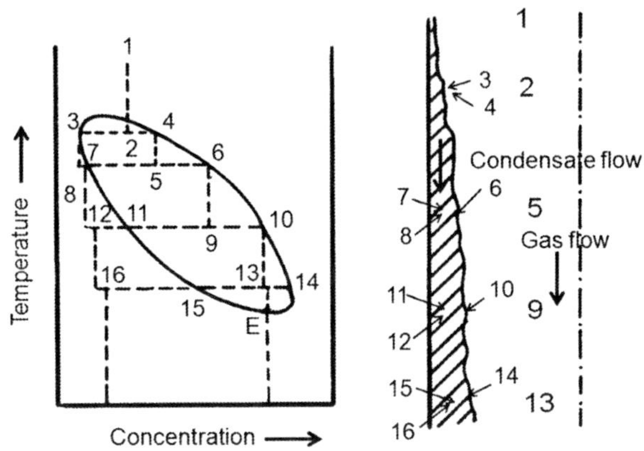

FIGURE 4.13 Schematic representation of a condensation process in a binary mixture. The gaseous mixture flows in a tube, gas and condensate flow relative to one another and the heat and mass transfer leads to concentration changes according to phase equilibrium in flow direction. *Modified from Ref. [44].*

transfer, these temperature differences can be neglected. Mass transfer from the bulk of the gaseous phase to the surface of the condensate film causes an increase of the concentration of the component that is chiefly liquefied. This boundary layer slows down mass transfer.

4.4.2.1.2 Condensation by Pressure Changes

Condensation of gaseous mixtures can also be caused by pressure changes. Basically, either a pressure reduction or a pressure enhancement can lead to the formation of a liquid phase, as illustrated in Figure 4.14 [44].

The left diagram of Figure 4.14 shows phase equilibrium for three temperatures. At T_1, both components are subcritical, at T_2 and T_3, the component with higher volatility is supercritical, while the other component is still subcritical. The two-phase area extends no longer to both concentration limits, meaning that two phases (as necessary for condensation) exist only in a limited range of concentrations.

In a binary gaseous mixture, condensation can be initiated by pressure reduction or pressure enhancement, depending on the conditions. This is illustrated in Figure 4.14 (right). Starting at low pressures, a gaseous mixture can be liquefied by pressure enhancement (from C via D and H to G in Figure 4.14). A liquid phase is formed that changes its composition along the phase boundary line during further pressure enhancement. The condensate at point G is actually a supercritical binary fluid with properties between a normal gas and a normal liquid. Corresponding to the composition change of the liquid phase, the remaining gaseous phase changes its condition along the other side of the two-phase area.

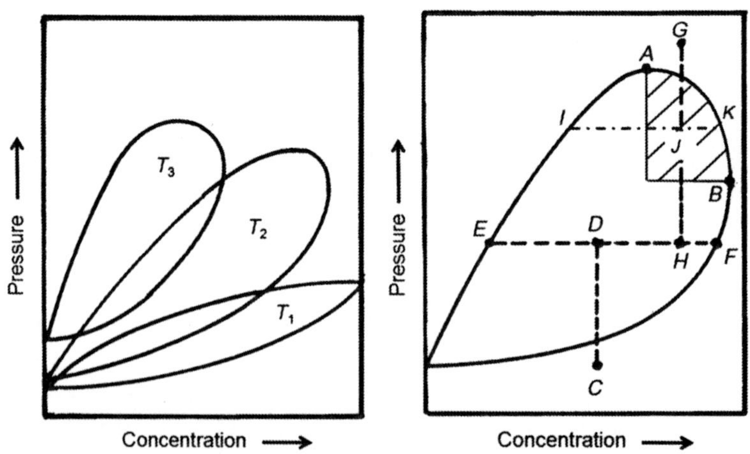

FIGURE 4.14 Pressure-concentration diagram of a binary mixture. See text. *Modified from Ref. [44].*

This is the type of condensation normally expected and occurs in all the two-phase areas with the exception of the area enclosed by AB and the phase boundary line (see in Figure 4.14, the hatched area). At conditions within this area, an increase in pressure leads to an increase of the gas phase (evaporation). The amount of the liquid phase, for example, represented by $\overline{JK}$ at total composition G, diminishes at increasing pressure. The inverse process, condensation by pressure reduction in this area is called retrograde condensation.

During condensation of a mixture, both variation of temperature and variation of pressure can occur at the same time. For high-pressure processes, this is common. Then, phenomena occur simultaneously and the resulting condensation is a combination of both cases.

In many processes, mixtures of more than two components are common. Therefore, the condensation in a ternary system is discussed next. Behavior of mixtures with more than three components can be qualitatively derived from the binary and ternary behavior.

4.4.2.2 Condensation in Ternary Mixtures

In Figure 4.15, the phase behavior of a ternary system is illustrated in a P,x- and a T,x-prism for a mixture of water, a hydrocarbon, and a gas, for example, water–toluene–hydrogen. Such types of mixtures are often the result of hydrothermal processes.

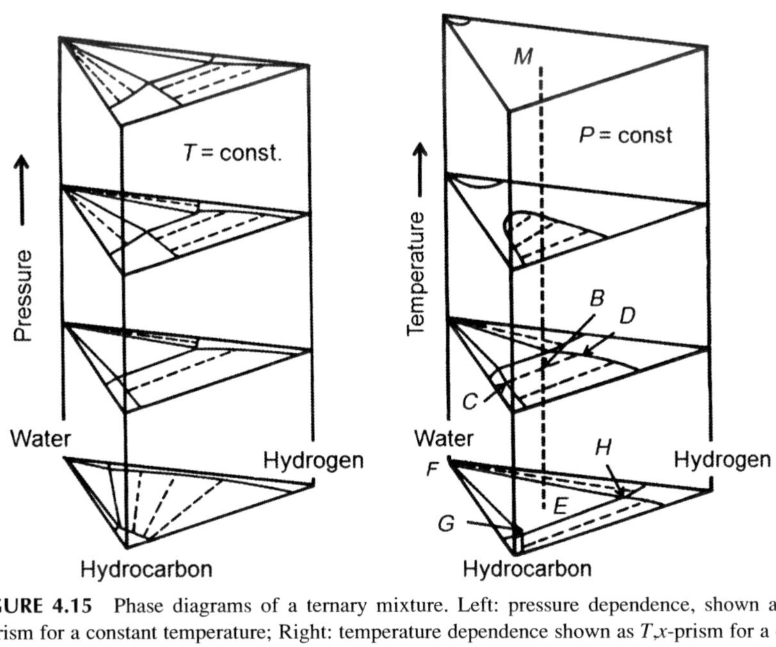

FIGURE 4.15 Phase diagrams of a ternary mixture. Left: pressure dependence, shown as P, x-prism for a constant temperature; Right: temperature dependence shown as T,x-prism for a constant pressure. *Modified from Ref. [44].*

Condensation in a ternary mixture is similar to that in binary mixtures and can be carried out by variation of pressure and temperature. The same phenomena as in a binary mixture can occur, including retrograde condensation. But the existence of extended three-phase areas in a ternary system is different to binary systems. (For the interpretation of the ternary diagrams, refer to Section 3.4.3.4 of Chapter 3.)

Condensation by variation of temperature at constant pressure is illustrated in Figure 4.15 (right). A homogeneous gaseous mixture, *M*, enters the condenser at high temperature and is cooled to lower temperatures. At lower temperature, the total composition of the initial gaseous mixture is represented by *B* in a two-phase area. Condensation into a liquid condensate *C* and a gaseous phase *D* occurs. All three components are present in these phases in appreciable amounts.

At even lower temperatures, the total composition of the original mixture is in a three-phase area (point *E* in Figure 4.15). At these conditions, the mixture separates into three phases, two liquid phases, one with a high concentration of water (point *F* in Figure 4.15), the other with a high concentration of the hydrocarbon (point *G* in Figure 4.15), and a gaseous phase (point *H* in Figure 4.15) with a high concentration of the very volatile component (e.g., hydrogen). In addition to the formation of the liquid phase by condensation, the three main components have been separated to a high degree.

The condensation process is similar to that of binary systems. The special feature of a ternary system is shown by the behavior of the equilibrium concentrations, as illustrated in Figure 4.16 for the influence of temperature on the composition of the condensate (and the gaseous phase) [44].

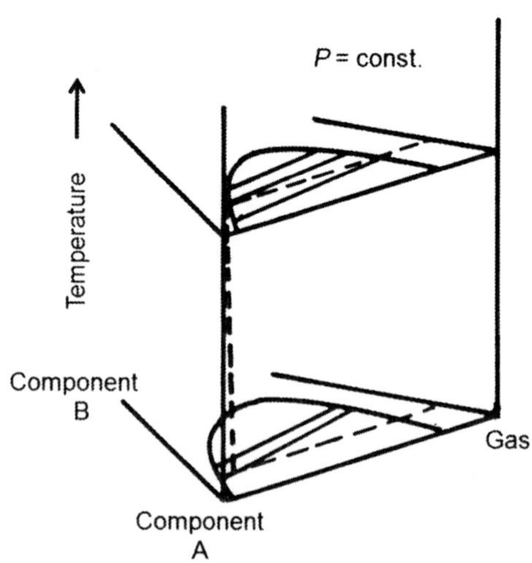

FIGURE 4.16 Influence of temperature on the composition of the condensate. See text. *Modified from Ref. [44].*

In a ternary system, as illustrated in Figure 4.16, the extension of the two-phase area changes with temperature. This initiates condensation or evaporation of the individual components but also changes the total amounts of condensate and gaseous phase. The composition of the equilibrium phases is also affected. In the ternary mixture shown in Figure 4.16, the phases (condensate and gas) in mutual equilibrium are connected by the broken line. The line is called "connode" and has only the function to visualize phases that are in mutual phase equilibrium. If temperature is changed, the inclination of the connode with respect to the basis line $\overline{A-\mathrm{Gas}}$ may also change, resulting in mass transfer processes that include liquefaction and evaporation of components, depending on the conditions. Condensation of the gaseous phase at constant composition of the phases can only be achieved along a connode. For this case, pressure and temperature must be kept constant. In all other cases, composition of both phases change and mass transfer is initiated in addition to heat transfer and condensation.

4.4.2.3 Example: Condensation of a Mixture of Water–Toluene–Hydrogen

For the condensation of a mixture of water, toluene, and hydrogen phase equilibrium is shown in Figure 4.17 for $P=10$ MPa and temperatures between $T=373.15$ and 673.15 K. Condensation by reduction of temperature occurs as discussed in the previous section.

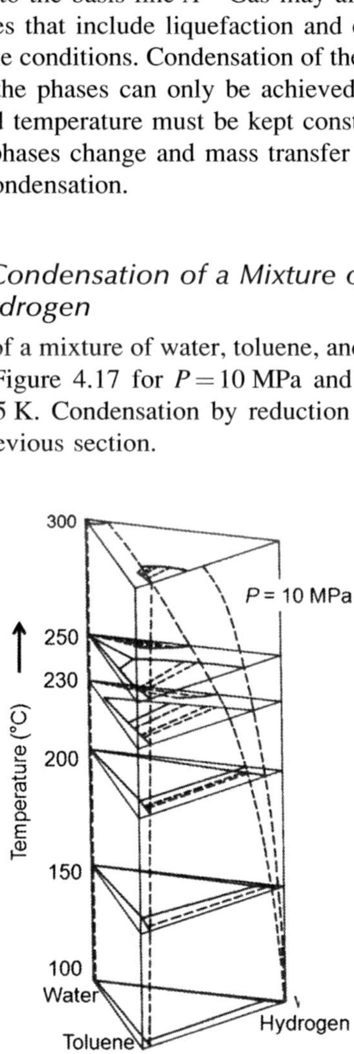

FIGURE 4.17 Condensation of a mixture of water, toluene, and hydrogen at 10 MPa and temperatures between 373.15 and 673.15 K. *Modified from Ref. [44].*

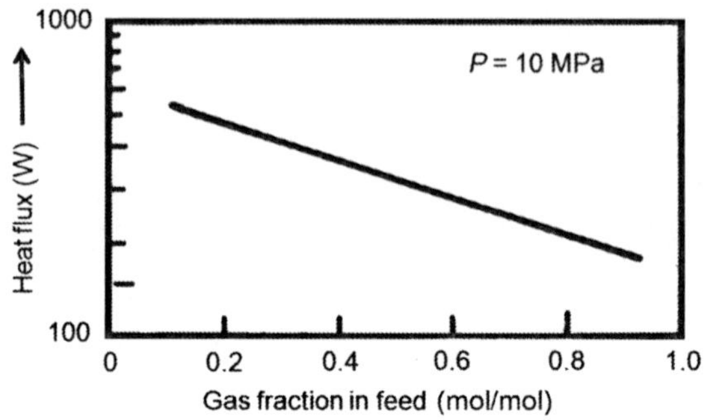

FIGURE 4.18 Heat flow in dependence on the hydrogen concentration. *Modified from Ref. [44].*

As an example for the influence of the various parameters, Figure 4.18 illustrates the heat flow in dependence on the hydrogen concentration.

Heat flow decreases with hydrogen concentration at constant pressure and increases with increasing water content. The lower heat flow at increased hydrogen concentration is due to the higher amount of hydrogen in the boundary layer gas-condensate film which hinders the transport of the other components to the surface. For a similar reason, the heat flow increases with increasing amount of water due to more water molecules present at the boundary layer.

Heat and mass transfer coefficients have been correlated with dimensionless equations, representing experimental data within the conditions given below [44]:

$$Nu = 26.237 Re^{0.35694} Pr^{1.26} \tag{4.84}$$

$$Sh = 53.294 Re^{0.3187} Sc^{0.34458} \tag{4.85}$$

Data were measured in the range

$60 < Nu < 300$, $400 < Sh < 5000$, $500 < Re < 1600$, $7 < Sc < 350$, $0.4 < Pr < 0.7$.

The influence of natural convection was not considered in the example since a calculation proved that the influence of natural convection is negligible. In any case, natural convection would enhance heat transfer [45].

4.5 HEAT TRANSFER FOR FLOW THROUGH BEDS OF PARTICLES

In hydrothermal processes, solids may be involved, for example, in conversion of wood chips. Heat and mass transfer in a bed of solids will then play a role.

Heat transfer coefficients between a fluid and a bed of particles are greater than that for heat transfer between a fluid and a single particle. According to

Gnielinski [46], the heat transfer coefficients in a bed of particles can be calculated from the heat transfer to a single particle [47]:

$$Nu = f\,Nu_{SP} \tag{4.86}$$

with

$$Nu_{SP} = 2 + \sqrt{Nu_{lam}^2 + Nu_{turb}^2} \tag{4.87}$$

$$Nu_{lam} = 0.664 Re^{0.5} Pr^{1/3} \tag{4.88}$$

$$Nu_{turb} = \frac{0.037 Re_{\varepsilon}^{0.8} Pr}{1 + 2.443 Re_{\varepsilon}^{-0.1}\left(Pr^{2/3} - 1\right)} \tag{4.89}$$

with

$$Nu = \frac{\alpha d_k}{\lambda}, \quad Re = \frac{u_f d_k}{\nu\varepsilon}, \quad Pr = \frac{\nu}{a} d_k = \sqrt{\frac{A_p}{\pi}}, \tag{4.90}$$

d_k, diameter of a sphere of the same surface area A_p as the particle; u_f, flow velocity of the fluid in the void cross section; ε, void volume fraction; $\varepsilon = (V - V_p)/V$; V, bulk volume; V_p, volume of all particles in the bulk volume. Values for the shape factor f are for spheres of the same size: $f = 1 + 1.5(1 - \varepsilon)$, for cubes and cylinders, $0.24 < l/d < 1.2$: $f = 1.6$. All properties of the fluid are applied at the average temperature between inlet (T_i) and outlet (T_o): $T_m = (T_o - T_i)/2$.

Equation (4.86) can be applied in the following limits for heat and mass transfer:

Spheres: $10^{-1} < Re_{\varepsilon} < 10^4$, $0.6 < Pr$ or $Sc < 10^4$,

Cylinders: $10^2 < Re_{\varepsilon} < 10^4$, $0.6 < Pr$ or $Sc < 1300$,

Cubes: $10^2 < Re_{\varepsilon} < 10^4$, $Pr = 0.7$.

Lower values observed than that obtained from Equation (4.86) are due to nonregular flow through the bulk of particles because of irregular void volumes.

4.6 HEAT TRANSFER COEFFICIENTS IN HEAT EXCHANGERS

Determination of the heat transfer coefficient may be tedious and needed data may not be available for preliminary design stages. Therefore, typical values of the heat transfer coefficient can be useful for estimating the required surface area. In Table 4.2, values for different applications and heat exchanger types are given. The values in the table are in the range of normal applications, with lower values for unfavorable conditions (lower flow velocities, higher viscosities, fouling resistances) and higher values for more favorable conditions [48].

TABLE 4.2 Typical Overall Heat Transfer Coefficients α in Heat Exchangers [48]

Type	Application and Conditions	α (W/(m^2 K))
Tubular, heating, or cooling	Gases at atmospheric pressure inside and outside tubes	5–35
	Gases at high pressure inside and outside tubes	150–500
	Liquid outside (inside) and gas at atmospheric pressure inside (outside) tubes	15–70
	Gas at high pressure inside and liquid outside tubes	200–400
	Liquids inside and outside tubes	150–1200
	Steam outside and liquid inside tubes	300–1200
Tubular, condensation	Steam outside and cooling water inside tubes	1500–4000
	Organic vapors or ammonia outside and cooling water inside tubes	300–1200
Tubular, evaporation	Steam outside and high-viscous liquid inside tubes, natural circulation	300–900
	Steam outside and low-viscous liquid inside tubes, natural circulation	600–1700
	Steam outside and liquid inside tubes, forced circulation	900–3000
Air-cooled heat exchangers	Cooling of water	600–750
	Cooling of liquid light hydrocarbons	400–550
	Cooling of tar	30–60
	Cooling of air or flue gas	60–180
	Cooling of hydrocarbon gas	200–450
	Condensation of low pressure steam	700–850
	Condensation of organic vapors	350–500
Plate heat exchanger	Liquid to liquid	1000–4000
Spiral heat exchanger	Liquid to liquid Condensing vapor to liquid	700–2500 900–3500

REFERENCES

[1] R.C. Hendricks, R.J. Simoneu, R.V. Smith, Survey of heat transfer to near-critical fluids, Lewis Research Center, Cleveland, Ohio 44135, NASA Technical note TN D-5886, NASA, Washington D.C., USA, 1970.

[2] VDI-Wärmeatlas, Verein Deutscher Ingenieure, Wärmeübertragung bei der Strömung durch Rohre, 10th ed., Springer, Berlin, Germany, 2006, Section Ga 1-Ga9. Engl. ed. VDI-Gesellschaft Verfahrenstechnik und Chemieingenieurwesen (Ed.), VDI Heat Atlas, 2nd Ed., Springer, Berlin, Heidelberg, New York, 2010.

[3] V. Gnielinski, Ein neues Berechnungsverfahren für die Wärmeübertragung im Übergangsbereich zwischen laminarer und turbulenter Rohrströmung, Forsch. im Ing.-Wes. 61 (1995) 240–248.

[4] F.W. Dittus, L.M.K. Boelter, Heat transfer in automobile radiators of the tubular type, Publ. Eng. 2 (1930) 443–461, University of California, Berkeley.

[5] VDI-Wärmeatlas, Verein Deutscher Ingenieure, Strömungssieden gesättigter Flüssigkeiten, 10th ed., Springer, Berlin, Germany, 2006, Section Hbb1-Hbb 33. Engl. ed. VDI-Gesellschaft Verfahrenstechnik und Chemieingenieurwesen (Ed.), VDI Heat Atlas, 2nd Ed., Springer, Berlin, Heidelberg, New York, 2010.

[6] K. Stephan, P. Preußler, Wärmeübergang und maximale Wärmestromdichte beim Behältersieden binärer und ternärer Flüssigkeitsgemische, Chem. Ingenieur Tech. 51 (1979) 37.

[7] Verein Deutscher Ingenieure, VDI-Wärmeatlas, Kritische Siedezustände, 10th ed., Springer, Berlin, Germany, 2006, Hbc1-Hbc 30. Engl. ed. VDI-Gesellschaft Verfahrenstechnik und Chemieingenieurwesen (Ed.), VDI Heat Atlas, 2nd Ed., Springer, Berlin, Heidelberg, New York, 2010.

[8] E. Schmidt, E.R.G. Eckert, U. Grigull, Wärmetransport durch Flüssigkeiten in der Nähe ihres kritischen Zustandes, Jahrbuch deutscher Luftfahrt-Forschung II (1939) 53–58. E. Schmidt, E.R.G. Eckert, U. Grigull, Heat Transfer by Liquids Near the Critical State. Trans. No. F-TS- 527-RE7 Air Material Command, Wright-Patterson AFB, Ohio, Apr. 26, 1946.

[9] M.E. Shitsman, Impairment of the heat transmission at supercritical pressures, High Temperature 1 (1963) 237–244.

[10] K. Yamagata, K. Nishikawa, S. Hasegawa, T. Fujii, Forced convective heat transfer in the critical region, in: Presented at the Japan Society of Mechanical Engineers 1967 Semi-International Symposium, Tokyo, Japan, Sept, 1967. K. Yamagata, K. Nishikawa, S. Hasegawa, T. Fujii, S. Yoshida, Forced convective heat transfer to supercritical water flowing in tubes, Int. J. Heat Mass Transf. 15 (1972) 2575–2593.

[11] Y.Y. Hsu, Discussion of an investigation of heat transfer of fluids flowing in pipes under supercritical conditions, Int. Dev. Heat Transf. ASME (1961) D-188–D-189. Cited from Ref. [1].

[12] F. Mayinger, M. Scheidt, Heat transfer in the supercritical region with vertical upflow, Wärme- und Stoffübertragung 18 (1984) 207–214.

[13] W.H. McAdams, Heat Transmission, third ed., McGraw-Hill Book Co., Inc., 1954.

[14] J.R. Larson, R.J. Schoenhals, Turbulent free convection in near-critical water, J. Heat Transf. 88 (1966) 407–414.

[15] J.P. Holman, J.H. Boggs, Heat transfer to Freon 12 near the critical state in a natural-circulation loop, J. Heat Transf. 82 (1960) 221–226.

[16] Z.L. Miropolskii, M.E. Shitsman, Heat transfer to water and steam at variable specific heat (in near-critical region), Soviet Phy. Tech. Phys. 2 (1957) 2196–2208.

[17] I.L. Pioro, R.B. Duffey, Heat Transfer and Hydraulic Resistance at Supercritical Pressures in Power-Engineering Applications, ASME Press, New York, N.Y., USA, 2007.

[18] I. Pioro, S. Mokry, S. Draper, Specifics of thermophysical properties and forced-convective heat transfer at supercritical pressures, Rev. Chem. Eng. 27 (2011) 191–214.
[19] P.L. Kirillov, V.V. Lozhkin, A.M. Smirnov, Investigation of borders of deteriorated regimes of a channel at supercritical pressures (In Russian). State Scientific Center of Russian Federation Institute of Physics and Power Engineering, A.I. Leypunskiy, FEI-2988, Obninsk, Russia, 2003.
[20] H.S. Swenson, J.R. Carver, C.R. Kakarala, Heat transfer to supercritical water in smooth-bore tubes, J. Heat Transf., Trans. ASME Ser. C 87 (1965) 477–484.
[21] E.N. Pis'mennyy, V.G. Razumovskiy, E.M. Maevskiy, et al., Experimental study on temperature regimes to supercritical water flowing in vertical tubes at low mass fluxes, in: Proceedings of the International Conference GLOBAL-2005, Tsukuba, Japan, October 9–13, 2005, Paper No. 519.
[22] Yu.V. Vikhrev, Yu.D. Barulin, A.S. Kon'kov, A study of heat transfer in vertical tubes at supercritical pressures, Thermal Eng. (Teploenergetika) 14 (1967) 116–119.
[23] Yu.V. Vikhrev, A.S. Kon'kov, V.A. Lokshin, Temperature regime of steam generating tubes at supercritical pressure (in Russian), in: Transactions of the IVth All-Union Conference on Heat Transfer and Hydraulics at Movement of Two-Phase Flow inside Elements of Power Engineering Machines and Apparatuses, Leningrad, Russia, 1971, pp. 21–40.
[24] M.A. Styrikovich, T.Kh. Margulova, Z.L. Miropol'skii, Problems in the development of designs of supercritical boilers, Thermal Eng. 14 (1967) 5–9.
[25] A.P. Ornatsky, L.P. Glushchenko, E.T. Siomin, et al., The research of temperature conditions of small diameter parallel tubes cooled by water under supercritical pressures, in: Proceedings of the 4th International Heat Transfer Conference, Paris-Versailles, France, Vol. VI, Elsevier Publishing company, 1970, Paper No. B 8.11.
[26] L.Yu. Krasyakova, I.I. Belyakov, N.D. Fefelova, Heat transfer with a downward flow of water at supercritical pressure, Thermal Eng. (Teploenergetika) 24 (1977) 9–14.
[27] J.W. Ackerman, Pseudoboiling heat transfer to supercritical pressure water in smooth and ribbed tubes, J. Heat Transf., Trans. ASME 92 (1970) 490–498.
[28] N.L. Kafengauz, About some peculiarities in fluid behavior at supercritical pressure in conditions of intensive heat transfer, Appl. Thermal Sci. (Prot. Teploenergetika) 8 (1986) 26–28.
[29] N.L. Kafengauz, The mechanism of pseudoboiling, Heat Transf. Soviet Res. 7 (1975) 94–100.
[30] P. Carles, Thermophysical Properties of SCFs, in: 9th International Symposium on Supercritical Fluids, May 18–20, 2009, Arcachon, France, Plenary lecture, 2009.
[31] N.M. Schnurr, V.S. Sastry, A.B. Shapiro, A numerical analysis of heat transfer to fluids near the thermodynamic critical point including the thermal entrance region, J. Heat Transf. Trans. ASME 98 (1976) 609–615.
[32] J.D. Jackson, Consideration of the heat transfer properties of supercritical pressure water in connection with the cooling of advanced nuclear reactors, in: Proceedings of the 13th Pacific Basin Nuclear Conference, Shenzhen City, China, October 21–25, 2002.
[33] E.A. Krasnoshchekov, V.S. Protopopov, F. Van, I.V. Kuraeva, Experimental investigation of heat transfer for carbon dioxide in the supercritical region, in: Proceedings of the 2nd All-Soviet Union conference on Heat and Mass Transfer, Minks, Belarus, May 1964, Published as Rand Report R-451-PR, C. Gazley, Jr., J.P. Hartnett, E.R.C. Ecker (Eds.) 1, (1967) 26–35, cited from [17].
[34] A.A. Bishop, R.O. Sandberg, L.S. Tong, Forced convection heat transfer to water at near-critical temperatures and supercritical pressures, AIChE Chem. Eng. Symp. Ser. 2 (1965) 77–85.

[35] S. Mokry, I. Pioro, A. Farah, K. King, S. Gupta, W. Peiman, P. Kirillov, Development of supercritical water heat-transfer correlation for vertical bare tubes, Nucl. Eng. Des. 241 (2011) 1126–1136.
[36] H. Zahlan, D.C. Groeneveld, S. Tavoularis, S. Mokry, I. Pioro, Assessment of supercritical heat transfer prediction methods, in: Proceedings 5th International Symposium SCWR (ISSCWR-5), Vancouver, British Columbia, Canada, March 13–16, 2011.
[37] Verein Deutscher Ingenieure, VDI-Wärmeatlas, Filmkondensation reiner Dämpfe, eighth ed., VDI-Verlag, Düsseldorf, Germany, 1997, Section Ja1–Ja16. Engl. ed. VDI-Gesellschaft Verfahrenstechnik und Chemieingenieurwesen (Ed.), VDI Heat Atlas, 2nd Ed., Springer, Berlin, Heidelberg, New York, 2010.
[38] W. Nusselt, Die Oberflächenkondensation des Wasserdampfes, Zeitschrift des Vereins Deutscher Ingenieure 60 (1916) 151.
[39] R. Stockfleth, G. Brunner, Hydrodynamics of a packed countercurrent column for the gas extraction, Ind. Eng. Chem. Res. 38 (1999) 4000–4006.
[40] R. Stockfleth, G. Brunner, Holdup, pressure drop, and flooding in packed countercurrent columns for the gas extraction, Ind. Eng. Chem. Res. 40 (2001) 347–356.
[41] G.B. Wallis, One-Dimensional Two-Phase Flow, Mc Graw Hill, New York, USA, 1969.
[42] Verein Deutscher Ingenieure, VDI-Wärmeatlas, Kondensation von Mehrstoffgemischen, eighth ed., VDI-Verlag, Düsseldorf, Germany, 1997, Section Jbb1–Jbb38. Engl. ed. VDI-Gesellschaft Verfahrenstechnik und Chemieingenieurwesen (Ed.), VDI Heat Atlas, 2nd Ed., Springer, Berlin, Heidelberg, New York, 2010.
[43] W.W. Ackers, M.M. Turner, Condensation of vapour of immiscible liquids, AIChE J. 8 (1962) 587–599.
[44] E. Liebold, Zur Abscheidung von Kohlenwasserstoffen aus wasserdampfhaltigen Gasgemischen bei höheren Temperaturen und Drücken, Ph. D. Dissertation, Hamburg University of Technology, Hamburg, Germany, 1987.
[45] Verein Deutscher Ingenieure, VDI-Wärmeatlas, Freie Konvektion in offenen Fluidschichten, eighth ed., VDI-Verlag, Düsseldorf, Germany, 1997, Section Fd1–Fd4. Engl. ed. VDI-Gesellschaft Verfahrenstechnik und Chemieingenieurwesen (Ed.), VDI Heat Atlas, 2nd ed., Springer, Berlin, Heidelberg, New York, 2010.
[46] V. Gnielinski, vt "verfahrenstechnik"12 (1978) 363–366; V. Gnielinski, vt "verfahrenstechnik"16 (1982) 36–39.
[47] Verein Deutscher Ingenieure, VDI-Wärmeatlas, Wärmeübertragung Partikel-Fluid in durchströmten Haufwerken, eighth ed., VDI-Verlag, Düsseldorf, Germany, 1997, Section Gj1-Gj2. Engl. ed. VDI-Gesellschaft Verfahrenstechnik und Chemieingenieurwesen (Ed.), VDI Heat Atlas, 2ndEd., Springer, Berlin, Heidelberg, New York, 2010.
[48] The Chemical Engineers' Resource Page, 1442 Goswick Ridge Road, Midlothian, VA 23114, 2011, www.cheresources.com.

Chapter 5

Reactions in Hydrothermal and Supercritical Water

5.1 INTRODUCTION

This chapter contains a discussion of chemical reactions of various compounds and materials in hydrothermal and supercritical water. Experimental methods for reaction mechanisms and reaction kinetics are discussed, since the type of experimental equipment (batch, semibatch, continuous) affects depth and precision of knowledge needed for modeling kinetics of reactions. Theory of reactions with hydrothermal and supercritical water (as well as for other compounds) is formulated on three levels of detailed description, molecular, mechanistic, and lumped level. Each level needs a different amount of knowledge and is useful for different purposes, for example, for fundamental research, conceptual process design, or process optimization. In this chapter, reactions of diverse components with hydrothermal and supercritical water are considered and representative examples are discussed.

The application of hydrothermal and supercritical water for chemical reactions is characterized by the wide range of conditions and the exceptional change of properties of the reaction medium "water." Conditions of state of water (and aqueous mixtures) can be adjusted without adding or removing compounds from the reaction mixture in the near-critical and supercritical region due to the compressibility of the reaction medium. Water can exert totally different actions to other compounds, depending on its thermodynamic state, in particular density and temperature. Therefore, water at high temperature and high pressure is a variable reaction medium [1]. In general, below the critical temperature of water, reaction rates usually increase with temperature. But, passing the critical point of water, the reaction rates decrease drastically [2], due to the change in physical properties.

Reactions with water are completely different in gaseous or liquid water, as is obvious from the different structure of water vapor (steam) and liquid water as discussed in Chapter 2. In the context of this book, only reactions in a dense state of water, considerably different from normal gases, are considered.

Supercritical Fluid Science and Technology, Vol. 5. http://dx.doi.org/10.1016/B978-0-444-59413-6.00005-4

For most reactions in the hydrothermal range, the reactions are kept in the liquid phase. Consequently, the operating pressure increases with the reaction temperature, according to the vapor pressure of water. The main reason for reactions in the liquid phase is that gaseous water, if temperatures are below the critical region, loses one of the most important properties for a reaction medium: solvent power.

Several properties of water change essentially with increasing temperature. The ionic product increases up to a temperature of about $T=250\ ^\circ C$ by three orders of magnitude, dielectric constant drops from about 80 D to about 2 D, and density decreases to about one-third of its value under ambient conditions, when the critical temperature is approached.

Nevertheless, up to relatively high temperatures, liquid water maintains essentially the familiar properties of a highly polar solvent. The dramatic change occurs in the neighborhood of the critical point, the so-called critical region. Within about 20 K, properties of water change in such a way that familiar properties of water are lost. Water loses its solvent power for ionic species and becomes a good solvent for nonpolar components. Most important, water in the critical region has a density that is easily changed by pressure and temperature. Although supercritical water largely loses its solvent power at low pressures, it is still a polar solvent. At higher pressures, a good solvent power and higher polarity can still be reached for supercritical water. Thus, the near-critical and supercritical region is a working field for realizing different reaction conditions and for recovering products by changing conditions. Adjusting temperature and pressure may change the selectivity of reactions in near- and supercritical water. For example, in supercritical water at low density (low pressure), the ionic product is low and therefore different reactions occur than at high density (high pressure) and a high ionic product.

A high relative dielectric constant lowers the activation energy of the reaction with a transition state of higher polarity than the initial state. By variation of the relative dielectric constant, the reaction rates may be controlled [2].

Furthermore, water at near-critical and supercritical conditions is fully miscible with many organic and inorganic compounds, enabling reactions in a homogeneous medium (see Chapter 3). Therefore, supercritical water can be a reaction medium for reactions usually carried out in organic solvents. Examples include reactions with organometallic complexes, which are unexpectedly stable in supercritical water [2]. Again, by changing conditions, a split in more than one phase can be induced, thus providing means for product recovery.

Supercritical water is an excellent reaction medium for reactions requiring heterogeneous catalysts, since high diffusivity avoids mass transfer limitation and high solvent power prevents coke formation and poisoning of the catalyst [2]. The reaction rates of small free radicals are increased due to the high collision frequency [3]. Reactions of high molecular mass free radicals, as they occur during pyrolysis, are slowed down by a cage effect caused by solvent molecules at high pressure [4]. This effect may be the reason why organometallic complexes exist and act as catalysts in supercritical water [2].

The ionic product of water increases slightly with temperature up to around 10^{-11} in the range $T = 200$–$300\,°C$. Under these conditions, acid- or base-catalyzed reactions have high reaction rates without addition of acids or bases. These reactions show a characteristic non-Arrhenius kinetic behavior near the critical point of water [5,6]. Water is assumed to participate in activated complexes during reaction, lowering the activation energy and acting thus as a catalyst [7,8].

5.2 CHEMICAL EQUILIBRIUM AND CHEMICAL KINETICS

Chemical reactions with hydrothermal and supercritical water are described by the general rules of chemical equilibrium and chemical kinetics. These are shortly repeated here. For detailed and further information, numerous textbooks are available discussing these fundamentals, for example [9].

5.2.1 Chemical Equilibrium

A chemical reaction of compounds A and B as reactants and C and D as products is considered as a forward reaction (→) and a backward reaction (←). At equilibrium, both reactions proceed to a certain degree.

The reaction $|\nu_A|A + |\nu_B|B \leftrightarrow |\nu_C|C + |\nu_D|D$ is written as

$$\nu_A A + \nu_B B + \nu_C C + \nu_D D = 0, \tag{5.1}$$

with ν_i as the stoichiometric coefficients. Reactant compounds A and B decrease and products C and D increase until equilibrium is attained, with ν_A and $\nu_B < 0$; ν_C and $\nu_D > 0$. The forward reaction rate is $r_{\rightarrow} = k_{\rightarrow} A^{\nu_A} B^{\nu_B}$, and the backward reaction rate is $r_{\leftarrow} = k_{\leftarrow} C^{\nu_C} D^{\nu_D}$. At equilibrium, forward and backward rates are equal:

$$k_{\rightarrow} a^{\nu_A} b^{\nu_B} = k_{\leftarrow} c^{\nu_C} d^{\nu_D} \tag{5.2}$$

with a^{ν_A} as the activity of compound A with $a^{\nu_A} = [A] \cdot \gamma_A$, where γ_A is the activity coefficient and [A] is the molar concentration of compound A. The ratio of the rate constants, the equilibrium constant K, is also a constant:

$$K = \frac{k_{\rightarrow}}{k_{\leftarrow}} = \frac{c^{\nu_C} \cdot d^{\nu_D}}{a^{\nu_A} \cdot b^{\nu_B}} \tag{5.3}$$

The law of mass action is valid only for one-step reactions that proceed through a single transition state, as rate equations do not, in general, follow the stoichiometry of more complex reactions. Equality of forward and backward reaction rates is a necessary condition for chemical equilibrium.

From a thermodynamic point of view, chemical equilibrium is reached (at constant temperature and pressure) for:

$$dG = \sum v_i \mu_i = 0(P, T = \text{const.}), \tag{5.4}$$

with G is the Gibbs free energy and μ_i is the chemical potential of compound i.

The expression for the equilibrium constant can be rewritten as the product of a concentration quotient (concentrations in square brackets [A]), the equilibrium constant K_c, and an activity coefficient quotient Γ.

$$K = \frac{[\mathrm{C}]^{\nu_\mathrm{C}} \cdot [\mathrm{D}]^{\nu_\mathrm{D}}}{[\mathrm{A}]^{\nu_\mathrm{A}} \cdot [\mathrm{B}]^{\nu_\mathrm{B}}} \cdot \frac{\gamma_\mathrm{C}^{\nu_\mathrm{C}} \cdot \gamma_D^{\nu_D}}{\gamma_\mathrm{A}^{\nu_\mathrm{A}} \cdot \gamma_\mathrm{B}^{\nu_\mathrm{B}}} = K_c \Gamma \tag{5.5}$$

or in fugacities, with $RT\ln f_i = \mu_i - \mu_i^0$ (where μ_i^0 is the chemical potential at a reference point).

$$\Delta g_\mathrm{R} = \sum v_i \, \Delta g_{\mathrm{B},i} = -RT\ln \prod \left(\frac{f_i}{f_i^0} \right)^{v_i} = -RT\ln K \tag{5.6}$$

with $\Delta g_{\mathrm{B},i} = \mu_i$ and K as the equilibrium constant.

The equilibrium constant K can then be written as

$$K = \left(\frac{f_1}{f_1^0} \right)^{v_1} \cdot \left(\frac{f_2}{f_2^0} \right)^{v_2} \cdots\cdots \left(\frac{f_n}{f_n^0} \right)^{v_n} = \prod \left(\frac{f_i}{f_i^0} \right)^{v_i} \tag{5.7}$$

with $f_i = y_i \varphi_i P = \varphi_i P_i$ and φ_i as the fugacity coefficient of compound i.

The influence of temperature and pressure on the equilibrium constant is expressed by Equations (5.8) and (5.9).

$$\frac{d(\ln K)}{dT} = \frac{\Delta h_R^0}{RT^2} \text{ (van't Hoff equation)} \tag{5.8}$$

The consequence of Equation (5.8) is that for exothermal reactions K decreases with increasing temperature, and for endothermic reactions K increases with increasing temperature.

$$\left(\frac{\partial \ln K}{\partial P} \right)_\mathrm{T} = \frac{-\Delta \overline{V}}{RT} \tag{5.9}$$

The partial molar volume $\overline{V}$ itself also depends on pressure. P, T, and x have influence not only on the chemical equilibrium, that is, on the maximum yield of a reaction under given conditions, but also on chemical kinetics, the velocity of a reaction, or the reaction rate.

5.2.2 Chemical Kinetics

Kinetics of reactions is usually described with models of different details: mechanistic models, pathway models, or lumped models. Mechanistic models include all the elementary steps in the reaction, pathway models consider only the observable species in the reaction process, and lumped or global models establish the overall rate law in terms of the numerical variables [10]. For the models discussed, it has to be borne in mind that chemical equilibrium as well as chemical kinetics is relevant. The reaction mechanism, which depends on the type of the model chosen, has to be regarded in terms of an equilibrium part and a kinetic part.

5.2.2.1 Mechanistic Models

Mechanistic models for kinetic rate laws describe each reaction in the overall process as a single elementary step. For the reaction $A+B \rightarrow D+E$, the rate law is given by the mass action rate, Equation (5.10), where r is the reaction rate, k is the rate constant, and C_i is the concentration.

$$r = kC_{\mathrm{A}}C_{\mathrm{B}} \tag{5.10}$$

Rate laws of mechanistic models are always first-order (unimolecular process) or second-order (bimolecular process) for homogeneous reactions. Due to the complex reaction mechanisms for reactions with water, only few and very simple hydrothermal reactions can be determined experimentally at the elementary level necessary for mechanistic models. Computational models at the quantum mechanical level may help to describe these elementary reactions [10,11]. With pathways models or global models other reaction rate orders are possible. Kinetics of elementary reactions in mechanistic models can, in many cases, be interpreted by the transition-state theory [12].

5.2.2.1.1 Transition-State Theory

The transition-state theory interprets the kinetics of elementary reactions as occurring through a transition-state species, or "activated complex" M^*. The transition state is defined as the state of maximum energy along the reaction coordinate, which is the minimum energy pathway between the reactants and the products on the potential energy surface. The transition-state theory for an elementary reaction is illustrated by Equation (5.11) [12].

$$\nu_{\mathrm{A}}\mathrm{A} + \nu_{\mathrm{B}}\mathrm{B} \leftrightarrow M^* \rightarrow \text{Products} \tag{5.11}$$

The rate constant k of the transition-state theory is given by Moore and Pearson [13]:

$$k = \kappa \frac{k_{\mathrm{B}}T}{h} K_{\mathrm{c}}^* \tag{5.12}$$

where κ is the transmission coefficient, k_{B} is the Boltzmann constant, h is Planck's constant, T is absolute temperature, and K_{c}^* is the concentration-based equilibrium constant for the reaction involving the reactants and the transition state.

The thermodynamic equilibrium constant, K_a^*, is related to K_c^* for molar concentrations by Equation (5.13) [14]:

$$K_c^* = \frac{K_a^*}{K_\gamma^*}, \tag{5.13}$$

with $K_\gamma^* = \prod \gamma_i^{\nu_i}$, γ_i is the activity coefficient of component i and ν_i is the stoichiometric coefficient for component i.

The rate constant in the transition-state theory can be written as

$$k_x = \kappa \frac{k_B T}{h} \frac{K_a^*}{K_\gamma^*}. \tag{5.14}$$

The rate constant in Equation (5.14) can be rewritten as

$$k_x = \kappa \frac{k_B T}{h} K_x^* \tag{5.15}$$

where $K_x^* = \prod_i x_i^{\nu_i}$.

Relating this equilibrium constant to the difference in Gibbs' free energy, ΔG^*, between the activated complex and the reactants, we can write the rate constant as

$$k_x = \kappa \frac{k_B T}{h} \exp\left(\frac{-\Delta G^*}{RT}\right) \tag{5.16}$$

where R is the gas constant. Thus, the rate constant for an elementary reaction step is a function of the difference in Gibb's free energies between the reactants and the transition state. This relationship between the kinetics and free energy is the basis for the linear free-energy relationships that are used to correlate kinetics for reaction families.

5.2.2.1.2 Pressure Effects—Activation Volume

Using the transition-state theory rate constant (Equation 5.15) and classical thermodynamics, the pressure dependence of the rate constant can be derived:

$$\left(\frac{\partial \ln k_x}{\partial P}\right)_T = \left(\frac{\partial \ln K_x^*}{\partial P}\right)_T + \left(\frac{\partial \ln \kappa}{\partial P}\right)_T = -\frac{\Delta v^*}{RT} + \left(\frac{\partial \ln \kappa}{\partial P}\right)_T \tag{5.17}$$

where Δv^* is the activation volume, that is the difference between the partial molar volumes $\overline{v}$ of the activated complex and the reactants:

$$\Delta v^* = \overline{v}^* - \nu_A \overline{v}_A - \nu_B \overline{v}_B - \cdots \tag{5.18}$$

The second term on the right-hand side of Equation (5.17), $(\partial ln\kappa/\partial P)_T$, is often neglected due to insufficient information and set to be equal to unity. Nevertheless, experimental measurements of the pressure dependence of the rate constant include the effect of pressure on the transmission coefficient. A discussion can be found in Ref. [15].

Assuming that kinetics is influenced by the variation of properties and phase behavior, the apparent activation volume for a reaction in a supercritical fluid can be related to the dramatic change in properties of supercritical fluids near the critical point.

Activation volumes for reactions in the liquid phase are typically between $\Delta v^* = -50$ and $30\ \mathrm{cm^3/mol}$. For reactions in fluids near their critical points, apparent activation volumes on the order of $\Delta v^* = 1000\ \mathrm{cm^3/mol}$ have been reported. Away from the critical point, activation volumes approach their liquid-phase values [12]. The activation volume for decomposition of α-chlorobenzyl methyl ether was $\Delta v^* = -6000\ \mathrm{cm^3/mol}$ at $T_r = 1.04$ and $P_r = 1.0$ but changed to $\Delta v^* = -71.6\ \mathrm{cm^3/mol}$ at $T_r = 1.09$ and $P_r = 6.1$ [16].

5.2.2.1.3 Reactions in Solution (Solvent Effects)

The transition-state theory provides a basis for exploring and explaining solvent effects on chemical reactions. Solvent effects are especially important for reactions in supercritical fluids. In some cases, it is of interest to manipulate these effects, whereas in other cases, these effects are used to infer molecular-level information. Throughout this section, only an elementary bimolecular reaction, $A + B \rightarrow C$, is considered, but the methods can be readily extended to other types of reactions.

The rate constant for a bimolecular reaction in a given solvent can be related to the rate constant, k_0, in an ideal fluid phase as

$$k = k_0 \frac{\gamma_A \gamma_B}{\gamma_{M^*}} \tag{5.19}$$

Activity coefficients can be derived from molecular thermodynamics to quantify solvent effects on chemical kinetics in appropriate cases. Regular solution theory gives the activity coefficient as

$$RT \ln \gamma_i = V_i \left(\delta_i - \overline{\delta}\right)^2 \tag{5.20}$$

where δ_i and V_i are the solubility parameter and molar volume of component i, and $\overline{\delta}$ is the solubility parameter of the solvent. Combining this activity coefficient model with the rate constant expression in Equation (5.19) gives

$$\ln \frac{k}{k_0} = \frac{V_A\left(\delta_A - \overline{\delta}\right)^2 + V_B\left(\delta_B - \overline{\delta}\right)^2 - V_{M^*}\left(\delta_{M^*} - \overline{\delta}\right)^2}{RT} \tag{5.21}$$

This expression allows correlation or prediction of the effect of solvents with different solubility parameters on the rate constant. The problem can be simplified by assuming that the molar volume of the activated complex is equal to the sum of the molar volumes of the reactants [12]:

$$V_{M^*} = V_A + V_B \tag{5.22}$$

According to Equation (5.21), the logarithm of the rate constant ratio varies linearly with the solvent solubility parameter $\overline{\delta}$. This has been verified experimentally for reactions in a series of different liquid solvents [13,17] and for hydrolysis reactions in supercritical water wherein the solubility parameter was changed by varying the density [18].

The dielectric constant can be used for correlating solution-phase reaction kinetics for reactions involving polar molecules. From electrostatics, Kirkwood derived that the free energy of a point dipole with dipole moment μ in a medium of dielectric constant ε varies according to Equation (5.23) [19].

$$\Delta G \sim \frac{\varepsilon - 1}{2\varepsilon + 1} \tag{5.23}$$

The combination of free energy with the transition-state expression for the rate constant leads to the consequence that the rate constant should vary linearly with the quantity $(\epsilon - 1)/(2\varepsilon + 1)$ on a semilogarithmic plot. The relation has been confirmed for reactions in the liquid phase. Kinetics can also be correlated using $1/\varepsilon$, which is linearly related to $(\epsilon - 1)/(2\varepsilon + 1)$ [12].

Ionic strength (I) is also a solvent property that can influences reactions involving ions or polar molecules. The influence of ionic strength on a chemical reaction involving ions at low concentrations can be derived from the transition-state theory and the Debye–Hückel expression for the activity coefficients, resulting in Equation (5.24):

$$\ln k \cong \ln k_0 + 2 z_A z_B \alpha \sqrt{I}, \tag{5.24}$$

where z_i is the charge on species i, α is a constant, and k is the rate constant at infinite dilution in a given solvent. According to Equation (5.24), the rate constant varies with the square root of the ionic strength, revealing the influence of salts [12].

Chemical reactions can also be influenced through cage effects from solvents. Solvent molecules can form a cage around reactive species. The reaction rate then depends on the rate at which the caged intermediates react with each other and the rate at which they escape from the cage [12]. High densities, low diffusivities, and high viscosities increase the probability that solvent cage effects may contribute to the observed kinetics.

By solute–solvent interactions, a solute's electronic energy levels can be affected by the solvent. The solvent-induced change in the electronic absorption spectrum of a solute is named solvatochromism. The solvent can influence the wavelength of maximum absorption (λ_{max}). This influence is expressed in terms of a transition energy, E_T:

$$E_T = \frac{h \cdot c \cdot N_{Av}}{\lambda_{max}}, \tag{5.25}$$

where c is the speed of light, h is Planck's constant, N_{Av} is Avogadro's number.

Solvatochromic scales can be used to correlate solvent effects on rate constants in the liquid state, but it can also be applied to supercritical fluids [16]. Rate constants can often be correlated as linear functions of the transition energy as in Equation (5.26), with a and b as pressure-independent parameters:

$$\ln k_x = -aE_T + b. \tag{5.26}$$

The activation volume can then be obtained as

$$\Delta v^* = aRT\left(\frac{\partial E_T}{\partial \rho}\right)_T \rho \kappa_T \tag{5.27}$$

Thus, solvatochromic shift data can be a route to determine activation volumes [12].

5.2.2.2 *Pathways Level*

Pathway models consider only the observable species in the reaction process. The rate law at the pathways level summarizes a multistep process in one expression. At the pathways level, the kinetic rate law for homogeneous processes can be determined from the reaction order or from stoichiometry of the reaction. Reactions in H_2O are frequently second-order but can be described as pseudo-first-order if H_2O is in large excess compared to the concentration of the other reactant [12]. For example, the rate law for the reaction $A + B \rightarrow D + E$ at the pathways level is given by Equation (5.28), where K is the equilibrium constant [12].

$$r = k(C_A C_B - C_D C_E / K) \tag{5.28}$$

Since the reaction pathway is a combination of several mechanistic reaction steps, the rate law at the pathways level is a simplification of the rate law at the mechanistic level [12]. For an example, the decarboxylation of propiolic acid in hydrothermal water (Equation 5.29), presented by Brill and Savage [10], is discussed. The decarboxylation rate of propiolic acid can be determined from the propiolic acid decomposed and from the CO_2 formed as shown in Figure 5.1.

Since propiolic acid is a moderately strong acid, its ionization has to be taken into account, as in Equation (5.29) (see also the discussion in Chapter 3). Even at $P = 27.5$ MPa and $T = 160$ °C, the acid is about 5% ionized [12].

$$HC{\equiv}CCO_2H + H_2O \leftarrow K_1 \rightarrow HC{\equiv}CCO_2^- + H_3O^+ \tag{5.29}$$

$$[HC{\equiv}CCO_2H]K_1 = [HC{\equiv}CCO_2^-][H^+]$$

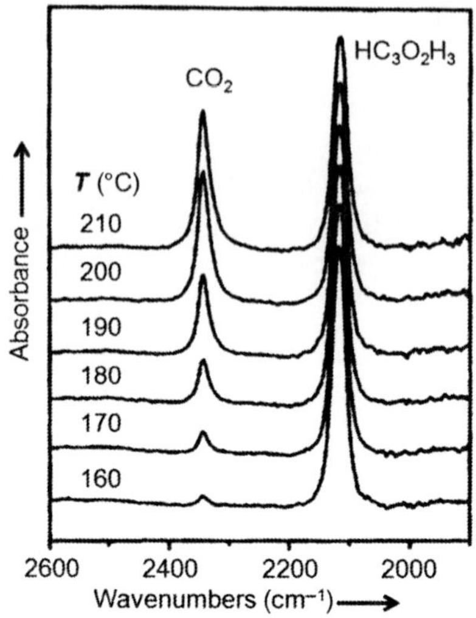

FIGURE 5.1 Transmission IR spectra for the decarboxylation of 0.5 molal propiolic acid as a function of temperature at a reaction time of 44 s at $P = 27.5$ MPa. *Modified from Ref. [12].*

Therefore, the total CO_2 released, $[CO_2]_t$, comes from both the neutral acid and its anions, Equation (5.30). The acid–base equilibrium is always maintained because the equilibration rate is very fast compared with the decarboxylation rate. In addition to the acid ionization, the mass balance and the charge balance (Equations 5.31 and 5.32) have to be considered.

$$[CO_2]_t = [CO_2]_{free} + [HCO_3^-] + [CO_3^{2-}] \quad (5.30)$$

$$[HC \equiv CCO_2H]_0 = [HC \equiv CCO_2H]_t + [HC \equiv CCO_2^-]_t + [CO_2]_t \quad (5.31)$$

$$[H^+] = [OH^-] + [HC \equiv CCO_2^-] \quad (5.32)$$

Hydrolysis equilibria of CO_2, as written in Equations (5.33) and (5.34), can be ignored at the pH of propiolic acid (pH = 1.91 at $T = 25$ °C). Then $[CO_2]_t = [CO_2]_{free}$.

$$CO_2 + H_2O \leftarrow K_{a1} \rightarrow H_3O^+ + HCO_3^- \quad (5.33)$$

$$HCO_3^- + H_2O \leftarrow K_{a2} \rightarrow H_3O^+ + CO_3^{2-} \quad (5.34)$$

Therefore, Equation (5.35) results and can be used when $[CO_2]$ is employed to indicate the reaction rate.

$$\ln([HC \equiv CCO_2H])_t = \ln([HC \equiv CCO_2H]_0 - [CO_2]_t) - \ln\left(1 + \frac{K_1}{[H^+]}\right) \quad (5.35)$$

By knowing the propiolic acid concentration as a function of time, the first-order rate equation (Equation 5.36) can be solved as a function of time to obtain the rate constant k.

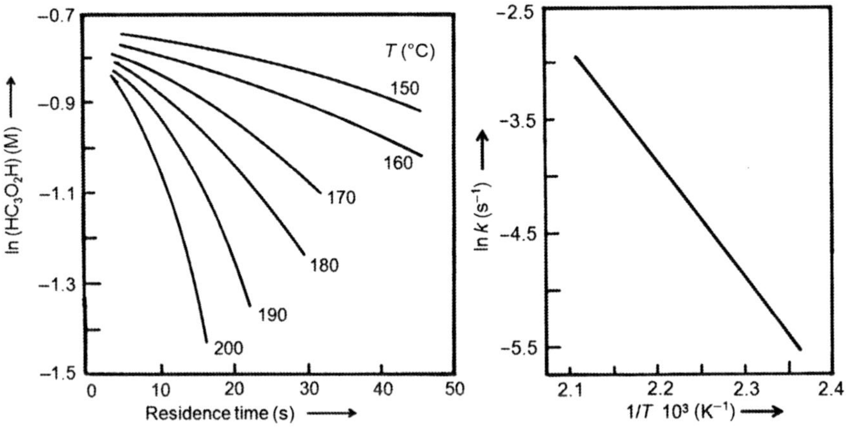

FIGURE 5.2 Left: Decarboxylation of 0.5 molal propiolic acid at $P=27.5$ MPa, derived from IR spectra for CO_2 formation and alkyne group concentration. Right: Arrhenius plot for the decarboxylation of propiolic acid at $P=27.5$ MPa. *Modified from Ref. [12].*

$$[HC\equiv CCO_2H]_t = [HC\equiv CCO_2H]_0 \exp(-kt) \quad (5.36)$$

In addition, since the experimental concentration of propiolic acid is also available from the area of C≡C infrared absorption, the reaction rate can also be calculated directly from Equation (5.36).

The variations of the logarithm of the acid concentration with residence time based on the decrease of the C≡C stretching mode intensity of the acid and CO_2 formation are shown in Figure 5.2 (left). The approximate linearity of these rate plots suggests that the decarboxylation of propiolic acid is of first-order or pseudo-first order. Figure 5.2 (right) shows the Arrhenius plot based on the average of the rate constants for CO_2 and propiolic acid. A comparison of the calculated and observed concentration profile of the species is shown in Figure 5.3.

The procedure used to calculate $[H^+]$ based on the observed $[CO_2]$ is more complicated than that based on the C≡C bond of the propiolate anion because Equations (5.30) and (5.31) become relevant. Therefore, the kinetic determination is more readily based on the intensity of the C≡C infrared absorption.

5.2.2.3 *Global or Lumped Level*

Lumped or global models establish the overall rate law in terms of the numerical variables [10]. Results of experiments are correlated with phenomenological kinetics models. This approach provides no immediate insight into the reaction mechanism but provides a mathematical model that can be used for engineering purposes. These phenomenological kinetics models are expressed in the form shown in Equation (5.37) [12].

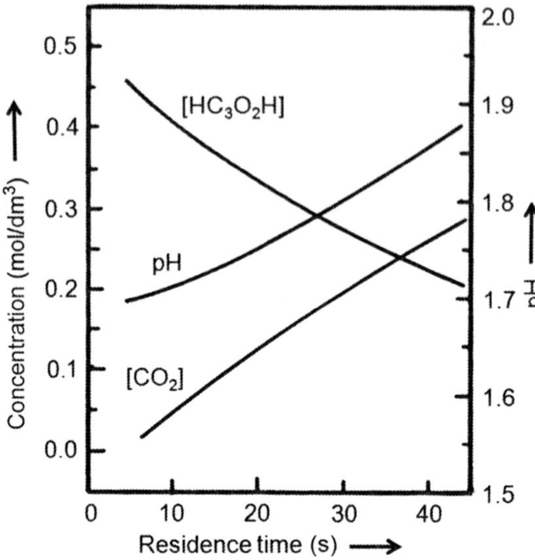

FIGURE 5.3 Concentration of species for the decarboxylaton of propiolic acid, $T=180\,°\text{C}$, $P=27.5$ MPa. *Modified from Ref. [12].*

$$\frac{\mathrm{d}C_i}{\mathrm{d}t}=r_i=-A\cdot\exp\left(\frac{-E_\mathrm{a}}{RT}\right)\prod C_i^{a_i} \tag{5.37}$$

The reaction order (a_i) and Arrhenius parameters (A, E_a) in the global rate constant are treated as adjustable parameters to fit the rate r_i. The rate equation is combined with the equation that describes the flow and reaction processes in the reactor used in the experiments.

The Arrhenius equation (Equation 5.38) provides the temperature dependence of the rate constant:

$$k=A\cdot\exp\left(\frac{-E_\mathrm{a}}{RT}\right), \tag{5.38}$$

where A is the preexponential factor and E_a is the activation energy. These quantities may be expressed in thermodynamic terms as the entropy of activation ΔS^A and the enthalpy of activation ΔH^A, respectively, by Equations (5.39) and (5.40):

$$\Delta S^\mathrm{A}=R\ln A-R\ln\left(\frac{kT}{h}\right)-R, \tag{5.39}$$

$$\Delta H^\mathrm{A}=E_\mathrm{a}-RT, \tag{5.40}$$

with k as the Boltzmann constant and h as the Planck's constant.

Knowledge about the temperature dependence of the rate constant makes possible limited extrapolation of the reaction rate, if it is assumed that the reaction mechanism does not change and that the effect of the variations in the properties of H_2O on the reaction rate is understood [12].

For reactions with rates that depend on the concentration of only one component (typically the organic compound), Equation (5.37) can be solved analytically for the reactant conversion, X, obtained in an isothermal reactor. Equation (5.41) results and is valid for reactions in liquid and in supercritical water, provided the reaction has a negligible effect on the total number of moles in the system [12].

$$X = 1 - \left(1 + (a-1)A\,e^{-E_a/RT}[\text{organic}]_0^{a-1}t\right)^{1/(1-a)} \quad \text{for } a \neq 1 \tag{5.41}$$

This equation can be applied also for reactions in which water participates in the reaction, provided the water concentration is high and can be taken to be essentially constant during the reaction. In these cases, the rate constant includes the concentration of water and any other reactants present in high amounts. The numerical values for the parameters in the rate equation can be determined by fitting a set of experimental data to the model equation via a nonlinear regression analysis [12].

5.2.2.4 *Detailed Chemical Kinetics Model*

A second engineering approach is to develop a detailed chemical kinetics model based on the governing reaction mechanism. This model provides information about reaction intermediates and rates of individual steps. An example is the study by Ederer *et al.* [4], who used the detailed chemical kinetics model approach for modeling *t*-butylbenzene pyrolysis in supercritical water. A set of elementary reaction steps was postulated to account for the underlying chemistry. The steps were drawn from well-studied thermal hydrocarbon chemistry. Assigning rate constants to each step allowed the mechanism to be coupled with the governing reactor design equation to model the reacting system. The model revealed the chemical causes for the differences observed experimentally for thermal pyrolysis and hydrolysis in supercritical water [12].

5.2.2.5 *Surface-Catalyzed Reactions*

Surface-catalyzed reactions are heterogeneous processes in which the rate law is much more complicated than for homogeneous systems. To date the most popular and successful means to describe catalytic kinetics is through the Langmuir–Hinshelwood–Hougan–Watson (LHHW) formalism [12,20]. Heterogeneous rate laws are discussed in terms of the classical LHHW formalism, given by Equation (5.42).

$$r = \frac{(\text{kinetic group})(\text{driving force group})}{(\text{adsorption group})^n} \tag{5.42}$$

The kinetic group contains the rate constant for the rate-determining step. The driving force group is related to the concentrations of the reactants and products and therefore to the equilibrium constant of the reaction, which is large initially and decreases as the reaction progresses. The adsorption group contains a term for each chemical component that adsorbs onto the surface [12].

Surface reactions under hydrothermal conditions are still under investigation. One effect is that species such as CO are able to extract metals (e.g., Ni) from steel and form highly toxic $Ni(CO)_4$ under hydrothermal conditions [12,21].

5.3 SPECIFIC REACTIONS IN SUPERCRITICAL WATER AS A REACTION MEDIUM

5.3.1 Introduction

In this section, chemical reactions of compounds of different chemical structures in high-temperature liquid and supercritical water are discussed with respect to chemical synthesis of specific compounds, where water is the reaction environment and water is included in the reactions as a reactant, as in hydrolysis. Water in the high-temperature liquid and supercritical state provides the reaction environment and can react with most molecules. The following paragraph will present first a short discussion of experimental methods for determination of reaction mechanisms and reaction kinetics. Then, hydrothermal reactions of hydrocarbons will be treated, followed by a discussion of hydrolysis reactions. Other reactions that are covered are reactions of acids and bases, some organic chemical reaction types, inorganic reactions, and reactions catalyzed by organometallic complexes, as well as partial oxidation reactions. Reactions of biomass, fuel sources, polymers, total oxidation, and inorganic reactions in an aqueous environment at high temperatures are treated separately in Chapters 7–12.

Water at higher temperatures and pressures can act as an aqueous environment, as a solvent, as a catalyst, and as a reactant [2,22]. As such its role has been discussed intensively in the literature. Quite a number of different names have been chosen to characterize the state of water throughout the numerous contributions about the role of high-temperature water and supercritical water with respect to chemical reactions and physical processes. Within this book, "high-temperature water" is used for water in the liquid state at "high" temperatures. In this context, "high temperature" begins at a temperature of $T = 100\ °C$ (373.15 K). Supercritical water is clearly defined as being in a state of a higher temperature than the critical temperature of water, $T = 647.096$ K (373.946 °C), and a pressure higher than the critical pressure of water ($P = 22.064$ MPa). Density of water beyond the critical temperature can be below the critical density ($\rho = 0.322\ g/cm^3$), a density not typical for a common liquid, or above the critical density with values approaching the density of common liquids. Variations of density influence reaction paths and reaction kinetics. Water in the neighborhood of the critical point is named "near-critical." The neighborhood is defined as the region where rapid changes in properties occur, compare also Chapter 2.4.3, last section. When the critical isotherm is crossed, either by temperature or by pressure variation, the properties of water change not abruptly, but rather in a smooth, but rapid change. The states of water, as they are named in this chapter, are illustrated in Figure 5.4.

Reactions with water are facilitated and can be directed by changes in the chemical and physical properties of water as temperature increases. Solvent properties of water at high temperature become similar to those of polar organic solvents at room temperature. Reactions that are normally carried out with organic compounds can be carried out in an environmentally friendly medium [23].

Advantages of reactions in hydrothermal and supercritical water include: due to the increase in the dissociation constant by three orders of magnitude, water at temperatures above $T=200\,°C$ acts as an acid, a base, or an acid–base catalyst. Thus, costly neutralization and catalyst regeneration steps can be avoided [23]. Furthermore, under supercritical conditions, the ionic product of water also increases with pressure, increasing reaction rates of acid- or base-catalyzed reactions. The addition of acids and bases may be avoided also in this case. Reaction rates of free radicals decrease with pressure, thus suppressing undesirable reactions during pyrolysis [2].

Pressure and density in that way are important parameters. Variations of these parameters may be applied to accelerate reaction rates or enhance

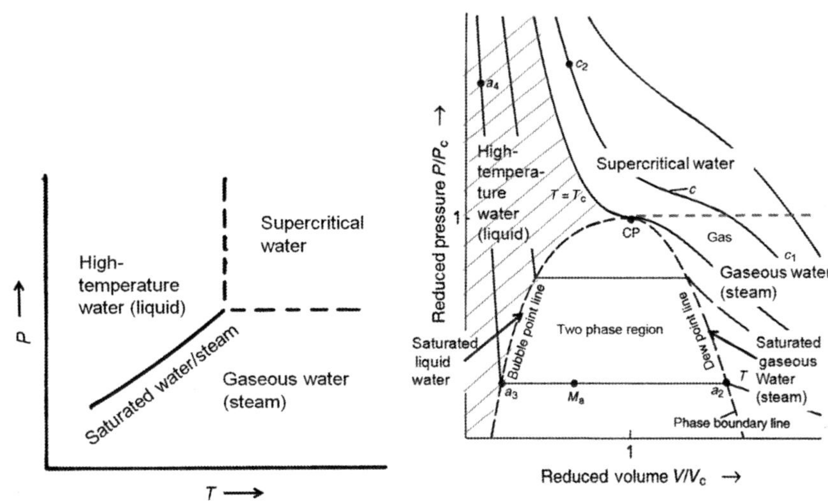

FIGURE 5.4 States of water, as used in this book.

selectivity at high temperatures, as applied for hydrothermal reactions, while at ambient temperature the rate enhancement due to high pressure may be limited. At a pressure of $P = 300$ MPa, which can be considered as an upper limit for large commercial pressure vessels, reaction rates may be 10- to 40-fold higher for pressure-sensitive reactions, as the Diels–Alder reaction ($\Delta V \approx -25$ to $-40\ \mathrm{cm^3/mol}$) [24,25]. Although the magnitude of these rate accelerations may be of interest for some applications, they are generally too low to efficiently activate reactions in general [25]. A particular advantage of carrying out reactions in high-temperature water or supercritical water is its high solvent power. After the reaction, water and organic compounds separate under ambient conditions [2].

The first measurement of a reaction under pressure in water was reported in 1892 by Röntgen in the acid-catalyzed inversion of sucrose [25,26]. Reactions of organic compounds with pure water at high temperatures, both sub- and supercritical, have been studied by many groups. Reviews have been published by Katritzky [23,28], Savage [10,12,22], Dinjus and Kruse [2], Jenner [25], Ikushima and Arai [27], Brunner [29], and others.

In the following, hydrothermal reactions of organic compounds are reviewed. A generally interesting topic is the stability and the reaction type of hydrocarbons in hydrothermal systems. The reactive action of water with organic compounds in many cases can be understood as a hydrolytic process. Other reactions, induced or assisted by an aqueous environment, are condensations, elimination reactions (e.g., dehydration), Diels–Alder reactions, Friedel–Crafts reactions, rearrangements, reactant ion reactions, oxidation reactions, organometallic reactions, hydroformylations, and others. Reactions of inorganic compounds with water play a dominant role in the so-called hydrothermal processes and are therefore discussed separately in Chapter 11.

5.3.2 Experimental Methods for Reaction Mechanisms and Reaction Kinetics

Reaction mechanisms and reaction kinetics have to be investigated in experiments. Laboratory reactors for determining reaction mechanism and reaction kinetics as basis for designing production processes are discussed in the following. Experiments must cover a wide variety of parameters under hydrothermal and supercritical water conditions. Often, temperature is around $T = 700$ K (427 °C), and pressure is about $P = 30$ MPa. The timescale of the reactions ranges from seconds to days. Corrosive conditions are common (see Chapter 12). Consequently, the number of experiments has to be reduced to as few as possible and experiments must be kept as simple as compatible with the expected results. Some properties of the different laboratory reactor types will be described below. Examples of reactors and experimental equipment are discussed in Chapter 13.

Construction of laboratory reactors and their operation are connected to the type of reaction and the method of data acquisition. Reaction yield and kinetic data are often determined by measuring the concentration of the reactants in dependence of reaction time. Data can be evaluated by integral or differential methods [55]. For the integral evaluation, data for the concentration $C_i = f(t)$ are fitted to the integral form of the equation for the course of the reaction yielding the respective parameters. For a differential evaluation, the measured concentration $C_i = f(t)$ is first differentiated to determine the reaction rates for all the reactants. Reaction rates are then related to the concentrations in order to determine the kinetic law and the respective parameters.

Reaction rates and reaction pathways are usually determined in the laboratory in batch reactors or continuous flow reactors. The characteristics of these reactors are discussed below and described in more detail in numerous textbooks, for example, in Refs. [9,55]. Reaction rates can be determined in differential reactors with controlled small transformation of the reactant. Due to experimental difficulties, these reactors are not common in hydrothermal and supercritical water research and will not be discussed. The reader is referred to the literature [55,56] for further information.

For the evaluation of reaction rates and reaction pathways, it is essential to determine the concentrations of all the components taking part in the reaction. Thus, the kinetic data can be directly related to the concentrations without time lag, which is essential for determining all the products occurring in a reaction and for the relevance of the data for process design purposes. Temperatures may vary during the reaction, in particular for those reactions with considerable reaction enthalpies. On the other hand, often an adiabatic operation of the reactor is maintained, since temperature can be correlated with time (batch reactor) or reactor length (flow reactor).

Reactions in fluid systems and in heterogeneous systems are mostly carried out in batch reactors, tubular flow reactors, and continuously operated stirred reactors.

Batch reactors are of a simple construction. No flow regulation but only a temperature control is necessary. In fluid systems, the reaction mixture must be thoroughly mixed to maintain constant conditions throughout the reactor. Sufficient mixing may be a problem at the onset of the reaction, since the reaction should begin in the whole reaction volume at a reaction time $t = 0$. It is obvious that the definition of $t = 0$ is prone to mistakes, which may be serious in cases in which the time for mixing and the reaction time are of the same order of magnitude.

Determination of the concentrations of the reactants in dependence on reaction time needs appropriate analytical methods to determine the concentrations at a certain time in the reaction volume. The best methods are those that determine the concentration *in situ*, for example, by measuring the electrical conductivity, or viscosity, or use spectroscopic methods like UV, IR, and others. In many cases, especially for multicomponent mixtures, samples have to be taken from the reactor. It must be ensured that the reaction is not proceeding in the samples after removing them from the reactor. For analyzing samples, an online chromatographic analysis is a good method. But off-line analysis is often applied due to the experimental complexity of directly coupling reactor and chromatograph.

Often batch reactors are used in a way that the reactants are introduced into the reaction volume. Then the reactor is closed and brought to reaction conditions. A certain time after reaching these conditions, the reactor is quenched and the resulting compounds of the mixture are analyzed. Such so-called bomb experiments have been widely used and are still used in investigations of chemical reactions. Results of such experiments are difficult to interpret with respect to kinetic data, reaction mechanism, and the time course of the reaction. Actually, those experiments have to be repeated if data for process design must be obtained.

Tubular flow reactors are convenient for determining the reaction yield and the reaction products after the reaction time which often is identical with the residence time of the reaction mixture in the tubular reactor. For determining relevant reaction data, a stationary situation has to be established in the flow reactor. Ideally, a turbulent flow regime exists in the tubular reactor during the reaction for a good mixing of the reactants. Analysis of the reaction products underlies the same conditions as with the batch reactor.

A semibatch reactor can overcome some problems of the tubular reactor. In a tubular reactor, the reaction time is limited by the flow rate of the reactants and the length of the reactor, usually to values of seconds to about several minutes. Reactor length in laboratory reactors is limited for practical reasons to a few meters and thus limits the extension of the residence time to higher values. At great reactor lengths, problems with pressure drop or fouling wipe out the advantages of the simple tubular reactor. To increase residence time, a continuous stirred reactor with a feed stream and an adequate removal of the product stream can be applied. In case of solid–fluid systems, a fixed bed consisting of the solid is contacted by the fluid reaction mixture. Residence times can such be regulated in a very wide range from minutes to days.

In addition to the statement about stationary conditions in flow reactors, it is possible to acquire more data about reaction rates and reaction mechanisms if the reactor is operated in a nonstationary way, provided the appropriate analytical and mathematical methods are available [55]. A discussion of methods for analyzing results of nonstationary experiments is beyond the scope of this book.

5.3.3 Hydrothermal Reactions of Hydrocarbons

The stability and the reactions of hydrocarbons in high-temperature aqueous systems are both of considerable interest for applications, for example, for transformations in synthetic fuels or in recycling of polymer compounds. As a result of numerous investigations, so far it can be concluded that reaction mechanisms for aromatic and aliphatic hydrocarbons (without heteroatoms) are essentially the same during thermal pyrolysis and in high-temperature liquid water. Water serves primarily as an inert solvent. Free-radical chemistry dominates and the rates and product distributions are often the same in both cases [10].

Although there are many cases supporting this conclusion, like, for example, bibenzyl [30], hexylbenzene and 1-decylnaphthalene [31], and *n*-hexadecane [32], there are also examples of differing behavior during thermal pyrolysis and reactivity in high-temperature water.

Aliphatic hydrocarbons can be reactive in hydrothermal systems, provided the temperature is sufficiently high. The chemistry is largely thermal, and the role of water is not central to the chemical conversions, which involve free-radical transformations. Hydrothermal reaction has been investigated, for example, for *n*-hexadecane and for polyethylene. The rates and product distribution from *n*-hexadecane pyrolysis at $T=400$–$450\ °C$ in supercritical water were almost indistinguishable from those observed from pyrolysis in argon at 0.1 MPa [32] (compare Figure 5.5, left). Pyrolysis of *n*-hexadecane was carried out at supercritical temperature for water, but in the low-density region of $\rho=0.1$ or $0.2\ g/cm^3$. Pyrolysis of polyethylene was carried out at low density of $\rho=0.13\ g/cm^3$, and at supercritical density of $\rho=0.42\ g/cm^3$. The reaction of polyethylene is much faster in water at supercritical temperatures than in argon, and the product distributions differ considerably (see Figure 5.5, right). These differences were attributed to water dissolving in the molten polyethylene phase and to pyrolysis products dissolving in water and thereby being removed from the reacting molten polyethylene phase [10,32]. Difference to argon pyrolysis is substantial and increases with

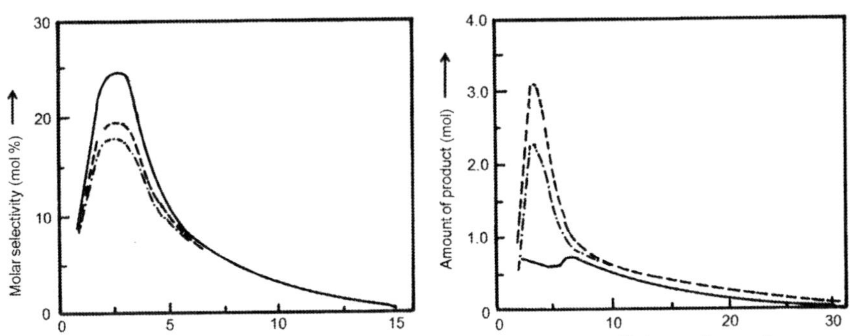

FIGURE 5.5 Product distribution of pyrolysis of *n*-hexadecane (left) and of polyethylene (right) Left: $T=723$ K, $t_R=30$ min, ——— in Ar, – – – in H_2O ($\rho_{H_2O}=0.1\,g/cm^3$), —●— in H_2O ($\rho_{H_2O}=0.2\,g/cm^3$), $x_{nC16}=56$–58%. Right: $T=693$ K, $t_R=30$ min, ——— in Ar, —●— in H_2O ($\rho_{H_2O}=0.13\,g/cm^3$), – – – in H_2O ($\rho_{H_2O}=0.42\,g/cm^3$). *Modified from Ref. [32].*

density. Argon pyrolysis results in a relative equal carbon number distribution of the pyrolysis products, while in water at supercritical temperature, a maximum, increasing with density, occurs around the carbon number 4 that is much higher than the concentration of Ar pyrolysis products. In addition, at supercritical density, the ratio of 1-alkene to *n*-alkane is much higher than at low density [32].

Ederer *et al.* [4] investigated *t*-butylbenzene pyrolysis and decomposition in high-temperature water. According to their results, the reaction rate for *t*-butylbenzene disappearance was about three orders of magnitude slower in water at supercritical temperature than in an inert gas (argon) at the same temperature and near-ambient pressure. Conversion was carried out in a flow reactor at $T=808$ K, $P=25$ MPa, $\rho_{H_2O}=83.656\ kg/m^3$, and less than 1 min reaction time. The differences between reaction in water at supercritical temperature and argon were attributed to the intrusion of cage effects on decomposition reactions in water and the promotion of substitution reactions.

Aromatic hydrocarbons are stable in high-temperature and supercritical water. For example, stability has been observed at $T=300$–$350\ °C$ for thousands of hours [33] and at $T=460\ °C$ for 1 h [31]. Chemical reactions that take place with aromatic compounds are limited to transformations of substituent groups on the ring. Toluene and benzene are nonreactive at $T=300\ °C$ for times on the order of $t=1000$ h [33]. Ethylbenzene, on the other hand, which can decompose thermally at $T=450\ °C$, only shows a 10% conversion after hydrothermal treatment at $T=450\ °C$ for 48 h [10,34].

The presence of additives in the hydrothermal system (minerals, salts, etc.) can dramatically alter the reactivity of hydrocarbons [10,28,33], see in the chapter on the conversion of fuel materials (Chapter 7).

Unsubstituted aromatic compounds are essentially stable in pure water, whereas aliphatic and aromatic compounds with thermally labile substituents can be reactive. The reactivity follows the classical free-radical mechanisms of thermal hydrocarbon chemistry, but the hydrothermal environment can alter rates and product distributions through the action of standard solvent effects [10]. In particular, the influence of density under supercritical conditions has to be taken into account and the mode of the experiments must be considered. Experiments, reported in the literature, have often been carried out at noncomparable conditions. Batch experiments carried out in batch reactors (bombs) are difficult to interpret because products cannot be related with some accuracy to process conditions, namely, residence time, temperature, and pressure (density). For example, conversion studies for diphenylether have been carried out for a total run time of $t=270$ min in a batch reactor, while the conversion of *t*-butylbenzene was carried out in a flow reactor with residence times of $t=10$–50 s. Therefore, batch reactor results can only be an indication of the parameters needed for designing a process.

Furthermore, results are often compared for experiments carried out at different densities. For example, the transformation of *n*-hexadecane was performed at low densities (see above), with the result that pyrolysis and hydrothermal reaction are not really different. The decomposition of polyethylene was carried out at low and high density, with the effect that the products at the higher density differed appreciably from that of the low-density experiments (all in the same investigation) [32].

Another example for competing pyrolysis and hydrolysis is molecules containing a saturated carbon attached to a heteroatom-containing leaving group. They undergo parallel pyrolysis and hydrolysis reactions in supercritical water. Selectivity and rate constant to hydrolysis increase with water density [5].

5.3.4 Hydrolysis/Cleavage

Hydrolysis is the chemical reaction in which a compound is decomposed by the action of water according to the formal equation:

$$A-B+H-OH \rightarrow A-H+B-OH \tag{5.43}$$

Hydrolysis processes of commercial importance are the cleavage of fats, of saccharose into glucose and fructose, of starch and cellulose into glucose, the reaction of esters and the cleavage of proteins into amino acids [37]. During hydrolysis, water acts simultaneously as solvent, reactant, and catalyst due to self-dissociation. The addition of a further catalyst may be needed to avoid undesirable side reactions [2]. Many hydrolysis reactions have been investigated like the hydrolysis of esters, ethers, amines, amides, nitriles, chlorocarbons, carbohydrates, proteins, and others. Carbohydrates and proteins are treated separately in the paragraph on biomass conversion in Chapter 8. For the other compounds, the basic reactions and an example are discussed in the following. Furthermore, specific reactions like condensation-, Diels–Alder-, organometallic-, and others will be addressed.

5.3.4.1 Esters

The hydrolysis of esters is of technical interest. Therefore, hydrolysis of many different esters was investigated. In the following, as examples, the hydrolysis of ethylacetate and triacylglycerides is presented in more detail.

5.3.4.1.1 Ethyl Acetate

The hydrolysis of ethyl acetate (Equation 5.44) was investigated in detail by Krammer and Vogel [38] in a tubular reactor, without the addition of a catalyst, at pressures of $P=23$–30 MPa, at a temperature range of $T=250$–450 °C, and at residence times of $t=4$–230 s. The degree of conversion and the selectivity of the reaction are shown in Figure 5.6.

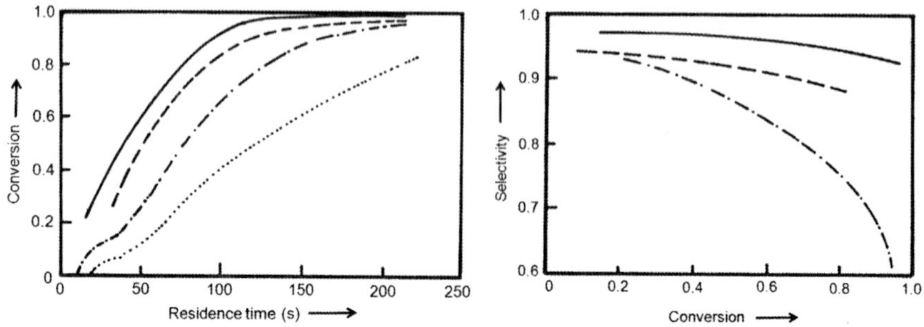

FIGURE 5.6 Left: Conversion of ethyl acetate for $P=28$ MPa. ——— 350 °C, $\rho_{H_2O}=0.64\,g/cm^3$, – – – 380 °C, $\rho_{H_2O}=0.51\,g/cm^3$, —●— 400 °C, $\rho_{H_2O}=0.26\,g/cm^3$, ●●●● 250 °C, $\rho_{H_2O}=0.82\,g/cm^3$. Right: Selectivity of the reaction for ethanol for $P=25$ MPa. ——— 300 °C, – – – 380 °C, —●—● 450 °C. *Modified from Ref. [38].*

$$H_3C-COOC_2H_5+H_2O \rightarrow CH_3COOH+C_2H_5OH \tag{5.44}$$

Conversion approaches a very high value at the temperature range of $T=350$–400 °C with residence times of about $t=100$–200 s, while at a temperature of $T=250$ °C, the conversion remains incomplete. The selectivity of the reaction for acetic acid was about 1 (100%) in all cases. The selectivity for ethanol (Figure 5.6, right) decreased with higher temperatures and residence times.

The stability of ethanol and acetic acid is of interest for choosing reaction conditions. Ethanol is stable under the reaction conditions of $P=25$ MPa, $T=250$–400 °C, and a residence time of $t=2$ min. Ethanol dehydrates to ethene at a reaction temperature of $T\geq 450$ °C [38]. According to a different investigation [39], 2.6% of 1 M ethanol is converted to ethylene at $T=400$ °C and $P=34.5$ MPa. At $T=500$ °C and $P=34.5$ MPa, the conversion decreases to 2%. Reaction products are a variety of gases: H_2, CO, CO_2, CH_4, C_2H_4, and C_2H_6 [39].

Acetic acid is stable up to a temperature of $T=400$ °C at pressures of $P=23$ and 28 MPa and at residence times of $t=0.5$–6 min. At $T=450$ °C, minor conversion of 2% can be determined [38]. A different investigation at the temperature range of $T=475$–600 °C at $P=24.6$ MPa and a residence time of $t=8$ s shows increasing conversion from 5% to 35% with decomposition products CH_4 and CO_2, and some H_2 and CO at higher reaction temperatures [40].

The reaction scheme for ethyl acetate, shown in Equations (5.45) and (5.46) [38], results in the equations for the conversion of ethyl acetate, Equations (5.47)–(5.50). The reaction rate is graphically shown in Figure 5.7.

$$-\frac{d[\mathrm{Etac}]}{dt}=k_3[\mathrm{EtacH^+}][H_2O]^2 \tag{5.45}$$

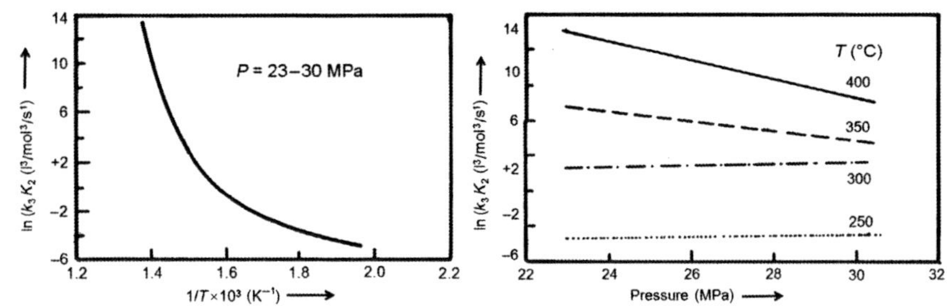

FIGURE 5.7 Left: Temperature dependence of the reaction rate of ethyl acetate hydrolysis at various pressures. Right: Pressure dependence of the reaction rate of ethyl acetate hydrolysis at various temperatures. *Modified from Ref. [38].*

$$[\mathrm{EtacH}^+] = K_2[\mathrm{Etac}][\mathrm{H}^+] \tag{5.46}$$

$$-\frac{\mathrm{d}[\mathrm{Etac}]}{\mathrm{d}t} = k_3K_2[\mathrm{Etac}][\mathrm{H}^+][\mathrm{H_2O}]^2 \tag{5.47}$$

$$-\frac{\mathrm{d}[\mathrm{Etac}]}{\mathrm{d}t} = k_3K_2[\mathrm{H_2O}]^2[\mathrm{Etac}]\sqrt{K_\mathrm{A}}\sqrt{[\mathrm{Etac}]_0 - [\mathrm{Etac}]} \tag{5.48}$$

$$[\mathrm{Etac}] = [\mathrm{Etac}]_0\left[1 - \tan h\left(\frac{a\sqrt{[\mathrm{Etac}]_0}t}{2}\right)^2\right] \tag{5.49}$$

with

$$a = k_3K_2[\mathrm{H_2O}]^2\sqrt{K_\mathrm{A}} \tag{5.50}$$

The temperature dependence of the reaction rate of ethyl acetate hydrolysis was correlated with Equation (5.51),

$$\ln(k_3K_2) = -\frac{1}{RT}\left(\Delta_\mathrm{R}H^0 + E_\mathrm{a}\right) + \mathrm{const.} \tag{5.51}$$

The reaction rate (Figure 5.7, left) shows a clear bend at temperatures above $T = 350\ °\mathrm{C}$ in the transition from the subcritical to the supercritical range. In the subcritical region, $\Delta_\mathrm{R}H^0 + E_\mathrm{a}$ is approximately 110 kJ/mol. In the supercritical region, the value is much higher, for example, 651 kJ/mol at $P = 25$ MPa [38], indicating a change in mechanism of two parallel ongoing reactions. In the subcritical region ($T = 250$–$350\ °\mathrm{C}$), the ester hydrolysis is autocatalyzed by the acetic acid formed during the reaction. In the supercritical region at $T = 400\ °\mathrm{C}$ and a pressure of $P = 25$ MPa, the pK_w value is 19.4, much smaller than at subcritical conditions. But the dissociation of water and acetic acid depend on density and temperature. The dissociation constant of water increases with pressure, for example, to a pK_w value of 12.5 at $P = 30$ MPa and $T = 380\ °\mathrm{C}$. Therefore, the change in the mechanism to the noncatalyzed mechanism should be minor at higher pressures. Nevertheless, the dominating mechanism in supercritical water is the noncatalyzed nucleophilic attack of water [38].

5.3.4.1.2 Triacylglycerides

Hydrolysis of fats, in particular triacylglycerides, is of general interest. As an example, hydrolysis of triacylglycerides from canola oil for formation of free fatty acids (FFA) is discussed in Ref. [41]. Reactions during hydrolysis in a mixture of triacylglycerides with water occur according to Equations (5.52)–(5.55).

$$\mathrm{TAG} + \mathrm{H_2O} \leftrightarrow \mathrm{DAG} + \mathrm{FFA} \tag{5.52}$$

$$\mathrm{DAG} + \mathrm{H_2O} \leftrightarrow \mathrm{MAG} + \mathrm{FFA} \tag{5.53}$$

$$\mathrm{MAG} + \mathrm{H_2O} \leftrightarrow \mathrm{Gly} + \mathrm{FFA} \tag{5.54}$$

$$\mathrm{TAG} + \mathrm{MAG} \leftrightarrow 2\mathrm{DAG} \tag{5.55}$$

with TAG, triacylglycerides; DAG, diacylglycerides; MAG, monoacylglycerides; FFA, free fatty acids; Gly, glycerol.

In equilibrium reactions, triacylglycerides are hydrolyzed to diacylglycerides, monoacylglycerides, FFA, and glycerol. It is supposed that the reaction is acid catalyzed by carbon dioxide dissolved in the aqueous phase. In order to determine the influence of CO_2 on the reactions, experiments were conducted at $T = 250$ °C, $P = 10$–30 MPa, using different mixtures of canola oil to water initial molar ratio (o/w: 1:3, 1:17, and 1:70) in a batch reactor with sampling during the experiments [41]. The concentrations of the reactant and the product compounds are illustrated in Figure 5.8, revealing that the reaction takes about 3 h with a selectivity for FFA of 93%. Products of the hydrolysis of triacylglycerides are mainly FFA. The influence of supercritical CO_2 dissolved in water on the hydrolysis is negligible at a temperature of

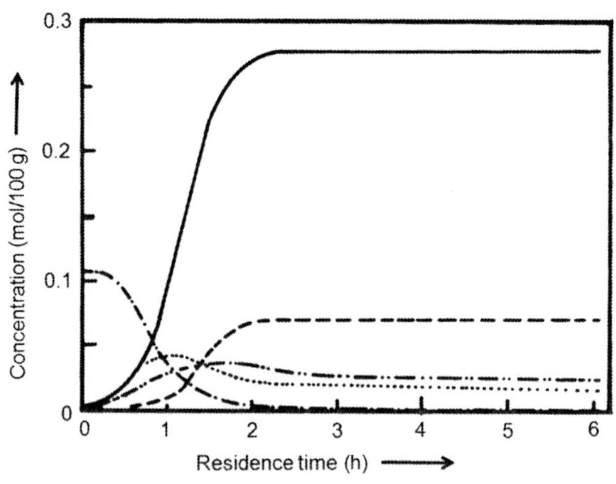

FIGURE 5.8 Hydrolysis of triacylglycerides of canola oil; $T = 250$ °C, $P = 10$ MPa CO_2, oil-to-water ratio = 1:17. ——— free fatty acids, — — — glycerol, —●●— monoacylglycerides, ●●●● diacylglycerides, —●—● triacylglycerides. *Modified from Ref. [41].*

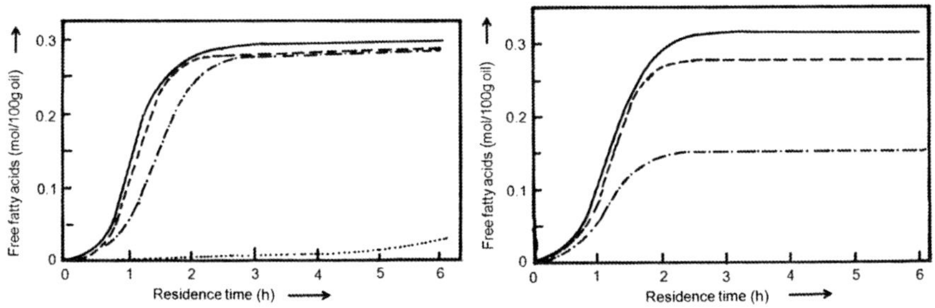

FIGURE 5.9 Left: Hydrolysis of triacylglycerides of canola oil. Oil-to-water ratio = 1:17, ●●●● 200 °C, 10 MPa CO_2, —●—● 250 °C, 30 MPa CO_2, ——— 250 °C, 10 MPa CO_2, ———250 °C, 10 MPa N_2. Right: Hydrolysis of triacylglycerides of canola oil, 250 °C, 10 MPa CO_2, oil-to-water ratio: —●—● 1:3, ——— 1:17, ——— 1:70. *Modified from Ref. [41].*

$T = 250$ °C because the hydrolysis with N_2 dissolved in the reaction mixture is practically the same (see Figure 5.9, left). The maximum rate of FFA production is not affected by pressure or added supercritical medium but is delayed about 30 min at $P = 30$ MPa. Hydrolysis increases significantly as the molar ratio of water is increased from 1:3 to 1:17 and 1:70 o/w (Figure 5.9, right).

The influence of increased water may be due to solubility effects. It is probable that at a ratio of 1:3 two condensed phases (oil and aqueous phase) are present, while at the higher ratios, only one homogeneous liquid phase exists. For comparison, the solubility of *n*-hexadecane in water may be considered (figure 3.11, Chapter 3). At $T = 200$ °C, the solubility of water in *n*-hexadecane is relatively small, while at $T = 300$ °C, the solubility has increased substantially. It is highly probable that triacylglyceride solubility in water at $T = 250$ °C is below the molar ratio of 1/3 o/w corresponding to 25 mol% solubility of water in triacylglycerides, while at 5.5 mol% corresponding to a molar ratio of 1/17 o/w, the concentration of the reaction mixture is in the one-phase region.

Adding CO_2 to the reaction mixture has no effect on the triacylglyceride conversion, as the reaction with dissolved N_2 reveals. Hydrolysis of triacylglycerides is minimal at $T = 200$ °C, $P = 10$ MPa compared to the hydrolytic conversion at $T = 250$ °C, $P = 10$ MPa. The low conversion of triacylglycerides cannot be attributed to solubility differences, and therefore not to acid-catalysis, induced by a pH-effect. The solubility of CO_2 in H_2O at $T = 200$ °C, $P = 10$ MPa is 0.013 mol/mol, at $T = 250$ °C, $P = 10$ MPa is 0.0123 mol/mol, and 0.042 mol/mol at $T = 250$ °C, $P = 30$ MPa (see Chapter 3). The similar results for $P = 10$ and 30 MPa can be explained by the small effect of the concentration of CO_2 in the aqueous phase. The essential decrease in pH (or of K_w), due to CO_2 dissolution in the aqueous phase, occurs at concentrations below 25% saturation of CO_2 in water (compare figure 3.49, Chapter 3) that are easily reached by the conditions at $T = 250$ °C. Although the concentration of H^+ is at its maximum for $T = 250$ °C, it is too low to compete with the nucleophilic attack of the much more abundant water molecule at this temperature.

The results for $T=250\,°C$ are in accordance with the hydrolysis of proteins and carbohydrates, where the influence of CO_2 also fades out at temperatures around $T=250\,°C$ (see Chapter 8 on biomass conversion). At this temperature, the ionic product of pure water is at a maximum and seems to be high enough to minimize the effect of additional ionization by CO_2. Nevertheless, there is an effect on hydrolysis reactions when CO_2 is added to an aqueous system at the temperature range of about $T=100$–$250\,°C$ (see reactions on biomass compounds, Chapter 8).

Under the conditions of temperature and pressure applied in the above discussed experiments ($T\geq 250\,°C$, $P=10$–30 MPa), hydrolysis of esters runs through an S_N2 mechanism (displacement of a leaving group by a nucleophile, see Figure 5.10) with the water molecule as a nucleophile, and not the hydroxyl group or the H^+ as the nucleophile. Similar results have been found by Klein *et al.* [5] for dibenzylether, phenyl ethyl phenyl ether, and guajacol, and by Harrell *et al.* [42] for benzonitrile hydrolysis (see below).

During hydrolysis of esters, the formation of soluble carboxylic acids creates the potential for autocatalysis. Autocatalytic effects have been used to form glycols by hydrolysis of the corresponding acetic acid diesters at $T=50$–$80\,°C$ [43], of formic acid from methyl formate at $T=90$–$140\,°C$ [44], and glycerol by hydrolysis of glycerol triacetate with water at 180–245 °C [45, cited from Ref. 28].

Hydrolysis of di-*n*-butylphthalate at the temperature range from $T=305$–$390\,°C$ and water densities up to $\rho=0.31$ g/cm^3 leads to *o*-phthalic acid, butanol, and butane, whereas thermolysis (pyrolysis without water) results in butane, benzoic acid, and benzene. Condensation of phenyl compounds to polymers is suppressed in supercritical H_2O [46].

Methyl 1-naphthoate is thermally relatively stable. It was hydrolyzed at $T=250\,°C$ after 5.5 days and at $T=343\,°C$ after 2 h. At $T=250\,°C$, the main product is naphthanoic acid which decarboxylates at $T=343\,°C$ and generates carbonic acid to catalyze the formation of naphthalene as the major product at the higher temperature [47]. Methyl benzoate and its 4-chloro, 4-methyl, and 4-methoxy derivatives are hydrolyzed to 50% within 30 min at 250 °C without decarboxylation [48]. Ethylacetoacetate is completely converted to acetone, ethanol, and CO_2 within 30 min at $T=250\,°C$. *t*-Butyl acetate degrades at the same conditions into a bright red, insoluble mixture resulting from polymerization of isobutylene [49].

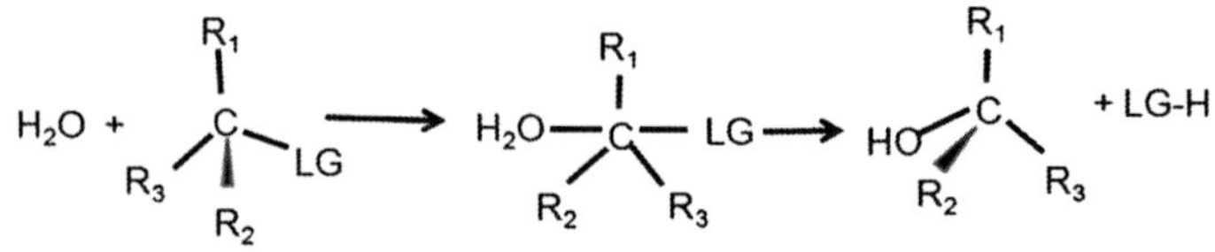

FIGURE 5.10 Hydrolysis pathway for high-temperature and supercritical water, prevailing at temperatures $T\geq 250\,°C$ [27]. (For color version of this figure, the reader is referred to the online version of this chapter.)

5.3.4.2 Ethers

The hydrolysis of ethers occurs in high-temperature and supercritical water similar to the hydrolysis of esters without the addition of acid catalysts. High density and the addition of NaCl improve reaction rate and selectivity of hydrolysis relative to other degradation reactions. The reaction leads at optimal conditions only to the respective alcohols [2]. Examples are the hydrolysis of guaiacol [5], dibenzylether [5,50], phenylethyl phenyl ether [5], diphenylether [35], methoxynaphthalenes [51], anisols [52], and from cellulose to glucose, fructose, and oligomers [53]. The hydrolysis of guaiacol, dibenzylether, phenylethyl phenyl ether, and diphenylether is discussed below in some detail. The hydrolysis of cellulose, glucose, fructose, and oligomers is discussed in Chapter 8 in the context of biomass conversion.

Hydrolysis of several ether compounds (guaiacol (2-methoxyphenol, brenzcatecholmonomethylether), phenylethyl phenylether, dibenzylether) under supercritical water conditions ($T = 373$–$400\ ^\circ C$) revealed parallel reactions of pyrolysis and hydrolysis (Equations 5.56 and 5.57), depending on water density, for example, for guaiacol [5]:

$$\text{Guaiacol} \rightarrow \frac{T}{k_P} \rightarrow CH_4 + \text{Catechol} + \text{Phenol} + \cdots + \text{Char} \tag{5.56}$$

$$\text{Guaiacol} \rightarrow \frac{+H_2O}{k_H} \rightarrow \text{Methanol} + \text{Catechol} \tag{5.57}$$

with k_P is the pyrolysis rate constant and k_H is the hydrolysis constant.

Experiments with $H_2{}^{18}O$-labeled water indicate that the H_2O molecules are incorporated into products in a nucleophilic substitution involving a saturated carbon with a heteroatom-containing leaving group as shown in Figure 5.10 [27]. For products of parallel free-radical pyrolysis decomposition, 1,2-diphenylethane and 1,3-diphenylpropane, no hydrolysis is observed as in the heteroatom-containing analogues [18]. These hydrocarbon molecules decompose by free-radical pathways at around $T = 400\ ^\circ C$ [27].

Selectivity of the reaction of guaiacol to the hydrolysis product methanol is shown in Figure 5.11 in dependence on reduced water density (ρ_r) under supercritical water conditions ($T = 373$–$400\ ^\circ C$).

Selectivity to hydrolysis increases with increasing density or concentration of the reactant water. The increase is nonlinear and not in accordance with the functional behavior expected from simple parallel reaction selectivity. The overall second-order hydrolysis rate constant k_H $l/(\text{mol s})$ is changing with water density due to a solvent effect that can be quantified by an application of Coulomb's law to kinetics (Kirkwood analysis) (Equation 5.58 and Figure 5.12) [5].

$$\ln k_H = \ln k_0 - \frac{N}{4\pi\varepsilon_0 RT} \frac{\varepsilon - 1}{\varepsilon} \cdot C \tag{5.58}$$

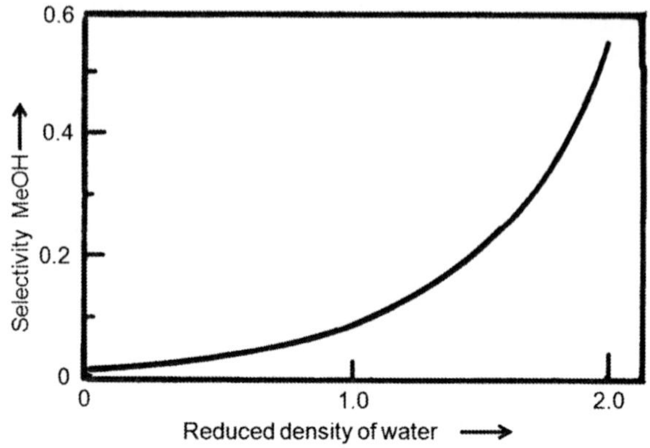

FIGURE 5.11 Selectivity of guaiacol hydrolysis to methanol with respect to density for supercriticial water conditions ($T=373$–$400\,°C$). *Modified from Ref. [5].*

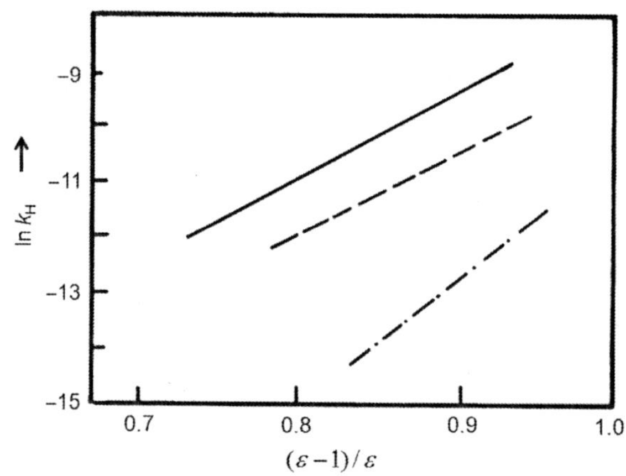

FIGURE 5.12 Correlation of the hydrolysis kinetics according to Equation (5.58). ——— polypropylene, – – – guaiacol, —●— dibenzylether (DBE). *Modified from Ref. [5].*

with k_H is the hydrolysis constant, k_0 is the rate constant in a condensed medium with $\epsilon=1$, ε_0 is the permittivity in vacuum, and C is the compound-specific constant.

Nevertheless, the substantial changes in k_H in Figure 5.12 are insignificant compared to the changes in K_w. At $\rho_r=0.3$, K_w is very small ($<10^{-20}$), while the hydrolysis rate is appreciable, meaning that the hydroxyl ion concentration, which is proportional to the magnitude of K_w, does not significantly influence the hydrolysis rate. Consequently, the S_N2 mechanism is a nucleophilic attack by a water molecule [5].

A further example is the reaction of diphenylether [35]. At low water density, from $\rho=0$ to approximately $\rho=0.3$ g/cm^3, that is at subcritical density

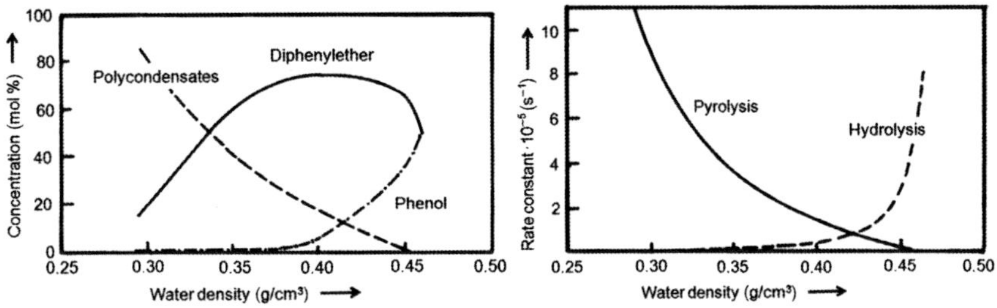

FIGURE 5.13 Left: Competition between pyrolysis and hydrolysis for the conversion of diphenylether and formation of phenol and polycondensates [35]. Right: Competition between pyrolysis and hydrolysis: rate constants of pyrolysis and hydrolysis in dependence of density. *Modified from Ref. [35].*

conditions, the conversion of diphenylether decreases. Products are typical for radical-type polycondensation reactions. At water densities greater than $\rho = 0.4$ g/cm^3, that is at supercritical density conditions, these products vanish and the conversion of diphenylether increases again and forms phenol as a sole reaction product, revealing ionic hydrolysis by attack through water molecules as the reaction mechanism (Figure 5.13) [35].

Dilute solutions of NaCl in supercritical water, in concentrations from $x = 0$–3.1 wt%, influence the rate of hydrolysis [54]. At $T = 430$ °C and a density of $\rho = 0.46$ g/cm^3, the rate decreased from the value at zero salt with addition of small amounts of salt and then increased with addition of salt to concentrations $x > 3.1$ wt.% to about twice the rate at zero salt content. Contrary to that observation, Klein *et al.* [5] observed at about the same density of water, an increase in hydrolysis rate for all salt concentrations applied. They attribute the increase to a higher polarity of the transition state in the salt solution. The capturing of H^+ by Cl^- ions, as stated by Penninger *et al.* [54], should not be of influence in view of the reaction mechanism of nucleophilic attack by H_2O molecules.

5.3.4.3 Carboxylic Compounds

Solution of carboxylic acids in water shows a small acidity due to the following reaction: $RCO_2H + H_2O \leftrightarrow RCO_2^- + H_3O^+$. In dilute solutions, the concentration of $[H_2O]$ is large and does not practically change. Therefore, this term is neglected and the equilibrium constant, the ionization constant of the acid K_A is given in Equation (5.59).

$$K_A = \frac{[RCO_2^-][H_3O^+]}{[RCO_2H]};\ pK_A = -\log K_A \qquad (5.59)$$

Decarboxylation of many carboxylic acids at elevated temperatures occurs with loss of CO_2,

$$RCO_2H \rightarrow RH + CO_2 \qquad (5.60)$$

Decarboxylation of carboxylic acids in hydrothermal and supercritical water is an important reaction since the carboxylate is the most abundant functional group of dissolved organic carbon in natural waters. A great number of studies of the decarboxylation rates of aliphatic acids (and aromatic acids, see below) have been reported. A compilation has been published by Brill and Savage [10]. Parameters that affect the rate at which an alkyl carboxylate loses CO_2 include the characteristics of the R group attached to the carboxylate, the surface in contact with the reacting fluid, the pH, and the counter ion [10].

The role of R for RCO_2H can be characterized by its electron-donating and electron-withdrawing properties. The electron-donating CH_3 group in acetic acid strengthens the $H_3C—CO_2H$ bond. As a result, acetic acid is the most stable and inert of the aliphatic acids toward decarboxylation [10]. The temperature at which acetic acid decomposes is in the range of $T=600$ °C. At this temperature, 35% of acetic acid is transformed in water in $t=8$ s at $P=240$ bar [40]. Acids with an electron-withdrawing R group decarboxylate at much lower temperatures, typically at $T=100$–250 °C [10,57].

The surface on which the acid reacts is of great influence when R is electron donating. For example, the rate of decomposition of CH_3CO_2H (to CH_4+CO_2) is a strong function of the substrate surface. Transformation rate at $T=100$ °C is more than 10 orders of magnitude lower on stainless steel than on titanium [10,58]. When R is electron withdrawing, changing R in a carboxylic acid is of much greater influence than by changing the construction metal of the reactor [57]. That means that acids with a weakened R—C bond decarboxylate without catalytic assistance from a surface, whereas decarboxylation of acids with a stronger R—C bond is promoted by surface catalysis [10].

The pH is important in the decarboxylation reaction because RCO_2H and RCO_2^- decarboxylate at different rates, as the example of malonic acid shows [59,60]. Between pH 1 and 4, the rate of decarboxylation is a competitive function of the rates for malonic acid and the monoanion, while above pH 5, the dianion decarboxylates much slower than either of the other two forms [10].

Hydrothermal stability of six aromatic carboxylic acids has been investigated by Dunn *et al.* [61]. The literature contains some information [62–64] on the reactivity of benzoic acid, the simplest aromatic acid, at elevated temperatures in water that is summarized in Figure 5.14. The stability of benzoic acid is expressed through the combined influence of temperature and residence time. The stability of benzoic acid was assumed for less than 3% conversion, represented by the full line in Figure 5.14. Above the line, benzoic acid is unstable, below it is stable.

Terephthalic acid and 2.6-naphthalene dicarboxylic acid are stable at $T=300$ °C for 1 h but are less stable than benzoic acid [61]. Conversion of terephthalic acid to benzoic acid is shown in Figure 5.15. 2.6-naphthalene

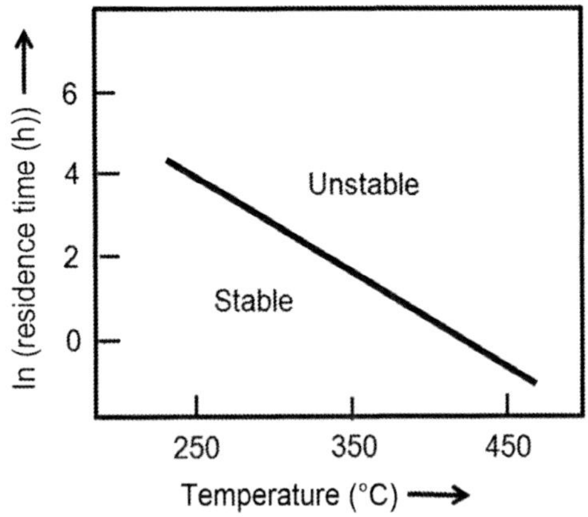

FIGURE 5.14 Stability of benzoic acid. *Data from Refs. [10,62–64].*

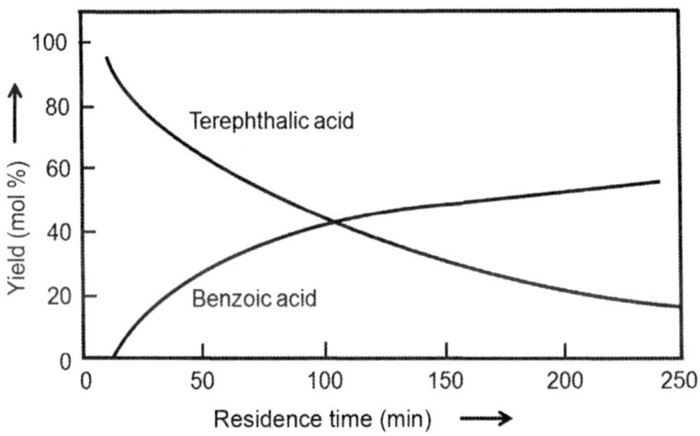

FIGURE 5.15 Decomposition of terephthalic acid to benzoic acid under supercritical water conditions ($T=400$ °C), [terephthalic acid]$_0=0.04$ M, [H_2O]$=14$ M. *Modified from Ref. [61].*

dicarboxylic acid is converted to naphthoic acid. Under hydrothermal and supercritical water conditions, 1- and 2-naphthoic acids are also more reactive than benzoic acid. According to Siskin *et al.* [47], 1-naphthoic acid is converted completely after 2 h at $T=343$ °C. 2-Naphthoic acid was converted to 42% at $T=450$ °C after 30 min [64].

During decarboxylation of aromatic acids, autocatalytic effects can occur due to production of CO_2, which forms carbonic acid to some degree under hydrothermal conditions and not too strong acidic conditions [10]. Such autocatalytic transformation was found for trimellitic anhydride. Under hydrothermal conditions, trimellitic anhydride forms a triacid and then is decarboxylated to di-acids, either phthalic or isophthalic acid, as shown in Figure 5.16 [61].

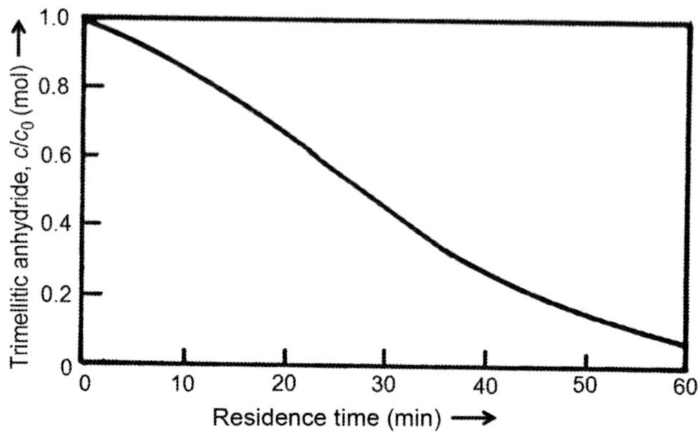

FIGURE 5.16 Hydrothermal conversion of trimellitic anhydride, $T=300\,°C$. *Modified from Ref. [61].*

According to Katritzky *et al.* [66,67], carboxylic acids lose CO_2 from the 2- or 4-position of a pyridine ring easily, less easily from the pyridine 3-position [68], and with difficulty when directly attached to benzene due to the stability of the carbon ion remaining after CO_2 loss from the acid anion [63].

5.3.4.4 Amines

Aliphatic amines are only modestly reactive in hydrothermal systems [10]. Aromatic amines are more reactive in hydrothermal and supercritical water but seem to need the assistance of a catalyst for hydrolysis [2]. For evaluation of experimental results for amines as well as other compounds, it has to be considered that many of the experiments on hydrolytic reactions have been carried out in batch reactors using the bomb-method, that is, filling the autoclave, heating it to reaction temperature, cooling it rapidly, removing the contents, and analyzing it. Consequently, the data for reaction mechanisms and reaction rates are only indications for the data needed to design a process.

The hydrolysis of aniline was investigated by Patat [70] in a silver-lined tube reactor at pressures between $P=40$ and 70 MPa, and at temperatures up to $T=450\,°C$ with phosphoric acid and its sodium salts as catalysts. The activation energy was lower in the subcritical than in the supercritical region. Under supercritical conditions, the reaction rate increases with pressure [2].

Hydrolysis of methylamine in hydrothermal and supercritical water was studied in a batch reactor (bomb-type experiment) at temperatures between $T=386$ and 500 °C [71]. The major products are ammonia and methanol. For water densities less than $\rho=0.28\,g/cm^3$ and pressures less than $P=25$ MPa, the effect of water on the reaction rate appears to be negligible, and there is little evidence of hydrolysis. Under these conditions, the reaction seems to be governed by pyrolysis. At water densities $\rho>0.28\,g/cm^3$, the rate of conversion increases and hydrolysis becomes more important. The methanol yield increases with water density [71].

Benzylamine is converted in 1 h at $T=400$ °C [30] with toluene as the major organic product but only after 5 days at $T=250$ °C [28]. Hydrolysis occurs in parallel with free-radical thermolysis [10].

Dibenzylamine is converted to 29% after 20 min at $T=300$ °C [69], a conversion that is much lower than by pyrolysis at comparable reaction conditions. Benzylalcohol and benzaldehyde are formed as hydrolysis products.

Hydrolysis of benzylphenyl amine leads to benzyl alcohol, benzaldehyde, and aniline [72,73] and in a later investigation [74] also to toluene as main products. The addition of salts increases hydrolysis due to increased ionic strength up to a maximum [50]. Beyond the maximum hydrolysis, yields are lower because the number of water molecules available for hydrolysis decrease due to more water molecules required to solvate the ions [10]. Aniline is converted to 20% at $T=450$ °C for 48 h and forms phenolic compounds and NH_3 [34].

5.3.4.5 *Phenols*

Phenol is stable in hydrothermal and supercritical water in the range of $T=300$–380 °C and at a pressure of $P=28$ MPa [33,75]. Cresols have been investigated for hydrolysis at lower temperature of $T=250$ °C and found to be stable. Only *p*-cresol undergoes some decomposition of 2.1% after 72 h. [10]. At $T=460$ °C, $P=25$ MPa and residence times up to about $t=30$ s, cresols are converted to <10%. Nitrophenols are much more reactive under these conditions [76]. Hydrolysis of substituted phenols produces phenol as a major product and shows *o*-phenol substitutes as more reactive than the other isomers. Nitrophenols react more rapidly than cresols [10].

5.3.4.6 *Acetals and Ketals*

Acetals and ketals are hydrolyzed readily at $T=200$–250 °C in about 30 min without side or secondary reactions [48]. More than 90% conversion occurred at $T=250$ °C in cyclopentanone ethylene ketal and 1,4-cyclohexanedionebis (ethylene ketal), benzaldehyde, and tolualdehyde diethyl acetals [23,28].

5.3.4.7 *Alcohols*

Alcohols like ethanol, propanol, *t*-butanol, and cyclohexanol can be hydrolyzed in hydrothermal and supercritical water. The reaction is acid-catalyzed but occurs to some extent without adding an additional catalyst because at the high temperatures involved, the concentration of H^+ from water is sufficient to cause the reaction. The dehydration mechanism depends on the structure of the alcohol and the reaction conditions [10]. Alcohol dehydration is the most important elimination reaction in which two groups are cleaved from a carbon structure.

Hydrolysis of *ethanol* in supercritical water without an additional catalyst is limited. In a flow reactor, 1.9–7.4% of ethanol is converted at temperatures from $T=433$–494 °C, a pressure of $P=24.6$ MPa, and residence times from $t=2$–12 s [79]. When a low concentration ($x<0.01$ mol dm^{-3}) of sulfuric acid is added, ethanol hydrolyses rapidly and selectively to ethene in supercritical water [80]. Addition of oxygen leads to a fast and total conversion of ethanol to acetaldehyde and formaldehyde in the liquid phase and carbon monoxide and carbon dioxide in the gas phase [79] (see Chapter 10).

Tertiary *butanol* is completely converted to isobutene in 30 s at subcritical temperatures without addition of acids [65].

Hydrolysis of *1-propanol* [77] and *cyclohexanol* [78] (see Figure 5.17) in pure, high-temperature water is characterized by an E2 mechanism. E2 stands for bimolecular elimination, a one-step mechanism that breaks carbon—hydrogen and carbon—halogen bonds to form a double bond. The E2 mechanism is very similar to the S_N2 reaction mechanism. Protonation of cyclohexanol produces an oxonium ion, which then reacts with water to form cyclohexene. The cyclohexene can be protonated to form a cyclohexyl carbocation, which rearranges to the more stable tertiary methylcyclopentyl cation and finally converts to methylcyclopentene [10].

Hydrolysis of several biomass-derived *polyols* in hydrothermal and supercritical water resulted in main products of 1,4-anhydroerythritol and propionaldehyde [81]. The hydrothermal conversion of *m*-erythritol is shown in Figure 5.18 as a function of temperature in the range of $T=300$–400 °C. The rate of conversion is low at $T=300$ °C and high at supercritical temperatures. Conversion is influenced by sulfuric acid and several salts, as shown in Figure 5.19 for $T=360$ °C, $P=34$ MPa, and residence times of $t=10$–180 s [81]. The dehydration of polyols is strongly influenced by the addition of salts. Bivalent transition metal cations increase both the conversion and the selectivity. Sodium sulfate inhibits conversion. The effect is due to pH shift for sodium sulfate into the alkaline region and the influence of ionic strength [81].

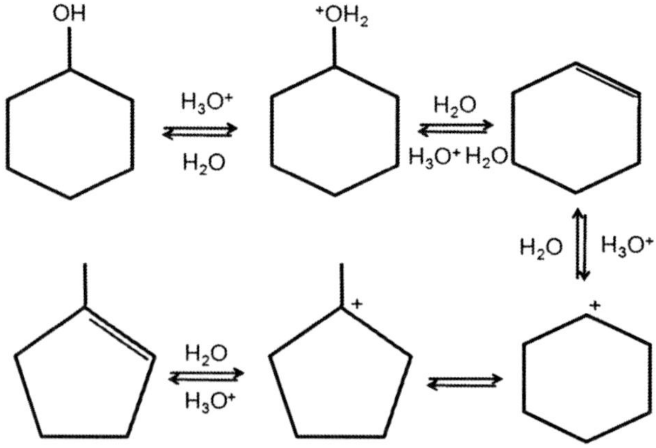

FIGURE 5.17 Hydrolysis of cyclohexanol in high-temperature liquid water [78].

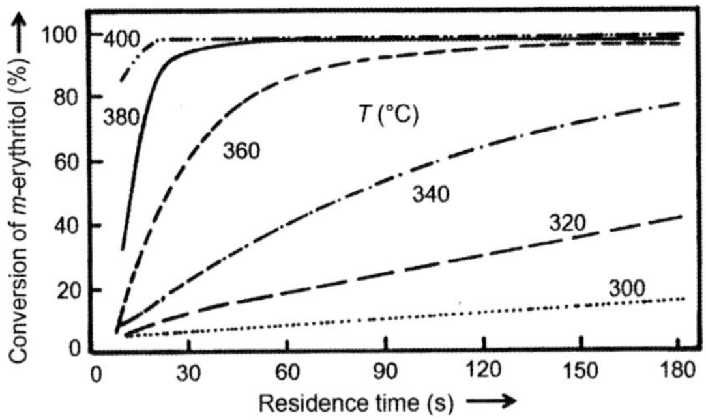

FIGURE 5.18 Conversion of *m*-erythritol at $P=34$ MPa, with 0.1 wt% $ZnSO_4$. *Modified from Ref. [81].*

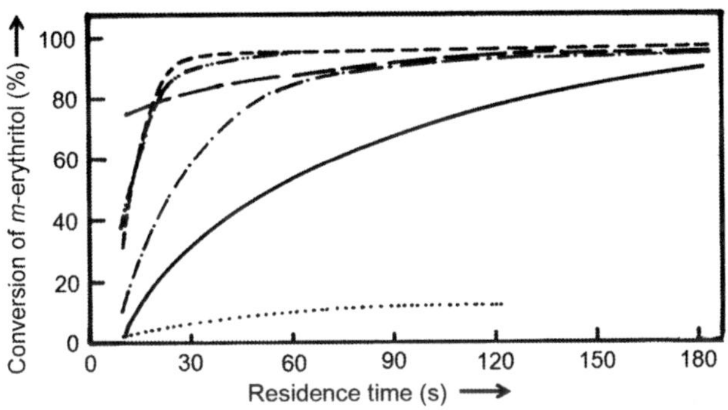

FIGURE 5.19 Conversion of *m*-erythritol with different catalysts at $P=34$ MPa, $T=360$ °C, ——— without additive, ●●●● 800 ppm sodium sulfate, —●—1000 ppm zinc sulfate, ——— 20 mmol/dm^3 H$^+$, —●●— 4000 ppm zinc sulfate, ——— 1000 ppm copper sulfate. *Modified from Ref. [81].*

Glycerol decomposes in hydrothermal and supercritical water to liquid and gaseous products as shown in Figures 5.20 and 5.21 [82]. The products of the glycerol degradation are (liquid) methanol, acetaldehyde, propionaldehyde, acrolein, allyl alcohol, ethanol, formaldehyde, (gaseous) carbon monoxide, carbon dioxide, and hydrogen. The overall degradation indicates two competing reaction pathways, one of ionic reactions, preferred at higher pressures and lower temperatures, the other a free-radical degradation, which dominates at lower pressures and higher temperatures as can be deducted from the experimental results in the Arrhenius plot of Figure 5.22 [82].

In *tert-butanol*, a complete conversion occurs at hydrothermal conditions without the addition of acid [65]. In other cases, like propanol [77,83], glycerol [84], glycol [85], and cyclohexanol [48,86], a mineral acid has to be added to obtain substantial yields and to prevent C—C bond breakage [2]. α-Ethyl-4-methoxy- and *d,l*-4-chloro-a-propylbenzyl alcohol react in high-temperature liquid water at $T=277$ °C [48]. Cyclohexanol and

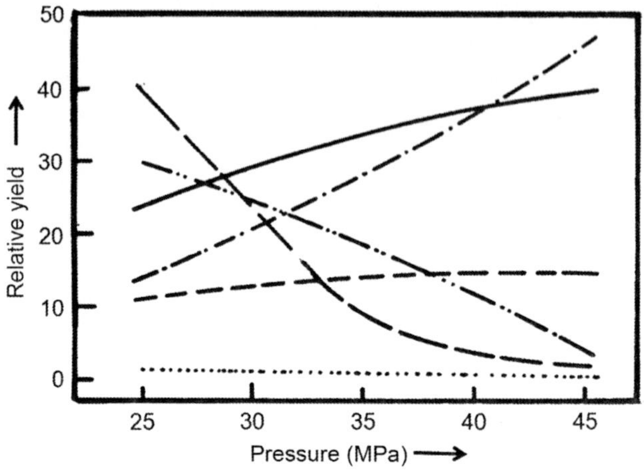

FIGURE 5.20 Liquid products from hydrolysis of glycerol at $T=667$ K. ——— formaldehyde, —●— acetaldehyde, — — — acrolein, —●●— methanol, ●———— allyl alcohol, ●●●● propionaldehyde. *Modified from Ref. [82].*

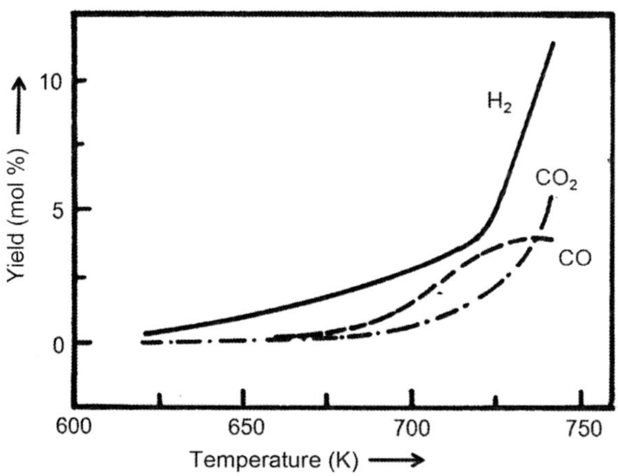

FIGURE 5.21 Gaseous products from glycerol hydrolysis at $P=45$ MPa and about $t=100$ s residence time, glycerol concentration 0.5%. *Modified from Ref. [82].*

methylcyclohexanol are dehydrated in water at temperatures of $T=250$–300 °C. Cyclohexanol is converted to cyclohexene, at $T=278$ °C to 85%, and at $T=300$ °C to 33%. Sulfuric or hydrochloric acid shifts the conversion to 99% at $T=250$ °C [87].

5.3.4.8 Aldehydes

Formaldehyde: Hydrolytic transformation of formaldehyde (HCHO) in supercritical water resulted in methanol (CH_3OH), formic acid (HCOOH),

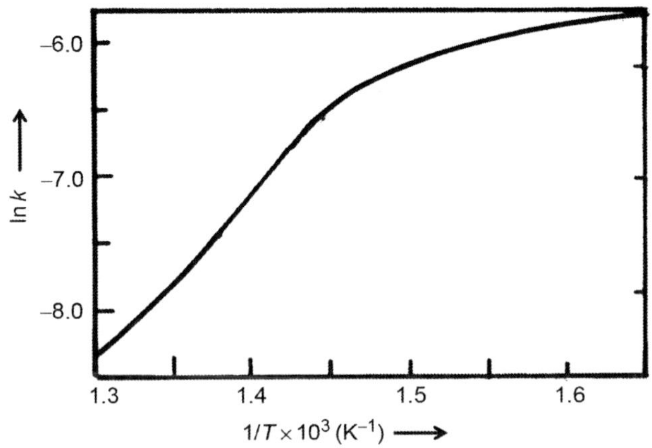

FIGURE 5.22 Global rate constant of glycerol hydrolysis in hydrothermal and supercritical water, at $P=45$ MPa. Glycerol concentration 0.5%. *Modified from Ref. [82].*

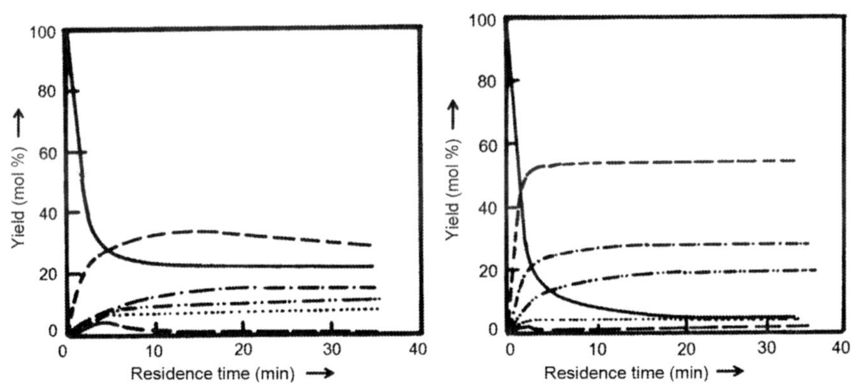

FIGURE 5.23 Hydrolysis of formaldehyde in supercritical water. $T=400\,°C$, $[HCHO]_0=0.56$ mol/dm^3, Left: $\rho_{H_2O}=0.17$ g/cm^3; Right: $\rho_{H_2O}=0.50$ g/cm^3. *Modified from Ref. [88].*

hydrogen (H_2), carbon monoxide (CO), and carbon dioxide (CO_2) as the major products [88]. Figure 5.23 shows the conversion for low density and high density.

Monomolecular decomposition of HCHO is the main reaction pathway at low water densities, indicated by increased CO and decreased CH_3OH yields. At higher water densities, methanol yields increase and the Cannizzaro reaction mechanism is dominant. Addition of base to the reacting mixtures promotes the Cannizzaro reaction path, whereas addition of acid promotes monomolecular decomposition [88], compare Figure 5.24.

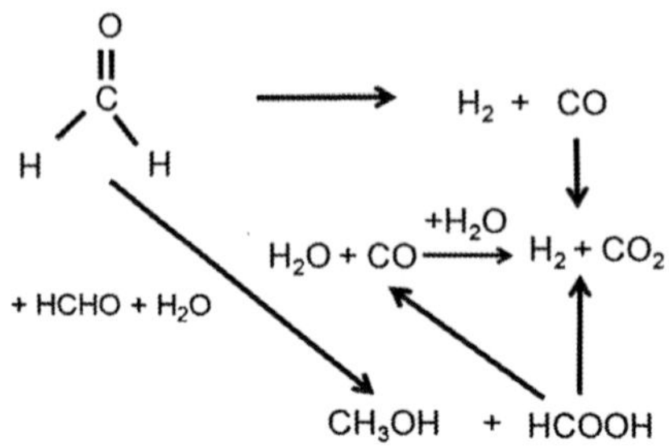

FIGURE 5.24 Reaction pathways for formaldehyde hydrolysis in supercritical water. *Modified from Ref. [88].*

5.3.4.9 *Nitriles*

Nitriles and amides are readily hydrolyzed in water at temperatures around $T=250\ °C$ but do not react substantially by pyrolysis. Hydrolysis and subsequent decarboxylation reactions are autocatalyzed by ammonia as a reaction product [23,89].

An explanation could be the increased concentration of $[H^+]$ and $[OH^-]$ in high-temperature liquid water caused by the increase in the ion product of water. A detailed study of the mechanism of noncatalytic hydrolysis of benzonitrile in high-temperature liquid water [42] revealed the nucleophilic attack at the nitrile carbon (RC*=N) as the rate-determining step and the H_2O molecule as the predominant nucleophile. $[H^+]$ or $[OH^-]$ concentration in high-temperature water is too low to promote catalytic hydrolysis, as was shown by adding acid or base catalysts. According to Harrell *et al.* [42], hydrolysis proceeds through a S_N2 reaction, as shown in Figure 5.25.

The influence of pressure on hydrolysis of nitriles seems to be substantial. Hydrolysis of butyronitrile at $T=330\ °C$ in high-temperature liquid, pressures from $P=12.8–260.0$ MPa, and residence times from $t=5–180$ min [91] in a bomb-type batch reactor confirmed this effect, as shown in Figures 5.26 and 5.27. Products of the hydrolysis are butyric acid, butanamide, and ammonia with negligible gas formation [91]. The transformation of butyronitrile was explained by a four-step autocatalytic rate model with butyric acid as the catalytic species [91]. The influence of reactor walls is of negligible influence for

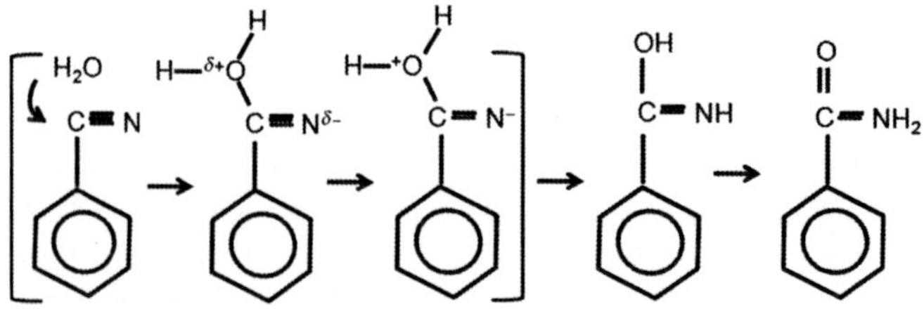

FIGURE 5.25 Hydrolysis of benzonitrile by nucleophilic attack of water [42,90].

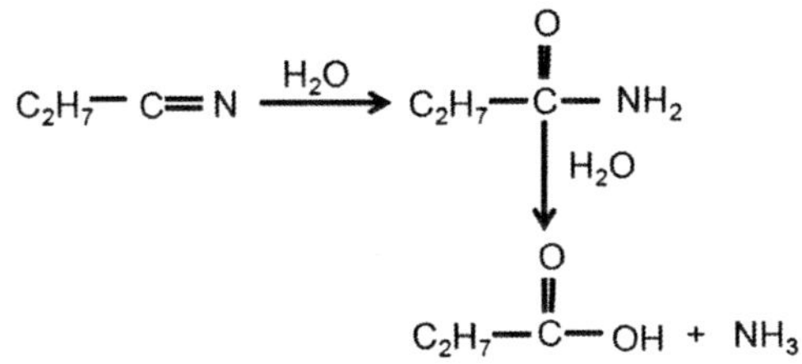

FIGURE 5.26 Butyronitrile hydrolysis in high-temperature water [91].

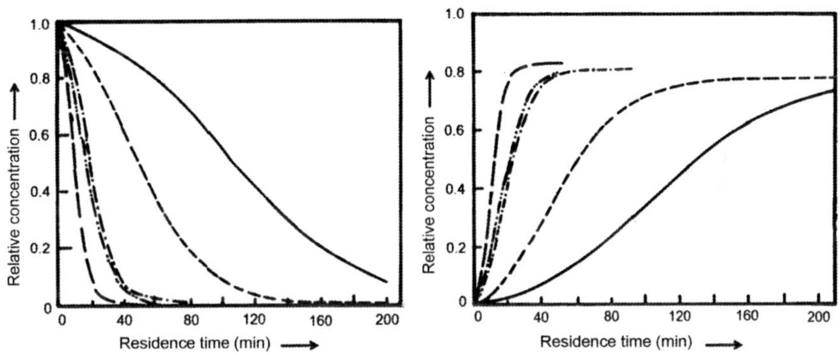

FIGURE 5.27 Left: Hydrolysis of butyronitrile in high-temperature liquid water at $T=330$ °C. Right: Production of butyric acid through the hydrolysis of butyronitrile. ——— $P=12.8$ MPa, ———$P=47.2$ MPa, —●— $P=87.1$ MPa, —●●— $P=146.9$ MPa, ——— $P=253.8$ MPa. *Modified from Ref. [91].*

the hydrolysis of nitriles. Reactor history, material of construction (stainless steel or titanium), and surface to volume ratio do not influence hydrolysis of benzonitrile [92].

The influence of temperature on hydrolysis of benzonitrile, acetonitrile, and acetamide under hydrothermal and supercritical water conditions, $T=350$–450 °C, $P=28$–32 MPa, residence times from $t=6$–400 s, was investigated in a tubular reactor without addition of catalysts [93,94]. Acetonitrile and benzonitrile decompose in hydrothermal and supercritical water according to Equations (5.61) and (5.62) via the amide to the carbonic acid. R stands for the methyl- or the benzyl group [93].

$$R-C\equiv N+H_2O \leftrightarrow R-CONH_2O \quad (5.61)$$

$$R-CONH_2+H_2O \leftrightarrow R-COOH+NH_3 \quad (5.62)$$

Hydrolysis of acetonitrile can be described by a simple first-order reaction model up to temperatures of $T=450$ °C, based on the experimental data shown in Figure 5.28. For a conversion, >30% substantial deviations from the results for the simple model occur, probably due to autocatalysis by the

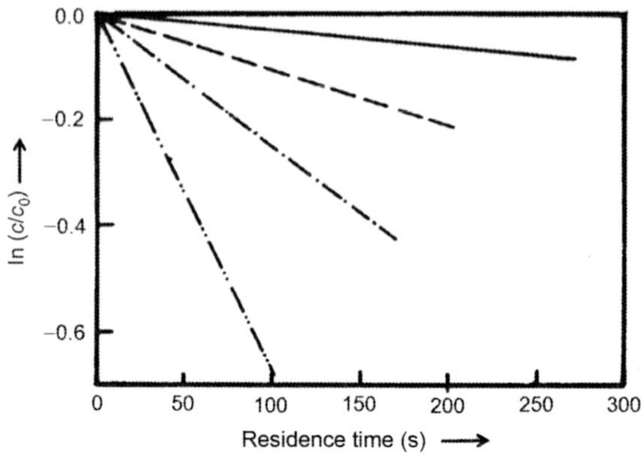

FIGURE 5.28 Decomposition of acetonitrile at $P = 28$ MPa. Initial concentration 5 wt%. ——— $T = 300$ °C, – – – – $T = 350$–380 °C, —●— $T = 400$ °C, —●●— T = 450 °C. *Modified from Ref. [93].*

degradation products acetic acid and ammonium acetate. Addition of acetic acid and ammonium acetate to the feed mixture leads to considerable corrosion inside the Inconel 625 reactor tube [93].

The reaction rate constant can be directly determined from the experimental results. The temperature dependence of the rate constant can be described by an Arrhenius equation (Equation 5.63).

$$\ln k_1 = -\frac{E_a}{RT} + \text{const.} \tag{5.63}$$

with k_1 as the reaction rate constant and E_a as the activation energy. From Equation (5.63), the activation energy can be derived. The activation energy E_a varies for acetonitrile from 128 kJ/mol at $P = 23$ MPa to 59 kJ/mol at $P = 32$ MPa, and for acetamide from 38 kJ/mol at $P = 23$ MPa to 50 kJ/mol at $P = 32$ MPa.

For the pressure dependence according to the transition-state theory, Equation (5.64) can be applied:

$$\ln k_1 = -\frac{V^*_{app}}{RT} P + \text{const.} \tag{5.64}$$

with k_1 as the reaction rate constant, V_{app} as the apparent activation volume, and P as the pressure.

From Equation (5.64), the apparent reaction volume can be calculated. V_{app} for acetonitrile varies from $V_{app} = -585$ cm^3/mol at $T = 300$ °C to $V_{app} = +793$ cm^3/mol at $T = 400$ °C, and for acetamide from $V_{app} = -11$ cm^3/mol at $T = 350$ °C to $V_{app} = -381$ cm^3/mol at $T = 400$ °C.

Acetamide is a product of the acetonitrile hydrolysis. Hydrolysis of acetamide leads to acetic acid, as shown in Figure 5.29, and in a small equilibrium amount also to acetonitrile. Acetic acid is stable under hydrothermal and supercritical water conditions at temperatures up to $T = 450$ °C.

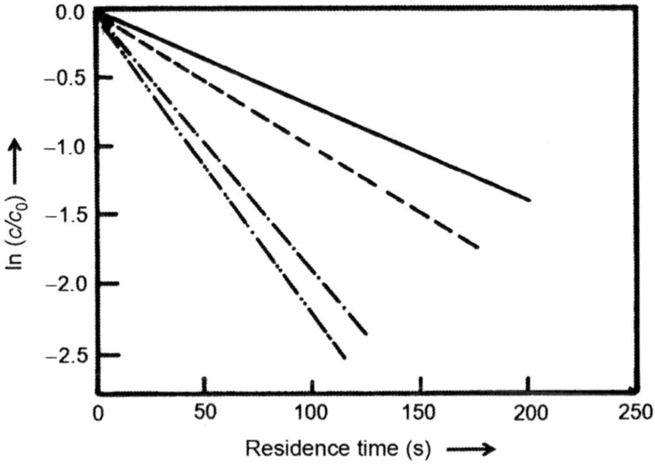

FIGURE 5.29 Hydrolysis of acetamide at $P=28$ MPa. Initial concentration of acetamide 5 wt%. ——— 350 °C, – – – 380 °C, —•— 400 °C, —••— 450 °C. *Modified from Ref. [93].*

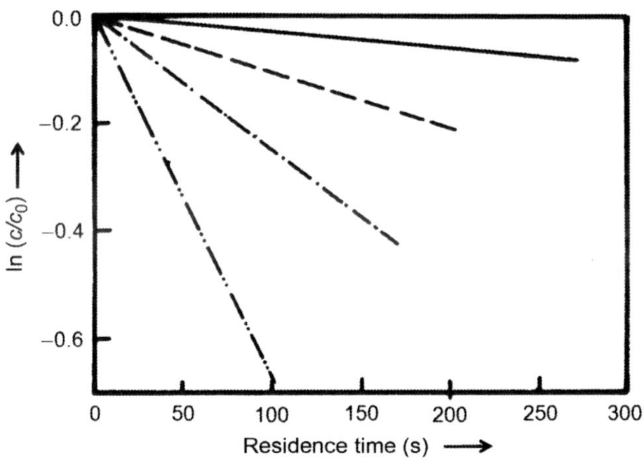

FIGURE 5.30 Hydrolysis of benzonitrile at 25 MPa. Initial concentration 5 wt%.——— $T=350$ °C, – – – $T=380$ °C, —•—• $T=400$ °C, —••— $T=450$ °C. *Modified from Ref. [93].*

Hydrolysis of benzonitrile leads to benzamide and then to benzoic acid. The rate of decomposition of benzonitrile is shown in Figure 5.30. Activation energy and apparent activation volume can be derived as for acetonitrile hydrolysis.

Hydrolysis of acetonitrile and benzonitrile leads selectively to the corresponding amides and carbonic acids. Both reactants react at ambient conditions only when strong acids or bases are added. Hydrothermal and supercritical water hydrolysis avoids the addition of compounds and consequently the separation and disposal of those compounds as salts [93].

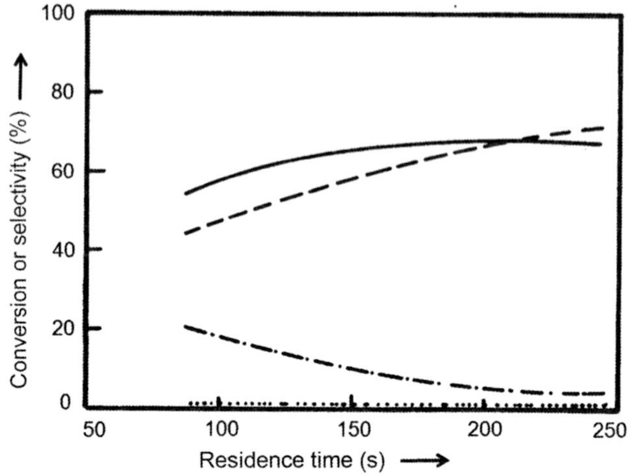

FIGURE 5.31 Hydrolysis of ε-aminocapronitrile at $T=350$ °C, $P=25$ MPa. Initial concentration of ε-aminocapronitrile 5 wt%. ——— conversion of ε-aminocapronitrile, ——— selectivity caprolactam, —●— selectivity aminocaproic acid amide, ●●●● selectivity dimers. *Modified from Ref. [93].*

An alternate synthesis path to ε-caprolactam runs via butadiene and ε-aminocapronitrile. The reaction of ε-aminocapronitrile in high-temperature liquid water leads to ε-caprolactam, which is stable. In Figure 5.31, yield and selectivity for the reaction compounds are shown for $T=350$ °C and $P=25$ MPa. Considerable yield and high selectivity can be achieved without the addition of catalysts. At temperatures $T>380$ °C, the selectivity decreases substantially [93].

5.3.4.10 Amides (Cyclic Azines)

Cyanamide, dicyandiamide, and related cyclic azines react in water at $T=100$–300 °C as shown for the first two compounds in Figure 5.32. Results are based on experiments in a sealed 316 SS tube [95]. The conversion of cyanamide to dicyandiamide dominates at $T=100$–175 °C (see Figure 5.32, left). At $T=175$–250 °C, for reaction times shorter than $t=15$ min, the major pathway is hydrolysis of the cyanamide–dicyandiamide mixture to CO_2 and NH_3. At higher temperatures than $T=225$ °C, hydrolysis occurs of these cyclic azines to aqueous NH_3 and CO_2. At $T=300$ °C, the conversion of all compounds to CO_2 and NH_3 is complete at $t=10$ min. Figure 5.32, right, shows the reaction of dicyanamide at $T=225$ °C and $P=27.5$ MPa [29].

5.3.4.11 Organic Chlorides

Hydrolysis of organic chlorides occurs in pure water and eliminates chlorine atoms (as Cl^-) under hydrothermal and supercritical water conditions. The formation of Cl^- ions leads to a corrosive fluid phase. These corrosion

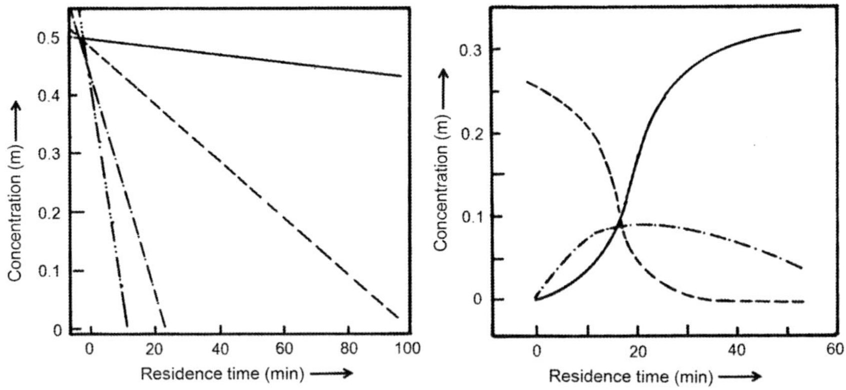

FIGURE 5.32 Hydrolysis of cyanamide. Left: Temperature dependence of the conversion of cyanamide, ——— $T = 100$ °C, – – – $T = 125$ °C, —•— $T = 150$ °C, —••— $T = 175$ °C. Right: Conversion of dicyanamide and reaction products at $T = 225$ °C and $P = 27.5$ MPa, – – – dicyandiamide, ——— $(NH_4)_2CO_3$, —•— cyclic azines (10^3). *Modified from Ref. [95].*

products may influence the reaction [10]. From the application of different reactor materials during the reactions of aliphatic and aromatic chlorides, it can be concluded that the chloride, or HCl as a secondary product, attacks the metal walls of the reactor and forms metal chlorides that catalyze the decomposition reactions of the organic chloride feed compound [96]. For example, decarboxylation of trichloroacetic acid is strongly influenced by reactor corrosion under hydrothermal conditions [57]. On the other hand, during the hydrothermal reaction of methylene chloride, no evidence was found for catalysis by metals or corrosion products [97,98].

Hydrolysis of methylene chloride in hydrothermal and supercritical water has been studied extensively [97–102, cited from 10]. The reaction network of the methylene chloride hydrolysis is given by Equations (5.65)–(5.67).

$$CH_2Cl_2 + H_2O \rightarrow CH_2O + 2HCl \quad (5.65)$$

$$CH_2O \rightarrow CO + H_2 \quad (5.66)$$

$$CO + H_2O \rightarrow CO_2 + H_2 \quad (5.67)$$

Reactions were carried out in a tubular reactor at $P = 24.6$ MPa, at temperatures from $T = 25$–600 °C, total residence times (preheater and reactor) of $t = 7$–23 s and CH_2Cl_2 concentrations from $x = 0.2$–0.6 mmol/dm^3, at the entrance to the reactor under supercritical conditions. Temperatures of the incoming feed streams were always $T = 25$ °C, while sand bath temperatures ranged from $T = 450$ to 600 °C [98].

Hydrolysis of CH_2Cl_2 leads to formaldehyde and HCl, followed by decomposition of formaldehyde to CO and H_2, and CO conversion to CO_2 and H_2 by the water-gas shift reaction. Reaction products for methylene chloride hydrolysis are shown in Figure 5.33.

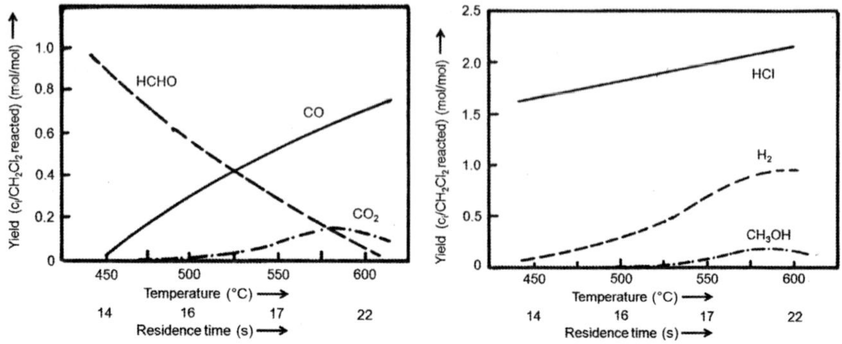

FIGURE 5.33 Products from hydrolysis of methylene chloride, $P=24.6$ MPa, residence time in the reactor $t=6$ s. *Modified from Ref. [98].*

The dominant reaction for CH_2Cl_2 is hydrolysis to HCHO and HCl, which begins at subcritical temperatures by a polar substitution mechanism. For HCHO, decomposition to CO and H_2 is the major reaction, followed by conversion of CO to CO_2 via the water-gas shift reaction. If catalytic reactions of CH_2Cl_2 occurred with the Hastelloy C-276 or Inconel 625 tubing used, they are less important than the homogeneous hydrolysis reaction [98]. This result is supported by Salvatierra *et al.* [97], who found no evidence for catalysis by metals or corrosion products during the hydrothermal reaction of methylene chloride.

Hydrolysis reaction of methylene chloride forming formaldehyde and HCl occurs readily in subcritical water but is slower in supercritical water [10]. This is due to the reduced bulk dielectric constant at supercritical conditions because the S_N2 reaction involves charged or polar species as reactants or intermediates which decrease considerably with increased temperature [10].

Results for reactions of 1-chloro-3-phenylpropane, 2-chlorotoluene, and 4-chlorophenol in supercritical water are available from experiments in batch reactors made from Inconel 600 for $T=400$–500 °C, $P=27$–45 MPa, reaction times from $t=30$–120 min, and reactant concentrations from $x=1.5$–7 mol% [96]. Organic chlorides are hydrolyzed at supercritical water conditions and react either directly or indirectly through HCl formed with the metal walls of the reactor to metal chlorides. Inserts from quartz or Vycor™ can protect the heated metal components, while inserts from Pyrex are not inert to supercritical water [96]. Data for the decarboxylation of trichloroacetic acid at hydrothermal conditions also indicate that this reaction is strongly influenced by reactor corrosion [57].

Further experimental data for hydrolytic dechlorination can be found [10] for 1,1,2-trichlorotri-fluoroethane [103], trichloroacetic acid and trichloroethylene [104], and polyvinylchloride [105].

5.3.4.12 Nitrocompounds

Aliphatic nitrocompounds belong to the so-called energetic materials and can be either the explosives as products or the byproducts of the production process. Decomposition of aliphatic nitrocompounds attracts attention because it is necessary to develop efficient methods for their conversion to simple compounds. The decomposition and oxidation of these compounds in supercritical water is a preferred method [106]. Oxidative destruction is treated in Chapter 10, and hydrolysis without addition of oxygen is discussed in the following.

Decomposition of aliphatic nitrocompounds (RNO_2) nitromethane, nitroethane, and 1-nitropropane shows a strong dependence of the rate constants for the decomposition reactions on density according to experiments in a tubular flow reactor at $T=663$–664 K and densities of $\rho=0.14$–0.51 g/cm^3 [106]. Parallel to the change in density from $\rho=0.14$–0.51 g/cm^3, the H_3O^+-concentration varies from $5.128\times10^{-11}1$ to 4.31×10^{-7} and rate constants of the overall decomposition reaction vary linearly with the H_3O^+-concentration. But the logarithm of the observed rate constant, ln k_{abs}, also varies linearly with density. Consequently, the rate constant depends more on density than on the concentration of the H_3O^+ ions, what may confirm the observation that the neutral water molecules act mainly as the nucleophile in the hydrolysis reaction. In addition to the hydrolytic reaction, a thermal (pyrolytic) decomposition of the reactants by catalysis on the metal walls of the tubular reactor was detected by the occurrence of products like H_2 and CH_4 [106].

In spite of the greater influence of density, a simple model that is based on the concentration of H_3O^+ ions is able to represent the experimental data (Equation 5.68) [106].

$$k_{obs}^{RNO_2} = k_0^{RNO_2} + k_1^{RNO_2}[H_3O^+], \tag{5.68}$$

with $k_{obs}^{RNO_2}$ as the observed rate constant, $k_0^{RNO_2}$ as the rate constant of the thermal decomposition, and $k_1^{RNO_2}$ as the rate constant of the H_3O^+ catalyzed reaction. Values for the rate constants are listed in Table 5.1.

As the previous discussion has shown, reaction of some classes of compounds in hydrothermal and supercritical water is promoted by the elevated

TABLE 5.1 Constants of Equation (5.68) and Confidence Intervals [106]

Reagent	$k_0^{RNO_2} \pm \Delta k_0\,(s^{-1})$	$k_1^{RNO_2} \pm \Delta k_1\,(s^{-1})$	σ (%)
CH_3NO_2	$6.6363\times10^{-3} \pm 6.4548\times10^{-4}$	$3.1290\times10^{+4} \pm 3.9989\times10^{+3}$	13.80
$C_2H_5NO_2$	$5.3108\times10^{-3} \pm 3.7586\times10^{-4}$	$2.5395\times10^{+4} \pm 2.2586\times10^{-3}$	13.04
$C_3H_7NO_2$	$5.5051\times10^{-3} \pm 3.9640\times10^{-4}$	$1.9054\times10^{+4} \pm 1.6409\times10^{+3}$	7.90

σ is the root-mean square relative deviation of experimental and calculated data.

temperature but more so by the nucleophilic activity of the water molecule or the water ions. Catalysis by H_3O^+ is found to be effective, for example, for alcohol dehydrogenation but often is not prevalent. In spite of the concentration increase of H_3O^+ of three orders of magnitude, the concentration may be still too low to effectively catalyze reactions. Furthermore, H_3O^+ concentration is at its maximum around $T=250$ °C. Most reactions are carried out at much higher temperatures in the near-critical and supercritical region of water, at which the ion product is again lower. Further information can be found in the reviews of chemical reactions with supercritical fluids [2,10,12,22,23,25,27–29,36,90].

5.3.5 Acids and Bases

Behavior of acids and bases in the neighborhood of the critical point has been determined by titration [107], which is useful for understanding some stable and prevalent species in hydrothermal systems, like ammonia, NaCl, HCl, and acetic acid. Acetic acid neutralizes both KOH and NH_3 at $T=350$ °C and $P=34$ MPa. At high temperatures, KOH remains a much stronger base than NH_4OH. The chloride ion increases the pH of a relatively concentrated solution of HCl at $T=380$ °C and pressures of $P=27$ and 34 MPa, but not to complete neutralization, as shown in Figure 5.34.

5.3.6 Miscellaneous Organic Reactions

In the following, some organic reactions carried out in high-temperature liquid water or in supercritical water will be mentioned as examples for the effect of an aqueous environment at high temperatures.

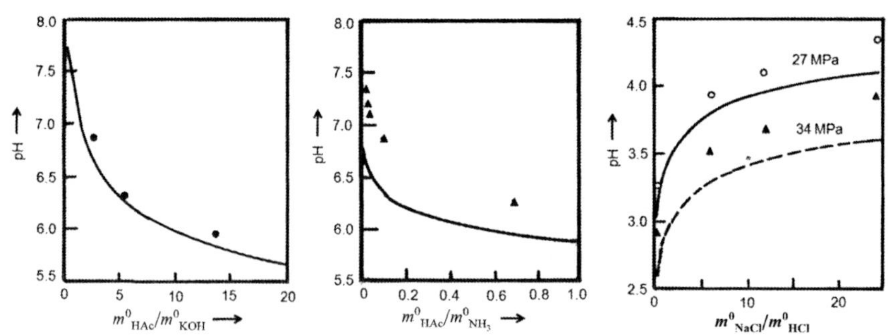

FIGURE 5.34 Titration curves for acid–base systems at hydrothermal conditions. Experimental results (symbols) compared to calculated results (lines). Left: KOH—acetic acid, $T=350$ °C, $P=34$ MPa, optical indicator: 2-Naphthoic acid; Middle: NH_3—acetic acid, $T=350$ °C, $P=34$ MPa, optical indicator: naphthoic acid; Right: HCl–NaCl, $T=380$ °C, $P=27$ and 34 MPa, optical indicator: acridine. *Modified from Ref. [107].*

5.3.6.1 Condensation Reactions

In condensation reactions, two or more molecules combine to a larger molecule and release a simple molecule like H_2O, HCl, or NH_3, as, for example, in the formation of esters. Esters are formed or hydrolyzed quite readily in aqueous solution [23]. Benzyl alcohol and phenethyl alcohol in aqueous acetic acid at hydrothermal conditions react to significant amounts of benzyl acetate and phenethyl acetate [23,63]. 2-Phenylethanol and benzyl alcohol react to considerable proportions to the corresponding ethers [63,108].

The formation of pyruvaldehyde from glyceraldehyde and dihydroxyacetone [36] and the formation of tetrahydrofuran [94] and dibenzylether [38,63] occur in hydrothermal and supercritical water without the addition of acids [2,109]. The reaction of *n*-butyraldehyde to 2-ethyl-3-hexanal occurs with complete conversion and 85% selectivity at $T = 275\ °C$ [2].

5.3.6.2 Cannizarro Reactions

The disproportionation of aldehydes to alcohol and carbonic acid is base catalyzed. Hydrolysis of benzaldehyde diethyl acetal in aqueous KOH solution is followed by a Cannizarro disproportionation, where benzyl alcohol and benzoic acid are predominantly formed [27,48] at $T = 254\ °C$ and an acetal to KOH molar ratio of 4. A Cannizarro reaction of this aldehyde in supercritical water can be catalyzed by NH_3 [110]. Benzaldehyde reacted completely in the supercritical region at $T = 397\ °C$ [111] without the addition of a base [2].

5.3.6.3 Diels–Alder Reactions

The Diels–Alder reaction is the most widely used synthetic method for the production of polycyclic ring compounds. For the [4+2]-cycloaddition of a conjugated diene and an alkene to cyclohexene derivatives, it has been proved that water is a favorable reaction solvent medium at low temperatures [27,112–115]. It is important that dienophile and dienes dissolve in the reaction solvent. That is a problem for nonpolar components in an aqueous environment at room temperature. But most of the reactants of a Diels–Alder reaction are soluble in hydrothermal and supercritical water. In addition, the elevated pressure increases the reaction rate since Diels–Alder reactions have a high activation volume of about -25 to $-50\ cm^3/mol$ [116]. Furthermore, Diels–Alder reactions show a more negative activation volume than the corresponding exo reaction of about $-2.5\ cm^3/mol$ [116]. Therefore, selectivity can be influenced by increased pressure and the ratio of the endocomponent to the exocomponent enhanced [2,117,118]. In most cases of different combinations of dienophiles and dienes, the reaction rate at near-critical hydrothermal conditions is faster than at low-temperature reaction conditions [2]. Reaction characteristics of Diels–Alder reactions in high-temperature liquid water and supercritical water are attributed to the unique properties of near-critical and supercritical water [27,119,120].

The use of water as the solvent for the Diels–Alder reaction results in higher endo selectivity in comparison with organic solvents [121,122]. This effect is general and illustrates the role of the high cohesive energy density of water. The acceleration of the Diels–Alder reaction in water has been attributed to the hydrophobic effect which is the tendency of nonpolar species to aggregate in water solution so as to decrease the hydrocarbon–water interfacial area [23]. Kinetics of the Diels–Alder reaction at low temperatures but high pressures is discussed in detail by Jenner [25].

Several other types of organic reaction are discussed in the literature like rearrangements, Friedel–Crafts reactions, reduction reactions, and more. The description of those reactions is considered to be outside the scope of this book. The interested reader is referred to the reviews on organic reactions in high-temperature liquid and supercritical water [2,22,23,25,27].

5.3.7 Catalyzed Reactions

So far, reactions with high-temperature liquid and supercritical water have been discussed essentially without considering catalysis. It was the intention of the previous Section 5.3.7 to highlight the reactions in pure water. The following section concentrates on the effect of catalysts in chemical synthesis reactions. Catalyzed reactions of conversion of biomass and other fuel sources as well as oxidative destruction of chemicals are discussed in Chapters 8 and 9, respectively.

High-temperature liquid and supercritical water as a reaction medium is considered mainly because water is present in the system and is a more environmentally friendly alternative than the solvent used in current practice. The catalysts are often acids and bases, and transition metal salts as homogeneous catalysts. Heterogeneous catalysts have also been employed [36,56,122].

For acid-catalyzed reactions, dissolved carbon dioxide should promote the reaction. CO_2 dissolves in water and forms carbonic acid which dissociates to some extent to form elevated levels of H^+, or H_3O^+ in high-temperature water solutions. The solution can be easily neutralized by releasing the CO_2 pressure after the reaction has been accomplished. Yet experiments revealed that the rates of many acid-catalyzed reactions are not very sensitive to the increased concentration of H_3O^+ [123–128]. The reason for this result is the increased activity of the water molecule at high temperatures and the relatively low concentration of the H_3O^+ ion. At lower temperatures up to about $T = 250\ °C$, the effect of dissolved CO_2 catalyzing reactions can be well demonstrated as is shown in the chapter on biomass conversion (Chapter 8).

5.3.7.1 *Organometallic Reactions*

Although organometallic complexes are usually assumed to be rather unstable with respect to high temperatures, organometallic-catalyzed reactions can be carried out in high-temperature liquid and supercritical water. An advantage of organometallic reactions in aqueous solution is the easy separation of products by cooling [2].

Organometallic catalysts that are used traditionally in organic solvents retain their activity in supercritical water. Hydrolysis reactions do not occur for some of the applications considered, for example, alkyne cyclotrimerization. Yield and selectivity for the substituted benzene product exceeds 95% for the $CpCo(CO)_2$-catalyzed cyclotrimerization of 2-butyne and phenylacetylene [129].

Cyclohexane derivatives have been used as a model system. Various reactions have been investigated like hydration and dehydration, hydrogenation and dehydrogenation, oxidation, and isomerization processes. As catalysts, different mineral acids, metal salts, and bases have been used as homogeneous catalysts and Pt and PtO_2 as heterogeneous catalysts. As a result, it was shown that these different classes of reactions can be carried out with some selectivity in supercritical water and an appropriate catalyst [36,86,130].

For the reaction of iodobenzene with alkenes, the precatalyst was $Pd(OAc)_2$ and a reduction agent (e.g., $N(Et)_3$), to form the Pd(O)complex *in situ*, which is the catalyst for the Heck reaction [2]. The reaction was carried out in high-temperature liquid water at $T=260$ °C and also in supercritical water at $T=400$ °C. The reducing agent NH_4HCO_3 is not needed at supercritical conditions, possibly due to the influence of the metal wall of the autoclave. The yield is 30% of both coupling products following the reaction of iodobenzene and styrene [2,131].

The cyclotrimerization of alkynes is a typical reaction catalyzed by organometallic complexes and is usually carried out in organic solvents [2]. This reaction can be carried out in supercritical water using cyclopentadienyl anion derivatives at $T=400$ °C [132]. Yield and selectivity of both benzene isomer products are comparable to results obtained from catalysis in organic solvents. With phenyl as the remaining R, conversion rates of >95% are achieved at $T=380$ °C, $P=25$ MPa with a yield of 24% of the symmetric and 71% of the second isomer [2].

Hydroformylation of hexene and cyclohexene in the presence of Co–carbonyl complexes in supercritical water is possible, but no organometallic products could be isolated after the reaction [2,133]. Up to 55% hydroformylated products are formed from cyclohexene and 1-hexene with hydrogenation as the main side reaction. Hydrogen required for the reaction was produced from carbon monoxide in the water-gas shift reaction, which seems to be catalyzed by the same species as the hydroformylation reaction [2].

5.3.7.2 Oxidation Reactions

Oxidation of compounds in supercritical water is of interest mainly for complete destruction of hazardous chemicals. This aspect is treated in Chapter 10.

In the following, partial oxidation of chemical compounds with the objective of obtaining useful chemicals is treated. There are mainly two groups of tested applications, the first one is partial oxidation of methane for production of methanol, and the second one concentrates on partial oxidation of aryl compounds.

The oxidation of methane may proceed through methanol as an intermediate as shown in Equations (5.69)–(5.71).

$$CH_4 + 3/2O_2 \rightarrow CH_3OH \tag{5.69}$$

$$CH_3OH + O_2 \rightarrow CO + 2H_2O \tag{5.70}$$

$$CO + 1/2O_2 \rightarrow CO_2 \tag{5.71}$$

Methane oxidation in supercritical water yields methanol with a methane conversion of 15–20% at a constant temperature of $T = 380$ °C and at pressures of $P = 30$ and 60 MPa [134]. In another investigation on the partial oxidation of methane to methanol without catalyst in near-critical and supercritical water, methanol selectivity was found from 4% to 75%, with a highest yield of 0.7%. The highest selectivity occurred at low conversions of <0.05%. Parameters of the investigation were: Temperature $T = 349$–481 °C, water density $\rho = 0.15$–0.359 g/cm^3, residence time in the batch reactor $t = 1$–9 min, with an estimated time for heating of about 1–2 min. Methanol, carbon monoxide, and carbon dioxide were the major products with trace amounts of acetaldehyde, ethanol, acetic acid, and formic acid [135].

Methane reacts with a Cr_2O_3 catalyst in a partial oxidation in supercritical water at $T = 450$ °C, $P = 33.2$ MPa partially to methanol. Results stem from experiments with a batch reactor. A highest yield of 4% is obtained after a residence time of $t = 40$ min, at a methanol selectivity of 40%. Catalytic oxidation is inhibited by supercritical water. The rate of reaction in supercritical water is only about one half of that for the gas-phase oxidation, but yield and selectivity for methanol increase by about one order of magnitude. The highest methanol yield is obtained at low oxygen supply [136].

Using Cr_2O_3/Al_2O_3 and MnO_2/CeO as heterogeneous catalysts leads to a maximum selectivity of 1.7% methanol, with a 6% conversion of methane and formic acid as the main product beside methanol, at $T = 400$–475 °C, $P = 24.1$ MPa. These results are from a flow reactor with residence times of $t = 13$–31 s [137]. For a technical application, the results of these investigations are not sufficient due to insufficient yield and selectivity and the lack of comprehensive experiments.

Much higher yields are found for the oxidation of aryl compounds to aldehydes, ketones, and carbon acids by oxygen in the presence of transition metal compounds ($MnBr_2$, $CoBr_2$, CuBr) as catalysts [2]. The oxidation of *p*-xylene to terephthalic acid is an important example of catalyzed partial oxidation in high-temperature liquid and supercritical water for synthesis of a bulk chemical product [36]. Carrying out the reaction in water would eliminate the distillation column for separating acetic acid and water in the conventional process. Furthermore, contamination by small amounts of methyl bromide could be avoided. A yield of terephthalic acid of 64% is obtained at $T = 375\ °C$, $P = 40$ MPa, and $t = 40$ min residence time in a batch reactor, with manganese bromide as the catalyst. Time for heating the reactor was in the range of the total reaction time. Therefore, subcritical conditions may have existed for most of the reaction time [36,138].

The method of supplying oxygen is of influence on the course of the oxidation reaction. A continuous oxygen supply leads to a slow reaction and lower yield and selectivity of terephthalic acid than the intermittent feeding of the same total amount of oxygen. With intermittent feeding, the rate is faster, selectivity reaches nearly 100%, and terephthalic acid yields exceed 80%. Furthermore, the formation of liquid and solid byproducts is prevented [139,140]. If water is used in the commercial terephthalic acid synthesis, environmental and potential economic advantages can be achieved [141]. For further information on this reaction, the reader is referred to the literature.

Hydrothermal and supercritical water provide a reaction atmosphere of wide varying parameters with respect to temperature, density, polarity, and ionic product. All the different states can be approached by controlling pressure and temperature. Therefore, the aqueous environment enables the variation of reaction paths within the same solvent. Although reaction conditions may be more severe for hydrothermal and supercritical water than for organic solvents at near-ambient conditions, the advantage of only one solvent that is compatible with the environment and the possibility for easy product separation may in many cases favor the process in water.

The reactions discussed in this chapter may be seen more as examples than as a compilation. Much more detailed treatises of reactions in hydrothermal and supercritical water are available in the literature, for example Refs. [2,10,12,23,25,27–29].

REFERENCES

[1] C.A. Eckert, D. Burk, J.S. Brown, C.L. Liotta, Tuning solvents for sustainable technology, Ind. Eng. Chem. Res. 39 (2000) 4615–4621.

[2] E. Dinjus, A. Kruse, Applications of supercritical water, in: R. van Eldik, F.-G. Klärner (Eds.), High Pressure Chemistry, Wiley-VCH, Weinheim, Germany, 2002, pp. 422–446, (Chapter 14).

[3] H. Hippler, Elementary reactions in supercritical fluids, in: E. Reverchon (Ed.), Proceedings of the 4th Italian Conference on Supercritical Fluids and Their Applications, CUES, Salerno, Italy, 1997, pp. 279–286.

[4] H.J. Ederer, A. Kruse, C. Mas, K.H. Ebert, Modelling of the pyrolysis of tert-butylbenzene in supercritical water, J. Supercrit. Fluids 15 (1999) 191–204.

[5] M.T. Klein, L.A. Torry, B.C. Wu, S.H. Townsend, S.C. Paspek, Hydrolysis in supercritical water: solvent effects as a probe of the reaction mechanism, J. Supercrit. Fluids 3 (1990) 222–227.

[6] M.T. Klein, B. Izzo, C.L. Harrell, T. Adschiri, in: K. Arai (Ed.), Proceedings of the 4th International Symposium on Supercritical Fluids, Tohoku University Press, Sendai, Japan, 1997, p. 543.

[7] N. Akiya, P.E. Savage, Effect of water density on hydrogen peroxide dissociation in supercritical water. 1. Reaction equilibrium, J. Phys. Chem. A 104 (2000) 4433–4440.

[8] N. Akiya, P.E. Savage, Effect of water density on hydrogen peroxide dissociation in supercritical water. 2. Reaction Kinetics, J. Phys. Chem. A 104 (2000) 4441–4448.

[9] O. Levenspiel, Chemical Reaction Engineering, third ed., John Wiley & Sons, Hoboken, NJ, USA, 1998.

[10] T.B. Brill, P.E. Savage, Kinetics and mechanisms of hydrothermal organic reactions, in: D.A. Palmer, R. Femandez-Prini, A.H. Harvey (Eds.), Aqueous Systems at Elevated Temperatures and Pressures: Physical Chemistry in Water, Steam and Hydrothermal Solutions, Elsevier, Amsterdam, The Netherlands, 2004, pp. 643–675, (Chapter 16).

[11] N. Akiya, P.E. Savage, Role of water in formic acid decomposition, AIChE J. 44 (1998) 405–415.

[12] P.E. Savage, S. Gopalan, T.I. Mizan, Ch.J. Martino, E.E. Brock, Reactions at supercritical conditions: applications and fundamentals, AIChE J. 41 (1995) 1723–1778.

[13] J.W. Moore, R.G. Pearson, Kinetics and Mechanism, third ed., Wiley, New York, 1981.

[14] S.I. Sandler, Chemical and Engineering Thermodynamics, second ed., Wiley, New York, 1989.

[15] R. Van Eldik, T. Asano, W.J. LeNoble, Activation and reaction volumes in solution 2, Chem. Rev. 89 (1989) 549–688.

[16] K.P. Johnston, C. Haynes, Extreme solvent effects on reaction rate constants at supercritical fluid conditions, AIChE J. 33 (1987) 2017–2026.

[17] C.A. Eckert, C.K. Hsieh, J.R. McCabe, Molecular thermodynamics for chemical reactor design, AIChE J. 20 (1974) 20–36.

[18] G.L. Huppert, B.C. Wu, S.H. Townsend, M.T. Klein, S.C. Paspek, Hydrolysis in supercritical water: identification and implication of a polar transition state, Ind. Eng. Chem. Res. 28 (1989) 161–165.

[19] K.A. Connors, Chemical Kinetics: The Study of Reaction Rates in Solution, VCH Publishers, New York, 1990.

[20] G.F. Froment, K.B. Bischoff, Chemical Reactor Analysis and Design, second ed., Wiley, New York, 1990.

[21] D. Miksa, T.B. Brill, Spectroscopy of hydrothermal reactions. 17. Kinetics of the surface-catalyzed water−gas shift reaction with inadvertent formation of $Ni(CO)_4$, Ind. Eng. Chem. Res. 40 (2001) 3098–3103.

[22] P.E. Savage, Organic chemical reactions in supercritical water, Chem. Rev. 99 (1999) 603–622.

[23] A.R. Katritzky, S.M. Allin, M. Siskin, Aquathermolysis: reactions of organic compounds with superheated water, Acc. Chem. Res. 29 (1996) 399–406.

[24] T. Asano, W.J. Le Noble, Activation and reaction volumes in solution, Chem. Rev. 78 (1978) 407–489.

[25] G. Jenner, Catalytic and solvophobic promotion of high pressure addition reactions, in: R. van Eldik, F.-G. Klärner (Eds.), High Pressure Chemistry, Wiley-VCH, Weinheim, Germany, 2002, pp. 305–347.

[26] W.C. Röntgen VIII, Ueber die Constitution des flüssigen Wassers, Ann. Phys. Chem. Neue Folge 45 (1892) 92–98.

[27] Y. Ikushima, M. Arai, Stoichiometric organic reactions, in: Ph.G. Jessop, W. Leitner (Eds.), Chemical Synthesis Using Supercritical Fluids, Wiley-VCH, Weinheim, Germany, 1999, pp. 267–275.
[28] A.R. Katritzky, D.A. Nichols, M. Siskin, R. Murugan, M. Balasubramanian, Reactions in high-temperature aqueous media, Chem. Rev. 101 (2001) 837–892.
[29] G. Brunner, Near critical and supercritical water. Part I. Hydrolytic and hydrothermal processes (Review), J. Supercrit. Fluids 47 (2009) 373–381.
[30] T.J. Houser, C.-C. Tsao, J.E. Dyla, M.K. Van Atten, M.E. McCarville, The reactivity of tetrahydroquinoline, benzylamine and bibenzyl with supercritical water, Fuel 68 (1989) 323–327.
[31] A.R. Katritzky, R.A. Barcock, M. Balasubramanian, J.V. Greenhill, M. Siskin, W.N. Olmstead, Aqueous high-temperature chemistry of carbo- and heterocycles. 20. Reactions of some benzenoid hydrocarbons and oxygen-containing derivatives in supercritical water at 460 °C, Energy Fuels 8 (1994) 487–497.
[32] M. Watanabe, H. Hirakoso, S. Sawamoto, T. Adschiri, K. Arai, Polyethylene conversion in supercritical water, J. Supercrit. Fluids 13 (1998) 247–252.
[33] T.M. McCollom, J.S. Seewald, B.R.T. Simoneit, Reactivity of monocyclic aromatic compounds under hydrothermal conditions, Geochim. Cosmochim. Acta 65 (2001) 455–468.
[34] T.J. Houser, D.M. Tiffany, Z. Li, M.E. McCarville, M.E. Houghton, Reactivity of some organic compounds with supercritical water, Fuel 65 (1986) 827–832.
[35] J.M.L. Penninger, R.J.A. Kersten, H.C.L. Baur, Reactions of diphenylether in supercritical water—mechanism and kinetics, J. Supercrit. Fluids 16 (1999) 119–132.
[36] P.E. Savage, A perspective on catalysis in sub- and supercritical water, J. Supercrit. Fluids 47 (2009) 407–414.
[37] J. Falbe, M. Regitz (Eds.), ninth ed., In: Römpp Chemie Lexikon, vol. 3, Georg Thieme Verlag, Stuttgart, Germany, 1995.
[38] P. Krammer, H. Vogel, Hydrolysis of esters in subcritical and supercritical water, J. Supercrit. Fluids 16 (2000) 189–206.
[39] M.J. Antal, A. Brittain, C. DeAlmeida, S. Ramayya, J.C. Roy, Heterolysis and homolysis in supercritical water, in: M.J. Comstock (Ed.), Supercritical fluids, ACS symposium series 329, Washington, DC, USA, 1989 7, pp. 77–86.
[40] J.C. Meyer, P.A. Marrone, J.W. Tester, Acetic acid oxidation and hydrolysis in supercritical water, AIChE J. 41 (1995) 2108–2121.
[41] P.H.L. Moquin, F. Temelli, Kinetic modeling of hydrolysis of canola oil in supercritical media, J. Supercrit. Fluids 45 (2008) 94–101.
[42] C.L. Harrell, M.T. Klein, T. Adschiri, Adv. Environ. Res. 1 (1993) 373.
[43] Y. Tanabe, J. Toriya, I. Kasahare, German Patent Application, Deutsche Offenlegungsschrift DE 2645030, April 14, 1977.
[44] J.D. Leonard, Preparation of formic acid by hydrolysis of methyl formate, U.S. Patent 4,299,981, November 10, 1981.
[45] W. Vogt, H. Glaser, German Patent Application, Deutsche Offenlegungsschrift DE 2062376, June 29, 1972, also GB1, 320,280.
[46] J.M.L. Penninger, Reactions of di-n-butylphthalate in water at near-critical temperature and pressure, Fuel 67 (1988) 490–496.
[47] M. Siskin, G. Brons, S.N. Vaughn, A.R. Katritzky, M. Balasubramanian, Aqueous organic chemistry. 3. Aqua thermolysis: reactivity of ethers and esters, Energy Fuels 4 (1990) 488–492.

[48] B. Kuhlmann, E.M. Arnett, M.J. Siskin, Classical organic reactions in pure superheated water, J. Org. Chem. 59 (1994) 3098–3101.
[49] T.P. Goldstein, Geocatalytic reactions in formation and maturation of petroleum, Am. Assoc. Pet. Geol. Bull. 67 (1983) 152–159.
[50] L.A. Torry, R. Kaminsky, M.T. Klein, M.R. Klotz, The effect of salts on hydrolysis in supercritical and near-critical water: reactivity and availability, J. Supercrit. Fluids 5 (1992) 163–168.
[51] J.M.L. Penninger, J.M.M. Kolmschate, Chemistry of methoxynaphthalene in supercritical water, in: K.P. Johnston, J.M.L. Penninger (Eds.), Supercritical fluid science and technology, ACS symposium series 406, American Chemical Society, Washington, DC, USA, 1989 16, pp. 242–258.
[52] M.T. Klein, Y.G. Mentha, L.A. Torry, Decoupling substituent and solvent effects during hydrolysis of substituted anisoles in supercritical water, Ind. Eng. Chem. Res. 31 (1992) 182–187.
[53] M. Sasaki, B. Kabyemela, R. Malaluan, S. Hirose, N. Takeda, T. Adschiri, K. Arai, Cellulose hydrolysis in subcritical and supercritical water, J. Supercrit. Fluids 13 (1998) 261–268.
[54] J.M.L. Penninger, R.J.A. Kersten, H.C.L. Baur, Hydrolysis of diphenylether in supercritical water. Effects of dissolved NaCl, J. Supercrit. Fluids 17 (2000) 215–226.
[55] M. Baerns, H. Hofmann, A. Renken, second ed., in: Chemische Reaktionstechnik, vol. 1, Georg Thieme Verlag, Stuttgart, Germany, 1992.
[56] A. Baiker, Supercritical fluids in heterogeneous catalysis, Chem. Rev. 99 (1999) 453–473.
[57] A.J. Belsky, P.G. Maiella, T.B. Brill, Spectroscopy of hydrothermal reactions 13. Kinetics and mechanisms of decarboxylation of acetic acid derivatives at 100−260 °C under 275 bar, J. Phys. Chem. A103 (1999) 4253–4260.
[58] D.A. Palmer, S.E. Drummond, Thermal decarboxylation of acetate, Part I. The kinetics and mechanism of reaction in aqueous solution, Geochim. Cosmochim. Acta 50 (1986) 813–823.
[59] G.A. Hall Jr., The kinetics of the decomposition of malonic acid in aqueous solution, J. Am. Chem. Soc. 71 (1949) 2691–2693.
[60] N.R. Gunawardena, T.B. Brill, Spectroscopy of hydrothermal reactions 15. The pH and counterion effects on the decarboxylation kinetics of the malonate system, J. Phys. Chem. A105 (2001) 1876–1881.
[61] J.B. Dunn, M.L. Burns, S.E. Hunter, Ph.E. Savage, Hydrothermal stability of aromaticcarboxylic acids, J. Supercrit. Fluids 27 (2003) 263–274.
[62] T.J. Houser, C.C. Tsao, in: W.W. Eckenfelder, A.R. Bowers, J.A. Roth (Eds.), Chemical Oxidation, Technologies for the Nineties, Technomic, Lancaster, PA, 1991, pp. 292–298.
[63] A.R. Katritzky, M. Balasubramanian, M. Siskin, Aqueous high-temperature chemistry of carbo- and heterocycles. 2. Monosubstituted benzenes: benzyl alcohol, benzaldehyde and benzoic acid, Energy Fuels 4 (1990) 499–505.
[64] L. Artok, H.H. Schobert, Reaction of carboxylic acids under coal liquefaction conditions: 1. Under nitrogen atmosphere, J. Anal. Appl. Pyrolysis 54 (2000) 215–233.
[65] X. Xu, M.J. Antal Jr., D.G.M. Anderson, Mechanism and temperature-dependent kinetics of the dehydration of tert-butyl alcohol in hot compressed liquid water, Ind. Eng. Chem. Res. 36 (1997) 23–41.
[66] A.R. Katritzky, A.R. Lapucha, M. Siskin, Aqueous high-temperature chemistry of carbo- and heterocycles. 3. 2-Substituted pyridines, Energy Fuels 4 (1990) 506–510.
[67] A.R. Katritzky, A.R. Lapucha, M. Siskin, Aqueous high-temperature chemistry of carbo- and heterocycles. 4. 4-Substituted pyridines, Energy Fuels 4 (1990) 510–514.

[68] A.R. Katritzky, A.R. Lapucha, R. Murugan, F.J. Luxem, M. Siskin, G. Brons, Aqueous high-temperature chemistry of carbo- and heterocycles. 1. Introduction and reaction of 3-pyridylmethanol, pyridine-3-carboxaldehyde, and pyridine-3-carboxylic acid, Energy Fuels 4 (1990) 493–498.
[69] M.A. Abraham, B.C. Wu, S.C. Paspek, M.T. Klein, Reactions of dibenzylamine neat and in supercritical fluid solvents, Fuel Sci. Tech. Int. 6 (1988) 557–568.
[70] F. Patat, Über die Hydrolyse von Anilin, Monatsh. Chem. 77 (1947) 352–375.
[71] K.M. Benjamin, Ph.E. Savage, Hydrothermal reactions of methylamine, J. Supercrit. Fluids 31 (2004) 301–311.
[72] M.A. Abraham, M.T. Klein, Pyrolysis of benzylphenylamine neat and with tetralin, methanol, and water solvents, Ind. Eng. Chem. Prod. Res. Dev. 24 (1985) 300–306.
[73] M.A. Abraham, M.T. Klein, in: T.G. Squires, M.E. Paulaitis (Eds.), Supercritical Fluids, Chemical and Engineering Principles and Applications, American Chemical Society, Washington, DC, 1987, pp. 67–76.
[74] S.H. Townsend, M.A. Abraham, G.L. Huppert, M.T. Klein, S.C. Paspek, Solvent effects during reactions in supercritical water, Ind. Eng. Chem. Res. 27 (1988) 143–149.
[75] T.D. Thornton, Ph.D. Dissertation, University of Michigan, 1991.
[76] C.J. Martino, P.E. Savage, Thermal decomposition of substituted phenols in supercritical water, Ind. Eng. Chem. Res. 36 (1997) 1385–1390.
[77] M.J. Antal, M. Carlsson, X. Xu, D.G.M. Anderson, Mechanism and kinetics of the acid-catalyzed dehydration of 1- and 2-propanol in hot compressed liquid water, Ind. Eng. Chem. Res. 37 (1998) 3820–3829.
[78] N. Akiya, P.E. Savage, Kinetics and mechanism of cyclohexanol dehydration in high-temperature water, Ind. Eng. Chem. Res. 40 (2001) 1822–1831.
[79] J. Schanzenbächer, J.D. Taylor, J.W. Tester, Ethanol oxidation and hydrolysis rates in supercritical water, J. Supercrit. Fluids 22 (2002) 139–147.
[80] X. Xu, C. De Almeida, M.J. Antal Jr., Mechanism and kinetics of the acid-catalyzed dehydration of ethanol in supercritical water in a flow reactor, J. Supercrit. Fluids 3 (1990) 228–232.
[81] L. Ott, V. Lehr, S. Urfels, M. Bicker, H. Vogel, Influence of salts on the dehydration of several biomass-derived polyols in sub- and supercritical water, J. Supercrit. Fluids 38 (2006) 80–93.
[82] W. Bühler, E. Dinjus, H.J. Ederer, A. Kruse, C. Mas, Ionic reactions and pyrolysis of glycerol as competing reaction pathways in near- and supercritical water, J. Supercrit. Fluids 22 (2002) 37–53.
[83] R. Narayan, J. Antal Jr., Influence of pressure on the acid-catalyzed rate constant for 1-propanol dehydration in supercritical water, J. Am. Chem. Soc. 112 (1990) 1927–1931.
[84] M.J. Antal Jr., W.S.L. Mok, J.C. Roy, A.T. Raissi, D.G.M. Anderson, Pyrolytic sources of hydrocarbons from biomass, J. Anal. Appl. Pyrolysis 8 (1985) 291–303.
[85] S. Ramayya, A. Brittain, C. DeAlmeida, W. Mok, Acid-catalysed dehydration of alcohols in supercritical water, Fuel 66 (1978) 1364–1371.
[86] R.C. Crittendon, E.J. Parsons, Transformations of cyclohexane derivatives in supercritical water, Organometallics 13 (1994) 2587–2591.
[87] E.L. Shock, H.C. Helgeson, Calculation of the thermodynamic and transport properties of aqueous species at high pressures and temperatures: correlation algorithms for ionic species and equation of state predictions to 5 kb and 1000 °C, Geochim. Cosmochim. Acta 52 (1988) 2009–2036.

[88] M. Osada, M. Watanabe, K. Sue, T. Adschiri, K. Arai, Water density dependence of formaldehyde reaction in supercritical water, J. Supercrit. Fluids 28 (2004) 219–224.
[89] A.R. Katritzky, A.R. Lapucha, M. Siskin, Aqueous high-temperature chemistry of carbo- and heterocycles. 12. Benzonitriles and pyridine carbonitriles, benzamides and pyridine carboxamides, and benzylamines and pyridylamines, Energy Fuels 4 (1990) 555–561.
[90] T. Adschiri, K. Arai, Y. Oshima, Oxidation and hydrolysis reactions in supercritical water, in: Y. Arai, T. Sako, Y. Takebayashi (Eds.), Supercritical Fluids, Springer, Berlin, Germany, 2002, pp. 364–381, (Chapter 6.3).
[91] S.D. Iyer, M.T. Klein, Effect of pressure on the rate of butyronitrile hydrolysis in high-temperature water, J. Supercrit. Fluids 10 (1997) 191–200.
[92] C.L. Harrell, J.S. Moscariello, M.T. Klein, The absence of wall effects during benzo-nitrile hydrolysis, J. Supercrit. Fluids 14 (1999) 219–224.
[93] A. Krammer, S. Mittelstädt, H. Vogel, Hydrolyse von Nitrilen in überkritischem Wasser, Chem. Ing. Tech. 71 (1999) 261–267.
[94] P. Krammer, S. Mittelstädt, H. Vogel, Untersuchungen zum Synthesepotential in überkritischem Wasser, Chem. Ing. Tech. 70 (1998) 1559–1563. P. Krammer, S. Mittelstädt, H. Vogel, Investigating the synthesis potential in supercritical water, Chem. Eng. Technol. 22 (1999) 126–130.
[95] A.J. Belsky, T.-J. Li, T.B. Brill, Reactions of cyanamide, dicyandiamide andrelated cyclic azines in high temperature water, J. Supercrit. Fluids 10 (1997) 201–208.
[96] T.J. Houser, X. Liu, Reactions of 1-chloro-3-phenylpropane, 2-chlorotoluene, and 4-chlorophenol in supercritical water, J. Supercrit. Fluids 9 (1996) 167–171.
[97] D. Salvatierra, J.D. Taylor, P.A. Marrone, J.W. Tester, Kinetic study of hydrolysis of methylene chloride from 100 to 500 °C, Ind. Eng. Chem. Res. 38 (1999) 4169–4174.
[98] P.A. Marrone, T.A. Arias, W.A. Peters, J.W. Tester, Solvation effects on kinetics of methylene chloride reactions in sub- and supercritical water: theory, experiment, and ab initio calculations, J. Phys. Chem. A 102 (1998) 7013–7028.
[99] P.A. Marrone, P.M. Gschwend, K.C. Swallow, W.A. Peters, J.W. Tester, Product distribution and reaction pathways for methylene chloride hydrolysis and oxidation under hydrothermal conditions, J. Supercrit. Fluids 12 (1998) 239–254.
[100] J.W. Tester, P.A. Marrone, M.M. DiPippo, K. Sako, M.T. Reagan, T. Arias, W.A. Peters, Chemical reactions and phase equilibria of model halocarbons and salts in sub- and supercritical water (200–300 bar, 100–600 °C), J. Supercrit. Fluids 13 (1998) 225–240.
[101] Y. Yamasaki, H. Enomoto, N. Yamasaki, M. Nakahara, NMR study of hydrothermal reactions of dichloromethane with and without alkali, Bull. Chem. Soc. Jpn. 73 (2000) 2687–2693.
[102] Y. Oshima, B. Bijanto, S. Koda, Kinetics of methylene chloride hydrolysis and the salt effect under hydrothermal conditions, Ind. Eng. Chem. Res. 40 (2001) 1026–1031.
[103] S.W. Park, J.H. Yoon, H. Lee, Destruction of CFC113 in supercritical and subcritical water, Korean J. Chem. Eng. 13 (1996) 640–641.
[104] B.R. Foy, K. Waldthausen, M.A. Sedillo, S.J. Buelow, Hydrothermal processing of chlorinated hydrocarbons in a titanium reactor, Environ. Sci. Technol. 30 (1996) 2790–2799.
[105] R.Y.Saleh, M. Siskin, G.A. Knudsen, Process for improving biodegradability of PVC, US Patent 5,324,817, 1994.
[106] V.I. Anikeev, A. Yermakova, V.A. Semikolenov, M. Goto, Effect of supercritical water density on the rate constant of aliphatic nitrocompounds decomposition, J. Supercrit. Fluids 33 (2005) 243–246.
[107] K.P. Johnston, J.B. Chlistunoff, Neutralization of acids and bases in subcritical and supercritical water: acetic acid and HCl, J. Supercrit. Fluids 12 (1998) 155–164.

[108] A.R. Katritzky, F.J. Luxem, M. Siskin, Aqueous high-temperature chemistry of carbo- and heterocycles. 7. Monosubstituted benzenes with two carbon atom side chains oxygenated at the alpha and beta positions, Energy Fuels 4 (1990) 525–531.
[109] K. Chandler, F. Deng, A.K. Dillow, C.L. Liotta, C.A. Eckert, Alkylation reactions in near-critical water in the absence of acid catalysts, Ind. Eng. Chem. Res. 36 (1997) 5175–5179.
[110] T.J. Houser, C.-C. Tsao, J.E. Dyla, M.K. Van Atten, M.E. McCarville, The reactivity of tetrahydroquinoline, benzylamine and bibenzyl with supercritical water, Fuel 68 (1989) 323–327.
[111] Y. Ikushima, K. Katakeda, O. Sato, T. Yokoyama, M. Arai, Structure and base catalysis of supercritical water in the noncatalytic benzaldehyde disproportionation using water at high temperatures and pressures, Angew. Chem. 113 (2001) 216–219.
[112] D. Rideout, R. Breslow, Hydrophobic acceleration of Diels-Alder reactions, J. Am. Chem. Soc. 102 (1980) 7816–7817.
[113] R. Breslow, Hydrophobic effects on simple organic reactions in water, Acc. Chem. Res. 24 (1991) 159–164.
[114] P.A. Grieco, P. Garner, Z.-M. He, 'Micellar' catalysis in the aqueous intermolecular Diels–Alder reaction: rate acceleration and enhanced selectivity, Tetrahedron Lett. 24 (1983) 1897–1900.
[115] P.A. Grieco, Organic chemistry in unconventional solvents, Aldrichim. Acta 24 (3) (1991) 58–66.
[116] F.-G. Klaerner, M.K. Diedrich, A.E. Wigger, Effect of pressure on organic reactions, in: R. Van Eldik, C.D. Hubbard (Eds.), Chemistry Under Extreme or Non-classical Conditions, Wiley, New York, 1997, pp. 103–162.
[117] J. Jurczak, D.T. Gryko, Organic synthesis at high pressure, in: R.R. Van Eldik, C.D. Hubbard (Eds.), Chemistry Under Extreme or Non-classical Conditions, Wiley, Weinheim, Germany, 1997, pp. 163–188.
[118] E. Dinjus, R. Fornika, M. Scholz, Organic chemistry in supercritical fluids, in: R. Van Eldik, C.D. Hubbard (Eds.), Chemistry Under Extreme or Non-classical Conditions, Wiley, Weinheim, Germany, 1997, pp. 219–272.
[119] M.B. Korzenski, J.W. Kolis, Diels-Alder reactions using supercritical water as an aqueous solvent medium, Tetrahedron Lett. 38 (1997) 5611–5614.
[120] J. Gao, Supercritical hydration of organic compounds. The potential of mean force for benzene dimer in supercritical water, J. Am. Chem. Soc. 115 (1993) 6893–6895.
[121] R. Breslow, U. Maitra, D. Rideout, Selective diels-alder reactions in aqueous solutions and suspensions, Tetrahedron Lett. 24 (1983) 1901–1904.
[122] P.E. Savage, Heterogeneous catalysis in supercritical water, Catal. Today 62 (2000) 167–173.
[123] S.E. Hunter, C.E. Ehrenberger, P.E. Savage, Kinetics and mechanism of tetrahydrofuran synthesis via 1,4-butanediol dehydration in high-temperature water, J. Org. Chem. 71 (2006) 6229–6239.
[124] C.M. Comisar, S.E. Hunter, A. Walton, P.E. Savage, Effect of pH on ether, ester, and carbonate hydrolysis in high-temperature water, Ind. Eng. Chem. Res. 47 (2008) 577–584.
[125] S.E. Hunter, P.E. Savage, Quantifying rate enhancements for acid catalysis in CO_2-enriched high-temperature water, AIChE J. 54 (2008) 516–528.
[126] C.M. Comisar, P.E. Savage, The benzil-benzilic acid rearrangement in high temperature water, Green Chem. 7 (2005) 800–806.
[127] S.E. Hunter, P.E. Savage, Recent advances in acid- and base-catalyzed organic synthesis in high-temperature liquid water, Chem. Eng. Sci. 59 (2004) 4903–4909.

[128] S.E. Hunter, P.E. Savage, Acid-catalyzed reactions in carbon dioxide-enriched high-temperature liquid water, Ind. Eng. Chem. Res. 42 (2003) 290–294.

[129] K.S. Jerome, E.J. Parsons, Metal-catalyzed alkyne cyclotrimerizations in supercritical water, Organometallics 12 (1993) 2991–2993.

[130] P. Reardon, S. Metts, C. Crittendon, P. Daugherity, E.J. Parsons, Palladium catalyzedcoupling reactions in superheated water, Organometallics 14 (1995) 3810–3816.

[131] J. Diminnie, S. Metts, E.J. Parsons, In situ generation and Heck coupling of alkenes in superheated water, Organometallics 14 (1995) 4023–4025.

[132] H. Borwieck, O. Walter, E. Dinjus, J. Ribizant, Organometallic chemistry in supercritical water: metallorganic products of the CpCo-catalyzed cyclotrimerization of acetylenes, J. Organomet. Chem. 570 (1998) 121–127.

[133] E. Dinjus, W. Riffel, H. Borwieck, German Patent DE-PS 19853371, 2000.

[134] E.U. Franck, High pressure combustion and flames in supercritical water, in: M.A. McHugh (Ed.), Proceedings of the 2nd International Symposium on Supercritical Fluids, Boston, MA, USA, 1991, pp. 91–96.

[135] P.E. Savage, R. Li, J.T. Santini Jr., Methane to methanol in supercritical water, J. Supercrit. Fluids 7 (1994) 135–144.

[136] C.N. Dixon, M.A. Abraham, Conversion of methane to methanol by catalytic supercritical water oxidation, J. Supercrit. Fluids 5 (1992) 269–273.

[137] S.N.V.K. Aki, M.A. Abraham, Catalytic partial oxidation of methane in supercritical water, J. Supercrit. Fluids 7 (1994) 259–263.

[138] R.L. Holliday, B.Y.M. Jong, J.W. Kolis, Organic synthesis in subcritical water—oxidation of alkyl aromatics, J. Supercrit. Fluids 12 (1998) 255–260.

[139] M. Osada, P.E. Savage, Terephthalic acid synthesis at higher concentrations in high-temperature liquid water. 1. Effect of oxygen feed method, AIChE J. 55 (2009) 710–716.

[140] M. Osada, P.E. Savage, Terephthalic acid synthesis at higher concentrations in high-temperature liquid water. 2. Eliminating undesired byproduct, AIChE J. 55 (2009) 1530–1537.

[141] J.B. Dunn, P.E. Savage, Economic and environmental assessment of high temperaturewater as a medium for terephthalic acid synthesis, Green Chem. 5 (2003) 649–655.

Chapter 6

Extraction Processes

6.1 INTRODUCTION

An extraction process in the context of this chapter is the separation of compounds from a substrate. In many cases, this is a physical process, where compounds are dissolved in a solvent, recovered with the solvent, and then separated from the solvent as product. But there are also many cases in which the solvent not only physically interacts with the substrate and the extract compounds but also reacts by breaking bonds between substrate and extract compounds or by reacting with the compounds itself. In particular, with hydrothermal and supercritical water as a solvent, such reactions occur at elevated temperatures and due to the properties and the increased reactivity of water. Useful products of this process can be the extract components, the residual substrate (raffinate), or both.

The extraction process is characterized by extraction kinetics and mass transfer. An important parameter is the solubility of the extract compounds in the solvent, in our context, in water at hydrothermal and supercritical conditions. At lower temperatures, ionic and polar species will be extracted by water, thus creating an alternative to weak-polar supercritical solvents like CO_2. At higher temperatures, in particular approaching the critical temperature, also nonpolar substances will be readily dissolved and extracted.

Extraction principles and extraction kinetics will be shortly discussed first. Then, examples are treated for cleaning of soil from unwanted compounds, for the extraction of bitumen from mineral matter, for cleaning of bone materials and retrieving protein fractions, and for the extraction of natural substances. Extractions related to fuel materials like coal, oil shale, and tar sand are discussed separately in Chapter 7.

6.2 FUNDAMENTALS OF THE EXTRACTION PROCESS

6.2.1 General Description

Extraction from solid substrates is carried out by continuously contacting the solid substrate with the solvent, in the context of this book with hydrothermal or supercritical water. The solid substrate in most cases forms a fixed bed. But a continuously operated extraction from solids may be of advantage. It can be

Supercritical Fluid Science and Technology, Vol. 5. http://dx.doi.org/10.1016/B978-0-444-59413-6.00006-6

carried out in a tubular reactor as cocurrent extraction or in a stirred reactor with continuous feeding and removal of the solid substrate and solvent. The stirred reactor at the high temperatures and pressures of hydrothermal and supercritical water, for example, can be realized in a fluidized-bed reactor.

Contacting a solid in a fixed bed is a one-stage process. There can be some advantages in contacting the solid substrate several times and under different conditions in the same pressure vessel. Countercurrent contacting, carried out, for example, in several different reactors, can be also of advantage.

The hydrothermal or supercritical water solvent contacts the solid substrate and extracts the product components until the substrate is depleted. For the solid as well as for the solvent, this is an unsteady process. The course of the extraction process can be followed by determining the amount of extract against time of extraction. From these data, more information on the process can be deducted, as is discussed below [1].

The amount of extract accumulating during the course of the extraction will, in principle, follow the schematic curve shown in Figure 6.1. The first part of the extraction curve may be a straight line, corresponding to a constant extraction rate. The second part forms a graph, approaching a limiting value which is given by the total amount of extractible substances.

The gradient of the first part of the graph can be given by the equilibrium solubility. Then, from this gradient the equilibrium solubility may be determined. On the other hand, a straight line can be caused by constant mass

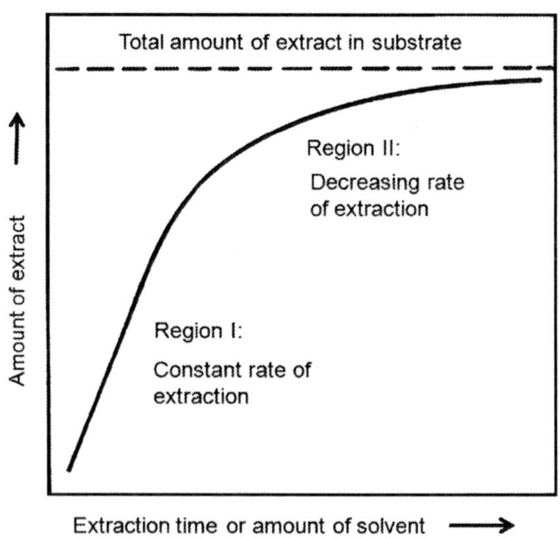

FIGURE 6.1 Integral extraction curve. Total amount of extract against time of extraction or amount of solvent. The graph in Region I is linear with a constant rate of extraction. The rate depends on the ratio of solvent to substrate. The graph in Region II is nonlinear due to mass transfer processes within the substrate and approaches a limiting value, given by the amount of extractible substances [1].

transfer resistance and there is no proof that equilibrium solubility is obtained during the extraction.

During the extraction from solid substrates, the distribution of the extract compounds throughout the substrate and the type of bonding of these to the substrate determine the extraction process. Furthermore, process properties influence the extraction, like axial and radial dispersion, back-mixing of the solvent, or nonhomogeneous distribution of the substrate in the reactor.

For the extraction in a fixed bed, it is characteristic that fresh solvent contacts a part of the solid substrate in a reactor that is already partially depleted. The solvent is further enriched in extract compounds along the reactor and can achieve a much higher concentration than would be possible by reaching only equilibrium at the entrance. Thus, complete extraction of the extractible components is possible, at least in some part of the reactor, while still a reasonable loading of the solvent can be achieved at the end of the reactor.

In a continuous cocurrent extraction, the maximum extract concentration is determined by equilibrium conditions that can be reached at the outlet of the reactor between the residual concentration of the extractible components in the substrate and the concentration of the extracted compounds in the solvent.

In a well-mixed stirred reactor, only equilibrium can be reached between the average concentration in the substrate and the concentration in the leaving solvent. But as fresh solvent is supplied to the stirred reactor, the concentration can be depleted to exhaustion, provided enough solvent is supplied. The concentration in the solvent will decrease with ongoing depletion in the substrate.

These different modes of operation have consequences for the modeling of the extraction process that is discussed in the following.

6.2.2 Influence of Process Parameters

The influence of the various process parameters is briefly discussed. Examples are further discussed in the following on cleaning of soil, cleaning of bone, bitumen extraction, and extraction of natural substances.

6.2.2.1 Pressure

In extraction processes where the density of the solvent depends substantially on pressure, an enhancement of pressure increases the extraction rate and yield, due to the increase of the solvent capacity with pressure at constant temperature. For high-temperature (liquid) water, this is not the case. Pressure has a negligible influence on the extraction in this case. Quite different is the extraction at near-critical and supercritical water conditions. In this region, the extraction rate and extraction yield strongly depend on pressure. Higher pressure and higher density increase solubility and therefore promote the extraction process.

6.2.2.2 Temperature

Temperature is of major influence on the extraction with high-temperature water. Rate and yield of the extraction are strongly enhanced with temperature. One reason is the increase of solvent power of high-temperature water for nonpolar compounds with temperature. Further reasons are the increase of mass transfer processes with temperature and the exponential rise of the vapor pressures of the extract compounds. In the near-critical and supercritical water region, increase of reaction rate and yield is still the case if density remains high. If density decreases too much, then solubility will drop and with it the amount of extracted substances. Increased temperature (in the range of the critical temperature) is, in particular, detrimental to the solubility of most inorganic salts. Their solubility decreases drastically at supercritical conditions.

6.2.2.3 Density

With increasing density, the extraction rate increases at constant temperature. Density is, in general, responsible for the capacity of a solvent, as the solubility of a compound rises with increasing density. As mass transfer is also of importance in the extraction process and increases with temperature, the extraction results will be different for the same density at different temperatures.

6.2.2.4 Solvent Ratio

Solvent ratio is the most important parameter for extraction, once approximate values of pressure and temperature are selected. Often, with increasing solvent ratio, the extraction rate can be enhanced more than with changing process parameters in a relatively narrow limit.

The influence of the solvent ratio cannot be discussed without considering economic consequences. A high solvent ratio causes enhanced operating costs if the loading of the solvent decreases with increasing solvent ratio and enhanced capital costs because the equipment for the solvent cycle is more expensive due to its larger size. On the other hand, a high solvent ratio increases the amount of extract and throughput of solid material, and thus production costs become lower.

If costs for cycling the supercritical solvent per unit quantity of product are dominant, then the solvent ratio will be more determined by the minimum quantity of solvent. The minimum quantity of solvent corresponds to a flow rate of the solvent with which equilibrium loading of the solvent at the exit of the extractor is just achieved. The actual optimum solvent ratio depends on the individual conditions and must be determined for each process and location (due to differing economic boundary conditions). In this optimization process, the possibilities of the regeneration of the solvent must be included, especially considering the residual concentration downstream to the regeneration. In addition to economic considerations, there may be further restrictions, for example, a negative influence of extraction time on the quality of the product [1].

6.2.3 Influence of the Condition of the Solid Substrate

6.2.3.1 Size of Solid Particles

In general, the extraction rate increases with decreasing particle size. The mass transfer area increases with smaller particle diameters, and the length of the transport path of compounds in the solid substrate is shortened if compounds are not only adsorbed at the surface but also present inside the solid substrate.

6.2.3.2 Pretreatment of Solid Substrate

Often the solid substrate is treated before the extraction for better extraction. This is, in particular, important for the extraction of plant materials. The pretreatment can consist of the addition of water, of flaking, of cell cracking, of heating, and of other measures.

6.2.4 Modeling of the Extraction Process

6.2.4.1 First Approximation Model

In Figure 6.2, the extraction of hydrocarbon contaminants from soil material with supercritical water is shown as an example. The extraction curve (shown as degree of extraction) consists of an initial nonlinear part, representing the initial time period of the experiment, wherein the equipment is started up.

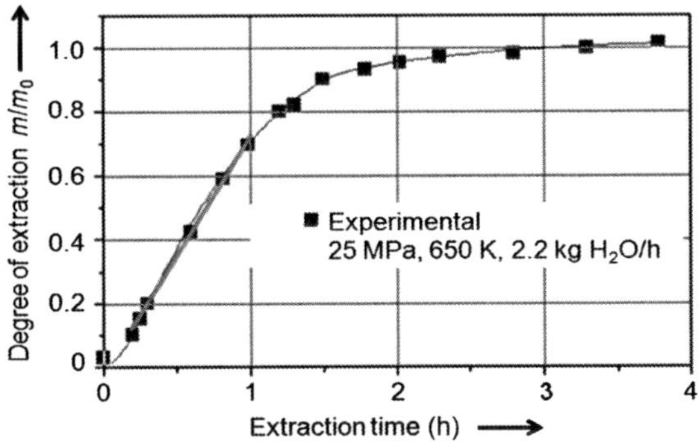

FIGURE 6.2 Course of an extraction of hydrocarbon contaminants from soil with supercritical water [2]. The full line is obtained by modeling the extraction (see below). The red line is a linear approximation of part of the extraction. For information on the extraction process, see Section 6.3. (For interpretation of the references to color in this figure legend, the reader is referred to the online version of this chapter.)

As mentioned above, the next part may be approximated by a straight line, representing the constant rate period of the extraction, where constant conditions of mass transfer are dominant. This part is followed by a section of the extraction, in which the rate of extraction is slowed down continuously by increasing mass transfer resistance, caused by depletion of the extract compounds and by lengthened transport paths.

The extraction process can be modeled using various relationships between the parameters. These relations contain models with specific coefficients, for example, mass transport models. The coefficients of these models may be obtained by independent methods, but can be calculated by fitting the result of the extraction to the model. The goal of the modeling procedure is to get a sufficient quantitative representation of the process with a simple system of equations and few and physically meaningful parameters.

The extraction of substances from solid material can be approximately modeled by assuming a steady-state process. This is a reasonable first approximation as the extraction processes are slow, residence time of the fluid is in the order of minutes, and extraction time for the solid is usually in the range from about 10 min to several hours. If these approximations are not applicable, a more detailed modeling has to be carried out as discussed below.

The amount of substances, m^*, extracted per unit of time, t, is given by Equation (6.1) [1]:

$$\frac{\mathrm{d}m}{\mathrm{d}t} = m^* = -m_\mathrm{s}\frac{\mathrm{d}c_\mathrm{m}}{\mathrm{d}t}, \tag{6.1}$$

with m as the mass of extract components, m_s as the mass of the solid substrate, and c_m as the mean concentration of extractible components in the solid.

The extract is transported within the solid to the interface solid–fluid and from there into the bulk of the fluid:

$$m^* = \beta_\mathrm{s} A(c_\mathrm{m} - c_0), \tag{6.2}$$

$$m^* = \beta_\mathrm{F} A(c_0 - c_\infty). \tag{6.3}$$

Total mass transfer resistance is given by Equation (6.4), if no other transport resistances must be taken into account and there is no phase transition at the interface solid–fluid, or the equilibrium distribution coefficient is set to 1

$$\frac{1}{k} = \frac{1}{\beta_\mathrm{s}} + \frac{1}{\beta_\mathrm{F}}, \tag{6.4}$$

with β_s as the mass transfer coefficient in the solid phase, β_F as the mass transfer coefficient in the fluid phase, k as the total mass transfer coefficient, c_0 as the initial mean concentration of extractible components in the solid, c_∞ as the concentration of extracted components in the bulk of the fluid, and A as the mass transfer area.

If a phase transition occurs, the equilibrium at the interface has to be taken into account in addition.

The differential equation can be integrated if the total mass transfer coefficient k is assumed to be constant. For $k = \text{constant}$, Equation (6.5) for the mean concentration of the extract components in the solid can be obtained:

$$\frac{c_\mathrm{m} - c_\infty}{c_0 - c_\infty} \approx \exp\left(-\frac{kA}{m_\mathrm{s} c_0} t\right). \tag{6.5}$$

If it can be assumed that mass transport resistance is dominant in the solid, then the result is

$$\frac{1}{\beta_\mathrm{F}} \ll \frac{1}{\beta_\mathrm{s}} \quad \text{and} \quad k \approx \beta_\mathrm{s}. \tag{6.6}$$

The approximation of Equation (6.5) is equivalent to a first-order reaction model for the extraction with constant reaction rate, as shown in Equation (6.7)

$$\frac{\partial X(z,t)}{\partial t} = -k_\mathrm{e} X(z,t) \tag{6.7}$$

with X—concentration of extract compound, z—extraction coordinate, t—time, and k_e—extraction rate constant.

If there is a first part of the extraction with a constant rate of extraction, then mass transfer in this part of the extraction can be calculated by dimensionless correlations, for example, Equation (6.8a) [1] or other similar equations

$$Sh = 2 + 1.1 Re^{0.6} Sc^{0.33} \quad \text{for } 3 < Re < 3000, \tag{6.8a}$$

with

$$Sh = \frac{\beta_\mathrm{F} d}{D_\mathrm{G}}, \quad Re = \frac{ud}{v}, \quad Sc = \frac{v}{D_\mathrm{G}}, \tag{6.8b}$$

where β_F is the mass transfer coefficient solid–fluid, d is the diameter of a volume equivalent sphere, D_G is the self-diffusion coefficient of the fluid, v is the viscosity of fluid, and u is the linear flow velocity of fluid.

According to this correlation, mass transport is proportional to the diffusion coefficient of the gas and the volumetric flow V of the gas with an exponent of 3/5, ($V^{3/5}$) [1].

Equations (6.5) and (6.7) are able to represent the extraction process if the rate constant k is adjusted to the different parts of the extraction. Using only the total amount of extract (at the end of the extraction) for determining k_e (i.e., taking into account only the long time solution) leads to a nonsufficient representation of the first part of the extraction (short time solution). Various models in the literature combine both parts in more or less similar ways. For a first approximation, Equations (6.5) and (6.7) are sufficient. A good modeling of the extraction process has to consider also various other parameters. Such a model is discussed in the following.

6.2.4.2 *Detailed Model for the Extraction from Solids (VTII Model)*

Extraction from solids with hydrothermal and supercritical water is a complicated process if the two-phase flow, the properties of the solid material, and the inhomogeneity of the fluid and the solid phases are considered. For most cases, a far simpler model, as indicated above, is sufficient. In the detailed model, the following parameters are considered sufficient to calculate course and result of an extraction [1]:

- equilibrium distribution between solid and supercritical solvent (adsorption isotherm),
- diffusion in the solid (effective diffusion coefficient or effective transport coefficient as defined by the applied transport model),
- mass transfer from the surface of the solid to the bulk of the fluid phase, and
- axial dispersion (effective dispersion coefficient, taking into account inhomogeneities of the fixed bed, the solvent distribution, and the influence of gravity).

The equations of the model are given in Equations (6.9)–(6.13):

Mass balance for the fluid phase:

$$\frac{\partial c_{\mathrm{F}}(z)}{\partial t} = D_{\mathrm{ax}} \frac{\partial^2 c_{\mathrm{F}}(z)}{\partial z^2} - \frac{u_z}{\varepsilon} \frac{\partial c_{\mathrm{F}}}{\partial z} - \frac{1-\varepsilon}{\varepsilon} \frac{\partial \bar{c}_{\mathrm{s}}(z)}{\partial t}. \quad (6.9)$$

Mass balance for the solid phase:

$$\frac{\partial \bar{c}_{\mathrm{s}}(z)}{\partial t} = a k_{\mathrm{oG}} \left(c_{\mathrm{F}}(z) - \bar{c}_{\mathrm{s}}(z) \frac{K(\bar{c}_{\mathrm{s}})}{\rho_{\mathrm{s}}} \right). \quad (6.10)$$

Equilibrium between fluid phase and solid phase:

$$K(\bar{c}_{\mathrm{s}}) = k_1 \bar{c}_{\mathrm{s}} \exp(-k_2). \quad (6.11)$$

Overall mass transfer coefficient:

$$\frac{\beta_{\mathrm{F}}}{k_{\mathrm{oG}}} = 1 + \frac{Bi K(\bar{c}_{\mathrm{s}})}{6}, \quad (6.12)$$

$$Bi = \frac{\beta_{\mathrm{F}} R K}{D_{\mathrm{es}}} \quad \text{Biot-number} \quad (6.13)$$

with c_{s} as mean concentration of extract components in the solid phase, c_{F} as the concentration of extract in the fluid (supercritical solvent), D_{ax} as the axial dispersion coefficient, u_z as the void volume linear velocity of supercritical solvent, and $K(\bar{c}_s)$ as the equilibrium distribution coefficient between solid and fluid phase,

$$K(\bar{c}_s) = c_F/\bar{c}_s;$$

D_{es} is the effective diffusion coefficient in the solid phase, k_{oG} is the overall mass transfer coefficient related to the fluid (gaseous) phase, z is the coordinate in the axial direction, ε is the void volume fraction (porosity of the fixed bed), t is the time of extraction, a is the specific surface of solid phase (mass transferring surface area), ρ_s is the density of solid, k_1 and k_2 are the coefficients of the sorption isotherm (e.g., Freundlich-isotherm), and β_F is the mass transfer coefficient for the fluid phase.

For the balance for the fluid phase, some assumptions have been made: Gradients of any kind are neglected in radial direction. It is assumed that gradients in radial direction can be taken into account by applying the model to several shells in axial direction, for which parameters can be considered constant. In the fluid phase, convection and axial dispersion cause dispersed plug flow. The extraction process is isothermal. The loading of the supercritical solvent is relatively low. Sorption enthalpy can therefore be neglected.

In the solid phase, the following assumptions have been made: Transport in the solid can be considered as one dimensional. The solid is uniform and the extract is evenly distributed over the solid material. Real transport phenomena, like membrane transition, pore diffusion, and diffusion in the solid, are summarized in an effective transport coefficient (effective diffusion coefficient). At the interface solid/fluid, phase equilibrium is assumed.

The equations have been written for the extract mixture. They may be used for individual components. Interactions between the individual components of the extract and the influence on the extraction process are neglected.

Experimental extraction curves have been represented by this model by fitting the kinetic coefficients (β_F, D_{es}, D_{ax}) to the experimental curves, obtained from a laboratory installation. The model can thus be used to determine the kinetic parameters from a laboratory experiment and then apply these parameters for scaling up the extraction [1].

In Figure 6.3, the simulation with such a detailed model of an extraction of contaminants from soil is shown [2]. The concentration of the extract in the solid material C_s is plotted against the dimensionless radius r/R of a single particle. Solubility of the extract compounds in the extracting solvent is assumed to be constant. The result shows the influence of different values of the extraction rate constant k_e on the extraction [2].

Case A, with the highest extraction rate constant, results in steep concentration profiles that vary with increasing extraction time through the particle. Equilibrium concentration for the extraction is reached fast and throughout the extraction. As discussed above, this equilibrium concentration may be the thermodynamic equilibrium concentration between the extract compounds

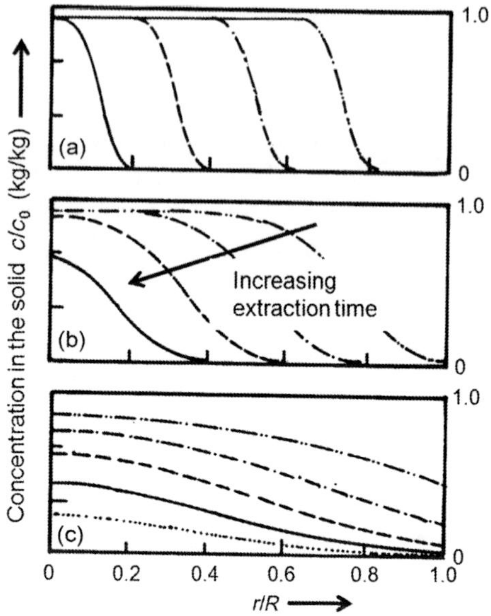

FIGURE 6.3 Simulation of an extraction from a single particle with radius R (contaminated soil) applying different values for the extraction rate: (a) $k_e = 4.5 \times 10^{-2}$/s; (b) $k_e = 4.5 \times 10^{-3}$/s; (c) $k_e = 4.5 \times 10^{-4}$/s. *Modified from Ref. [2].*

and the solvent. But in many cases, where interaction forces between extract compounds and the solid material exist, the equilibrium is different to pure component equilibrium and determined by the extract compounds in the solid substrate, the solid substrate itself, and the solvent. This fast extraction is analogue to a shrinking core behavior, where mass transfer from the solid to the fluid phase is fast.

In case B, the concentration profiles become less steep, representing decreased mass transfer rates, caused by the assumed extraction rate constant, which is one order of magnitude lower than in case A. Equilibrium concentration in the fluid is not reached at the end of the extraction. This last part of the extraction refers to the second part of the extraction curve in Figure 6.1, showing a decreasing rate of extraction due to mass transfer limitations.

In case C, mass transfer is dominant throughout the extraction. This is due to the further reduction of the extraction rate constant by one order of magnitude. The equilibrium concentration in the solvent phase is not approached at any extraction time.

The detailed model seems to be applicable for all cases of extraction rates. The most important parameters are the solubility of the extract in the solvent, which was assumed to be constant in the discussion above, and the extraction rate constant. The equilibrium solubility can be determined from separate phase equilibrium measurements or calculations. The extraction rate constant may be known from experiments or can be used as an adjustable parameter for representing experimental results of an extraction.

The simulation method can be applied to an extraction from a single particle with r as the radial coordinate as discussed above and shown in Figure 6.3. But it can also be applied to an extraction in a well-mixed stirred reactor, if the one-particle solution is assumed to be valid for all the particles in the reactor and for an extraction from a fixed bed, if the coordinate r is replaced by the coordinate in the direction of the flow of the solvent in the reactor. It can even be applied to a continuous cocurrent reaction, if mass transfer is limited to reaching equilibrium between the solid and solvent phase at the outlet.

The detailed single particle model can also describe the temperature dependence of an extraction. For an example, the results of the simulation of the extraction of contaminants from soil with hydrothermal and supercritical water are shown in Figure 6.4 and compared to experimental values [3].

Limits for the simulation may be set by the uneven distribution of the solid substrate. As mentioned, a fixed bed can be simulated in different parts, with different parameters for each section. In a continuously operated extractor, the effect of gravity may be of importance. In a vertical operated extraction, gravity may induce separation of solid particles according to size. In a horizontal continuously operated extractor, gravity effects may result in a layering of the solid substrate.

This has been demonstrated for the continuous extraction of contaminated soil. At low solid concentrations, the flow is nearly evenly distributed, but at higher concentrations the solid accumulates in the lower part of the tube. Therefore, the extraction conditions are different in the upper and the lower

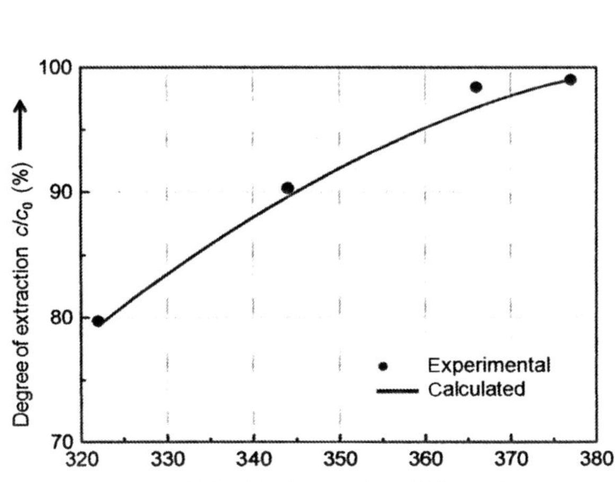

FIGURE 6.4 Simulation of the temperature dependence for an extraction of contaminants from soil with water. $P = 24$ MPa, solid concentration 1 wt% [3].

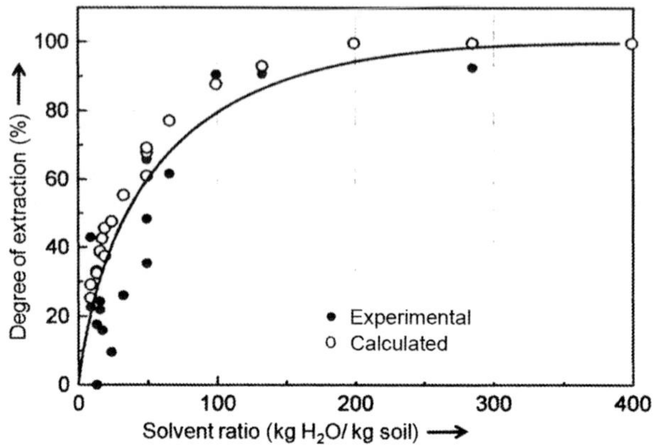

FIGURE 6.5 Simulation of an extraction of contaminants from soil in a continuous cocurrent horizontal reactor. $P = 24$ MPa, $T = 378$–393 °C, $\rho = 167$–249 kg/m^3, residence time $\tau = 23$–49 s [3,4].

part of the tube. The extraction can then be simulated by assuming that the total amount of solid moves in layers along the tubular extractor. Experimental results can be modeled with the detailed extraction model, Equations (6.9)–(6.13), using an empirical parameter representing the part of the layered solid relative to the total flow. This empirical parameter can be directly derived from the ratio of feed rate to solvent rate. In addition, it is assumed that mass transfer from the solid to the solvent takes place only on the outside of the layered fraction [3,4]. The results of the modeling are shown in Figure 6.5.

For more detailed information on the simulation procedure and the applied programs, the reader is referred to the original publications [2–5].

In the following, the application of hydrothermal and supercritical water for the extraction of substances from solid substrates is discussed for some examples: cleaning of soil material from contaminations, bitumen extraction, cleaning of bone material from organic matter, and the extraction of valuable compounds from natural materials.

6.3 SOIL CLEANING

The incentive for the cleaning of soil is the reuse of the contaminated sites and the restricted possibility for dumping the contaminated soil. Therefore, contaminants are removed and the cleaned soil material can be reused.

Soils with a high content of fine particles ($d_p < 63$ μm) are especially difficult to clean. A first step is washing the soil with water to separate the larger particles from the fines. Since by far the greatest part of the contamination is adsorbed to the fraction of fines, processing of the fines from soil material is now the goal of the cleaning process. If that process can be carried out successfully, other soil materials can be treated in an analogue way.

Water at hydrothermal and supercritical conditions has been proposed for the cleaning of soil, removing polycyclic aromatic hydrocarbons (PAH), hydrocarbons, and metals [5–7]. Experimental results prove that hydrothermal and supercritical water is well applicable for recovering these contaminants. The results are based on cleaning of different soil materials that are typical for old industrial sites and contain high levels of weathered hydrocarbons and heavy metals [2–5].

Cleaning of soil materials contaminated with hydrocarbons was effectively carried out with hydrothermal and supercritical water in a semi-continuous fixed-bed extraction and in a continuous extraction in tubular reactors [7]. Extraction time needed for cleaning could be reduced drastically, from $t=6$ h for the semi-continuous extraction to less than $t=1$ min in the continuous extraction. At a temperature of $T=380$ °C and a residence time of $t=45$ s, 98% of the contaminants could be removed. But solvent-to-feed ratio must be kept high for the continuous extraction, S/F > 100, corresponding to a concentration of about 1 wt% of soil material in the aqueous feed suspension [5,7]. Alternatively, a longer tubular reactor (increased residence time), multiple treatment, or a reactor with feeding and removal of the solid phase can be applied. Mixed contaminations of heavy metals and hydrocarbons can be cleaned with water at temperatures of $T=250$–350 °C at a pressure of $P=25$ MPa. For destruction of the contaminants, supercritical water oxidation (SCWO) proved to be a good alternative, as discussed in Chapter 10.

6.3.1 Materials

The soil materials used for the cleaning experiments were contaminated with a high content of weathered hydrocarbons ($\approx$100,000 mg/kg_{ds}, ds = dry substance) and contained up to 85% fine mineral particles with $d_p < 63$ μm.

The extraction was carried out in a fixed-bed reactor. The experimental conditions and the results are listed in Table 6.1. Residual contaminations are below $x=100$ ppm. During the treatment with supercritical water, an effluent emulsion of oil in water is produced, with concentrations of about $x=1000$–2000 mg/dm^3 total organic carbon (TOC) content. The emulsion is stable and can be treated with microorganisms, with limited success, or totally destroyed by SCWO, as described in Chapter 10 [5].

6.3.2 Extraction Results

In Figure 6.2, results of an extraction experiment are shown for a temperature of $T=380$ °C and a pressure of $P=25$ MPa. At these conditions, a high degree of extraction, or cleaning of the soil material, of >99% can be obtained. It is supposed that only a small amount of the contamination is adsorbed or chemically attached to the soil material. Artificial contaminations, or short-lived contaminations, can be removed from the soil four to

TABLE 6.1 Extraction Results of Real Contaminated Soil Materials [2,5]

Soil material from:	Petrol station	Industrial site	Old barrel depot
Sauter diameter (μm)	30	7	14
Contamination	Lubricating oil aged 1 year	Lubricating oil aged >20 years	PAH aged >45 years
Initial concentration (mg hydrocarbons/kg dry substance)	20,200	103,500	11,050
Extraction			
Final concentration (mg hydrocarbons/kg dry substance)	<20	<100	<10
Amount extracted (%)	>99.9	>99.9	>99.9
Temperature (°C)	390	382	392
Pressure (MPa)	24	24	25
Solvent/soil ratio (kg water/kg dry matter)	6	6	12

six times faster than weathered contaminations. The greatest part of the contamination is removed independent of the age of the contamination. But to achieve a high value of separation, supercritical water is essential.

Experiments with supercritical CO_2 as solvent, carried out for comparison, reveal that supercritical water is a much better extracting agent for weathered contaminations from soil [3,8] than supercritical carbon dioxide, as shown in Figure 6.6. At comparable densities of the two solvents—for water ($T = 380\,°C$, $P = 25$ MPa) $\rho_{H_2O} = 445\,\mathrm{kg/m^3}$, for CO_2 ($T = 100\,°C$, $P = 20$ MPa) $\rho_{CO_2} = 482\,\mathrm{kg/m^3}$—soil material can be cleaned with water but not with carbon dioxide. The polar structure of water and the high temperature at supercritical conditions break any bonds between contaminants and soil material. The total degree of extraction with supercritical carbon dioxide is low, total extract is about 10–20% from the total contamination, while the total degree of extraction for supercritical water as solvent is very high, >99%. Addition of water as a modifier has no significant effect on the result of the extraction. For nonweathered soil material (artificially contaminated), the cleaning effect of CO_2 was found to be sufficient.

Although a semi-batch extraction with supercritical water is able to remove all of the initial hydrocarbon contaminations from soil material, the 6 h of extraction time is too long. To improve the extraction, the process

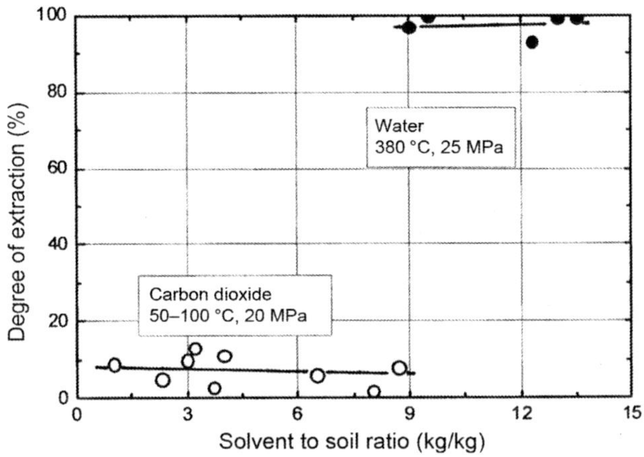

FIGURE 6.6 Extraction results (total degree of extraction) for a weathered soil material. Comparison of supercritical water and carbon dioxide as solvents [5].

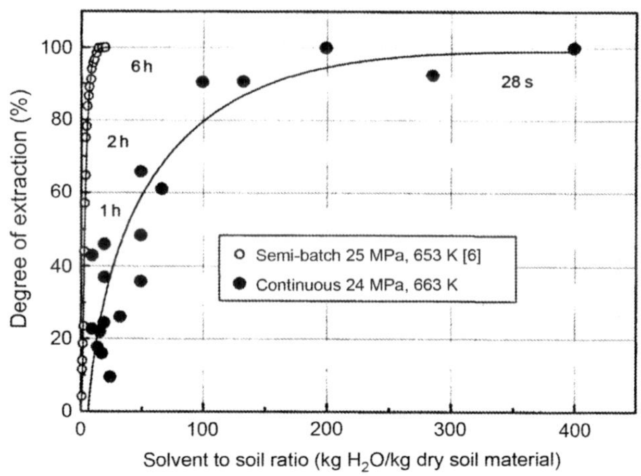

FIGURE 6.7 Extraction results for the continuous and semi-batch extraction [3,5].

can be operated continuously for both solvent and soil material. The flow scheme of the apparatus and details are discussed in Chapter 13.

The soil material for the experiments was a clayey loam with an equivalent particle diameter $d_p < 355$ μm, weathered for more than 20 years and containing 19 wt% hydrocarbons (37% long chain alkanes, 34% monoaromatic, 16% diaromatic, 12% PAH). Conditions of the experiments were $T = 250$–400 °C, $P = 24$ MPa, and concentration of the contaminated soil material in water $x \leq 10$ wt%. Representative results are shown in Figure 6.7 [3,5].

Continuous operation reduces the necessary residence time for cleaning to about $t = 28$ s compared to $t = 6$ h for semi-batch extraction. But a high solvent-to-soil ratio is needed for the continuous process. A solvent ratio of $S/F > 200 kg_{H_2O}/kg_{soil}$ (<0.5 wt% soil in water) is needed to extract more than 92% of the contamination in the tubular reactor of given length in one run. With increasing concentration of soil in water, the degree of extraction is reduced, as shown in Figure 6.8. For a higher cleaning effect, the soil material must be treated more than one time in the given reactor, or the length of the reactor and the residence time must be increased.

The degree of extraction increases with increasing temperature, as shown in Figure 6.9. Although hydrocarbon contaminations are already miscible with water at $T = 350$ °C and $P = 25$ MPa, the best extraction result is obtained at

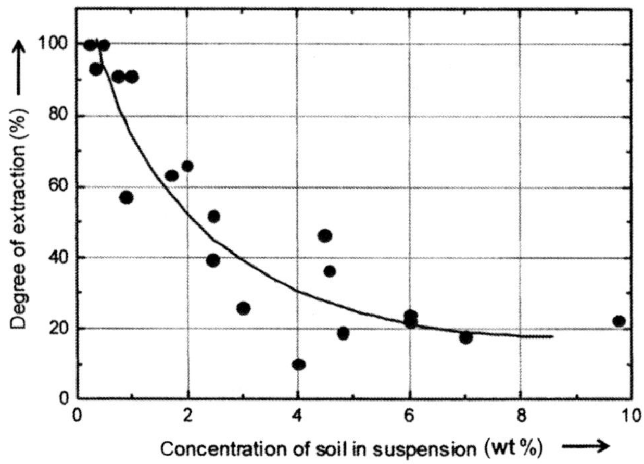

FIGURE 6.8 Degree of extraction in dependence on solvent-to-soil ratio. $T = 390$ °C, $P = 24$ MPa, residence time: $t = 28$ s [3,5].

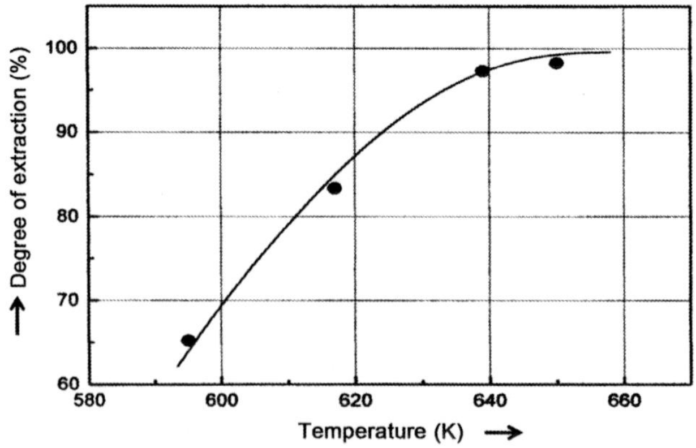

FIGURE 6.9 Temperature dependence of extraction results. $P = 24$ MPa, residence time: $t = 45$ s; $x = 1$ wt% soil in water [3,5].

$T=377$ °C and $P=24$ MPa. The extraction result can be improved by about 8% by extending the residence time from $t=28$ to 45 s with a concentration of 1 wt% of soil in water. But due to the cocurrent extraction mode, this effect is limited. More effective is a repeated treatment. With a suspension of $x=2$ wt% of soil material, the degree of extraction was improved up to 94% at a total extraction time of $t=90$ s. Due to thermal decomposition and hydrolytic reactions, carbon dioxide and methane are present in the gas phase at conditions above $T=387$ °C and $P=24$ MPa. The amount of CO_2 in the gas phase increases proportional with the amount of solids in suspension.

In many cases, heavy metals can be found as contaminating compounds, as mixtures of hydrocarbons and heavy metals occur frequently in contaminated soil of industrial sites.

6.3.3 Extraction of Heavy Metals and Organic Compounds

The soil materials investigated contained lead, copper, antimony, and zinc, in addition to the contaminating hydrocarbons [4,9]. The extraction results presented are considered to represent the lower limit for the effect of metal removal. The reason is that methods for the analyses of the metal concentrations vary substantially in their results, in particular, for lower concentrations. Therefore, the higher values of the metal analyses have been assumed as valid.

Heavy metals are fixed to the soil material by different bonding mechanisms. Various methods can lead to a mobilization. In a neutral environment, heavy metals are least mobile. At higher pH values than neutral, a mobilization of humid compounds can lead to an increased mobilization of heavy metals. At lower pH values than neutral, dissolution of hydroxides, carbonates, and oxides occurs. Therefore, dissolution of supercritical carbon dioxide in water to lower the pH value was used to improve the removal of heavy metal compounds.

In Figure 6.10, the extraction yield of lead from soil is shown as a function of residence time at different pH values and a temperature of $T=100$ °C. The extraction yield increases with residence time. The highest yield is obtained with pH 3, adding acetic acid. With dissolved carbon dioxide, only a pH value of >4 is obtained.

Results of the extraction of antimony are presented in Figure 6.11. Extraction yields of 80% are obtained. The extraction yield decreases with increasing concentration of soil, although the absolute level is low.

The organic content of the soil material can be extracted at comparable conditions as for the hydrocarbon contamination discussed above (about $T=400$ °C, $P=25$ MPa). The extraction of mixed contaminants is more difficult than that of hydrocarbons [4]. Cleaning of contaminated soil with near-critical and supercritical water from the organic content is possible. Heavy metals can be removed at lower temperatures and lowered pH values. Dissolved carbon dioxide can be used, but also organic acids in low concentration [5].

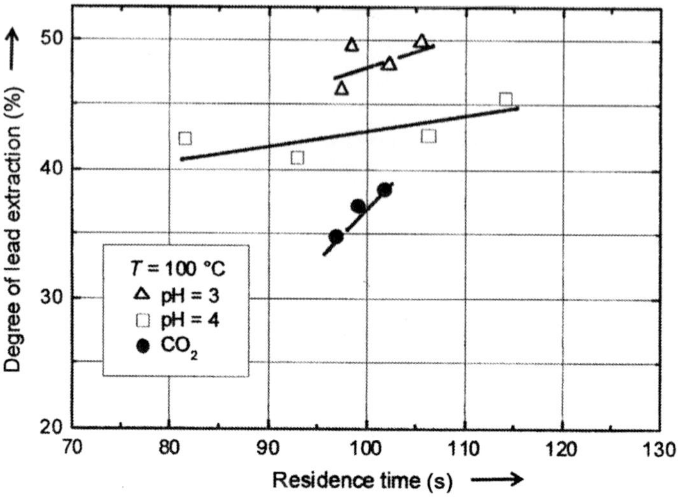

FIGURE 6.10 Degree of extraction of lead with water from contaminated soil material in dependence on residence time. $P = 25$ MPa, $T = 100$ °C [4,5].

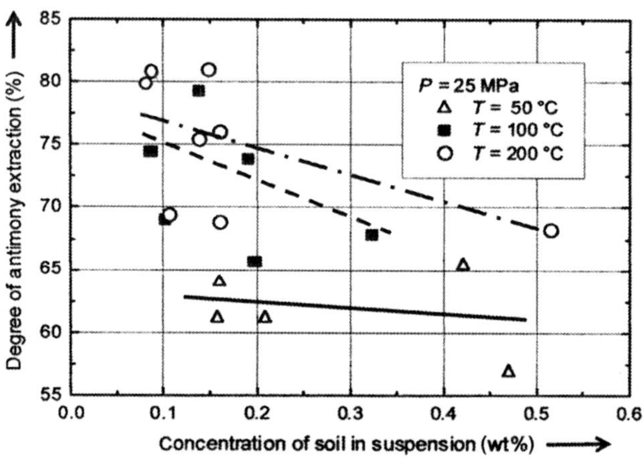

FIGURE 6.11 Degree of extraction of antimony with water from contaminated soil material in dependence on concentration of soil material. $P = 25$ MPa [4,5].

Results of this investigation prove the feasibility for the cleaning process of contaminated soil with hydrothermal and supercritical water. Nevertheless, the process is intensive in capital and energy. The amount of soil to be cleaned is very high, even if only small areas have to be remediated. Therefore, the separation of heavily contaminated soil fraction from the noncontaminated soil fraction is essential before treating the contaminated part with water at hydrothermal and supercritical conditions. In particular, the fraction of fines, being only a relatively small part of the total amount of soil that often causes heavy problems during processing, can be treated with advantage with hydrothermal and supercritical water, without causing further environmental problems. Many investigations on the cleaning of soil apply oxidation of the extract in supercritical water. This part is discussed in Chapter 10.

6.4 VARIOUS EXTRACTIONS FROM MINERAL SUBSTRATES

High-temperature and supercritical water can also be used for other separations of organic compounds from mineral matter. Examples are the extraction of hydrocarbons from fly ash, the total separation of bitumen and minerals for asphalt analysis, and the regeneration of activated carbon.

6.4.1 Extraction of PAH from Fly Ashes

Toxic PAH are frequently found in incineration fly ash. Analytical procedures are hindered by the fact that conventional extraction results only in a low yield of the hydrocarbons. Supercritical water at $T=400$ °C and $P=24$ MPa extracts the hydrocarbons better by a factor of more than an order of magnitude from an initial yield in the range of $x=0.1$–21% and gives a more detailed fingerprint of the hydrocarbons [10].

6.4.2 Bitumen Extraction from Asphalt

Asphalt is a mixture of bitumen and mineral particles used for paving road surfaces. For analytical purposes, the bitumen fraction with a medium molecular weight of $M_W=300$–500 must be separated from the mineral fraction. The analytical method aims to determine both the quantity and the quality of bitumen. Because of degradation at higher temperature, the temperature is limited to about $T=335$ °C.

Hydrocarbons dissolve in H_2O with increasing temperature. They show a Type III behavior (according to van Konynenburg and Scott) with interrupted critical curves. The solution for the separation of bitumen and mineral matter is a Soxhlet-type extraction with near-critical water ($T\approx608$ K (335 °C), $P\approx15$ MPa, $t\leq3$ h). Asphalt, the main structure of bitumen, and the experimental Soxhlet extraction are shown in Figure 6.12. The Soxhlet extraction ensures an automatic cycle for the solvent (H_2O). Increase of temperature of only 5 °C to $T=613$ K (340 °C) results in an extraction time of $t\leq1$ h, but also in an unacceptable degradation of the bitumen [11,12].

6.4.3 Regeneration of Activated Carbon

Activated carbon, saturated with phenol, can be efficiently regenerated with supercritical water at $T=400$–500 °C and $P=26$ MPa [13]. Residence times of about $t=10$ min lead to a 100% regeneration. A temperature-programmed

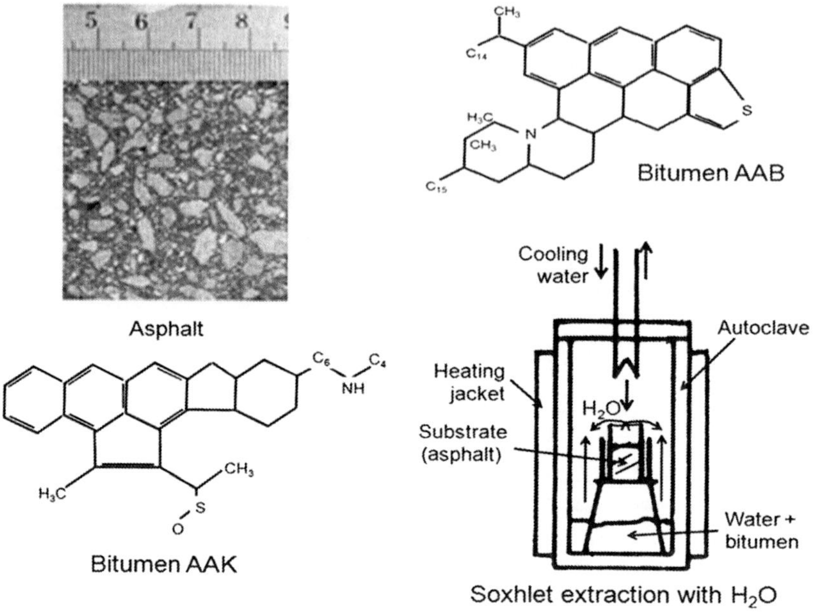

FIGURE 6.12 Asphalt and bitumen, left; Soxhlet extraction with near-critical H_2O, right [11,12].

desorption using high-temperature water and supercritical water shows that a fraction of the phenol is physically adsorbed, while another fraction is chemically bonded to the activated carbon surface. For an extraction time $t \geq 3$ min and temperatures $T \geq 450$ °C, the adsorption capacity of regenerated activated carbon is fully recovered or even increased. The mass losses during the regeneration are <6%. The chemisorbed part of phenol increases with aging time ($t = 0.5$–60 days). Full regeneration of activated carbon with aged phenol requires a longer extraction time and a higher temperature than a fresh saturated activated carbon [13].

6.5 SEPARATION OF ORGANIC COMPOUNDS FROM BONE MATERIAL

Subcritical and supercritical water can be successfully applied to remove organic compounds completely from bone materials for the purpose to use the cleaned bone material for human implants. Bones contain lipid and protein compounds that have to be removed from the hydroxyl apatite, the structural material of bones, before it can be used for implants. With hydrothermal and near-critical water, lipids and proteins can be extracted completely.

For the experiments, cited in the following, the bone material was prepared from horse femur bones. The residual protein and lipid content of milled bone material and larger bone fragments are below analytical detection limits after treatment at temperatures between $T = 200$ and 350 °C at a pressure of $P = 25$ MPa and a residence time of $t \leq 1$ h [14–16].

Treatment of the bone fragments was carried out in various modes: batch, semi-continuous, continuous, and in a combination of semi-continuous and continuous mode. The experimental setup consists of a stirred tank reactor ($V = 1000$ ml) operated with continuous solvent flow and a tubular reactor ($V = 400$ ml). Both could be connected in series. The reactor setup is described in detail in Chapter 13.

The study comprises four parts: (1) extraction of proteins from defatted bone material with high-temperature water, (2) extraction of fat and proteins with high-temperature water, (3) extraction of larger bone structures, (4) investigation of the influence of CO_2 on the hydrolysis and extraction, and (5) characterization and separation of the protein extract.

6.5.1 Extraction of Proteins from Defatted Bone Material with High-Temperature Water

Extraction in an unsteady batch extraction was of interest to determine the protein degradation and protein removal during the time needed for the extraction autoclave to reach the end temperature. In these batch experiments, 2 g of previously defatted bone material ($d_p \approx 1$ mm) was extracted with water at a concentration of $x = 0.5$ wt%. The materials were placed in one of two autoclaves ($V = 250$ ml, $T = 250$ °C, $P = 15$ MPa and $V = 1000$ ml, $T = 500$ °C, $P = 30$ MPa) and then the autoclave was heated. After reaching the end temperature, the autoclave was cooled in a water bath. The time needed for heating ranged from $t = 20$ min to reach $T = 130$ °C up to $t = 90$ min to reach $T = 250$ °C. The extraction yield of proteins obtained during heating and cooling the autoclave is shown in Figure 6.13. The yield is equal to the degree of protein removal from the bone material.

The reduction during the heating and cooling of a batch autoclave is substantial, as shown in Figure 6.13. At a low temperature of $T = 130$ °C, already 47 wt% of the total bone protein is removed. At temperatures above $T = 200$°C, a high degree of extraction of the protein material is reached. But the necessary total removal of the protein material from the bones cannot be achieved by heating the material to temperatures in the range of up to $T = 300$°C.

The ratio of solvent (water) to bone material is of importance for the protein extraction. Variation of the solvent ratio from S/F = 0.2 to 2 wt% in batch experiments carried out with an end temperature of $T = 330$ °C and $P = 10$ MPa initial nitrogen pressure in the autoclave shows that with a higher solvent ratio, the remaining protein content in the bones can be reduced, as shown in Figure 6.14.

The best extraction results are obtained at high ratios of the solvent water to the amount of bone material. The protein content after the batch experiments is reduced from $x = 2$ wt% at a solvent ratio of S/F = 50 g/g to about

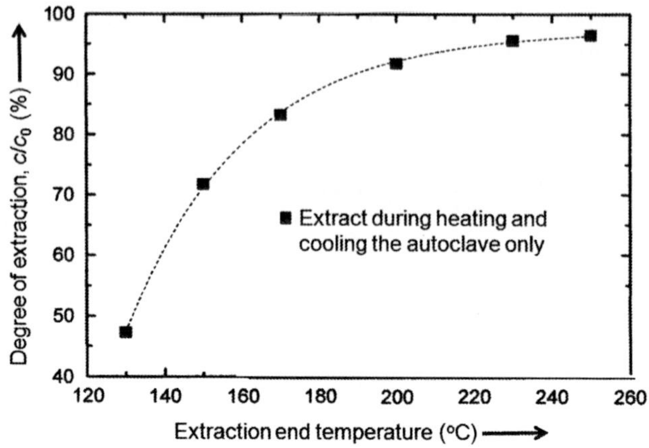

FIGURE 6.13 Protein extraction during heating to the end temperature and immediate cooling in a batch reactor. The pressure corresponds to the vapor pressure of water [16].

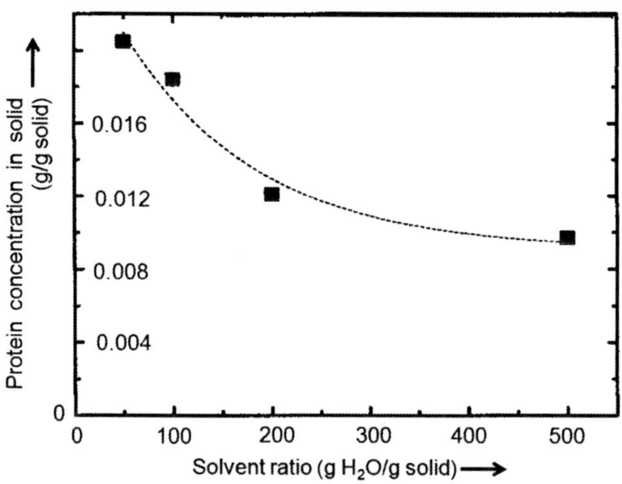

FIGURE 6.14 Remaining protein content in the bone material after heating to $T=330\,°C$ and immediate cooling. Initial pressure: $P=10$ MPa nitrogen [16].

$x=1$ wt% at a solvent ratio of S/F = 500 g/g. Still the extraction yield is not sufficient to use the cleaned bone material for implant purposes.

6.5.2 Extraction of Fat and Proteins with High-Temperature Water

Extraction of fat from bones is usually carried out with a chloroform and methanol mixture. An equivalent extraction can be achieved with supercritical carbon dioxide. For simplifying the cleaning process, defatting and protein

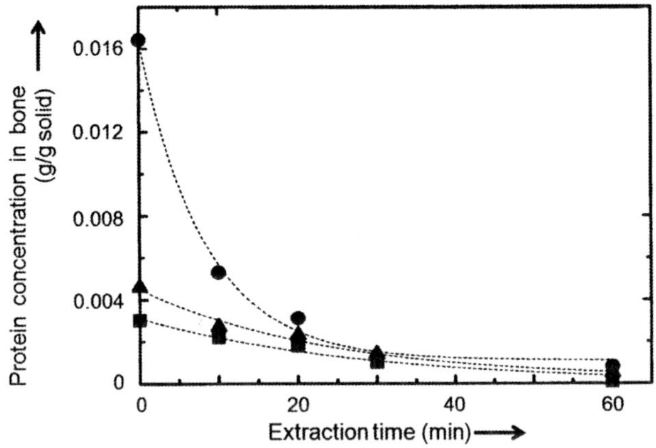

FIGURE 6.15 Cleaning of bone material from fat and proteins with different methods, $P=22$ MPa. ▲: Defatting with supercritical CO_2, protein extraction with H_2O, $T=270$ °C; ■: defatting with supercritical CO_2, protein extraction with H_2O, $T=300$ °C; ●: defatting and protein extraction with H_2O, $T=300$ °C [15,16].

extraction using only water are of interest. Such experiments were carried out in the larger autoclave in a semi-continuous mode with continuous flow of water at temperatures between $T=270$ and 300 °C and residence times at the end temperature between $t=0$ and 60 min. Results are plotted in Figure 6.15. Defatting with supercritical carbon dioxide and subsequent extraction with water lead to the same results (residual concentration $x=0.004$–0.002 mg protein/mg bone material) as defatting with chloroform/methanol and subsequent extraction with water. Extension of the residence times to $t=30$–60 min leads to residual protein concentrations below the analytical detection limit.

Bone material can be cleaned simultaneously from fat and proteins with high-temperature water. The cleaning of the non-pretreated material is slower at first, as fat compounds have to be extracted too. But after a residence time of $t=30$ min at the end temperature, no difference can be seen within the analytical uncertainties.

Extraction kinetics and extraction time are determined in continuous experiments as discussed in the following.

In order to get information on extraction kinetics, the extraction of protein compounds from bone material was investigated in a continuous extraction. This continuous extraction with water was carried out in a tubular reactor in cocurrent flow. The bone material was previously defatted with CO_2 and milled to a medium particle diameter of $d_p=100$ μm. The extraction was carried out at $T=250$–300 °C with residence times between $t=0$ and 120 s. Pressure was kept constant at $P=25$ MPa, concentration of bone material in water $x=1$ wt%. The degree of extraction is substantial and increases with residence time and temperature (Figure 6.16). At $T=300$ °C, about 90 wt%

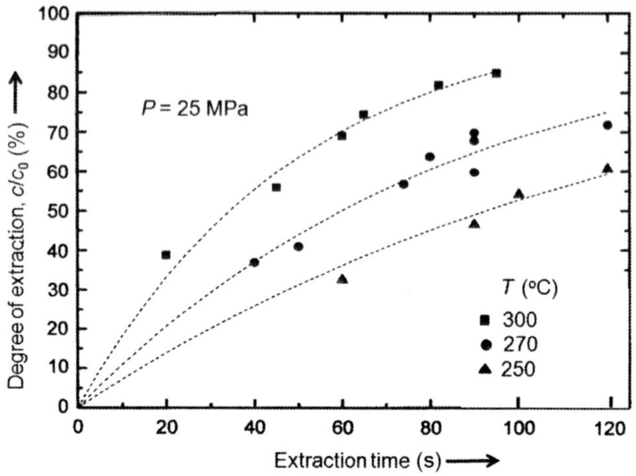

FIGURE 6.16 Continuous cocurrent extraction of protein compounds with water from bone material, $d_p = 0.1$ mm, $P = 25$ MPa, $c_0 = 1$ wt% bone material [15,16].

of the organic compounds in the bone material is degraded and extracted in $t = 90$ s [15,16].

From the experimental data, parameter for an Arrhenius equation can be derived (Equation 6.14) and the activation enthalpy determined to 49.3 kJ/mol.

$$-\ln k = -5930.5\frac{1}{T} + 6.4502 \tag{6.14}$$

6.5.3 Extraction of Larger Bone Structures

Removal of the organic compounds of bone material was also determined for larger particles in the semi-continuous mode. In this mode, the bone material forms a fixed bed in the reactor and water flows continuously through the reactor. Conditions for the experiments were particle size up to $10 \times 10 \times 10$ mm with a weight of about 10 g, $P = 25$ MPa, $T = 200$–350 °C, and solvent ratio S/F = 240–480 g H_2O/g bone material (for 10 g samples) at a solvent rate of $\dot{m} = 20 - 40\,\mathrm{g}H_2O/\mathrm{min}$. The results are shown in Figure 6.17.

The solvent ratio is apparently of no influence in these experiments. This may be due to the relatively large amounts of water and the limitation by the mass transport from the bone particles to the solvent. An increase of the solvent ratio leads to a dilution of the extract at these high solvent ratios. Residual protein concentrations for temperatures from $T = 300$ to 350 °C are at the limit of the analytical procedure.

The semi-continuous procedure has some advantages over the continuous processing. If a non-pretreated material is to be processed, the lipid content inhibits the milling of the bone material. In a fixed bed, there is no problem

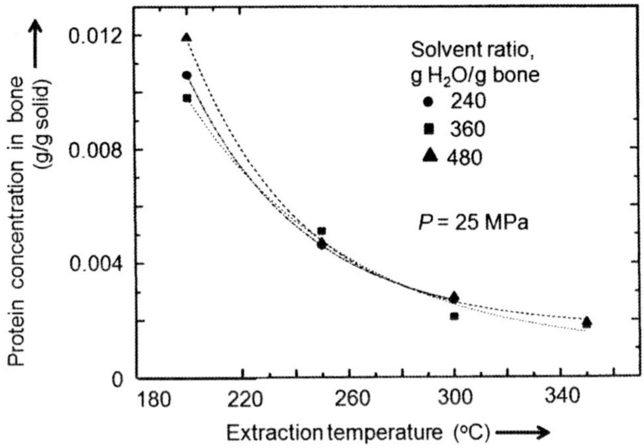

FIGURE 6.17 Residual protein content of large particles of bone material [6,15].

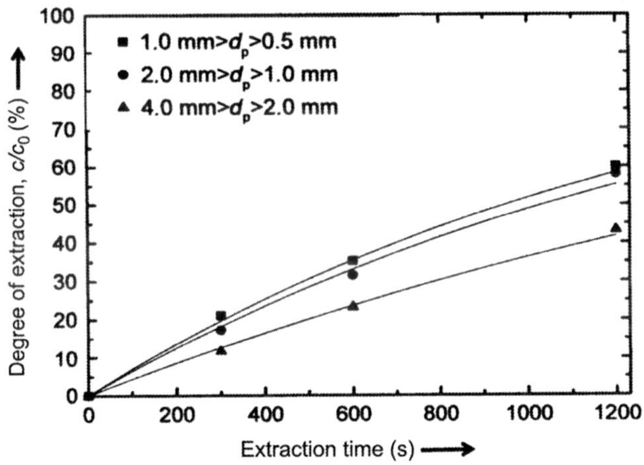

FIGURE 6.18 Influence of particle size on degree of extraction of bone material with H_2O [16].

with larger particles, and the cleaning process can be carried out in one step. Furthermore, a semi-continuous procedure does not limit the extraction time. The process can be extended until all the organic compounds are removed from the bone material.

The influence of the particle diameter on the extraction was investigated at a temperature of $T = 170\ °C$, at $P = 25$ MPa, $m = 5$ g bone material, and a flow rate of the high-temperature water of $\dot{m} = 40\,\text{g/min}$. The results are shown in Figure 6.18. As expected, for small particles up to $d_p = 2$ mm diameter, the extraction is faster than for larger particles of $d_p = 4$ mm. In the range of particle diameters from $d_p = 0.5$ mm up to $d_p = 2$ mm, no significant differences of the extraction rate can be detected.

The influence of temperature on the extraction of organic compounds from bone material was investigated in the range from $T=110$ to $300\,°C$ with extraction times from $t=1$ to 60 min, at a pressure $P=25$ MPa, a flow rate of water of $\dot{m}=40\text{g}/\text{min}$, and $m=5$ g bone material. The amount extracted based on the TOC content is shown in Figure 6.19.

As the results show, the extraction is strongly enhanced by rising temperatures. While at $T=110\,°C$, the degree of extraction after $t=60$ min is only 15%, it is more than 80% after $t=40$ min at $T=170\,°C$. At temperatures higher than $T=200\,°C$, a total extraction is obtained, at $T=230\,°C$ after $t=20$ min and at $T=300\,°C$ after $t=5$ min. The temperature dependence of the extraction can be described with an Arrhenius equation. In Figure 6.20, the function $\ln(k(T))=f(1/T)$ is plotted and can be used to determine the parameters of the Arrhenius equation: $k_0=6.6\times 10^3\ s^{-1}$ and the activation energy $E_A=59.8$ kJ/mol.

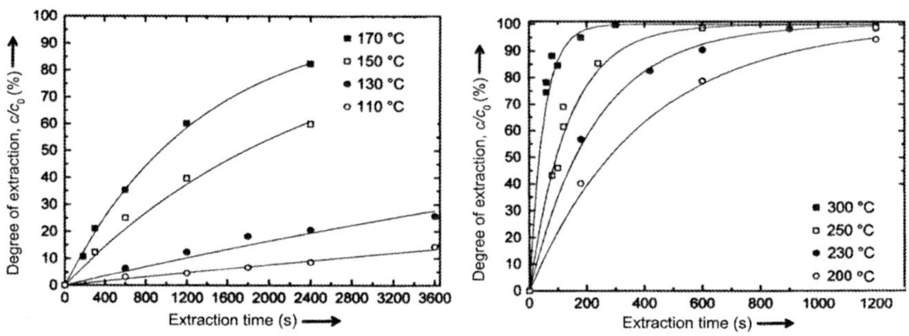

FIGURE 6.19 Extraction of bone material with high-temperature water. Semi-continuous mode. $P=25$ MPa. Left: $T=110$–$170\,°C$. Right: $T=200$–$300\,°C$ [15,16].

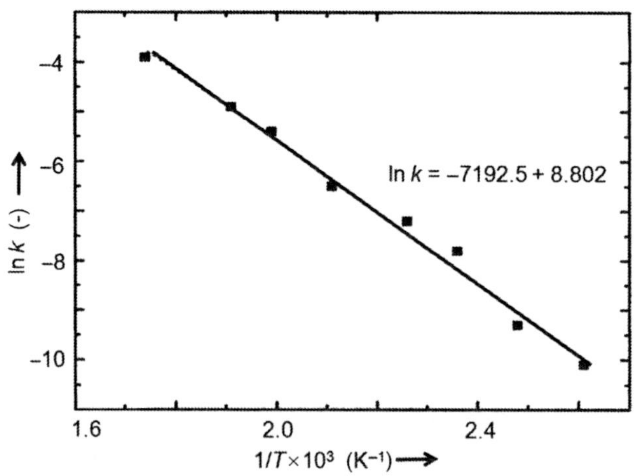

FIGURE 6.20 Arrhenius plot of the extraction of organic compounds from bone material with high-temperature water [16].

Cleaning of bone material from organic compounds with high-temperature water seems to be a good alternative to other cleaning procedures.

6.5.4 Investigation of the Influence of CO_2 on Hydrolysis and Extraction

The addition of carbon dioxide to high-temperature water during the extraction results in a decrease of the pH value. For an acid-catalyzed hydrolysis, as well as for the protein degradation, the extraction process should be faster if carbon dioxide is dissolved in the water phase. Experiments were carried out in the temperature range from $T=110$ to 300 °C, at $P=25$ MPa, $m=5$ g bone material, a flow rate of water of $\dot{m}=40\,\mathrm{g/min}$, and particle diameters from $d_p=0.5$ to 1.0 mm. Experimental results for varying the concentration of CO_2 are shown in Figure 6.21 for a temperature of $T=170$ °C. As the results prove, the concentration of CO_2 has some influence on the reaction rate. Without CO_2, after $t=10$ min about 30% of the organic material is extracted. Water, saturated with CO_2, extracts about 60% after $t=10$ min. The influence of the extraction temperature is shown in Figure 6.22.

The catalytic effect of CO_2 increases from $T=110$ to 130 °C. The effect can also be seen at the higher temperatures of $T=150$ and 170 °C. At 170 °C, without CO_2, about 75% of the initial organic content is extracted. Addition of CO_2 enhances the extraction to more than 90%, at the same conditions. At higher temperatures, the influence of CO_2 becomes smaller.

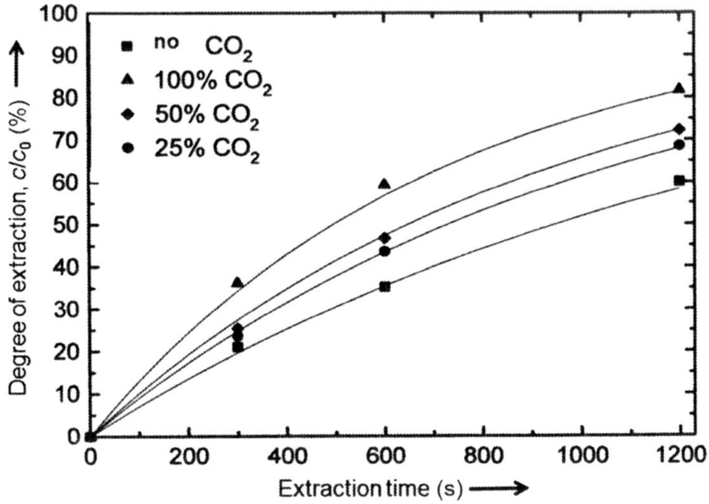

FIGURE 6.21 Influence of dissolved CO_2 concentration on the extraction rate of organic compounds from bone material. $T=170$ °C, $P=25$ MPa, initial amount of solid: $m=5$ g bone material, particle size $d_p=0.5$–1.0 mm, H_2O flow rate $\dot{m}=40\,\mathrm{g/min}$ [16].

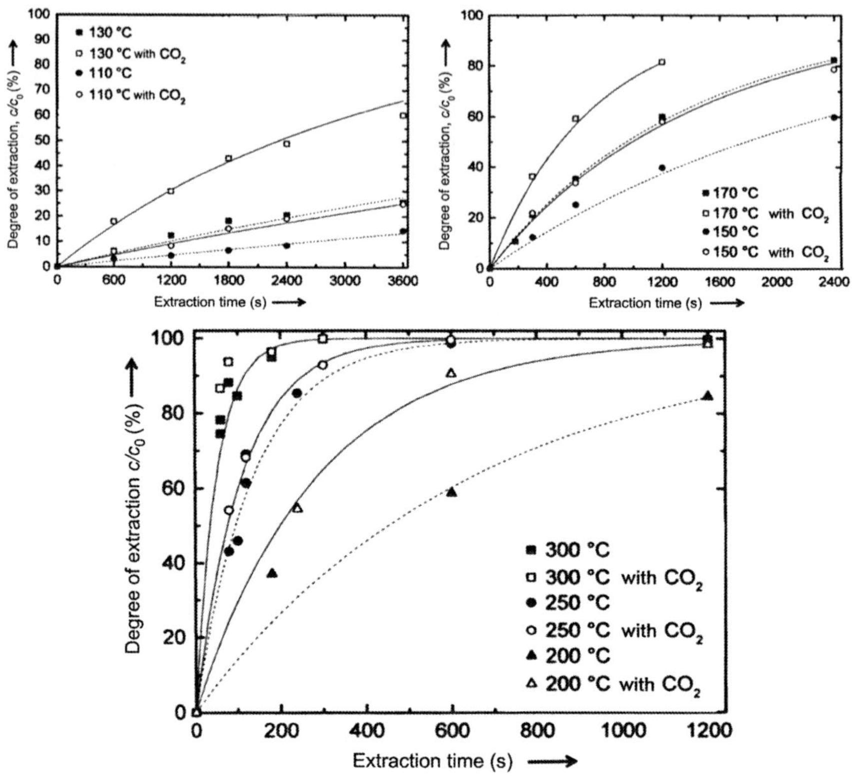

FIGURE 6.22 Influence of temperature and dissolved CO_2 on the extraction of organic compounds from bone material. $P = 25$ MPa, initial amount of solid $m = 5$ g bone material, particle size $d_p = 0.5$–1.0 mm, H_2O flow rate $\dot{m} = 40$ g/min [15,16].

At $T = 250$ °C, the difference is 5%, which is about the range of the experimental uncertainty. At $T = 300$ °C, no difference can be seen.

This result is also illustrated by the Arrhenius plot for CO_2-assisted extraction and pure water extraction as shown in Figure 6.23.

From this investigation, it can be concluded that the total extraction of fat and protein compounds from bone material with high-temperature water is feasible. Water and CO_2 are environmentally friendly substances to treat bone materials, not only for implants but also for producing valuable compounds from waste bone material. Therefore, the extract from the bone treatment is analyzed regarding type and molecular weight of the extract compounds in the following.

6.5.5 Characterization and Separation of the Protein Extract

The liquid extract from the high-temperature water treatment of bone material was analyzed for dissolved organic compounds. Proteins and peptides were

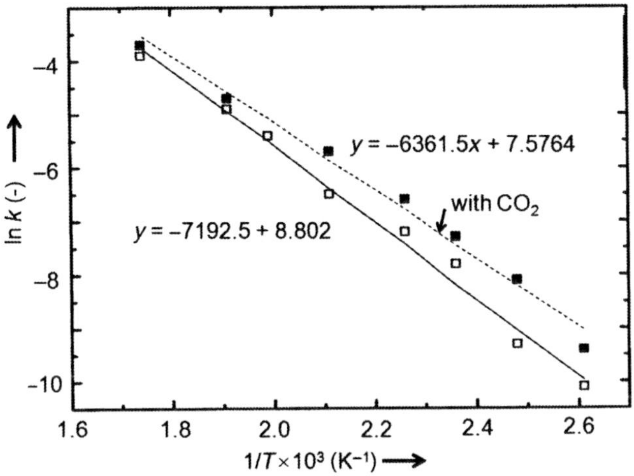

FIGURE 6.23 Reaction rate for the extraction of organic compounds from bone material for pure water extraction and CO_2-assisted extraction [15,16].

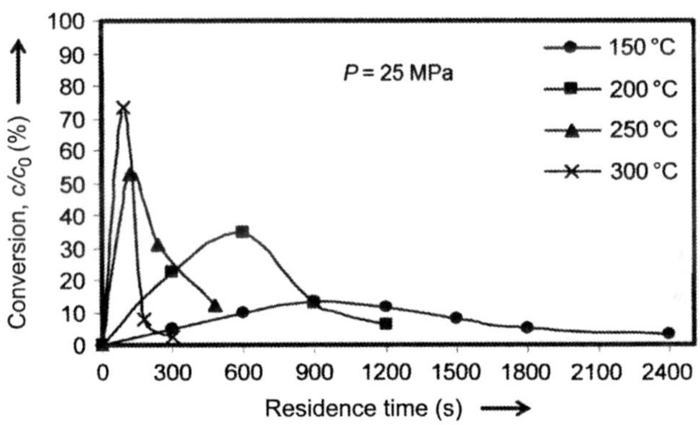

FIGURE 6.24 Conversion rate of organic compounds of the bone material into dissolved compounds. $P = 25$ MPa, $m = 5$ g bone material, flow rate of water $\dot{m} = 40\,\mathrm{g/min}$, particle diameter $d_p = 0.5$–1.0 mm [16].

determined with the Kjeldahl method [16] with respect to their concentration and for a mass balance. Determination of the nitrogen content covers all nitrogen-containing substances, such as dipeptides, monomers of proteins—amino acids—and further products of the hydrolytic degradation.

The rate of conversion of the organic bone content into dissolved organic compounds after the extraction, as obtained from semi-continuous experiments, is shown in Figure 6.24.

The conversion rate of the organic bone material increases strongly with temperature, shifting the maximum rate to lower extraction times. At $T = 300$ °C, the maximum is reached at about $t = 1.5$ min, whereas at $T = 150$ °C, the much flatter

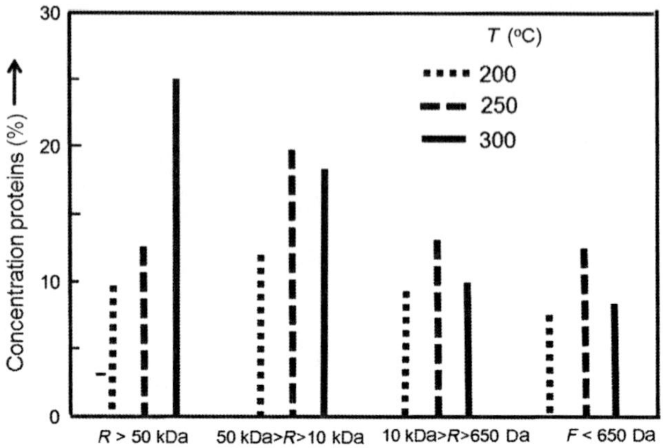

FIGURE 6.25 Distribution of proteins in the aqueous extract from bone material for different extraction temperatures. Experimental conditions as in Figure 6.24 [16].

and lower maximum is approached at about $t = 17$ min (see Figure 6.24). The mass balance could be matched at $T = 150$ °C to 97%. At higher temperatures of $T = 200$ and 250 °C, the mass balance matched to only about 60%, indicating the formation of gaseous products containing nitrogen. At $T = 300$ °C, the mass balance was close to about 73% which may be due to the reduced time available for degradation [15,16].

The extract was separated by membrane filtration into fractions of molecular weight of $M_W > 50$ kDa, M_W: 50–10 kDa, M_W:10 kDa > R > 650 Da, and $M_W < 650$ Da. Protein yield is determined from the total protein content in the extract in relation to the initial yield of total proteins in the bone material. The result is presented in Figure 6.25. At $T = 200$ °C, the protein products are distributed relatively even over the various fractions. At higher temperatures and shorter extraction times, the formation of protein products with lower molecular weight is favored.

Cleaning of bone materials from organic compounds with high-temperature water is possible. Dissolved carbon dioxide enhances the extraction up to temperatures of $T = 250$ °C. The materials in the original state and after extraction are shown in Figure 6.26 for small particles and for bigger structures. The cleaned bone material was applied in cell-cultivating experiments. Growth of the cells revealed an inhibition by extracted metals from the stainless steel autoclave. After substitution of the autoclave by an autoclave manufactured from titanium, much lower metal contents were obtained and the growth experiments were successful [15,16].

6.6 EXTRACTION OF NATURAL SUBSTANCES

Extraction of natural substances, mostly from plant materials, with high-temperature water is mainly aimed at the recovery of polar and water-soluble

FIGURE 6.26 Bone material used for the extraction of organic compounds with high-temperature water. Upper left: Milled bone material untreated, lower left: Milled bone material after the extraction. Upper right: Bone material (≈10 × 10 × 10 mm) untreated, lower right: Bone material (≈10 × 10 × 10 mm) after the extraction [15,16]. (For color version of this figure, the reader is referred to the online version of this chapter.)

substances. Water-insoluble nonpolar substances that are usually recovered by hydrodistillation can also be favorably extracted by high-temperature water. This is due to several effects.

- Lowering of the boiling point in immiscible mixtures, analogue to hydrodistillation, but extended to higher pressures. Liquid–gas transition occurs at a lower temperature than given by the vapor pressure curves, as the total pressure is composed of the vapor pressures of both components and the sum of both adds up to the total pressure at a lower temperature than the liquid–gas transition temperature for each of both components.
- The higher temperature, applied in high-temperature water extraction, favors higher solubility of the target compounds in water and increases the rate of mass transport.
- Hydrolysis in high-temperature water makes available compounds for extraction.

The extraction process is carried out according to the principles discussed at the beginning of this chapter. Examples of experimental equipment are discussed in Chapter 13.

Extraction with high-temperature water is widely used for analytical purposes. In a review [17], numerous examples are cited for the (analytical) extraction of organic contaminants from foodstuff and plant materials, for the extraction of flavors, fragrances, and bioactive compounds, and the extraction of contaminations from soils or sediments [17]. High-temperature water extraction follows the traditional herbal preparations which usually involve sequential steps with boiling in water.

Some examples are listed in the following: Extraction of volatile essential oil from *Cuminum cyminum* L. at a temperature of $T = 150\,°C$ with water produces yields comparable to Soxhlet extraction and steam distillation [18]. Similar results are reported for Borneol [19] and Pulegone [20]. Capsaicinoids present in peppers are successfully extracted with water at $T = 200\,°C$ [21]. Organic contaminants can be removed from foodstuff, in particular pesticides, veterinary drugs, persistent environmental chemicals, and naturally occurring toxicants, as described in reviews [17,22]. High-temperature water is applied to obtain shikimic acid from Chinese star anise in high yields and at short extraction time [23,24] (see Figure 6.28).

Extraction of anthraquinones from dried roots of *Morinda citrifolia* (Noni) with high-temperature water at a temperature of $T = 110–220\,°C$ and $P = 7$ MPa was found to be able to recover all the anthraquinones present in the roots within an extraction time of $t = 3$ h, as shown in Figure 6.27. Conventionally, the compound is extracted with ethanol, followed by evaporation to separate the solvent from the product. This procedure leaves residual solvent in the product. Figure 6.27 shows that yield and rate of the extraction increase with higher temperatures [25].

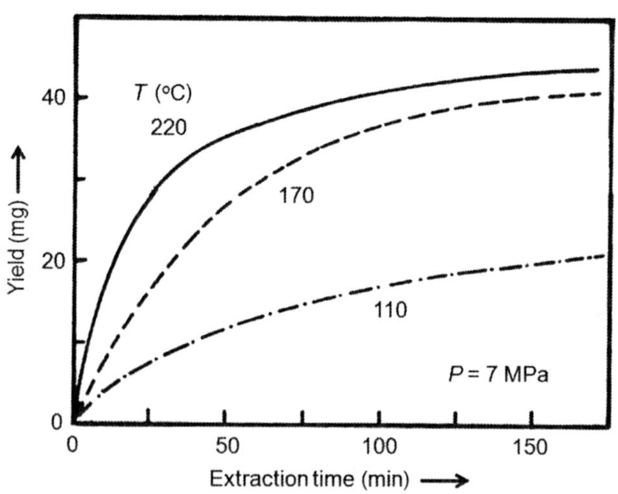

FIGURE 6.27 Extraction of anthraquinones from roots of *Morinda citrifolia* (Noni) with water, $\dot{m} = 4\,cm^3/min$. *Modified from Ref.* [25].

TABLE 6.2 Comparison of Different Extraction Methods for Extraction of Anthraquinones [26]

Extraction Method	Temperature (°C)	Extraction Time (h)	Recovery (%)
Batch solvent extraction, magnetic stirrer	25	72	81.2
Ultrasonic-assisted extraction	60	2	79.6
Soxhlet extraction	78.3	1	93.4
		2	94.8
		3	97.0
		4	97.9
Extraction with high-temperature water	150	1	81.1
	170	2	92.6
	200	1	80.5
		2	94.2

Results of the extraction of anthraquinones with high-temperature water, according to different extraction methods, are shown in Table 6.2 and Figure 6.28 [26]. For about 80% recovery of anthraquinones, extraction in a stirred vessel requires an extraction time of $t=72$ h, ultrasound-assisted extraction at $T=60$ °C requires an extraction time of $t=2$ h, Soxhlet extraction at $T=78$ °C needs less than $t=1$ h, and water extraction at temperatures of $T=150$–200 °C also requires an extraction time of $t=1$ h. Extraction yield of the extraction with water after $t=2$ h is the same as with Soxhlet extraction. Remarkably, the extract from ultrasonic-assisted extraction has a lower antioxidant activity than that from the other extraction methods. It can be concluded that extraction with high-temperature water is more effective than the other extraction techniques for recovery of anthraquinones [26].

Antioxidants from rosemary leaves were extracted with high-temperature water at $T=200$ °C. At this temperature, the maximum antioxidant activity was obtained. The extract was spray-dried to a powder by means of supercritical carbon dioxide [27].

Extraction of phenolic compounds and flavonoids from grape pomace (winery waste) with high-temperature water at $T=100$–140 °C and $P=8$–15 MPa resulted in significantly higher polyphenols, flavonoids, and antioxidant activity compared to extraction with water and ethanol at atmospheric pressure [28]. The results are shown in Table 6.3.

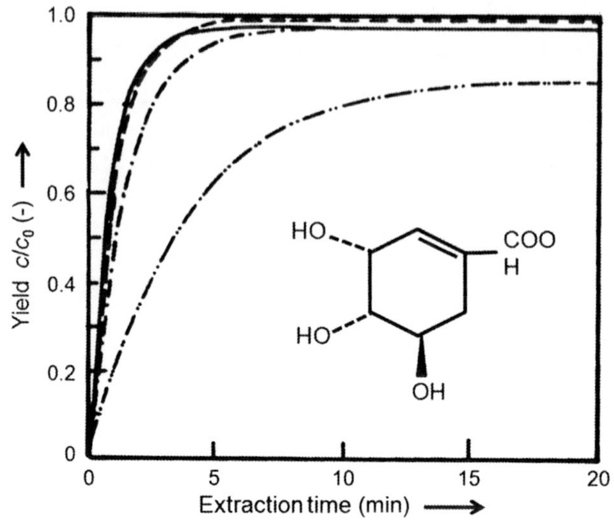

FIGURE 6.28 Extraction of shikimic acid from Chinese star anise with water. *Modified from Ref.* [24].

TABLE 6.3 Extraction of Grape Pomace with High-Temperature Water [28]

Extraction Procedure	Temperature (°C)	Pressure (MPa)	Total Polyphenols (mg_{GAE}/g_{DP})	Total Flavonoids (mg_{CE}/g_{DP})	Antioxidant Power ($\mu g_{DPPH}/\mu l_{Extr}$)
High-temperature water	100	8	12.8	8.3	5.9
		11.5	16.7	10.0	7.1
		15	16.5	10.0	7.1
	120	8	20.2	12.3	9.1
		11.5	23.9	14.4	9.3
		15	23.2	11.5	7.3
	140	8	32.5	15.1	12.6
		11.5	30.8	15.3	12.9
		15	28.5	15.3	13.9
Conventional, water	25	0.1	1.7	1.3	4.2
Conventional, ethanol	25	0.1	7.9	14.5	22.6

DPPH—2,2-diphenyl-1-picryl hydrazyl; DP—dried pomace, GAE—gallic acid equivalent; CE—catechin equivalent.

Extraction of shikimic acid from Chinese star anise carried out in a semi-continuous flow apparatus enabled 100% recovery of the 8% content of shikimic acid in star anise, particle size $d_p = 355–600\ \mu m$ with water at a temperature of $T = 150\ °C$ and a pressure of $P = 15$ MPa and an extraction time of $t = 4$ min. In Figure 6.28, the extraction curves are shown [24].

Extraction of asiatic acid and asiaticoside (Figure 6.29) from *Centella asiaticais* is carried out with high-temperature water at temperatures from $T = 100$ to $250\ °C$ and pressures from $P = 10$ to 40 MPa. At the optimal extraction conditions of $P = 40$ MPa and $T = 250\ °C$, the extraction yield of asiatic acid is 7.8 mg/g and the extraction yield of asiaticoside is 10.0 mg/g. The extraction yield increases with temperature and pressure, as is illustrated in Figure 6.30. The extracted products, asiatic acid and asiaticoside, precipitate when the extract is depressurized and cooled to room temperature. Particles containing asiatic acid are larger than particles containing asiaticoside. Thus, separation of asiatic acid and asiaticoside can be achieved by choosing appropriately sized filters [29].

Further examples for the extraction with hydrothermal (hot liquid) water are Glucan extraction from waxy barley [30], extraction of a natural insecticide from neem seed kernels (*Azadirachta indica* A. Juss) [31], and extraction of natural sweetener from licorice (*Glycyrrhiza glabra*) roots [32]. Other

FIGURE 6.29 Structure of asiatic acid (left) and asiaticoside (right) [29].

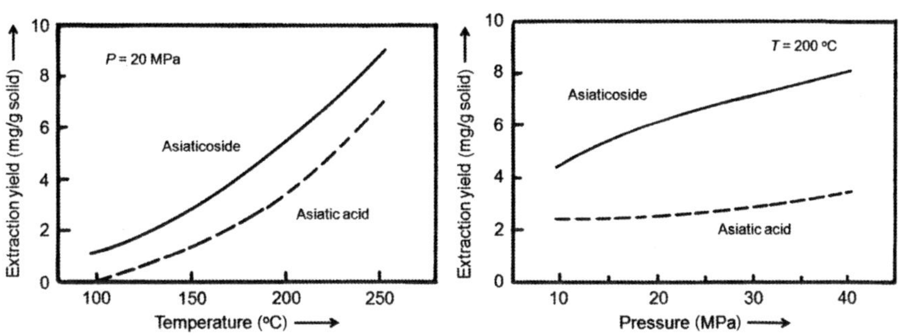

FIGURE 6.30 Extraction of asiatic acid and asiaticoside with water. Extraction time: $t = 5$ h. Influence of temperature (left) and of pressure (right). *Modified from Ref. [29].*

extractions reported are [25] extraction of essential oils from marjoram [33], extraction of oxygenates from savory and peppermint [34], essential oils from oregano [35], extraction of hypericin and pseudo-hypericin from St. John's wort [36], extraction of iridoid glycosides from *Veronica longifolia* [37] and kava lactones from kava roots [38], extraction of thermal labile compounds, such as berberine, glycyrrhizin, liquorice, and baicalein [39], and extraction of sage, *Salvia officinalis* L. [40].

High-temperature water can be effectively used as an environmentally friendly extraction medium, also in combination with supercritical carbon dioxide, for a sustainable technology. It might be suspected that high-temperature water processes are costly with respect to energy needs. However, this is not the case because water stays liquid and the latent heat of evaporation is not required. For high-temperature water extraction, only 505 kJ/kg is required to heat liquid water from $T=30$ to 150°C, compared with 2550 kJ/kg required to convert water at $T=30$ °C to steam at $T=100$°C for steam distillation. A greater mass of high-temperature water may be needed for a given mass of material to be extracted. However, up to a fivefold higher solvent ratio, the energy needed is lower for high-temperature water extraction. Moreover, it is relatively easy to recycle the heat in a high-temperature water process by passing the water leaving the extraction through a heat exchanger to heat the water flowing to the extraction. More than 80% of the energy can be realistically recovered in this way [41].

REFERENCES

[1] G. Brunner, Gas Extraction. An Introduction to Fundamentals of Supercritical Fluids and the Application to Separation Processes, Steinkopff, Springer, Darmstadt, New York, 1994 (Chapter 7).

[2] K. Nowak, Reinigung kontaminierter Bodenmaterialien mit überkritischem Wasser, Ph.D. Dissertation, Hamburg University of Technology, Hamburg, Germany, 1995.

[3] A. Firus, Reinigung von Bodenmaterial durch Extraktion und Reaktion mit überkritischem Wasser und Kohlendioxid, Ph.D. Dissertation, Hamburg University of Technology, Hamburg, Germany, 1996.

[4] B. Misch, Reinigung mischkontaminierten Bodenmaterials und kontinuierliche Extraktion von Feststoffen mit überkritischen Fluiden, Ph.D. Dissertation, Hamburg University of Technology, Hamburg, Germany, 2001.

[5] G. Brunner, B. Misch, A. Firus, K. Nowak, Cleaning of soil with supercritical water and supercritical carbon dioxide, in: R. Stegmann, G. Brunner, W. Calmano, G. Matz (Eds.), Treatment of Soil—Fundamentals, Analysis, Applications, Springer, Berlin, Germany, 2001, pp. 491–517.

[6] J. Kronholm, J. Kalpala, K. Hartonen, M.-L. Riekkola, Pressurized hot water extraction coupled with supercritical water oxidation in remediation of sand and soil containing PAHs, J. Supercrit. Fluids 23 (2002) 123–134.

[7] G. Brunner, Near critical and supercritical water. Part I. Hydrolytic and hydrothermal processes, J. Supercrit. Fluids 47 (2009) 373–381.

[8] A. Firus, W. Weber, G. Brunner, Supercritical carbon dioxide for the removal of hydrocarbons from contaminated soil, Sep. Sci. Technol. 32 (1997) 1403–1414.
[9] B. Misch, G. Brunner, Hydrolytische und thermische Spaltung organischer Verunreinigungen kontaminierter Böden mit überkritischem Wasser zur Erzeugung biologisch abbaubarer Produkte, Deutsche Forschungsgemeinschaft, Sonderforschungsbereich 188, TP A10. Final Report, 2000.
[10] K.-A. Hsueh, F.-C. Chang, H. Paul Wang, H.-C. Wang, Y.-J. Huang, H.-L. Huang, Y.-J. Tuan, Enhanced extraction of PAHs hindered in fly ashes with supercritical water, J. Supercrit. Fluids 73 (2013) 116–119.
[11] G. Brunner, M. Budich, S. Jung, Verfahren und Vorrichtung zum Abtrennen von Bitumen aus Asphalt. European Patent EP 0 768 362 B1, May 6, 1999; (Appl.) October 17, 1995.
[12] G. Brunner, Supercritical Fluids for Effective Separation Processes, in: Plenary Lecture, ISSF 2012: 10th International Symposium on Supercritical Fluids, San Francisco, USA, 2012. www.ISSF2012.com.
[13] F. Salvador, N. Martin-Sanchez, M.J. Sanchez-Montero, J. Montero, C. Izquierdo, Regeneration of activated carbons contaminated by phenol using supercritical water, J. Supercrit. Fluids 74 (2013) 1–7.
[14] D. Doncheva, G. Brunner, Cleaning of bone xenografts from organic compounds with supercritical fluids, in: Proceedings of the 8th Italian Conference on Supercritical Fluids, Ischia, Italy, May 28–31, 2006.
[15] D. Doncheva, G. Brunner, Cleaning of animal-derived bone material for implantation by combined extraction/reaction process of organic matrix with subcritical water and characterisation of hydrolysates, in: Proceedings of the 5th International Symposium on High Pressure Process Technology and Chemical Engineering, Segovia, Spain, June 24–27, 2007.
[16] D.A. Doncheva-Albrecht, Untersuchung der vollständigen Extraktion organischer Stoffe aus Knochenmaterial mit Wasser und überkritischen Fluiden, Ph.D. DissertationHamburg University of Technology, Hamburg, Germany, 2009.
[17] C.C. Teo, S.N. Tana, J.W.H. Yong, C.S. Hew, E.S. Ong, Pressurized hot water extraction (PHWE), J. Chromatogr. A 1217 (2010) 2484–2494.
[18] M.H. Eikani, F.G. Mohammad, M. Mirza, S. Rowshanzamir, Subcritical water extraction of essential oils from coriander seeds (Coriandrum sativum L.), J. Food Eng. 80 (2007) 735–740.
[19] M.Z. Özel, H. Kaymaz, Superheated water extraction, steam distillation and Soxhlet extraction of essential oils of Origanum onites, Anal. Bioanal. Chem. 379 (2004) 1127–1133.
[20] M.Z. Özel, F. Gogus, J.F. Hamilton, A.C. Lewis, Analysis of volatile components from Ziziphora taurica subsp taurica by steam distillation, superheated-water extraction, and direct thermal desorption with GCxGC-TOFMS, Anal. Bioanal. Chem. 382 (2005) 115–119.
[21] G.F. Barbero, M. Palma, C. Barroso, Pressurized liquid extraction of capsaicinoids from peppers, J. Agric. Food Chem. 54 (2006) 3231–3236.
[22] S. Bogialli, A. Corcia, Matrix solid-phase dispersion as a valuable tool for extracting contaminants from foodstuffs, J. Biochem. Biophys. Methods 70 (2007) 163–179.
[23] H. Machida, M. Takesue, R.L. Smith Jr., Green chemical processes with supercritical fluids: properties, materials, separations and energy, J. Supercrit. Fluids 60 (2011) 2–15.
[24] H. Ohira, N. Torii, T.M. Aida, M. Watanabe, R.L. Smith Jr., Rapid separation of shikimic acid from Chinese star anise (Illicium verum Hook. f.) with hot water extraction, Sep. Purif. Technol. 69 (2009) 102–108.
[25] A. Shotipruk, J. Kiatsongserm, P. Pavasant, M. Goto, M. Sasaki, Pressurized hot water extraction of anthraquinones from the roots of Morinda citrifolia, Biotechnol. Prog. 20 (2004) 1872–1875.

[26] B. Pongnarayane, M. Goto, M. Sasaki, T. Anekpankul, P. Payasant, A. Shotipruk, Extraction of anthraquinones from roots of Morinda citrifolia by pressurized hot water: antioxidant activity of extracts, J. Supercrit. Fluids 37 (2006) 390–396.

[27] I. Rodríguez-Meizoso, M. Castro-Puyana, P. Börjesson, J.A. Mendiola, C. Turner, E. Ibáñez, Life cycle assessment of green pilot-scale extraction processes to obtain potent antioxidants from rosemary leaves, J. Supercrit. Fluids 72 (2012) 205–212.

[28] B. Aliakbarian, A. Fathi, P. Perego, F. Dehghani, Extraction of antioxidants from winery wastes using subcritical water, J. Supercrit. Fluids 65 (2012) 18–24.

[29] W.-J. Kim, J. Kim, B. Veriansyah, J.-D. Kim, Y.-W. Lee, S.-G. Oh, R.R. Tjandra-winata, Extraction of bioactive components from Centella asiatica using subcritical water, J. Supercrit. Fluids 48 (2009) 211–216.

[30] Ó. Benito-Román, E. Alonso, M.J. Cocero, Pressurized hot water extraction of β-glucans from waxy barley, J. Supercrit. Fluids 73 (2013) 120–125.

[31] G.C. Jadeja, R.C. Maheshwari, S.N. Naik, Extraction of natural insecticide azadirachtin from neem (Azadirachta indica A. Juss) seed kernels using pressurized hot solvent, J. Supercrit. Fluids 56 (2011) 253–258.

[32] M. Mukhopadhyay, P. Panja, A novel process for extraction of natural sweetener from licorice (Glycyrrhiza glabra) roots. Department of Chemical Engineering, I.I.T. Bombay, Mumbai 400076, India, Journal unknown.

[33] M.M. Jiménez-Carmona, J.L. Ubera, L.D. Castro, Comparison of continuous subcritical water extraction and hydrodistillation of marjoram essential oil, J. Chromatogr. A 855 (2002) 625–632.

[34] A. Kubátová, A.J.M. Lagadec, J.D. Miller, S.B. Hawthorne, Selective extraction of oxygenates from savory and peppermint using subcritical water, Flavour Fragr. J. 16 (2001) 64–73.

[35] R.G. Ayala, M.D.L. Castro, Continuous subcritical water extraction as a useful tool for isolation of edible essential oils, Food Chem. 75 (2001) 109–113.

[36] M.H. Mannila, H. Kim, C.M. Wai, Supercritical carbon dioxide and high-pressure water extraction of bioactive compounds in St. John's Wort, in: Proceedings of Supergreen 2002, Kyung Hee University, Korea, 2002.

[37] J. Suomi, H. Sire´n, K. Hartonen, M.L. Riekkola, Extraction of iridoid glycosides and their determination by micellar electrokinetic capillary chromatography, J. Chromatogr. A 868 (2000) 73–83.

[38] A. Kubátová, D.J. Miller, B.S. Hawthorne, Comparison of subcritical water and organic solvents for extracting kava lactones from kava root, J. Chromatogr. A 923 (2001) 187–194.

[39] E.S. Ong, S.M. Len, Pressurized hot water extraction of berberine, baicalein, and glycyrrhizin in medicinal plants, Anal. Chim. Acta 482 (2003) 81–89.

[40] M. Ollanketo, A. Peltoketo, K. Hartonen, R. Hiltunen, M.L. Riekkola, Extraction of Sage (Salvia officialis L.) by pressurized hot water and conventional methods: antioxidant activity of the extracts, Eur. Food Res. Technol. 215 (2002) 158–163.

[41] A.A. Clifford, Separations using superheated water, in: M. Goto, T. Adschiri, B. Han, Y.-W. Lee (Eds.), Proceedings 8th International Symposium on Supercritical Fluids, Kyoto, Japan, Presentation OC 2 10, 2006.

Chapter 7

Processing of Fuel Materials with Hydrothermal and Supercritical Water

7.1 INTRODUCTION

Processing fuel materials as coal, oil shale (kerogen), tar sands, heavy bitumen, and similar materials with hydrothermal and supercritical water involves the aspects of phase equilibrium between different hydrocarbons and reactions of these components with water. The fundamentals of these aspects have been treated in Chapter 3 for the physical behavior of mixtures with water and in Chapter 5 for the reactive behavior of hydrocarbons with water.

In general, the physical behavior of hydrocarbon compounds with water can be illustrated by their very different solubility behavior. At low temperatures, hydrocarbons are nearly immiscible with water, while at high temperatures, the solubility increases until around the critical temperature, often total miscibility occurs. In parallel, the reactivity of hydrocarbons with water increases with temperature and hydrolytic and thermolytic transformations are observed. Only simple hydrocarbons are relatively stable at near-critical and supercritical water conditions.

With increasing temperature, also other properties of hydrocarbons are modified. At higher temperatures, viscosity is drastically reduced and surface tension decreases with enhanced mutual solubility.

Nonhydrocarbon components of a hydrocarbon mixture are also affected by high temperature water and may react to compounds that can be removed more readily from the hydrocarbon mixture.

The hydrocarbon product of a treatment with near-critical or supercritical water can be removed from water by lowering the temperature to a region where water again is nearly immiscible with hydrocarbon compounds, compare Chapter 3. Thus, separation of the product streams is simplified.

The resulting process water can be recycled into the process, while the hydrocarbon mixture can be treated and upgraded to conventional hydrocarbon mixtures. Handling of solids poses no fundamental problem in treatment

Supercritical Fluid Science and Technology, Vol. 5. http://dx.doi.org/10.1016/B978-0-444-59413-6.00007-8

with near-critical and supercritical water. Fine particles are readily processed, as has been discussed in Chapter 6 for the example of cleaning of fine particles from soil.

Processing of water-containing materials is inherently no problem for a water-based process. Therefore, fuel material sources containing substantial amounts of water like biomass, brown coals, or peat can be processed without previous drying.

Energy for compression of the feed materials is relatively low since only condensed materials are compressed (pumped). Heat recovery from the relatively high temperature level can be achieved since all process flows are fluid, either liquid, gaseous, or a slurry.

The whole treatment with near-critical and supercritical water can be confined to a closed process plant. Gases, formed during the process, can be separated and treated also in a confined plant and only released after sufficient conversion or cleaning.

Although the application of high temperature water seems to have many advantages, so far no commercial plant using near-critical or supercritical water has gone into operation. Nevertheless, the potential is given and may be exploited in the future, in particular, for more environmentally friendly processes.

In the following, examples for treating coal, oil shale, and tar sands and the upgrading of bitumen with hydrothermal and supercritical water are presented, while biomass is treated separately in Chapter 8. Characteristic results of experiments are presented for selected examples and further investigations reported in the literature are annotated. Information on the experimental facilities for coal, oil shale, and tar sands are given in Chapter 13 since they are similar to those of other areas of application like reaction, extraction, biomass conversion, and polymer recycling. The properties of the initial fuel materials are not discussed since information on resources and properties are readily accessible in the literature.

7.2 PROCESSING OF COAL WITH WATER

The interest of processing coals for alternative fuels was at its height end of the 1970s of past century. Then, the so-called first and second oil crisis spurred research, development, and demonstration of several coal conversion processes like gasification, liquefaction, and extraction, including extraction with supercritical fluids. Due to the decline of the oil price during the 1980s of past century, most of the interest vanished and most projects were terminated. While for gasification and liquefaction of coal demonstration plants have been built and operated, this was not the case for extraction, neither for solvent extraction nor for extraction with supercritical fluids. Nevertheless, it is of interest to review the findings of that time and the few developments on laboratory scale that have been published since then, because of the long-time availability of coal and the similarity of the problems to processing other fuel materials and biomass.

As with all natural feedstocks, coal is not a material that is more or less equal, independent of its source. On the contrary, coal is a general name for carbon-containing solids derived from natural materials and transformed in the carbonization process. Coals differ widely from one another in their rank (degree of carbonization), from lignite to anthracite, in the content of bituminous compounds, and in the content of mineral matter. Therefore, results from investigations must be interpreted with care and for design of processes, experiments with the expected feed material are necessary.

Processing of coal with water was investigated in parallel with the use of hydrothermal and supercritical water in conversion of biomass, see Chapter 8. In both fields, the same general types of problems are addressed, such as to improve yields of liquid products, suppress gas formation, and suppress retrogressive reactions [1, 2].

In the following, first examples for investigations on the extraction and liquefaction of coal and then examples for the experimental gasification of coal with water are discussed.

7.2.1 Extraction and Liquefaction of Coal in Hydrothermal and Supercritical Water

The approach of supercritical water for the production of liquid fuels from coals is illustrated in Figure 7.1 [2].

The general process for coal conversion is as follows: Coal is introduced together with process water to the processing unit. There, high temperature or supercritical water reacts with the coal and extracts soluble compounds and reaction products. The product stream is cooled and products and process water are separated. The different coal conversion product streams are separated into gas, char, and conversion products. The conversion products can be further processed, for example, by hydrogenation into transport fuels and chemical feedstocks.

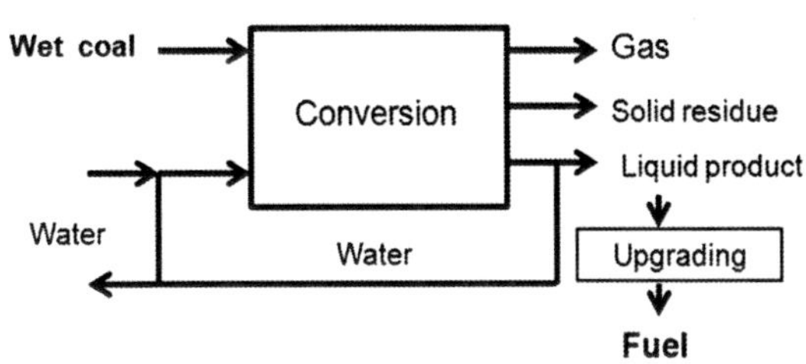

FIGURE 7.1 Simplified scheme for processing of coals with sub- and supercritical water. *Modified from Ref. [2].*

The processing of coal with water is assumed to comprise two steps:

(i) High temperature or supercritical water penetrates the micropore structure of the coal and extracts the material physically held in the micropores [2].

(ii) Pyrolysis and hydrolytic reactions occur and transform part of the coal material. Products are extracted at temperatures at which these products would not be volatile. This limits secondary reactions of the products. The lower the rank of the coal, the more the pathway (ii) can be expected to dominate [2].

Beside the supercritical fluid extraction of coal investigated by the National Coal Board of Britain in a $\dot{m} = 5\,\mathrm{kg\,h^{-1}}$ pilot plant in the 1970s, conversion of coal with supercritical fluids, including water, has been limited to laboratory studies, mainly using enclosed autoclaved and semi-continuous reactors [2].

Since water is an excellent solvent for organic compounds at subcritical and supercritical conditions, it can be expected that at sufficiently high solvent density supercritical water is a good solvent for coal extraction. Relatively high conversions are obtained for some low-rank coals [2]: German brown coal is converted into 70–75% extractible compounds at $T = 375$ °C and $P = 23$ MPa [3]. Conversions of 42–68% are obtained for Australian brown coals with supercritical water at $T = 380$ °C and $P = 22$ MPa (see Figure 7.2) [4–6], and a conversion of 43% is reported for North Dakota lignites under similar conditions [7].

For Australian brown coals, conversion to extractible compounds are considerably higher with water at $T = 380$ °C and $P = 22$ MPa than with

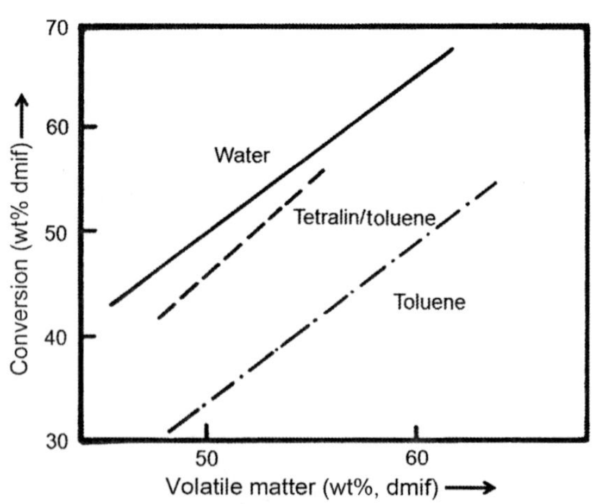

FIGURE 7.2 Conversion of Australian brown coals with water, dry, free of mineral matter (dmif). ——: water, $T = 380$ °C, $P = 22$ MPa, — • — • —: toluene with 5% tetralin, $T = 400$ °C, $P = 10$ MPa, – – –: toluene, $T = 400$ °C, $P = 10$ MPa. *Modified from Ref. [2].*

toluene. For black coals conversions are lower. For bituminous coal, a conversion of 58% is reported (Bruceton coal) at $T = 380\,°C$ and $P = 30\,MPa$ after a residence time of $t = 60$ min, but for two other bituminous coals much lower conversions are obtained: 19% for Powhatan and 29% for Illinois No. 6. It can be concluded that yields are reasonable for supercritical water extraction of bituminous coals, but generally offer no improvement on supercritical toluene extraction under similar conditions of temperature and pressure [2].

In summary, advantages of water compared to organic solvents are:

- Low cost and availability,
- easy separation of solvent and product due to the immiscibility of water and extract at lower temperature and pressure,
- drying of coal is not necessary.

To avoid drying is particularly significant for lignites (brown coals) with their high moisture contents. Higher conversions were reported for lignites than for sub-bituminous coals [7]. But lignites from various sources behave differently and some are less reactive with supercritical water [3,8].

Processing history is also of influence on coal conversion with water. Sub-bituminous coal treated with steam at $T = 240\,°C$ prior to extraction at $T = 400\,°C$ increased conversion from 30.5% to 40.3% [9]. The heating rate is important for the conversion also. Conversion of a German brown coal is 58% when the coal is present during the heating-up period, compared with 70–75% when the coal is injected into the hot reactor [3]. In experimental investigations, a higher conversion is obtained using a semi-continuous reactor with a fast heating rate of about $100\,°C\,min^{-1}$ than the conversion obtained with a slow heating rate of about $7\,°C\,min^{-1}$ [2]. Rapid heating and product removal from the reactor minimize retrogressive reactions leading to insoluble products [2]. Since in most investigations on conversion of coal with water semi-continuous reactors with a slow heating rate have been employed, it can be expected that the conversions obtained for coal conversion with supercritical water in continuous plants will be higher than most of the published values.

Increasing the density of supercritical water increases the conversion, as shown in Figure 7.3 [2,3]. The effect of pressure on the conversion vanishes at high pressure [7,11].

Since solvent density decreases rapidly when water is heated above its critical temperature, a lower conversion may be obtained than at near-critical, high density conditions. But similar conversions are found for the extraction of an Australian brown coal at $T = 380\,°C$, $T = 420\,°C$, and $T = 460\,°C$ [5,10], and also for extraction of a North Dakota lignite at $T = 380\,°C$ and $T = 440\,°C$ [2]. Increased pyrolysis and hydrolysis at higher densities before reaching the final temperatures ($T = 420–460\,°C$) may contribute to the results of these studies carried out in slow heating semi-continuous reactors.

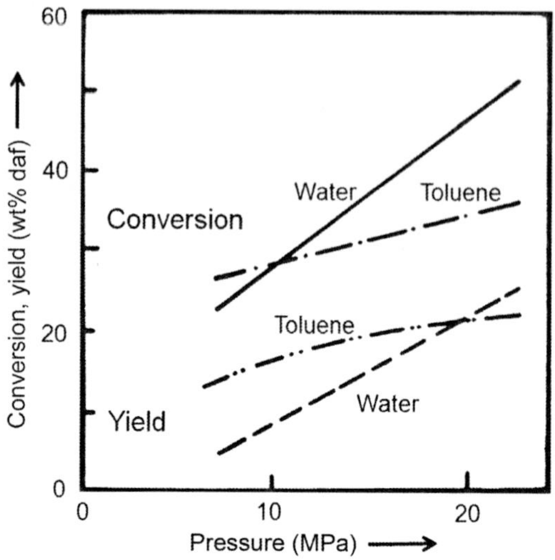

FIGURE 7.3 Influence of pressure on conversion and extract yield of brown coal (Loy Yang, Australia), dry ash free (daf), at $T = 380\ °C$. *Modified from Ref. [10].*

TABLE 7.1 Enhancement of Extraction Yield from Coals with Supercritical Water by Added Bases [2]

				Conversion (%)	
Coal	*P* (MPa)	*T* (°C)	Base	H_2O	H_2O+base
Morwell (brown)	22	380	0.5 M NaOH	51.7	82.2
Yalourn (brown)	22	380	2 M Pentylamine	56.7	95.8
Wyodak (sub-bituminous)	5.2	400	2.5% Ammonia	28.7	42.5

Supercritical water extraction is considerably enhanced for lignites and sub-bituminous coals when bases are present (see Table 7.1). The effect may be attributed to cleavage of ester bonds in the coal occurring during processing [5,12]. Yield is also enhanced when hydrogen donors, like tetralin or tetrahydroquinoline, are present in the aqueous phase [5,6,13].

Supercritical water is a better medium for oxygen removal and coal conversion than toluene due to hydrolysis reactions [14]. Supercritical water and carbon monoxide make an effective medium for liquefaction. The effectiveness seems not due to the *in situ* generation of hydrogen via the water–gas shift reaction. This topic has been reviewed and discussed by Ross [15] and Savage *et al.* [1]. High coal conversions are obtained in supercritical CO/H_2O/solvent systems with or without an added catalyst [16]. For example, in one case, supercritical water leads to 25% coal conversion and 11% desulfurization, but adding CO increases the coal conversion to 33% and the desulfurization to 20% [1]. Adding a hydrogen donor organic solvent (e.g., tetrahydroquinoline) increases coal conversion further, supposedly because it can disperse and dissolve the coal fragments [16].

Extraction from coal with the CO/H_2O binary gives oil yields higher than H_2/H_2O, N_2/H_2O, and CO_2/H_2O. Oils from aqueous-phase extraction have a lower oxygen content and higher carbon content than oils from pyrolysis [17,18]. It is postulated that the enhancement of liquefaction due to the addition of CO is the influence of the formate ion formation on the ionic strength of the solution [1].

In a comparative investigation on the conversion of Taiheiyo coals, a considerable portion of coal is converted into light oils with supercritical water ($T=653$ K, $P=35$ MPa), supercritical toluene ($T=653$ K, $P=20$ MPa), or a supercritical-formic acid mixture ($T=653$ K, $P=35$ MPa). The liquid yield for the $HCOOH–H_2O$ mixture is higher than the yields in supercritical water and supercritical toluene, probably through hydrolysis and hydrogenation in $HCOOH–H_2O$, see Figure 7.4 [19]. It is interesting that residence time has no influence in the range from several seconds to about 1 h for supercritical water and toluene. But conversion decreases at a prolonged residence time for the $HCOOH–H_2O$ mixture [19].

The complex structure and composition of coal and other hydrocarbon fuel sources make it difficult to define the fundamental chemical reactions that occur during treatment with hydrothermal and supercritical water. The investigation of the behavior of model compounds offers a way to overcome the

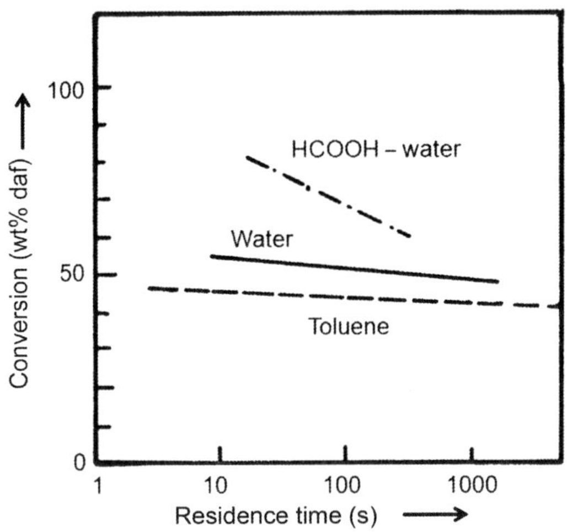

FIGURE 7.4 Coal conversion at various residence times (dry ash free, daf) at $T=653$ K and $P=20$ MPa (supercritical toluene), $P=35$ MPa (water, $HCOOH–H_2O$). *Modified from Ref. [19].*

variability of working with coal. Results for the model compounds must be related to real conversions of coal with water. Several detailed investigations on model compounds are discussed in Chapter 5. An overview can be found in Ref. [1].

Supercritical water has been investigated as a vehicle for coal hydrogenation with high-pressure hydrogen in the presence of stannous chloride or molybdenum trisulfide catalyst, or with carbon monoxide, making use of the water–gas shift reaction [2].

The results of the investigations may be summarized [2]:

Advantages

- Separation of the products from the residual solid is relatively simple,
- separation of the products from water is achieved readily,
- no high-pressure gas (hydrogen) is required. The energy requirement for compression is low because the extraction medium is compressed as a liquid,
- the extract is relatively hydrogen rich which should be advantageous in the further hydroprocessing step,
- the residual solid (char) can be gasified to provide process heat,
- gas production is minimal, at least for bituminous coals.

Disadvantages

- Conversions are low in comparison to hydrogenation processes,
- extraction with hydrothermal and supercritical water has not been developed to the same extent as solvent hydrogenation processes. Engineering problems may arise during scale-up, for example, for removal of the solid residue from the high-pressure system.

More recently, results of the group of Vostrikov has reported on several investigations on the kinetics of coal conversion in supercritical water and on the conversion of brown coal [20–23].

Conversion of the coal used for the experiments with supercritical water in a semi-batch reactor at $P=30$ MPa, $T=500$–750 °C, and reaction times of $t=60$–720 s can be described as a first-order reaction showing Arrhenius dependence with an activation energy of conversion of $E=103$ kJ mol^{-1} and an pre-exponential factor of $A_0=1.3\times10^{3.1}$ s^{-1}.The addition of CO_2 into supercritical water decreases the conversion rate and increases the CO yield [20].

The organic matter of brown coal is converted in a continuously operated tubular reactor, with a supply of a water-coal suspension to a vertical flow reactor, to 44–63% in supercritical water at $T=390$–760 °C and $P=30$ MPa. Reaction time corresponds to the time needed for the solids to fall through the reactor. In a subsequent batch conversion, the residue consisting of a

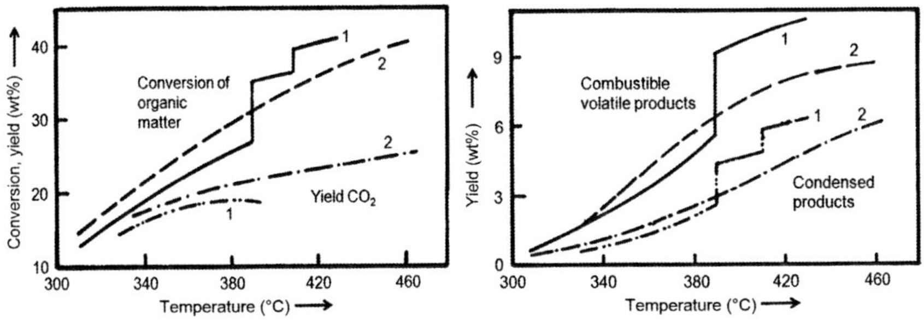

FIGURE 7.5 Temperature dependence of the degree of conversion α of organic matter of coal and the yield of CO_2 (left), combustible volatile and condensed products (right). $P=29$ MPa; 1 and 2 refer to different experiments. *Modified from Ref. [23].*

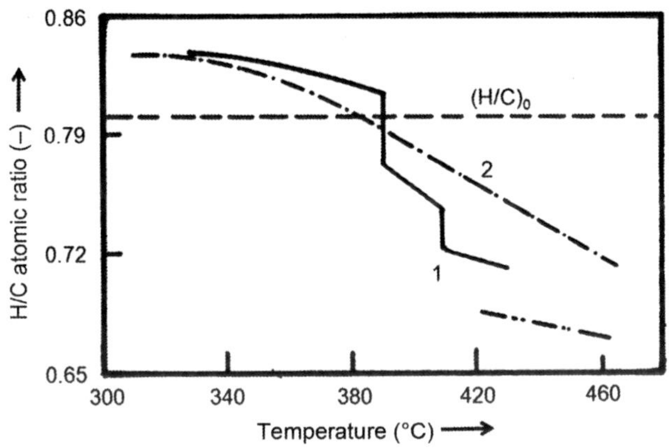

FIGURE 7.6 Variation of the H/C ratio of the residues of the organic matter of coal. —•••—: H/C ratio of the conversion residues, – – –: H/C ratio of the initial organic matter in coal. $P=29$ MPa; 1 and 2 refer to different experiments. *Modified from Ref. [23].*

nonagglomerated layer of coal particles was further reacted at $T=550$–760 °C. Total volatile products comprised CH_4, C_2H_6, C_6H_6, C_7H_8, C_8H_{10}, H_2, CO, and CO_2 [22,23].

In a new technique, the conversion products of the reaction in the tubular reactor are fast discharged and depressurized. Temperature ($T=310$–460 °C) and pressure ($P=9$–31 MPa) are raised stepwise. Conversion of coal and yield of products is shown in Figure 7.5. The heating value of combustible products and solid residue is higher than that of the initial coal (per unit of mass) since most of the native oxygen (up to 85.2%) is removed as CO_2 (Figures 7.5 and 7.6). Sulfur is removed up to 77.2% as H_2S and nitrogen is concentrated up to 67.2% in the solid conversion residue (Figure 7.7) [23].

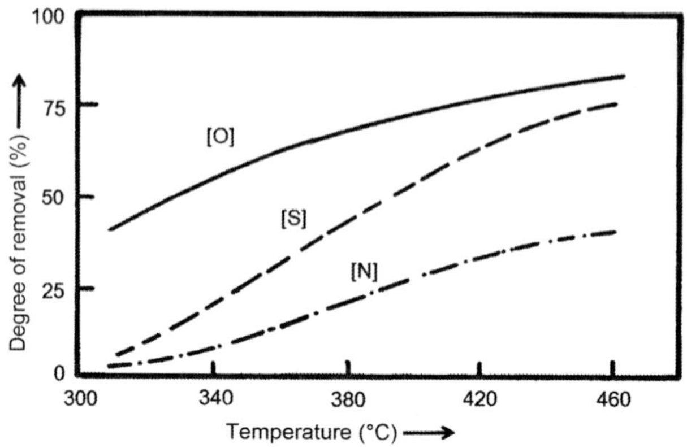

FIGURE 7.7 Temperature dependence of the degree of removal of oxygen [O], sulfur [S], and nitrogen [N] from the organic matter of coal. *Modified from Ref. [23].*

Extraction of low-rank coals with supercritical water in a semi-continuous laboratory extraction at $T = 380\,°C$ results in conversions of 40–50% with extract yields from 10% to 30%. Analyses of the extracts show that supercritical water extracts the volatile, hydrogen-rich fraction of the coal. The extract is highly polar with significant quantities of phenols and long-chain aliphatic fatty acids [24].

7.2.2 Gasification of Coal in Supercritical Water

Gasification of solid coal in supercritical water is another way to achieve a high conversion of coal to a product of high heating value. The idea was brought forward, among others, by Modell *et al.* [25] in context with the gasification of liquid or solid organic materials in general. Coal, or other organic liquid and solid material, is contacted with water at the critical temperature of water or at a higher temperature and at or above the critical pressure of water to achieve the critical density of water. A catalyst, in addition, can promote hydrogenation, reforming, and cracking of the organic material. A high conversion to a gas with high heating value is obtained without accompanying the formation of undesired char. Contrary to that, at conditions below the critical for water, substantially increased conversion to char occurs and conversion to gas is reduced. The carbon from the organic material reacts with water to form synthesis gas ($CO + H_2$), which then reacts to form methane and carbon dioxide. The gas product contains different concentrations of carbon monoxide, methane, hydrogen, carbon dioxide, and other low-molecular-weight hydrocarbon gases such as ethane and ethylene. From a slurry of bituminous coal particles ($d \approx 1$ mm) and water at $P = 233$ bar, $t = 60$ min, $T = 647$ K results a conversion to 8% gas and 20% liquid without the formation of char. 80% of the sulfur in the coal feed is precipitated in the water as elemental sulfur or removed as H_2S [25]. For a higher amount of gas, the temperature must be increased and the pressure adjusted to a density higher than the critical.

Because a high density at high temperatures can only be achieved at relatively high pressures (at $T = 700\ ^\circ C$, a pressure of about $P = 100$ MPa is necessary to achieve the critical density, see Chapter 2), various means are tried to get a good conversion at lower density, for example, by adding bases or dissolved salts.

For gasification of coals, calcium hydroxide was used both as a catalyst and an absorbent for CO_2 at $T = 700\ ^\circ C$ and pressures up to 30 MPa [26]. Without the addition of Ca, the carbon conversion of Yallourn coal is about 44% at $T = 690\ ^\circ C$ and $P = 28$ MPa. The addition of Ca up to a Ca/C molar ratio of 0.6 increases the carbon conversion to about 54%. Hydrogen yield is increased from 0.23 to 0.37 dm^3(STP)/g-coal (daf: *d*ry and *a*sh *f*ree), methane yield from 0.16 to 0.19 dm^3(STP)/g-coal (daf), and ethane yield from 0.015 to 0.03 dm^3(STP)/g-coal (daf). At a Ca/C ratio of 0.6, almost no CO_2 remained in the gas phase and calcium was utilized to 50–60% [26].

Another approach is to dissolve salts in supercritical water. Although the solvent power for salts is drastically reduced in supercritical water, the catalyzing salt is precipitated at the coal surface in a higher concentration and is more active than for initially catalyst-impregnated coals with the effect that the reaction is considerably promoted [27].

A slurry of 16 wt% coal was continuously gasified in supercritical water at $T = 650$–$800\ ^\circ C$, $P = 23$–27 MPa and a slurry flow rate from $\dot{m} = 3$–$7\ kg\,h^{-1}$ with K_2CO_3 and Raney-Ni as catalysts, and H_2O_2 as oxidant. The hydrogen fraction reached up to 73% [28].

In a similar investigation, lignite was converted in a coal–water slurry at $T = 500$–$650\ ^\circ C$, $P = 20.0$–30.0 MPa, Ca/C = 0–0.45 mol mol^{-1}, and O/C = 0–0.35 mol mol^{-1}. Product yields are shown in Figure 7.8. There is a notable increase of the hydrogen content and the yield with higher reaction temperature, see Figure 7.9.

Conversion of coals with high temperature and supercritical water can produce a high amount of gaseous and liquid fuel products from the feed material, which can be coal of minor quality. Processes for coal conversion have been carried out to production size during World War II and later in South Africa. Similar and new coal conversion processes have been in operation in demonstration size during the 1970s and 1980s. They were rendered not profitable due to the drastic decline of oil prices in the 1980s. Since then, no major efforts have been initiated to convert coals to liquid fuels. Although the application of high temperature and supercritical water proved that it can be an interesting conversion route, no major conversion project has been initiated, in spite of steadily increasing oil prices. One reason is probably the concern about carbon dioxide emissions, but the major reason can be the trend to technologically simple and short-time return of investment processes.

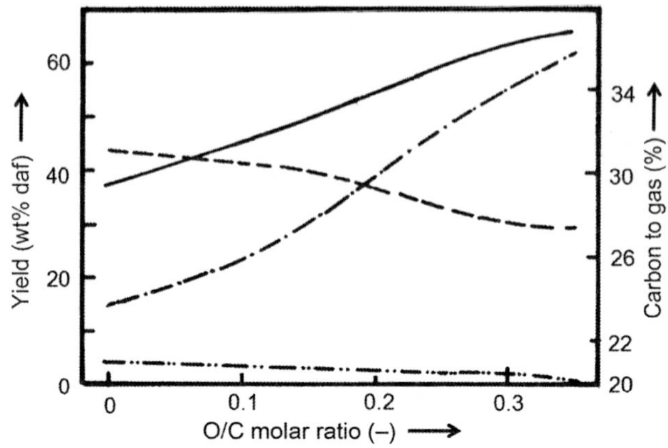

FIGURE 7.8 Conversion of lignite in water. Product yields at various O/C ratios; $T=600$ °C, $P=25$ MPa, coal slurry concentration 20 wt%. ———: gas, – – –: residue, — • • —: liquid, — • —: carbon to gas. *Modified from Ref. [29].*

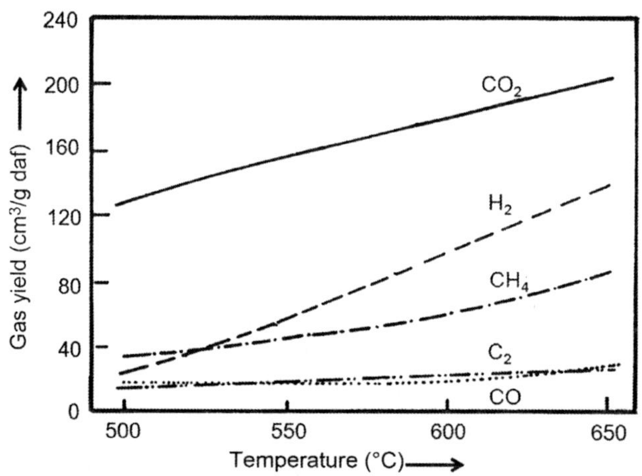

FIGURE 7.9 Conversion of lignite in water. Yield of gaseous products at various temperatures; $P=25$ MPa, coal slurry concentration 40 wt%. *Modified from Ref. [29].*

7.3 OIL SHALE CONVERSION WITH HYDROTHERMAL AND SUPERCRITICAL WATER

Typical oil shales are nonporous, impermeable rocks containing 80–95 wt% minerals, and only 5–20 wt% organics. The organic part consists of bitumen and kerogen. Bitumen can be extracted with organic solvents. Kerogen is an insoluble solid that is finely dispersed throughout the mineral matrix. Heating kerogen to high enough temperatures converts the kerogen and the products that are obtained so that they can be used as fuels or sources of chemicals.

Supercritical water is applicable for kerogen conversion. For a Green River oil shale up to 92 wt% of the organic carbon content can be converted. The yield in liquid oil products with a boiling point of $T \geq 150\ ^\circ C$ is 13.8 wt%. CO_2 is the main component of the gaseous products ($x_{CO_2} \leq 80$ vol%). The soluble (bituminous) part of the oil shale—about 8 wt%—can be extracted below $T = 350\ ^\circ C$. Kerogen conversion starts at $T = 350\ ^\circ C$. Asphaltenes and preasphaltenes react to oil components at temperatures $T > 400\ ^\circ C$. At $T = 420\ ^\circ C$ the maximum rate of conversion is observed. Carbonates react at temperatures $T > 500\ ^\circ C$ and release carbon dioxide. The best conditions for kerogen conversion with supercritical water are temperatures from $T = 400$ to $450\ ^\circ C$. From an oil shale with a kerogen content of 18.4 wt%, the following products could be obtained: 11 wt% oil, 2 wt% low-boiling oil components, 2.5 wt% asphaltenes, 0.3 wt% preasphaltenes, and about 0.7 wt% hydrocarbon gases. The oil shale residue still contains about 1.9 wt% coke.

Products from the kerogen conversion with supercritical water are significantly different in product yield and composition, depending on the feedstock origin and its chemical composition. Several studies on kerogen conversion of different oil shales with water and added compounds, mainly carbon monoxide, are discussed in the following and the individual differences are addressed.

Kerogen conversion processes using water at supercritical conditions are operated at pressures $P > 22.0$ MPa and at temperatures $T > 380\ ^\circ C$. At these conditions, kerogen conversion and liquid product yield are at the same level as for other kerogen conversion processes using supercritical solvents or pressurized hydrogen. Advantages of using supercritical water result from the high availability of the reaction medium and from the drastic changes of properties of water between the supercritical state and the state at ambient conditions. At supercritical conditions, water is a good solvent for hydrocarbons, while at more ambient conditions, it is not a solvent for hydrocarbons at all. These changes allow a relatively simple separation of reaction solvent and reaction products. Water can be handled at ambient conditions safely and in a simple manner.

In a kerogen conversion process with water the raw shale is crushed, can be mixed with a recycle stream of hydrocarbons derived from the conversion products and with supercritical water, and is introduced into the conversion reactor. Conversion products are separated from the product stream by cooling, separating the product stream into an aqueous phase, lean in hydrocarbons, and an organic phase, rich in hydrocarbons, and gaseous products. From the gaseous products, the remaining hydrocarbon low-boiling liquids and acid gases are removed in a scrubber. The washed gas stream can be used for energy supply. Energy recovery from the spent shale can be achieved by sequentially quenching the hot shale with water, as has been proposed for hydrogen retorting. The liquid product phase is processed for hydrocarbon fractions. The aqueous phase is further depressurized; emerging gaseous and liquid products are separated and introduced into the process. The residual water is pressurized again, heated, and sent as recycle stream to the conversion reactor. Dissolved salts can enhance the conversion reaction but also increase the risk of corrosion. The kerogen conversion process with supercritical water is the least developed one of all the supercritical kerogen conversion processes.

7.3.1 Kerogen Conversion with Supercritical Water

The use of water allows extraction of the organics not incorporated into the kerogen matrix, similar to the extraction of nonpolar organic solvents. The extraction yield and composition from Kukersite oil shale at $T=200\ °C$ with water is quite similar to that obtained with methanol-modified carbon dioxide [30]. At supercritical conditions, water can be used for degradation of kerogen, and thus can be used to produce oil from shale. At $T=250\ °C$ the oil yields are low (5–10% of the total organic matter). At an increased temperature of up to $T=300\ °C$ oil yields of up to 35% in water and 58% in benzene are obtained. The extracts are characterized by a high content of aliphatic compounds and especially of olefins with a double bond in the middle of the chain. The addition of an organic solvent to water increases extraction yields. For benzene–water (1:1) at $T=350\ °C$ the extraction yield increases up to 74%. The extracts consist of high-molecular-weight oils containing a great deal of asphaltenes, with a yield comparable to Fischer assay results (Fischer assay is a standardized method to determine pyrolysis yield of carbon-containing materials). Addition of water to carbon dioxide at a 1:5 ratio results at $T=340\ °C$ in yields of extract that are 27% larger than with pure CO_2 and 63% larger than for the extraction with pure water, with an overall recovery as high as 94% of the kerogen. At these conditions, the presence of minerals can lead to the formation of additional products. Also, using modifiers to carbon dioxide having a hydrogen-donor capability, like alcohols, lead to hydrogenation reactions [30].

These laboratory investigations, directed to analytical use, can give indications on the use of supercritical fluids in conversion processes. In the following, conversion of kerogen with supercritical water is covered.

Water at near-critical and supercritical conditions ($T_c=373.946\ °C$, 647.096 K, $P_c=22.064$ MPa) is a good solvent for hydrocarbons, contrary to its properties at near-ambient conditions. But water at supercritical conditions is still a polar solvent. Thus it can dissolve polar compounds. Changing the conditions from critical to lower temperatures provides the means to separate organic products from water (compare Chapter 3, the section on solubility, in particular, phase equilibrium of three components). Furthermore, water is a component of the kerogen conversion process anyway and it is readily available compared to other process components. In the following, some investigations using water as a processing medium for kerogen conversion are reviewed.

7.3.2 Investigation of Oil Shale Conversion on a Green River Oil Shale

In a systematic study, Missal [31] and Missal and Hedden [32] investigated kerogen conversion with sub- and supercritical water on a Green River oil shale. Experiments were carried out in a semi-continuous apparatus in which water at constant pressure flows through a fixed bed of oil shale particles and is heated during the experiment from subcritical to supercritical temperatures. For the experimental setup, see Chapter 13.

The course of the experiments is shown in Figure 7.10 for the evolution of the liquid and gaseous products [31]. The liquid product emerges in the temperature range of $T=400$–450 °C in about $t=30$ min after reaching the conversion temperature, pressure is about $P=30$ MPa. The gaseous products show up from about $T=350$ °C and increase steadily until the end of the experiment after $t=150$ min, with a maximum production rate at $T=450$ °C after $t=100$ min. At that time, the production of liquid products already approaches zero.

The influence of the reaction temperature is shown in Figure 7.11a for the temperature range $T=350$–550 °C, and in Figure 7.11b, the influence of residence time is shown at different tempertures.

The degree of conversion increases from 9 wt% at $T=350$ °C to 30 wt% at $T=550$ °C. These temperatures are the end-temperatures of the experiment. The evolution of the products is shown in Figure 7.11 in dependence on the end-temperature. At an end-temperature of $T=350$ °C, the conversion of the kerogen is low. Conversion increases with temperature and reaches at about $T=400$ °C, the maximum yield after $t=150$ min. At $T=450$ °C, the highest amount of liquid product is reached in about $t=30$ min. It is of interest to look at the production of the individual products in dependence on the end-temperature as shown in Figure 7.12.

For oil formation, a maximum occurs at $T=430$ °C, for the formation of preasphaltenes at $T=400$ °C, and for asphaltenes at $T=410$ °C. Carbon dioxide formation is at a maximum at $T=530$ °C, caused by the decomposition of the carbonate part of the oil shale. For lower end-temperatures of the kerogen conversion, the maximum of CO_2 formation is shifted to lower temperatures of $T=410$–430 °C, determined by the kerogen conversion. The maximum of formation of other gases, CH_4, C_2- to C_4-hydrocarbons, and CO lies at about $T=430$ °C. Hydrogen formation has a maximum at $T=460$ °C. Conversion of the organic part of the oil shale reaches 20 wt% at $T=350$ °C, 83 wt% at $T=400$ °C, and 88 wt% at an end-temperature of $T=550$ °C. At temperatures above $T=350$ °C, appreciable amounts of N_2 are produed by the oil shale conversion.

Influence of solvent ratio and of particle size and of pressure are shown in Figure 7.13.

An increase of solvent ratio leads to a somewhat better conversion of the kerogen. The influence of the heating rate leads to fast kerogen conversion

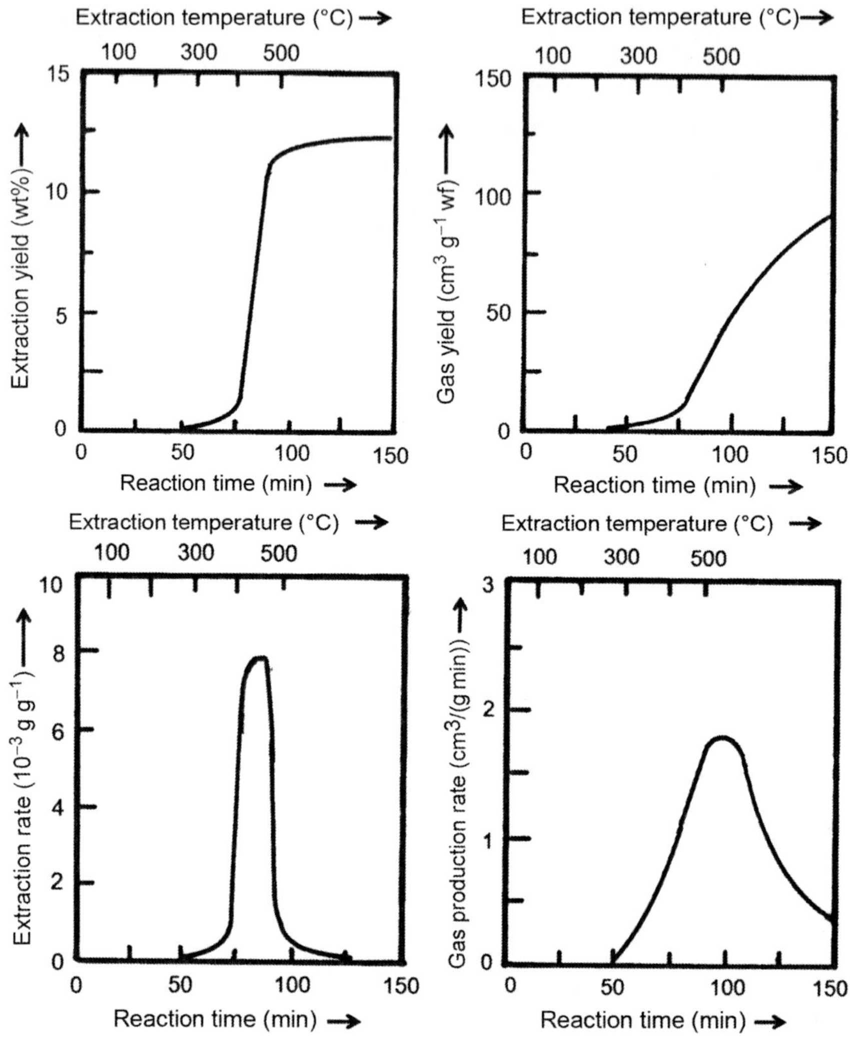

FIGURE 7.10 Conversion of Green River oil shale in water. Development of liquid and gaseous products during the experiments. $P = 30$ MPa. *Modified from Ref. [31].*

at high heating rates due to the fact that temperatures are approached faster and products then evolve at higher temperatures. At lower heating reates, kerogen conversion proceeds at lower temperatures and already decreases when higher temperatures are reached. In general, the maximal formation rates occur at lower temperatures for lower heating rates. Degree of conversion, liquid yield, and amount of gaseous products decrease with increasing particle diameter. For particle diameters from $d_p = 0.6$ to 1.2 mm about 89% of the

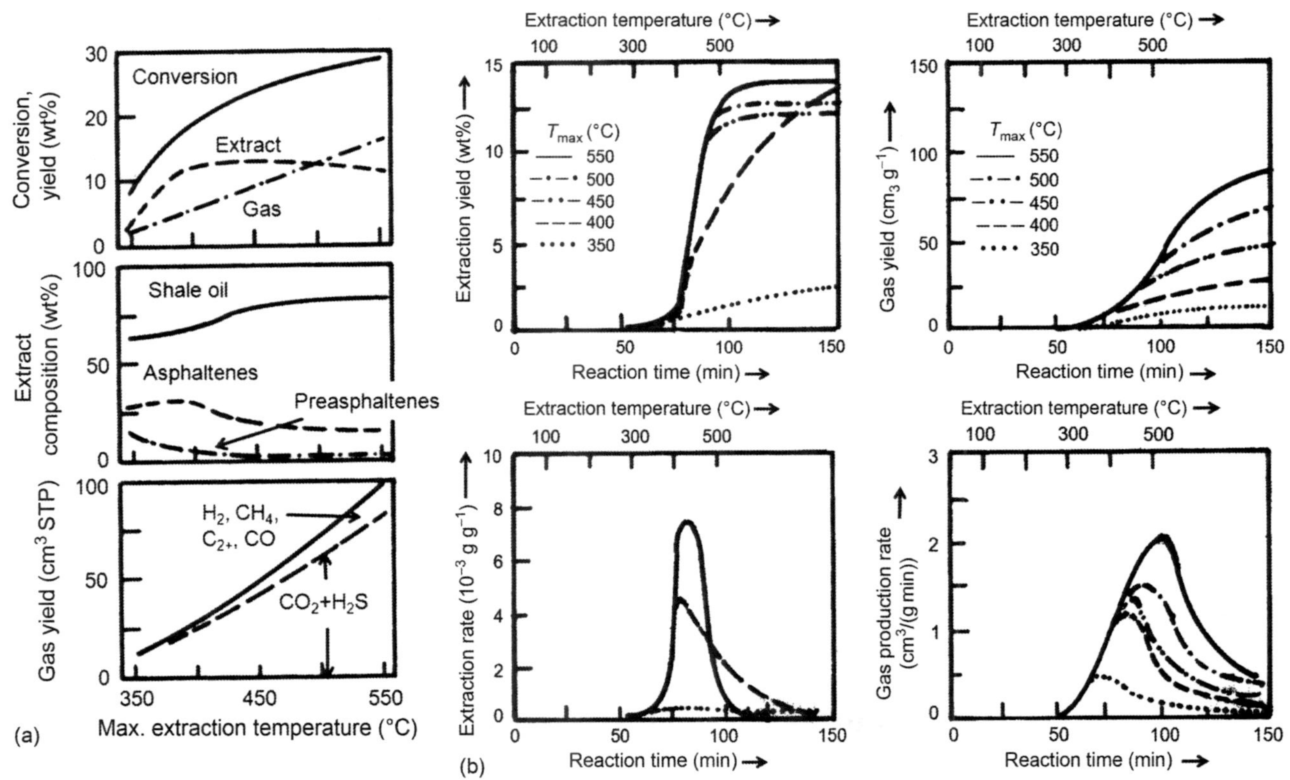

FIGURE 7.11 (a) Influence of temperature on the kerogen conversion. $P = 30$ MPa. (b) Formation of products during kerogen conversion of a Green River oil shale as a function of the end-temperature of the experiment. $P = 30$ MPa. *(a) Modified from Ref. [31] and (b) modified from Ref. [31].*

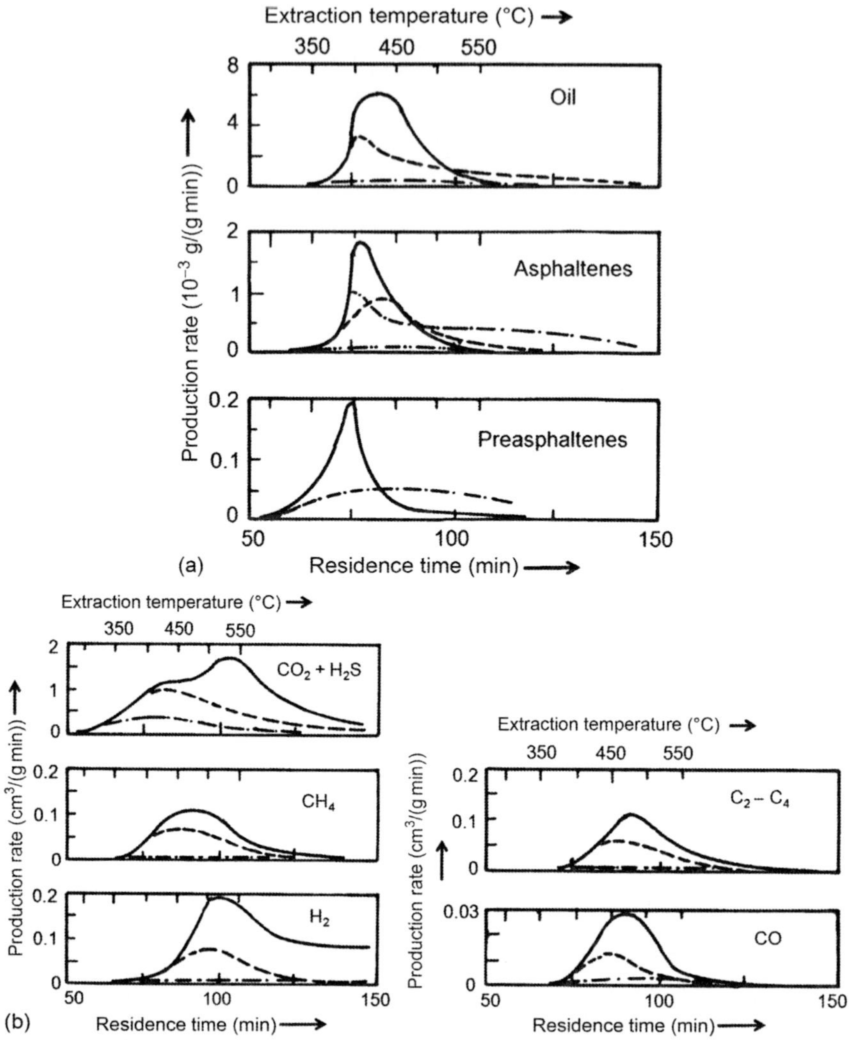

FIGURE 7.12 (a) Formation of products during kerogen conversion with water, liquid products, T_{max}: ———, 550 °C; – – –, 450 °C; — • —, 350 °C. $P = 30$ MPa. (b) Formation of products during kerogen conversion with water, gaseous products, T_{max}: ———, 550 °C; – – –, 450 °C; — • —, 350 °C. $P = 30$ MPa. *(a) Modified from Ref. [31] and (b) modified from Ref. [31].*

organic matter is transformed, while for particle diameters from $d_p = 8$–12 mm only 84% are transferred to products. If the conversion at 0.1 MPa with water as a sweep gas is compared to the conversion at supercritical conditions, the amount of products is enhanced. Also enhanced are the amount of asphaltenes. For gases, the conversion at 0.1 MPa shows the production of hydrogen as a major component, while at the higher pressures

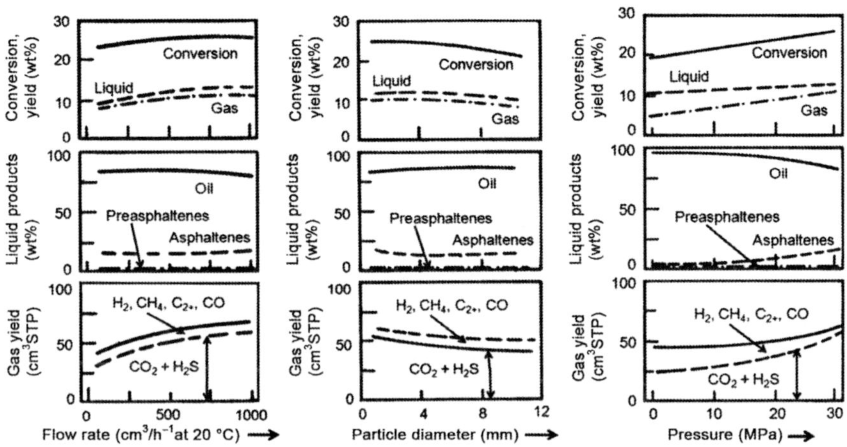

FIGURE 7.13 Kerogen conversion of a Green River oil shale with water. Left: influence of flow rate, middle: influence of particle size, right: influence of pressure. Conditions (if not varied): amount of oil shale: 100 g (wf), particle diameter $d_p = 2\text{–}2.5$ mm, $T_{max} = 550$ °C, $dT/dt = 5$ K min^{-1}, $P = 30$ MPa, flow rate $m_{H_2O} = 500$ml/min^{-1} at $T = 20$ °C, extraction time $t = 106$ min. *Modified from Ref. [31].*

TABLE 7.2 Conversion of Green River Oil Shale with Water

Temperature (°C)	C (wt%)	H (wt%)	N (wt%)	S (wt%)	O (wt%)	H/C
390–410	82.5	10.9	1.2	1.5	3.9	1.6
410–430	84.2	11.2	1.3	1.3	2.0	1.6
430–450	84.5	11.2	1.6	0.9	1.8	1.6
450–470	84.5	11.3	2.1	0.6	2.6	1.6
470–490	83.5	10.7	2.8	0.8	2.2	1.5

Elemental analysis of the liquid products [31].

carbon dioxide is the main component. The other gas components are of minor importance. An elemental analysis of the liquid products is shown in Table 7.2 and the boiling curve of the liquid products in Figure 7.14.

All the liquid products show the same boiling behavior. The boiling curve begins at $T = 190$ °C and ends at $T = 500$ °C. 50 wt% of the liquid products is middle distillate with a boiling end at $T = 360$ °C, 50 wt% is a heavy oil fraction with boiling temperatures $T > 360$ °C [31].

The reaction of kerogen in oil shale in water is assumed to follow a simple scheme (Figure 7.15) with a parallel physical extraction and a chemical transformation of the kerogen.

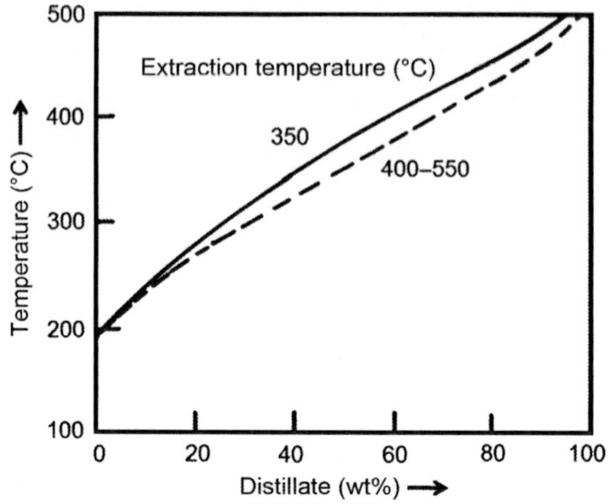

FIGURE 7.14 Boiling curves of liquid products from kerogen conversion of Green River oil shale with water. *Modified from Ref. [31].*

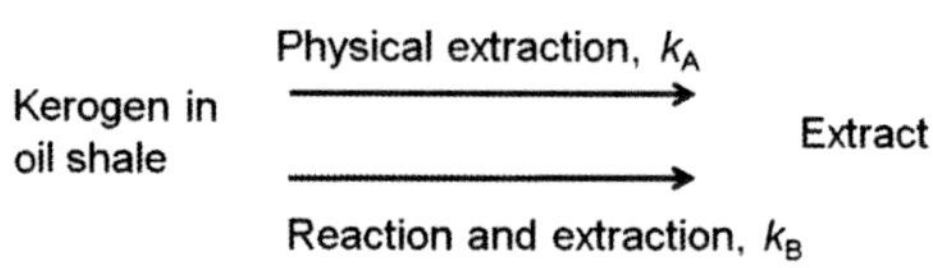

FIGURE 7.15 Reaction scheme for the kerogen conversion in water [31].

Rate equations for the reactions of the kerogen conversions are as follows, with A for the physical extraction and B for the transformation reaction of kerogen:

$$\left(\frac{dm_E}{dt}\right)_A = k_A m_A \tag{7.1}$$

$$\left(\frac{dm_E}{dt}\right)_B = k_B m_B \tag{7.2}$$

Rate constants are described by an Arrhenius equation. Due to the nonisothermal experimental procedure, the integration of the reaction equations is somewhat complicated. It is documented in the literature [31]. The results are shown in Figure 7.16. The physical extraction has an activation energy of $E_A = 46$–63 kJ mol^{-1}, the kerogen conversion reaction has an activation energy of $E_A = 145$–196 kJ mol^{-1}. These values compare well with other literature values [33].

From the results, it can be concluded that water is applicable for kerogen conversion. For the specific oil shale, used in the investigation, up to 92 wt% of the organic carbon content is converted and the yield in liquid oil products with a boiling point $T \geq 150$ °C is up to 13.8 wt%. CO_2 with up to 80 vol% is

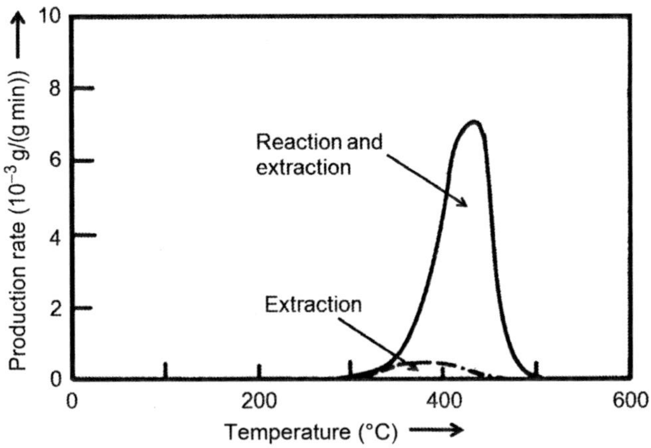

	Extraction	Reaction and extraction
E_a [kJ/mol]	61	171
k_0 [1/s]	93	$1.7 \cdot 10^{10}$
m_E [g]	1.4	10.3

FIGURE 7.16 Representation of the kerogen conversion with a simplified reaction model. Conditions as for Figure 7.13. *Modified from Ref. [31].*

the main component of the gaseous products. Detailed information is available on the kinetics of the kerogen conversion. The soluble (bituminous) part of the oil shale—about 8 wt%—can be extracted below $T = 350$ °C. Kerogen conversion starts at $T = 350$ °C, asphaltenes and preasphaltenes further react to oil components at temperatures $T > 400$ °C. At $T = 420$ °C the maximum rate of conversion is observed. The overall reaction can be described by a first-order model. Carbonates react at temperatures $T > 500$ °C and release carbon dioxide. The best conditions for kerogen conversion with supercritical water are temperatures from $T = 400$ to 450 °C. From the oil shale with a kerogen content of 18.4 wt% the following products can be obtained: 11 wt% oil, 2 wt% low-boiling oil components, 2.5 wt% asphaltenes, 0.3 wt% preasphaltenes, and about 0.7 wt% hydrocarbon gases. The oil shale residue still contains about 1.9 wt% coke.

7.3.3 Further Investigations on Kerogen Conversion with Supercritical Water

Supercritical water, methanol, and mixtures were used for supercritical fluid extraction of both western and eastern oil shales [34–39]. Batch-type extractions were conducted at $T = 400$ °C and under pressure up to $P = 34.5$ MPa.

About 90% of the organic matter from Green River oil shale is converted. Methanol is incorporated into the product as ester liquid. The extracts have a high H/C atomic ratio of about 1.6 and a number average molecular weight of $M_W = 500$–800, more than twice the amount of the shale-oil pyrolysis product. The high oil yield with water–methanol mixture could not be reproduced by others [40]. But, by using NH_4OH in place of methanol, 90% yield in 1 h of extraction is obtained at $T = 390$ °C and 37.3 MPa pressure [41].

In another study, thermal conversion of Green River oil shale was investigated in carbon monoxide and steam at $T = 300$–450 °C under a pressure of $P = 1.4$–41.4 MPa [42–45]. The intent of these studies was to use the water–gas shift reaction: $CO + H_2O \leftrightarrow H_2 + CO_2$ for hydrogen production. The conversion of kerogen to liquid and gaseous products also in this case increases with temperature, 30% at $T = 350$ °C, 50% at $T = 375$ °C, 78% at $T = 400$ °C, and 98% at $T = 450$ °C, in 1 h or less reaction time. Increasing the CO pressure above 1.4 MPa reduces the yield of the liquids. At lower pressure and higher temperature ($T > 400$ °C), the liquid product contained a volatile oil fraction [41].

In water alone (supercritical in most cases), comparable total yields of liquids and gases were observed. However, the evolution of large amounts of gases (three to nine times the amount with CO and steam) at all temperatures and the appearance of carbon residue at the highest temperature used ($T = 450$ °C) bear an indirect evidence that, in CO and steam, some hydrogen is added to the products formed, minimizing gas and coke formation. Although the laboratory results of the high-pressure CO and steam process were encouraging, subsequent cost analysis of a simulated commercial process showed that the process is uneconomical. High-pressure operation and insufficient gas production for CO makeup are cited as the major factors raising the cost of oil [41,46].

Extraction of eastern oil shale was studied [36,47] at $T = 400$ °C and pressures up to $P = 34.5$ MPa. By using binary mixtures of water with methanol, toluene, octanol, and tetralin, an increase in oil yield from 25% to 92% for 1 h extraction of Chattanooga oil shale was observed, with an average molecular weight of the liquid products of $M_W = 500$–700 [41].

Due to the comparatively high pressure for supercritical water ($P > 22$ MPa), some investigations use water at supercritical temperatures but subcritical pressures. In those regions, the high temperature is sufficient for thermal kerogen conversion but the solvent power of water may not be optimal.

In such an investigation on oil shale in the range of $T = 400$–450 °C and $P = 14$–21 MPa, extracted kerogen conversion products account for up to 70% of the organic matter in the shale, up to 75% higher than oil yields obtained from shale retorting (without water at high pressures) [48]. H/C ratio is about 1.57; carbon aromaticities are as low as 0.35–40. Extraction in the presence of CO also removes most of the sulfur in the source material as H_2S. Kerogen conversion products contain higher proportions of asphaltenes and resins than oils produced by pyrolysis of the shale at the same temperature and pressure conditions, and are of significantly higher molecular weight.

For the conversion of Estonian oil shales it was found [49] that extracts are significantly different in product yield and composition, depending on the feedstock origin and its chemical composition. The majority of hydrocarbons are represented by polycyclic aromatic compounds. Aliphatic hydrocarbons, making 6–11% of the benzene-soluble oil, were represented by *n*-alkanes up to C_{33}. Conversion gases, especially those of peat and willow are characterized by high carbon dioxide content. The majority of hydrocarbons are polycyclic aromatic compounds. Aliphatic hydrocarbons, making 6–11% of the benzene-soluble oil, are represented by *n*-alkanes up to C_{33}. Conversion gases are characterized by high carbon dioxide content [49].

Water conversion was applied for Kukersite oil shale and Dictyonema shale (Estonia) in batch experiments for 4 h in a 0.5 l autoclave at $T=380$ °C [50]. Oil yield for Kukersite shale is 62.7% of organic matter, 84.9% of it is benzene soluble. 30.8% of the organic matter is transferred to gas and 6.5% remains in the solid residue. From Dictyonema shale, only 20.5% of organic matter is transferred to oil and 46.6% to gas, 32.9% remains in the solid residue. The content of benzene solubles is 59.3% of the total oil. The dominating gas from both oil shales is CO_2 with about 70 vol% of total gas amount from Kukersite and 40 vol% from Dictyonema shale. In coprocessing of carbonaceous Kukersite and argillaceous Dictyonema shale, no synergy in total oil as well as its group yield could be found [50].

Water extraction experiments with Huadian oil shale (Jilin, China) carried out in a semi-continuous operation apparatus shows that the extract is obtained mainly in the temperature range between $T=300$ and 500 °C. Gas formation is only observed at temperatures higher than $T=350$ °C with a maximum formation rate for extract and gas at about $T=390$ °C [51]. The major extract component is asphaltene and the major gas formed is CO_2. At high temperature, the formation rate of H_2 and CH_4 increases and causes a secondary increase of gas formation starting at $T=460$ °C. From isothermal experiments it was found that a maximum extract yield occurs at $T=400$ °C, and a conversion plateau exists for temperatures $T>400$ °C. Experimental results correspond to the variation of hydrocarbon solubility in water and the degree of pyrolysis of oil shale. Compared to extraction with toluene, extraction with water shows a higher conversion, slightly lower extract yield, and a higher oil content [51].

A more recent investigation on the same type of shale (Huadian, China) by subcritical water proved that water at elevated temperatures can crack kerogen into smaller hydrocarbon compounds at relatively low temperatures. At $T=260$ °C, the extract yield reaches 7 wt% ($t=2.5$ h at $P=15$ MPa). With increasing pressure and temperature, large amounts of high-molecular-weight hydrocarbons are gradually decomposed and increase levels of low-molecular-weight hydrocarbons, polycyclic, and heterocyclic compounds [52].

In the supercritical region from $T = 380$ to 400 °C, extraction of Moroccan (Timahdit) oil shale with water [53] showed that the yield and the fraction of paraffins and aromatics increase while the percentage of asphaltenes decreases with increasing temperature. From Moroccan oil shales new pitches are obtained by supercritical extraction with water as solvent in a batch autoclave at a temperature of $T = 390$ °C, $t = 2$ h treatment, and a maximum pressure of $P = 28$ MPa [54].

In a comparative study, oil shales from Mishor Rotem, Israel, Kentucky, and Green River, USA, were investigated. The conversion was carried out batchwise, in autoclaves, and in a continuous flow system. The best solvents were found to be isopropanol–water and water–CO mixtures [55].

Extraction of oil from Australian Condor oil shale, carried out in a batch autoclave, using water with CO and Na_2CO_3 as additional components, results in a maximum oil yield of 46 wt% at a water density of $\rho = 200\ \mathrm{kg\ m^{-3}}$ (at $T = 673$ K) and $P = 1$ MPa partial-pressure of CO. The yield is higher than the yield of the Fischer assay of 33 wt% [56]. Due to the experimental procedure in a batch autoclave, no data can be obtained that would be useful for real production process. Such data can be derived from investigations, in which kinetics of kerogen conversion with water are measured, for example, from the above-cited investigation of Missal [31] on Green River oil shale, or a more recent one on a Turkish oil shale [57].

A comparative investigation with respect to different conversion procedures was carried out on Göynük oil shale comprising slow pyrolysis, flash pyrolysis, and super- and subcritical fluid extraction with water. It reveals considerable differences in yields and compositions of the oils obtained under the different conditions [58].

Another investigation was recently reported on the coprocessing of Himmetoglu (Turkey) oil shale and specific berries in supercritical water [59]. The yields of gas, hexane and benzene extracts, and organic residue formed in the supercritical water conversion consist of partial contributions of the yields from the initial objects. There is no synergistic effect at supercritical water pyrolysis of shale and terebinth berries. Although supercritical water leads to a decrease in the amount of organic residue from oil shale, the extract contains more polar hetero-compounds and less nonaromatic hydrocarbon than those of dry pyrolysis [59].

7.3.4 Conversion with Supercritical Brine

The conversion of fuel materials with supercritical brine aims at the effect of salts on the conversion and considers the often occurring salinity of deposits with organic compounds. Using water with dissolved salts (brine), Gregoli *et al.* [60] proposed to convert heavy crudes, tars, and bitumens to lighter products in the presence of brine at supercritical conditions for surface and subsurface operations. The proposed process is mentioned here, due to the aspect of *in situ* conversion and the effect of supercritical water with dissolved salts on the conversion products.

According to the proposal, heavy hydrocarbon crude is reacted with a minor amount of brine at supercritical temperature and pressure for the brine, in order to upgrade and convert the heavy hydrocarbon crude into a lighter hydrocarbon crude of higher API gravity.

The upgrading and conversion is accomplished in a continuous reactor system and may be accomplished in a subterranean petroleum reservoir at supercritical temperature and pressure. In order to provide the necessary temperature, heat is added to the system prior to the reaction. For an *in situ* application, a combustion operation may be utilized to provide the necessary temperature, and is initiated using an oxidizing gas injected through an injection well. After a predetermined amount of time, injection of the oxidizing gas is terminated and the injection well is shut in for a predetermined period of time to permit the petroleum reservoir to undergo a soak period in order to increase the temperature and decrease the viscosity of the viscous heavy hydrocarbonaceous crude oil. As the viscosity of the heavy hydrocarbonaceous crude oil decreases, the oil flows downwardly into the combustion zone steam. Brine is injected, which is at supercritical conditions in order to upgrade the heavy hydrocarbonaceous crude oil into lighter fractions.

The proposed process can in the same way be applied for an oil shale reservoir. Reaction time and conditions will have to be adjusted to the ones needed for kerogen conversion. But the principle of hydrocarbon conversion with supercritical brine (water + salts) remains the same.

The reason for subjecting brine and heavy crude at or above supercritical condition is to provide intimate mixing of the reactants and the desired conditions for converting heavy crude into a lighter crude. Brine and crude are totally miscible and provide the necessary residence time not obtainable under subcritical conditions in a conventional reactor system.

Conditions of the process for conversion of kerogen and upgrading of the conversion products are: A heavy crude oil and brine mixture comprising from about 2 wt% to less than about 50 wt% brine is reacted in a vessel at $P = 23.45$–27.6 MPa and $T = 377$–482 °C for $t = 0.5$–6 h. From 20 to 90 wt% of the original heavy crude oil is converted into a lighter crude oil. Some of the unconverted original heavy crude oil of lower API gravity is recycled back and mixed with the original heavy hydrocarbon crude feed. Products of the process are shown in Figure 7.17.

An 11° API, heavy crude oil was reacted with approximately 20 wt% brine in a reactor having a temperature of $T = 426.7$ °C and a pressure of about $P = 24.1$ MPa. The recovered oil reaches a maximum API gravity (approximately 30° API) after a residence time of $t = 1.0$–1.5 h (see Figure7.17). Coke and gas are produced as by-products. About 25–28 wt% of light crude

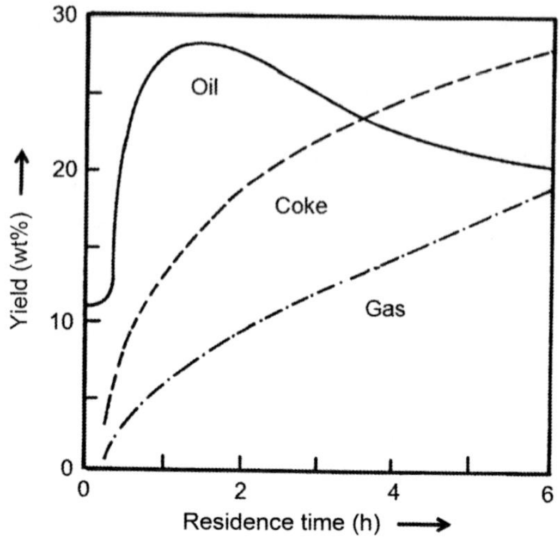

FIGURE 7.17 Conversion of heavy crude with supercritical brine. See text. *Modified from Ref. [60].*

(25–28° API) from the original amount of 11° API heavy crude oil, 12–14 wt % coke and 5–7 wt% gas are produced. As an unexpected result, a high ratio of saturates to unsaturates is found in the product in the presence of the produced H_2S and CO.

Other results are that coke formation and production of unsaturates are lower than for thermal cracking and useful conversion product yields are far greater than those obtained from conventional coking. Furthermore, the ratio of saturates to unsaturates in light ends is significantly greater with supercritical brine conversion than obtained from delayed coking. The sensitivity to temperature shows the far greater ratio of saturates/unsaturates at $T=399$ and 426.7 °C compared to delayed coking. For example, the ratio of saturated to unsaturated C_4 compounds is >100, compared to 1.5 for coking. The same is valid for C_3- and C_2-compounds.

In order to obtain the benefits of *in situ* upgrading in bitumen or heavy crude reservoirs, it is necessary to devise a method of raising the reservoir temperature for the time required to allow the desired cracking reactions to take place. The minimum temperature is in the $T=260$–371 °C range with corresponding residence times of a few hours to several weeks. One method is to perform a fireflood operation in the lower part of a thick reservoir, followed by a soaking period.

Oil shale processes using water at supercritical conditions are operated at pressures $P>22$ MPa and temperatures $T>380$ °C. At these conditions, kerogen conversion and liquid product yield are at the same level as for the other oil shale processes using supercritical solvents or pressurized hydrogen. Advantages of using supercritical water result from the availability of the reaction medium and from the drastic changes of properties of water between the supercritical state and the state at ambient conditions. At supercritical conditions water is a good solvent for hydrocarbons, while at more ambient conditions it is not a solvent for hydrocarbons at all. These changes allow a relatively simple separation of reaction solvent and reaction products. In addition, water can be handled at ambient conditions safely and in a simple manner. Nevertheless, water has not been investigated as intensively for use in oil shale processes as have been liquid solvents and hydrogen.

7.4 OIL SANDS PROCESSING WITH HYDROTHERMAL AND SUPERCRITICAL WATER

Oil sands are a mixture of sand grains, water, and bitumen that dissolves in organic solvents and decreases in viscosity upon heating [61]. The bitumen is present in a porous sand matrix, with porosities of typically 25–35% (Athabasca deposits, Alberta, Canada). The bitumen and water content is generally about 17 wt%. The sand fraction of about 83 wt% consists predominantly of fine quartz grains about 75–250 μm in size, with small amounts of clay minerals attached to the grain surfaces. The sand grains are in direct contact with one another, and wet with a thin layer of water of a few micrometers thick. Bitumen fills the void volume among them and forms a continuous phase through the pores. In this type of oil sand, the bitumen is not in direct contact with the sand grains but is separated from them by the water layer. Other oil sand deposits, as, for example, in the United States, do not have the water layer around the sand grain. The bitumen adheres directly to the grains, which is important for selection of the extraction method [61]. Oil sand bitumen has a higher concentration of asphaltenes, resins, and heteroatomic species containing nitrogen, sulfur, and heavy metals than conventional crude oils [62].

Bitumen from oil sands can be recovered after the mining by physical separation of bitumen and sand in a hot water process. In processes such as the steam-assisted gravity drainage process and the hot water extraction process, the heated oil with steam condensate is obtained in the form of water-in-bitumen emulsions. After the lighter components from the bitumen are removed, the heavier components of the bitumen can be upgraded. Conventional processes for deasphalting of crude oils may be not applicable to treat the bitumen emulsion [62]. The advantage of using hydrothermal and supercritical water for bitumen recovery in oil sand processing is probably that both types of oil sands can be processed.

The description of those processes is beyond the scope of this book. More detailed information on oil sand processing is readily available in the literature, for example, on general information on synthetic fuel processes in Ref. [61] and specific properties and problems with the water-based recovery of bitumen from oil sands in Ref. [63]. In the following, some results of investigations on bitumen recovery from oil sands are discussed.

7.4.1 Extraction of Bitumen from Oil Sands with Sub- and Supercritical Water

Oil sands are treated mostly in a hot water process. The effluents of such processes seem to be not environmentally friendly. Processing with hydrothermal and supercritical water bears the possibility to keep the process contained and products can be released in a more controlled way. The extract can be processed for fuel applications and the residue cleaned from all of the hydrocarbons from the mineral matter, as has been shown in Chapter 6 for the cleaning of soil material. Mineral matter can be removed from the extract, and emulsions can be effectively split, applying supercritical fluid technology.

Oil sands can be extracted with sub- and supercritical water to more than 80% of the initial bitumen content [64,65]. Conditions vary from about $T=390$ to 410 °C at pressures of $P=14$–30 MPa. For Northern Alberta's Athabasca oil sand samples, only little toluene-insoluble preasphaltic and heavier material was generated. The extracts showed significant composition shifts to lower molecular weight [64].

Oil sand samples (Fort McMurray, "high grade") containing 12–18 wt% bitumen were extracted in a semi-continuous reactor [64]. Most of the bitumen of the oil sand samples is soluble in toluene. Only a minor fraction of $x<0.5$ wt % consists of preasphaltenes. The extraction of an oil sand sample containing about 13 wt% bitumen at temperature of $T=400$ °C and a pressure of $P=21$ MPa, at a solvent ratio of water to oil sand of about 6 $dm^3\ kg^{-1}\ h^{-1}$ and a residence time of about $t=3$ min (deduced from the flow rate of 3 $ml\ min^{-1}$ and the sample size of 30 g), results in a total yield of 83.4% extract and 4.0% coke. The extraction with toluene yields 88.9% extract at $T=350$ °C, $P=21$ MPa [64]. Higher pressures (with water) increase the amount of coke without affecting the yield of the extraction. Some asphaltenes are transformed to lower molecular weight polar and aromatic hydrocarbons.

Extraction with water is considered less effective than with toluene, regarding yield only. However, for a production process, the oil sand residue has to be considered also. Residue containing water may be handled somewhat easier than one containing toluene.

A more recent investigation considered the extraction of oil sand from Inner Mongolia (Tumuji) [65]. The bitumen content of this oil sand sample was $x_B=13.55$ wt%, with 55.8 wt% saturates, 16.7% aromatics, 12.9% resin, 7% asphaltenes, and 7.7% preasphaltenes. The extract yield at a final temperature of $T=500$ °C, approached with a heating rate of 4 $K\ min^{-1}$, is 73.8 wt% daf at $P=20$ MPa, 76.2 wt% daf at $P=25$ MPa, and 81.1 wt% daf at $P=30$ MPa.

The hydrothermal pyrolysis of oil sand occurs in two steps. In the first step, at around $T=350$ °C, relatively light organic substances are removed.

In the second step, at temperatures of $T = 350$–$500\ °C$, the heavy organic substances begin to crack and the products are removed. The course of the formation of the extract and the different fractions is illustrated in Figure 7.18 for a pressure of $P = 30$ MPa.

The conversion and extraction of oil sand bitumen components begins at a temperature of about $T = 300\ °C$, steeply increases from $T = 350$ to $\approx 390\ °C$, at which temperature the maximum extraction rate occurs. At higher temperatures, the conversion rate decreases and drops to zero just below $T = 500\ °C$. The influence of pressure is shown in Figure 7.19. The maximum conversion and extraction rate is pressure dependent. Increasing pressure from $P = 20$ to

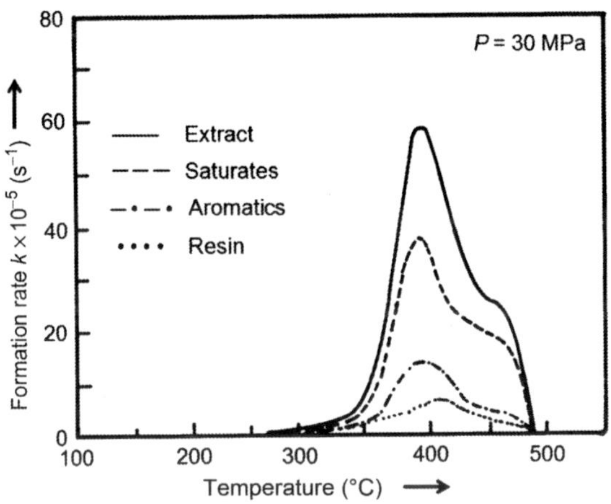

FIGURE 7.18 Extraction of oil sand with water at $P = 30$ MPa. Formation rate of products. *Modified from Ref. [65].*

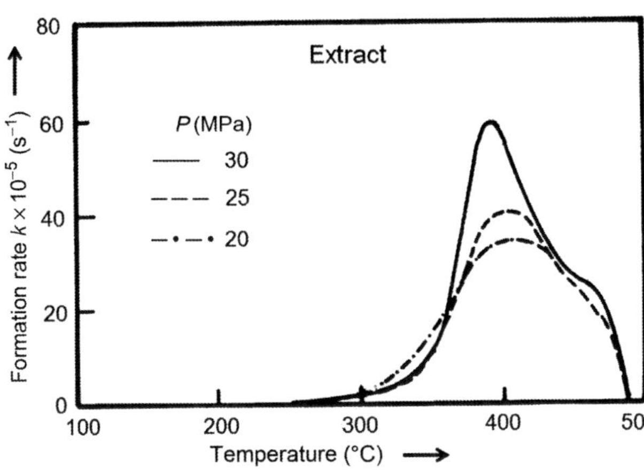

FIGURE 7.19 Extraction of oil sand with water. Formation rate in dependence on pressure [65].

30 MPa, enhances the extraction rate from $33 \times 10^{-5}\ s^{-1}$ at $T=412\ °C$ to $58 \times 10^{-5}\ s^{-1}$ at $T=390\ °C$, see Figure 7.19. Pressure also affects the extract composition. More of saturated compounds are in the extract at the lower extraction pressure of $P=20$ MPa than that at the higher extraction pressures of $P=25$–30 MPa. The hydrogen-rich components are extracted during extraction of oil sand with water. Extracts have a higher H/C ratio than the original oil sand sample [65].

The formation rates for the saturated extract components, for the aromatic extract components, and for the resin components are shown in Figure 7.20.

The formation of gaseous compounds is of minor importance and is not much influenced by pressure. Total gas yield is $4.85 \times 10^{-2}\ cm^3\ g^{-1}$ at a pressure of $P=20$ MPa, $4.48 \times 10^{-2}\ cm^3\ g^{-1}$ at a pressure of $P=25$ MPa, and $4.38 \times 10^{-2}\ cm^3\ g^{-1}$ at a pressure of $P=30$ MPa. Main gas components are CO_2 with about 40%, H_2 with about 18%, and CH_4 with about 25%.

From these experiments, it can be concluded that a high liquid extract yield of oil sand can be obtained by using sub- and supercritical water extraction.

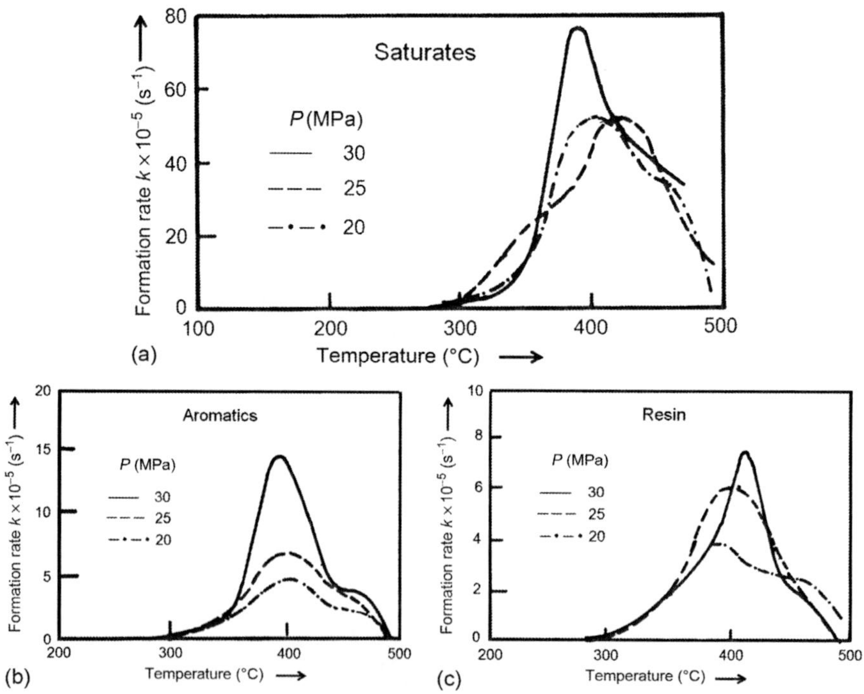

FIGURE 7.20 Extraction of oil sand with water. Formation rate of (a) saturates, (b) aromatics, and (c) resin in dependence on pressure. *Modified from Ref. [65].*

7.4.2 Upgrading of Bitumen with Hydrothermal and Supercritical Water

Hydrothermal and supercritical water can be used for upgrading the oil sand bitumen. At densities above the critical density of water, a substantial amount of the bitumen may be extracted. The extraction process is determined by the equilibrium phase behavior which can be estimated from the general behavior of hydrocarbons with water discussed in Chapter 3. According to the experimental data available, a phase separation into at least a water-rich phase and an oil-rich phase is probable for processes with supercritical water. In addition, water reacts with the hydrocarbons as discussed in Chapter 5. According to the result of numerous investigations, the reaction mechanisms for aromatic and aliphatic hydrocarbons (without heteroatoms) are essentially the same during thermal pyrolysis and in high-temperature liquid water. Water serves primarily as an inert solvent.

Reaction of oil sand bitumen leads to reported conversion yields in the range of 20% but with ill-defined conversion. It can be concluded that some reaction of water with the bitumen components takes place, although most of the upgrading effect seems to be physical due to the extraction of lighter and aromatic compounds. Sulfur and asphaltene content of the products from the water phase are lower. The oil-rich phase on the other hand has a higher asphaltene content. Coke formation in supercritical water seems to be suppressed to some extent, but not total. Still about 10% coke is formed at supercritical water conditions and reaction times of up to $t=2$ h. The results stem from batch reactor experiments that cannot be taken as a basis for evaluating an upgrading process with supercritical water. Nevertheless, it is of advantage that with supercritical water, oil sand bitumen containing a substantial amount of solids can be processed [66,67].

Results of an extraction of Athabasca oil sand bitumen with sub- and supercritical water in a batch reactor of 43 cm^3 at $T=360$–380 °C, $P=15$–30 MPa, $\rho_{H_2O}=0.07-0.65 g/cm^3$, and a reaction time of up to $t=2$ h showed a maximum conversion of 24% after $t=90$ min extraction at supercritical conditions. About 18% of initial sulfur in bitumen can be removed at maximum conversion condition. The asphaltene content of the residue is significantly higher than that of the original bitumen due to preferential extraction of aromatic compounds in supercritical condition [62].

REFERENCES

[1] P.E. Savage, S. Gopalan, T.I. Mizan, Ch.J. Martino, E.E. Brock, Reactions at supercritical conditions: applications and fundamentals, AIChE J. 41 (1995) 1723–1778.

[2] J.R. Kershaw, Supercritical fluids in coal processing, J. Supercrit. Fluids 2 (1989) 35–45.

[3] G.V. Deshpande, G.D. Holder, A.A. Bishop, J. Gopal, I. Wender, Extraction of coal using supercritical water, Fuel 63 (1984) 956–960.

[4] J.R. Kershaw, Proceedings of the International Conference on Coal Science, 1985, p. 242.
[5] J.R. Kershaw, L.J. Bagnall, Extraction of Australian coals with supercritical aqueous solvents, in: Th.G. Squires, M.E. Paulaitis (Eds.), Supercritical Fluids, Chemical and Engineering Principles and Applications, American Chemical Society Symposium Series 329, American Chemical Society, Washington, DC, 1987, pp. 266–280.
[6] J.R. Kershaw, L.J. Bagnall, Preprint Paper Am. Chem. Soc. Div. Fuel Chem. 30 (1985) 101.
[7] M.L. Swanson, E.S. Olson, J.W. Diehl, S.A. Farnum, Preprint Paper Am. Chem. Soc. Div. Fuel Chem. 31 (1986) 43.
[8] W.P. Scarrah, W.P. Scarrah, in: M.E. Paulaitis, J.M.L. Penninger, R.D. Gray, P. Davidson (Eds.), Chemical Engineering at Supercritical Fluid Conditions, Ann Arbor Science, Ann Arbor, MI, 1983, pp. 395–407.
[9] P.R. Bienkowski, R. Naruyan, R.A. Greenkorn, K.C. Choa, Enhanced coal liquefaction with steam pretreatment, Ind. Eng. Chem. Res. 26 (1987) 202–205.
[10] J.R. Kershaw, Extraction of Victorian brown coals with supercritical water, Fuel Process. Technol. 13 (1986) 111–124.
[11] B. Slomka, T. Aida, T. Squires, Preprint Paper Am. Chem. Soc. Div. Fuel Chem. 30 (1985) 368.
[12] P.R. Bienkowski, R. Narayan, R.A. Greenkorn, K.C. Chao, Liquefaction of a subbituminous coal with steam and ammonia, Ind. Eng. Chem. Res. 26 (1987) 206–208.
[13] S.E. Towne, Y.T. Shah, G.D. Holder, G.V. Deshpande, D.C. Cronauer, Liquefaction of coal using supercritical fluid mixtures, Fuel 64 (1985) 883–889.
[14] L. Li, N.O. Egiebor, Oxygen removal from coal during supercritical water and toluene extraction, Energy Fuels 6 (1992) 35–40.
[15] D.S. Ross, Coal conversion in carbon monoxide-water systems, Coal Sci. 3 (1984) 301–337.
[16] L.A. Amestica, E.E. Wolf, Catalytic liquefaction of coal with supercritical water-CO-solvent media, Fuel 65 (1986) 1226–1232.
[17] J.M.L. Penninger, Extraction of oil from Wyoming coal with aqueous solvents at elevated pressure, in: J.M.L. Penninger, M. Radosz, M.A. McHugh, V.J. Krukonis (Eds.), Supercritical Fluid Technology, Elsevier, Amsterdam, 1985.
[18] J.M.L. Penninger, Selectivity effects in aqueous supercritical fluid extraction of subbituminous coal, Fuel 68 (1989) 983–989.
[19] O. Cazalla, E. Sebastian, G. Cultrone, M. Nechar, M.G. Bagur, T. Adschiri, T. Sato, H. Shibuichi, Z. Fang, S. Okazaki, K. Arai, Extraction of Taiheiyo coal with supercritical water-HCOOH mixture, Fuel 79 (2000) 243–248.
[20] A.A. Vostrikov, S.A. Psarov, D.Y. Dubov, O.N. Fedyaeva, M.Y. Sokol, Kinetics of coal conversion in supercritical water, Energy Fuels 21 (2007) 2840–2845.
[21] A.A. Vostrikov, O.N. Fedyaeva, S.A. Psarov, D.Yu. Dubov, M.Y. Sokol, Brown coal conversion under the action of supercritical water, Solid Fuel Chem. 41 (2007) 280–289.
[22] A.A. Vostrikov, O.N. Fedyaeva, D.Y. Dubov, S.A. Psarov, M.Y. Sokol, Conversion of brown coal in supercritical water without and with addition of oxygen at continuous supply of coal–water slurry, Energy 36 (2011) 1948–1955.
[23] O.N. Fedyaeva, A.A. Vostrikov, A.V. Shishkin, M.Ya. Sokol, N.I. Fedorova, V.A. Kashirtsev, Hydrothermolysis of brown coal in cyclic pressurization-depressurization mode, J. Supercrit. Fluids 62 (2012) 155–164.
[24] M.L. Swanson, E.S. Olson, J.W. Diehl, S.A. Farnum, Extraction of Low-Rank Coals with Supercritical Water, University of North Dakota Energy Research Center, University Station, Grand Forks, NO, 1986.
[25] M. Modell, R.C. Reid, S.I. Amin, Gasification process, US Patent No. 4,113,446, September 12, 1978.

[26] J. Wang, T. Takarada, Supercritical Water Gasification of Coal Using $Ca(OH)_2$ as an Additive, Department of Biological and Chemical Engineering, Gunma University, Kiryu, Japan.
[27] M. Nywlt, S. Peter, Improvement of coal gasification, Sci. Technol. 45 (1992) 349.
[28] Y. Li, L. Guo, X. Zhang, H. Jin, Y. Lu, Hydrogen production from coal gasification in supercritical water with a continuous flowing system, Int. J. Hydrogen Energy 35 (2010) 3036–3045.
[29] R. Zhang, W. Jiang, L. Cheng, B. Sun, D. Sun, J. Bi, Hydrogen production from lignite via supercritical water in flow-type reactor, Int. J. Hydrogen Energy 35 (2010) 11810–11815.
[30] M. Koel, S. Ljovin, K. Hollis, J. Rubin, Using neoteric solvents in oil shale studies, Pure Appl. Chem. 73 (2001) 153–159.
[31] P. Missal, Extraktion eines Colorado-Ölschiefers mit Wasser in unter- und überkritischer Phase, Ph.D. Dissertation, University of Karlsruhe, Fortschritts-Berichte VDI Series 3, No. 164, VDI-Verlag, Düsseldorf, Germany, 1988.
[32] P. Missal, K. Hedden, Extraction of a Colorado oil shale by water in the sub-and supercritical phases, Erdöl Erdgas Kohle 42 (1989) 346–352.
[33] R.L. Braun, A.J. Rothmann, Oil shale pyrolysis: kinetics and mechanism of oil production, Fuel 54 (1975) 129–131.
[34] J.F. McKay, S.-L. Chong, G.W. Gardner, Recovery of organic matter from Green River oil shale at temperatures of 400 °C and below, Liquid Fuels Technol. 1 (1983) 259–287.
[35] J.F. McKay, S.-L. Chong, Characterization of organic matter recovered from Green River oil shale at temperatures of 400 °C and below, Liquid Fuels Technol. 1 (1983) 289–324.
[36] J.F. McKay, S.-L. Chong, Recovery of oil from eastern shales using supercritical extraction techniques, in: Proceedings of the Synthetic Fuels from Oil Shale and Tar Sands, Third Symposium, Chicago, USA, 1983, pp. 389–393.
[37] J.F. McKay, M.S. Blanche, A comparison of transport process allowing organic matter to be recovered from Green River shale at 400 °C, Fuel Sci. Technol. Int. 4 (1986) 19–43.
[38] J.F. McKay, M.S. Blanche, Effect of time and pressure on recovery of oil from Green River oil shale in extraction experiments conducted at 400 °C, Fuel Sci. Technol. Int. 4 (1986) 413–431.
[39] S.-L. Chong, J.F. McKay, Interaction of supercritical methanol and water with organic material in Green River oil shale, Fuel Sci. Technol. Int. 5 (1987) 513–541.
[40] R.T. Ellington, D.E. Baugh, Oil shale: a new process approach, Energy Process. 7 (1987) 208–211.
[41] K. Das, Solvent and supercritical fluid extraction of oil shale: a literature survey, Technical Note, U.S. Department of Energy, DOE/METC-89/4092 (DE 89011708), 1989.
[42] J.J. Cummins, W.E. Robinson, Thermal conversion of oil shale kerogen in the presence of carbon monoxide and water, Preprint Paper Am. Chem. Soc. Div. Fuel Chem. 21 (1976) 94–109.
[43] S.-L. Chong, J.J. Cummins, W.E. Robinson, Fractionation of soluble extracts obtained from kerogen thermal degradation with CO and H_2O, Preprint Paper Am. Chem. Soc. Div. Fuel Chem. 21 (1976) 265–277.
[44] J.J. Cummins, W.E. Robinson, Thermal conversion of oil shale kerogen using CO and water at elevated pressures, U.S. DOE Laramie Energy Research Center, LERC/RI-78/1, 1978, p. 19.
[45] J.J. Cummins, D.A. Sanchez, W.E. Robinson, Conversion of oil shale kerogen in CO-steam at low pressure: a preliminary report, Energy Commun. 6 (1980) 117–135.
[46] W.J. Parkinson, T.J. Merson, Cost analysis of the carbon monoxide-steam process, Los Alamos National Laboratory, LA-9463-MS, NTIS/DE82021834, 1982, p. 15.
[47] J.F. McKay, S.-L. Chong, J.F. McKay, S.-L. Chong, Supercritical fluid extraction of oil from eastern oil shales, 1983, in: Eastern Oil Shale Symposium Proceedings, Lexington, Kentucky, USA, 1983, pp. 235–239.

[48] O.M. Ogunsola, N. Berkowitz, Extraction of oil shales with sub- and near-critical water, Fuel Process. Technol. 45 (1995) 95–107.
[49] L. Luik, H. Luik, V. Palu, K. Kruusement, H. Tamvelius, Conversion of the Estonian fossil and renewable feedstocks in the medium of supercritical water, J. Anal. Appl. Pyrol. 85 (2009) 492–496.
[50] K. Kruusement, H. Luik, L. Luik, V. Palu, N. Vink, F. Vogel, Liquefaction and gasification of Estonian oil shales in the medium of hot compressed water, in: Oil Shale Symposium Tallinn, 2009.
[51] H. Hu, J. Zhang, S. Guo, G. Chen, Extraction of Huadian oil shale with water in sub- and supercritical states, Fuel 78 (1999) 645–651.
[52] S. Deng, Z. Wang, Q. Gu, F. Meng, J. Li, H. Wang, Extracting hydrocarbons from Huadian oil shale by sub-critical water, Fuel Process. Technol. 92 (2011) 1062–1067.
[53] K. El harfi, C. Bennouna, A. Mokhlisse, M. Ben chanâa, L. Lemée, J. Joffre, A. Amblès, Supercritical fluid extraction of Moroccan (Timahdit) oil shale with water, J. Anal. Appl. Pyrol. 50 (1999) 163–174.
[54] A. Abourriche, A. Adil, M. Oumam, H. Hannache, R. Pailler, R. Naslain, M. Birot, J.-P. Pillot, New pitches with very significant maturation degree obtained by supercritical extraction of Moroccan oil shales, J. Supercrit. Fluids 47 (2008) 195–199.
[55] R. Kramer, M. Levy, Extraction of oil shales under supercritical conditions, Fuel 68 (1989) 702–709.
[56] T. Funazukuri, K. Mizuta, N. Wakao, Oil extraction from Australian Condor oil shale with water and CO in the presence of Na_2CO_3, Fuel 67 (1988) 1510–1515.
[57] M. Canel, P. Missal, Extraction of solid fuels with sub- and supercritical water, Fuel 73 (1994) 1776–1780.
[58] J. Yanik, M. Yüksel, M. Sağlam, N. Olukçu, K. Bartle, B. Frere, Characterization of the oil fractions of shale oil obtained by pyrolysis and supercritical water extraction, Fuel 74 (1995) 46–50.
[59] J. Yanik, P. Secim, S. Karakaya, L. Tiikma, H. Luik, J. Krasulina, P. Raik, V. Palu, Low-temperature pyrolysis and co-pyrolysis of Goynuk oil shale and terebinth berries (Turkey) in an autoclave, Oil Shale 28 (2011) 469–486.
[60] A.A. Gregoli, U.M. Oko, F. Leder, Process for converting heavy crudes, tars, and bitumens to lighter products in the presence of brine at supercritical conditions, US Patent No. 4,818,370, April 4, 1989.
[61] R.R. Probstein, R.E. Hicks, Synthetic Fuels, McGraw-Hill Book Co., New York, NY, 1982.
[62] J.H. Park, S.H. Son, Extraction of bitumen with sub- and supercritical water, Korean J. Chem. Eng. 28 (2011) 455–460.
[63] J. Masliyah, Z. Zhou, Z. Xu, J. Czarnecki, H. Hamza, Understanding water-based bitumen extraction from Athabasca oil sands, Can. J. Chem. Eng. 82 (2004) 628–654.
[64] N. Berkowitz, J. Calderon, Extraction of oil sand bitumens with supercritical water, Fuel Process. Technol. 25 (1990) 33–44.
[65] M.H. Hu, Q. Zhang, M. Ding, Extraction of Tumuji oil sand with sub- and supercritical water, Energy Fuels 20 (2006) 1157–1160.
[66] M. Watanabe, S.-N. Kato, S. Ishizeki, H. Inomata, R.L. Smith Jr., Heavy oil upgrading in supercritical water: basic study, in: M. Goto, T. Adschiri, B. Han, Y.-W. Lee, M. Goto, T. Adschiri, B. Han, Y.-W. Lee (Eds.), Proceedings of the 8th International Symposium on Supercritical Fluids, Kyoto, Japan, November 5–8, 2006, 2006.
[67] M. Morimoto, Y. Sugimoto, Y. Saotome, S. Sato, T. Takanohashi, Effect of supercritical water on upgrading reaction of oil sand bitumen, J. Supercrit. Fluids 55 (2010) 223–231.

Chapter 8

Processing of Biomass with Hydrothermal and Supercritical Water

This chapter covers the processing of biomass and its compounds with high-temperature and supercritical water. Extraction with water is shortly addressed since it is inherently connected to processing of biomass. The main discussion regarding extraction can be found in Chapter 6. Treatment of biomass with high-temperature water and the initiated hydrolytic reactions comprise the major part of Chapter 8. The discussion on reactive transformations concentrates on biomass and its compounds. Fundamentals of reactions with high-temperature and supercritical water are treated in Chapter 6, with the main attention on nonbiomass compounds.

In Chapter 8, the conversion with high-temperature and supercritical water will be treated first for single components or fractions of biomass, such as sugars, starch, cellulose, hemicellulose, lignin, and proteins. Then, conversion of natural mixtures of these compounds will be discussed. The chapter is concluded with the discussion of biomass gasification and liquefaction.

8.1 INTRODUCTION

Processing of biomass is an extensive subject, even for the narrowed aspect of processing with high-temperature and supercritical water. In Chapter 8, these aspects are discussed relatively short but as comprehensive as possible. The reader may be aware of some overlapping of the chapters on extraction (Chapter 6), fuel material processing (Chapter 7), and oxidation with supercritical water (Chapter 10).

Probstein and Reid have given a useful definition for biomass [1]: "Biomass is any material that is directly or indirectly derived from plant life and that is renewable in time periods of less than about 100 years. ... Animal wastes are also biomass materials in that they are derived, either directly or via the food chain, from plants which have been consumed as food."

Biomass as feedstock consists of the polymers cellulose, hemicelluloses, starch, lignin, proteins, and many other compounds in varying concentrations

Supercritical Fluid Science and Technology, Vol. 5. http://dx.doi.org/10.1016/B978-0-444-59413-6.00008-X

like triacylglycerides, terpenes, and others. Major intermediate monomeric products of the depolymerization of biomass compounds are glucose, xylose, amino acids, fatty acids, 5-hydroxymethylfurfural (5-HMF), furfural, and phenolic compounds.

Biomass and its use will remain an important topic because biomass is a renewable resource. Beside the direct use for food purposes, alternate ways to exploit biomass are explored. Biomass is a complex mixture of diverse compounds, but in nearly all cases, a substantial amount of water is inherent to the material. Therefore, processing of biomass with hydrothermal and supercritical water can play an important role in the further exploitation of this resource. Biomass may be processed with high-temperature and supercritical water for different purposes, such as to produce food, derive useful compounds for chemical synthesis, to enable fermentation, or make biomass available for energy purposes. Further advantage is that biomass is a renewable resource with low-sulfur content and near-to-zero emission of CO_2 can be realized in the cycle of growing and conversion of biomass.

Most effort has been put into the conversion of simple biomass compounds like sugars and starch. Cellulosic materials have gained attention more recently due to the dual use of the simple biomass compounds for food and nonfood purposes. The mixture of cellulose, hemicellulose, and lignin as occurring in cellulosic materials has been extensively investigated, but so far has not been processed in a commercial large-scale plant, partly due to the challenges of separating this mixture efficiently. Real biomass feedstock contains additional compounds like proteins, and others [2]. Separation of proteins from the biomass mixture has not attracted much attention, but is important, in particular, for processing agricultural and animal wastes.

Biomass can be used in different ways and for producing various products. Therefore, the processing of biomass will be different. It can be assumed that it is the intention to make use of as many compounds as possible from biomass. This idea recently led to the vision of a biorefinery, in which the biomass feedstock is separated into all its valuable compounds. These compounds can be extracted, modified by hydrolysis or thermal pyrolysis and then separated, transformed by high-temperature treatment into fractions of gaseous, liquid, and solid compounds (liquefaction), or totally gasified to a directly used gaseous energy carrier, or further processed to synthesis gas or liquid products.

Conversion of biomass with high-temperature and supercritical water can be carried out as partial liquefaction, total gasification, oxidation with supercritical water, and even incineration. A possible scheme for biomass conversion is shown in Figure 8.1 in which a block diagram illustrates the various ways for utilizing biomass.

A multiunit processing scheme based on hydrothermal and supercritical water for the treatment of a biomass substrate is shown in Figure 8.2 [2]. The target substrate is pretreated with pressurized water to prepare it for extraction or reaction using high-temperature water, for example, by reducing

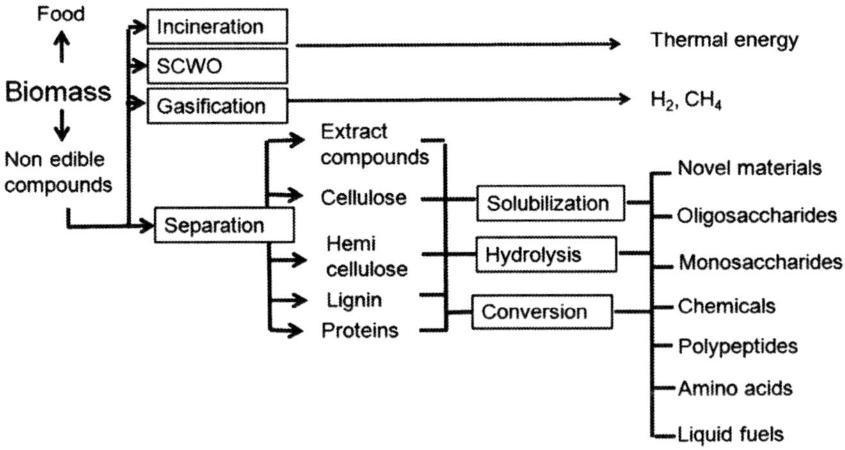

FIGURE 8.1 Ways for biomass processing. *Modified from Ref. [3].*

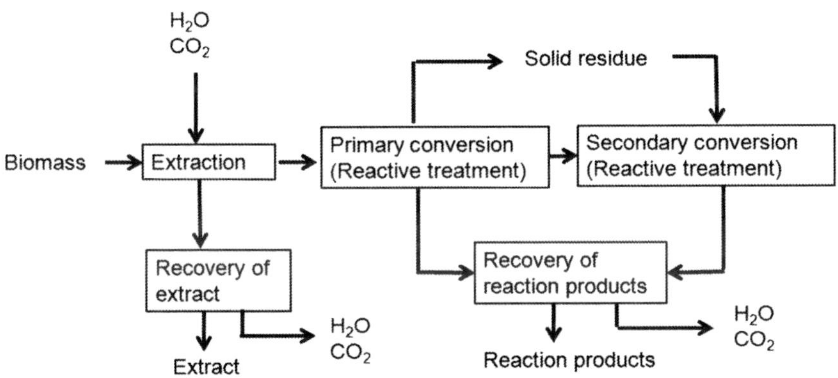

FIGURE 8.2 Generalized processing scheme of biomass. The different process steps can be carried out in one reactor subsequently.

the size or by swelling. Extracts are separated from the substrate prior to applying subcritical water as a reaction solvent which is then performed at higher temperatures to hydrolyze the biomass and further convert it to a lower molecular weight fraction [2]. A residual, mostly solid fraction can then be treated by supercritical water oxidation. The advantage of such a processing scheme is that it can be potentially conducted in one reactor of an integrated processing plant and therefore will save cost [2].

Increasingly it is accepted that for a sustainable society, the design philosophy of production processes need to be changed from only large-scale mass production systems to decentralized local and small- to medium-scale production systems, with the effect that products can be supplied from distributed renewable resources and biomass can be exploited economically.

High-temperature and supercritical water allows fast reaction rates, high selectivity, and high-yield conversions of much biomass-related feedstock and allows processing in compact devices. In combination with supercritical carbon dioxide that allows selective separations, a sustainable technology can be designed [3].

8.2 PROCESSING OF PLANT-BASED MONOMERS WITH HIGH-TEMPERATURE AND SUPERCRITICAL WATER

The majority of the polysaccharides found in plant cell walls belongs to the cellulose, hemicellulose, or pectin classes. Cellulose, the dominant structural polysaccharide of plant cell walls, is a linear β-(1 → 4)-D-glucopyranoside polymer. Cellulose functions as the rigid, load-bearing component of the cell wall. The rigidity of the cellulose microfibril is strengthened within a matrix of hemicelluloses and pectins. Pectins are noncellulosic acidic cell wall polysaccharides with multiple purposes [4].

Plant biomass consists of a number of compounds that can be processed by high-temperature and supercritical water, like lipids, sugars, starch, cellulose, hemicellulose, lignin, proteins, and others more. Starch, pectin, cellulose, and hemicellulose are sugar-polymers and can be transformed to sugar monomers. From starch, monosugars are produced straightforward by microorganisms or by hydrolysis and can be further transformed to ethanol, which is not only of importance for bioethanol production but also of some concern because of the competitive situation to food. Cellulose and hemicellulose, being available from agricultural sources without competition to food, can be transformed to monosugars. Most of these resources also contain considerable amounts of lignin. So far, scientific and technological development has not led to a commercial process of such materials. Nevertheless, successful and economic processing of lignocellulosic material will be the future for bioethanol production and for biomass utilization [5].

Conversion of monosaccharides, hemicellulose, cellulose, lignin, and proteins are discussed in the following. After this discussion, mixtures of the different compounds, as they occur in nature, are treated. Conversion to intermediate compounds is the topic of the transformations in high-temperature and supercritical water, followed by the discussion of transformations aiming at liquid products that can be used for fuels, and by gasification of biomass to produce gas with a high heating value, hydrogen, or synthesis gas.

8.2.1 Conversion of Sugars to Liquid Products

Glucose, fructose, and xylose are the main sugars in biomass. D-Glucose dissolved in water exists as an open chain, or a pyranose ring, or a furanose ring. In the same way, D-fructose dissolved in water can exist as an open chain, or a pyranose ring, or a furanose ring. Consequently, glucose or fructose are present in at least six forms of monosaccharide in water and follow the same general reaction pathways in high-temperature water. Glucose reversibly isomerizes into fructose via the LBAE (Lobry de Bruyn–Alberda van Ekenstein) transformation [4]. The rate of isomerization is slow compared to the rates of degradation of glucose and fructose. For example, during the conversion of glucose, the amount of fructose is quite small compared to the amounts of other degradation products, and vice versa [6]. D-Xylose, a monosaccharide found in hemicelluloses, is important for bioethanol production and as a source for chemicals, such as furfural and organic acids.

8.2.1.1 Glucose Conversion

The hydrolysis of glucose is a matter of interest since more than a century within a range of temperature $T=150$–$300\ °C$ [7–10]. The hydrolysis of glucose in high-temperature and supercritical water was examined by Modell [11,12] and later at Tohoku University [13,14]. During hydrolysis of glucose in high-temperature water, gaseous products can be neglected up to a temperature of $T=300\ °C$, at a pressure of $P=40$ MPa, and a residence time of up to $t=5$ min. The main hydrolysis products are 1,6-anhydroglucose, fructose, erythrose, glycolaldehyde, pyruvaldehyde, and dihydroxyacetone. All products are soluble in water. Hydrolytic reactions of glucose at temperatures $T=300$–$400\ °C$, pressures $P=25$–40 MPa, and residence times $t=0.02$–2 s reveal a number of parallel and secondary reactions. Eventually, the hydrolysis products react to acids, aldehydes, and alcohols. A possible reaction scheme is shown in Figure 8.3 [15–17].

Pressure has no major influence on the rate of conversion of glucose in the subcritical region. In the supercritical region, the rate of conversion of glucose decreases with increasing pressure. At temperatures higher than $T=300\ °C$, gaseous products are formed as discussed also in Section 8.6.

Some investigations carried out on the hydrolysis of glucose in high-temperature water provide detailed information. Results of experiments carried out in a continuously operated tubular reactor in the temperature range

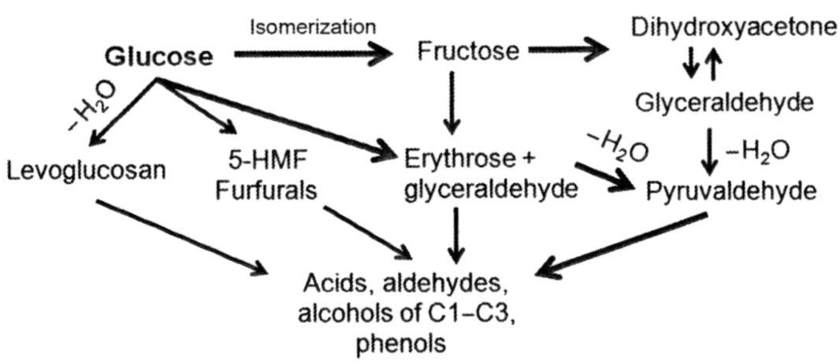

FIGURE 8.3 Reaction paths for the decomposition of glucose under hydrothermal conditions. *Modified from Refs. [15–17].*

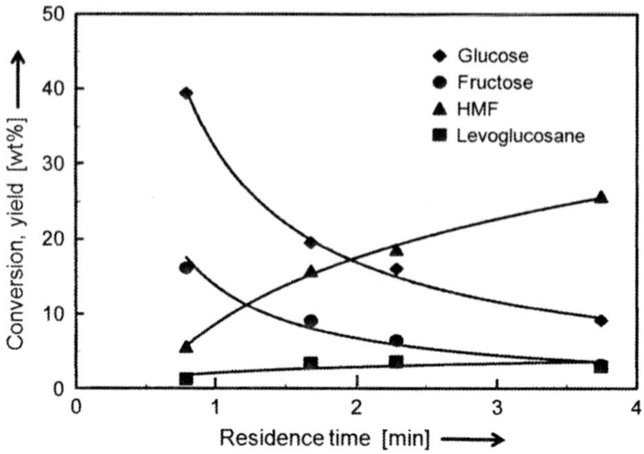

FIGURE 8.4 Conversion of glucose in high-temperature water. $T=250\ °C$, $P=24$ MPa. Initial glucose concentration $x_{G,0}=0.264$ wt% [18].

of $T=140$–$270\ °C$, at pressures of $P=6$–24 MPa, with glucose concentrations of $x_G=0.2$–1 wt%, and residence times $t=0.8$–4 min are shown in Figure 8.4 for the conversion of glucose and the formation of some main degradation products [18–20].

Glucose is hydrolyzed and fructose, 5-HMF, and levoglucosan are formed. Fructose, formed first from glucose, is rapidly hydrolyzed due to its low-temperature stability. Yields are calculated as fractions of the stoichiometrically maximal quantity, for fructose:

$$Y_F = \frac{\frac{C_F}{180}}{\frac{C_{G,0}}{180}} \times 100$$

with Y_F, yield of fructose; C_F, quantity of fructose; $C_{G,0}$, initial quantity of glucose.

Temperature is of major influence on the hydrolysis of glucose as shown in Figure 8.5. Pressure is of negligible influence, as mentioned above, and acidification by dissolving CO_2 in water is only of minor consequence, quite different to the findings for the hydrolysis of starch (see section 8.3.1) [18].

The conversion reaction of glucose is assumed to be of first order (Equation 8.1), with the temperature dependence represented by an Arrhenius equation (Equation 8.2):

$$x_G = x_{G,0} \cdot \exp(-k_G \cdot t) \tag{8.1}$$

$$k_G = k_{0,G} \cdot \exp\left(-\frac{E_{a,G}}{RT}\right) \tag{8.2}$$

with x_G, $x_{G,0}$, concentration of compound A; k_G, rate constant; $k_{0,G}$, preexponential factor; t, time; $E_{a,G}$, activation energy; R, gas constant; T, absolute temperature.

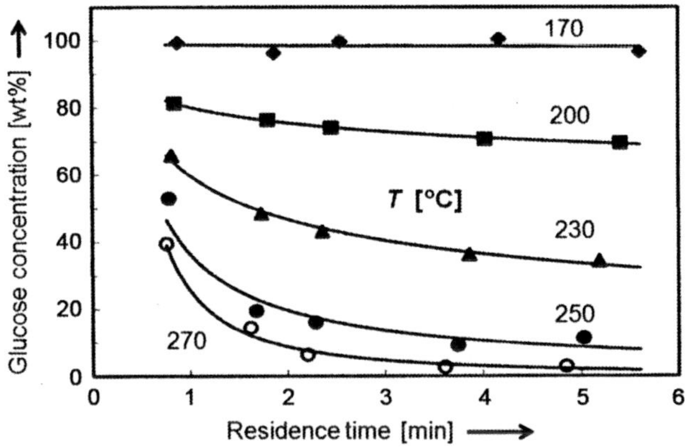

FIGURE 8.5 Hydrolysis of glucose in high-temperature water. Influence of temperature. $P=24$ MPa, initial glucose concentration $x_{G,0}=0.264$ wt% [18].

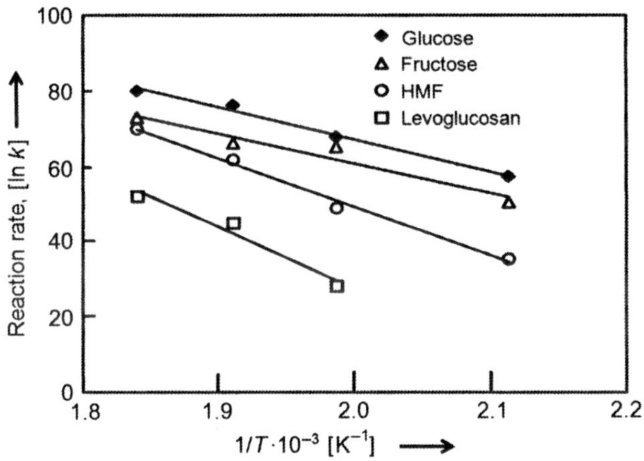

FIGURE 8.6 Reaction rates of the hydrolysis of glucose in high-temperature water and the formation of products [18].

The temperature dependence of the rate constant for the hydrolysis of glucose and the formation of the products is shown in Figure 8.6. The kinetic parameters of the Arrhenius equation are listed in Table 8.1.

A first-order reaction for glucose degradation is the simplest way to represent reaction behavior of glucose degradation. More detailed models include the formation and degradation of intermediate products. Such models can be found in Refs. [16,17,21,22].

According to results published, the conversion of glucose in high-temperature water already begins at temperatures below $T=160$ °C. At a temperature of $T=200$ °C about 20% and at $T=240$ °C about 50% of glucose is converted, as shown in Figure 8.7 [15]. Up to feed concentrations of $x_G=6$ wt%

TABLE 8.1 Hydrolysis of Glucose in High-Temperature Water and Formation of Products

Reaction	Activation Energy ($E_{G,A}$) (J mol^{-1})	Preexponential Factor ($k_{0,G,A}$) (s^{-1})
Glucose conversion	7.249×10^4	1.830×10^5
Fructose formation	6.683×10^4	2.493×10^4
HMF formation	1.060×10^5	1.131×10^8
Levoglucosan formation	1.920×10^5	8.450×10^{15}
Formation of unknown compounds	4.137×10^4	1.830×10^1

Parameters of the Arrhenius equation [18].

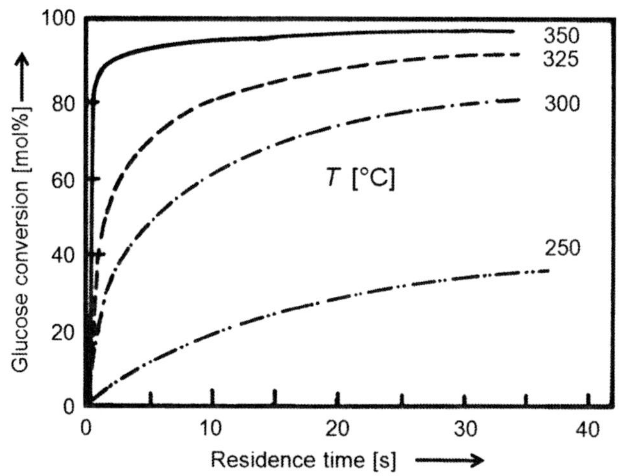

FIGURE 8.7 Temperature dependence of glucose conversion in subcritical water at temperatures of $T=250$–350 °C, $P=24$ MPa. Feed concentration $x_G=0.3$ wt% glucose. *Modified from Ref. [16].*

glucose in water, the concentration is not of significant influence. At temperatures beyond $T=300$ °C, glucose destruction proceeds fast. 55% of the initial amount of glucose is converted at $T=300$ °C after $t=2$ s and 90% at $T=350$ °C after $t=1$ s [21]. According to another investigation, at $T=350$ °C, a conversion of more than 90% is reached in $t=6$ s residence time (Figure 8.7) and glucose conversion is nearly complete in $t=4$ s at the supercritical temperature of $T=420$ °C [16]. Generally, most investigations come to very similar results. Nevertheless, experimental details, as for example, the heating and cooling rate, flow regimes, analytical procedures, and typical experimental errors lead to somewhat differing results.

Figure 8.8 shows the glucose degradation rate as a function of temperature from a number of studies.

Decomposition products were mainly obtained via keto-enol tautomerism to fructose, via dehydration to 1,6-anhydro-β-D-glucose and 5-HMF, and organic acids (lactic acid, formic acid, acetic acid, and glycolic acid) which further decomposed to aldehydes via retro-aldol condensation [15]. The product distribution is shown in Figure 8.9 in dependence of temperature [15].

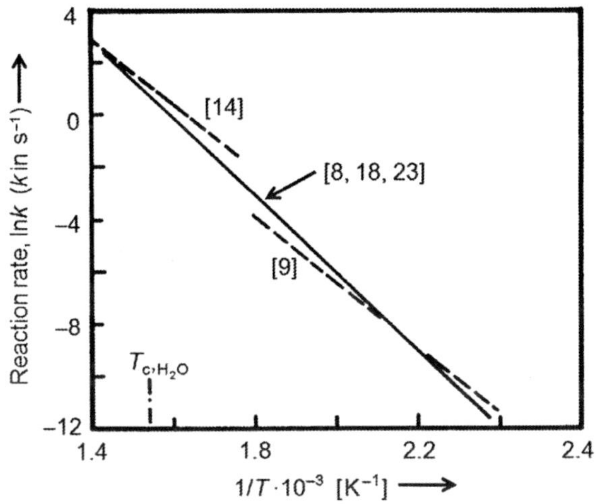

FIGURE 8.8 Glucose degradation according to an assumed first-order reaction. *Data from Refs. [8,9,14,18,23]. Modified from Ref. [4].*

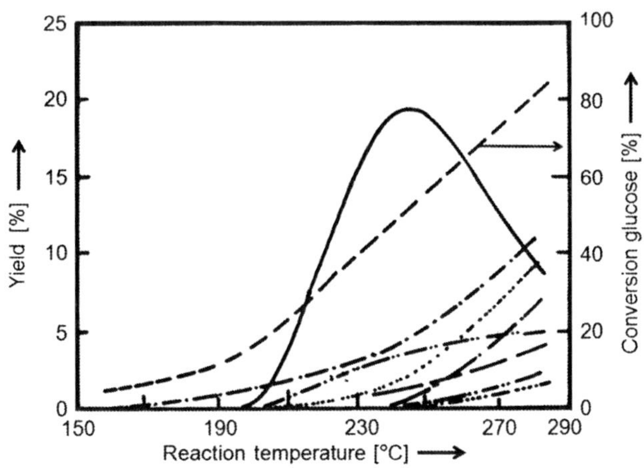

FIGURE 8.9 Conversion products of glucose with subcritical water in dependence on temperature. Residence time $t=60$ s, $P=10$–25 MPa. Conversion: ———, glucose; Yield: ———, fructose; —•—•, 5-HMF; —••—, 1,6 anhydroglucose; ••••, formic acid; ——•——•, 2-furfural; —— ——, glycolic acid; ——••——, lactic acid; —••••—, acetic acid. *Modified from Ref. [15].*

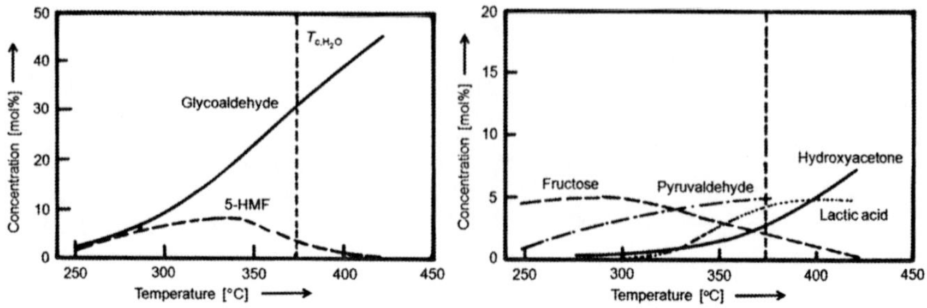

FIGURE 8.10 Conversion products of glucose reaction in high-temperature and supercritical water. Residence time $t=5$–8 s, $P=24$ MPa for $T<374$ °C, $P=34$ MPa for $T>374$°C, feed concentration $x_G=0.3$ wt% glucose. *Modified from Ref. [16].*

In general, with increasing residence time and temperature, higher molecular products (e.g., fructose) are decomposed to smaller molecules. Product distribution for temperatures from $T=250$ to 420 °C is shown in Figure 8.10. At $T=300$ °C, the main products are glycolaldehyde and 5-HMF (Figure 8.10, left). With increasing residence time, more 5-HMF is formed, whereas fructose, mannose, 1,6-anhydroglucose, and pyruvaldehyde yields decrease [16].

Glycolaldehyde concentration increases substantially with temperature (Figure 8.10), while 5-HMF yield first increases with temperature but decreases above $T=350$ °C. Fructose according to these experiments begins to decrease around $T=300$ °C, somewhat higher in temperature than found in another investigation [15]. Pyruvaldehyde is only formed at subcritical temperatures and is not observed at $T=420$ °C. Hydroxyacetone and glycolic acid yields (not shown) both increase notably at supercritical temperature. A reaction network is proposed but is considered out of scope for this book. The proposed reaction scheme is discussed in some detail in the literature [16].

At subcritical temperatures, neither increasing the reaction pressure nor increasing the reactant concentration has any significant influence on conversions and yields [16]. Feed concentration for these results was relatively low, too low for industrial applications. An increase in feed concentration changes the relative concentration of the products. In particular, formation of brown humic polymers will increase at higher feed concentrations [16].

Adding oxygen (in stoichiometric quantity) to the reaction and increasing the feed concentration of glucose has no effect on conversion, but leads to higher concentrations of glycolaldehyde, glycolic acid, and CO_2 and oxidation of intermediately formed CO. Addition of small amounts of zinc sulfate accelerates the conversion of glucose and catalyzes the formation of lactic acid.

The influence of the heating rate on glucose conversion and product distribution is considerable. When the glucose solution is slowly heated at a rate of $\Delta T=0.18$ K s^{-1} to a temperature of $T=350$ °C at saturation pressures,

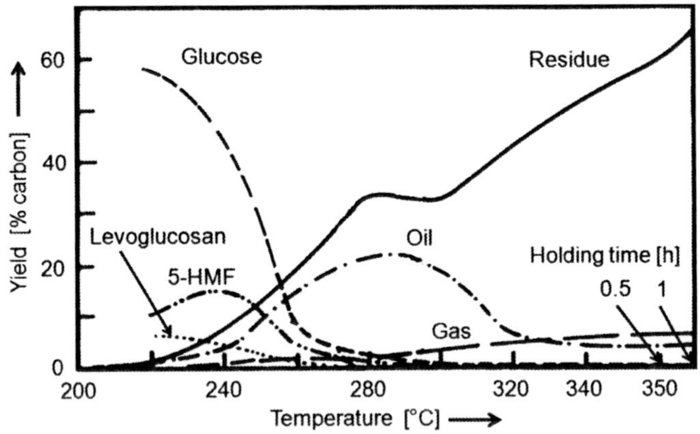

FIGURE 8.11 Product yields (carbon%) for decomposition of glucose in high-temperature water at saturation conditions. Initial glucose concentration $x_{G,0}=0.9$ M, low heating rate: $\Delta T=0.18$ K s^{-1}. Time to reach $T=350$ °C: $t=30$ min. *Modified from Ref. [17].*

glucose decomposition starts at $T=220$ °C and is finished at $T=280$ °C (Figure 8.11) [17]. The initial decomposition products 5-HMF and levoglucosan are converted into yellow and orange materials, including a solid residue, and finally to nanoparticles at 65 wt% yield at $T=350$ °C [17]. Gaseous products form to about 5% (carbon), oil is formed to less than 5%, while the solid residue increases linearly with temperature to about 57% (carbon) and increases further to more than 65% if the temperature is kept at $T=350$ °C for $t=1$ h. Rapid heating of the glucose solution to $T=410$ °C at a rate of $\Delta T=9.5$–17 K s^{-1} yields brownish sugar-like materials and a homogeneous phase with yellow color [17].

8.2.1.2 Fructose Conversion

Fructose, as glucose, reacts in high-temperature and supercritical water to several products. For the hydrolysis of fructose to 5-(hydroxymethyl)-furfural (HMF), Antal and coworkers [6,24] found a sequence of reactions commencing with the fructofuranose ring and retaining it intact. The fructofuranosyl-cation intermediate, necessary in this hypothesis, is produced directly by the hydrolysis of sucrose and reacts to produce HMF in high yields [6].

Glucose isomerization to fructose is important in high-temperature water, while the reverse reaction of fructose to glucose is not [14]. Despite the isomerization, fructose reacts much faster than glucose, at least in the presence of phosphoric acid. After 2 min at $T=340$ °C, 98% of fructose is reacted, but only 52% of glucose [25]. At near-critical temperatures (and high pressures), fructose is converted in short-reaction times below $t=1$ s, as shown in Figure 8.12. At a temperature of $T=400$ °C and a pressure of

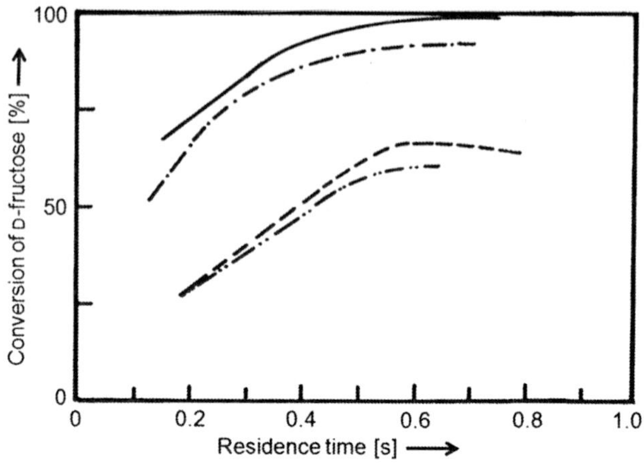

FIGURE 8.12 D-Fructose conversion in high-temperature and supercritical water. —••—, $T = 350$ °C, $P = 40$ MPa; ———, $T = 350$ °C, $P = 80$–100 MPa; —•—•, $T = 400$ °C, $P = 40$–80 MPa; ———, $T = 400$ °C, $P = 100$ MPa. *Modified from Ref. [26].*

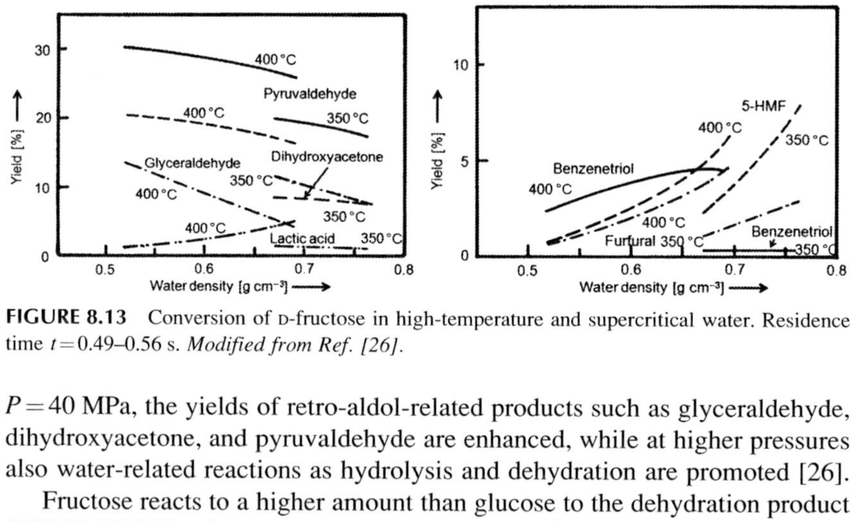

FIGURE 8.13 Conversion of D-fructose in high-temperature and supercritical water. Residence time $t = 0.49$–0.56 s. *Modified from Ref. [26].*

$P = 40$ MPa, the yields of retro-aldol-related products such as glyceraldehyde, dihydroxyacetone, and pyruvaldehyde are enhanced, while at higher pressures also water-related reactions as hydrolysis and dehydration are promoted [26].

Fructose reacts to a higher amount than glucose to the dehydration product 5-HMF. HMF has been proposed as a building-block chemical for bio-based products [27]. HMF is the starting material for biomass-derived dimethylfuran, proposed to replace gasoline [28]. Compared to ethanol, 2,5-dimethylfuran has a higher energy density (40%), a higher boiling point (20 K), and is not soluble in water [4].

The conversion of fructose is shown in Figure 8.12, and the yield of degradation products is shown in Figure 8.13. At $T = 400$ °C and $P = 100$ MPa, fructose is completely converted in $t < 1$ s. Increasing density, with increasing pressure from $P = 40$ to 100 MPa, shifts the converted amount of fructose from about 90% to about 100%.

At these conditions, the main degradation product is pyruvaldehyde, followed by dihydroxyacetone. An aromatic compound, 1,2,4-benzenetriol, is formed at supercritical temperature and high density [26]. This is of interest because it confirms previous observations during the conversion of lignocellulosic material. Usually, it is assumed that aromatic compounds originate from lignin [29,30].

The conversion of fructose in high-temperature water is significantly affected by the addition of small amounts of acids. 2 mM H_2SO_4 caused increased yields of 5-HMF and furfural and decreased yields of pyruvaldehyde and lactic acid, but had no measurable effect on the isomerization of fructose to glucose [6]. Phosphoric acid gives an optimal yield of 65% HMF at $T=240$ °C after $t=120$ s, with a feed of $x_F=0.05$ M fructose in a phosphoric acid solution with an initial pH = 2. Reaction rates decrease with increasing fructose concentration, while the solid residue increases [25].

8.2.1.3 Xylose Conversion

D-Xylose is a monosaccharide that is found in hemicelluloses and is considered to be one of the key compounds for bioethanol production. Future biomass industries may rely on chemicals like furfural and organic acids, derived from D-xylose.

Conversion of xylose in high-temperature and supercritical water begins at temperatures below $T=160$ °C, as results show from a continuously operated flow reactor for $t=60$ s residence time [15]. In the temperature range $T=160$–280 °C and pressures $P=10$–25 MPa, conversion of xylose is 20% at $T=200$ °C and 50% at $T=240$ °C, about the same values as for glucose conversion (see Figure 8.14) [15]. Up to 6 wt% glucose in water, the feed concentration is of not of significant influence. For 10 wt% feed concentration, the conversion of xylose is substantially higher in the temperature range up to $T=220$ °C than for the lower feed concentrations [15].

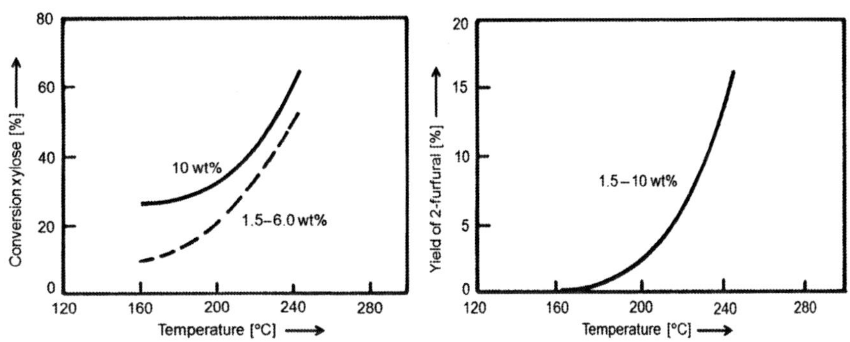

FIGURE 8.14 Conversion of xylose in high-temperature water. $P=10$–25 MPa, residence time $t=60$ s. Left: Conversion of xylose. Right: Yield of 2-furfural. *Modified from Ref. [15].*

Products of the decomposition of D-xylose are similar to those of D-glucose decomposition, that is, 2-furfural by dehydration (see Figure 8.14), some aldehydes, and organic acids produced by further decomposition [15].

Conversion of D-xylose is low up to temperatures of $T=200\ °C$ at short-reaction times of $t=1$ min. At the higher temperature of $T=240\ °C$, conversion of xylose and furfural concentration are substantially higher but still not within the target range for a process. Both, conversion and yield increase with extended residence time and by simultaneous extraction with supercritical CO_2. At $T=230\ °C$ and a CO_2 pressure of $P=12$ MPa, conversion approaches 97% of the initial xylose amount, and the yield in furfural is nearly 60% of the product, as shown in Figure 8.15. The enhancement is due to the extraction of furfural from the liquid reaction mixture and the catalytic effect of carbonic acid formed by dissolved CO_2 and water. The optimal initial concentration of xylose is between $x_X=4$ and 10 wt%, depending whether the main focus is on maximizing yield or furfural concentration [31].

Model 1 consists of a first-order reaction scheme, where part of the pentose reacts to a degradation product (DP1) and another part to furfural (Figure 8.15). Furfural further reacts to a second degradation product (DP2). According to this model, xylose degradation is much more pronounced than furfural degradation. Thus, the impact of furfural extraction is limited. [31].

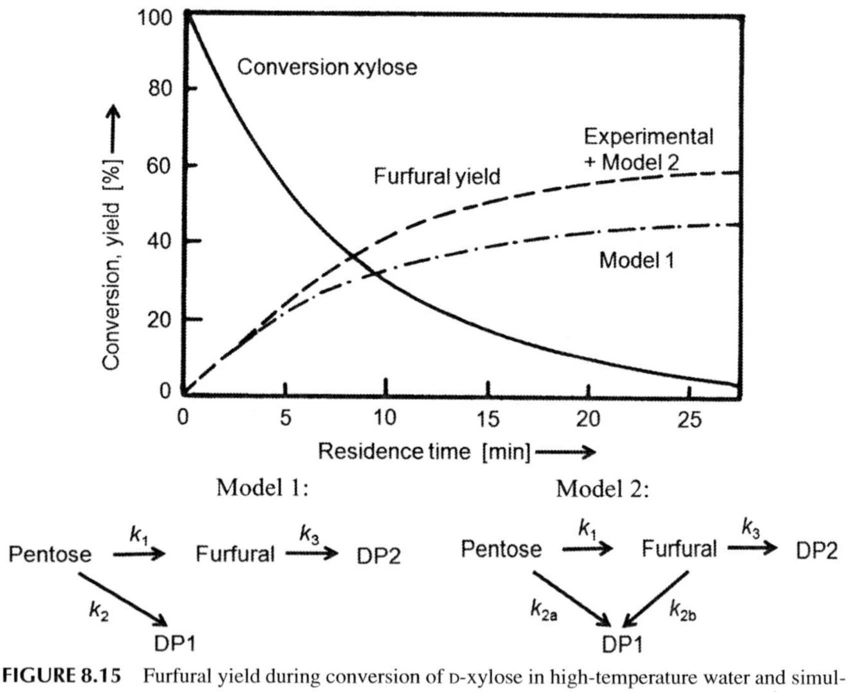

FIGURE 8.15 Furfural yield during conversion of D-xylose in high-temperature water and simultaneous extraction with CO_2. $T=230\ °C$, $P=12$ MPa, CO_2 solvent ratio$\approx$0.18 min^{-1}, initial xylose concentration $x_X=4$ wt%. Models 1 and 2, see text. *Modified from Ref. [31].*

Model 2 considers the reaction of furfural with the intermediates, and xylose degradation is a function of furfural concentration. Extraction of furfural maintains a low furfural concentration in the reactor and influences yield substantially (Figure 8.15) [31].

At higher temperatures of $T=350$ and 400 °C and pressures of $P=40$–100 MPa conversion rate of xylose is much higher. Results are based on experiments with an average concentration of *o*-xylose of $x_X=0.049$ M. Heating of the xylose solution occurred rapidly in a mixing tee. Details of the experimental set-up can be found in Chapter 13. Residence time for a nearly complete conversion is in the range of a few seconds [32]. The products obtained from the reaction of D-xylose were furfural, D-xylulose, glyceraldehyde, glycolaldehyde, dihydroxyacetone, pyruvaldehyde, lactic acid, and formaldehyde (see Figure 8.16).

Figure 8.17 shows the Arrhenius plot for xylose decomposition in high-temperature and supercritical water for pressures up to $P=100$ MPa. Results reasonably agree. The main difference with the investigation of Sasaki *et al.* [33] may be due to different experimental procedures. The influence of pressure on the rate constant seems to be not important. But a more detailed analysis showed evidence for dependence of the kinetic constants on water density and therefore on pressure in the range of $P=8$–100 MPa. Details can be found in Ref. [32].

In Figure 8.18, a possible reaction scheme is shown for the conversion of xylose. Experimental results showed evidence of a dehydration reaction pathway, a retro-aldol reaction pathway and an LBAE pathway from D-xylulose (see Figure 8.18). Details can be found in [34].

8.2.1.4 L-Arabinose Conversion

L-arabinose can be a significant fraction of hemicellulose and hemicellulose hydrolysates. Conversion of L-arabinose in high-temperature water and furfural yield is shown in Figure 8.19 for temperatures of $T=180$–260 °C [31].

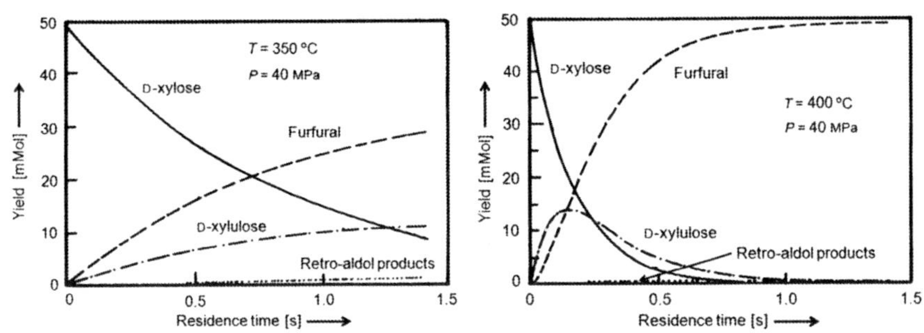

FIGURE 8.16 Conversion of D-xylose in high-temperature water. Left: $T=350$ °C, $P=40$ MPa, Right: $T=400$ °C, $P=40$ MPa. *Modified from Ref. [32].*

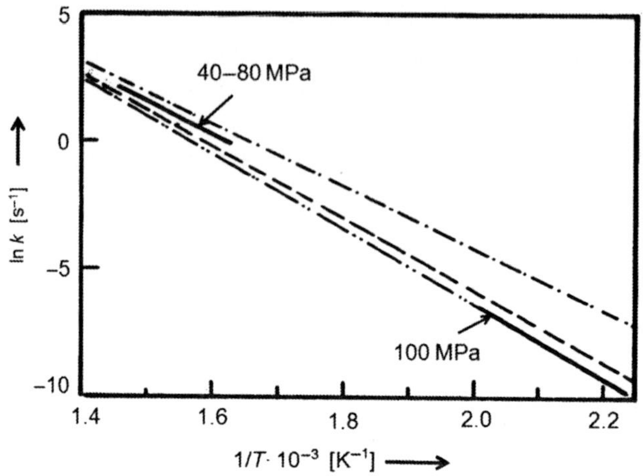

FIGURE 8.17 Arrhenius plot for the decomposition of D-xylose in high-temperature and supercritical water. ———, $P=40$–80 MPa, 100 MPa [11]. — — —, [12]; —•—•, [13]; —••—, [14]. *Modified from Ref. [32].*

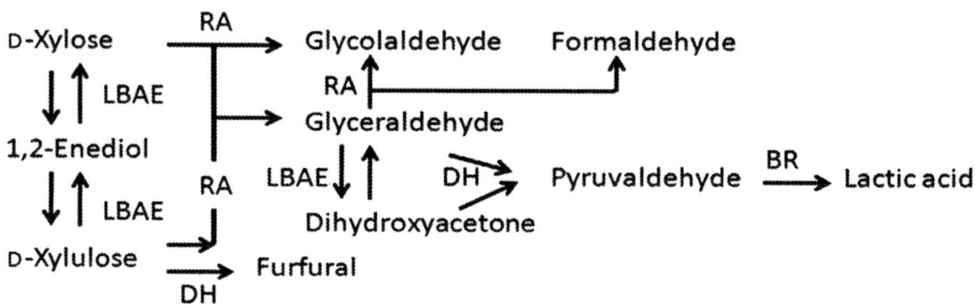

FIGURE 8.18 Reaction pathway of D-xylose decomposition in high-temperature and supercritical water. LBAE, Lobry de Bruyn–Alberda van Ekenstein transformation; RA, retro aldol; DH, dehydration; BR, bencilic acid rearrangement. *Modified from Ref. [32].*

A temperature of higher than $T=220\,^{\circ}\mathrm{C}$ is necessary to reach an almost complete disappearance of arabinose within less than $t=60$ min. The activation energy (about 120 kJ mol^{-1}) and the preexponential factor ($\ln k_1 \approx 25$ min^{-1}) are similar to the values obtained for D-xylose [35–37] and are close to the values obtained for L-arabinose degradation [36], but differ from kinetic parameters obtained by Usuki *et al.* [38].

Conversion of monomers like the ones treated above is of great interest for knowledge on the pathways of decomposition reactions. This knowledge indicates the major products and intermediates and on the reaction rate of specific transformations. It is therefore important for designing experiments on, but cannot replace the investigations for process parameters on real feedstock.

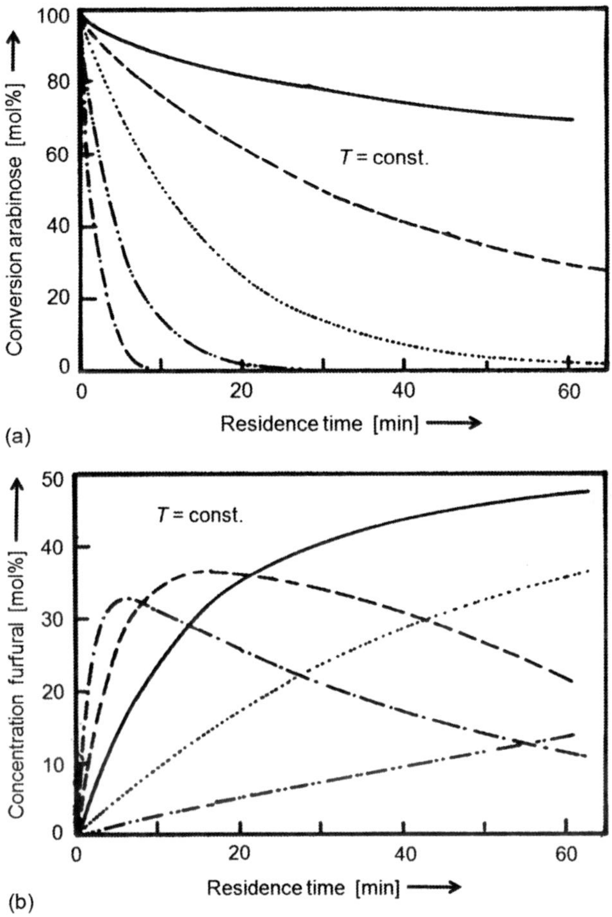

FIGURE 8.19 Left: Conversion of arabinose in high-temperature water. $P=3$ MPa. ———, $T=180$ °C; — — —, $T=200$ °C; ••••, $T=220$ °C; —••—, $T=240$ °C; —•—•, $T=260$ °C. (b) Concentration of furfural in high-temperature water (in presence of arabinose and arabinose degradation products). $P=3$ MPa. —••—, $T=180$ °C; ••••, $T=200$ °C; ———, $T=220$ °C; — — —, $T=240$ °C; —•—•, $T=260$ °C. *Modified from Ref. [31].*

8.3 PROCESSING OF NATURAL POLYMERS WITH HIGH-TEMPERATURE AND SUPERCRITICAL WATER

Conversion of monosaccharides like glucose, fructose, and xylose is important as an intermediate step in processing of lignocellulosic plant material to make use of biomass. The polymers of the monosaccharides, like starch, hemicelluloses, and cellulose are among the main components in plant materials, beside lignin and proteins. The conversion of these polymers in high-temperature and supercritical water is discussed in the following for each polymer before the discussion of natural mixtures of these polymers.

The unique properties of water at elevated temperatures and pressures, in particular, the variability of ionic product, dielectric constant, and density, render it a promising reaction medium. Biopolymers react with high-temperature water and supercritical water in very short residence times and with high rates of conversion. The achievable products are manifold and can be varied to a large extent by changing the operating conditions. Adding carbon dioxide to the reaction mixture at subcritical temperatures up to about $T = 250\ °C$ expands the possibilities of sub- and supercritical water significantly due to catalytic activity of the hydrolysis products of dissolved CO_2 [5].

8.3.1 Starch

Starch is an alpha-linked polysaccharide, composed of two components with different molecular weights: linear amylose (20–30%) and branched amylopectin (70–80%), see Figure 8.20. The hydrolysis of starch is an acid-catalyzed hydrolytic decomposition where a C—O—C linkage is cracked between two glycopyranose units and a water molecule is inserted. Carried to the limit, starch will be hydrolyzed to the basic building block, D-glucose.

About 70% of the annual 17 million tons world production of starch is converted to glucose-containing sweeteners by using either an acid-catalyzed or an enzymatic process. The initial suspension for the acid-catalyzed process is a 40 wt% starch suspension which is acidified to a pH of 4–5, usually using hydrochloric acid. After the hydrolysis, the solution is neutralized and the resulting salt is removed using an ion-exchanger.

Hydrolysis of starch to monosugars is an old industrial process. The hydrolysis reaction, originally catalyzed with mineral acids, is now replaced by a biotransformation. Microorganisms for this reaction are selective but slow. Hydrolysis using CO_2, dissolved in hot pressurized water, without mineral acids can achieve excellent conversion of starch and may be able to replace the biotransformation since reaction times are much shorter [18–20]. From corn starch, high yields of glucose of about 65% can be obtained with

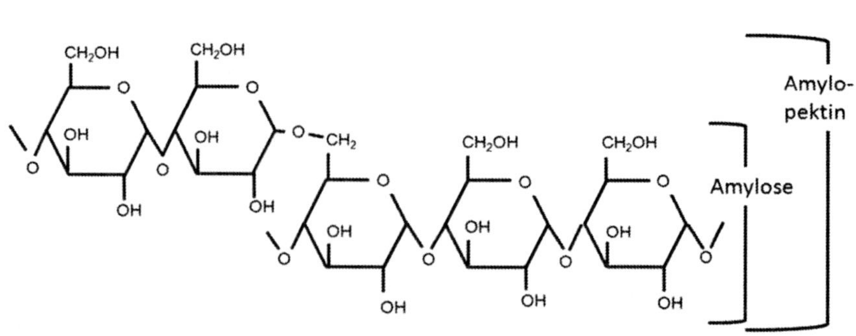

FIGURE 8.20 Structure of a starch molecule [18–20].

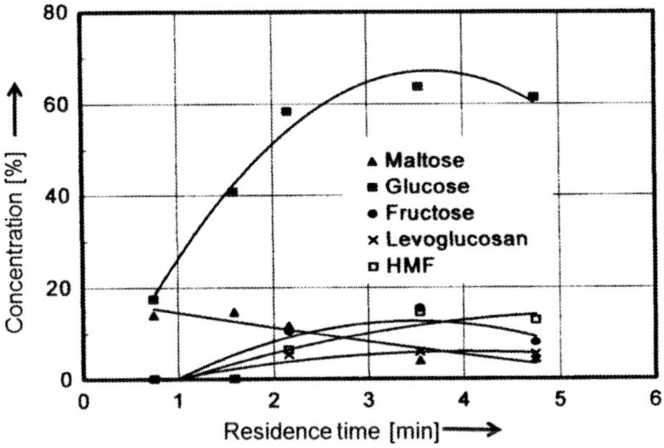

FIGURE 8.21 Conversion of corn starch with high-temperature water saturated with CO_2. Concentration of hydrolysis products. $T = 230$ °C, $P = 24$ MPa, feed concentration $x_S = 0.2$ wt% [18–20].

high-temperature water in which carbon dioxide is dissolved. CO_2 dissociates and acts as a catalyst [18–20]. The reaction is carried out continuously in a tubular reactor, residence time $t < 180$ s, starch concentrations $x_S = 0.2$–10 wt% in the aqueous feed slurry, pressure $P = 60$–240 bar, temperature $T = 170$–380° C. There are only few by-products, and their concentration can be kept low by short-residence time. Figure 8.21 shows the results of the hydrolysis of starch at $T = 230$° C and $P = 240$ bar, for an example. Glucose yield is increased from 5% to 60% by adding carbon dioxide. Details are discussed below after shortly reviewing some contributions from literature.

8.3.1.1 Batch Processing of Starch Conversion

The following conversions of starch have been carried out in a closed autoclave. That implies a duration of the experiments of at least several 10 min as compared to a continuously operated reactor where the reaction time can be reduced to several minutes, that is, by a factor of 10. Nevertheless, the conversion at the prolonged reaction times runs to a glucose yield comparable to that of the continuously operated reaction.

In an investigation covering the lower temperature range, carried out in the early 1970s of past century, starch was hydrolyzed in a closed system, without the addition of CO_2 [39]. Concentration of starch was $x_S = 3$ wt%, the reaction was carried out at $T = 120$–180 °C, $P = 0.2$–1.2 MPa, and a residence time of $t = 1$ h after a heating time of also about 1 h. Total sugar after the hydrolysis is 11.2 mg per 100 ml at $T = 120$ °C and 265 mg per 100 ml at $T = 180$ °C, corresponding to 0.37% and 8.83% sugar production, respectively, with a glucose content of 3.1–41.5% of the total sugar [39].

In a recent investigation, starch was hydrolyzed in a 250-ml batch reactor with high-temperature water under the addition of CO_2 [40]. Concentration of

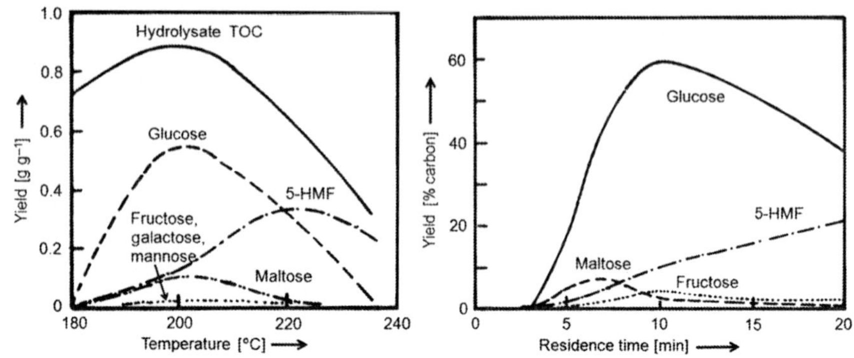

FIGURE 8.22 Starch hydrolysis in a batch reactor. Left: Influence of temperature. $P = 3$ MPa, residence time $t = 15$ min after 20–30 min heating. Right: Influence of residence time. $T = 220$ °C, P not specified. *Left: Modified from Ref. [49]; right: modified from Ref. [34].*

starch was $x_S = 4$–20 wt% at $T = 180$–235 °C, $P = 3$ MPa, at a reaction time of $t = 15$ min after a heating period of 20–30 min [40]. The maximum yield of glucose is 54.8 wt% of starch carbon at $T = 200$ °C. The degree of conversion and the conversion products are shown in Figure 8.22 in dependence on temperature [40].

Similar results were obtained in a small batch reactor [34] at $T = 180$–240 °C, see Figure 8.22, right. The major product is glucose. Further products are oligomers with various degrees of polymerization, maltose, fructose, 5-HMF, and furfural. The maximum yield of glucose is 63 wt% on carbon basis at $T = 200$ °C and $t = 30$ min residence time. Production of char and gaseous products is negligible. Glucose production increases drastically with CO_2 addition at the ratio of 0.1 g CO_2 per g_{H_2O} [41], a result that has been found before in continuous processing experiments.

8.3.1.2 Continuous Processing of Starch

The hydrothermal conversion of starch is discussed in the following based on experiments in a continuously operated tubular reactor (Inconel 600) of 6-mm inner diameter and 4 m length [18–20]. Details of the reactor and the experimental procedure are presented in Chapter 13.

The hydrolysis of corn starch was carried out in water and CO_2 at $T = 170$ and 300 °C and $P = 6$–24 MPa at a starch concentration of $x_S = 0.2$–10 wt% and residence time $t = 0.4$–20 min [18–20]. If carbon dioxide is added, carbon dioxide concentration is expressed in degree of saturation which is the ratio of the added carbon dioxide divided by the maximum quantity of carbon dioxide soluble in water at given conditions. A degree of saturation of more than 100% means that carbon dioxide is in excess. The dissolved carbon dioxide considerably increases with an increase in pressure, thus creating an acid environment inside the reactor, which catalyzes the hydrolysis of corn starch.

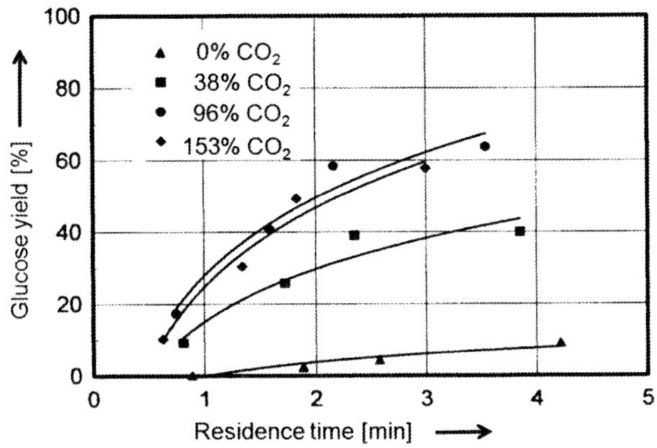

FIGURE 8.23 Conversion of corn starch in high-temperature water. $T = 230\ °C$, $P = 24$ MPa. Yield of glucose at different carbon dioxide concentrations [18–20].

Exemplary results of the hydrolysis of starch at $T = 230\ °C$ and $P = 24$ MPa are shown in Figure 8.23.

Product yields reach a maximum during the reaction. From a certain point on, the degradation becomes faster than the formation. This maximum is different for each substance and depends on the temperature. During starch hydrolysis, many reactions occur in parallel or in sequence. According to the mechanism of the acid-catalyzed hydrolysis, the probability of H^+ attacking the C—O—C bonds of starch is almost the same for every C—O—C bonds. Therefore, the direct hydrolysis of starch to glucose is possible, but starch can also be hydrolyzed first to oligomers or maltose and then to glucose.

The influence of the concentration of carbon dioxide can be demonstrated clearly. In Figure 8.23, the influence of the concentration of carbon dioxide on the formation of glucose is shown [18–20].

The glucose yield is significantly higher when carbon dioxide is dissolved in the reaction medium than without the use of carbon dioxide [18–20,42]. The yield can be increased from 5% to 60% by adding carbon dioxide by enhancing the carbon dioxide partial pressure. At higher CO_2 partial pressures than the saturation pressure for the given temperature, the influence of pressure is negligible. The hydrolysis reaction is catalyzed by the H^+ (or H_3O^+) ions. Therefore, only the dissociated part of dissolved carbon dioxide is of influence on the hydrolysis of starch. A detailed discussion of the carbon dioxide equilibrium in high-temperature water is given in Chapter 3. The influence of carbon dioxide concentration on the total sugar yield is shown in Figure 8.24. Total sugar yield means the sum of maltose, glucose, and fructose yield. Adding carbon dioxide can increase the total sugar yield from 20% to more than 80% [18–20].

Temperature has the strongest influence on the hydrolysis of starch in water. This is shown in Figure 8.25 for the relatively small temperature range

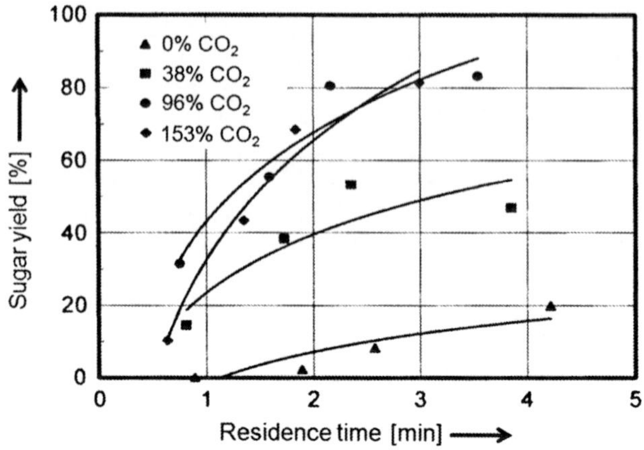

FIGURE 8.24 Conversion of corn starch in high-temperature water. $T = 230\ °C$, $P = 24$ MPa. Total sugar yield at different carbon dioxide concentrations [18–20].

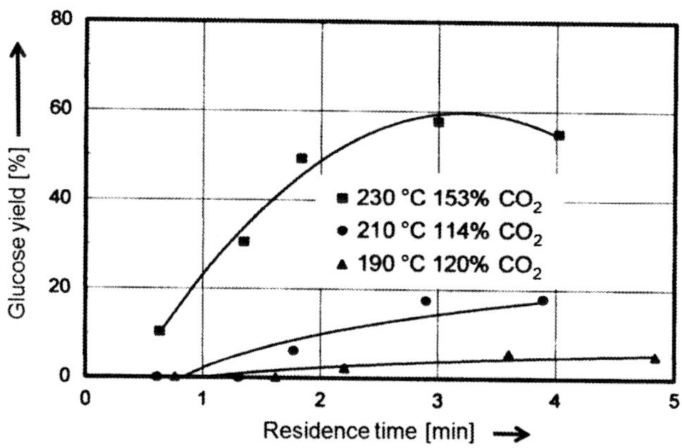

FIGURE 8.25 Conversion of corn starch in high-temperature water. Influence of temperature on the glucose yield under an excess of carbon dioxide. $P = 24$ MPa [18–20].

of $T = 190$–$230\ °C$. The higher temperature accelerates both formation and further decomposition of glucose. Therefore, the maximum glucose concentration occurs at lower residence time at higher temperatures. This effect increases with increasing temperature.

For industrial applications, the space-time yield of the product is of major importance. Up to an initial starch concentration of $x_S = 10$ wt%, no influence of the concentration occurs, as demonstrated by the results shown in Figure 8.26.

For the kinetic analysis of the reaction of starch with high-temperature water, it is assumed that all reactions are of the first order. The reaction scheme is considered as follows, see Figure 8.27.

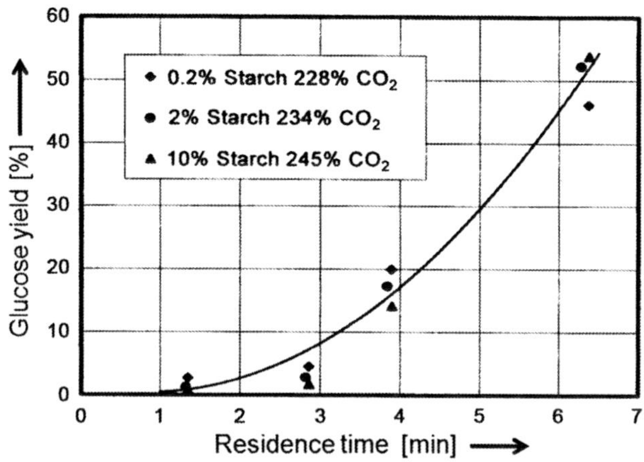

FIGURE 8.26 Conversion of corn starch in high-temperature water. $T = 230\,°C$, $P = 24$ MPa. Glucose yields at different initial starch concentrations [18–20].

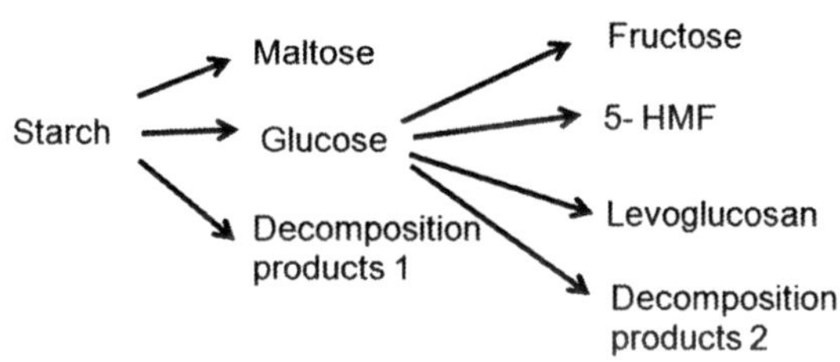

FIGURE 8.27 Conversion of starch in high-temperature water [18–20].

Glucose, maltose, and some unknown decomposition products I are produced through parallel reactions. Glucose further hydrolyzes into fructose, 5-HMF, levoglucosan, and some unknown decomposition products II. Concentration of starch is evaluated as the concentration of glucose units, since the molecular size of starch is variable. In Figures 8.28 and 8.29, the Arrhenius plots for the hydrolysis of starch and further decomposition of glucose are shown [18–20].

The activation energy E_a and frequency factor k_0 (Equations 8.1 and 8.2) for the different hydrolysis reactions are shown in Table 8.2. Compared to other biopolymers, the hydrolysis of starch in high-temperature water starts at lower temperatures and is faster than, for example, that of cellulose.

Conversion of starch to monosaccharides is a fast and clean process in high-temperature water that is saturated with CO_2. By depressurizing the product stream, CO_2 can be removed and the product stream neutralized. Compared with acid-catalyzed process, the hydrolysis of starch with carbon dioxide as catalyst has the advantage that the process steps of neutralizing the solution and removing the resulting salt with ion exchangers can be eliminated [18–20].

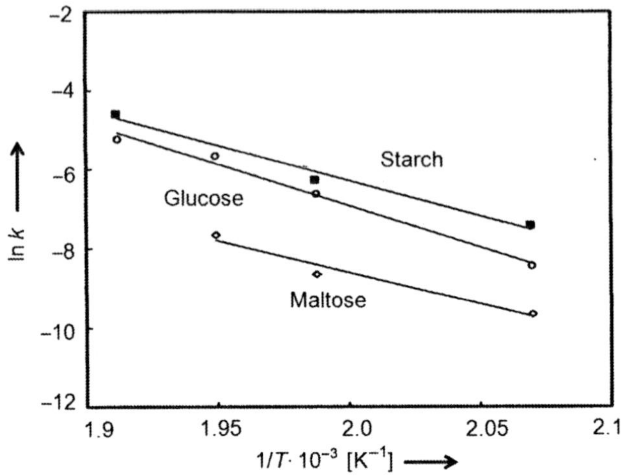

FIGURE 8.28 Arrhenius plot of the hydrolysis of starch in high-temperature water saturated with CO_2. Pressure $P=24$ MPa; CO_2 pressure higher than the saturation pressure at given temperature [18–20].

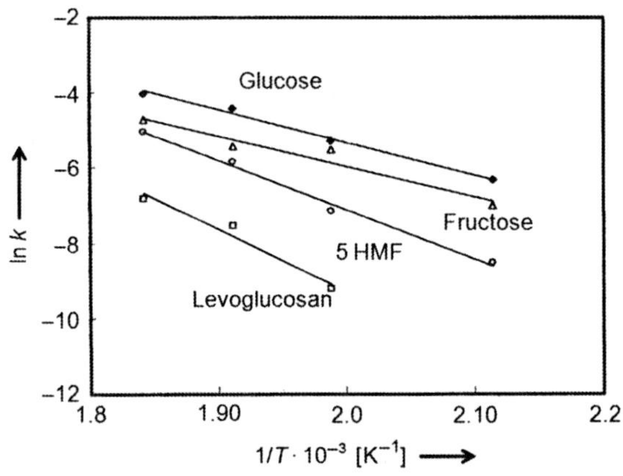

FIGURE 8.29 Arrhenius plot for the hydrolysis of glucose produced from the hydrolysis of starch in high-temperature water saturated with CO_2. Pressure $P=24$ MPa; CO_2 pressure higher than the saturation pressure at given temperature [18–20].

8.3.2 Pectin

Pectin is a structural heteropolysaccharide. The main chain consists of α-1,4-linked D-galacturonic acid. Pectin is contained in the primary cell walls of terrestrial plants. For example, citrus peels consist of about 30% pectin. It is produced commercially as a white to light brown powder, mainly extracted from citrus fruits, and is used in food as a gelling agent particularly in jams and jellies, as a stabilizer in fruit juices and milk drinks, and as a source of dietary fiber. Isolated pectin has a molecular weight of typically

TABLE 8.2 Activation Energy E_a and Frequency Factor k_0 for the Different Hydrolysis Reactions

Reaction	Activation Energy (E_a) (J mol^{-1})	Frequency Factor (k_0) (s^{-1})
Starch conversion	1.48×10^{5}	5.29×10^{-12}
Maltose formation	1.32×10^{5}	1.05×10^{10}
Glucose formation	1.35×10^{5}	1.12×10^{11}
Composition products I formation	1.67×10^{5}	2.07×10^{14}
Glucose conversion	1.830×10^{5}	7.249×10^{4}
Fructose formation	2.493×10^{4}	6.683×10^{4}
HMF formation	1.131×10^{8}	1.060×10^{5}
Levoglucosan formation	8.450×10^{15}	1.920×10^{5}
Composition products II formation	1.830×10^{1}	4.137×10^{4}

$P=24$ MPa [18–20].

$M_W=60$–130,000 g mol^{-1}, depending on origin and extraction conditions. 80% of carboxyl groups of galacturonic acid are esterified with methanol. Pectins are classified as high- or low-ester pectins, with more or less than half of all the galacturonic acid esterified [43].

Recovery of pectin from citrus fruits requires separation of pectin from other polysaccharides, like hemicellulose and cellulose. Treatment of residues from fruit juice processing with high-temperature water makes possible to separate pectin and hemicelluloses. The by-product of processing *Citrus junos*, an orange from China, is considered as an example in the following. Results are based on experimental investigations in a semi-batch flow reactor at temperatures of $T=110$–320 °C, at pressures of $P=4$–30 MPa [44–46]. Results are shown in Figure 8.30.

Pectin yield is more than 75% at pressures $P=4$–30 MPa in the temperature range $T=120$–140 °C. In this temperature range, pectin has almost the same molecular weight as commercial pectin. Extracts produced at the higher temperature of $T=160$ °C contain about 78% of pectin of lower molecular weight. Extraction and conversion at higher temperatures of $T=160$–320 °C at a pressure of $P=20$ MPa, with water to feed solvent ratios of $S/F=6.3$–21 $g_{H_2O}\,g_{Feed}{}^{-1}\,h^{-1}$ lead to a nearly complete separation of hemicelluloses from cellulose at $T=200$ °C. Cellulose in the residue reached about 80% after hydrothermal treatment at $T=200$ °C (Figure 8.31). At temperatures $T>200$ °C, cellulose is converted to secondary products as discussed further in Section 8.3.3.

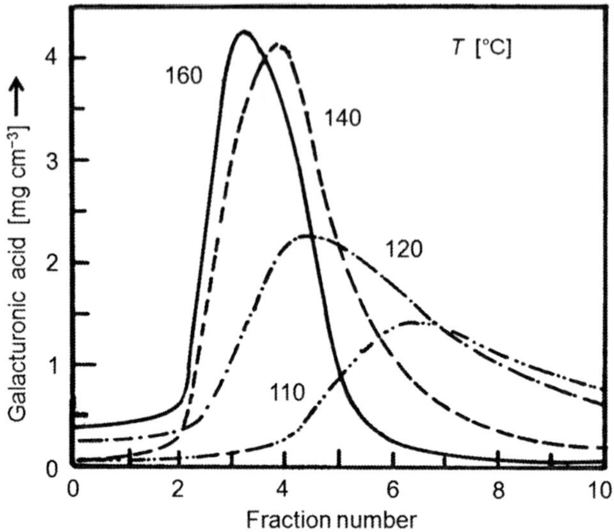

FIGURE 8.30 Pectin extraction with high-temperature water. Effect of temperature. $P = 4$ MPa, flow rate $\dot{m} = 0.5$ ml min^{-1}. *Modified from Ref. [45].*

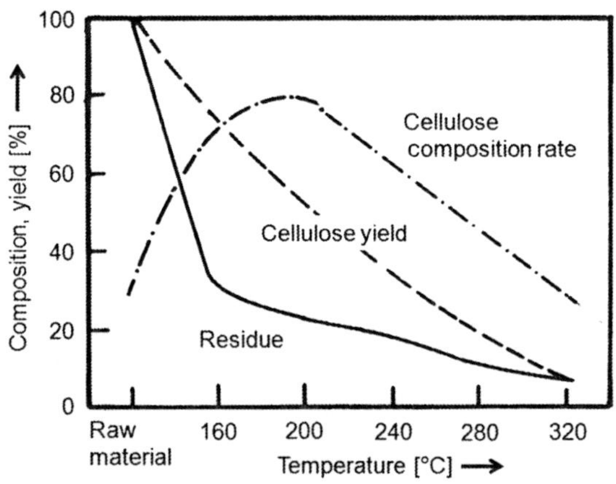

FIGURE 8.31 Extraction of citrus fruit residue with high-temperature water. Influence of temperature on concentration and yield of cellulose and yield of residue. $P = 20$ MPa, solvent ratio $S/F \approx 21\ g_{H_2O} g_{Feed}{}^{-1} h^{-1}$. *Modified from Ref. [44].*

8.3.3 Cellulose

Cellulose is a main component of lignocellulosic materials which constitutes the bulk of the dry mass of plant materials and is therefore one of the most abundant biochemicals [4]. Cellulose is a polysaccharide composed of units of glucose that are connected via β-(1→4)-glycosidic bonds and form strong intra- and intermolecular hydrogen bonds, leading to a crystalline structure. This structure makes cellulose resistant to swelling in water and against attack by enzymes [4].

8.3.3.1 Kinetics of Conversion of Cellulose

In high-temperature and supercritical water, cellulose reacts to oligomer sugars, monomer sugars, and various degradation products as pyruvaldehyde and HMF, and others, Equation (8.3):

$$\begin{aligned} &\text{Cellulose} \rightarrow \text{oligomer sugars} \rightarrow \text{monomer sugars} \rightarrow \\ &\qquad \text{secondary degradation products} \end{aligned} \tag{8.3}$$

Cellulose is derived from different biological sources. The consequence is a different physical and chemical structure which affects its behavior. In due course, experimental results from cellulose conversion vary considerably, as is shown in Figure 8.32 for the conversion rates of cellulose.

Results of several investigations on the conversion of cellulose in high-temperature water are in reasonable coincidence [47–58]. At near-critical temperatures, the reaction is probably accelerated and the energy of activation is enhanced. So far, there is no agreement on these data. Due to the scattering of experimental results, the enhancement of the reaction rate may well be in the range of experimental deviations, which may be attributed to different sources of cellulose (and different degrees of crystallization) or to normal scattering of experimental results. A reason for the enhancement near the critical temperature of water may be that cellulose is dissolved in water at these

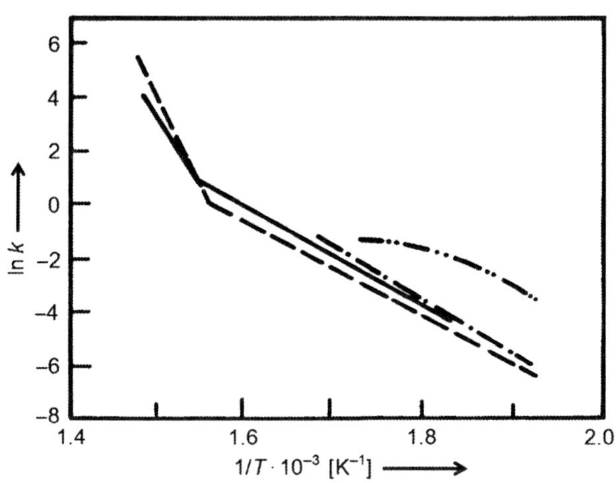

FIGURE 8.32 Rate constant for the conversion of microcrystalline cellulose in high-temperature water, according to different authors. Particle size $d_p = 20$–130 μm. —•—•, [47]; ——, [48]; — — —, [49]; —••—, [50]. *Modified from Ref. [48].*

temperatures and a homogeneous phase is formed and mass-transfer restrictions omitted [48].

Kinetics of cellulose conversion in water is usually modeled as a first-order reaction with a characteristic activation energy E_a and a preexponential factor k_0 (Equation 8.4):

$$k_T = k_0 \exp\left(-\frac{E_{a,C}}{RT}\right) \tag{8.4}$$

For temperatures $T = 200$–$350\ °C$, fairly consistent values for the activation energy are reported. For cotton cellulose degradation, an activation energy of $E_a = 129.1\ kJ\ mol^{-1}$ was determined [51]. Significantly higher activation energy of $E_a = 165\ kJ\ mol^{-1}$ was found for powdered cellulose of unspecified plant origin [52], for unspecified cellulose $E_a = 146\ kJ\ mol^{-1}$ [54,55], for microcrystalline cellulose (Avicel) with a purity of 99% (Merck, Darmstadt, Germany) $E_a = 163.8\ kJ\ mol^{-1}$ [56–58], and for microcrystalline cellulose with a particle size $d_P = 20$–$130\ \mu m$ (VWR Chemical Company) $E_a = 154.4 \pm 9.5\ kJ\ mol^{-1}$ [48]. At temperatures higher than $T = 350$–$370\ °C$, a drastic acceleration of the reaction kinetics occurs to $E_a = 548\ kJ\ mol^{-1}$ for cellulose from an unspecified plant source [54,55], or $E_a = 430.3 \pm 6.3\ kJ\ mol^{-1}$ for microcrystalline cellulose with a particle size $d_P = 20$–$130\ \mu m$ [48].

In the following, results for the conversion of microcrystalline cellulose are presented in some detail and the formation of glucose and secondary degradation products is shown as well as the influence of dissolved carbon dioxide on the acid-catalyzed reactions. Results are derived from continuous experiments on the conversion of pure cellulose with an initial cellulose concentration of $x_C = 0.1$–1 wt% at the reactor inlet.

Cellulose conversion in high-temperature water in dependence of temperature and residence time is shown in Figure 8.33, expressed as the water-soluble reaction products. As mentioned before, in the temperature range

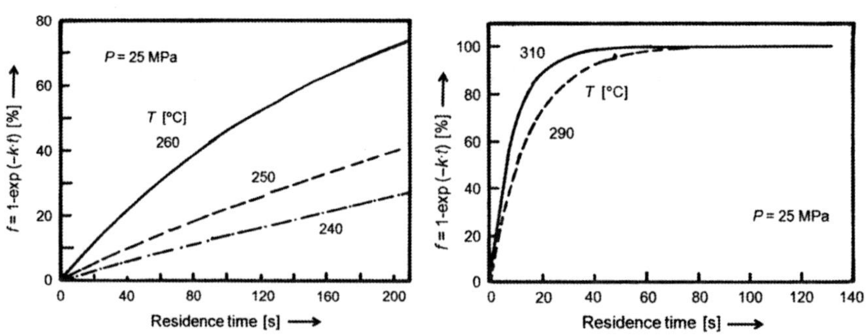

FIGURE 8.33 Conversion of cellulose on a carbon basis, plotted as the degree of liquefaction (Equation 8.6). $P = 25$ MPa; initial cellulose concentration: $x_C = 0.5$ wt%. The lines were calculated according to a first-order kinetics approach [56].

below about $T = 320\ °C$, the formation of gaseous products is negligible. Due to the low compressibility of water and the low cellulose concentration, the fluid density can be assumed as constant along the reactor and the influence of the cellulose concentration can be neglected. For these conditions, the degree of cellulose liquefaction f in an ideal plug flow reactor can be expressed as:

$$f = 1 - \exp(-k \cdot \tau) \tag{8.5}$$

and after rearrangement:

$$-\ln(1 - f) = k \cdot \tau \tag{8.6}$$

The degree of liquefaction progressively increases with temperature. At a temperature of $T = 280\ °C$, essentially all carbon is liquefied at reaction times of $t = 100$–120 s. A further increase in temperature to $T = 290$ and $310\ °C$ (Figure 8.33, right) leads to even more rapid conversion [56]. Pressure is of negligible influence, as long as it is higher than the vapor pressure of water. Reaction rate constants for the conversion of cellulose in high-temperature water are given in Table 8.3.

Products of the conversion of cellulose are glucose, the monomeric building block, fructose, which is formed from glucose by isomerization, pyruvaldehyde, levoglucosan, and HMF, and carboxylic acids. At lower temperatures and short-residence times, water-soluble oligosaccharides occur as hydrolysis products [56]. Figure 8.34 shows the formation of the main products glucose and HMF.

Glucose and HMF yield follow the pattern of a consecutive reaction. At lower temperatures, the maximum yield for both components is beyond the experimental residence time. For temperatures between $T = 270$ and $310\ °C$, the maximum of glucose concentration is shifted to lower residence times. At $T = 310\ °C$, also for HMF, a maximum occurs, indicating that this product is itself further converted. The same pattern can be seen for pyruvaldehyde.

The scattering experimental data can be represented by the lines in Figure 8.34, which were calculated under the assumption of a consecutive reaction with a first-order rate law for formation and decomposition [56]. The values for the rate constants of formation and decomposition are listed in Table 8.4.

TABLE 8.3 Reaction Rate Constants of Cellulose Liquefaction at Different Temperatures [56]

T (°C)	240	250	260	270	280	290	310
$k(T)$ (s^{-1})	0.0015	0.0025	0.0062	0.0214	0.0263	0.0647	0.1086

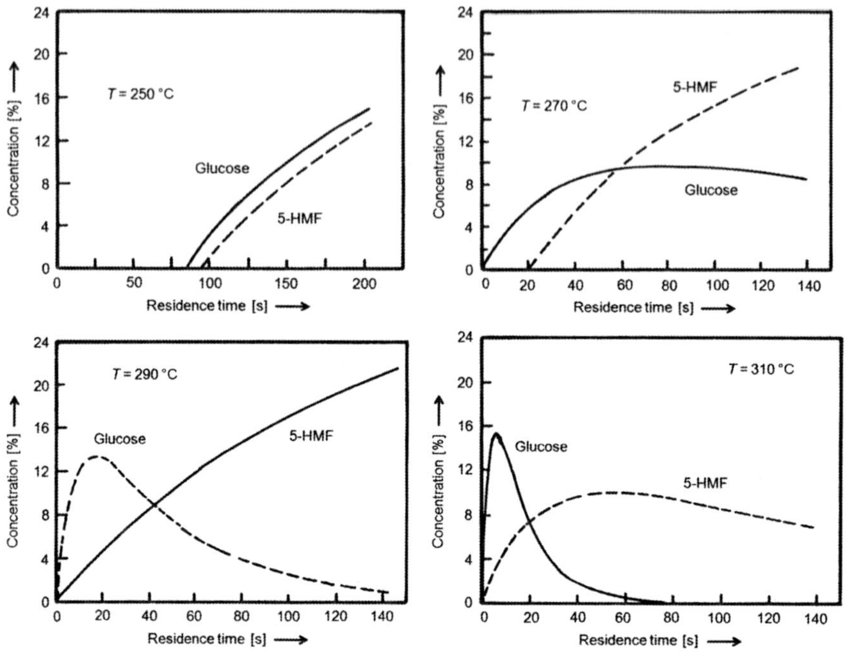

FIGURE 8.34 Conversion of cellulose in high-temperature water. Yield of glucose and 5-HMF. $T = 250–310\,°C$, $P = 25$ MPa [56].

TABLE 8.4 Conversion of Cellulose in High-Temperature Water

T (°C)	Glucose		HMF		Pyruvaldehyde	
	k_f (s^{-1})	k_d (s^{-1})	k_f (s^{-1})	k_d (s^{-1})	k_f (s^{-1})	k_d (s^{-1})
250	0.002	0.006	0.0018	0.005	0.001	0.007
270	0.004	0.03	0.003	0.008	0.0017	0.01
290	0.024	0.12	0.0025	0.005	0.0036	0.013
310	0.075	0.31	0.006	0.043	0.015	0.045

Rate constants of formation (k_f) and decomposition (k_d) for main degradation products [56].

Selectivity of main products can be calculated as the amount of carbon transformed to product i related to the overall degree of conversion, expressed as dissolved organic conversion products (DOC):

$$S_i = \frac{y_i}{f} = \frac{c_i}{DOC} \tag{8.7}$$

with y_i, amount of compound i in liquid product; f, amount of liquefied feed; c_i, amount of compound i in the feed; DOC, total amount of converted feed.

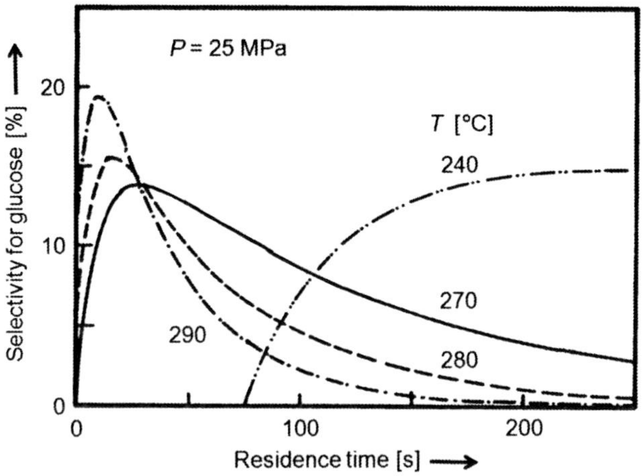

FIGURE 8.35 Conversion of cellulose in high-temperature water. Calculated selectivity of glucose formation as a function of reaction time for different temperatures [56].

The calculated selectivity of glucose formation as a function of residence time for different temperatures is shown in Figure 8.35.

Selectivity for the formation of glucose and maximum glucose concentration increase with higher temperature and are shifted to lower residence time. For the temperature range of $T = 320–400$ °C, this trend is also valid for the highest glucose concentrations occurring at very short-residence times. Glucose yield is maximal 50 wt% at $T = 400$ °C, $P = 25$ MPa, at a residence time of $t = 0.0025$ s. After a reaction time of $t = 0.15$ s, glucose is completely reacted to secondary products [54,55].

Beside the so far discussed products, during conversion of cellulose in high-temperature water also carboxylic acids like acetic acid, propionic acid, isobutyric acid, and *n*-butyric acid are formed in small quantity. However, only acetic acid contributes significantly to the total carbon of the products, as shown in Figure 8.36.

For the interpretation of the results, it is important to consider that the carbon mass balance is not closed by the mentioned hydrolysis products. The missing portion of the carbon balance can be attributed to partial hydrolysis products at lower temperature and the formation of various secondary reaction products at higher temperature, both of which have not been analyzed.

8.3.3.2 Catalytic Influence of CO_2 Addition

Carbon dioxide promotes the acid-catalyzed reaction pathway of cellulose hydrolysis by the formation and dissociation of carbonic acid, as shown for the hydrolysis of starch. The addition of carbon dioxide as catalyst has the advantage of an easy recovery by expansion and subsequent reutilization in

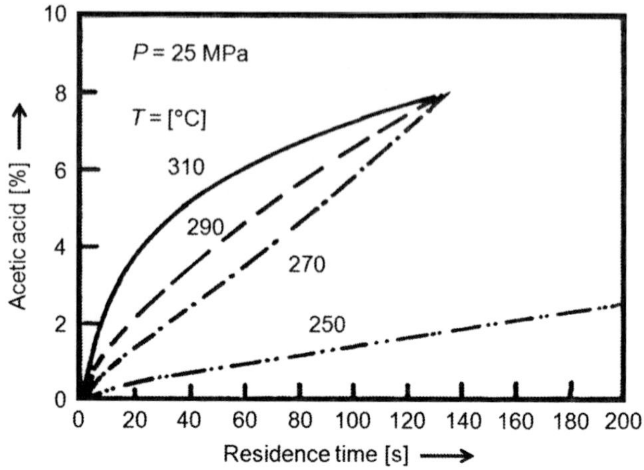

FIGURE 8.36 Conversion of cellulose in high-temperature water. Yield of acetic acid as a product of cellulose hydrolysis. $P=25$ MPa. [56].

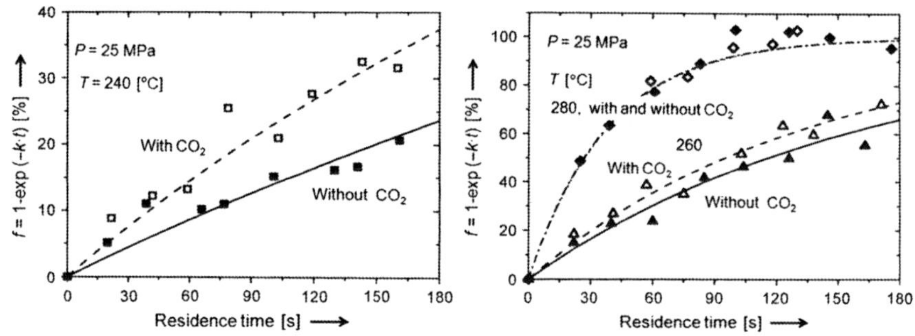

FIGURE 8.37 Conversion of cellulose in high-temperature water. Influence of CO_2 addition. CO_2 concentration $x_{CO_2}=75$–100% of saturation. Initial cellulose concentration $x_C=0.5$ wt% [56].

a closed cycle. The effect of carbon dioxide on the liquefaction of cellulose is shown in Figure 8.37 and on glucose yield in Figure 8.38. [56].

The addition of carbon dioxide enhances cellulose hydrolysis at temperatures below $T=280$ °C. Around this temperature, the reactivity of the water molecule is high enough not to be influenced by the relatively small amount of H_3O^+ ions formed by addition of CO_2 as is discussed in more detail in Chapter 5.

The hydrolysis of crystalline cellulose in high-temperature water results in important conclusions about the conversion of lignocellulosic biomass. Pure cellulose can readily be converted to water-soluble products at reaction times in the order of seconds to minutes without the addition of a catalyst. Increasing temperature increases the reaction rate considerably. A marked gas formation is not observed for temperatures below $T=300$ °C. Cellulose is degraded to saccharides, mainly glucose, and secondary reaction products like HMF,

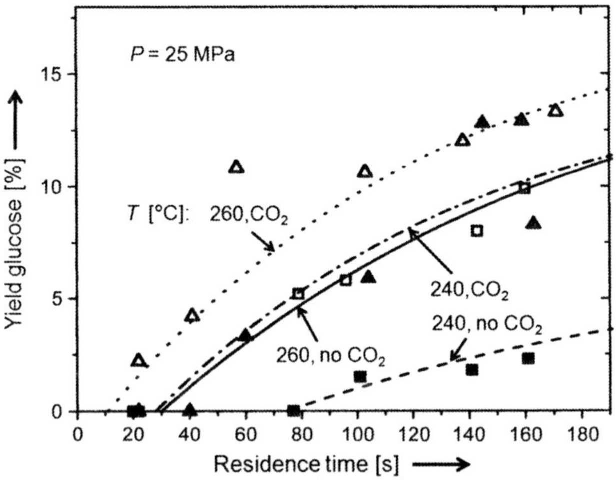

FIGURE 8.38 Conversion of cellulose in high-temperature water. Influence of CO_2 addition on the glucose yield. CO_2 concentration 90% saturation. Initial cellulose concentration $x_C = 0.5$ wt%, $P = 25$ MPa [56].

pyruvaldehyde, and carboxylic acids. Kinetics of conversion and of product formation can be described by global first-order reactions. The addition of carbon dioxide leads to a rate enhancement of cellulose hydrolysis in temperature ranges below $T = 280$ °C due to partly formation and dissociation of carbonic acid [56].

At temperatures $T = 350$–400 °C and a pressure of $P = 25$ MPa, cellulose is hydrolyzed very fast (Figure 8.39). In supercritical water, at $T = 400$ °C, cellulose is completely converted in $t = 0.05$ s [59]. Most of the products are hydrolysis products, such as glucose, fructose, and oligomers of glucose, with a total yield of the hydrolysis products of up to 75% [59].

8.3.4 Hemicellulose

Occurrence and chemical composition of hemicelluloses is shortly addressed. Conversion of hemicellulose and product formation is shown in examples and kinetics of hemicelluloses hydrolysis is derived according to simple degradation models. Finally, examples for hemicelluloses extraction and hydrolysis are discussed for illustration.

8.3.4.1 What is Hemicellulose?

Hemicellulose is a heteropolymer with side chains and is composed of pentoses (xylans), alternating units of mannose and glucose (mannans or glucomannans) or galactose units (galactans). The amount of the sugar monomers can change quite dramatically for different feedstock sources.

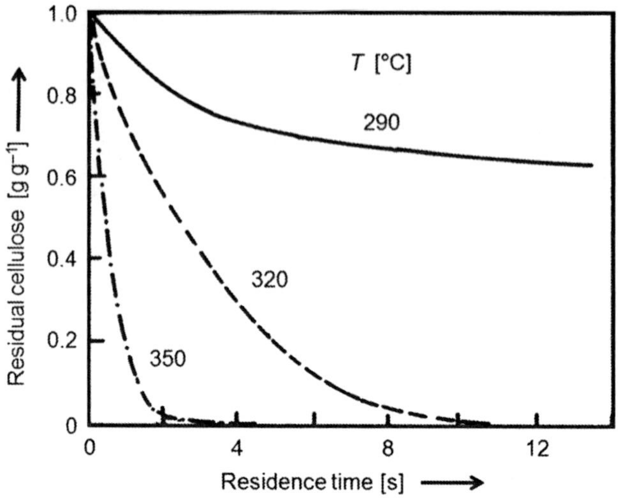

FIGURE 8.39 Conversion of cellulose in high-temperature and near-critical water. $P = 25$ MPa. *Modified from Ref. [59].*

Unlike cellulose, hemicellulose consists of 50–3000 sugar units as opposed to 7000–15,000 glucose molecules per polymer in cellulose. Hemicelluloses are classified according to the main sugar residue in the backbone as xylans, mannans, and glucans. Depending on the plant species, developmental stage, and tissue type, various subclasses of hemicellulose may be found [60]. These different subtypes can be grouped into two general categories based on the hydration of the fibers. Low hydration polysaccharides function primarily to stabilize the cell wall through hydrogen-bonding interactions with cellulose and covalent interaction with lignin. They are water soluble due to their branched structure. The second type is hemicelluloses composed mainly of hydrocolloids, which function primarily as an extracellular energy and raw materials storage system and as a water-retention mechanism in seeds. Further information on the structure and functionality of the various building blocks of hemicelluloses can be found in Ref. [60].

Hemicellulose is amorph, not crystalline as cellulose, and therefore more susceptible to hydrothermal extraction and hydrolysis [4]. It readily dissolves in water at temperatures above about $T = 180$ °C [61]. High amounts of the hemicelluloses can be easily extracted with high-temperature water. Mok and Antal [62] could extract an average of 95% of hemicellulose as monomeric sugars at $P = 34.5$ MPa and $T = 200$–230 °C in a few minutes. Garrote *et al.* [63] report in a review that in most studies yields of 65–82% are reported for hemicellulose extraction and recovery as sugars or oligomers. They review reaction kinetics for hemicellulose degradation into sugars and subsequent degradation of the sugars into furfurals and other degradation compounds [63].

Conversion kinetics of hemicelluloses is discussed in the following. Extraction and conversion of hemicellulose is then treated in an example. Further information and examples are also presented in Section 8.6.2.

8.3.4.2 Kinetics of Hemicelluloses Hydrolysis

Kinetic models to describe hemicellulose hydrolysis, in general, assume a homogeneous first-order series reaction mechanism. In simple models, it is further assumed that oligomers react so fast to monomers that oligomers can be omitted from consideration [64]. Thus, hemicellulose is supposed to react fast to xylose and further to degradation products. Since these assumptions proved to be too simple for modeling hemicelluloses conversion in high-temperature water, the kinetic model has been modified by considering two arbitrarily defined fractions of hemicelluloses to account for fast hydrolysis during the first part and slow hydrolysis in the second part of the conversion [65]. Furthermore, oligomers of various degrees of polymerization are present during the hydrolysis reaction and must be considered. The hydrolysis reaction of oligomers may lead in parallel to the formation of xylose directly to degradation products like furfural.

Reaction rate can be expressed in the form of first-order reactions (Equation 8.1). Rate constants k_i are assumed to follow an Arrhenius temperature relationship (Equation 8.2).

Despite significant effort to describe the kinetics of hemicellulose hydrolysis, the models do not predict consistent results, but predict different hemicellulose removal rates and yields [66]. In most studies, the conversion of hemicelluloses is observed after its extraction from a certain substrate without considering the extraction as a separate process step. This extraction process underlies the same influences as any fluid-solid extraction, and for example, the influence of solvent ratio can be explained in terms of extraction kinetics as discussed in Chapter 6. Furthermore, solubility of intermediate oligomers in high-temperature water probably is a determining parameter in the subsequent hydrolysis reactions.

The influence of temperature and residence time can be taken into account in a so-called severity approach, defining a severity factor R_0 as given in Equation (8.8) [67]:

$$R_0 = t \cdot \exp((T_H - T_R)/14.75) \tag{8.8}$$

where t is reaction time in minutes; T_H is hydrolysis temperature; and T_R is reference temperature, most often 100 °C.

Hydrolysis of hemicelluloses-containing substrates like corn stover and corn cobs was analyzed [68] covering the results of several investigations carried out as batch experiments at saturation pressure [69–72]. In Figure 8.40, the fraction of potential xylose remaining in the solids, and in Figure 8.41, yield of monomers and oligomers are shown.

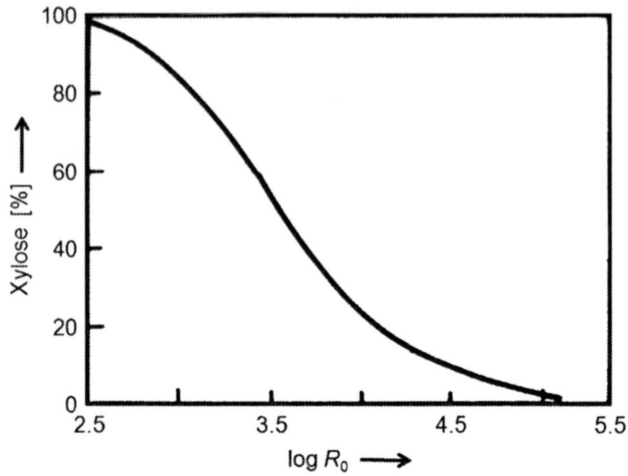

FIGURE 8.40 Potential xylose remaining in the solid residue during hydrolysis of hemicelluloses with high-temperature water from corn stover and corn cobs. *Data from different authors. Modified from Ref. [68].*

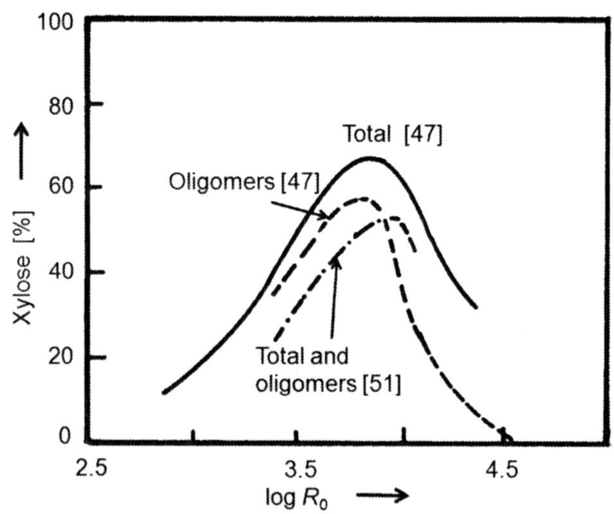

FIGURE 8.41 Potential xylose as monomers plus oligomers (total) and as only oligomers from hydrolysis of hemicelluloses with high-temperature water from corn stover and corn cobs. *Modified from Ref. [68].*

The maximum yield of xylose monomers and oligomers in the hydrolysate is about 60% and occurs at a severity factor log $R_0 = 3.8$–4.0. Xylose oligomers are the main product compounds, accounting at lower severities for up to 90% of the solubilized species and for about 80% at severity conditions corresponding to the maximum yield [68].

Figure 8.41 shows that the maximum yield (monomers plus oligomers) occurs at about $t = 14$ min with a severity factor of about $\log R_0 = 3.8$ for corn stover-derived hemicellulose, and at about $t = 100$ min, and the same a severity factor of about $\log R_0 = 3.8$ for corn cobs.

In a different investigation using corn cobs, Garrote *et al.* [72] were successful with modeling residue and products by including parallel degradation reactions of oligomers directly to furfural, see Figure 8.42. Corn cobs were treated with high-temperature water at temperatures $T = 145$–$190\,^{\circ}$C for up to $t = 12.3$ h at a water to solid ratio of $S/F = 8$ or 10 kg kg^{-1}. Concentrations of xylan and xylan-degradation products, xylose oligomers, xylose, furfural, and other degradation products are shown in Figure 8.43.

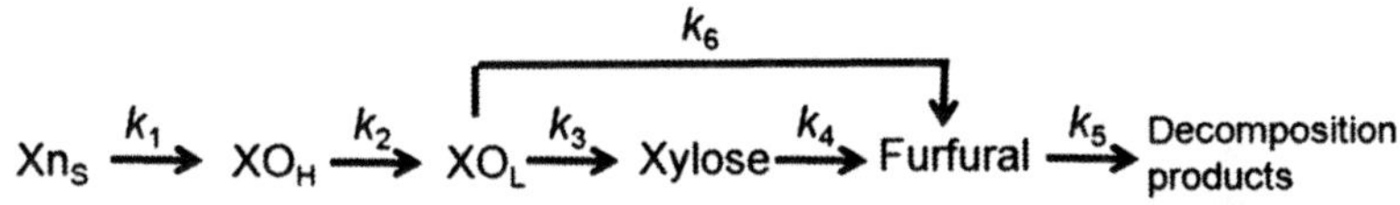

FIGURE 8.42 Model for the conversion of hemicelluloses (xylan) derived from corn cobs in high-temperature water [72].

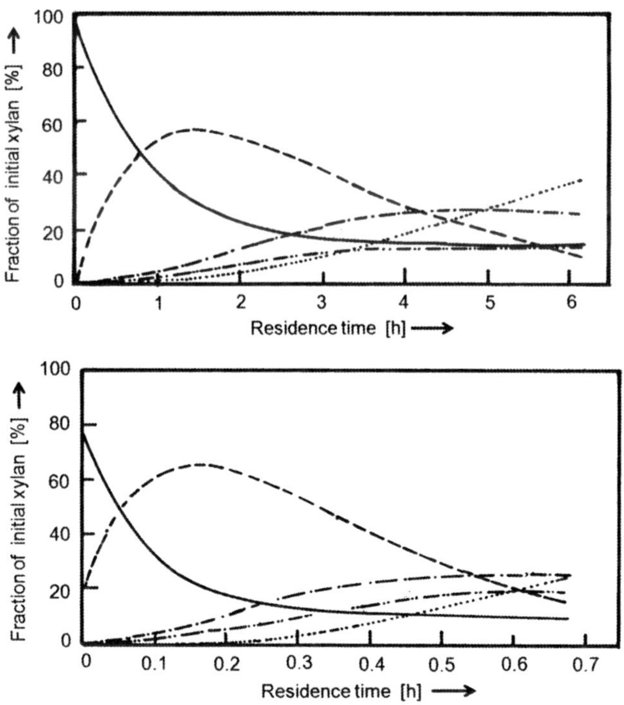

FIGURE 8.43 Conversion of corn cob-derived hemicelluloses with high-temperature water. Concentrations of residual xylan and xylan-degradation products. ———, Residual xylan; — — —, oligomers; —•—•, xylose; —••—, furfural; ••••, decomposition products. *Modified from Ref. [72].*

The kinetics of xylan degradation was modeled by means of a mechanism involving sequential, first-order, pseudo-homogeneous kinetics, see Figure 8.42. The model provides a good correlation of the experimental data, as shown in Figures 8.43 and 8.44.

In semicontinuous experiments, sweet sorghum bagasse was hydrolyzed with high-temperature water in a flow-through reactor [73] at reaction temperatures of $T = 15–235\ °C$, a pressure of $P = 5\ MPa$, and flow rates from $\dot{m} = 8$ to $30\ ml\ min^{-1}$. The effect of temperature on the concentration of xylose, glucose, and arabinose is shown in Figure 8.45, and the effect on the degradation products furfural, 5-HMF, and formic acid is shown in Figure 8.46. The concentration of xylose, including xylose monomer and xylose oligomers, drops dramatically for temperatures higher than $T = 195\ °C$.

The influence of flow rate on the concentration of xylose and on the degradation product 5-HMF is shown in Figure 8.47. At higher flow rates, the concentration of xylose in the liquid product is increased and the maximum is shifted to lower residence times. This clearly indicates the influence of the extraction of hemicelluloses from the substrate as a first step before the conversion of hemicelluloses due to an increased mass transfer. At a low flow rate of $\dot{m} = 5\ ml\ min^{-1}$, a higher 5-HMF yield occurs due to the longer residence time for xylose in the reactor than at a higher flow rate of $\dot{m} = 30\ ml\ min^{-1}$.

It is supposed that the major xylan of hemicelluloses derived from sweet sorghum bagasse is *O*-acetyl-4-*O*-methyl-glucuronoarabionoxylan. Xylose, arabinose, glucose, and their oligosaccharides are the main products. Glucuronic acid and acetic acid are produced from the decomposition of glucuronic and O-acetyl groups in the branched chain of the hemicellulose. Furfural

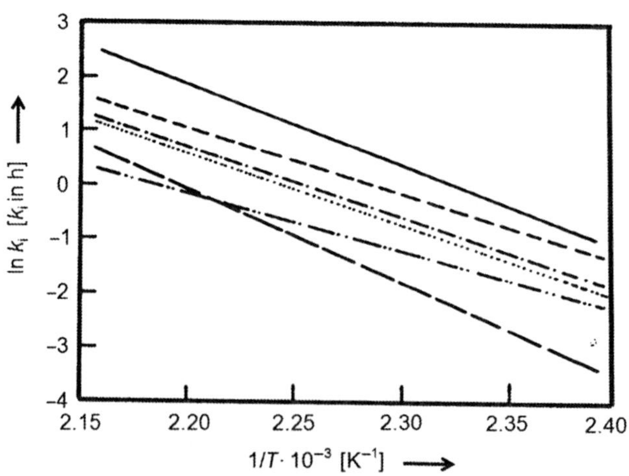

FIGURE 8.44 Conversion of corn cob-derived hemicelluloses with high-temperature water. Arrhenius plot for k_1 to k_6. ———, ln k_1; — — —, ln k_2; —•—•, ln k_3; —— ——, ln k_4; ••••, ln k_5; —••—; ln k_6. *Modified from Ref. [72].*

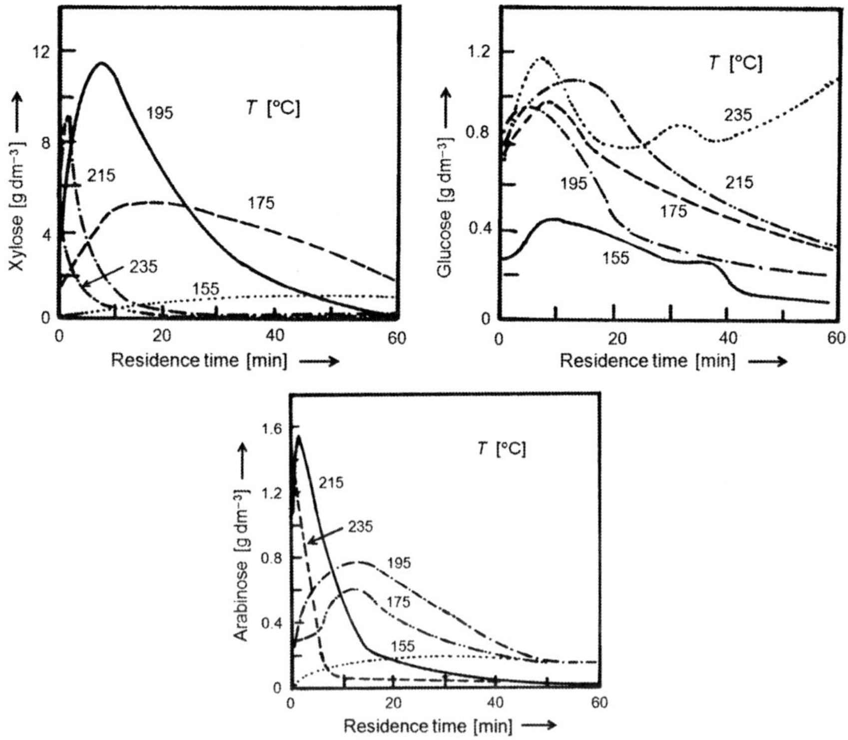

FIGURE 8.45 Conversion of hemicelluloses derived from sweet sorghum bagasse with high-temperature water. Concentration of total xylose, glucose, and arabinose. *Modified from Ref. [73].*

and 5-HMF are the direct degradation products. Glucuronic acid is oxidatively degraded into glycolic acid with low-molecular weight organic acids, as formic acid, as a further degradation product. Figure 8.48 shows a chromatogram of the xylose oligomers stemming from the conversion of hemicelluloses of sweet sorghum bagasse. It is clear that the decomposition of hemicelluloses is a complex process with quite a number of degradation products [73]. It is of interest to look at the comparison of the results of flow-through (semicontinuous) and batch experiments as listed in Table 8.5. In this case, the results of both modes of operation are quite similar.

8.3.5 Lignin

Lignin is a complex high-molecular-weight compound with a highly random structure [4]. A simplified structure of lignin is shown in Figure 8.49. Lignin is the most stable component of biomass. Its conversion is of great importance for the development of processes that make use of renewable resources. The chemical composition is the basis for evaluation of applications of lignin.

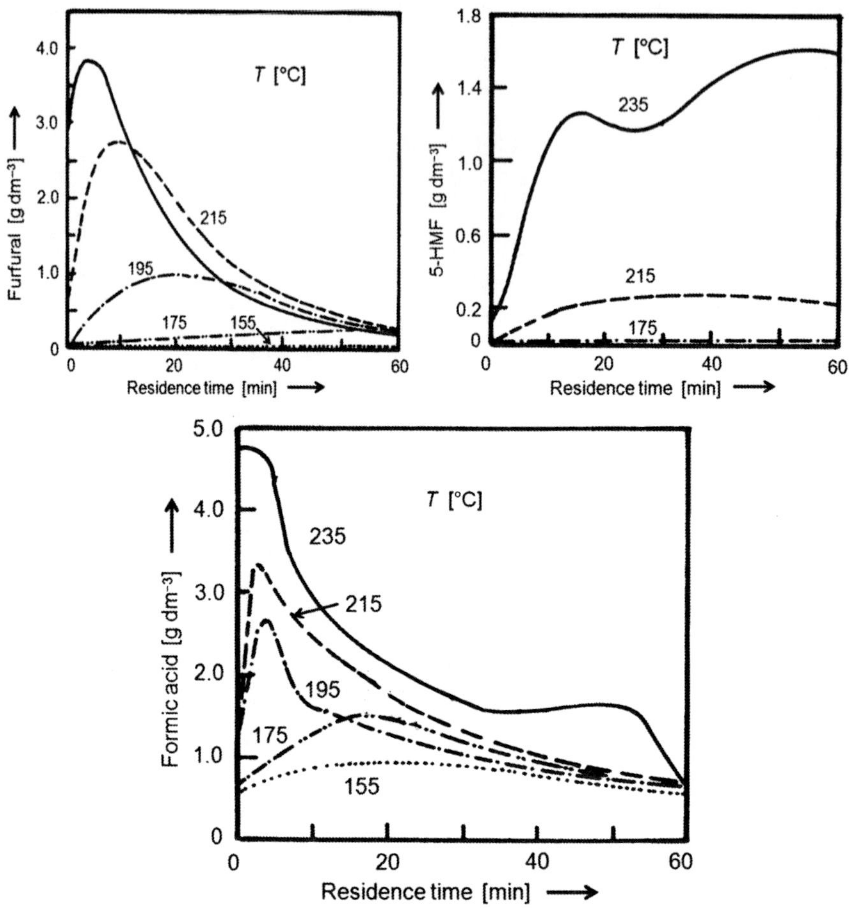

FIGURE 8.46 Conversion of hemicelluloses derived from sweet sorghum bagasse with high-temperature water. Concentration of degradation products furfural, 5-HMF, and formic acid [73].

The conversion in high-temperature and supercritical water, as the topic of this book, is the main content of Section 8.3.5. Dissolution, conversion, the various products and the reaction pathways are addressed, as is kinetic modeling, although the complex nature of lignin and the many pathways of lignin conversion make it not possible to go into many details. Therefore, simple models and the conversion of model compounds for lignin, like guaiacol and catechol are considered.

8.3.5.1 Chemical Composition

Lignin is a complex primary component of plant biomass (Figure 8.49). It contains many oxygen functional groups as hydroxyl, carboxyl, and carbonyl groups. Hydroxyphenylpropane units such as *trans-p*-coumaryl alcohol, coniferyl

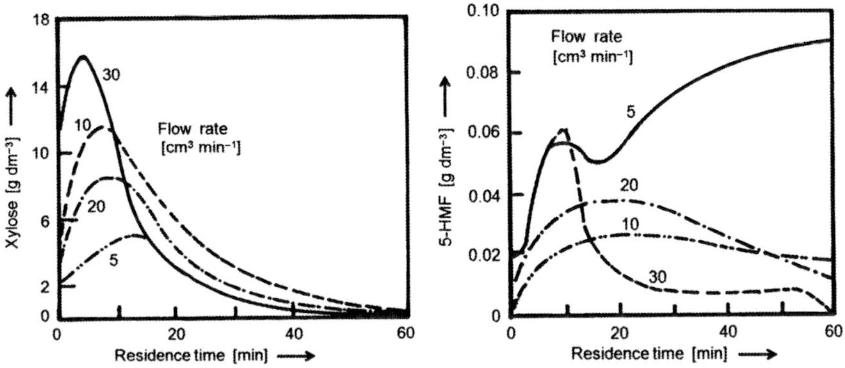

FIGURE 8.47 Conversion of hemicelluloses derived from sweet sorghum bagasse with high-temperature water. Effect of flow rate on concentrations of glucose (left) and 5-HMF (right). *Modified from Ref. [73].*

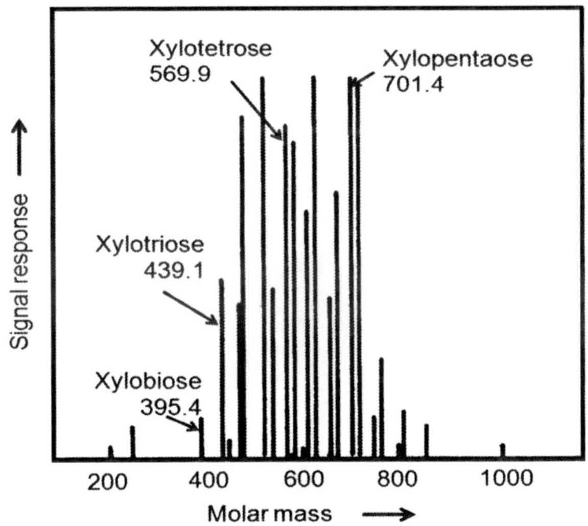

FIGURE 8.48 Conversion of hemicelluloses derived from sweet sorghum bagasse with high-temperature water. Chromatogram of xylose oligomers in the liquid fraction. *Modified from Ref. [73].*

alcohol, and sinapyl alcohol are connected by ether and carbon—carbon bonds [4,74]. The ether bond in lignin between one propyl side chain of a hydroxyphenylpropane unit and the hydroxy group attached to a benzene ring of another hydroxyphenylpropene unit is a major feature of the structure. Hence, the conversion of ether bonds and carbon—carbon bonds of alkylphenol are important factors in lignin conversion [75].

Due to its complex composition and structure, the thermal and hydrothermal decomposition of lignin occurs over a broad temperature range. Lignin

TABLE 8.5 Composition of the Liquid Hydrolysate and the Remaining Solid

	Residual Solid	Glucan (%)		Xylan (%)		Klason Lignin (%)
		Solid	*Liquid*	*Solid*	*Liquid*	
Untreated	100	45.2	–	25.2	–	16.6
Batch	77.5	42.6	2.2	4.9	17.0	15.3
Flowthrough	75.9	41.0	3.0	3.5	19.1	14.5

T=195 °C, P=5 MPa, water flow $\dot{m}$=30 ml min^{-1}, residence time t=8 min, liquid to solid ratio S/F=12.8 [73].

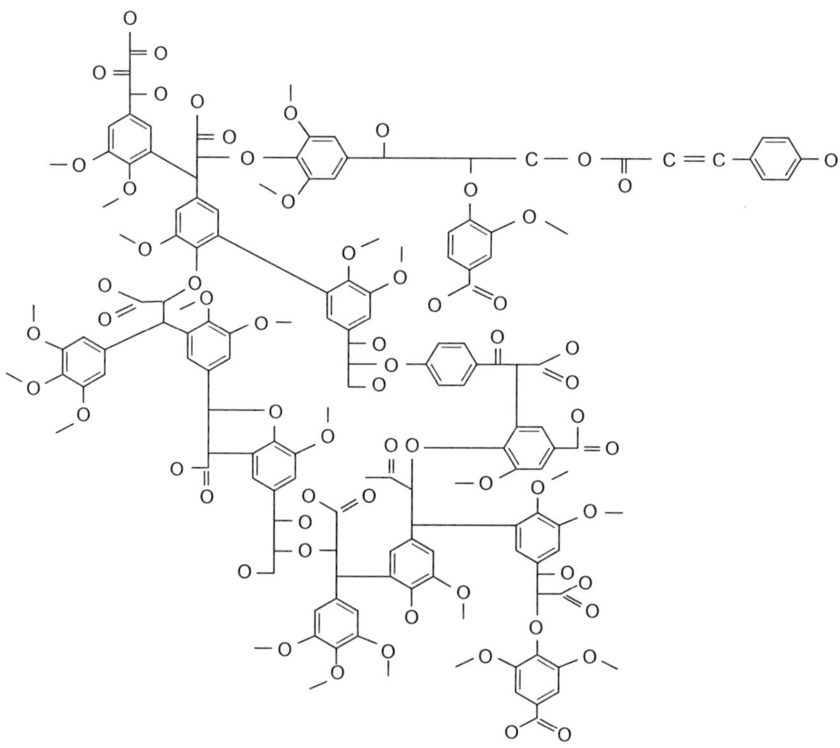

FIGURE 8.49 Simplified structure of lignin [4].

composition varies with the plant source and the pretreatment. Live lignin cannot be recovered, since during the recovery process the lignin structure is modified. Table 8.6 shows some examples for the composition of different lignin materials [76]. The structure of lignin shown in Figure 8.49 is simplified [4].

TABLE 8.6 Composition of Lignin from Different Sources

	Kraft Pine Lignin	Organosolv Lignin from Oat Hulls	Hardwood Residue	Switchgrass Residue
Glucan		11.1	47.5	22.5
Xylan		30.9	7.6	2.3
Galactan		3.6	1.8	2.1
Arabinan		5.5	0.8	1.8
Mannan		–	0.8	2.9
Klason lignin	92.3	39.3	32.2	41.1
Sulfur	2.1	–	–	–
Moisture	5.6	6.1	5.2	3.1
Ash	–	1.4	0.94	23.2

Air dry, wt%.
Modified from Ref. [76].

8.3.5.2 Conversion of Lignin in Water

8.3.5.2.1 Dissolution of Lignin

Isolated lignin is more cross-linked than native lignin. Hence, it can be assumed that isolated lignin is more resistant to hydrothermal degradation than native lignin [56] and is a reasonable material for representing lignin conversion. Experiments in continuous and semicontinuous modes show that organosolv lignin, which is essentially water insoluble at ambient conditions, can be completely dissolved in high-temperature and supercritical water [56]. Yet, the hydrolysis products are only partly soluble in water at ambient conditions.

The amount of insoluble (at ambient conditions) reaction products decreases rapidly at temperatures of $T > 285$ °C and within less than $t = 60$ s residence time, as has been shown by Bobleter and Concin [77] for the hydrothermal treatment of poplar wood lignin, see Figure 8.50. At a constant reaction temperature, a minimum of insoluble reaction products occurs which is shifted to a lower amount of insoluble products and shorter residence times with increasing temperature. At extended reaction times beyond $t = 60$ s, the amount of insoluble reaction products increases equally fast with time. At near-critical temperatures, the amount of residue reaches about 35% after $t = 2$ min.

Similar results were obtained with the conversion of organosolv lignin [56]. At conversion temperatures of $T = 370$–380 °C, a pressure of $P = 25$ MPa, and a residence time of about $t = 30$–40 s, 20–25% of the original carbon is dissolved in water at ambient temperature. About 75% of the carbon is present in form of a water-insoluble and very finely suspended phase at

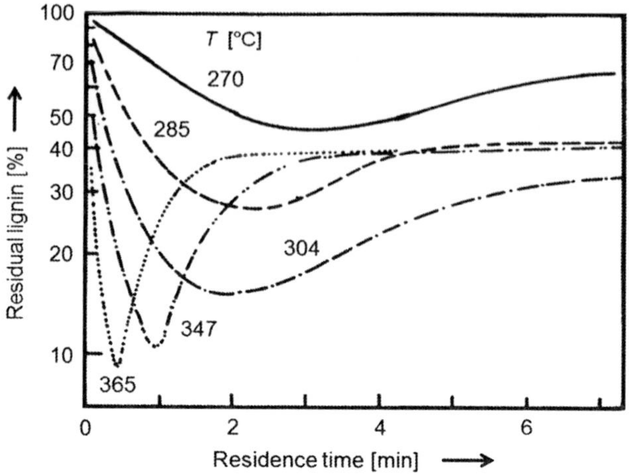

FIGURE 8.50 Hydrothermal degradation of lignin. Water-insoluble reaction products at ambient temperature. *Modified from Ref. [77].*

ambient conditions [56]. The amount of water-insoluble products is decreasing up to a reaction temperature of about $T=450$ °C, where 40% soluble products have been found. At even higher temperatures, the amount of water-insoluble products decreases to 14% at $T=480$ °C and 5% at $T=490$ °C. The reduction in water-soluble carbon can be partly attributed to the formation of gaseous species like carbon dioxide and methane. Another reason is the recombination of reactive lignin fragments to water-insoluble residues. The low water density at higher temperatures presumably leads to free-radical reactions producing insoluble pyrolysis products [56].

These results lead to the conclusion that a complete conversion to ambient temperature-soluble products from isolated lignin is not feasible in pure high-temperature or supercritical water. Furthermore, the treatment in pure water leads to a broad product spectrum and not to the selective production of few compounds, as can be seen in Figure 8.51 from the GC/MS analysis of soluble reaction products [56]. The chromatogram shows the existence of numerous degradation products in considerable concentrations. Although some major classes of degradation products can be identified, for example, phenols, phenol derivatives, and other aromatic substances, a complete identification and quantification of all reaction products is not of practical importance. Nevertheless, researchers have explored the conversion of lignin using high-temperature and supercritical water to extract potentially valuable chemicals [78–81]. At extended reaction times, the number of products is drastically reduced due to the disappearance of instable intermediate products. This can be seen in Figure 8.52, where several chromatograms show the development of major product compounds with time [76].

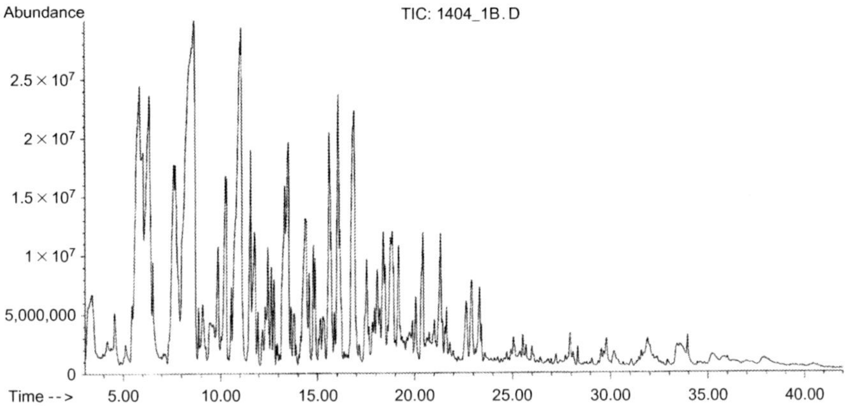

FIGURE 8.51 GC/MS analysis of lignin degradation products after hydrolysis. $T = 417$ °C, $P = 25$ MPa, residence time $t = 29$ s [56].

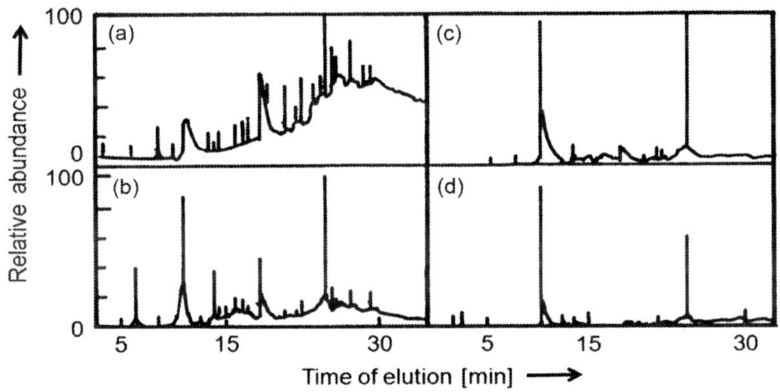

FIGURE 8.52 GC chromatograms of the liquid products from Kraft pine lignin. $T = 300$ °C, $P = 10$ MPa, residence time: (a) 2 min, (b) 4 min, (c) 10 min, (d) 30 min. *Modified from Ref. [76].*

8.3.5.2.2 Conversion of Lignin and Its Pathways

Lignin decomposition in supercritical water starts by hydrolysis to form phenolic compounds. Ether bonds in lignin occur between one propyl side chain of a hydroxylphenylpropane unit and the hydroxyl group attached to the benzene ring of another hydroxylphenylpropane unit. Therefore, disassembly of ether bonds and carbon—carbon bonds is important in the degradation of lignin. Hydrolysis occurs at ether and ester bonds, cleaving them by the addition of one molecule of water for every broken linkage, a process which is accelerated by the high ion product of water [61,74,82,83].

Figure 8.53 shows a simplified reaction scheme for lignin decomposition under high-temperature and supercritical water conditions. A detailed

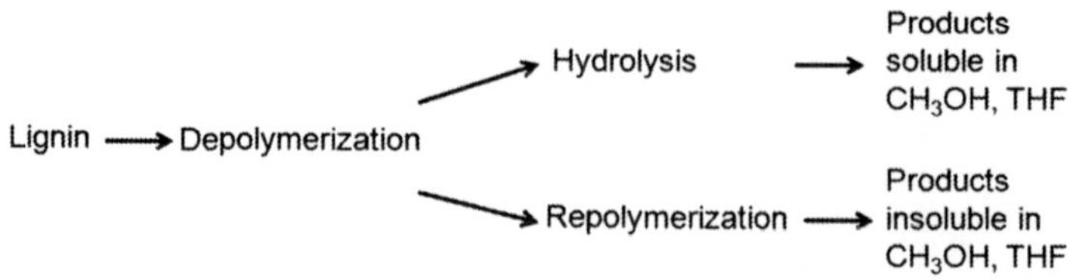

FIGURE 8.53 Simplified reaction scheme for lignin conversion in high-temperature and supercritical water.

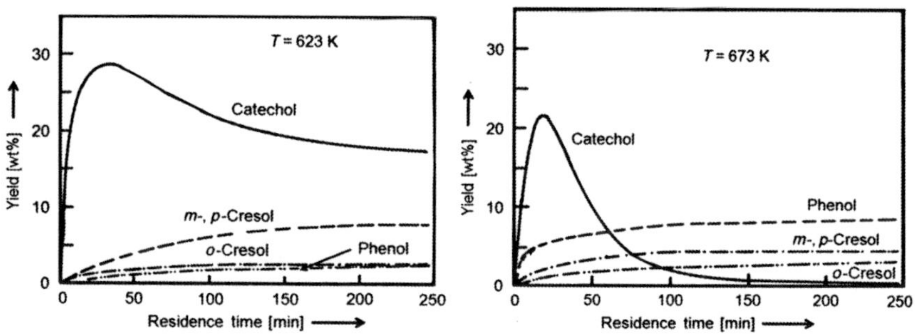

FIGURE 8.54 Yield of products for lignin decomposition in high-temperature and supercritical water at $P=25$ MPa. *Modified from Ref. [84].*

mechanism is complicated by the fact that every compound can be obtained through different reaction pathways. Degradation of lignin in near- and supercritical water strongly depends on water density. Hydrolysis is a key reaction that promotes the degradation of natural polymers. In parallel, formation of char due to repolymerization occurs at the same time, which may be the main reason why the yield of chemicals is low for lignin conversion [74].

Degradation of lignin was studied in high-temperature and supercritical water at temperatures of $T=623$ and 673 K in batch experiments. Results show the typical differences between conversion of lignin in near-critical and supercritical water. Main methanol-soluble products obtained in batch mode at a water density $\rho_{H_2O}=0.16$–$0.67\ g\ cm^{-3}$ were catechol, phenol, *m*-, *p*-cresol, and *o*-cresol [84]. Yield of products is shown in Figure 8.54, and the influence of density on product yields for phenol and *p*-, *m*-cresol in Figure 8.55.

Reaction times of the batch experiments are two orders of magnitude longer than for continuous or semicontinuous experiments. Within short-reaction time, the catechol yield is high with a high selectivity at the near-critical water temperature of $T=623$ K. At the supercritical temperature of $T=673$ K, the catechol yield is initially still high, but catechol rapidly degrades after about $t=10$ min reaction time. At near-critical conditions, *m*-, *p*-cresol is the main other product, while at supercritical conditions, phenol is the main product compound. Density is of influence on product formation (Figure 8.55).

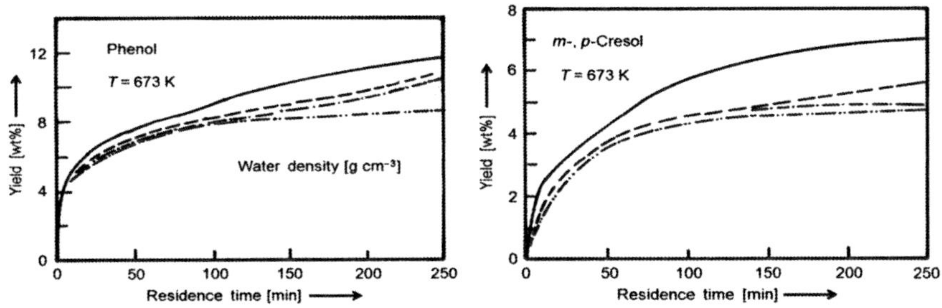

FIGURE 8.55 Influence of water density on the yield of phenol and *m*-, *p*-cresol for lignin decomposition in supercritical water. Water density ρ_{H_2O} (g cm^{-3}): ———, 0.16; —•—•, 0.35; — — —, 0.47; —••—, 0.52. *Modified from Ref. [84].*

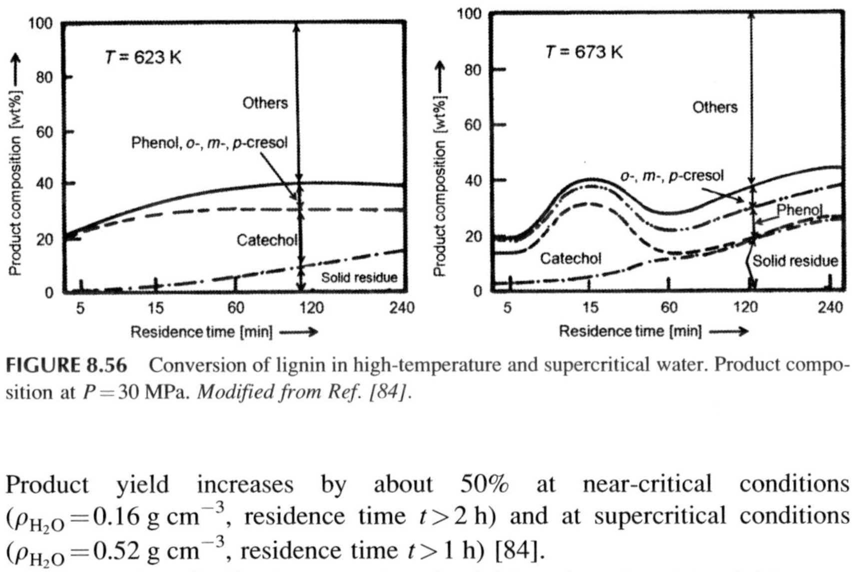

FIGURE 8.56 Conversion of lignin in high-temperature and supercritical water. Product composition at $P = 30$ MPa. *Modified from Ref. [84].*

Product yield increases by about 50% at near-critical conditions ($\rho_{H_2O} = 0.16$ g cm^{-3}, residence time $t > 2$ h) and at supercritical conditions ($\rho_{H_2O} = 0.52$ g cm^{-3}, residence time $t > 1$ h) [84].

The product distribution to methanol-soluble and methanol-insoluble compounds is shown in Figure 8.56. The methanol-insoluble (solid) part of the products increases with reaction time. At supercritical conditions, this part of the products is higher than at subcritical conditions and reaches nearly 20% after $t = 2$ h. The identified methanol-soluble compounds catechol, phenol, *o*-, *m*-, *p*-cresol are only a minor part of the total amount of methanol-soluble compounds [84].

Lignin is degraded and liquefied in water at temperatures of $T = 623$ and 673 K and pressures of $P = 25$–40 MPa, as discussed before. Higher temperatures favor the conversion of lignin and the formation of heavier components. The compounds identified as catechol, phenol, *m*-, *p*-cresol, and *o*-cresol are only a part of the total methanol-soluble derivative compounds. With reaction time, the amount of the lower molecular weight compounds increases. At the same time, recombination to solid, methanol-insoluble compounds also increases with temperature and reaction time [84].

Conversion of lignin produces a substantial amount of solid residue which is often found as being resistant to further reaction. A pathway to total conversion of lignin is oxidation, which is treated in Chapter 10. Another way can be the addition of compounds to promote the formation of soluble products and reduce the amount of solid residue. The addition of phenol, as a compound that occurs anyway in the reaction mixture, has considerable success [75]. The organosolv lignin used was completely soluble in tetrahydrofuran (THF). Decomposition of this organosolv lignin in supercritical water at $T=673$ K and a water density of up to $\rho_{H_2O}=0.5$ g cm^{-3} proves that the addition of phenol reduces the amount of solid residue, insoluble in THF. THF-soluble compounds found during the conversion of lignin are syringols, guaiacols, and catechols. Figure 8.57 shows the influence of phenol on the formation of solid residue in lignin conversion in supercritical water [75]. A pronounced effect is obtained by adding phenol, but at a high phenol-to-lignin ratio. The influence of density, shown in Figure 8.57 on the right-hand side for a reaction time of $t=69$ min, is great for densities $\rho_{H_2O}<0.1$ g cm^{-3} and becomes smaller with higher densities. Molecular weight of the products becomes lower with increasing density and with the addition of phenol, see Figure 8.58. With respect to a low yield in insoluble products, the reaction time should be kept short, around $t<10$ min.

The presence of phenol prevents cross-linking by reacting with reactive sites of the decomposed fragments and suppresses the formation of char significantly by employing a sufficiently high phenol-to-water ratio [85,86]. A similar effect occurs with the addition of p-cresol [81,87].

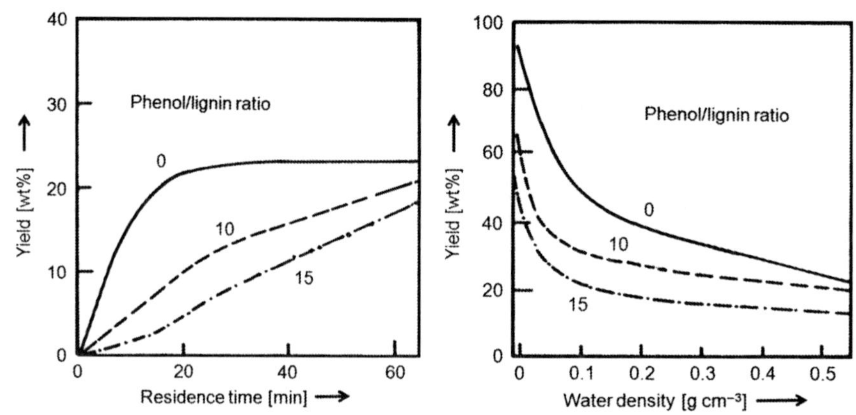

FIGURE 8.57 Conversion of organosolv lignin in supercritical water. Formation of THF-insoluble residue at $T=673$ K. Left: Influence of reaction time at a water density of $\rho_{H_2O}=0.5$ g cm^{-3}. Right: Influence of water density at a reaction time of $t=60$ min. *Modified from Ref. [75].*

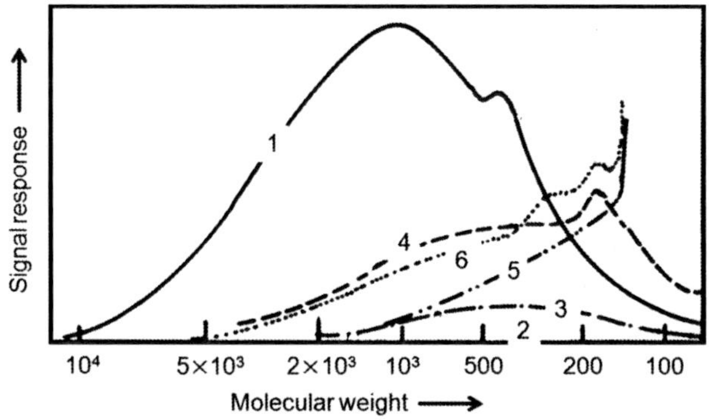

FIGURE 8.58 Conversion of organosolv lignin in supercritical water. GPC analysis of raw lignin (1) and THF-soluble products at a water density of $\rho_{H_2O}=0$ g cm^{-3} (2), $\rho_{H_2O}=0.1$ g cm^{-3}(3), and $\rho_{H_2O}=0.5$ g cm^{-3} (4) and at a water density of $\rho_{H_2O}=0.1$ g cm^{-3}, phenol/lignin ratio 10 (5), and at a water density of $\rho_{H_2O}=0.5$ g cm^{-3}, phenol/lignin ratio 10 (6). *Modified from Ref. [75].*

There is a great number of further and similar investigations on the conversion of lignin, for example [84,88,89]. Conditions of the experiments vary and so do results. A detailed discussion is beyond the scope of this book.

Generally, the reaction pathways that take place during decomposition of lignin consist of many steps and involve a large number of compounds. A detailed modeling of the kinetics is of fundamental interest, but of minor practical importance, since many of the intermediate and product compounds can be produced in various ways. The conversion of lignin, the formation of solid residue, and the formation of main conversion compounds is a good guide for considering conversion processes and for formulation of a simplified model for the lignin conversion in water.

8.3.5.2.3 Kinetic Modeling

A simple model for the lignin conversion in high-temperature and supercritical water is presented in the following. The model is based on experiments on the conversion of lignin from different sources in water, carried out in batch mode at $T=300$ °C at saturation pressure, and at $T=374$ °C and a density of about $\rho_{H_2O}=0.6$ g cm^{-3}, with a residence time of $t=10$ min, additional to the time for heating of about 2–3 min [76].

The yields depend on the composition and structure of the raw materials. At least the structure is determined by the pretreatment process. Therefore, each raw material needs to be evaluated individually [76]. The yields for some materials are presented in Table 8.7.

The simplified model is based on a fast and a slow reaction phase. Lignin is converted fast in about $t=1$ min into low-molecular components, monomers, and higher molecular components, mainly oligomers [77]. Then, in a

TABLE 8.7 Products of Treatment of Lignin Materials with Supercritical Water at $T=374$ °C [76]

	Yield (wt%)				Gas yield (wt% of gas)			
	Residue	*Liquid*	*Gas*	*Dried Liquid*	H_2	CO	CH_4	CO_2
Kraft pine lignin	37.6	57.8	4.6	33.9	3.0	7.4	12.7	76.9
Organosolv lignin	12.1	79.1	8.8	42.8	2.1	19.7	0.3	77.9
Hardwood residue	24.2	56.3	19.4	17.6	3.8	22.9	8.0	65.3
Switchgrass residue	48.8	32.6	18.6	15.2	2.7	25.3	7.9	64.1
Aspen lignin	3.8	90.2	5.9	78.0	7.6	7.7	13.3	71.4

slower reaction, insoluble polymers and char are produced by condensation between soluble components and insoluble polymers. For the lignin concentration during the fast reaction:

$$\frac{d[L]}{dt} = -k_1[L] \tag{8.9}$$

with [L], lignin concentration at time t in wt% of the initial amount of lignin; k_1, reaction rate constant.

In the slow degradation phase, a part of the soluble compounds reacts with the other compound in solution to form the insoluble polymers and char (P_L):

$$\frac{d[P_L]}{dt} = k_2[L_s]^2 \tag{8.10}$$

with [L_S], sum of the low- and high-molecular weight compounds soluble in acetone-water; [P_L], concentration of insoluble components and char; k_2, reaction rate constant.

For the first reaction phase, $k_1=2.95\ \text{min}^{-1}$ for both temperatures. For the second reaction phase, $k_2=0.030\ \text{min}^{-1}$ for $T=300$ °C and $k_2=0.072\ \text{min}^{-1}$ for $T=374$ °C. Lignin conversion follows an Arrhenius equation with the pre-exponential factor of $k_0=70.2\ \text{min}^{-1}$ and an activation energy of $E_A=37\ \text{kJ mol}^{-1}$ [76]. Comparison of modeling results and experimental data is shown in Figure 8.59. The simple model is able to describe the overall formation of residue with reasonable coincidence to experimental data.

Beside the conversion of lignin which varies in composition according to source and pretreatment, model compounds for this conversion may give hints

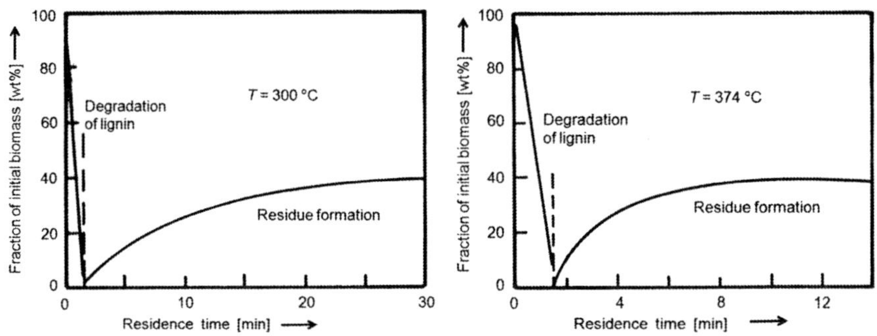

FIGURE 8.59 Conversion of lignin in high-temperature water at $T=300$ °C (left) and in supercritical water at $T=374$ °C. *Modified from Ref. [76].*

on the underlying degradation schemes. Therefore, the degradation of guaiacol and catechol as model compounds are of interest, since they belong to the most abundant phenolic compounds in nature and contain representative functional groups [74].

8.3.5.3 Conversion of Model Compounds for Lignin

8.3.5.3.1 Guaiacol

Reaction of guaiacol in high-temperature and supercritical water leads to products like catechol, phenol, and *o*-cresol through primary hydrolysis and to high-molecular weight compounds as the major products of pyrolysis [74]. The selectivity of guaiacol hydrolysis is sensitive to density of water, as shown in experiments with supercritical water at $T=383$ °C and reduced densities of water of $\rho_R=0$–1.6 in batch mode. Guaiacol pyrolysis yields catechol and char as major and minor products including phenol and *o*-cresol [90].

The product spectrum of guaiacol pyrolysis and hydrolysis is shown in Figure 8.60. The decrease in reactant concentration during pyrolysis corresponds to a pseudo-first-order rate constant of $k=0.0198$ min^{-1}.

The principal primary product of the pyrolysis of guaiacol is catechol, reaching a molar yield of about 20% at a residence time of $t=90$ min. Hydrolysis of guaiacol at a reduced density of $\rho_R=1.6$ leads to a decrease in guaiacol concentration with a pseudo-first-order rate constant of $k=0.0242$ min^{-1}. In hydrolysis, yields of catechol increase over those of pyrolysis to about 55%, while the yields of *o*-cresol and char decrease and the concentrations of guaiacol and phenol remain essentially unchanged [90].

The influence of density of product yields is shown in Figure 8.61 for reduced densities of $\rho_R=0.0$–1.6.

The conversion of guaiacol is insensitive to water density in this range. At low values of $\rho_{R,w}$ ($0<\rho_{R,w}<0.10$–0.15), the addition of water even decreases guaiacol conversion relative to pyrolysis. Only at higher reduced

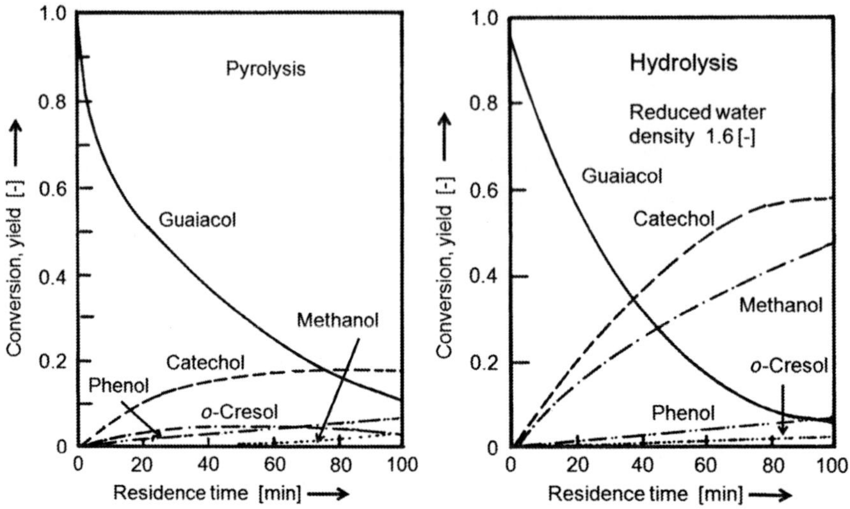

FIGURE 8.60 Conversion of guaiacol by pyrolysis (left) and hydrolysis in supercritical water (right) at $T=383\ °C$, reduced density of water $\rho_R=1.6$. *Modified from Ref. [90].*

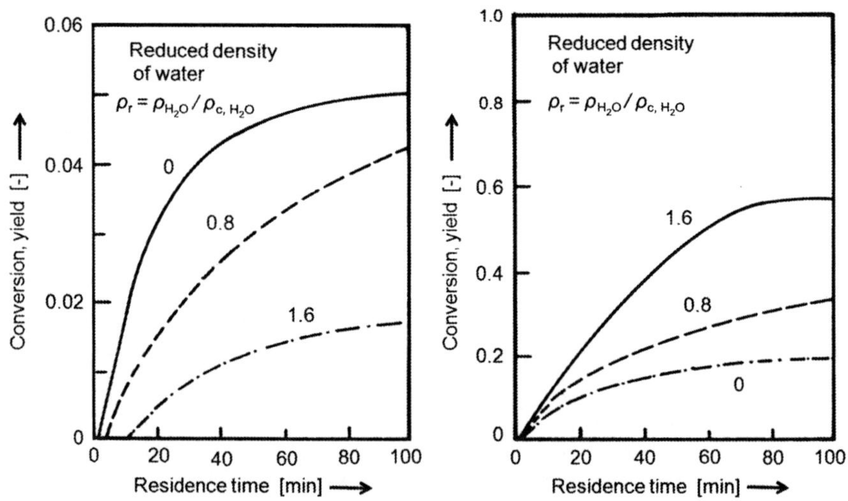

FIGURE 8.61 Conversion of guaiacol by hydrolysis in supercritical water at $T=383\ °C$. Influence of reduced density of water on yield of catechol (left) and *o*-cresol (right). *Modified from Ref. [90].*

densities >1, guaiacol conversion increases. At a water density of $\rho_{R,w}=2.1$, the conversion of guaiacol is greater than that for neat pyrolysis at a reaction time of $t=90$ min [90].

Two competing reactions occur during guaiacol reaction in water. The first is neat pyrolysis that leads to catechol and high-molecular weight material as major products. The second is hydrolysis of guaiacol to catechol and methanol, the selectivity to which is a continuous and increasing function of water density [90].

8.3.5.3.2 Catechol

Catechol hydrolyses in supercritical water to phenol as the main decomposition product and a number of other compounds as can be seen from the chromatogram shown in Figure 8.62. Other identified compounds are cyclopentanone, 2-cyclopentenone, 1,2-benzenedicarboxylic acid, nonylphenol, 1,4-dipropylbenzene, acetophenone, 2,6-di-*tert*-butylnaphthalene, 4-butoxyphenol, *o*-ethoxyphenol, 2-methylterephthalaldehyde, *p*-diethoxybenzene, and 5-methoxy-2,3,4-trimethylphenol [91].

The conversion of catechol and the formation of phenol increases with temperature and water density (Figures 8.63 and 8.64). First-order kinetics are appropriate to describe the catechol decomposition in supercritical water with a global rate constant value for the catechol decomposition of $k = 3.0 \times 10^{-4}$ to $11.0 \times 10^{-4}\ \text{min}^{-1}$ (see Figure 8.65).

High-temperature and supercritical water is a promising method to treat lignin and its derived compounds. Decomposition of lignin is a complex process with different pathways, including competitive and consecutive reactions. Lignin thermally decomposes over a broad temperature range due to the existence of various oxygen functional groups, with their scission occurring at different temperatures. Under high-temperature and supercritical water conditions, lignin

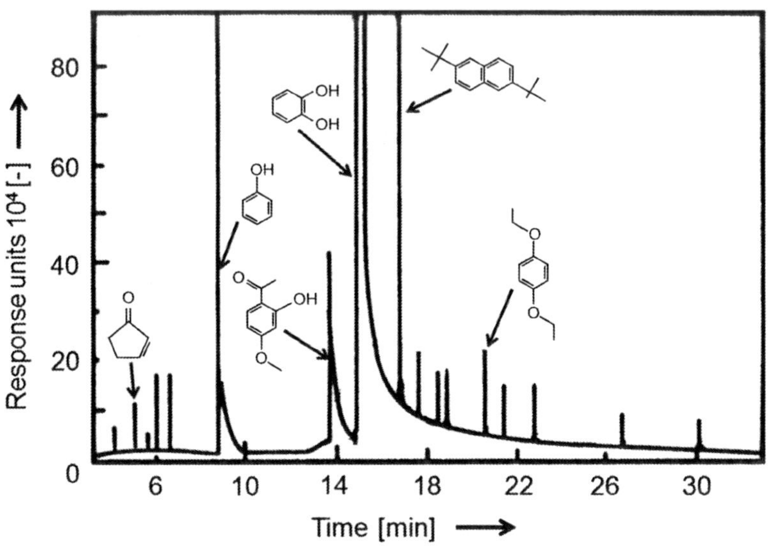

FIGURE 8.62 Conversion of catechol in supercritical water. GC/MS chromatogram of the liquid phase recovered after treatment at $T = 673$ K, $P = 30$ MPa, $t = 240$ min. *Modified from Ref. [91].*

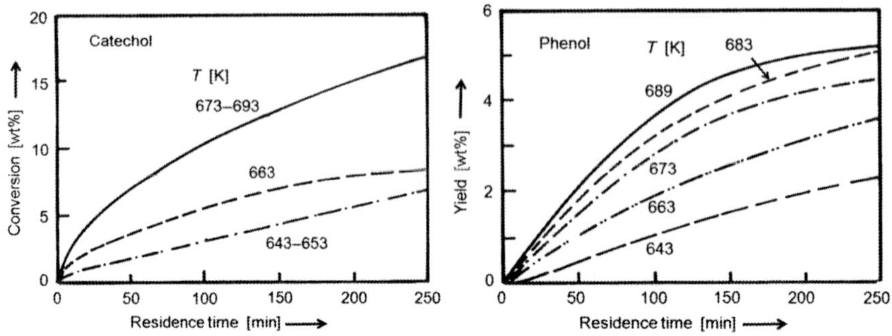

FIGURE 8.63 Conversion of catechol in supercritical water. Influence of temperature on catechol conversion (left) and phenol formation (right). $P = 30$ MPa. *Modified from Ref. [91].*

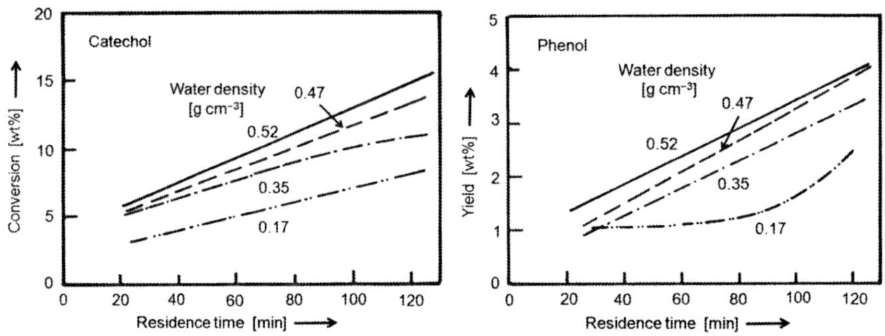

FIGURE 8.64 Conversion of catechol in supercritical water. Influence of water density on catechol conversion (left) and phenol formation (right). $T = 673$ K [91].

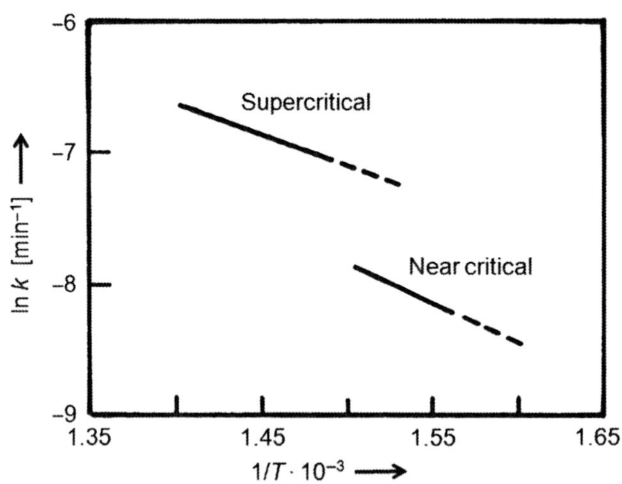

FIGURE 8.65 Arrhenius plot of the rate constants for the decomposition of catechol in near- and supercritical water. *Modified from Ref. [91].*

decomposes into phenolic compounds as the main products. In parallel, formation of char due to repolymerization occurs. In supercritical water, rapid and direct decomposition pathways favor the decomposition of guaiacol and catechol, though catechol is a very stable compound not easily cleaved under thermal conditions.

8.3.6 Proteins

Proteins are fundamental building blocks of living cells. To understand the action of high-temperature water on proteins, their chemical structure is addressed, followed by the discussion of the conversion of amino acids and albumin as a model protein. The kinetics of the conversions and the role of adding CO_2 and NaCl are discussed.

8.3.6.1 Structure of Amino Acids and Proteins

Proteins are macromolecules consisting of basically carbon, hydrogen, oxygen, nitrogen, and sulfur and are building blocks of biological cells. Proteins are constructed of 20 amino acids which are connected by peptide bonding to polymers. The peptide bonds can be hydrolyzed in aqueous environment (see Figure 8.66). In high-temperature water, the hydrolysis of proteins can be carried out without catalysts, a processing pathway that will be discussed in this chapter.

The protein structure is characterized by peptide-bonded amino acids which form a nonbranched poly-peptide chain. For more than about 100 amino acids in a chain, the polymer is called "protein." Some of the proteins contain disulfide bridges that originate from the oxidation of cysteine and form the disulfide cysteine (CYSS). Proteins with a high content in cysteine act as mechanical stabilizers in protein structures, for example, for keratin in feathers and hair.

The three-dimensional structure of a protein consists of a primary one, which is the sequence of amino acids, of a secondary one, which comprises the folding scheme of the amino acid chains, like the α-helix, of a tertiary one comprising the three-dimensional folding of the secondary protein structure,

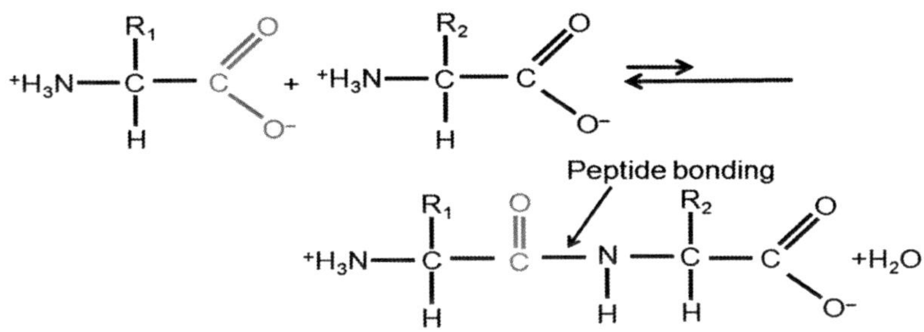

FIGURE 8.66 Formation of a peptide bond. *Modified from Ref. [92].* (For color version of this figure, the reader is referred to the online version of this chapter.)

and of a quartery one, which comprises the three-dimensional structure of more than one protein. The formation of the peptide bonding is shown in Figure 8.66.

Hydrolytic decomposition of proteins is discussed using albumin as an example. Albumin for the experiments is produced from bovine serum and named bovine serum albumin, BSA. The elliptical shaped BSA-protein molecule consists of a chain of 607 amino acids and has a molecular weight of about 69,300 g mol^{-1}. All amino acids are part of the BSA-molecule. For the hydrolytic conversion of proteins, the thermal stability of the amino acids is important information and is therefore addressed in the following.

8.3.6.2 Conversion of Amino Acids

Stability and degradation of amino acids are of fundamental importance for the conversion of proteins. Qualitatively, a great number of degradation products is detected after the hydrolysis of various amino acids in supercritical water at $T=375$ °C after $t=3.75$ h [93]. The products consist of gaseous products like CO, CO_2, H_2, CH_4, low-molecular alkanes and alkenes, alcohols ($\leq$C5), amines, aldehydes, and carbonic acids [92].

According to other investigations, products of amino acid hydrolysis are mainly organic acids such as acetic, formic, and lactic acid [94–96]. Yield increases with temperature and reaction time. Further degradation at higher temperatures leads to CO, CO_2, and H_2O. Parallel to degradation reactions, polymerization of glycine can occur to oligomers at $T=250$–400 °C, $P=25$–44 MPa [97,98]. Formation polymers from glycine comprising 8–10 monomers is reported at relatively mild conditions of $T=270$ °C, $P=10$ MPa, and a residence time of $t=27$ s [99]. Hydrothermal degradation in continuous mode in a tubular reactor at $T=200$–340 °C and $P=20$ MPa leads to the conclusion that two main pathways are followed during degradation of amino acids: deamination to organic acids and ammonia and decarboxylation to carbonic acids and amides [100]. Conversion rates of serine and asparagic acid are higher than those of leucine and phenylalanine.

Stability of amino acids has also been discussed in context with the hypothesis concerning the relationship between submarine hot springs and the origin of life. Stability of amino acids under these conditions is supposed to end around $T=250$ °C [101]. In the following, the limits of stability and the conversion reactions are discussed for some of the more important amino acids. The conversion of glycine and alanine, two of the simplest amino acids, confirm the already mentioned conclusion that the primary mechanisms of degradation of these amino acids are decarboxylation and deamination [102]. The amino acids alanine and glycine are contacted with high-temperature water at temperatures of $T=250$–450 °C, a pressure of $P=24$ MPa, residence times of $t=2.5$–30 s, and feed concentrations of $x_F=1$–2 wt% in a continuously operated reactor. The conversion of alanine is shown in Figure 8.67, and the concentration of the conversion products in Figure 8.68. For the conversion of glycine, results are shown in Figure 8.69.

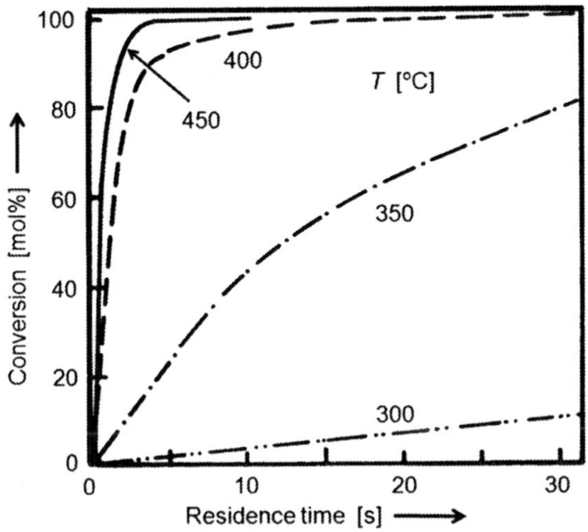

FIGURE 8.67 Conversion of alanine in high-temperature water at $P = 34$ MPa, feed concentration $x_F = 1$ wt%. *Modified from Ref. [102].*

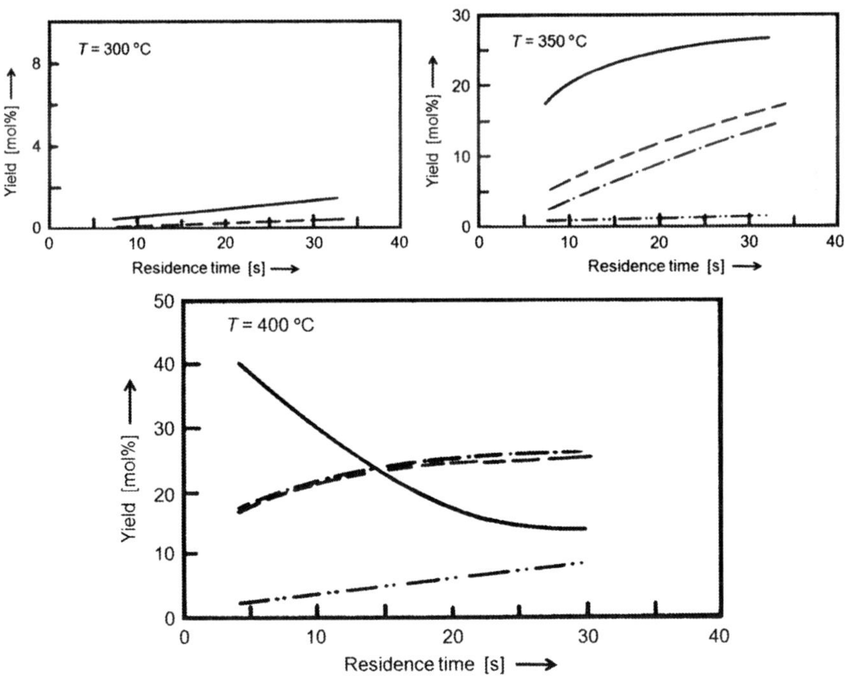

FIGURE 8.68 Product yield of alanine decomposition at hydrothermal conditions at $P = 34$ MPa. Feed concentration of $x_F = 1$ wt%, ———, lactic acid; – – –, ethylamine; —•—, CO_2; —••—, acetaldehyde. *Modified from Ref. [102].*

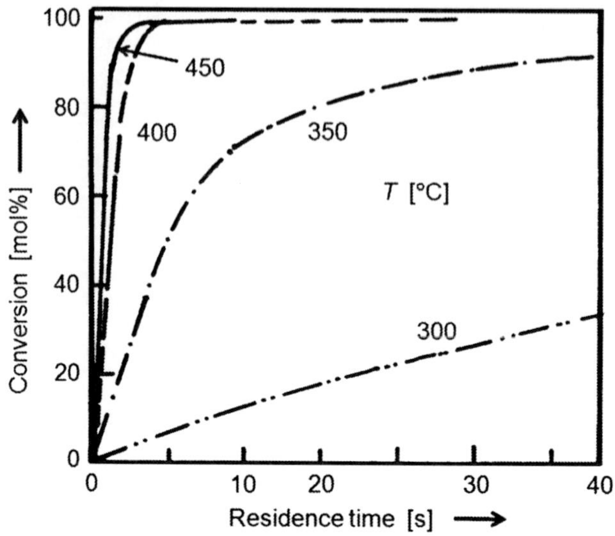

FIGURE 8.69 Conversion of glycine in high-temperature water at $P=34$ MPa and a feed concentration of $x_F=1$ wt%. *Modified from Ref. [102].*

Conversion of alanine and glycine is strongly temperature dependent. At $T=250$ °C, the hydrolysis reaction is still slow. At $T=350$ °C, the reaction is much faster and is further accelerated for $T=400$ °C, while a further temperature increase is of minor effect on the conversion reaction rate. Within $P=24$–34 MPa, pressure has no significant influence. Increase of the concentration from $x_F=1$ to 2 wt% increases the reaction rate.

Decomposition of alanine and glycine can be described by simple rate laws with the order of $n=0.47$ for alanine and $n=0.78$ for glycine [102]. For alanine, the preexponential factor is $k_0=3.6\times10^{11}$ $mol^{0.53}\,L^{-0.53}\,s^{-1}$and the activation energy $E_a=160$ kJ mol^{-1}. For glycine decomposition, the preexponential factor is $k_0=1.4\times10^{12}$ $mol^{0.22}\,L^{-0.22}\,s^{-1}$ and the activation energy is $E_a=156$ kJ mol^{-1} [102]. Sato *et al.* [100] assumed for the decomposition of alanine and glycine at a concentration of $x_F=0.1$ wt% a reaction order of 1 and obtained different values for the Arrhenius equation. For alanine decomposition, the preexponential factor is $k_0=2.7\times10^{12}$ and the activation energy is $E_a=154$ kJ mol^{-1}. For glycine decomposition, the preexponential factor is $k_0=3.5\times10^{13}$ with an activation energy of $E_a=166$ kJ mol^{-1}.

The stability of the amino acids glycine, proline, serine, and valine in high-temperature water was investigated by Rogalinski *et al.* [92] at $T=250$–350 °C, $P=25$ MPa, feed concentration of $x_F=0.1$ wt% in a continuous reactor for single amino acids and for a mixture of amino acids. In preliminary experiments, the onset of the degradation of amino acids was determined at about $T=180$ °C by visual observation of a brownish hue of the reaction solution. Therefore, preheating of the reaction solution was limited to this temperature. Results are shown in Figures 8.70 and 8.71.

Glycine is stable at $T=250$ °C. At $T=300$ °C, a substantial conversion has taken place after $t=50$ s. At a temperature of $T=350$ °C, glycine is no more detectable after that reaction time. The conversion of glycine is nearly the same for the pure component as in the mixture. For valine, the results are comparable to that of glycine. Serine degrades faster than glycine and valine. In the mixture, degradation of serine is even faster. At $T=250$ °C, nearly no serine can be detected. A similar trend is observed for proline. But for the pure component, even at $T=350$ °C, about 43% are nonreacted, very different

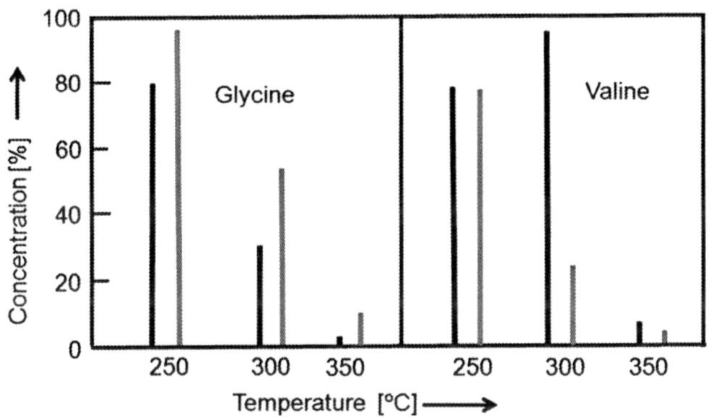

FIGURE 8.70 Stability of glycine and valine in high-temperature water. ———, pure component; ———, component in the mixture. $P=25$ MPa, residence time $t=50$ s. *Modified from Ref. [92].* (For interpretation of the references to color in this figure legend, the reader is referred to the online version of this chapter.)

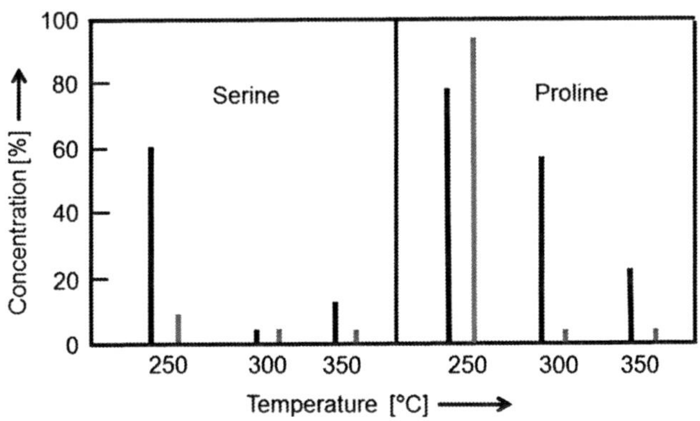

FIGURE 8.71 Stability of serine and proline in high-temperature water. ———, pure component; ———, component in the mixture. $P=25$ MPa, residence time $t=50$ s. *Modified from Ref. [92].* (For interpretation of the references to color in this figure legend, the reader is referred to the online version of this chapter.)

to the behavior in the mixture. Stability of serine and proline is negatively influenced by the presence of other amino acids.

The experimental values at $T = 300$ °C for valine and at $T = 350$ °C for serine are probably due to experimental error, although yields, including higher ones than 100% for simple amino acids in mixtures may stem from decomposition reactions of more complex amino acids. Also the analysis of amino acids, in particular, at low concentrations, may produce scattering values.

8.3.6.3 Degradation Products of Protein and Amino Acid Conversion

Amino acids at high temperatures react to low-molecular degradation products. Products soluble in the aqueous solution are shown as an example in Figure 8.72. Acetic acid is the main degradation product at short-residence times, followed at $t = 3$–5 min reaction time by several other components like iso- and *n*-butyric acid and ethanolamine. Degradation products are formed in the same order of magnitude as amino acids.

8.3.6.4 Conversion of Proteins

Proteins are converted in high-temperature and supercritical water to amino acids, other low-molecular degradation products, and gases. Yield of amino acids is, in general, below 20%. Beside the produced gas, a substantial amount of the feed is found after the reaction as oligomers and polymers, and as solid residue. At least part of these fractions is formed by polymerization of amino acids. Experimental results for the degradation of a model protein (albumin) illustrate this. But before presenting these results, a method for kinetic modeling of the degradation reaction of proteins is recapitulated, which has been already discussed in Chapter 5.

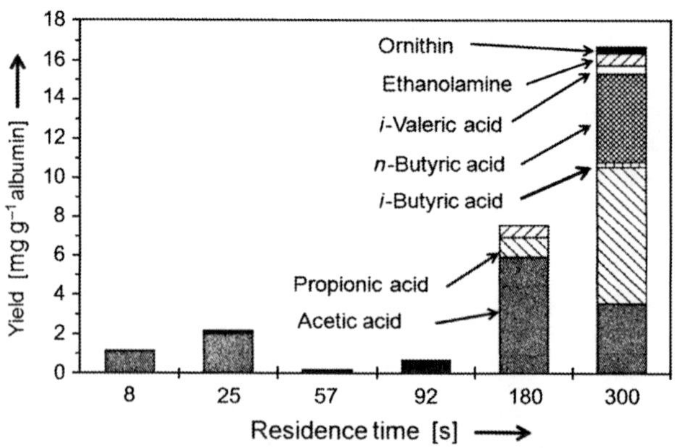

FIGURE 8.72 Low-molecular degradation products of hydrolysis of albumin in high-temperature water. $T = 250$ °C, $P = 25$ MPa [92].

8.3.6.4.1 Kinetic Modeling

The conversion of proteins, in the following albumin as a model compound for proteins, is formulated as a simplified model as shown in Equation (8.11), which does not take into account parallel reactions and repolymerization reactions. The justification for that is the interest in the formation of amino acids.

$$\text{Albumin} \xrightarrow{k_1} \text{Amino acids} \xrightarrow{k_2} \text{Degradation products} \tag{8.11}$$

Albumin in high-temperature water is hydrolyzed by cleavage of the peptide bonds to amino acids such as alanine and glycine, which may further react to degradation products like acetic acid, propionic acid, ethanolamine, and others. For kinetic modeling of the albumin conversion, stationary conditions, plug flow, isothermal and isobaric conditions are assumed for the reaction occurring in an ideal tubular reactor. The conversion of albumin to amino acids, as well as the formation and degradation of amino acids is lumped in a pseudo-first-order reaction, Equation (8.12):

$$\frac{\mathrm{d}c_{\mathrm{Pr}}}{\mathrm{d}t} = -k_1 c_{\mathrm{Pr}}, \tag{8.12}$$

with c_{Pr} as product concentration.

The concentration of the free amino acids is determined by the formation of the amino acids from albumin and the further reaction of amino acids to degradation products, Equation (8.13):

$$\frac{\mathrm{d}c_{\mathrm{AS}}}{\mathrm{d}t} = k_1 c_{\mathrm{Pr}} - k_2 c_{\mathrm{AS}}, \tag{8.13}$$

with c_{AS} for the concentration of amino acids.

The concentration of the degradation products is represented by Equation (8.14):

$$\frac{\mathrm{d}c_{\mathrm{Abb}}}{\mathrm{d}t} = k_2 c_{\mathrm{AS}}, \tag{8.14}$$

with c_{Abb} as the concentration of the degradation products.

Integration and combination of Equations (8.13) and (8.14) results in Equation (8.15), valid for the concentration of amino acids in the assumed sequential reactions.

$$\frac{c_{\mathrm{AS}}}{c_{\mathrm{Pr},0}} = \frac{k_1}{k_1 - k_2}[\exp(-k_1 t) - \exp(-k_2 t)] \tag{8.15}$$

Rate constants k_1 and k_2 are determined by fitting to the experimental data. Temperature dependence of the rate constants can be described with an Arrhenius equation (8.16).

$$k_{\mathrm{i}} = k_{0,\mathrm{i}} \exp\left(-\frac{E_{\mathrm{a}}}{RT}\right), \quad \text{or } \ln k_{\mathrm{i}} = \ln k_{\mathrm{i},0} - \frac{E_{\mathrm{a}}}{RT} \tag{8.16}$$

From Equation (8.16), the activation energy and the preexponential factor can be determined.

8.3.6.4.2 Conversion of Albumin as a Model Protein

The conversion of albumin was investigated in continuous mode in a tubular reactor at temperatures of $T = 250$–350 °C, pressures of $P = 5$–35 MPa, residence times of $t = 5$–6000 s, and initial feed concentrations of $x_F = 1$–6 wt% [92,103]. Pressure is not of significant influence on the conversion and the formation of individual amino acids, opposed to temperature and reaction time. In Figure 8.73, influence of these parameters on the conversion of albumin and in Figure 8.74 on the yield of amino acids is shown.

Yield of total amino acids increases at constant temperature with increasing residence time. At $T = 250$ °C, no maximum occurs within the experimental time limit of about $t = 1.5$ h and yield approaches 17 wt%. This yield is not reached again at higher temperatures. The conversion rate of albumin increases drastically

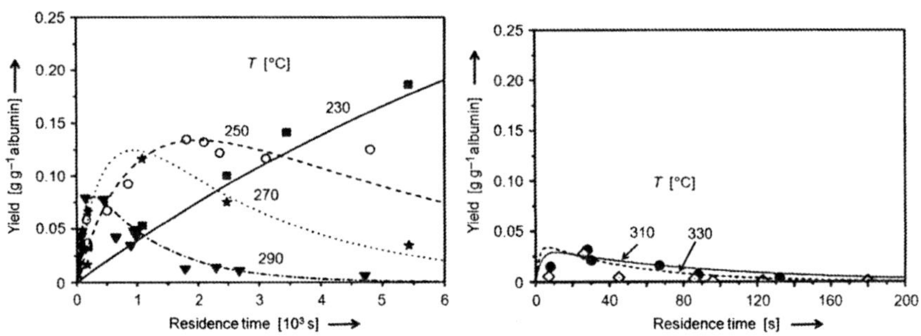

FIGURE 8.73 Yield of total amino acids during hydrolysis of albumin in high-temperature water. $P = 25$ MPa, feed concentration $x_F = 1$ wt% [92].

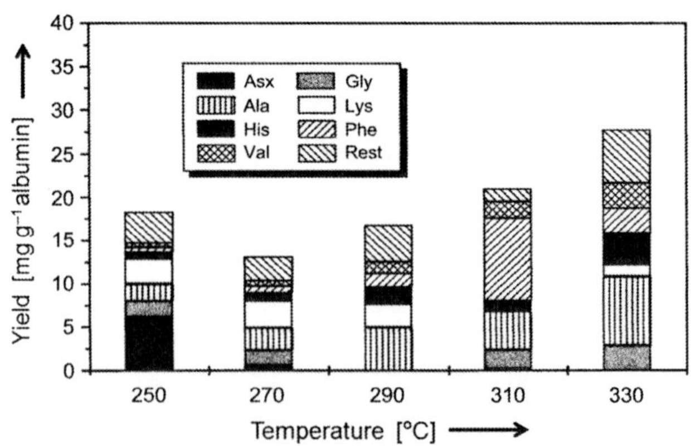

FIGURE 8.74 Yield of individual amino acids during hydrolysis of albumin in high-temperature water. $P = 25$ MPa, feed concentration $x_F = 1$ wt%, residence time $t \approx 30$ s [92].

with temperature but so does also the degradation of amino acids. A maximum in yield is reached at constant temperature and that maximum is shifted to shorter residence times with temperature. At $T=250$ °C, the maximum reaches 12.5 wt% at a residence time of $t=1.5$ h, while at $T=310$–330 °C, the maximum is reached in about $t=30$ s reaction time. Total yield of amino acids is steadily decreasing with temperature and is only 2.5 wt% above $T=310$ °C. The amino acid ASX (asparagic acid) is the major one at $T=250$ °C, but is present only in traces at $T=290$ °C. Alanine and glycine are more stable at high-temperature conditions and can be formed from other more complex amino acids. Reaction rate constants are summarized in Table 8.8. Degradation reactions of amino acids are faster than the formation of amino acids, but the activation energy is higher.

Conversion of albumin in high-temperature water, carried out in batch mode, showed that even in $t=9$ h the yield of amino acids was in the range of 20 wt%, as shown in Figure 8.75. For $T=230$ °C, the conversion is also shown for the continuous experiment. Even after $t=9$ h reaction time, the yield of amino acids does not increase beyond the value of about 20 wt%. For $T=190$ °C, yield of amino acids has still a positive slope, indicating that no decomposition of amino acids occurs, while for $T=210$ °C, the maximum yield seems to be approached at $t=9$ h residence time.

For the amino acid formation and degradation, phenylalanine and glycine are chosen as examples and their behavior in high-temperature water is shown in Figures 8.76 and 8.77.

The amino acids tyrosine and phenylalanine behave similar during the hydrolysis of albumin in high-temperature water as the sum of all amino

TABLE 8.8 Reaction Rate Constants and Constants of the Arrhenius Equation for the Formation and the Degradation of Amino Acids During the Hydrolysis of Albumin in High-Temperature Water

T (°C)	k_1 (s^{-1})	k_2 (s^{-1})
230	0.000041	0.000044
250	0.000200	0.001000
270	0.000400	0.002200
290	0.000890	0.008500
310	0.011000	0.335000
330	0.020000	0.510000

$k_{0,1}$ (s^{-1})	$E_{a,1}$ (kJ mol^{-1})	$k_{0,2}$ (s^{-1})	$E_{a,2}$ (kJ mol^{-1})
7.36×10^{11}	157.17	1.87×10^{20}	236.46

$P=25$ MPa, feed concentration $x_F=1$ wt% [92].

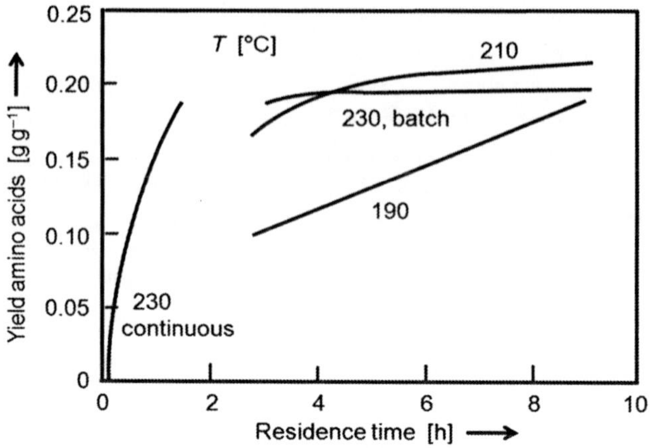

FIGURE 8.75 Hydrolysis of albumin in high-temperature water. Batch experiments: $P=10$ MPa. Continuous experiments: $P=25$ MPa. *Modified from Ref. [92].*

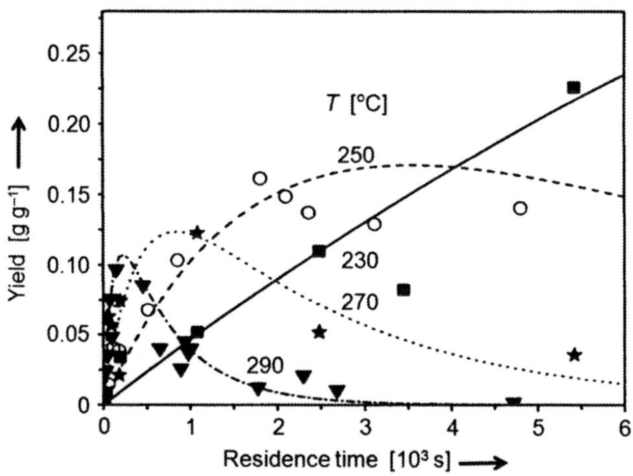

FIGURE 8.76 Yield of phenylalanine during hydrolysis of albumin in high-temperature water. $P=25$ MPa, feed concentration $x_F=1$ wt%. Symbols: experimental results, lines are calculated [92].

acids. At $T=230$ °C and $t=1.5$ h residence time, a yield of 25 wt% related to the original content of the specific amino acid is reached. Asparagic acid and serine are obtained in quantities of less than 10 wt%. Alanine and glycine yields reach values beyond 100 wt%, an indication of the formation of these amino acids as intermediate product from other, more complex amino acids. The rate constants for the formation of the individual amino acids are listed in Table 8.9, those for the degradation in Table 8.10. Activation energy and preexponential factor of the Arrhenius equation are listed in Table 8.11.

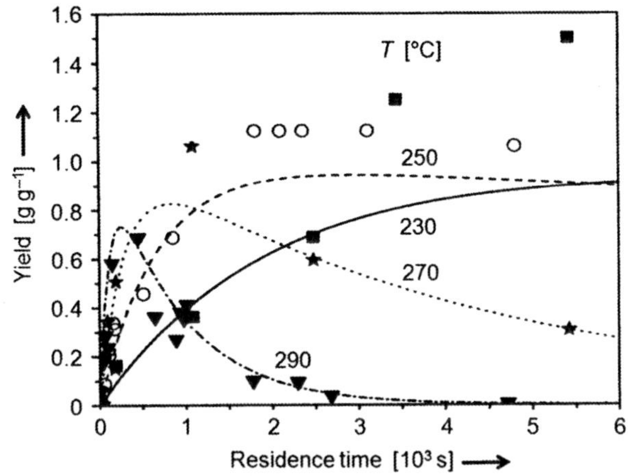

FIGURE 8.77 Yield of glycine during hydrolysis of albumin in high-temperature water. $P=25$ MPa, feed concentration 1 wt%. Symbols: experimental results, lines are calculated [92].

TABLE 8.9 Reaction Rate Constants k_1 for the Formation of Individual Amino Acids During the Hydrolysis of Albumin in High-Temperature Water

	Temperature (°C)					
Amino acid	230	250	270	290	310	330
ASX	0.00040	0.00065	0.00140	0.00300	0.00500	0.01000
GLX	–	–	–	–	–	–
SER	0.00032	0.00080	0.00100	0.00350	0.01500	0.02200
LYS	0.00005	0.00015	0.00020	0.00060	0.00400	0.00800
LEU	0.00006	0.00015	0.00025	0.00100	0.06000	0.10000
HIS	0.00012	0.00025	0.00110	0.00250	0.01500	0.03000
GLY	0.00050	0.00150	0.00350	0.00900	0.06000	0.06500
THR	–		–	–	–	–
ARG	0.00060	0.00150	0.00200	0.00400	0.01000	–
ALA	0.00050	0.00100	0.00400	0.00700	0.04000	0.06000
TYR	0.00006	0.00015	0.00050	0.00150	0.01000	0.05000
TRP	0.00004	0.00015	0.00025	0.00200	–	–
MET	0.00005	0.00015	0.00035	0.00140	–	–
VAL	0.00007	0.00023	0.00040	0.00180	0.00900	0.02000
ILE	0.00002	0.00008	0.00016	0.00100	0.01000	–
PHE	0.00005	0.00014	0.00045	0.00150	0.02000	0.04000

$P=25$ MPa, feed concentration $x_F=1$ wt% [92].

TABLE 8.10 Reaction Rate Constants k_2 for the Degradation of Individual Amino Acids During the Hydrolysis of Albumin in High-Temperature Water

	Temperature (°C)					
Amino acid	230	250	270	290	310	330
ASX	0.00520	0.01000	0.02200	0.07000	0.80000	1.50000
GLX	–	–	–	–	–	–
SER	0.00900	0.01900	0.05000	0.20000	0.80000	0.90000
LYS	0.00008	0.00090	0.00160	0.00600	0.20000	0.50000
LEU	0.00002	0.00050	0.00075	0.00650	0.50000	0.90000
HIS	0.00020	0.00060	0.00220	0.00630	0.07000	0.10000
GLY	0.00001	0.00002	0.00023	0.00120	0.09000	0.18000
THR	–	–	–	–	–	–
ARG	0.01500	0.02500	0.10000	0.30000	0.60000	–
ALA	0.00001	0.00008	0.00038	0.00220	0.06000	0.12000
TYR	0.00003	0.00050	0.00200	0.01000	0.20000	0.50000
TRP	0.00003	0.00100	0.00350	0.02500	–	–
MET	0.00012	0.00120	0.00500	0.02500	–	–
VAL	0.00002	0.00050	0.00090	0.00600	0.15000	0.19000
ILE	0.00002	0.00080	0.00190	0.01500	0.10000	–
PHE	0.00002	0.00050	0.00250	0.01000	0.13000	0.30000

$P=25$ MPa, feed concentration $x_F=1$ wt% [92].

8.3.6.4.3 Influence of Additional Compounds on the Conversion of Proteins

Conversion of proteins in pure water leads not to very attractive yields of amino acids. The addition of acids and bases, which is known to increase the yield, makes the conversion process more complex. Simple additives are CO_2 and NaCl. Experiments with CO_2 lead to an increase in conversion for temperatures below $T=250$ °C. At $T=250$ °C, the effect is negligible, analogue to the effect on hydrolysis of starch and other polysugars. At the lower temperature of $T=200$ °C, the addition of CO_2 (saturation of the liquid phase) and of NaCl, as well as a combination of CO_2 and NaCl as additives, lead to a substantial increase in yield of amino acids (Figure 8.78). Nevertheless, the total yield is not in the range of being attractive for a production process.

TABLE 8.11 Activation Energy and Preexponential Factor for the Formation and the Degradation of Individual Amino Acids During the Hydrolysis of Albumin in High-Temperature Water

Amino Acid	$k_{0.1}$ (s^{-1})	$E_{a,1}$ (kJ mol^{-1})	$k_{0.2}$ (s^{-1})	$E_{a,2}$ (kJ mol^{-1})
ASX	1.35×10^{5}	82.71	1.85×10^{13}	152.30
GLX	–	–	–	–
SER	1.05×10^{8}	111.90	1.57×10^{11}	128.49
LYS	1.49×10^{9}	131.17	4.32×10^{18}	219.73
LEU	1.76×10^{16}	201.05	5.90×10^{23}	274.78
HIS	1.37×10^{11}	146.36	3.15×10^{13}	166.89
GLY	1.75×10^{10}	130.98	5.68×10^{22}	271.78
THR	–	–	–	–
ARG	1.32×10^{5}	80.34	3.53×10^{10}	120.11
ALA	7.75×10^{9}	128.02	3.26×10^{20}	246.85
TYR	1.37×10^{13}	169.29	1.17×10^{21}	246.03
TRP	8.18×10^{10}	148.33	9.87×10^{21}	253.59
MET	1.08×10^{9}	128.86	3.35×10^{17}	205.93
VAL	8.53×10^{10}	146.40	5.23×10^{19}	234.06
ILE	5.04×10^{13}	178.49	7.92×10^{20}	244.39
PHE	9.87×10^{13}	178.9	2.37×10^{20}	239.57

P=25 MPa [92].

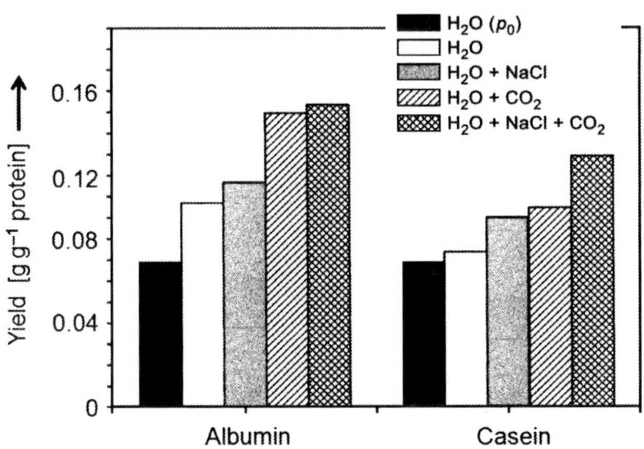

FIGURE 8.78 Conversion of proteins in high-temperature water. Influence of CO_2 and NaCl on the total yield of amino acids at T=200 °C, P=10 MPa, residence time t=1 h [92].

The conversion of pure proteins in high-temperature water at temperatures $T = 200$–$350\ ^\circ C$ does not lead to high conversion of proteins to amino acids. A substantial amount of the initial protein is transformed to low-molecular degradation products, remains as solid residue, or as repolymerizate. The conversion of structural proteins, contained in hairs and feathers is discussed in Chapter 8.4.2.

8.4 PROCESSING OF MIXTURES OF NATURAL COMPOUNDS OF BIOMASS IN HIGH-TEMPERATURE AND SUPERCRITICAL WATER

The behavior of natural compounds of biomass in high-temperature and supercritical water is treated in Section 8.2 for separate compounds like sugars, starch, hemicellulose, cellulose, lignin, amino acids, and proteins. In natural materials, some of these compounds, or even all, occur together as mixtures in natural materials like straw, bagasse, wood, and various forms of biowaste. During a conversion process with water, they generally behave similar in the mixture as in pure-component processing. But in the mixture, the lignocellulose compounds are bound together by chemical and physical forces. Any conversion process for separate compounds must first break these bonds before removal and further conversion can take place. Furthermore, the natural materials are very different, not only in composition but much more in their structure. A process that works for one material may not be applicable for another one. Nevertheless, the principles of such processes are similar and can be transferred from one material to another. But for a process design, it is necessary to carry out experiments on the conversion of the specific material and on the recovery of the target compounds. It has also to be borne in mind that natural materials vary according to their source, harvesting time, storage, and according to various influences on the materials from harvesting to processing. Experimental results will therefore be different for the same type of material, beyond the normal scattering of experimental results. Due to this wide variability in materials and the possible processing results, the following discussion can only consider examples. The discussion of the conversion processes has to further take into account the fundamentals of reactions, as discussed in Chapter 5, and of extraction, as discussed in Chapter 6.

Treatment of natural material in high-temperature water may not lead to a sufficient yield of the target compounds in one processing step. In separate process steps, compounds may be added, for example, to change the pH value of the solution, catalysts may be used to enhance the reaction rate, or other means may increase the yield of specific compounds. Furthermore, often mixtures will be obtained. Target compounds can then be separated from the product mixture by well-known separation techniques that are beyond the scope of this book.

8.4.1 Lignocellulosic Materials

Lignocellulosic materials are mainly composed of cellulose, hemicellulose, lignin, and other minor components, such as ash and proteins. Valuable components may have been separated from the lignocellulosic materials before they are considered as feedstock for conversion processes with hydrothermal and supercritical water. Those materials include agricultural by-products, like wheat and rye straw, corn stover, rice straw, bagasse, grass, forest residues, and others. Table 8.12 lists the composition of some materials. In Figure 8.79, a simplified structure of lignocellulosic material is shown [104].

TABLE 8.12 Composition of Selected Lignocellulosic Materials [104]

Material	Hemicellulose	Cellulose	Lignin	Ash	Sum
	wt%, dry basis				
Spruce	26.5	42.0	28.6	0.4	97.5
Aspen	28.7	50.8	15.5	0.2	95.2
Red maple	33.0	29.0	23.0	0.2	85.2
Wheat straw	27.6	24.0	18.0	1.3	70.9
Cornstalk	32.6	33.5	11.0	1.0	78.1
Bagasse	34.0	38.0	11.0	–	83.0
Tall fescue	24.5	30.0	3.1	–	57.6

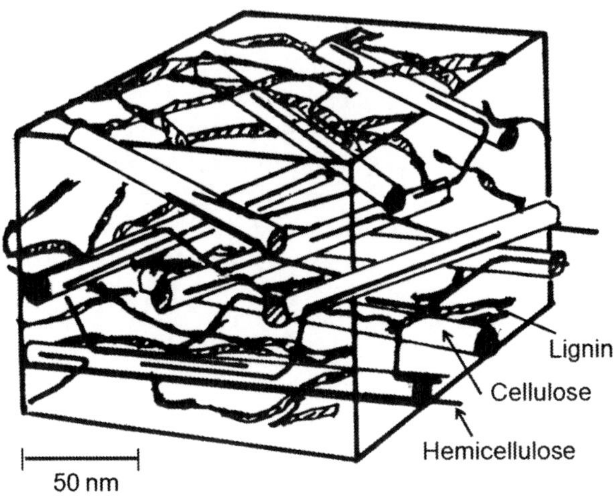

FIGURE 8.79 Structure of lignocellulosic materials (schematic). *Modified from Ref. [40].*

8.4.1.1 Straw

Straw, an agricultural by-product, consists of the dry stalks of cereal plants, after the grain and chaff have been removed. Straw makes up about half of the yield of cereal crops such as barley, oats, rice, rye, and wheat [105]. The conversion of straw is of special interest since a successful conversion in high-temperature water opens new potentials for the use of lignocellulosic materials as a feedstock for valuable chemicals. Hydrolysis of straw in high-temperature water to obtain saccharides that can be converted to ethanol by fermentation is only one example. The composition of rye straw is listed in Table 8.13.

Glucose is mainly bound in cellulose, and xylose is mainly bound in hemicelluloses. Other sugars occurring are arabinose, mannose, and galactose. Other compounds comprise oils, proteins, and waxes.

Utilization of straw by processing with high-temperature or supercritical water follows several pathways. Conversion of straw into water-soluble products at lower temperatures $T = 150$–$250\ ^\circ C$ can be used to separate hemicelluloses, cellulose, and lignin or to produce a liquid solution that can further be treated with enzymes. The conversion of lignocellulosic material at higher temperatures leads to valuable compounds from the constituents of the biomass. Conversion at high temperatures into gaseous and liquid products can make available base chemical compounds for production of specific chemicals, or lead to a basic material that can be processed for fuel purposes. Furthermore, total conversion is relevant for total recycling of organic materials.

8.4.1.1.1 Solubilization of Straw at Temperatures up to About $T = 250\ ^\circ C$

Solubilization of straw at moderate temperatures up to about $T = 250\ ^\circ C$ is the pathway that is the most promising one for recovering the constituents of lignocellulosic materials and for further conversion of the products to chemicals. The influence of processing conditions for the conversion of lignocellulosic materials (and others) in high-temperature water can be expressed with the so-called severity factor which combines the effect of temperature and reaction time. It is assumed, as has been shown before, that pressure has a negligible influence on conversion, if temperatures are well below the critical temperature of water and pressures are above the saturation pressure. The severity factor is defined by Equation (8.17) [67]:

TABLE 8.13 Composition of Rye Straw [104]

Glucose	Xylose	Other Sugars	Ash	Kraft Lignin	Others
		wt%, dry matter			
43.6	20	3.5	3.4	18.6	10.9

$$R_0 = t \cdot \exp\left(\frac{T - 100}{14.75}\right), \tag{8.17}$$

or for a temperature range [104]:

$$R_0 = \int_0^t \exp\left(\frac{T(t) - 100}{14.75}\right) dt \tag{8.18}$$

The solubilization of rye straw in a continuous-flow reactor at $P = 10$ MPa is shown in Figure 8.80 for severity factors of log $R_0 = 1$–6.5, standing for temperatures $T = 120$–310 °C, residence times of $t = 1.0$–14.5 min, and for feed concentrations of $x_F = 1$–5 wt%. In some experiments, carbon dioxide is added until saturation.

The degree of solubilization increases almost linearly with log R_0, the increasing severity factor, and reaches values of about 65% at log $R_0 = 6.5$. The high degree of solubilization is due to the enhanced cleavage of the glycosidic linkages of hemicellulose and cellulose at higher temperatures and longer residence times. The influence of carbon dioxide, added to saturation at $P = 10$ MPa, is negligible, in coincidence with literature. Although enhanced degradation of some substrates like starch, hemicellulose, and others is observed (see Sections 8.2 and 8.3), an increased degradation rate of lignocellulosic biomass has not yet been reported. As determined in batch experiments, even a decrease of pH to a value of pH = 2.0 by the addition of sulfuric acid does not affect solubilization [47,104].

High severity factors that lead to high values of solubilization are not the optimum since monomeric sugars degrade further to by-products that reduce the yield of sugars and interfere with further processing pathways. Formation of the main degradation products furfural and 5-HMF is shown in Figure 8.81.

These results are based on experiments in a continuously operated tubular reactor. Such a continuous flow apparatus is restricted with respect to residence time, in the case of rye straw solubilization to $t < 8$ min. Furthermore,

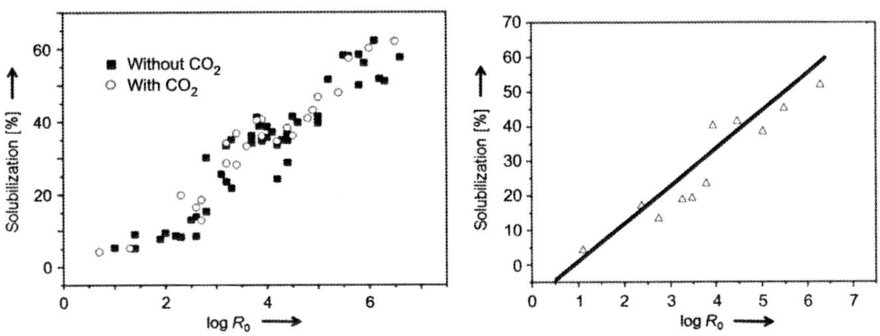

FIGURE 8.80 Solubilization of rye straw in a continuous-flow reactor at $P = 10$ MPa. Influence of the addition of CO_2 and of temperature ($T = 120$–310 °C) and residence time ($t = 1.0$–14.5 min), feed concentration $x_F = 1$–5 wt% in dependence on the severity factor [47,104,106,107].

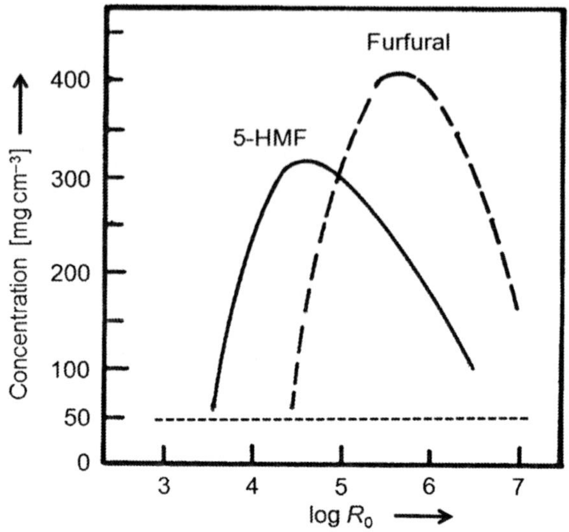

FIGURE 8.81 Concentration of sugar degradation products. Rye straw hydrolysis in a continuous-flow reactor at $P = 10$ MPa. ---, analytical limit [47,104,106,107].

feed and reaction medium (high-temperature water) cannot be handled separately due to the cocurrent mode of the conversion process. Residue and dissolved compounds have the same residence time. A fixed-bed reactor, in which the solid biomass forms the fixed bed, solves these problems at the cost that the solid biomass in the reactor is exposed to the heating, before the reaction medium is introduced. Therefore, a fixed-bed reactor is not useful for short-residence times. Residence time should be one order of magnitude longer than heating time.

The conversion of rye straw in a fixed-bed reactor is shown in Figure 8.82 as the elution curve of dissolved compounds. Rye straw was percolated with water and the temperature increased stepwise, with a residence time of water at each step necessary to replace the water once.

Elution of compounds from rye straw with water begins at low temperatures (Figure 8.82) due to extractible compounds such as starch. The hemicellulose fraction of rye straw is hydrolyzed at $T = 170$–$200\,°C$, while the cellulose fraction degrades at $T = 250$–$280\,°C$, as reported in literature [61,62,66,108–110]. Amorphous parts of cellulose degrade at considerably lower temperatures. The cross-linked, heterogeneous structure of hemicellulose and covalent cross-linkages to lignin are the reason that hemicellulose is solubilized at temperatures of $T = 160$–$230\,°C$. Sample A (Figure 8.82) contains the monomeric sugars glucose and fructose, present in the biomass as free sugars, or originating from starch. Sample B contains xylose and various xylans, originating from hemicelluloses. Sample C contains monomeric glucose, HMF, and furfural [104]. Up to 50% of the initial Klason lignin is also

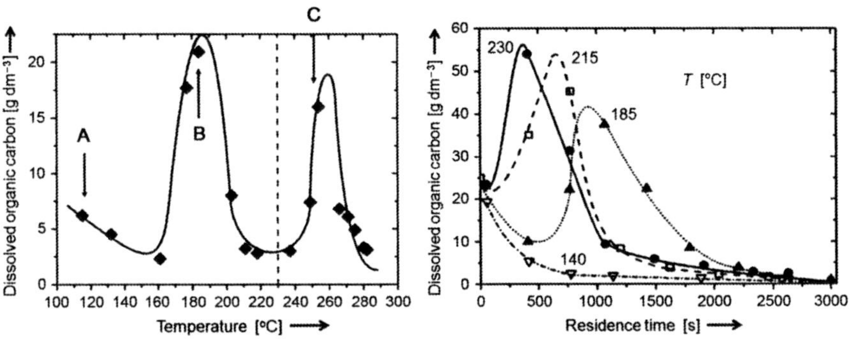

FIGURE 8.82 Thermal elution profiles of the conversion of rye straw in high-temperature water. Continuous flow of water through a fixed bed of straw. Left: Total dissolved organic carbon (DOC) in dependence of temperature. $P=10$ MPa, residence time $t_{H_2O}=10$ min. Right: Time dependence of total dissolved organic carbon (DOC) at different temperature. $P=50$ bar, residence time $t_{H_2O}=9$ min [47,104,106].

dissolved. 10–15% of the initial dry matter is recovered as solid residue, which consists of up to 99% of lignin and ash.

Elution profiles for different temperatures, shown in Figure 8.82, right, illustrate the time dependence of the conversion of rye straw in high-temperature water. At lower temperatures, a certain amount of biomass is dissolved at short-reaction times but the main part is converted and dissolved at extended reaction times, as can be seen in Figure 8.82, right, for $T=185$ °C. With increasing temperatures, hydrolysis reactions begin earlier, peaks increase, and are shifted to shorter reaction times. Elution of conversion products ends at about $t=50$ min. Even for reaction times $t>4$ h, biomass was not converted further [104].

Composition of the reaction products are shown in Figure 8.83 for $T=230$ °C. Glucose and xylose stand not only for the monomer sugars but also for the respective oligomer sugars, which are the greater part of the sugars dissolved. The glucose concentration is high in the beginning. The fraction at $t=12.5$ min consists mostly of xylose and furfural, while at about $t=32$ min, the product flow consists of HMF, glucose, and small amounts of furfural and xylose.

8.4.1.1.2 Kinetics of Solubilization of Rye Straw in High-Temperature Water

The conversion of rye straw is assumed to follow a first-order reaction,

$$\frac{dc_S}{dt} = -k_S \cdot c_S \tag{8.19}$$

with c_S, concentration of solid straw (kg m^{-3}) in the reactor and k_S, reaction rate constant (s^{-1}). Temperature dependence of the reaction rate constant can be described with the Arrhenius equation:

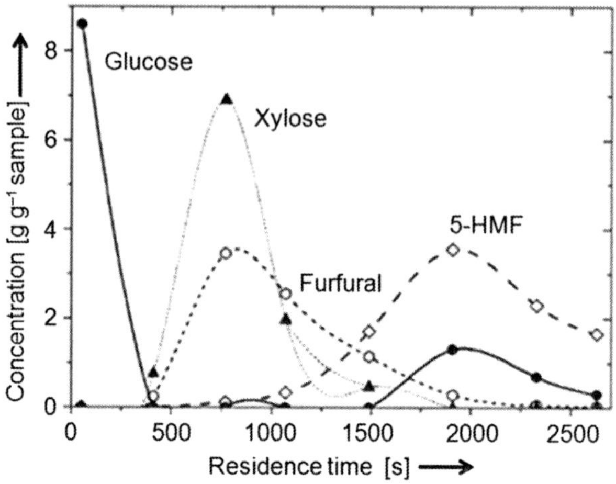

FIGURE 8.83 Composition of the hydrolysate from the conversion of rye straw in high-temperature water. $P=5$ MPa, $T=230$ °C, residence time $t_{H_2O}=9$ min, c_{Sa} (kg m^{-3}) concentration in the sample, c_{to} (kg m^{-3}) concentration in the total hydrolysate [47,104,106].

$$k_s(T)=k_0\cdot\exp\left(-\frac{E_a}{RT}\right), \tag{8.20}$$

with E_a, activation energy and k_0, preexponential factor.

Part of the solid feed (rye straw) is converted by hydrolytic reactions at a certain temperature, while the other part remains intact, as has been shown by the experimental results. The maximum conversion $\lambda_{max}(T)$ for a specified temperature is defined in Equation (8.21) [47,104]:

$$\lambda_{max}(T)=\frac{c_{s,l}(T)}{c_{s,0}} \tag{8.21}$$

with $c_{s,l}(T)$, maximum amount of feed converted to liquid solution; $c_{s,0}$, initial amount of feed.

A mass balance in combination with Equation (8.19) (rate equation) leads to the amount of liquefied feed, Equation (8.22):

$$f=\frac{M_l}{M_0}=\lambda_{max}(1-\exp(-k_s t)) \tag{8.22}$$

with M_l, total dissolved mass, M_0, total initial mass of solid.

Conversion of rye straw in high-temperature water is shown in Figure 8.84 for different reaction temperatures. Experimental data are correlated with Equation (8.22). Reaction parameters are listed in Table 8.14. Activation energy derived from Equation (8.20) is $E_a=11$ kJ mol^{-1} and the preexponential factor is $k_0=0.027$ s^{-1}. The correlation represents data for the conversion part and ends for zero conversion at a value of the time scale t_0, which is taken as the time for the starting of the

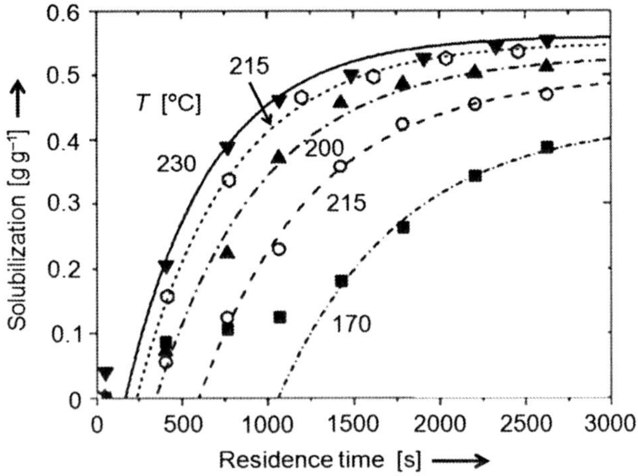

FIGURE 8.84 Conversion of rye straw in high-temperature water. P = 5 M, t = 9 min [47,104,106].

TABLE 8.14 Reaction Parameters for the Degradation of Rye Straw in High-Temperature Water [47,104,106]

T (°C)	λ_{max}	k_s (s^{-1})	t_0 (s)
170	0.40	0.0012	1060
185	0.43	0.0014	580
200	0.45	0.0016	375
215	0.46	0.0018	231
230	0.47	0.0019	176

hydrolysis reactions. The initial part of the liquefaction is not covered by this model.

The maximum dissolved amount of feed for the conversion of rye straw in high-temperature water is shown in Figure 8.85 for a temperature range of T = 25–270 °C. At about T = 165 °C, hemicellulose dissolves, and at T = 250 °C, cellulose is dissolved. Data represent results from two semicontinuous fixed-bed reactors of 50 and 1000 ml volume and from a continuously operated tubular reactor [104].

Degradation products of the conversion of rye straw and the composition of the solid residue are shown in Figure 8.86. The amounts of 5-HMF and furfural are related to the initial fraction of glucose and xylose in rye straw.

The amount of not dissolved xylose and lignin in the solid residue decreases with temperature. The entire content of xylose is dissolved at T = 230 °C. Bound glucose, as cellulose, has a maximum conversion of about

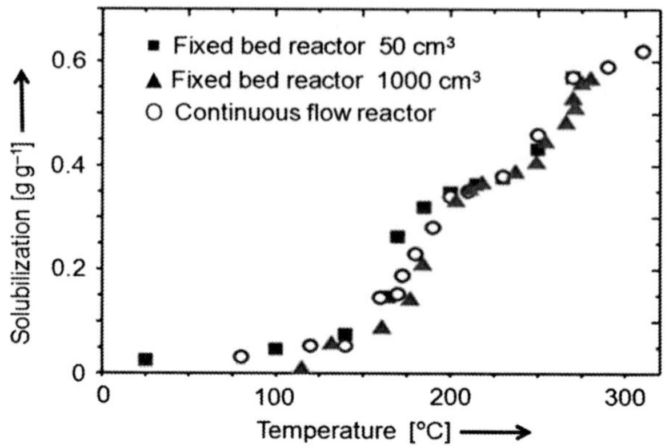

FIGURE 8.85 Maximum conversion (λ_{max}) of rye straw in high-temperature water [47,104].

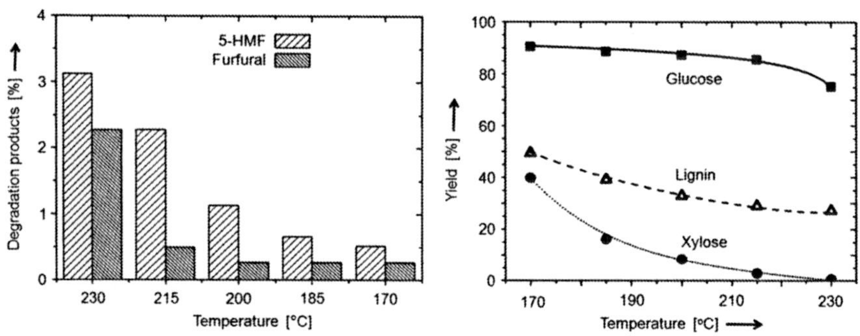

FIGURE 8.86 Conversion of rye straw in high-temperature water. Left: Liquid degradation products. Right: Compounds of the solid residue. $P=5$ MPa, $t_{H_2O}=9$ min [47,104,106].

80 wt% at $T=215$ °C. At higher temperatures, parts of the cellulose are converted. The lignin fraction, representing Klason lignin and ash, reaches a value of 35% at $T=230$ °C. The cellulosic residue is hydrolyzed at values for the severity factor $R_0>5$ and reaches 65% at $R_0=5.4$. Pure cellulose is converted to 90 wt% into degradation products [56–58]. The residual lignin restricts the degradation of cellulose [47,61,104].

Treatment of rye straw with water up to temperatures of about $T=250$ °C yields two products, a cellulose-rich solid residue and a liquid hydrolysate, which contains the dissolved degradation products.

Treatment of wheat straw, carried out in context with experiments on the total recycling of organic waste, shows that the conversion is limited to about 80 wt%, Figure 8.87 [56–58]. CO_2 has no influence on the conversion. The degree of liquefaction f is calculated as the ratio of dissolved carbon in the reaction products to the total initial carbon.

On the average, from straw, 20–30% of the initial carbon cannot be converted to water-soluble carbon with pure water or with the addition of CO_2 within the residence time of about $t=90$ s, even at supercritical temperatures.

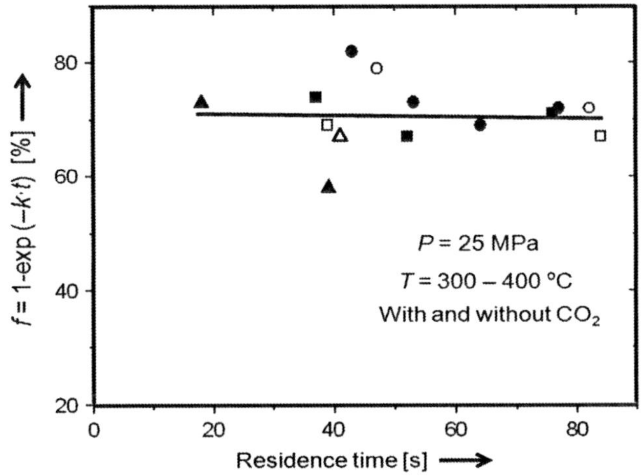

FIGURE 8.87 Conversion of wheat straw to water-soluble carbon in high-temperature and supercritical water. $f = DOC_{out}/C_{in}$ (DOC—dissolved organic carbon). CO_2 concentration ≈ 70% of saturation, initial concentration of solid $x_F = 1$ wt% [56–58].

One reason is the formation of carbon-rich, insoluble compounds by repolymerization of reactive radicals. The formation of product gases is of minor influence for the hydrolysis of wheat straw at these conditions in a flow reactor. Experiments in a fixed-bed reactor show that the carbon is successively converted to water-soluble substances during the heating-up phase. If products are instantaneously cooled down in the heat exchanger, a degree of liquefaction close to 100% can be obtained at temperatures of $T = 370–380$ °C [56–58].

The addition of oxygen leads to a conversion of 100% into water-soluble and gaseous products, as shown in Figure 8.88 in terms of the residual dissolved carbon content. The carbon that is not contained in the liquid products (Figures 8.88 and 8.89) is converted to the gaseous fraction, which has not been specified.

The oxidative conversion of wheat straw in high-temperature and supercritical water leads to solid-free effluents for temperatures $T > 300$ °C. About 40% of the total initial carbon content is liquefied without oxygen in $t = 90$ s. At oxygen concentration of twice the stoichiometric amount, about 90% of the initial carbon is transformed into gaseous substances within a residence time of less than $t = 20$ s. The other part is converted to gaseous species. The amount of acetic acid in the liquid product fraction is shown in Figure 8.89.

It can be concluded that lignocellulosic wastes can be completely converted in high-temperature and supercritical water into water-soluble and gaseous products. A total conversion of the lignin part and the repolymerized compounds is only possible at high temperatures and short-reaction times, or with the addition of oxygen. Similar investigations on conversion of lignocellulosic material are reported in the literature, for example, on the conversion of rice straw [111], corn stover [112], brewery waste [31], and bagasse [113–115].

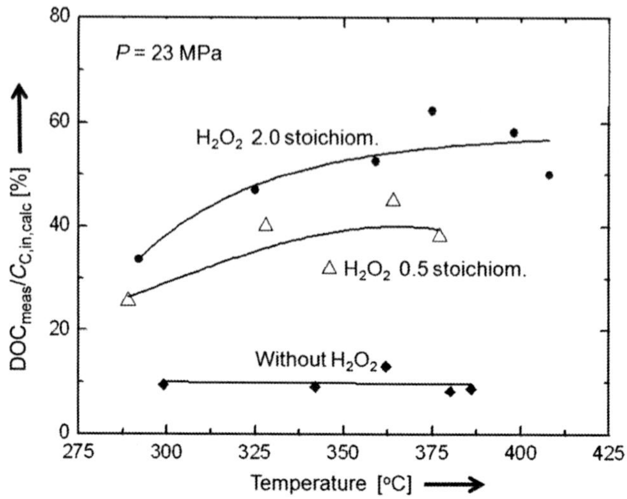

FIGURE 8.88 Oxidative conversion of wheat straw to water-soluble substances. Total dissolved organic carbon in the liquid product. $P=23$ MPa, initial solid concentration $x_F=1$ wt%. Residence time: With H_2O_2: $t=4$–17 s, without H_2O_2: $t=13$–84 s [56–58].

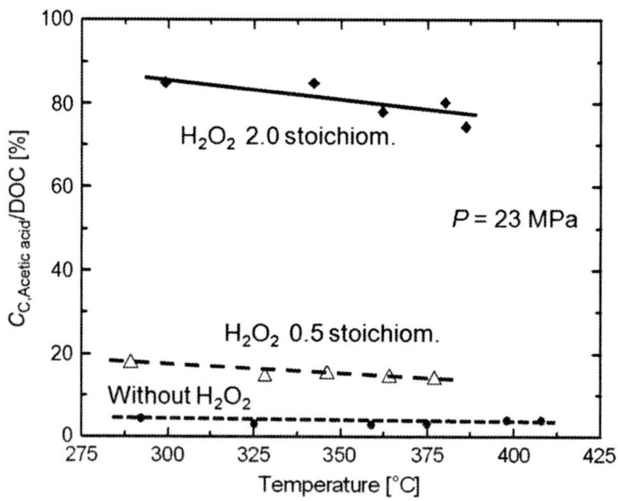

FIGURE 8.89 Conversion of wheat straw in high-temperature and supercritical water. Acetic acid in the liquid product. $P=23$ MPa, initial solid concentration $x_F=1$ wt%. Residence time: With H_2O_2: $t=4$–17 s, without H_2O_2: $t=13$–84 s [56–58].

8.4.1.2 Wood

Wood is a hard, fibrous tissue composed of cellulose, hemicelluloses, lignin, and extractives. Cellulose, hemicelluloses, and lignin have been described previously in this chapter. Extractives are compounds that can be extracted with water and organic solvents. The composition of some wood species is

TABLE 8.15 Composition of Some Wood Species [61,116]

Plant	Cellulose	Hemicellulose	Lignin	Ash	Sum
		wt%, dry matter			
White poplar	49.0	25.6	23.1	0.2	97.9
Trembling aspen	49.4	30.9	18.1	0.4	98.8
European birch	48.5	25.1	19.4	0.3	93.3
European spruce	40.4	31.1	28.2	0.3	100.0
Pinus densiflora	45	22	27	0.2	94.2

listed in Table 8.15. Comprehensive information on the composition of many species of wood and other biomass materials is reported in Refs. [61,116].

Hydrothermal treatment of wood (and straw) was reported by Bobleter *et al.* as early as 1968 [61,117]. They found that at a reaction temperature of $T = 180–190$ °C, hemicellulose is converted almost completely and the lignin is converted to some part and removed from the initial substrate, leaving behind 60% of the initial material as residue. Only at temperatures above $T = 240$ °C, the attack on cellulose becomes more severe and the residue amounts only to 40%. Antal *et al.* [117] treated different species of hardwood and grasses. At $T = 230$ °C, conversion products are 40–60% of the feed material. Hemicellulose is converted and removed as oligomer, monomer sugars to 90–100%, lignin to about 50%, and cellulose to 5–20% [117].

In the following years, numerous investigations on the conversion of wood with water were published. Most of them deal with steam for treatment of wood, known as "steam explosion," a processing step that is used mainly for pretreatment purposes rather than for recovering the different fractions of wood. Treatment of steam explosion is out of the scope of this book. Information can be derived from Ref. [63].

Treatment of wood with high-temperature and supercritical water is very similar to that of straw which is discussed in this chapter in section 8.4.1.1. In the following, three examples of more recent reports on the conversion of pine wood with high-temperature water are cited, two for the conversion of Japanese species of wood and one on the conversion of European pine wood.

Japanese red pine wood was treated with high-temperature water at $T = 270$ °C in a batch reactor [118]. Sugar products dissolved in the aqueous phase are shown in Figure 8.90 together with the total dissolved organic matter (DOC). The main amount of products occurs within $t = 5$ min reaction time, with a peak around $t = 3$ min. Only glucose and cellobiose, products of cellulose conversion, are found after $t = 7$ min in substantial quantity. Consequently, DOC approaches the conversion limit after about $t = 5$ min. Degradation products reach high levels after $t = 3–5$ min. HMF reaches a

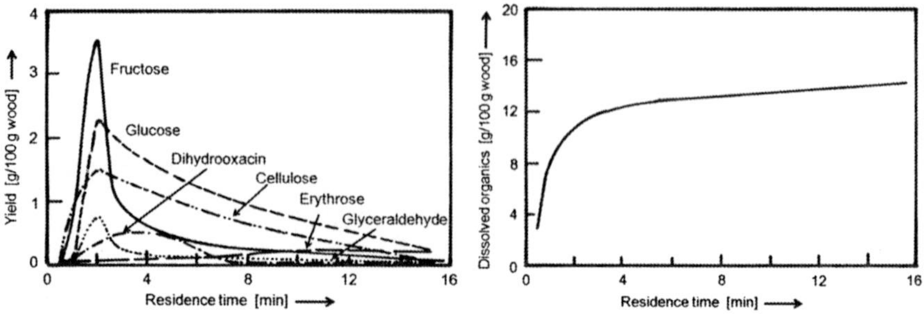

FIGURE 8.90 Conversion of red pine wood with high-temperature water in a batch reactor. Yield of sugar products and total dissolved organics (DOC). $T=270\,°C$. *Modified from Ref. [118].*

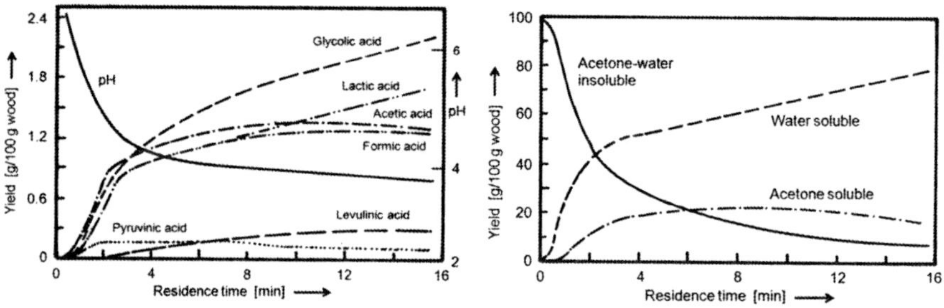

FIGURE 8.91 Conversion of red pine wood with high-temperature water in a batch reactor. Yield of organic acids (left) and of acetone-soluble, acetone-water-insoluble, and water-soluble products (right). $T=270\,°C$. *Modified from Ref. [118].*

concentration of about 15 wt% of the initial feed, and furfural reaches a level of about 9 wt%. The amount of organic acids is represented in Figure 8.91, left. Organic acids reach levels up to 2 wt%. In Figure 8.91 on the right-hand side, water-soluble products, acetone-soluble products, and water-acetone-insoluble products, which can be attributed mainly to the degraded lignin, tar, and unreacted wood chips, respectively, are shown. Water-soluble products reach about 70 wt% after $t=15$ min of treatment, acetone-soluble products are mainly at a level of about 20 wt%, and acetone-water-insoluble products decrease to less than 10 wt% after $t=15$ min at $T=270\,°C$ [118].

Treatment of sugi (*Cryptomeria japonica D. Don*) wood in high-temperature water at $T=310$–$320\,°C$, $P=25$ MPa, and a solvent ratio of $S/F=32.5\,g_{H_2O}(g_{Wood}\,min)^{-1}$ in a semicontinuous reactor results in a conversion of 70–90 wt% of the sugi wood meal into water-soluble compounds, mainly monosaccharides and oligosaccharides, as a result of hydrolysis of cellulose and hemicelluloses [119]. Conversion is completed within about $t=20$ min. Precipitates, identified as crystallized cellulose, appear after $t=48$ h in the samples. Yield of the conversion can be improved by 60% with pretreating the wood meal with a surfactant to increase wettability. Conversion products are shown in Figure 8.92 and listed in Table 8.16 as monosaccharides,

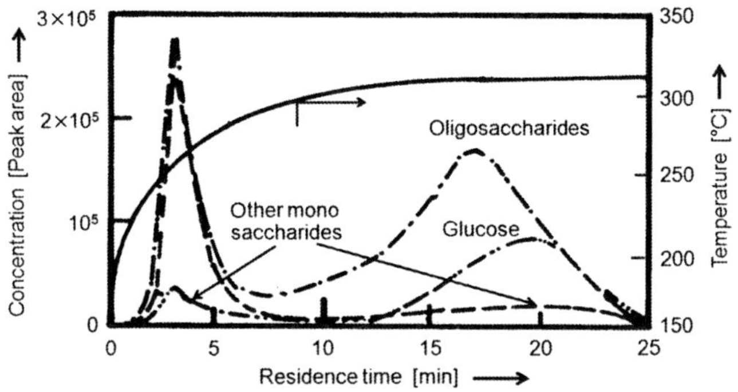

FIGURE 8.92 Conversion of sugi wood in high-temperature water in a semicontinuous reactor. $T=310$–$320\,°C$, $P=25$ MPa, solvent ratio $S/F=32.5\,g_{H_2O}(g_{Wood}\,min)^{-1}$. *Modified from Ref. [119].*

TABLE 8.16 Conversion of Sugi Wood in High-Temperature Water in a Semicontinuous Reactor

T (°C)	Glucose	Xylose	Galactose	Mannose	Residue
			wt%		
250–260	20.7	8.7	6.4	14.3	33.2
310–320	33.6	7.5	5.8	8.7	16.6
350–360	30.4	5.9	5.1	7.4	11.1

Yield of saccharides. $P=25$ MPa, solvent ratio $S/F=32.5\,g_{H_2O}(g_{Wood}\,min)^{-1}$ [119].

glucose, and oligosaccharides. The conversion yield drops by only 11% when the solvent ratio is reduced to $S/F=2.2\,g_{H_2O}(g_{Wood}\,min)^{-1}$ [119].

In a different investigation, pine wood meal is contacted in a semicontinuous reactor with steam at $T=435\,°C$, $P=9$ MPa, and a solvent ratio for water as reaction medium of $S/F=0.036\,g_{H_2O}(g_{Wood}\,min)^{-1}$. The influence of the water content in hydrogen as the basic reaction medium is shown in Figure 8.93 [120].

Results are shown as conversion products such as gaseous phase, aqueous phase, water-insoluble liquid organic phase, and as total conversion of pine wood. Total conversion is nearly constant for pure hydrogen treatment and pure water treatment at 65 wt% of dry matter. Water-insoluble liquid product increases from 10 wt% with pure hydrogen to about 23 wt% in pure water and the amount of gaseous products decreases slightly from 20 to 16.5 wt% [120].

Conversion in high-temperature water at $T=320\,°C$ and $P=23$ MPa results in a conversion of pine wood meal of 88 wt% of dry matter. Cellulose treated at the same conditions is converted to about 80 wt%. Treatment of the

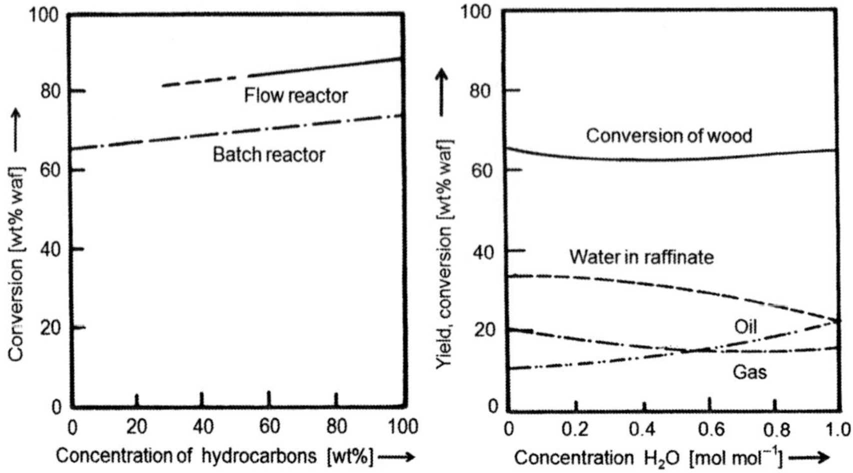

FIGURE 8.93 Conversion of pine wood in water. Left: hydrogen–steam mixtures, semicontinuous. $T=435$ °C, $P=9$ MPa. Right: Pure water, $T=325$ °C, $P=23$ MPa (batch), $T=380$ °C, $P=23$ MPa semicontinuous. Water in raffinate means water in the aqueous product phase. *Modified from Ref. [120].*

pine wood meal with supercritical water at $T=380$ °C and $P=23$ MPa in a batch reactor reduces the overall conversion of wood and cellulose to about 70 wt%. Organosolv lignin is converted to about 65 wt% [120]. More information on the conversion of wood and other biomass feedstock to liquid products is presented in Section 8.5.

8.4.2 Materials Containing Proteins

The role of high-temperature and supercritical water for processing of skleroproteins and other proteins from natural materials is considered in this section. Structural proteins are discussed on the example of the decomposition of keratin of duck feathers into water-soluble compounds. Treatment of fish meat is discussed as an example for other protein-containing natural materials. Furthermore, some investigations reported in the literature are reviewed.

The structural bond that links amino acids together into proteins is the peptide bond, a C—N bond between the carboxyl and amine groups present in all amino acids [4]. This bond hydrolyzes easily in high-temperature water. But yields of amino acids are significantly lower than those obtained by conventional acid hydrolysis due to degradation of amino acids. Optimal yields are below 10 wt% and are obtained at temperatures of about $T=250$ °C [4]. See also the discussion on protein conversion in Section 8.3.6.

8.4.2.1 Skleroproteins

Structural proteins, or skleroproterins, are a group of water-insoluble fibrous proteins from animal sources. Structural proteins act as supporting and constructive compounds and comprise keratin, collagen, fibroin, elastin, and resilin. Skleroproteins consist of amino acids with special properties, for example, α-keratin consists mainly of amino acids with side chains like cysteine rests, while β-structures (feathers) consist mainly of the simple amino acids glycine, alanine, and serine [121].

Keratins are skleroproteins produced from higher vertebrae like feathers, hairs, and others. Keratins have a high mechanical strength and are chemically stable. They are insoluble in water at ambient conditions and in dilute acids, but swell in basic media [92].

Keratin-based natural materials are a commercial source of L-cysteine, L-tyrosine, and some other amino acids. The keratin-materials are hydrolyzed with hydrochloric acid and neutralized with ammonia. Amino acids are then separated by ion-exchange chromatography. The amino acid composition of feathers is shown in Figure 8.94.

8.4.2.2 Conversion of Keratin from Feathers

The results of the conversion of keratin from feathers is based on experiments in a continuous flow reactor at temperatures of $T = 190–310$ °C and pressures of $P = 10–25$ MPa [92]. In Figure 8.95, the yield of amino acids is shown in dependence on temperature and residence time [92]. Treatment at $T = 250$ °C is the best with yields of more than 10 wt% after a reaction time of $t = 15$ min.

The influence of particle size is negligible. No significant difference occurs for particle sizes of $d_P = 112–1000$ μm or whole feathers. Pressure also has no significant effect on the yield of amino acids, as well as CO_2 or NaCl, as is shown in Figure 8.96 [92] for short-residence times. At long-reaction times, additives enhance the yield, with the most effective additive being NaOH.

The influence of the conversion conditions on the yield of amino acids is determined by a subsequent acid hydrolysis of the hydrolysates. The acid hydrolysis shows the content of all amino acids. If this value is higher than the

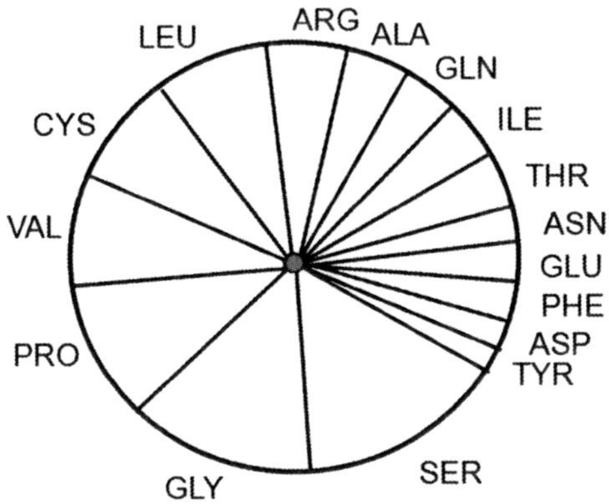

FIGURE 8.94 Amino acid composition of duck feathers. *Modified from Ref. [92].* (For color version of this figure, the reader is referred to the online version of this chapter.)

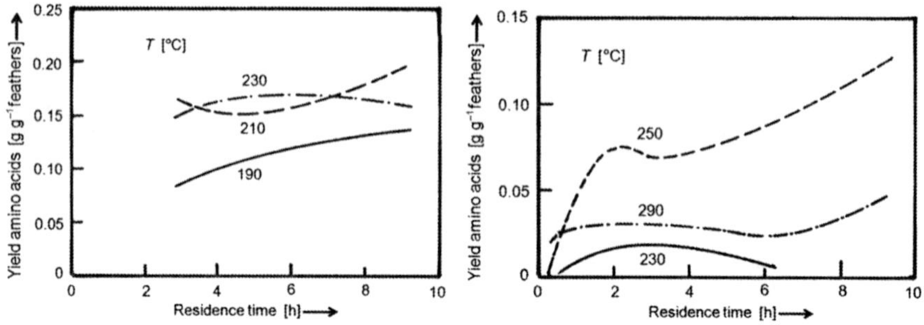

FIGURE 8.95 Conversion of duck feather keratin in high-temperature water. Left: Batch reactor, $P = 10$ MPa. Right: Continuous flow reactor, $P = 25$ MPa. *Modified from Ref. [92].*

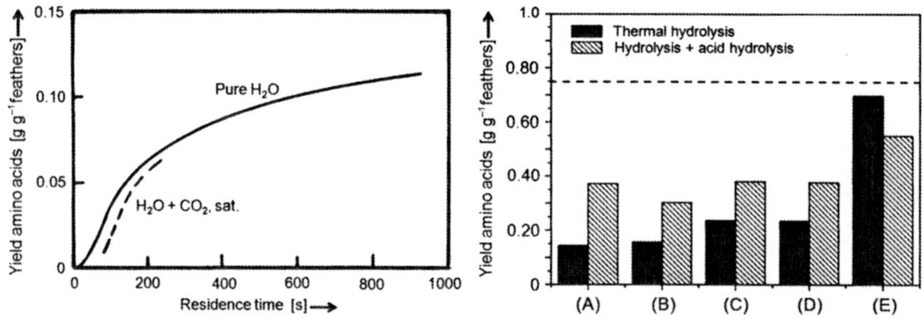

FIGURE 8.96 Conversion of duck feather keratin in high-temperature water. Feed $x_F = 1$ wt%. Left: Influence of CO_2. Continuous flow reactor, $T = 250$ °C, $P = 25$ MPa. Right: Batch reactor, $T = 250$ °C, $P = 25$ MPa. A, H_2O; B, $H_2O + CO_2$; C, $H_2O + NaCl$; D, $H_2O + CO_2 + NaCl$; E, $H_2O + NaOH$ at $T = 235$ °C, $P = 10$ MPa, residence time $t = 1$ h. CO_2 is added to saturation, NaCl, NaOH: 5 mol%. *Modified from Ref. [92].*

hydrothermal treatment yield, hydrolysates contain some unconverted compounds. If the acid yield is higher than the yield obtained by hydrothermal hydrolysis, then part of the amino acids was destroyed by the treatment at high temperatures. The difference of both analyses is the amount of nonconverted peptides.

Conversion of feather keratin in high-temperature produces up to 20 wt% of hydrolyzate. For production of specific amino acids, the hydrolysate must be further processed.

In the literature, various investigations are reported on other feedstock of skleroprotein materials that are converted with high-temperature and supercritical water. An example is the decomposition of silk fibroin to amino acids that was carried out at temperatures of $T = 210$–310 °C and saturation pressures of $P = 1.4$–9 MPa with residence times $t = 2$–62 min. The addition of acids or bases increase the yield of amino acids, also in this case [122].

8.4.2.3 *Other Protein-Based Natural Materials*

8.4.2.3.1 Fish Meat

Fish meat is a waste product that comprises about 40 wt% of the total fish weight. Waste fish meat is easily liquefied by hydrolysis in high-temperature water and forms an aqueous phase containing the water-soluble conversion products and a water-insoluble phase containing oil and solid fat-like compounds [123].

Hydrothermal conversion of fish meat (horse mackerel) was carried out in a batch reactor of $V=7\ cm^3$ in a temperature range of $T=200$–$400\ °C$. For subcritical conditions, 1 g of fish meat (water content 70 wt%), corresponding to about 30 wt% of the feed suspension, was charged and the conversion carried out at saturation pressure. For supercritical conditions, feed suspension was 30–55 wt% and pressure was set to $P=30$ MPa. Results for the total dissolved organic carbon are shown in Figure 8.97. Conversion of fish meat is completed after $t=10$ min and reaches 60 wt% at an optimum reaction temperature of about $T=570$–600 K.

The amount of the solid phase, representing nonreacted fish meat, decreases with reaction temperature, see Figure 8.98. The water-insoluble phase increases with the reaction temperature and with conversion of the fish meat up to $T=580$ K and then decreases due to decomposition of compounds to further degradation products. The oil phase contains fatty acids such as arachidonic acid, eicosapentanoic acid, and docosahexanoic acid. The solids are nonreacted fish meat. Figure 8.98 also shows that the water-insoluble phase was produced from a part of the solids as the reaction temperature increased. Then, the water-insoluble phase is degraded to other organic compounds in the high-reaction temperatures. Analysis with a GC/MS shows that the oil extracted with hexane contains many useful compounds [123].

Products that can be found are as follows (see Figure 8.98): Lactic acid, stable up to $T=513$ K, about $0.03\ g\ g^{-1}$ dry meat; pyroglutamic acid, $0.095\ kg\ kg^{-1}$ of dry meat (30 min reaction time, $T=553$ K, $P=6.42$ MPa);

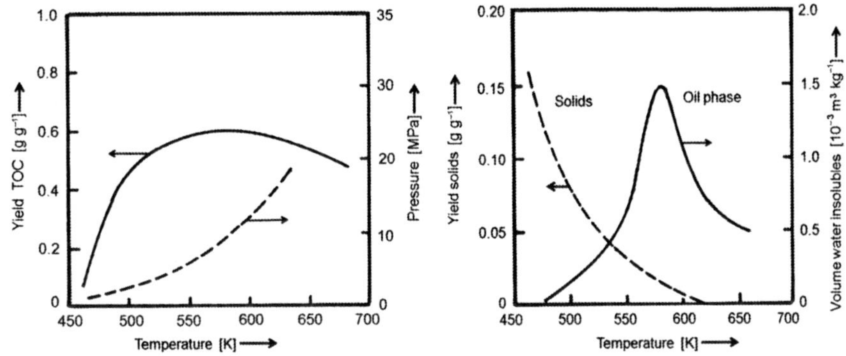

FIGURE 8.97 Conversion of fish meat in water. Reaction time $t=5$ min. Left: Effect of reaction temperature. Right: Amount of solids and of water-insoluble phase. *Modified from Ref. [123].*

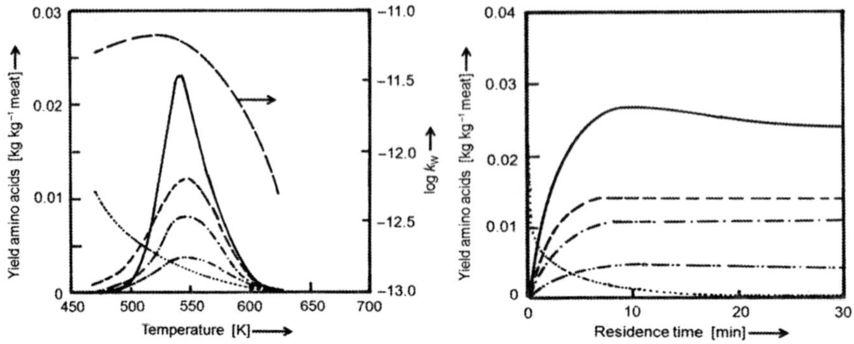

FIGURE 8.98 Conversion of fish meat in water. Yield of amino acids. Left: Effect of reaction temperature, residence time $t=5$ min. Right: Effect of residence time, $T=543$ K, $P=5.51$ MPa. ———, Cystine; — — —, alanine; —•—•, glycine; —••—, leucine; ••••, histidine. *Modified from Ref. [123].*

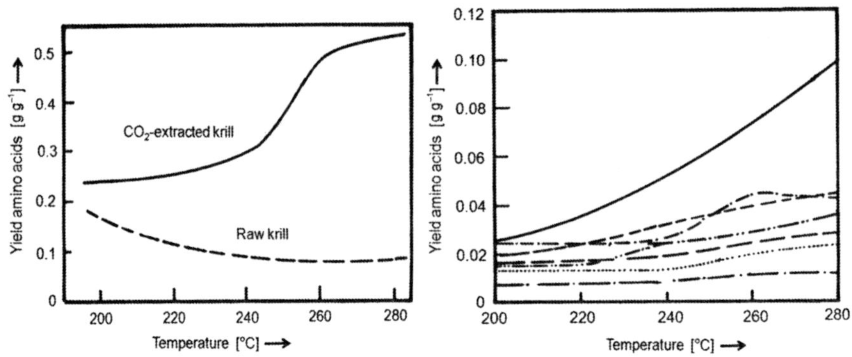

FIGURE 8.99 Conversion of krill to amino acids. $P=0.1$–6.4 MPa (saturation pressure) Left: Total amino acid yield. Right: Yield of individual amino acids for deoiled krill. ———, Arginine; — — —, glycine; —•—•, leucine; —— ——, lysine; ——•——, methionine; ••••, alanine; —••—, phenylalanine. *Modified from Ref. [124].*

cystine 0.024; alanine 0.013; glycine 0.009; and leucine 0.004 kg kg^{-1} of dry meat, reaction time 5 min at $T=543$ K, $P=5.51$ MPa.

8.4.2.3.2 Krill

Treatment of Krill residues with high-temperature water is more effective for deoiled substrate than for the original one. Deoiling with supercritical carbon dioxide enhances the yield in amino acids from <100 mg g^{-1} to about 650 mg g^{-1} for the subsequent conversion in water in a batch reactor at $T=280$ °C and $t=5$ min reaction time, as shown in Figure 8.99 [124]. The yields glycine, arginine, and leucine increase with water temperature, with the highest yield of amino acids in deoiled krill hydrolysate at 280 °C, opposed to the highest yield of amino acids from raw krill occurs at the lower temperature of 200 °C. Content of reducing sugar also is higher in hydrolysate from deoiled krill than from raw krill [124].

Similar investigations on various types of marine waste also show similar results, for example, on shrimp shells [125], fish entrails [95], and scallop viscera waste [94], approaching approximately 15 wt% yield in amino acids at optimal conditions.

Treatment with high-temperature water is also a good pathway to recover and separate the different components of materials that are mixtures of lignocellulose and proteins, like bean residues [126], rice bran [127], or brewery wastes. Utilization of such materials with high-temperature water was so far less intensively investigated compared to materials that contain only one group of compounds. Usually, only one type of components is the target of the investigations; rice bran is discussed in the following.

8.4.2.3.3 Rice Bran

Treatment of deoiled rice bran with high-temperature water recovers protein, amino acids, and reduced sugars [127]. Results are based on experiments in a batch reactor with 20 wt% solids content. The highest yield of protein is 219 mg g^{-1} and of amino acids 8.0 mg g^{-1} of dry bran at $T = 200$ °C and a reaction time of $t = 30$ min. The amount of protein and amino acids produced is higher than that obtained from alkali hydrolysis. Generally, yields increase with temperature and reaction time. At higher temperatures ($T > 200$ °C) and extended reaction times, thermal degradation of the product occurs [127]. Composition of the rice bran is listed in Table 8.17, and the results of the hydrolysis in high-temperature water are shown in Figures 8.100 and 8.101.

High-temperature and supercritical water is able to convert natural materials into fractions and compounds without addition of acids, bases, or catalysts. In some cases, such an addition may be worth the effort, but in general, the advantage of processes that use only water is overwhelming, in particular, if all costs for the additives, the recovery, and the decontamination of products are considered.

TABLE 8.17 Composition of Rice Bran [127]

	Raw	Deoiled
Moisture (wt%)	9.70	10.90
Oil (wt%)	16.75	1.90
Protein (wt%)	12.25	18.56
Fiber (wt%)	6.64	10.48
Nonprotein (wt%)	54.66	58.16

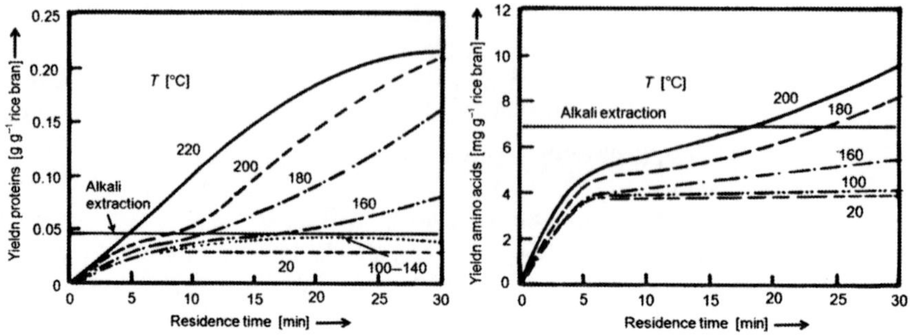

FIGURE 8.100 Conversion of rice bran in high-temperature water. Left: Yield of protein. Right: Yield of amino acids. *Modified from Ref. [127].*

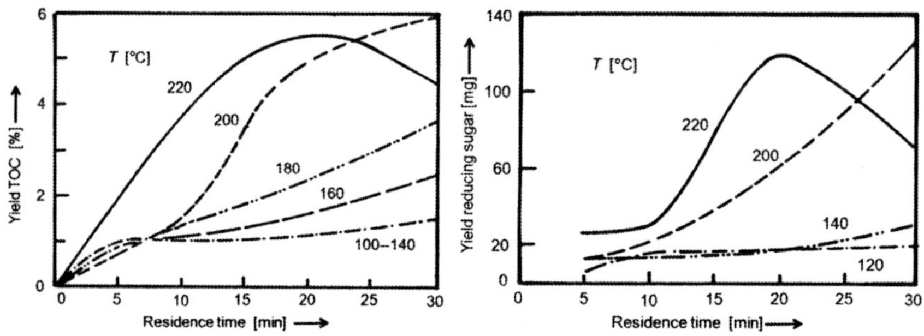

FIGURE 8.101 Conversion of rice bran in high-temperature water. Left: Total organic carbon of products. Right: Yield of reduced sugars. *Modified from Ref. [127].*

The most important issue for utilizing biomass, as for any other process, is to find and define products that are needed and can be sold to cover the costs of processing and financing, and that produce some profit.

So far, the goals have been set too much on the recovery of singular types of compounds or even special compounds. Often, oligomers as products of natural polymers are not considered as valuable products. Since natural materials are mixtures of several types of biopolymers and depolymerization cannot be achieved to a high degree without producing degradation products, the intermediate oligomers must get much more attention.

Investigations mostly are carried out on laboratory scale with small samples in batch mode, semicontinuous mode or in flow reactors. Scale up is one of the issues to be addressed and problems solved before any commercial process can be designed.

The application of chemical engineering techniques can improve yield and selectivity to a high degree. Such techniques can be applied to all conversions with high-temperature and supercritical water.

8.5 BIOMASS LIQUEFACTION

Liquefaction of biomass is aimed to obtain as much as possible of a liquid fuel-like product. Liquefaction of biomass with high-temperature or supercritical water places the reactions in an aqueous atmosphere. Products are a liquid phase, often named "oil," containing the water-insoluble compounds, an aqueous phase, containing the water-soluble compounds, and a gaseous phase, containing the gaseous products. The oil generally needs further processing to be acceptable as fuel. The gaseous product can provide the necessary energy for processing. Compounds in the aqueous phase can be removed and considered as separate and valuable products. Even, if this phase is recycled, some separation from the aqueous phase must be achieved.

Throughout the liquefaction of biomass, the same reactions occur as have been discussed previously in context with conversion of the different components and mixtures of biomass. But the goal, to produce a liquid phase that shall be used for fuel purposes leads to process conditions that maximize the yield of the water-insoluble phase, considering only the composition relevant to fuel. The liquefaction process is relatively simple when individual components are not recovered. This anticipated simplicity may have contributed to the widespread idea of converting biomass into a product that can be used as replacement for fuel based on mineral oil. As will be discussed further, the properties of biomass, such as being mostly a solid with a high water content and a low energy density, and other peculiarities, lead to the fact that commercial biomass processes are still in development, and it is not guaranteed that they will ever be profitable, if the intention is to produce fuel products only. With this intention, biomass conversion processes will have to compete with the products of mineral oil exploration for a long time, before the advantage of being a renewable resource may change the issue.

Liquefaction in high-temperature water is a medium-temperature, high-pressure process in which the macromolecules of the biomass are hydrolyzed and degraded into smaller molecules. Some intermediate products are unstable and reactive, and recombine into larger molecules, often insoluble in the reaction medium, or are cracked to gases. A substantial part of the oxygen in the biomass is removed by dehydration or decarboxylation [128]. By selection of appropriate reaction conditions, thermal conversion processes can be optimized for the production of a variety of liquid, gaseous, or solid fuels from biomass [129].

At the end of the 1980s, it was accepted that rapid pyrolysis processes that are simple and can be carried out in a variety of reactor types can be used to produce the liquid product at a high yield (up to 80%) and at low cost. Such processes are carried out at temperatures of $T > 500\ °C$, at ambient pressure, and with a short-residence time of seconds [129].

High-pressure systems are more costly, and the overall efficiency of conversion with conversions investigated at that time was less. For both cases, production costs for the biomass gasoline product were estimated to be about the same as producing gasoline from petroleum and roughly the same as producing oxygenates such as ethanol or methanol from biomass.

A typical investigation on wood conversion—a continuous slurry process (Lawrence Berkeley Laboratory) was carried out in a stirred reactor through which an aqueous slurry of biomass is pumped. Reaction temperature is maintained at about $T = 350$ °C, at a pressure of $P = 20$ MPa, and a residence time of $t = 10$–30 min. In addition to the water of the slurry, the reaction medium consisted of sodium carbonate (2.5–5%) and gas composed of 60% CO and 40% H_2 [129]. For preparation of the slurry, which contained 20 wt% of solids, the wood is subjected to an acid hydrolysis prior to the liquefaction. Wood chips are heated to $T = 180$ °C for about $t = 45$ min in an aqueous solution with an initial pH = 1.9 (about 0.075% sulfuric acid).

Liquefaction, according to this investigation, produces an oil phase with yields up to 35 wt% of the feedstock. For comparison, fast pyrolysis achieves yields up to 53 wt%. Oxygen content of the liquefaction process is 12–18 wt%. More than 70% of the total oil consists of products with one or more phenolic groups, with an average molecular weight of $M_W = 200$–300 D. The aqueous product contains about 25 wt% dissolved organics, with about 50% organic acids such as formic, acetic, glycolic, and others, and about 50% other organics such as ketones and phenolics. It is assumed that the compounds in the aqueous phase are difficult to recover, without referring to relevant procedures [129].

Similar liquefaction results are achieved in high-temperature and supercritical water, as has been presented in context with processing of wood (Section 8.4.1.2). Conversion in high-temperature water at $T = 320$ °C and $P = 23$ MPa results in a conversion of pine wood meal of 88 wt% of dry matter. Treatment of the pine wood meal with supercritical water at 380 °C and $P = 23$ MPa in a batch reactor reduces the overall conversion of wood and cellulose to about 70 wt%. Oil yield is about 30–35 wt% in pure water [120].

In a more recent investigation, wood (silver birch) was liquefied by water in a batch autoclave at temperatures of $T = 280$–420 °C with sodium carbonate as catalyst and a hydrogen atmosphere of initial $P = 8$ MPa pressure. The maximum yield of oil of about 54 wt% was obtained at a reaction temperature of $T = 380$ °C, as shown in Figure 8.102. The oil consists of hydrocarbons, aldehydes, ketones, hydroxybenzenes, and esters [130]. Although batch reactors do not yield results that can be directly applied to continuous processes, this result indicates that with an inexpensive catalyst the oil yield can be enhanced substantially. As reported from other investigations, the effect and necessity of reducing gases (H_2 and CO) is unclear [4]. Either no reducing gas is consumed [131] or results indicate that reducing gases or inert gases (CO_2 and N_2) may be useful [132].

These investigations indicate that conversion of wood is possible in neat water. Yield enhancement can be achieved with simple catalysts, if wanted.

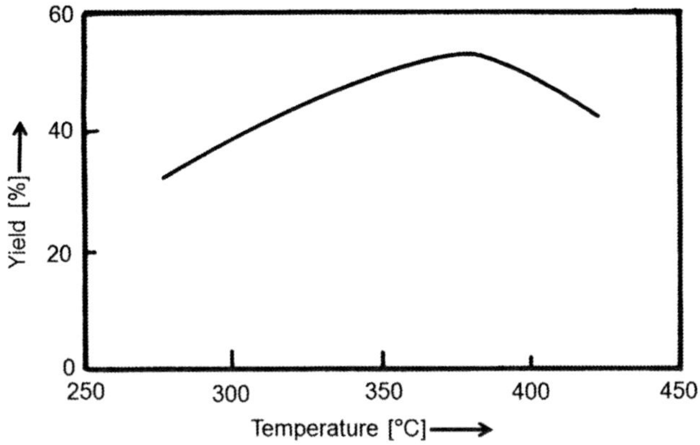

FIGURE 8.102 Conversion of wood in water with Na_2CO_3 as catalyst, initial H_2 pressure $P=8$ MPa. *Modified from Ref. [130].*

Although it is now accepted that no enhancement of reaction rates at critical conditions takes place for water, observed rates are significantly enhanced by the lack of mass-transfer limitations, because most organic species become miscible with near-critical water, as well as by its ability to sustain ionic and free-radical reactions [4].

For producing fuels from biomass, one main objective is to remove oxygen, since biomass feedstock often contain 40–60 wt% oxygen, while conventional oils only have an oxygen content of <1%. Removal of the oxygen heteroatom occurs by dehydration in the form of water, and by decarboxylation, in the form of carbon dioxide. Dehydration reactions occur in hydrothermal systems at elevated temperatures and pressures although an excess of water is present. Decarboxylation reactions decrease not only the oxygen content of the feedstock but also increase the H/C ratio. In mixtures containing multiple functional groups, interactions between these groups change chemical pathways and reactions. Conditions of liquefaction and results of investigations reported in the literature are reviewed by Peterson *et al.* [4].

For the conversion of biomass at hydrothermal conditions to liquid products, several approaches exist. Conditions with liquid water range from temperatures of $T=280$ to 380 °C, pressures of $P=7$ to 30 MPa, and residence times of $t=10$ to 60 min. Catalysts may be used, generally alkaline, and also reducing gases, such as CO and H_2, although their role is at least unclear.

The oil produced has a heating value of 30–36 MJ kg^{-1} and an oxygen content of 10–20%. Biomass feedstocks used for liquefaction have heating values of 10–20 MJ kg^{-1} and oxygen contents of 30–50%. Various biomass substrates have been investigated, for example, sugar beet pulp [133], swine manure [134], bagasse, coconut shells as model garbage [135,136], algae [137], and various animal- and plant-derived wastes [138].

The total conversion and the quality of the products may be enhanced by two-step liquefaction. In the conversion of switch grass, as an example, the first step is carried out at $T = 200\ °C$ in acidic subcritical water to liquefy hemicelluloses and avoid repolymerization reactions. In the second step, the remaining nonliquefied biomass reacts in supercritical water at $T = 380\ °C$ with $Ca(OH)_2$ as catalyst. Thus, the formation of char is minimized, lignin solubilization is enhanced, and liquefaction of the remaining polysaccharides is increased. The yield of the two-step process is 40 wt% and 67% on energy basis of the feedstock [139].

8.6 BIOMASS GASIFICATION

Several techniques to thermally convert biomass into useful gaseous compounds have been investigated and some have been developed. Problems remaining are the formation of significant amounts of char, the high water content of many materials. Supercritical water as the medium for gasification leads to hydrolysis reactions in parallel to pyrolysis and avoids the need for drying the feed material. The water–gas shift reaction becomes an important tool to accomplish CO elimination and the gaseous compounds produced are at high pressure, often necessary for subsequent application. Gasification in supercritical water requires lower temperatures than other gasification techniques. Achieving the high pressure is not energy intensive since it is carried out in the liquid state of water. Furthermore, high-pressure processes, even for commodity chemical, are well known and used routinely [140].

Therefore, gasification of biomass feedstock in high-temperature and supercritical water promises a rapid, direct route to gases like H_2, CO, CO_2, CH_4, and light hydrocarbons without char formation [4], and without the variety of products that have to be recovered from an aqueous phase. Unfortunately, experiments revealed that gasification is not simple, but has to overcome its special problems. High yield of valuable gases without char formation is difficult to achieve. Nevertheless, hydrothermal gasification is still a promising pathway to utilize biomass feedstock, if the intention is to recover mainly gaseous products. Since the work of Modell and coworkers [9,12,141,142], research in hydrothermal gasification has three target products: (1) hydrogen-rich gas via high temperature ($T > 500\ °C$) without catalysts or with nonmetal catalysts, (2) methane-rich gas via near-critical temperatures of $T \approx 500\ °C$, usually with catalysts, and (3) subcritical catalytic processing to a gaseous product [4].

Hydrothermal gasification of wet biomass can be useful for processing biomass residues of high water content. A great part of such residues is lignocellulosic biomass. For experimental purposes, this type of biomass can be modeled with phenol as substitute for lignin and with glucose as a substitute for cellulose. Protein-containing biomass has been modeled by the gasification of glycine.

Gasification of model compounds and of separate natural polymers is discussed in the following. Then, examples for gasification of biomass feedstock are treated, an overview on gasification research is presented and finally, additional examples are discussed, covering important features of gasification. Concluding remarks sum up Section 8.6.

8.6.1 Hydrothermal Gasification of Model Compounds and Separate Natural Polymers

8.6.1.1 Glucose

Conversion of glucose at temperatures above $T = 300$ °C produces a substantial amount of gaseous products, beside the liquid fraction, as discussed above. For conversion at temperatures in the supercritical range of water of $T = 425$–600 °C ($P = 24.6$ MPa, residence time $t = 4$–11 s), three steps of product formation occur. In the first one, relatively stable compounds are formed like acetic acid, acetaldehyde, propenic acid, and acetonylacetane. In the second step, these compounds are transformed into stable gases like methane, ethane, and ethylene. These are further transformed in a third step into CO_2, CO, H_2, and H_2O [143]. Similar results, were obtained by Antal *et al.* [144] at similar conditions (temperature $T = 450$–650 °C, pressure $P = 34.5$ MPa, residence time $t \approx 20$ s). According to these results, glucose can be totally transformed to gases in pure water if the temperature is higher than $T = 600$ °C, with the main products being H_2 and CO_2.

Gasification of glucose is enhanced in presence of an alkaline catalyst, as shown in Figure 8.103, while small quantities of phenol in the feed solution inhibit the conversion. It is assumed that phenol acts as a free-radical scavenger [146]. Salts that are added have to be kept in solution to avoid plugging [145]. By addition of K_2CO_3, carbon gasification efficiency is improved up to 94% at $T = 600$ °C and $P = 20$ MPa [147].

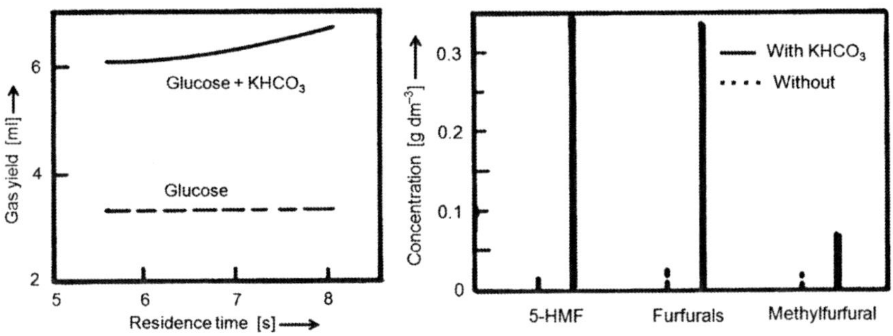

FIGURE 8.103 The influence of $KHCO_3$ on the gas yield (left) and the formation of furfurals (right) during glucose gasification with supercritical water. Feed solution: 1.5–2 wt% glucose, 0.2 wt% $KHCO_3$ in water. $P = 25$–26 MPa, $T = 500$ °C. HMF, hydroxymethylfurfural; FU, furfural; MF, methylfurfural. *Modified from Ref. [145].*

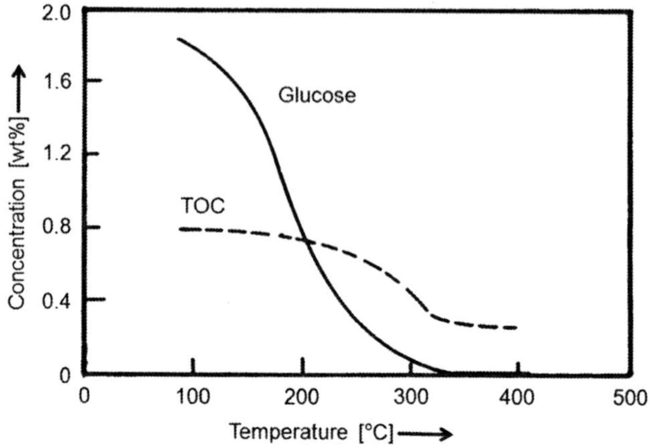

FIGURE 8.104 Conversion of glucose catalyzed by a Pt-Pt-γ-alumina catalyst in a flow reactor. $P=10$–30 MPa, residence time $t \approx 3.75$ min. *Modified from Ref. [149].*

The presence of transition metal chelates (organometallic compounds) slightly improves the decomposition of glucose at $T=310$–350 °C, $P=10$–21 MPa, while Raney-nickel particles enhance the gas yield by a factor of 3–5. The yield also increases with concentration (0.06 → 0.65 M) and almost doubles with a longer reaction time (3 → 30 min) [148].

Glucose can be converted at relatively low temperatures in a flow reactor and short-residence times of a few minutes, if the reaction is catalyzed with a Pt catalyst. In this case, glucose concentration decreases rapidly at temperatures above $T=100$ °C and completely disappears at $T=325$ °C (Figure 8.104). TOC in the aqueous phase also decreases with temperature. At $T=360$ °C, about 67% of the carbon content of the feed is converted to gas. Main components in the gas are CO_2, H_2, and CH_4, with up to 44 mol% H_2 at $T=290$ °C [149].

8.6.1.2 Cellulose

Gasification of cellulose in hydrothermal water occurs according to the following reactions [149]:

$$\text{Cellulose}: \quad (C_6H_{10}O_5)_n + nH_2O \rightarrow nC_6H_{12}O_6 \tag{8.23}$$

$$\text{Glucose}: \quad C_6H_{10}O_5 \rightarrow 6H_2 + 6CO \tag{8.24}$$

$$\text{Water-gas shift reaction}: \quad CO + H_2O \rightarrow H_2 + CO_2 \tag{8.25}$$

$$\text{Overall}: \quad (C_6H_{10}O_5)_n + 7H_2O \rightarrow 12nH_2 + 6nCO_2 \tag{8.26}$$

According to Equations (8.23)–(8.26), the gas produced from cellulose gasification can contain up to 66.7 mol% H_2. In Figure 8.105, carbon yields for the products gas, oil, aqueous phase, and residue are shown for the

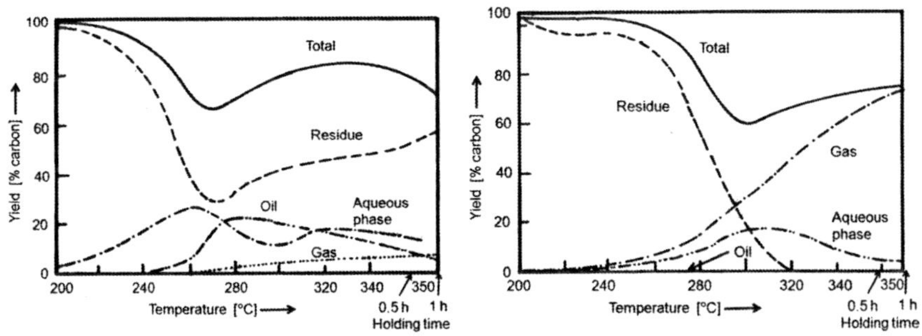

FIGURE 8.105 Conversion of cellulose in high-temperature water in a stirred batch autoclave. Heating rate $\Delta T = 0.18\ °C\ s^{-1}$, $P = 4.3$ MPa ($T = 200\ °C$) to 16.5 MPa ($T = 350\ °C$) Left: No catalyst. Right: Pt-γ-alumina catalyst. *Modified from Ref. [149].*

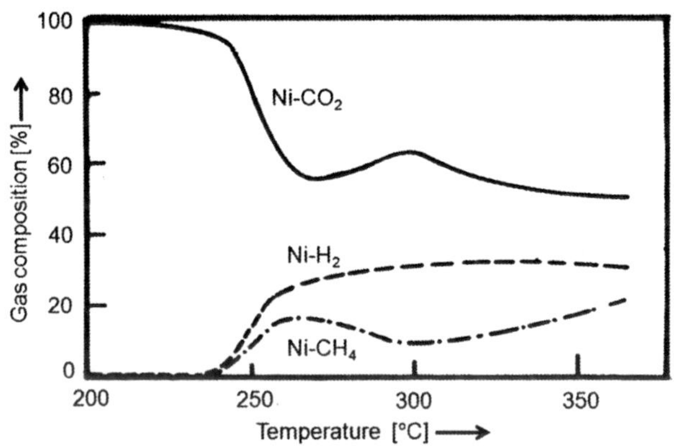

FIGURE 8.106 Composition of the gas product during conversion of cellulose in high-temperature water in a stirred batch autoclave. Heating rate $\Delta T = 0.18\ °C\ s^{-1}$, $P = 4.3$ MPa ($T = 200\ °C$)–16 MPa ($T = 350\ °C$). *Modified from Ref. [149].*

decomposition of cellulose in a stirred autoclave. The autoclave was heated with $\Delta T = 0.18\ °C\ s^{-1}$ and cooled after reaching the end temperature, or temperature was held for a specified time, as indicated in Figure 8.105. The influence of a Ni catalyst is shown in the right-hand part of Figure 8.105.

The composition of the gas product of the cellulose conversion is plotted in Figure 8.106 [149]. At higher, near-critical temperatures, the concentration of CO_2 is lower than at temperatures $T = 200$–$250\ °C$. H_2 concentration reaches about 30% and methane concentration about 15%. An extended reaction time by $\Delta t = 30$ min at $T = 350\ °C$ changes these concentrations only to a minor extent, see Figure 8.105 [149].

Cellulose can be gasified up to 96 wt% with a Ni-silica/alumina catalyst and by good mixing in an autoclave [149]. Cellulose completely dissolves in water at $T = 318\ °C$ and then is hydrolyzed to glucose and oligomers, when heated rapidly. According to these results, a three-step conversion can be carried out for the gasification of cellulose: Cellulose is completely dissolved in a first step. In the second step, the dissolved product compounds are hydrolyzed, and in a third step, the hydrolysis products are catalytically converted to gaseous products [149].

Other investigations confirm these results in principle. At temperatures of $T=400$–$450\ °C$ and reaction times or around $t=2$ h in a batch reactor, 97% cellulose is gasified with RuO_2 catalyst [150]. With a Ni catalyst at $T=374\ °C$ and $t=30$ min reaction time in a batch reactor, a gasification rate of 18% is obtained [12], and at $T=350\ °C$, a gasification rate of 84% [151].

8.6.1.3 Lignin

Gasification of lignin usually is investigated with isolated lignin. Such lignin products are modified during extraction from the plant material, leading to more highly condensed, cross-linked materials. Therefore, degradation of isolated lignin is probably more difficult compared to processing lignin in whole-plant materials [61,140,152].

Lignin decomposition in high-temperature and supercritical water starts by hydrolysis and forms phenol compounds. Connections between the building blocks often are ether bridges that are cleaved by hydrolysis and one molecule of water for every broken linkage is added. At hydrothermal conditions, the low-molecular-weight fragments of lignin hydrolysis undergo cross-linking reactions to heavier compounds that are solid [140]. The gas forming reactions compete with these cross-linking reactions. At temperatures around $T=400\ °C$ and without a catalyst, total gas yields are low (<5–$10\ mmol\ g^{-1}$) [153]. Gasification in supercritical water, without catalysts, at $T=600\ °C$ and 5.0 wt% lignin loading results in the highest total gas yield of 90 wt%, while the highest H_2 yield of $7.1\ mmol\ g^{-1}$ is obtained at $T=725\ °C$ and $t=60$ min residence time [140]. As the extended residence time indicates, these results are based on experiments in a quartz batch reactor (volume $0.58\ cm^3$) [140]. Gasification in a continuous reactor will most probably lead to similar results at far shorter residence times.

Results for the gasification of lignin in microbatch reactors are shown in Figure 8.107. Visual observation indicates that char is formed in all cases [140].

Density of water is of major importance for the gasification of lignin which proceeds through the decomposition of lignin to low-molecular weight compounds and the gasification of the low-molecular weight compounds, often over metal catalysts. Higher water densities increase the first step, that is, the breakdown of lignin [154]. Total gasification efficiency decreases when the density is $\rho_{H_2O}>0.4\ g\ cm^{-3}$, probably due to Le Chatelier's principle. In mixtures like biomass, lignin seems to inhibit the rates of gas formation [155].

Gas production from lignin gasification in supercritical water with Ni/MgO catalysts, carried out in a stainless steel batch reactor is shown in Figure 8.108. Carbon yield of the gas products increases with surface area

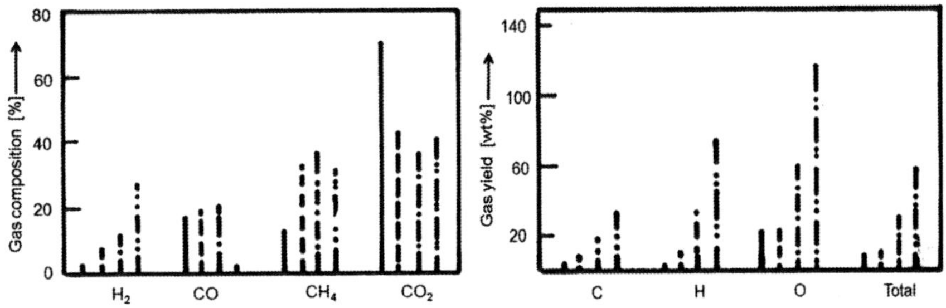

FIGURE 8.107 Gasification of lignin in supercritical water. Residence time $t=45$ min. Left: Gas composition. Right: Gas yield. Temperature: ———, $T=365$ °C; — — —, $T=500$ °C; —•—•, $T=600$ °C; —••—, $T=725$ °C. *Modified from Ref. [140].*

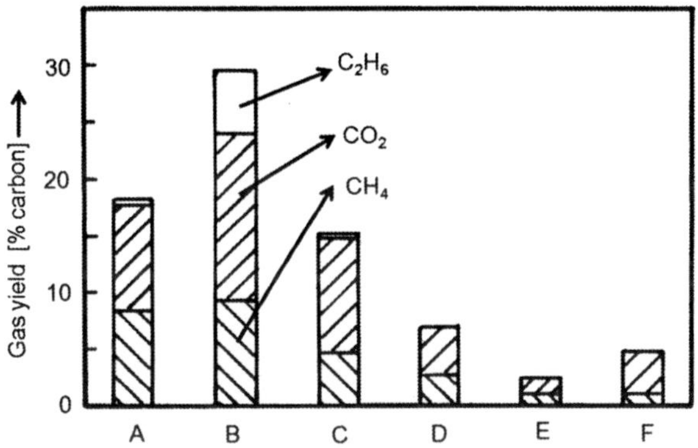

FIGURE 8.108 Gas yield for the gasification of lignin in supercritical water with various catalyst. A, 10 wt% Ni/MgO (500 °C); B, Ni (600 °C); C, Ni (700 °C); D, Ni (800 °C); E, Ni (900 °C); F, 10 wt% Co/MgO (600 °C). *Modified from Ref. [156].*

of the Ni catalyst, with the exception of the 10 wt% Ni/MgO catalyst at $T=600$ °C. Carbon efficiency for the gas production is relatively low, notwithstanding the comparatively high amounts of catalyst [156]. Detailed discussion of the effect of various catalysts is beyond the scope of this book. Information on the effect of various catalysts for the gasification of lignin is available from the literature [153,154,156–163].

8.6.1.4 Glycine

Glycine gasification in supercritical water serves as a model compound for protein. In a flow reactor at temperatures of $T=380$–500 °C and residence times of $t=1$–1.5 min, with sodium carbonate as catalyst, glycine is almost completely gasified. The gas products include H_2, CO_2, N_2, CH_4, C_2–C_3, with hydrogen up to >50% by volume and a yield of 1.8 $dm^3\ g^{-1}$. Sodium carbonate accelerates the decarboxylation reaction of formic acid, which is a hydrolysis product of glycine [164]. Results are shown in Figure 8.109.

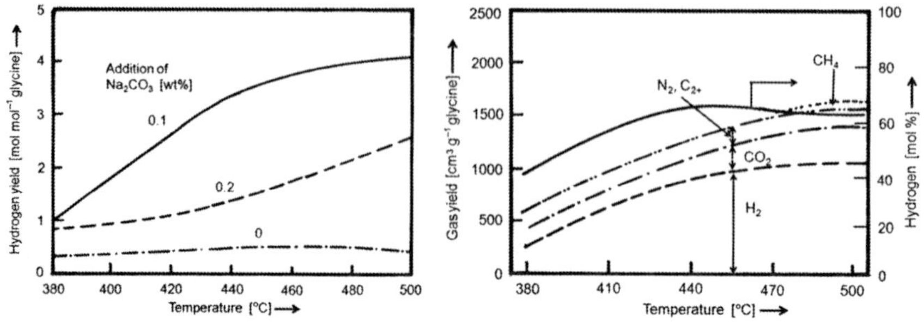

FIGURE 8.109 Gasification of glycine in supercritical water. Left: Influence of Na_2CO_2 concentration. Right: Gas composition and yield. *Modified from Ref. [164].*

Gasification of glycine without catalyst or with a Ni catalyst is difficult [165]. 20–90% of the initial carbon remained in the aqueous phase even after $t=1$ h for noncatalyzed and Ni-catalyzed reactions. CO was most abundant after the noncatalyzed reaction, while hydrogen was the major gas product in presence of a nickel catalyst [165].

8.6.2 Hydrothermal Gasification of Real Biomass

During gasification of biomass, mainly hydrogen and carbon dioxide are formed. In addition, methane, CO, and some higher hydrocarbons are formed. At high temperatures, the conversion is fast, intermediate products are difficult to determine, and CO is converted by the water–gas shift reaction. At lower temperatures, biomass first reacts to intermediate products that are subsequently gasified.

8.6.2.1 Introductory Example

An example for real biomass is the gasification of corn silage [166] in a flow reactor of $L=1$–1.7 m length, with $d=8$ mm inner diameter, made of Inconel 625, at $T=773$–973 K, $P=25$ MPa. The gasification yield reaches >90% at $T=973$ K after about $t=25$ s and increases about 60% from $T=773$ to 973 K (Figure 8.110). The amount of gas produced does not vary with changing process pressure [166]. Hydrogen production increases from a small amount at $T=773$ K to >30 vol% at $T=973$ K. At the lower temperature, the concentration of CO is around 15 vol%, while at $T=973$ K, the CO concentration is <1 vol%, as shown in Figure 8.111 [166].

It can be concluded that the gasification of corn silage as an example for a real biomass is possible in supercritical water. Only reduction of the solid particle size is necessary. Potassium enhances the gasification of corn silage biomass up to a concentration of 0.05 wt%.

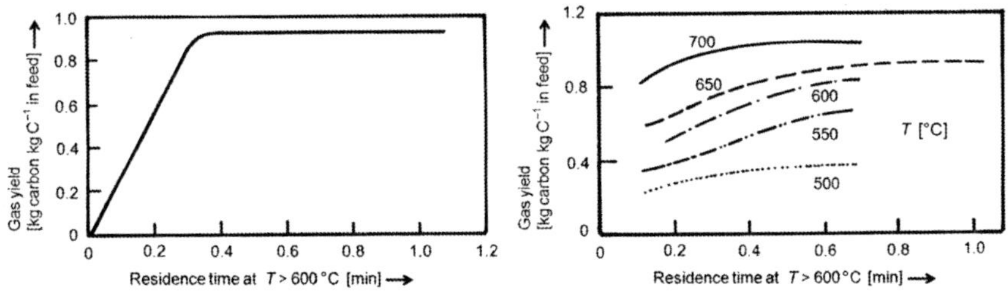

FIGURE 8.110 Gasification of corn silage in supercritical water in a continuous flow reactor. $P = 25$ MPa, $x_F = 5$ wt% corn silage as dry matter. *Modified from Ref. [166].*

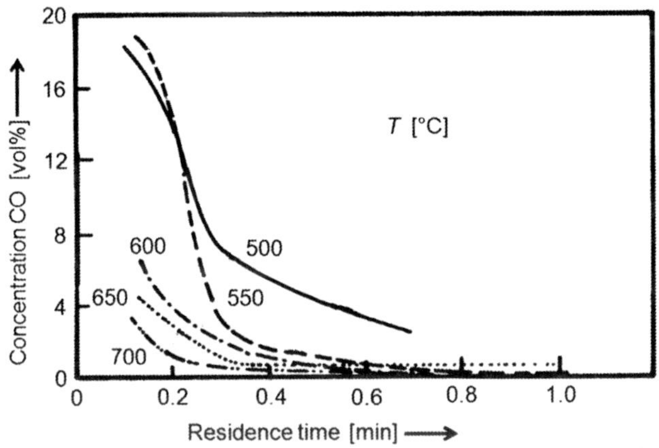

FIGURE 8.111 CO concentration during the gasification of corn silage in supercritical water in a continuous flow reactor. $P = 25$ MPa, $x_F = 5$ wt% corn silage dry matter. *Modified from Ref. [166].*

8.6.2.2 Mixtures of Natural Biopolymers

Much information is available on the gasification of components of natural biopolymers in high-temperature and supercritical water. The influence of such components in mixtures of natural biopolymers is addressed in a review that claims to represent the status of gasification in high-temperature and supercritical water up to the year 2005 [167]. Artificially mixing the components of lignocellulose, the major influence of lignin on gasification of cellulose and xylan can be seen, as shown in Figure 8.112. Lignin inhibits the gasification of these compounds of cellulosic material, indicated by the steep decrease in gas yield, if lignin is added to the mixture. Gas yield of mixtures of cellulose and xylan, on the other hand, behave like ideal mixtures, that is, the gas yield changes linearly with concentration. Although the graphs for mixtures containing lignin are interesting and confirm a common belief, it is not clear whether they have been artificially produced, or whether the

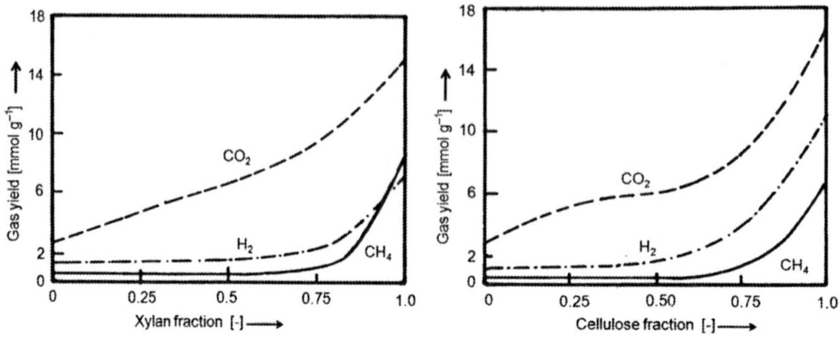

FIGURE 8.112 Influence of lignin on gasification of lignin–cellulose and lignin–xylan mixtures. *Modified from Ref. [167].*

results are on gasification of a real cellulosic material. Furthermore, conditions have not been specified.

8.6.2.3 Overview on Gasification of Biomass with Water

Several overviews on gasification with high-temperature and supercritical water have been published, for example [4,168–172]. The following summary is retrieved from the overview of Peterson *et al.* [4], if not otherwise stated.

In the temperatures range of about $T = 500–700$ °C, a variety of biomass feedstock can be converted to a very high degree into gases in supercritical water, as described above in the introductory example. The product gas is rich in H_2, while most of the carbon is converted to CO_2. Residence time for continuous operation is in the range of minutes. Reaction temperature has a strong effect on yields and gas compositions. On the other hand, pressure above $P = 21$ MPa is of little effect on gasification of the feed or composition of the product gases. Catalysts are not necessarily needed, but activated carbon and alkali salts do have positive effects. Investigations on the conversion of a number of different sources have been carried out in sizes from small-scale batch rectors, in continuous reactors, to pilot plant scale installations. So far, no commercial plant is in operation. Research is still in progress in many laboratories, many more than cited by Peterson *et al.* [4], as can be concluded from the list of references from this chapter and Chapter 9.

Experimental evidence shows that the addition of alkali salts leads to higher hydrogen yields under supercritical water conditions. In parallel, the interaction of metal reactor walls with alkali salt solutions leads to hydrogen, compare Equations (8.27) and (8.28), obscuring the reason for higher H_2 yields.

$$\text{Dissolution: } Cr_2O_3(s) + 2\,OH^-(aq) \leftrightarrow 2\,CrO_2^-(aq) + H_2O(l) \tag{8.27}$$

$$\text{Oxidation: } 2\,Cr(s) + 3\,H_2O(l) \leftrightarrow Cr_2O_3(s) + 3\,H_2(g) \tag{8.28}$$

Corrosion products occurring during experiments in supercritical water with alkali salts indicate that the alkali environment dissolves the protective metal oxide layer on the reactor walls. Fresh, temporarily reduced metal is oxidized in supercritical water, producing hydrogen.

Protein-containing biomass reduces the gas yield compared to materials containing only lignocellulose, as demonstrated on the example of the amino acid alanine during gasification of animal waste [173,174]. Minerals contained in the feed can be concentrated in the char produced by two-step processing. In the first step, the feed is flash pyrolized, and in the second step, the condensate is fed to the reactor, in which the gasification in supercritical water is carried out [175,176]. At $T=600$–$650\ °C$, $P=28$ MPa, a gasification efficiency of 60–80% is obtained in a tubular reactor made of Incoloy with H_2, CH_4, CO, and CO_2 as major products.

At lower temperatures of $T \approx 500\ °C$, the production of methane is favored in gasification reactions and catalysts are generally needed to achieve high-gasification rates and high selectivity. The equilibrium gas composition has been calculated for $x_F=5$ wt% biomass (wood sawdust, approximate formula $CH_{1.35}H_{0.617}$) in water at $P=25$ MPa [177,178]. At the moderate temperatures of $T=400$–$600\ °C$, CH_4, CO, H_2, and CO_2 are the major products. Around $T=400\ °C$, concentration of these compounds is similar. At temperatures approaching $T=600\ °C$, CH_4 decreases substantially, CO and CO_2 increase to >30 mol kg_{Feed}^{-1}, and H_2 increases to nearly 80 mol kg_{Feed}^{-1} from as low as 20 mol kg_{Feed}^{-1} at $T=400\ °C$. In experiments, it is not possible to approach these equilibrium values due to formation of nonreactive polymers from intermediates and the slow methane-formation reaction. Catalytic effect of the reactor walls may be important in experiments, but below $T=500\ °C$, these effects are less important and the product distribution is affected mainly by the catalyst. Catalysts achieve complete conversion by gasifying reactive intermediates. This gasification step must be fast enough to avoid the formation of polymers and char. The intermediates are mostly water-soluble products that react along two competing pathways, one leading to gaseous products such as CO, CO_2, H_2 and the other leading to oils and char [151]. A detailed discussion of catalysis for the production of methane from biomass in high-temperature and supercritical water is reported in Refs. [157,163,167,179].

Without a catalyst, a slurry of $x_F=10$ wt% wood sawdust is converted to gases to about 21% at $T=409\ °C$ [180]. At the somewhat lower temperature of $T=350\ °C$ and a residence time of $t=60$ min, cellulose is converted to gases to only about 10% carbon. 67% of the carbon in the feed was recovered as char, 5% as oil, and 13% dissolved in the aqueous phase [181].

8.6.2.4 Gasification of Wood Sawdust

Sawdust can be gasified in supercritical water fast and with high gas yield and high efficiency. The results are summarized in Figures 8.113–8.115 for

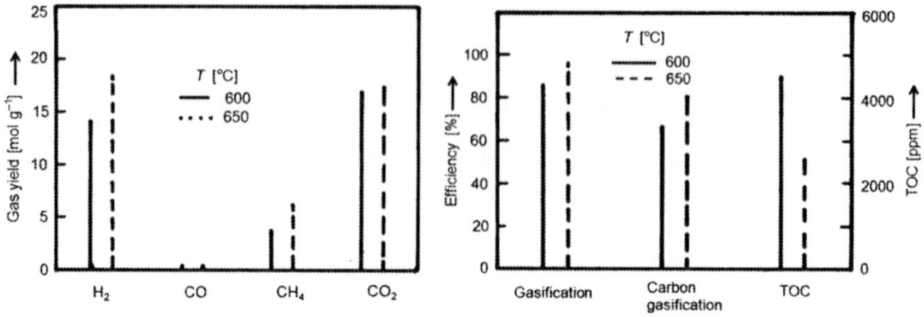

FIGURE 8.113 Influence of temperature on wood sawdust gasification in supercritical water. Left: Gas yield. Right: Gasification efficiency, carbon gasification efficiency, and TOC. $P = 30$ MPa, residence time $t = 27$ s, feed $x_F = 2$ wt% wood sawdust. *Modified from Ref. [177].*

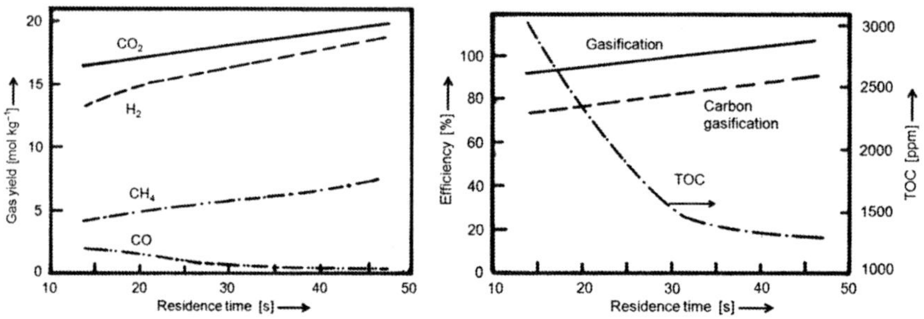

FIGURE 8.114 Influence of residence time on wood sawdust gasification in supercritical water. Left: Gas yield. Right: Gasification efficiency, carbon gasification efficiency, and TOC. $T = 923$ K; $P = 30$ MPa, feed $x_F = 2$ wt% wood sawdust. *Modified from Ref. [177].*

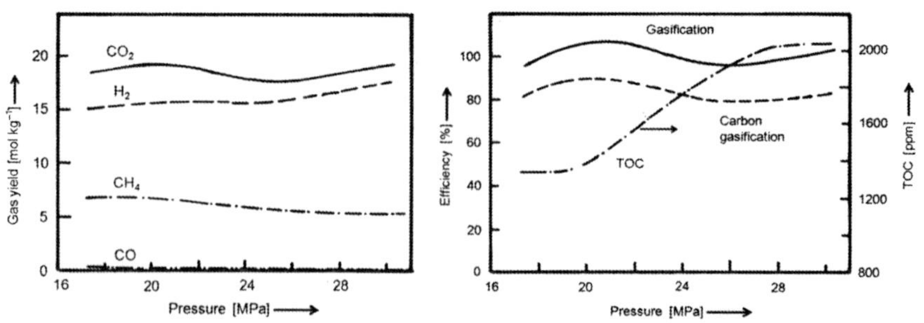

FIGURE 8.115 Influence of pressure on wood sawdust gasification in supercritical water. Left: Gas yield. Right: GE, gasification efficiency; CE, carbon gasification efficiency; and TOC. $T = 923$ K, residence time $t = 27$ s, feed $x_F = 2$ wt% wood sawdust. *Modified from Ref. [177].*

temperatures of $T=873$ K (600 °C) and $T=923$ K (650 °C), pressures above $P=22$ MPa, and short-residence times of around $t=30$ s. In Figure 8.113, gas yield and efficiencies are shown in dependence on temperature, in Figure 8.114 in dependence on residence time, and in Figure 8.115 in dependence on pressure. The results have been obtained in a continuous flow reactor for feed concentration of $x_F=2$ wt% [177].

A higher temperature, in general, leads to higher gas concentrations and higher efficiencies, although the temperature difference of these experiments is rather small. Gas concentrations rise with residence time, with the exception of CO, which decreases from 2 mol kg^{-1} to less than 1 mol kg^{-1} after $t=45$ s. TOC decreases rapidly with residence time, but reaches a lower level around $t=30$ s, while efficiencies seem to increase steadily with residence time. Pressure has no significant influence on gas concentrations and on efficiencies, but TOC increases steadily from a pressure of about $P=20$–30 MPa, with a probable maximum value for TOC of 2000 ppm.

It is of interest to compare the gasification behavior of some not too different materials as shown in Figure 8.116. Gas composition and efficiency are essentially the same for most of the chosen materials. Only for corn stalks, efficiency values are remarkably lower than for the other materials. TOC varies more, with a lower value of about 1800 ppm for rice shells and the highest value for wood sawdust of about 2700 ppm [177].

8.6.2.5 Gasification of Black Liquor

Black liquor is a waste stream with a high content of water and mainly contains lignin, its derivatives, and alkali salt. The alkali salt is an effective catalyst for gasification in supercritical water. It promotes the water–gas shift reaction, increases the H_2 content, and reduces the CO content of the gaseous product. The following results are based on continuous experiments with black liquor, derived from the wheat straw soda-pulping process.

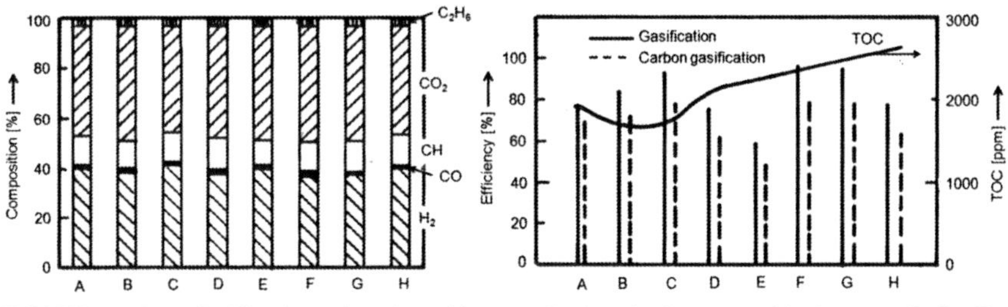

FIGURE 8.116 Gasification of various biomass feedstocks in supercritical water. Left: Gas composition. Right: Gasification efficiency, carbon gasification efficiency, and TOC. $T=923$ K; residence time 27 s; $P=25$ MPa, feed $x_F=2$ wt% biomass. A, rice straw; B, rice shell; C, wheat stalk; D, peanut shell; E, corn stalk; F, corn cob; G, sorghum stalk; H, wood sawdust. *Modified from Ref. [177].*

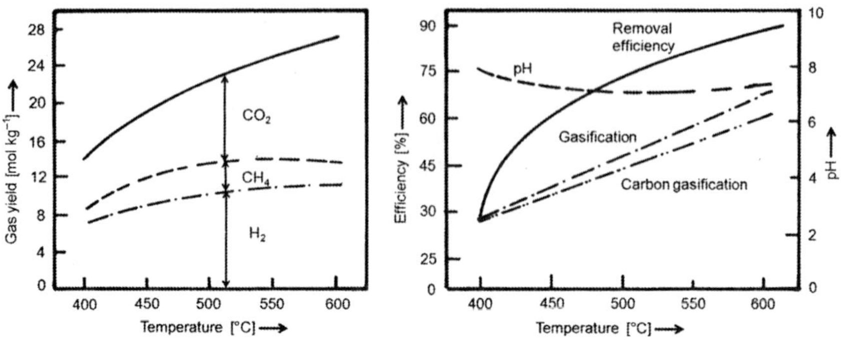

FIGURE 8.117 Temperature dependence of gasification of black liquor in supercritical water. $P = 25$ MPa, flow rate $\dot{m} = 5$ kg h^{-1}, feed concentration (nonaqueous) $x_F = 9.5$ wt%. Left: Gas yield. Right: Gasification efficiencies, wt%. *Modified from Ref. [182].*

Experimental conditions were $T = 400$–600 °C, $P = 19$–28 MPa, a flow rate of $\dot{m} = 5$ kg h^{-1} of black liquor with $x_F = 9.5$ wt% waste components in water. Dependence of gas yield and gasification efficiency is shown in Figure 8.117 [182].

With increasing temperature, the total gas yield increases from 14 mol kg^{-1} at $T = 400$ °C to 28 mol kg^{-1} at $T = 600$ °C. The chemical oxygen demand (COD) is reduced with temperature from 26% at $T = 400$ °C to 88.7% at $T = 600$ °C. Gasification efficiency is raised from about 27% at $T = 400$ °C to 61–68% at $T = 600$ °C (Figure 8.117).

Assuming a pseudo first-order reaction for the COD removal in gasification with supercritical water leads to an activation energy of $E_A = 74.38 \pm 6.02$ kJ mol^{-1} and a preexponential factor of $k_0 = 10^{4.05 \pm 0.41}$ s^{-1} (Equation 8.29) [182].

$$k = 10^{4.05 \pm 0.41} \exp(74.38 \pm 6.02/RT) \left[\mathrm{s}^{-1}\right] \tag{8.29}$$

Pressure in the range of $P = 19$–28 MPa is of minor importance for gasification efficiency and COD removal for black liquor gasification in supercritical water. Total gas yield increases from 20 mol kg^{-1} at $P = 19$ MPa to 27 mol kg^{-1} at $P = 28$ MPa. Of similar minor importance is the flow rate on gas composition, while total gas yield and gasification efficiencies drop slightly with increasing flow rate.

Concentration of the nonaqueous components, on the other hand, has a major influence on gasification. Total gas yield drops substantially with increasing concentration, as shown in Figure 8.118, as well as the heating value of the gaseous product (CgE) and hydrogen gasification efficiency.

More examples have been reported in the literature, but are beyond the scope of this book. For examples, the reader is referred to the following publications [163,167,175,176,178,183–185].

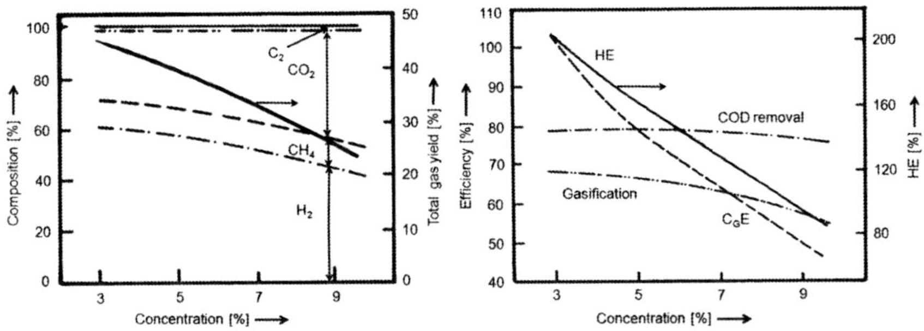

FIGURE 8.118 Concentration dependence of gasification of black liquor in supercritical water. $T = 550\,°C$, $P = 25$ MPa, flow rate $\dot{m} = 5\ kg\ h^{-1}$. Left: Gas yield. Right: Gasification efficiencies, wt%: GE, gasification efficiency, wt%; CgE, heating value efficiency; LHV of product gas/LHV of BL, %; COD removal efficiency, wt%. *Modified from Ref. [182].*

Organic materials can be gasified efficiently in supercritical water. For a nearly total gasification, catalysts are needed at lower temperatures, while at higher temperatures, well above the critical temperature of water, gasification without catalysts can lead to gases with a high heating value. Technical problems are mainly feeding, heat recovery, coke formation, and plugging by precipitated salts. If the organic biomass is not a waste that must be processed anyway, the product gas has to compete with similar gaseous products from other sources, which may be decisive for the economic feasibility of gasification. Furthermore, for full-scale applications, a continuous process design is required. Batch experiments are not a viable basis for this purpose.

REFERENCES

[1] R.R. Probstein, R.E. Hicks, Synthetic Fuels, McGraw-Hill Book Co., New York, USA, 1982, p.381.

[2] J.W. King, K. Srinivas, Multiple unit processing using sub- and supercritical fluids, J. Supercrit. Fluids 47 (2009) 598–610.

[3] K. Arai, R.L. Smith Jr., T.M. Aida, Decentralized chemical processes with supercritical fluid technology for sustainable society, J. Supercrit. Fluids 47 (2009) 628–636.

[4] A.A. Peterson, F. Vogel, R.P. Lachance, M. Fröling, M.J. Antal Jr., J.W. Tester, Thermochemical biofuel production in hydrothermal media: a review of sub- and supercritical water technologies, Energy Environ. Sci. 1 (2008) 32–65.

[5] G. Brunner, Near critical and supercritical water. Part I. Hydrolytic and hydrothermal processes, J. Supercrit. Fluids 47 (2009) 373–381.

[6] M.J. Antal Jr., W.S.L. Mok, G.N. Richards, Kinetic studies of the reactions of ketoses and aldoses in water at high temperature. Mechanism of formation of 5-(hydroxymethyl)-2-furaldehyde from D-fructose and sucrose, Carbohydr. Res. 199 (1990) 91–109.

[7] G. Düll, Chem. Ztg. 19 (1895) 216.

[8] O. Bobleter, G. Pape, Hydrothermal decomposition of glucose, Monatsh. Chem. 99 (1968) 1560–1567.

[9] S. Amin, R.C. Reid, M. Modell, Reforming and decomposition of glucose in an aqueous phase, in: ASME Paper #75-ENAs-21, Intersociety Conference on Environmental Systems, San Francisco, CA, USA, July 21–24, 1975.

[10] G. Bonn, W. Schwald, O. Bobleter, Hydrothermolysis of short rotation forestry plants, in: W. Palz, J. Coombs, D.O. Hall (Eds.), Energy from Biomass, Elsevier, London, UK, 1985, pp. 953–958.

[11] M. Modell, Reforming of organic substances in supercritical water, in: ACS Spring Meeting, St. Louis, Missouri, USA, 1980.

[12] M. Modell, Gasification and liquefaction of forest products in supercritical water, in: R.P. Overend, T.A. Milne, L.K. Mudge (Eds.), Fundamentals of Thermochemical Biomass Conversion, Elsevier, New York, USA, 1985, pp. 95–119.

[13] R.M. Malaluan, A study of cellulose decomposition in subcritical and supercritical water, Ph.D. Dissertation, Tohoku University, Sendai, Japan, 1995.

[14] B.M. Kabyemela, T. Adschiri, R.M. Malaluan, K. Arai, Kinetics of glucose epimerization and decomposition in subcritical and supercritical water, Ind. Eng. Chem. Res. 36 (1997) 1552–1558.

[15] S. Noguchi, R. Uehara, M. Sasaki, M. Goto, Thermal stability of monosaccharides in subcritical water, in: M. Goto, T. Adschiri, B. Han, Y.-W. Lee (Eds.), Proceedings 8th International Symposium on Supercritical Fluids, November 5–8, Kyoto, Japan; Paper PB 2-47, 2006.

[16] D. Klingler, H. Vogel, Influence of process parameters on the hydrothermal decomposition and oxidation of glucose in sub- and supercritical water, J. Supercrit. Fluids 55 (2010) 259–270.

[17] Z. Fang, R.L. Smith Jr., J.A. Kozinski, T. Minowa, K. Arai, Reaction of d-glucose in water at high temperatures (410 °C) and pressures (180 MPa) for the production of dyes and nano-particles, J. Supercrit. Fluids 56 (2011) 41–47.

[18] K. Liu, Zur Hydrolyse von Biopolymeren in Wasser und Kohlendioxid unter erhöhten Drücken und Temperaturen, Ph.D. Dissertation, Hamburg University of Technology, Hamburg, Germany, 2000.

[19] K. Liu, G. Brunner, Hydrolysis of some biopolymers with water and carbon dioxide at high pressures and temperatures, in: Proceedings of International Meeting of the GVC-Fachausschuß Hochdruckverfahrenstechnik, Karlsruhe, Germany, March 3–5, 1999, pp. 123–126, Wissenschaftliche Berichte FZKA 6271.

[20] K. Liu, G. Brunner, Hydrolysis of some biopolymers with water and carbon dioxide at high pressures and temperatures, in: G. Brunner (Ed.), Reaction Media, Proceedings of the AIChE Annual Meeting, Miami Beach, FL, USA, November 15–20, 1998, Elsevier B.V, Amsterdam, 2004, pp. 147–162, Chapter 1.6.

[21] B. Kabyemela, T. Adschiri, R. Malaluan, K. Arai, Glucose and fructose decomposition in subcritical and supercritical water: detailed reaction pathway, mechanisms, and kinetics, Ind. Eng. Chem. Res. 38 (1999) 2888–2895.

[22] Q. Jing, X. Lü, Kinetics of non-catalyzed decomposition of glucose in high-temperature liquid water, Chin. J. Chem. Eng. 16 (2008) 890–894.

[23] Y. Matsumura, S. Yanachi, T. Yoshida, Glucose decomposition kinetics in water at 25 MPa in the temperature range of 448–673 K, Ind. Eng. Chem. Res. 45 (2006) 1875–1879.

[24] M.J. Antal Jr., W.S.L. Mok, A study of the acid catalyzed dehydration of fructose in near-critical water, in: A.V. Bridgwater, J.L. Kuester (Eds.), Research in Thermochemical Biomass Conversion, Elsevier, New York, USA, 1988, p. 464.

[25] F.S. Asghari, H. Yoshida, Acid-catalyzed production of 5-hydroxymethyl furfural from D-fructose in subcritical water, Ind. Eng. Chem. Res. 45 (2006) 2163–2173.
[26] T.M. Aida, K. Tajima, M. Watanabe, Y. Saito, K. Kuroda, T. Nonaka, H. Hattori, R.L. Smith Jr., K. Arai, Reactions of d-fructose in water at temperatures up to 400°C and pressures up to 100MPa, J. Supercrit. Fluids 42 (2007) 110–119.
[27] M. Kunz, Hydroxymethylfurfural-a possible basic chemical for industrial intermediates, in: A. Fuchs (Ed.), Studies on Plant Science, vol. 3, Inulin and Inulin Containing Crops, Elsevier, Amsterdam, New York, USA, 1993, pp. 149–160.
[28] Y. Roman-Leshkov, C.J. Barrett, Z.Y. Liu, J.A. Dumesic, Production of dimethylfuran for liquid fuels from biomass-derived carbohydrates, Nature 447 (2007) 982–985.
[29] G.C.A. Luijkx, F. van Rantwijk, H. van Bekkum, Hydrothermal formation of 1,2,4-benzenetriol from 5-hydroxymethyl-2-furaldehyde and D-fructose, Carbohydr. Res. 242 (1993) 131–139.
[30] D.A. Nelson, P.M. Molton, J.A. Russell, R.T. Hallen, Application of direct thermal liquefaction for the conversion of cellulosic biomass, Ind. Eng. Chem. Prod. Res. Dev. 23 (1984) 471–475.
[31] K. Gairola, I. Smirnova, Hydrothermal pentose to furfural conversion and simultaneous extraction with SC-CO_2—kinetics and application to biomass hydrolysates, Bioresour. Technol. 123 (2012) 592–598.
[32] T.M. Aida, N. Shiraishi, M. Kubo, M. Watanabe, R.L. Smith Jr., Reaction kinetics of d-xylose in sub- and supercritical water, J. Supercrit. Fluids 55 (2010) 208–216.
[33] M. Sasaki, T. Hayakawa, K. Arai, T. Adschiri, Measurement of the Rate of Retro-Aldol Condensation of D-Xylose in Subcritical and Supercritical Water, World Scientific, Singapore, 2003, pp. 169–176.
[34] M. Nagamori, T. Funazukuri, Glucose production by hydrolysis of starch under hydrothermal conditions, J. Chem. Technol. Biotechnol. 79 (2004) 229–233. http://onlinelibrary.wiley.com/doi/10.1002/jctb.976/abstract.
[35] Q. Jing, L. Xiuyang, Kinetics of non-catalyzed decomposition of D-xylose in high-temperature liquid water, Chin. J. Chem. Eng. 15 (2007) 666–669.
[36] F. Carvalheiro, G. Garrote, J.C. Parajo, H. Pereira, F.M. Girio, Kinetic modeling of Brewery's spent grain autohydrolysis, Biotechnol. Prog. 21 (2005) 233–243.
[37] P.J. Oefner, A.H. Lanziner, G. Bonn, O. Bobleter, Quantitative studies on furfural and organic acid formation during hydrothermal, acidic and alkaline degradation of D-xylose, Monatsh. Chem. 123 (1992) 547–556.
[38] C. Usuki, Y. Kimura, S. Adachi, Degradation of pentoses and hexouronic acids in subcritical water, Chem. Eng. Technol. 31 (2008) 133–137.
[39] K. Lorenz, J.A. Johnson, Starch hydrolysis under high temperatures and pressures, American Association of Cereal Chemists, Contribution No. 736. http://www.aaccnet.org/publications/cc/backissues/1972/Documents/Chem49_616.pdf, 1972 (accessed January 10, 2013).
[40] R.L. Orozco, M.D. Redwood, G.A. Leeke, A. Bahari, R.C.D. Santos, L.E. Macaskie, Hydrothermal hydrolysis of starch with CO_2 and detoxification of the hydrolysates with activated carbon for bio-hydrogen fermentation, Int. J Hydrogen Energy 37 (2012) 6545–6553. Manuscript accepted 11/1/12—from: http://eprints.bham.ac.uk/997/1/Orozco_et_al_2012_-_Int_J_Hydrog_Energy_-_Starch_hydrolysis.pdf.
[41] T. Miyazawa, T. Funazukuri, Polysaccharide hydrolysis accelerated by adding carbon dioxide under hydrothermal conditions, Biotechnol. Prog. 21 (2005) 1782–1785.
[42] M. Goto, M. Tanaka, A.T. Quitain, M. Sasaki, R. Fukuzato, Recovery of phytochemicals by hybrid extraction process using supercritical CO_2 and water. http://issf2012.com/handouts/documents/425_004.pdf.

[43] http://en.wikipedia.org/wiki/Pectin.
[44] M. Tanaka, A. Takamizu, M. Hoshino, M. Sasaki, M. Goto, Extraction of dietary fiber from citrus junos peel with subcritical water, Food Bioprod. Process. 90 (2012) 180–186.
[45] M. Hoshino, M. Tanaka, A. Terada, M. Sasaki, M. Goto, Separation and characterization of pectin from juice processing residue extracted by sub-critical water. http://staff.aist.go.jp/tohru-kamo/FSRJ/output/7_nenkai/05IFSR/Proceeding-IFSR05/R-41.pdf.
[46] M. Goto, S. Machmudah, M. Sasaki, M. Tanaka, Utilization of citrus peel by sub- and supercritical fluid technology. http://www.icef11.org/content/papers/fpe/FPE908.pdf.
[47] T. Rogalinski, T. Ingram, G. Brunner, Hydrolysis of lignocellulosic biomass in water under elevated temperatures and pressures, J. Supercrit. Fluids 47 (2008) 54–63.
[48] D.A. Cantero, M.D. Bermejo, M.J. Cocero, Kinetic analysis of cellulose depolymerization reactions in near critical water, J. Supercrit. Fluids 75 (2013) 48–57.
[49] K. Ehara, S. Saka, Decomposition behavior of cellulose in supercritical water, subcritical water, and their combined treatments, J. Wood Sci. 51 (2005) 148–153.
[50] X. Lü, A. Sakoda, M. Suzuki, Decomposition of cellulose by continuous near-critical water reactions, Chin. J. Chem. Eng. 8 (2000) 321–325.
[51] W. Schwald, O. Bobleter, Hydrothermolysis of cellulose under static and dynamic conditions at high temperatures, J. Carbohydr. Chem. 8 (1989) 565–578.
[52] T. Adschiri, S. Hirose, R. Malaluan, K. Arai, Non catalytic conversion of cellulose in supercritical and subcritical water, J. Chem. Eng. Jpn. (1993) 676–680.
[53] K. Mochidzuki, A. Sakoda, M. Suzuki, Measurement of the hydrothermal reaction rate of cellulose using novel liquid-phase thermogravimetry, Thermochim. Acta 348 (2000) 69–76.
[54] M. Sasaki, Z. Fang, Y. Fukushima, T. Adschiri, K. Arai, Dissolution and hydrolysis of cellulose in subcritical and supercritical water, Ind. Eng. Chem. Res. 39 (2000) 2883–2890.
[55] M. Sasaki, T. Adschiri, K. Arai, Kinetics of cellulose conversion at 25 MPa in sub- and supercritical water, AIChE J. 50 (2004) 192–202.
[56] T. Albrecht, Conversion of lignocellulosic materials and model compounds in sub- and supercritical water, Ph.D. Dissertation, University of Technology Hamburg, Hamburg, Germany, 2006.
[57] T. Albrecht, G. Brunner, Production of valuable organic compounds by hydrolyzing biopolymers in water at elevated temperatures and pressures, in: G. Brunner, I. Kikic, M. Perrut (Eds.), Proceedings 6th International Symposium on Supercritical Fluids, ISBN 2-905-267-37-02, Versailles, France, vol. II, 2003, pp. 1411–1416.
[58] T. Rogalinski, K. Liu, T. Albrecht, G. Brunner, Hydrolysis kinetics of biopolymers in subcritical water, J. Supercrit. Fluids 46 (2008) 335–341.
[59] M. Sasaki, B. Kabyemela, R. Malaluan, S. Hirose, N. Takeda, T. Adschiri, K. Arai, Cellulose hydrolysis in subcritical and supercritical water, J. Supercrit. Fluids 13 (1998) 261–268.
[60] C.E. Wyman, S.R. Decker, M.E. Himmel, J.W. Brady, C.E. Skopec, L. Viikari, Hydrolysis of cellulose and hemicellulose, in: S. Dumitriu (Ed.), Polysaccharides: Structural Diversity and Functional Versatility, Marcel Dekker, Inc., New York, 2005. nsm1.nsm.iup.edu/jford/projects/misc/hydrolysisofcelluloseandhemicellulose_review.pdf.
[61] O. Bobleter, Hydrothermal degradation of polymers derived from plants, Prog. Polym. Sci. 19 (1994) 797–841.
[62] W.S.L. Mok, M.J. Antal Jr., Uncatalyzed solvolysis of whole biomass hemicellulose by hot compressed liquid water, Ind. Eng. Chem. Res. 31 (1992) 1157–1161.
[63] G. Garrote, H. Dominguez, J.C. Paraj, Hydrothermal processing of lignocellulosic materials, Holz Roh Werkst. 57 (1999) 191–202.

[64] J. Saeman, Kinetics of wood saccharification—hydrolysis of cellulose and decomposition of sugars in dilute acid at high temperature, Ind. Eng. Chem. 37 (1945) 42–52.

[65] T. Kobayashi, Y. Sakai, Hydrolysis rate of pentosan of hardwood in dilute sulfuric acid, Bull. Agric. Chem. Soc. Jpn. 20 (1956) 1–7.

[66] S.E. Jacobsen, C.E. Wyman, Cellulose and hemicelluloses hydrolysis models for application to current and novel pretreatment processes, Appl. Biochem. Biotechnol. 81 (2000) 84–86.

[67] R.P. Overend, E. Chornet, Fractionation of lignocellulosics by steam-aqueous pretreatments, Phil. Trans. R. Soc. Lond. A321 (1987) 523–536.

[68] T. Lloyd, C.E. Wyman, Application of a depolymerization model for predicting thermochemical hydrolysis of hemicelluloses, Appl. Biochem. Biotechnol. 105–108 (2003) 53–67.

[69] M. Rubio, J.F. Tortosa, J. Quesada, D. Gómez, Fractionation of lignocellulosics. Solubilization of corn stalk hemicelluloses by autohydrolysis in aqueous medium, Biomass Bioenergy 15 (1998) 483–491.

[70] J.F. Tortosa, M. Rubio, D. Gomez, Afinidad 457 (1995) 181–188.

[71] T.P. Schultz, M.C. Templeton, C.J. Biermann, G.D. McGinnis, Steam explosion of mixed hardwood chips, rice hulls, corn stalks, and sugar cane bagasse, J. Agric. Food Chem. 32 (1984) 1166–1172.

[72] G. Garrote, H. Dominguez, J.C. Parajo, Kinetic modeling of corncob autohydrolysis, Process Biochem. 36 (2001) 571–578.

[73] Q. Yu, X. Zhuang, Z. Yuan, W. Wang, W. Qi, Q. Wang, X. Tan, Hydrolysis of sweet sorghum bagasse hemicelluloses with liquid hot water and its mechanism, J. Chem. Ind. Eng. 63 (2012) 605–610.

[74] Wahyudiono, S. Machmudah, M. Goto, Utilization of sub and supercritical water reactions in resource recovery of biomass wastes, Eng. J. 17 (1) (2013) 1–12. http://www.engj.org/.

[75] M. Saisu, T. Sato, M. Watanabe, T. Adschiri, K. Arai, Conversion of lignin with supercritical water-phenol mixtures, Energ. Fuel 17 (2003) 922–928.

[76] B. Zhang, H.-J. Huang, S. Ramaswamy, Reaction kinetics of the hydrothermal treatment of lignin, Appl. Biochem. Biotechnol. 147 (2008) 119–131.

[77] O. Bobleter, R. Concin, Degradation of poplar lignin by hydrothermal treatment, Cellul. Chem. Technol. 13 (1979) 583–593.

[78] G. Wu, M. Heitz, E. Chornet, Improved alkaline oxidation process for the production of aldehydes (vanillin and syringaldehyde) from steam-explosion hardwood lignin, Ind. Eng. Chem. Res. 33 (1994) 718–723.

[79] Q. Xiang, Y. Lee, Production of oxychemicals from precipitated hardwood lignin, Appl. Biochem. Biotechnol. 91–93 (2001) 71–80.

[80] A. Quitain, N. Sato, H. Daimon, K. Fujie, Qualitative investigation on hydrothermal treatment of hinoki (Chamaecyparisobtusa) bark for production of useful chemicals, J. Agric. Food Chem. 51 (2003) 7926–7929.

[81] K. Okuda, X. Man, M. Umetsu, S. Takami, T. Adschiri, Efficient conversion of lignin into single chemical species by solvothermal reaction in water-p-cresol solvent, J. Phys. Condens. Matter 16 (2004) 1325–1330.

[82] M.J. Negro, P. Manzanares, I. Ballesteros, J.M. Olivia, A. Cabanas, M. Ballesteros, Hydrothermal pretreatment conditions to enhance ethanol production from poplar biomass, Appl. Biochem. Biotechnol. 105 (2003) 87–100.

[83] J. Weil, M. Brewer, R. Hendrickson, A. Sarikaya, M.R. Ladisc, Continuous pH monitoring during pretreatment of yellow poplar wood sawdust by pressure cooking in water, Appl. Biochem. Biotechnol. 70–72 (1998) 99–111.

[84] Wahyudiono, M. Goto, M. Sasaki, Recovery of phenolic compounds through the decomposition of lignin in near and supercritical water, Chem. Eng. Process. 47 (2008) 1609–1619.
[85] K. Okuda, M. Umetsu, S. Takami, T. Adschiri, Disassembly of lignin and chemical recovery-rapid depolymerization of lignin without char formation in water-phenol mixtures, Fuel Process. Technol. 85 (2004) 803–813.
[86] Z. Fang, T. Sato, R.L. Smith Jr., H. Inomata, K. Arai, J.A. Kozinski, Reaction chemistry and phase behavior of lignin in high-temperature and supercritical water, Bioresour. Technol. 99 (2008) 3424–3430.
[87] K. Okuda, S. Ohara, M. Umetsu, S. Takami, T. Adschiri, Disassembly of lignin and chemical recovery in supercritical water and p-cresol mixture. Studies on lignin model compounds, Bioresour. Technol. 99 (2008) 1846–1852.
[88] R. Pecina, P. Burtscher, G. Bonn, O. Bobleter, GC-MS and HPLC analyses of lignin degradation products in biomass hydrolyzates, Fresen. Z. Anal. Chem. 325 (1986) 461–465.
[89] S. Karagoz, T. Bhaskar, A. Muto, Y. Sakata, Comparative studies of oil compositions produced from sawdust, rice husk, lignin and cellulose by hydrothermal treatment, Fuel 84 (2005) 875–884.
[90] J.R. Lawson, M.T. Klein, Influence of water on guaiacol pyrolysis, Ind. Eng. Chem. Fundam. 24 (1985) 203–208.
[91] M. Wahyudiono, M. Sasaki, Goto, Conversion of biomass model compound under hydrothermal conditions using batch reactor, Fuel 88 (2009) 1656–1664.
[92] T. Rogalinski, Hydrolyse von Proteinen in Wasser und Wasser-Kohlendioxid-Systemen unter erhöhten Drücken und Temperaturen, Ph.D. Dissertation, Hamburg University of Technology, Hamburg, Germany, 2006.
[93] W. Walter, H.-P. Harke, R. Polchow, Das Verhalten von Glykokoll, Alanin, α-Aminobuttersäure, Leucin, Phenylalanin und Asparaginsäure unter hydrothermalen Bedingungen, Z. Naturforsch. 22 (1967) 931–937.
[94] H. Yoshida, O. Tavakoli, Sub-critical water hydrolysis treatment for waste squid entrails and production of amino acids, organic acids, and fatty acids, J. Chem. Eng. Jpn. 37 (2004) 253–260.
[95] K. Kang, A.T. Quitain, H. Daimon, R. Noda, N. Goto, H.-Y. Hu, K. Fujie, Optimization of amino acids production from waste fish entrails by hydrolysis in sub- and supercritical water, Can. J. Chem. Eng. 79 (2001) 65–70.
[96] K. Kang, A.T. Quitain, S.-Y. Urano, H. Daimon, K. Fujie, Rapid sample injection in semibatch hydrothermal treatment of solid wastes, Ind. Eng. Chem. Res. 40 (2001) 3717–3720.
[97] D.K. Alargov, S. Deguchi, K. Tsujii, K. Horikoshi, Reaction behaviors of glycine under super- and subcritical water conditions, Orig. Life Evol. Biosph. 32 (2002) 1–12.
[98] M.N. Islam, T. Kaneko, K. Kobayashi, Reaction of amino acids in a supercritical waterflow reactor simulating submarine hydrothermal systems, Bull. Chem. Soc. Jpn. 76 (2003) 1171–1178.
[99] T. Goto, Y. Futamura, Y. Yamaguchi, K. Yamamoto, Condensation reactions of amino acids under hydrothermal conditions with adiabatic expansion cooling, J. Chem. Eng. Jpn. 38 (2005) 295–299.
[100] N. Sato, A.T. Quitain, K. Kang, H. Daimon, K. Fujie, Reaction kinetics of amino acid decomposition in high-temperature and high-pressure water, Ind. Eng. Chem. Res. 43 (2004) 3217–3222.
[101] J. Corliss, J. Baross, S. Hoffman, An hypothesis concerning the relationship between submarine hot springs and the origin of life on earth, Oceanol. Acta 4 (Suppl.) (1981) 59–69.

[102] D. Klingler, J. Berg, H. Vogel, Hydrothermal reactions of alanine and glycine in sub- and supercritical water, J. Supercrit. Fluids 43 (2007) 112–119.
[103] T. Rogalinski, S. Herrmann, G. Brunner, Production of amino acids from bovine serum albumin by continuous sub-critical water hydrolysis, J. Supercrit. Fluids 36 (2005) 49–58.
[104] T. Ingram, Thermal and thermal-enzymatic hydrolysis of lignocellulosic biomass for the production of bioethanol, Diploma Thesis, Hamburg University of Technology, Hamburg, Germany, 2008.
[105] http://en.wikipedia.org/wiki/Straw.
[106] T. Ingram, T. Rogalinski, V. Bockemühl, G. Antranikian, G. Brunner, Semi-continuous liquid hot water pretreatment of rye straw, J. Supercrit. Fluids 48 (2009) 238–246.
[107] A. Wöhlk, Thermischer Aufschluss nachwachsender Rohstoffe zur Gewinnung von Bioethanol, Master Thesis, Hamburg University of Technology, Hamburg, Germany, 2007.
[108] O. Bobleter, H. Binder, R. Concin, E. Burtscher, The conversion of biomass to fuel raw material by hydrothermal treatment, in: W. Palz, P. Chartier, D.O. Hall (Eds.), Energy from Biomass, Applied Science Publishers, London, 1981, pp. 554–562.
[109] J. Kim, Y. Lee, R. Torget, Cellulose hydrolysis under extremely low sulfuric acid and high-temperature conditions, Appl. Biochem. Biotechnol. 91 (2001) 331–340.
[110] M. Sasaki, T. Adschiri, K. Arai, Fractionation of sugarcane bagasse by hydrothermal treatment, Bioresour. Technol. 86 (2003) 301–304.
[111] N. Yoswathana, P. Phuriphipat, P. Treyawutthiwat, M.N. Eshtiaghi, Bioethanol production from rice straw, Energ. Res. J. 1 (2010) 26–31.
[112] C. Liu, C.E. Wyman, The effect of flow rate of compressed hot water on xylan, lignin, and total mass removal from corn stover, Ind. Eng. Chem. Res. 42 (2003) 5409–5416.
[113] M.E. Vallejos, M.D. Zambon, M.C. Area, A.A. da Silva Curvelo, Low liquid–solid ratio (LSR) hot water pretreatment of sugarcane bagasse, Green Chem. 14 (2012) 1982–1989.
[114] D. Abril, M. Medina, A. Abril, Sugar cane bagasse prehydrolysis using hot water, Braz. J. Chem. Eng. 29 (2012) 31–38.
[115] G. Zhu, Z. Xiao, X. Zhu, F. Yi, X. Wan, Reducing sugars production from sugarcane bagasse wastes by hydrolysis in sub-critical water, Clean Technol. Environ. Policy 15 (2013) 55–61.
[116] R.C. Pettersen, The chemical composition of wood, in: R.M. Rowell (Ed.), The Chemistry of Solid Wood, in: American Chemical Society, Washington, DC, USA, 1984, pp. 57–126.
[117] O. Bobleter, G. Pape, Austrian Patent 263,661 (1968).
[118] F.S. Asghari, H. Yoshida, Conversion of Japanese red pine wood (Pinus densiflora) into valuable chemicals under subcritical water conditions, Carbohydr. Res. 345 (2010) 124–131.
[119] M. Matsunaga, H. Matsui, Y. Otsuka, S. Yamamoto, Chemical conversion of wood by treatment in a semi-batch reactor with subcritical water, J. Supercrit. Fluids 44 (2008) 364–369.
[120] Th. Willner, Thermische Holzumwandlung unter dem Einfluß von Wasserstoff und Wasser, Ph.D. Dissertation, Hamburg University of Technology, Hamburg, Germany,1993.
[121] http://www.wissenschaft-online.de/abo/lexikon/biok/10872, Spektrum Akademischer Verlag.
[122] K. Kang, B.-S. Chun, Behavior of hydrothermal decomposition of silk fibroin to amino acids in near-critical water, Korean J. Chem. Eng. 21 (2004) 654–659.

[123] H. Yoshida, M. Terashima, Y. Takahashi, Production of organic acids and amino acids from fish meat by sub-critical water hydrolysis, Biotechnol. Prog. 15 (1999) 1090–1094.

[124] A. Ali-Nehari, S.-B. Kim, Y.-B. Lee, B.-S. Chun, Production of value added materials by subcritical water hydrolysis from krill residues extracted by supercritical carbon dioxide, Afr. J. Biotechnol. 10 (2011) 18450–18457.

[125] A. Quitain, N. Sato, H. Daimon, K. Fujie, Production of valuable materials by hydrothermal treatment of shrimp shells, Ind. Eng. Chem. Res. 40 (2001) 5885–5888.

[126] G. Zhu, X. Zhu, Q. Fan, X. Wan, Kinetics of amino acid production from bean dregs by hydrolysis in sub-critical water, Amino Acids 40 (2011) 1107–1113.

[127] I. Sereewatthanawut, S. Prapintip, K. Watchiraruji, M. Goto, M. Sasaki, A. Shotipruk, Extraction of protein and amino acids from de-oiled rice bran by subcritical water hydrolysis, Bioresour. Technol. 99 (2008) 555–561.

[128] S.S. Toor, L. Rosendahl, A. Rudolf, Hydrothermal liquefaction of biomass: a review of subcritical water technologies, Energy 36 (2011) 2328–2342.

[129] D.J. Stevens, Review and analysis of the 1980–1989 biomass thermochemical conversion program, Technical Report, U.S. Department of Energy, NREL/TP-421-7501, 1994.

[130] Y. Qian, C. Zuo, J. Tan, J. He, Structural analysis of bio-oils from sub-and supercritical water liquefaction of woody biomass, Energy 32 (2007) 196–202.

[131] H. Davis, C. Figueroa, L. Schaleger, Hydrogen or carbon monoxide in the liquefaction of biomass, in: World Hydrogen Energy Conference IV, 1982.

[132] B.J. He, Y. Zhang, Y. Yin, T.L. Funk, G.L. Riskowski, Effects of alternative process gases on the thermochemical conversion process of swine manure, Trans. Am. Soc. Agric. Eng. 44 (2001) 1873–1880.

[133] F. Goudriaan, B. van de Beld, F. Boerefijn, G. Bos, J. Naber, S. van der Wal, J. Zeevalkink, Thermal efficiency of the HTU process for biomass liquefaction, in: A. Bridgwater (Ed.), Progress in Thermochemical Biomass Conversion, Blackwell Science, Tyrol, Austria, 2001, pp. 1312–1325.

[134] B.J. He, Y. Zhang, Y. Yin, T.L. Funk, G.L. Riskowski, Operating temperature and retention time effects on the thermochemical conversion process of swine manure, Trans. Am. Soc. Agric. Eng. 43 (2000) 1821–1825.

[135] T. Minowa, M. Murakami, Y. Dote, T. Ogi, S. Yokoyama, Oil production from garbage by thermochemical liquefaction, Biomass Bioenergy 8 (1995) 117–120.

[136] T. Minowa, S. Yokoyama, M. Kishimoto, T. Okakura, Oil production from algal cells of dunaliella tertiolecta by direct thermochemical liquefaction, Fuel 74 (1995) 1735–1738.

[137] Y. Dote, S. Sawayama, S. Inoue, T. Minowa, S. Yokoyama, Recovery of liquid fuel from hydrocarbon-rich microalgae by thermochemical liquefaction, Fuel 73 (1994) 1855–1857.

[138] A. Sinağ, A. Kruse, P. Maniam, Hydrothermal conversion of biomass and different model compounds, J. Supercrit. Fluids 71 (2012) 80–85.

[139] H. Ramsurn, R.B. Gupta, Two-step biomass hydrothermal liquefaction process through acidic subcritical- followed by alkaline supercritical-water treatment, in: ISSF 2012: 10th International Symposium on Supercritical Fluids, May 13–16, 2012, San Francisco, CA, USA, 2012. http://issf2012.com/handouts/documents/418_004.pdf.

[140] F.L.P. Resende, S.A. Fraley, M.J. Berger, P.E. Savage, Noncatalytic gasification of lignin in supercritical water, Energy Fuels 22 (2008) 1328–1334.

[141] M. Modell, R. Reid, S. Amin, Gasification process, US Patent 4 113 446 (1978).

[142] M. Modell, Reforming of organic-substances in supercritical water, J. Electrochem. Soc. 127 (1980) C139.

[143] H.R. Holgate, Oxidation chemistry and kinetics in supercritical water: hydrogen, carbon monoxide and glucose, Ph.D. Dissertation, Department of Chemical Engineering, Massachusetts Institute of Technology, Cambridge, USA, 1993.

[144] M.J. Antal Jr., S. Manarungson, W.S.L. Mok, Hydrogen production by steam reforming glucose in supercritical water, in: A.V. Bridgewater (Ed.), Advances in Thermochemical Biomass Conversion, Blackie Academic & Professional, London, 1993, pp. 1367–1377.

[145] A. Kruse, D. Forchheim, M. Gloede, F. Ottinger, J. Zimmermann, Brines in supercritical biomass gasification: 1. Salt extraction by salts and the influence on glucose conversion, J. Supercrit. Fluids 53 (2010) 64–71.

[146] E. Weiss-Hortala, A. Kruse, Ch. Ceccarelli, R. Barna, Influence of phenol on glucose degradation during supercritical water gasification, J. Supercrit. Fluids 53 (2010) 42–47.

[147] T.G. Madenoğlu, M. Sağlam, M. Yüksel, L. Ballice, Simultaneous effect of temperature and pressure on catalytic hydrothermal gasification of glucose, J. Supercrit. Fluids 73 (2013) 151–160.

[148] P. Azadi, A.A. Khodadadi, Y. Mortazavi, R. Farnood, Hydrothermal gasification of glucose using Raney nickel and homogeneous organometallic catalysts, Fuel Process. Technol. 90 (2009) 145–151.

[149] Z. Fang, T. Minowa, C. Fang, R.L. Smith Jr., H. Inomata, J.A. Kozinski, Catalytic hydrothermal gasification of cellulose and glucose, Int. J. Hydrogen Energy 33 (2008) 981–990.

[150] Y. Izumizaki, K.C. Park, Y. Tachibana, H. Tomiyasu, Y. Fujii, Organic decomposition in supercritical water by an aid of ruthenium (IV) oxide as a catalyst-exploitation of biomass resources for hydrogen production, Prog. Nucl. Energy 47 (2005) 544–552.

[151] T. Minowa, T. Ogi, Hydrogen production from cellulose using a reduced nickel catalyst, Catal. Today 45 (1998) 411–416.

[152] F.L.P. Resende, P.E. Savage, Expanded and updated results for supercritical water gasification of cellulose and lignin in metal-free reactors, Energy Fuels 23 (2009) 6213–6221.

[153] M. Watanabe, H. Inomata, M. Osada, T. Sato, T. Adschiri, K. Arai, Catalytic effects of NaOH and ZrO_2 for partial oxidative gasification of n-hexadecane and lignin in supercritical water, Fuel 82 (2003) 545–552.

[154] M. Osada, O. Sato, M. Watanabe, K. Arai, M. Shirai, Water density effect on lignin gasification over supported noble metal catalysts in supercritical water, Energy Fuels 20 (2006) 930–935.

[155] T. Yoshida, Y. Matsumura, Gasification of cellulose, xylan, and lignin mixtures in supercritical water, Ind. Eng. Chem. Res. 40 (2001) 5469–5474.

[156] T. Furusawa, T. Sato, H. Sugito, Y. Miura, Y. Ishiyama, M. Sato, N. Itoh, N. Suzuki, Hydrogen production from the gasification of lignin with nickel catalysts in supercritical water, Int. J. Hydrogen Energy 32 (2007) 699–704.

[157] M. Osada, T. Sato, M. Watanabe, M. Shirai, K. Arai, Catalytic gasification of wood biomass in subcritical and supercritical water, Combust. Sci. Technol. 178 (2006) 537–552.

[158] M. Osada, T. Sato, M. Watanabe, T. Adschiri, K. Arai, Low-temperature catalytic gasification of lignin and cellulose with a ruthenium catalyst in supercritical water, Energy Fuels 18 (2004) 327–333.

[159] T. Sato, T. Furusawa, Y. Ishiyama, H. Sugito, Y. Miura, M. Sato, N. Suzuki, N. Itoh, Effect of water density on the gasification of lignin with magnesium oxide supported nickel catalysts in supercritical water, Ind. Eng. Chem. Res. 45 (2006) 615–622.

[160] M. Osada, N. Hiyoshi, O. Sato, K. Arai, M. Shirai, Reaction pathway for catalytic gasification of lignin in presence of sulfur in supercritical water, Energy Fuels 21 (2007) 1854–1858.

[161] M. Osada, N. Hiyoshi, O. Sato, K. Arai, M. Shirai, Effect of sulfur on catalytic gasification of lignin in supercritical water, Energy Fuels 21 (2007) 1400–1405.
[162] M. Osada, O. Sato, K. Arai, M. Shirai, Stability of supported ruthenium catalysts for lignin gasification in supercritical water, Energy Fuels 20 (2006) 2337–2343.
[163] D.C. Elliott, Catalytic hydrothermal gasification of biomass, Biofuels Bioprod. Biorefin. 2 (2008) 254–265.
[164] D. Xu, S. Wang, Q. Zhang, Ch. Chen, Y. Li, L. Zheng, Glycine gasification in supercritical water for hydrogen production, J. Chem. Ind. Eng. 99 (2008) 735–742.
[165] G.J. DiLeo, M.E. Neff, S. Kim, P.E. Savage, Supercritical water gasification of phenol and glycine as models for plant and protein biomass, Energy Fuels 22 (2008) 871–877.
[166] P. D'Jesús, N. Boukis, B. Kraushaar-Czarnetzki, E. Dinjus, Gasification of corn and clover grass in supercritical water, Fuel 85 (2006) 1032–1038.
[167] Y. Matsumura, T. Minowa, B. Potic, S.R.A. Kersten, W. Prins, W.P.M. van Swaaij, B. van de Beld, D.C. Elliott, G.G. Neuenschwander, A. Kruse, M.J. Antal Jr., Biomass gasification in near- and super-critical water: status and prospects, Biomass Bioenergy 29 (2005) 269–292.
[168] H. Schmieder, J. Abeln, N. Boukis, E. Dinjus, A. Kruse, M. Kluth, G. Petrich, E. Sadri, M. Schacht, Hydrothermal gasification of biomass and organic wastes, J. Supercrit. Fluids 17 (2000) 145–153.
[169] A. Kruse, Hydrothermal biomass gasification, J. Supercrit. Fluids 47 (2009) 391–399.
[170] L. Guo, C. Cao, Y. Lu, Supercritical water gasification of biomass and organic wastes, in: M.N.B. Momba (Ed.), Biomass, ISBN: 978-953-307-113-8, 2010. InTech, http://www.intechopen.com/books/biomass/supercritical-water-gasification-of-biomass-and-organic-wastes.
[171] Y. Calzavara, C. Joussot-Dubien, G. Boissonnet, S. Sarrade, Evaluation of biomass gasification in supercritical water process for hydrogen production, Energy Convers. Manage. 46 (2005) 615–631.
[172] Y. Matsumura, M. Sasaki, K. Okuda, S. Takami, S. Ohara, M. Umetsu, T. Adschiri, Supercritical water treatment of biomass for energy and material recovery, Combust. Sci. Technol. 178 (2006) 509–536.
[173] A. Kruse, A. Krupka, V. Schwarzkopf, C. Gamard, T. Henningsen, Influence of proteins on the hydrothermal gasification and liquefaction of biomass. 1. Comparison of different feedstocks, Ind. Eng. Chem. Res. 44 (2005) 3013–3020.
[174] A. Kruse, P. Maniam, F. Spieler, Influence of proteins on the hydrothermal gasification and liquefaction of biomass. 2. Model compounds, Ind. Eng. Chem. Res. 46 (2007) 87–96.
[175] J. Penninger, M. Rep, Reforming of aqueous wood pyrolysis condensate in supercritical water, Int. J. Hydrogen Energy 31 (2006) 1597–1606.
[176] J. Penninger, G. Maass, M. Rep, Compressed hydrogen-rich fuel gas (CHFG) from wet biomass by reforming in supercritical water, Int. J. Hydrogen Energy 32 (2007) 1472–1476.
[177] Y. Lu, L. Guo, C. Ji, X. Zhang, X. Hao, Q. Yan, Hydrogen production by biomass gasification in supercritical water: a parametric study, Int. J. Hydrogen Energy 31 (2006) 822–831.
[178] Y. Lu, L. Guo, X. Zhang, Q. Yan, Thermodynamic modeling and analysis of biomass gasification for hydrogen production in supercritical water, Chem. Eng. J. 131 (2007) 233–244.
[179] F. Vogel, Catalytic conversion of high-moisture biomass to synthetic natural gas in supercritical water, in: P. Anastas (Ed.), Handbook of Green Chemistry, in: R. Crabtree (Vol. Ed.), Heterogeneous Catalysis, vol. 3, Wiley-VCH, Weinheim, Germany, 2009, ISBN: 978-3-527-32498-9.

[180] M. Waldner, F. Vogel, Renewable production of methane from woody biomass by catalytic hydrothermal gasification, Ind. Eng. Chem. Res. 44 (2005) 4543–4551.

[181] T. Minowa, F. Zhen, T. Ogi, Cellulose decomposition in hot compressed water with alkali or nickel catalyst, J. Supercrit. Fluids 13 (1998) 253–259.

[182] C. Cao, L. Guo, Y. Chen, Y. Lu, Hydrogen production by supercritical water gasification of alkaline black liquor, in: D. Stolten, T. Grube (Eds.), Proceedings of the WHEC, May 16–21, 2010, Essen, Germany, 18th World Hydrogen Energy Conference 2010—WHEC 2010, Parallel Sessions Book 3: Hydrogen Production Technologies—Part 2.

[183] T.G. Madenoğlu, S. Kurt, M. Sağlam, M. Yüksel, D. Gökkaya, L. Ballice, Hydrogen production from some agricultural residues by catalytic subcritical and supercritical water gasification, J. Supercrit. Fluids 67 (2012) 22–28.

[184] E. Kipçak, M. Akgün, Oxidative gasification of olive mill wastewater as a biomass source in supercritical water: effects on gasification yield and biofuel composition, J. Supercrit. Fluids 69 (2012) 57–63.

[185] Y. Guo, S.Z. Wang, D.H. Xu, Y.M. Gong, H.H. Ma, X.Y. Tang, Review of catalytic supercritical water gasification for hydrogen production from biomass, Renew. Sustain. Energy Rev. 14 (2010) 334–343.

Chapter 9

Reactions of Synthetic Polymers with Water

The production of synthetic polymers has reached more than 200 million tons per year. Development of effective processes for recycling the polymers or of useful products from a decomposition process is an important issue. High-temperature and supercritical water is an excellent reaction medium for decomposition of synthetic polymers [1], since it does not introduce compounds that contaminate the products, and water is easy to remove from the decomposition products. In this chapter, reactions of synthetic polymers with water are discussed and several examples illustrate the principles of depolymerization reactions with water. Proposed schemes for commercial plants are presented in Chapter 13.

Polymers from condensation polymerization processes, such as polyethylene terephthalate (PET), nylon, and polyurethane are readily depolymerized to their monomers in supercritical water [1]. Polymers from addition polymerization processes, such as phenol resin, epoxy resin, and polyethylene, can also be depolymerized in high-temperature and supercritical water. Composite polymer materials can be separated into useful fractions. Chlorinated polymers can be treated in water to remove halogens, beside the conversion of the polymers to useful degradation products. From biodegradable polymers, such as poly(lactic acid), the monomer can be recovered in high purity and yield.

9.1 CONDENSATION POLYMERS

Polymers from condensation polymerization with ether, ester, or acid amide linkages can be hydrolyzed in high-temperature water. An important example is PET. PET is a semi-crystalline thermoplastic polyester that is widely used for high-strength fibers, photographic films, and soft-drink bottles [2]. Various chemical recycling methods have been investigated for PET [3].

In subcritical water, without a catalyst, nearly complete depolymerization of PET can be achieved at $T = 265\ ^\circ C$, with a ratio of water/PET (w/w) $S/F > 5$ [3].

Supercritical Fluid Science and Technology, Vol. 5. http://dx.doi.org/10.1016/B978-0-444-59413-6.00009-1

In supercritical water, at $T=673$ K, $P=40$ MPa, PET decomposes to terephthalic acid and ethylene glycol within $t=12.5$ min, with a yield of terephthalic acid of up to 91% and a concentration of up to >97%. The reaction rate is reduced at lower temperatures. At $T=573$ K, decomposition of PET to 90% terephthalic acid needs $t=90$ min. Increasing reaction pressure suppresses char and carbon dioxide formation [4]. The yield for ethylene glycol is lower due to further decomposition catalyzed by terephthalic acid. Secondary products observed are benzoic acid, diethylene glycol, 1,4-dioxane, acetaldehyde, and crotonic acid. Formation of further decomposition is negligible [1]. Results for the decomposition of PET in supercritical water are shown in Figure 9.1.

Most systems of PET–water become homogeneous in a concentration range of 12–59 wt% [2]. For slow heating, homogenization temperature is around the PET melting point ($T=241$ °C). At rapid heating and a solid–liquid transition during heating, the homogeneous state occurs at a temperature of $T=297$–318 °C [2]. Terephthalic acid–water become miscible at $T=356$ °C. Although the water/PET ratio seems to play a major role in the hydrolysis of PET, no general conclusions on the phase behavior can be drawn on the effect of initial PET concentration. This may be due to other prevailing factors, such as heating rate or heat transfer to the solid PET. If water is completely absent, PET decomposition is different to that in an aqueous environment [2].

Kobe Steel, Japan, developed a recycling process for PET using supercritical water with a recycling efficiency for PET of about 99 wt%. Compared to methanolysis and glycolysis, the process is simpler and has a reduced reaction time [6], see Chapter 13.

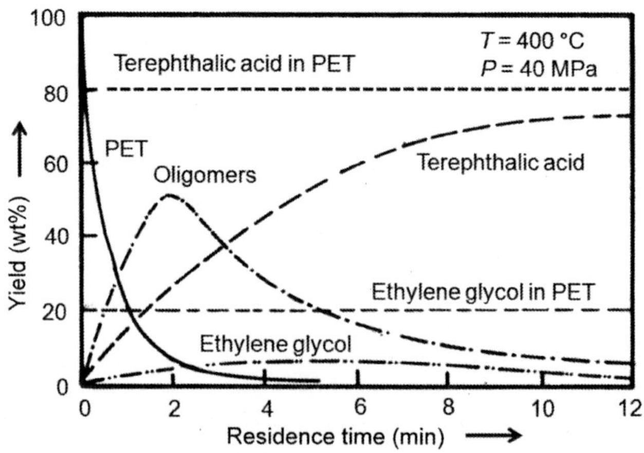

FIGURE 9.1 Decomposition products of polyethylene terephthalate (PET) in supercritical water. $T=400$ °C, $P=40$ MPa. *Modified from Refs. [4,5].*

The same process is applied to recycling of polyurethanes [6]. Polyurethanes are produced by the reaction of a polyisocyanate with a polyalcohol (polyol) in a catalyzed process. Hydrolysis of polyurethane in high-temperature water results in polyol and diamine corresponding to the starting isocyanate [7]. The optimum temperature for the decomposition of polyurethane is in the range of $T=523$–573 K at $P=10$–20 MPa. At temperatures above $T=543$ K, polyurethane is totally converted, while at $T=523$ K and $t=30$ min reaction time, the yield is only 72% [8]. The product splits into two liquid phases. The upper layer is an aqueous solution of tolylenediamine and the lower oil phase consists of polyols. Polyol and diamine are almost completely recovered [7], but the yield of recovered monomers decreases at temperatures $T>603$ K [8].

Nylon 6, a polymer synthesized by polymerization of ε-caprolactam, is hydrolyzed in high-temperature and supercritical water to ε-aminocaproic acid, followed by cyclodehydration to ε-caprolactam or decomposition further to smaller molecules [9]. According to batch experiments, the total yield of monomers approaches 100% at a temperature of $T=573$ K in $t=60$ min, or in $t=30$ min at $T=603$ K. At longer reaction times, the yield of ε-aminocaproic acid decreases rapidly [9].

Polycarbonates, widely used because of their temperature resistance, impact resistance, and optical properties, is decomposed in high-temperature and supercritical water to bisphenol A, phenol, *p*-isopropenylphenol, and *p*-isopropylphenol in a temperature range of $T=250$–430 °C, as shown in Figure 9.2 [10].

The reaction temperature is of major influence on the hydrolysis of polycarbonate. At supercritical temperature, decomposition reaches >90 wt% in

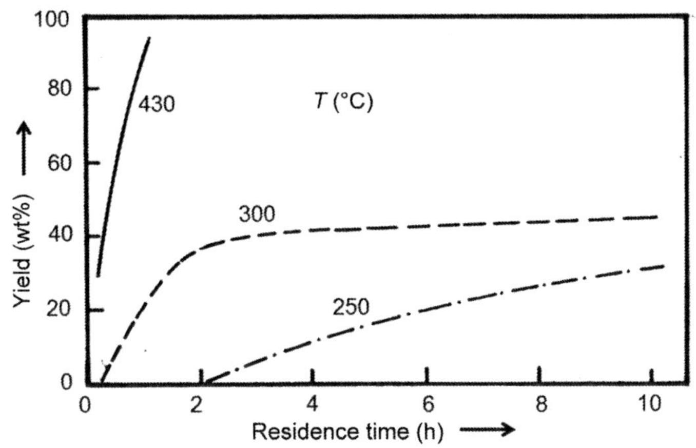

FIGURE 9.2 Decomposition of polycarbonate in high-temperature and supercritical water. *Modified from Refs. [4,5].*

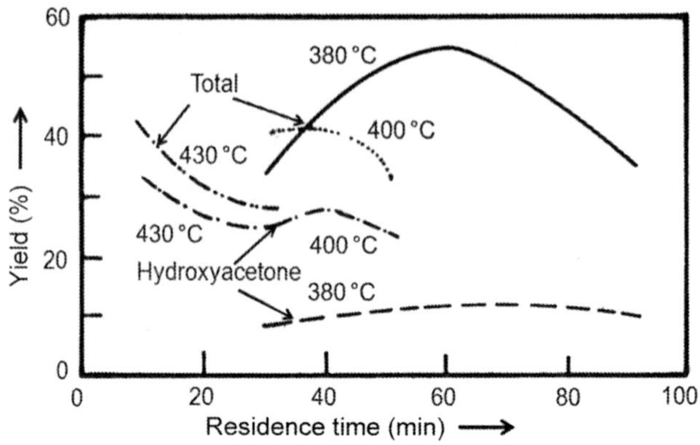

FIGURE 9.3 Decomposition of polypropylene glycol in supercritical water. *Modified from Ref. [12].*

less than $t=1$ h. The reaction is accelerated by the addition of Na_2CO_3, with a yield of identified products of 68 wt% at $T=523$ K in $t=1$ h [11].

Polypropylene glycol, as an example for the hydrolysis of polyols, is decomposed in supercritical water to hydroxyacetone and diols (propylene glycol, dipropylene glycol, and tripropylene glycol) [12]. The yield of hydroxyacetone is 33 wt% at $T=430$ °C and $t=10$ min reaction time, while the total yield of hydroxyacetone and diols is about 57 wt% at $T=380$ °C after $t=60$ min (Figure 9.3). Yield for both hydrolysis products increases with increasing ratio of water to polypropylene glycol. Hydroxyacetone is the end product of polypropylene glycol decomposition [12].

9.2 POLYETHYLENE

Polyethylene is a common plastic mostly used for packaging. It is a thermoplastic polymer consisting of long hydrocarbon chains, $(C_2H_4)_nH_2$, that differ in chain length. The melting point for high-density polyethylene is typically in the range of $T=120$–130 °C, and for low-density polyethylene about $T=105$–115 °C. Polyethylene polymers have excellent chemical resistance. They are not attacked by strong acids or strong bases and are resistant to gentle oxidants and reducing agents. Polyethylene (except cross-linked polyethylene) can be dissolved at elevated temperatures in aromatic hydrocarbons or in chlorinated solvents [13].

Polyethylene can be hydrolyzed in high-temperature and supercritical water. The distribution of reaction products can be controlled by the operating conditions. In comparison to pyrolysis in argon, higher yields of shorter chain hydrocarbons, higher 1-alkene/*n*-alkane ratio, and higher conversions are obtained in supercritical water [14]. The enhancement of polyethylene decomposition in supercritical water is at least partially due to the dissolution of high-molecular weight hydrocarbons in supercritical water and diffusion of water into the molten polyethylene phase [14].

The product from conversion in supercritical water at $T = 698$ K (425 °C) at a reaction time of $t = 120$ min is a yellow-brown grease, comprising about 90 wt%, and a gaseous phase containing 6.5 wt% of the initial polyethylene. In contrast, the product from thermal cracking is a blackish brown liquid with a conversion of 72 wt% to an oily phase and 12 wt% to a gaseous phase. Coke production is small compared to thermal cracking [15]. Hydrogen is integrated from supercritical water into the product oil, an effect that increases with higher density [16]. Conversion of polyethylene leads to 2-propanol and to 2-propanone.

In batch experiments on the conversion of low-density polyethylene with a mean molecular weight of 68,000 D in supercritical water, about 30 wt% of the initial amount of polymer is converted [17]. The mean molecular weight of tetrahydrofuran-insoluble products is about 3800 D for a density of $\rho_{H_2O} = 0.13\,g/cm^3$ and 2000 D for a density of $\rho_{H_2O} = 0.42\,g/cm^3$, which is lower than the mean molecular weight of 6000 D for pyrolysis in Ar [17]. Clearly, hydrolysis of polyethylene differs in mechanism from the pyrolysis in Ar. The product distribution for gaseous and products that are soluble in tetrahydrofuran is shown in Figure 9.4 and the ratio of 1-alkenes/*n*-alkanes for these products in Figure 9.5 [17].

Pyrolysis of polyethylene in an Ar atmosphere occurs mainly in a molten polyethylene phase as free-radical bimolecular reaction and leads to the relatively flat product distribution and low 1-alkene/*n*-alkane ratio (Figures 9.4 and 9.5). In supercritical water, the molten polyethylene phase is diluted with dissolved water and the contribution of the p-scission (unimolecular) reaction

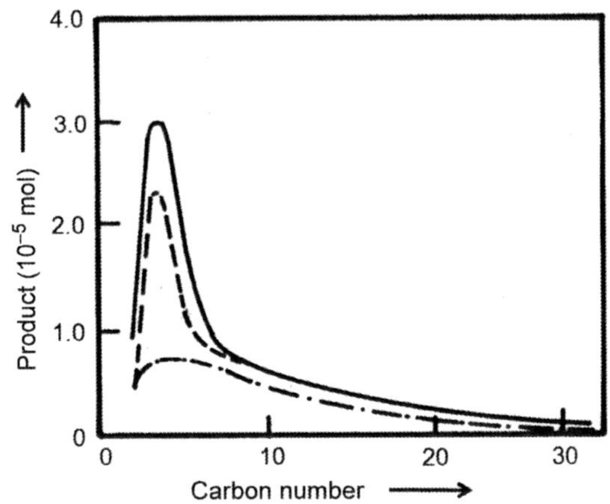

FIGURE 9.4 Product distribution (gas + tetrahydrofuran soluble) for the conversion of polyethylene. $T = 693$ K (420 °C), reaction time $t = 30$ min. — — —, water $\rho_{H_2O} = 0.13\,g/cm^3$; ———, water $\rho_{H_2O} = 0.42\,g/cm^3$; — • —, Ar. *Modified from Ref. [17].*

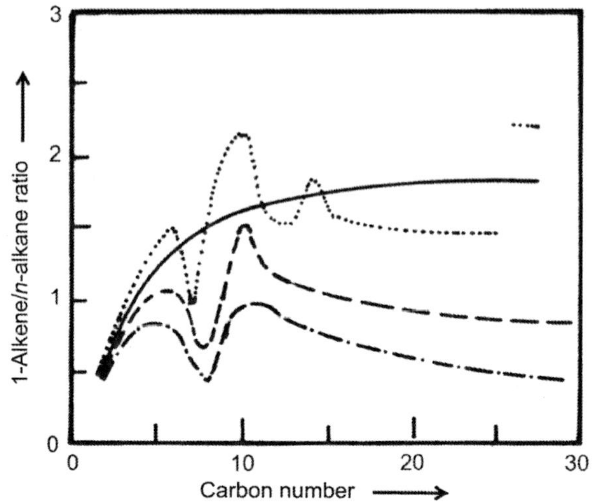

FIGURE 9.5 1-Alkene/*n*-alkane distribution (gas + tetrahydrofuran soluble products) for the conversion of polyethylene. $T = 693$ K (420 °C), reaction time $t = 30$ min. — — —, water $\rho_{H_2O} = 0.13\,g/cm^3$; ——, (averaged experimental values); • • • •, (individual experimental values) water $\rho_{H_2O} = 0.42\,g/cm^3$; — • — •, Ar. *Modified from Ref. [17].*

increases leading to more l-alkene and shorter chain hydrocarbons. Some of the products dissolve in supercritical water, where at a relatively low concentration also p-scission (unimolecular) reactions are predominant and convert produced hydrocarbons further [17]. Phase behavior of polyethylene–water mixtures with a concentration of 12–30 wt% polyethylene at high pressures ($P = 110$–690 MPa) and rapid heating reveal formation of a liquid spherule polyethylene phase. Polyethylene and water remain as a heterogeneous system. Only after polyethylene decomposes above about $T = 565\,°C$ to lower molecular weight hydrocarbons, homogeneous reaction conditions result [2].

9.3 ADDITION POLYMERIZATION POLYMERS

9.3.1 Phenol Resins

Thermosetting resins are widely used for electronic applications. Phenol resin, one of the thermosetting resins, has a high thermal stability due to aromatic units that are connected by methylene bonds. Decomposition of prepolymers of phenol resin in high-temperature and supercritical water into their monomers such as phenol, cresols, and *p*-isopropylphenol leads at $T = 703$ K to a maximum yield of 78 wt% after $t = 30$ min. The decomposition is accelerated by the addition of Na_2CO_3 with a yield of identified monomers of >90 wt% [13]. Similar, molding materials of phenol resin with a phenol resin content of <50 wt% are decomposed into their monomers (phenol and cresols) by reaction in supercritical water, and the decomposition is accelerated by the addition of polycarbonate [18].

Printed circuit board waste has been converted in supercritical water in a batch reactor into a gaseous, liquid, and solid phase [19]. About 80% of the feed is transformed into liquid or gaseous products. The liquid phase consists of small molecules dissolved in water. Total organic carbon of liquid phase products amounts to about 48 wt% of carbon in the feed, with a maximum at $T=673$ K, which decreases with increasing temperature due to conversion to gaseous products. At $T=723$ K, the conversion is >60 wt% after $t=20$ min.

The yield of phenol reached 5 wt% at $T=733$ K in $t=80$ min, corresponding to 18 wt% of the phenol resin. The yield of cresols is similar with a highest value of around 3 wt%, corresponding to 11 wt% of the phenol resin. Together, about 30 wt% of the phenol resin is converted to phenol and cresols [19].

Epoxy resin and polyether-ether-ketone (PEEK) resin are decomposed into their monomers, such as phenol, cresols, and their analogues, by hydrolysis in high-temperature and supercritical water. In a 10-ml batch reactor, the yield of epoxy resin hydrolysis of identified products reaches 10 wt% at $T=703$ K in $t=60$ min (Figure 9.6). For hydrolysis of PEEK resin, the total yield of phenol and dibenzofuran reaches 88 wt% at $T=703$ K in $t=180$ min [20].

Decomposition of a brominated epoxy resin (used as flame-resistant polymer) at $T=300$ °C, $P=25$ MPa, leads in $t=30$ min reaction time to water-soluble products and tetrahydrofuran soluble products with a high-molecular weight and transfers bromine to >98% into water [21].

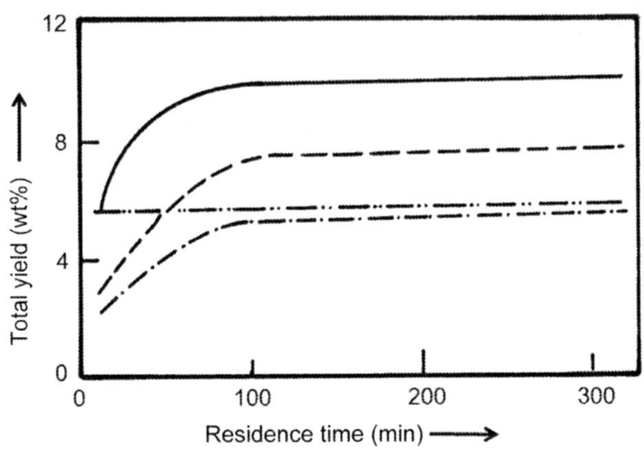

FIGURE 9.6 Conversion of epoxy resin in water. $T=703$ K. — • — •, H_2O, 1 ml injection volume; — • • —, H_2O+3 wt% Na_2CO_3, 1 ml injection volume (corresponding to a density of 0.1 g/cm^3); ———, H_2O+3 wt% Na_2CO_3, 3 ml injection volume (corresponding to a density of 0.3 g/cm^3); — — —, 1-methylnaphthalene, 1 ml injection volume (density 0.1 g/cm^3). *Modified from Ref. [20].*

9.3.2 Fiber-Reinforced Plastics

Fiber-reinforced plastics (FRPs) are widely used as high-strength materials. Glass fiber-reinforced plastics are composite materials manufactured by laminating unsaturated polyester resin with glass fiber and filler. Carbon fiber-reinforced plastics consist of carbon fiber and epoxy resin.

FRPs can be decomposed in sub- and supercritical water, see Figure 9.7. In a batch reactor, FRPs are almost completely converted at $T=653$ K in $t=5$ min [23,24]. From carbon fiber-reinforced epoxy resin, about 80 wt% of the resin is removed at $T=673$ K, $P=28$ MPa. The tensile strength of the reclaimed fibers is 90–98% of the original fibers [22]. Potassium hydroxide as an alkali catalyst improves the recovery to >95 wt%.

At the lower temperature of $T=230$ °C, thermosetting polyester resin of FRPs is completely dissolved with subcritical water and an alkali catalyst after a reaction time of $t=4$ h at saturation pressure ($P=2.8$ MPa). In a downstream filter press, inorganic filler and milled glass fiber are separated from the reaction liquid with a recovery ratio of 95% [25]. A solid, identified as a styrene-fumaric acid copolymer, corresponding to 75 wt% of the initial resin can be recovered. From the initial content of glycol and fumaric acid, 71 wt% and 22 wt%, respectively, are in the reaction liquid. Up to 96 wt% of the initial resin is recovered in the reaction liquid [25].

At the higher reaction temperature of $T=360$ °C, the highest recovery ratio of the glycol and of fumaric acid is observed with calcium carbonate as an alkali catalyst after the short-reaction time of $t=20$ min. 65 wt% of

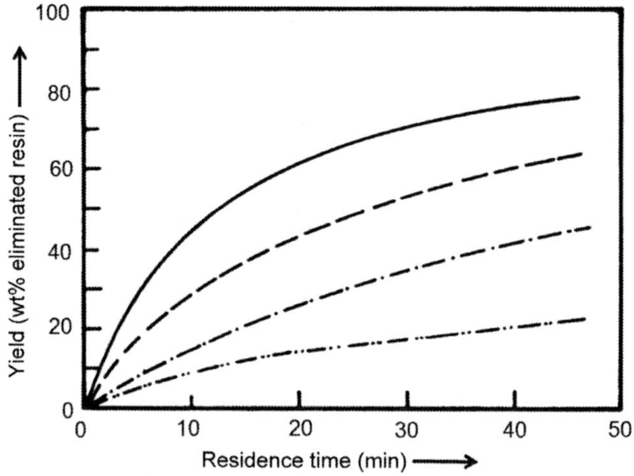

FIGURE 9.7 Conversion of epoxy resin from carbon fiber-reinforced plastics in supercritical water in a batch reactor, reaction volume 10 cm^3. Kinetic curves expressed at different operating pressures (MPa) and resin/solvent weight ratios (g/g) placed initially in the reactor system. — • • —, 523 K, 0.196 g/g, 4 MPa; — — —, 673 K, 0.196 g/g, 14 MPa; — • — •, 573 K, 0.065 g/g, 10 MPa; ———, 673 K, 0.065 g/g, 27 MPa. *Modified from Ref. [22].*

TABLE 9.1 Conversion of Fiber-Reinforced Plastics (Thermosetting Polyester Resin) in Subcritical Water with Different Catalysts

Catalyst	Conversion (wt%)	Glycol Recovery (wt%)	Fumaric Acid Recovery (wt%)
No catalyst	15.9	33.7	9.0
$CaCO_3$	30.8	32.2	1.2
K_3PO_4	41.5	58.6	3.6
KOH	96.9	70.7	21.8

$T=230\ °C$, $t=4$ h, saturation pressure [25].

the decomposing organic material is chloroform soluble. Since it is known that pyrolysis of the cross-linking species of the thermosetting polyester resin starts at a temperature of $T=230\ °C$, this temperature is selected as reaction temperature. After $t=4$ h reaction time, no chloroform soluble reaction products other than initial glycols are detected [25,26]. The influence of alkali catalysts is listed in Table 9.1.

9.4 VARIOUS POLYMERS

9.4.1 Cross-Linked Polyethylene

Cross-linked polyethylene (XLPE) is a thermosetting resin with three types of cross-linking: peroxide cross-linking, radiation cross-linking, and silane cross-linking. The polymer can be effectively recycled if only the cross-linking points are decomposed. From peroxide cross-linked polyethylene, the cross-links are selectively removed by using supercritical water without severe decomposition of the backbone chains [27]. For silane cross-linked polyethylene, treatment in subcritical water for temperatures $T>340\ °C$ leads to zero cross-linking, with a reduction of the molecular weight. For comparison, in methanol, zero cross-linking is achieved at $T>300\ °C$, without reduction of the molecular weight, see Figure 9.8 [1].

9.4.2 Chlorinated Polymers

From polyvinylchloride polymers, chlorine is removed with high-temperature water at saturation pressure. At $T=315\ °C$ in a small batch reactor, 90 wt% of the chlorine is removed after $t=1$ h, and at $T=250\ °C$, 86 wt% of chlorine is removed after $t=2$ h [28]. Butyl rubbers are depolymerized to lower molecular weight organic compounds in high-temperature liquid water. For rubbers

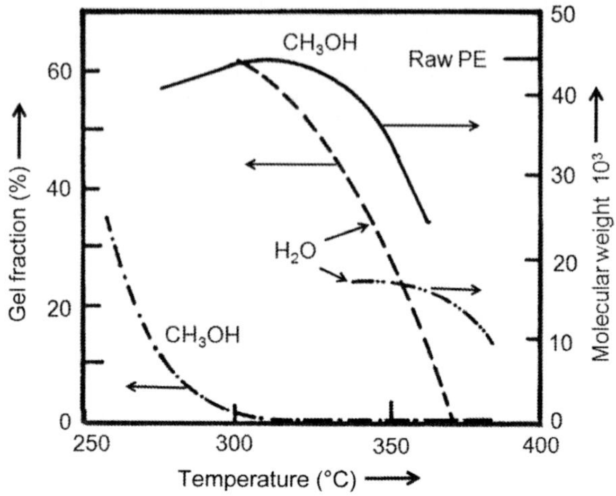

FIGURE 9.8 Conversion of silane cross-linked polyethylene. *Modified from Ref. [1].*

containing halogen groups, the amount of halogen is reduced. At $T=350$ °C after $t=1$ h, the chlorine content is reduced by 63 wt%. Bromobutyl rubber is depolymerized at $T=350$ °C in water to a liquid product and about 30 wt% of the bromine is removed [29].

9.4.3 Polystyrene

The degradation of polystyrene at pressures of $P=240$–320 bar at a temperature of $T=370$ °C leads to a conversion of about 80 wt% in $t=5$ min. At supercritical conditions, conversion is completed (100%) after $t=15$ min at $T=380$ °C, and at $T=390$ °C after $t=3$ min. Selectivity for the degradation products, styrene monomer, styrene dimer, styrene trimer, decreased with reaction time, while that for toluene, ethyl benzene, and isopropyl benzene increased at $T=400$ °C and $P=28$ MPa. Selectivity is only slightly influenced by temperature in the range from $T=370$ to 400 °C. Degradation of polystyrene in high-temperature and supercritical water can be formulated by the first-order kinetic law with an activation energy of $E_A=157$ kJ/mol [30].

9.4.4 Poly(Lactic Acid)

Poly(lactic acid) is polymerized from lactic acid that can be obtained from starch containing renewable resources. It can be hydrolyzed in high-temperature water at $T=200$–300 °C with high efficiency [31]. At lower temperatures, the recovered monomer of lactic acid is predominant, while at higher temperatures, lactic acid is degraded into other low-molecular weight

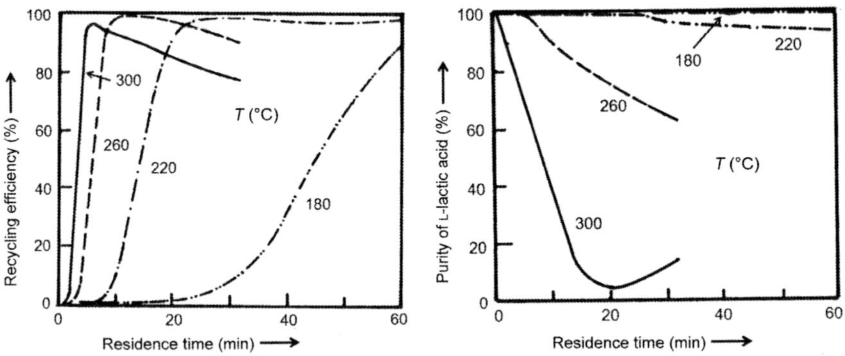

FIGURE 9.9 Hydrolysis of poly(lactic acid) in high-temperature water at saturation pressure. Left: Decomposition of poly(lactic acid). Right: Concentration of L-lactic acid in the hydrolysis product. *Modified from Ref. [31].*

organic compounds. High temperatures, such as $T > 300$ °C, induce dramatic racemization, compared to lower temperatures, at which L-lactic acid is recovered at a yield of 98%. The effect of reaction temperature and time on the recycling efficiency to lactic acid in high-temperature water is shown in Figure 9.9.

The products from poly(lactic acid) hydrolysis are mostly lactic acid. The amount of lactic acid increases with reaction temperature up to $T = 300$ °C. At short-reaction times of $t = 0$–3 min, depolymerization results only in the reduction of molecular weight. At reaction times of $t = 3$–5 min, a significant increase of lactic acid is observed. The yield of lactic acid reaches around 97 wt% at $t = 5$ min and $T = 300$ °C. The optimum yield of lactic acid is obtained at $t = 20$–30 min at $T = 220$ °C, with more than 99% of poly(lactic acid) converted into lactic acid. At $T = 350$ °C, the degradation rate of poly(lactic acid) is high and small amounts of propionic acid and acetic acid occur.

The optimum conditions for recovery of L-lactic acid ($T = 220$ °C, $t = 30$ min, $P \geq P^0_{H_2O}$) applied to commercial products of poly(lactic acid) (brister bag, body of a personal computer) resulted in about 96 wt% recovery for the brister bag and only 36 wt% for the body of a personal computer, as shown in Figure 9.10. Probably, the composition of the body of the personal computer is different from standard poly(lactic acid) [31].

Hydrolysis without any acid or base catalyst in high-temperature water is a promising technology for the chemical recycling of polymers. For the further development of the process and its application to various fields, fundamental information on kinetics and mechanism of the reactions and on the reaction phase is required. In combination with unit operation processes in high-temperature and supercritical water, this will lead to a wider application of hydrolysis in industrial processes [4].

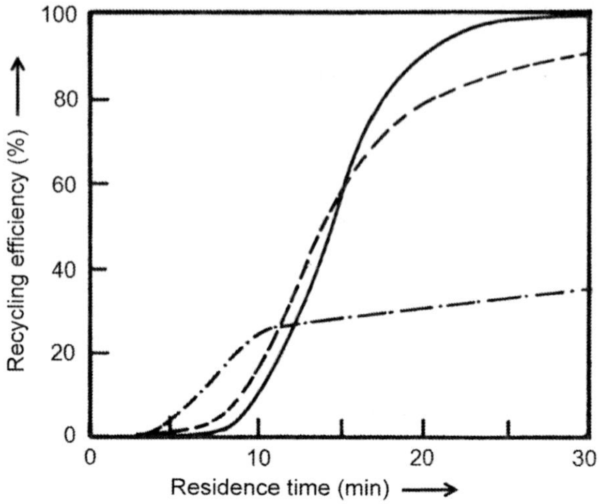

FIGURE 9.10 Hydrolysis of poly(lactic acid) and commercial products in high-temperature water at saturation pressure, $T = 220\ °C$, $t = 30$ min reaction time.——, poly(lactic acid) sample; — — —, brister bag; — • — •, body of personal computer. *Modified from Ref. [31].*

REFERENCES

[1] M. Goto, Chemical recycling of plastics using sub- and supercritical fluids, J. Supercrit. Fluids 47 (2009) 500–507.

[2] Z. Fang, R.L. Smith Jr., H. Inomata, K. Arai, Phase behavior and reaction of polyethylene terephthalate-water systems at pressures up to 173 MPa and temperatures up to 490 °C, J. Supercrit. Fluids 15 (1999) 229–243.

[3] J.R. Campanelli, M.R. Kamal, D.G. Cooper, A kinetic study of the hydrolytic degradation of polyethylene terephthalate at high temperatures, J. Appl. Polym. Sci. 48 (1993) 443–451.

[4] K. Arai, R.L. Smith Jr., T.M. Aida, Decentralized chemical processes with supercritical fluid technology for sustainable society, J. Supercrit. Fluids 47 (2009) 628–636.

[5] T. Adschiri, O. Sato, K. Machida, N. Saito, K. Arai, Recovery of terephthalic acid by decomposition of PET in supercritical water, Kagaku Kogaku Ronbun. 23 (1997) 505–511.

[6] Y. Nagase, M. Yamagata, R. Fukuzato, Development of a chemical recycling process for waste plastics using supercritical water, Kobelco Technol. Rev. 22 (1999) 11–14.

[7] Y. Nagase, M. Yamagata, T. Matsuda, T. Naito, K. Kodama, R. Fukuzato, Chemical recycling process for scrap polyurethane plastics using supercritical water, in: Proceedings of the 5th Meeting on Supercritical Fluids, 1998, pp. 127–132.

[8] Z. Dai, B. Hatano, J. Kadokawa, H. Tagaya, Effect of diaminotoluene on the decomposition of polyurethane foam waste in superheated water, Polym. Degrad. Stab. 76 (2002) 179–184.

[9] M. Goto, M. Umeda, A. Kodama, T. Hirose, S. Nagaoka, Monomerization of nylon 6 in sub- and supercritical water, Kobunshi Ronbun. 58 (2001) 548–551.

[10] H. Tagaya, K. Katoh, J.-I. Kadokawa, K. Chiba, Decomposition of polycarbonate in subcritical and supercritical water, Polym. Degrad. Stab. 64 (1999) 289–292.

[11] A. Ikeda, K. Katoh, H. Tagaya, Monomer recovery of waste plastics by liquid phase decomposition and polymer synthesis, J. Mater. Sci. 43 (2008) 2437–2441.

[12] Z. Dai, B. Hatano, H. Tagaya, Decomposition of a polyol in supercritical water, Polym. Degrad. Stab. 80 (2003) 353–356.

[13] http://en.wikipedia.org/wiki/Polyethylene.

[14] M. Watanabe, T. Adschiri, K. Arai, Polyethylene decomposition via pyrolysis and partial oxidation in supercritical water, Kobunshi Ronbun. 58 (2001) 631–641.

[15] T. Moriya, H. Enomoto, Characteristics of polyethylene cracking in supercritical water compared to thermal cracking, Polym. Degrad. Stab. 65 (1999) 373–386.

[16] T. Moriya, H. Enomoto, Supercritical fluid in polymer science and technology. II. Conversion of polyethylene to oil using supercritical water and donation of hydrogen in supercritical water, Kobunshi Ronbun. 58 (2001) 661–673.

[17] M. Watanabe, H. Hirakoso, S. Sawamoto, T. Adschiri, K. Arai, Polyethylene conversion in supercritical water, J. Supercrit. Fluids 13 (1998) 247–252.

[18] Y. Suzuki, H. Tagaya, T. Asou, J. Kadokawa, K. Chiba, Decomposition of prepolymers and molding materials of phenol resin in subcritical and supercritical water under an Ar atmosphere, Ind. Eng. Chem. Res. 38 (1999) 1391–1395.

[19] M. Goto, M. Kitamura, T. Hirose, K. Shibata, Depolymerization of printed circuit board in near-critical water, in: S.H. Feng, J.S. Chen, Z. Shi (Eds.), Hydrothermal Reactions and Techniques, Proceedings of the Seventh International Symposium on Hydrothermal Reactions, Changchun, China, 14–18 December 2003, World Scientific, New Jersey, 2003, pp. 201–208.

[20] H. Tagaya, Y. Shibasaki, C. Kato, J.I. Kadokawa, B. Hatano, Decomposition reactions of epoxy resin and polyether ether ketone resin in sub- and supercritical water, J. Mater. Cycles Waste Manage. 6 (2004) 1–5.

[21] I. Okajima, T. Sugeta, T. Sako, Supercritical fluid in polymer science and technology. II. Decomposition and debromination of flame-resistant polymers containing bromine atoms with subcritical water, Kobunshi Ronbun. 58 (2001) 692–696.

[22] R. Pinero-Hernanz, C. Dodds, J. Hyde, J. Garcia-Serna, M. Poliakoff, E. Lester, M.J. Cocero, S. Kingman, S. Pickering, K.H. Wong, Chemical recycling of carbon fibre reinforced composites in near critical and supercritical water, Compos. Appl. Sci. Manuf. 39 (2008) 454–461.

[23] T. Sugeta, S. Nagaoka, K. Otake, T. Sako, Decomposition of fiber reinforced plastics using fluid at high-temperature and pressure, Kobunshi Ronbun. 58 (2001) 557–563.

[24] I. Okajima, K. Yamada, T. Sugeta, T. Sako, Decomposition of epoxy resin and recycling of CFRP with sub- and supercritical water, Kagaku Kogaku Ronbun. 28 (2002) 553–558.

[25] T. Nakagawa, T. Urabe, T. Miyazaki, M. Hidaka, K. Oka, T. Maekawa, T. Yoshimura, H. Yoshida, FRP recycling technology using sub-critical water hydrolysis, in: 10th International Symposium on Supercritical Fluids, ISSF 2010, Kyoto, Japan, 2006.

[26] T. Nakagawa, FRP recycling technology using sub-critical water hydrolysis, JEC Composites Magazine (March–April) (2008) 40–43.

[27] S. Watanabe, K. Komura, S. Nagaya, H. Morita, T. Nakamoto, S. Hirai, F. Aida, Development of cross-linked polymer material recycling technology by supercritical water, in: Proceedings of the International Conference on Properties and Applications of Dielectric Materials, vol. 7, 2003, pp. 595–598.

[28] R.Y. Saleh, M. Siskin, G.A. Knudsen, Process for improving biodegradability of PVC, U.S. Patent 5,324,817, June 28, 1994.

[29] R.Y. Saleh, M. Siskin, G.A. Knudsen, Process for depolymerization of butyl rubbers and halobutyl rubbers using hot liquid water, U.S. Patent 5,283,318, February 1, 1994.

[30] H. Kwak, H.-Y. Shin, S.-Y. Bae, H. Kumazawa, Characteristics and kinetics of degradation of polystyrene in supercritical water, J. Appl. Polym. Sci. 101 (2006) 695–700.

[31] M. Faisal, Y. Atsuta, H. Daimon, K. Fujie, Chemical recycling of poly (L-lactic) acid with condensed fluids, in: M. Goto, T. Adschiri, B. Han, Y.-W. Lee (Eds.), Proceedings of the 8th International Symposium on Supercritical Fluids, Kyoto, Japan, November 5/8, OC-2-07, 2009.

Chapter 10

Oxidation in High-Temperature and Supercritical Water

10.1 GENERAL ASPECTS OF OXIDATION IN WATER

Oxidation in high temperature and supercritical water refers to the oxidation of organic compounds in water at high temperatures and elevated pressures. It comprises (i) oxidation in a liquid aqueous environment in the temperature range from $T=100$–374 °C, known as “wet air oxidation” (WAO), since in most cases air is used as the oxidation agent; (ii) oxidation in supercritical water, known as “supercritical water oxidation” (SCWO) at temperatures above the critical temperature of water in the range of up to about $T=550$ °C; and (iii) the oxidation of organic compounds in water in the so-called hydrothermal flames at temperatures higher than $T=1000$ °C. All oxidation reactions aim at the conversion of the organic compounds, mostly to totally oxidized end products, but also in some cases as partial oxidation, for the purpose of forming interesting intermediate products. Nevertheless, the main goal of such oxidation processes is the conversion of waste compounds to harmless oxidation products.

In WAO, the oxidation is carried out usually at moderate pressures that can reach up to $P=20$ MPa, if the temperature needed for the conversion approaches the critical temperature of water. The oxidation is moderately fast and takes about $t=1$–3 h for the oxidation reaction. In SCWO, pressures are higher than in WAO, but not necessarily higher than the critical pressure of water. The oxidation reaction is fast and takes typically less than $t=60$ s. The special properties of water at supercritical conditions not only make it possible to control the reaction via density but also make it necessary to take care of precipitating salts. In hydrothermal flames, pressures are high, typically higher than the critical pressure of water, and the reaction is very fast in the range of $t=1$ ms to $t=1$ s. In all three cases, corrosion is a major issue.

In this chapter, the fundamentals and applications of WAO, SCWO, and hydrothermal flames are described with examples given. Process conditions as derived from basic experiments and process verifications are discussed. The reactors for the processes and some processes are discussed in more detail

Supercritical Fluid Science and Technology, Vol. 5. http://dx.doi.org/10.1016/B978-0-444-59413-6.00010-8

and separately in Chapter 13. Corrosion in high temperature and supercritical water is discussed in Chapter 12.

10.2 WET AIR OXIDATION

10.2.1 Introduction

WAO is the oxidation of organic compounds by molecular oxygen in the liquid phase at high temperatures $T \geq 100$–$374\ °C$ and elevated pressures to maintain a liquid phase and to dissolve sufficient oxygen in the liquid phase. Typical conditions are for the temperature $T = 200$–$325\ °C$, for the pressure 5–18 MPa, and for the residence time around $t = 1$ h. The high temperature and pressure conditions convert the organic matter to carbon dioxide and water. WAO can be applied to treat chemical and municipal wastes. The load of waste components, expressed as chemical oxygen demand (COD), typically is in the range from $x_{COD} = 10$–$80\ kg/m^3$. WAO is an enclosed process with limited interaction with the environment [1]. It can be applied at mild conditions to alter molecular structures and improve subsequent biological treatment, or at more severe conditions to achieve almost entire oxidation of the organic matter. The oxidation reactions are exothermic. Energy self-sufficiency is obtained for COD loads that exceed $x_{COD} = 12$–$15\ kg/m^3$.

At temperatures of about $T = 250\ °C$, nearly all compounds can be completely transformed except acetic and propionic acid. At such conditions, the final product is not only carbon dioxide and water but also various other compounds, in particular carboxylic acids with the main component acetic acid. Acetic acid can be oxidized to $\geq 90\%$ at a temperature of $T = 310\ °C$ in $t = 1$ h. The oxidation proceeds in a chain reaction mechanism shown in Figure 10.1.

Organic carbon is oxidized to carbon dioxide, organic nitrogen is reacted to ammonia, nitrogen is converted to nitrous compounds, sulfur compounds react to sulfuric acid or sulfates, phosphorus compounds to phosphates, and chlorine compounds react to hydrochloric acid. The general material balance for the WAO process can be described with the general reaction, Equation (10.1). The overall heat of reaction is about $\Delta H_{WAO} = 435$ kJ/mol for any mole of O_2 reacted [1].

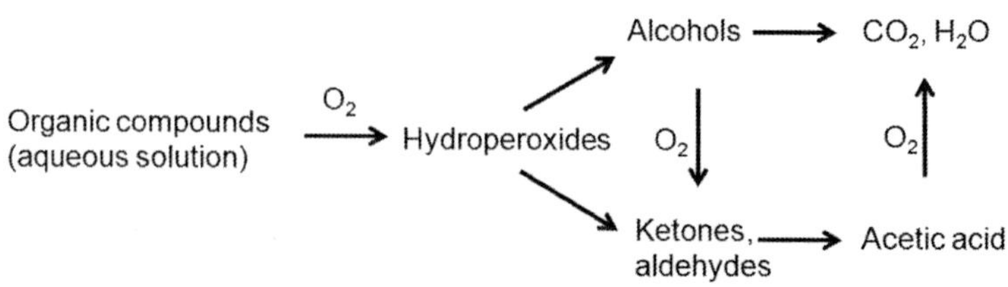

FIGURE 10.1 Chain reaction mechanism of wet air oxidation [1].

$$C_mH_nO_kCl_wN_xS_yP_z + (m + 0.25(n - 3x) - 0.5k + 2(y + z))O_2 \rightarrow mCO_2 + 0.5(n - 3x)H_2O + xNH_3 + wCl^- + ySO_4^{2-} + zPO_4^{3-} + \Delta H_R \quad (10.1)$$

Mechanism of the reactions:

$$O{-}O + R^* \quad (10.2)$$

$$RH + O_2 \rightarrow R^* + HO_2^* \quad (10.3)$$

$$RH + HO_2^* \rightarrow R^* + H_2O_2 \quad (10.4)$$

$$H_2O_2 + M \rightarrow 2HO^* + M \quad (10.5)$$

$$ROO^* + RH \rightarrow R^* + ROOH \quad (10.6)$$

The reaction series, Equations (10.2)–(10.6), shows that the reaction is propagated by an organic radical R* during a coupling with molecular oxygen, Equation (10.2). The radical comes from a reaction between oxygen and the weakest C—H bond, Equation (10.3), or from the reaction with the radical HO_2^*, Equation (10.4). Hydrogen peroxide, obtained in Equation (10.4), decomposes rapidly according to Equation (10.5), to hydroxyl radicals. A propagation reaction leads to hydroperoxides and other oxidized species. The initiation step, Equation (10.3), is limiting and is strongly temperature dependent, with an activation energy of $E_a \geq 100$–200 kJ/mol. The propagation step, Equation (10.2), is usually rapid with rate constants of about $k = 10^7$–10^9 dm^3 mol^{-1} s^{-1} [1]. To overcome the thermal limitation of reaction Equation (10.3), metal oxides and precious metals are proposed for heterogeneous catalysis [2,3], and ions like Fe^{2+}, Fe^{3+}, Cu^{2+}, Ni^{2+}, Ag^+, and Cr^{3+}, in combination with hydrogen peroxide or organic peroxides, for homogeneous catalysis [4].

Regarding the kinetic aspects of the chemical reactions, partial order for the organic compound is almost always 1, and close to 0.4 for dissolved oxygen [5,6]. For a complex mixture of compounds as found in real wastes, some of the organic compounds are destroyed to the final oxidation products, while others are transformed to intermediates, having a lower reactivity. The global rate of oxidation depends on the final formation rates of products as well as of the intermediates [1]. In a generalized model, shown in Figure 10.2, group A includes initial and relatively unstable

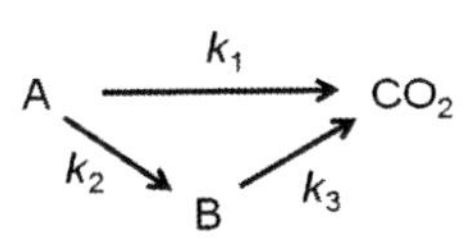

FIGURE 10.2 Generalized kinetic model for wet air oxidation of organic compounds [7].

intermediates and group B contains the more stable intermediates represented mainly by acetic acid [7]. Both groups can lead to oxidation end products, like carbon dioxide. Equation (10.7) is used for any of the elemental reactions.

For oxygen in excess, the reactions can be described with the following equations:

$$-\frac{dC_A}{dt} = k_1 C_A + k_2 C_A \tag{10.7}$$

$$\frac{dC_B}{dt} = k_2 C_A - k_3 C_B \tag{10.8}$$

$$k_1 = k_{0,1} \exp\left(\frac{-E_1}{RT}\right) [O_2]^{n_1} \tag{10.9}$$

$$k_2 = k_{0,2} \exp\left(\frac{-E_2}{RT}\right) [O_2]^{n_2} \tag{10.10}$$

$$k_3 = k_{0,3} \exp\left(\frac{-E_3}{RT}\right) [O_2]^{n_3} \tag{10.11}$$

$$[C_A + C_B]/[C_A + C_B]_0 = \frac{k_2}{[k_1 + k_2 - k_3]} \exp(-k_3 t) + \frac{[k_1 - k_3]}{[k_1 + k_2 - k_3]} \exp(-[k_1 + k_2] t) \tag{10.12}$$

The evolution of the concentration of organic compounds over the time is given in Equation (10.8). Parameter $[C_A + C_B]$ is a representation of the actual value of the organic load, usually expressed as COD or total oxygen demand (TOC). k_1, k_2, and k_3 are apparent constants for the chemical rates. They depend on the temperature value and the dissolved oxygen concentration according to Equations (10.9)–(10.11). A high value of k_2 indicates accumulation of low reactivity intermediates, like acetic acid, and a low value indicates a high global reactivity [1].

Experimental results on the WAO of a real waste from chemical industry are shown in Figure 10.3 together with the modeling according to Equations (10.7)–(10.12) after fitting k_i values to the experimental data [1].

The basic flow diagram of a WAO plant, operated continuously, is shown in Figure 10.4. The feed waste is pumped to the lower part of a bubble column reactor through a series of preheaters. The oxidation reaction takes place in the liquid phase with a residence time of 0.5–3 h. Air or oxygen is compressed and flows upward in the reactor.

The typical removal efficiency for the COD usually is below 95%. The reaction effluent cannot be discharged into the environment but is treated in a further cleaning facility, mostly of biological type. An example for a

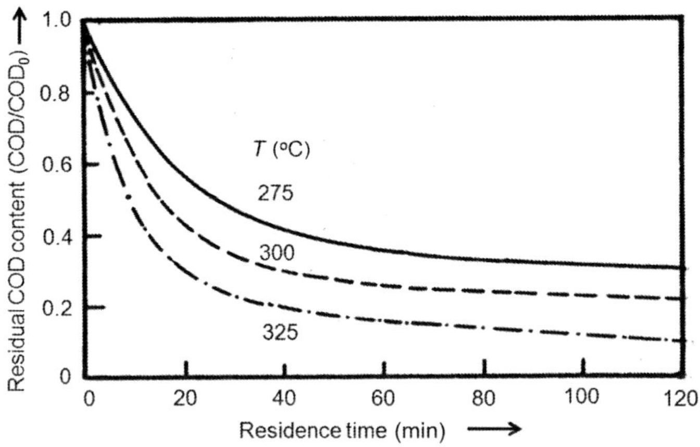

FIGURE 10.3 Reduction of COD during wet air oxidation of a waste from chemical industry, ——— $T=275\,°C$, — — — $T=300\,°C$, —•— $T=325\,°C$, $[O_2]=0.75\ g/dm^3$. *Modified from Ref. [1].*

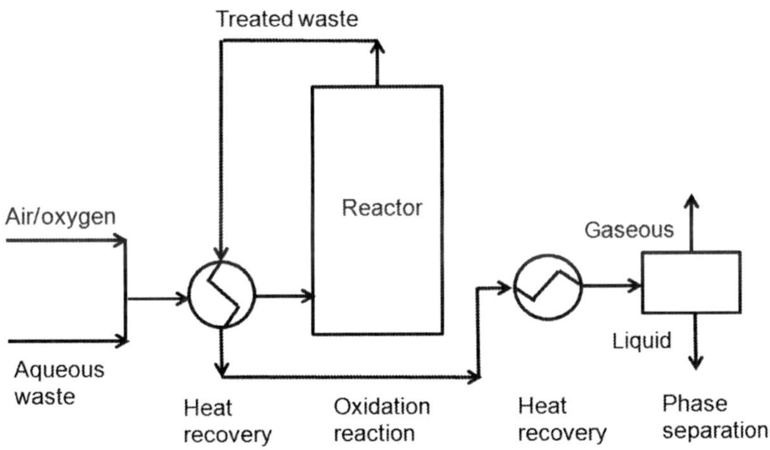

FIGURE 10.4 Generalized flow diagram of a wet air oxidation plant. *Modified from Ref. [1].*

technical WAO plant is discussed in Chapter 13. In the following, some results of experimental investigations on WAOs are presented for specific compounds or wastes.

10.2.2 Examples

10.2.2.1 WAO of Phenol

Phenol is a common component of industrial wastewaters. Therefore, attention has been directed to the oxidative reactions of phenol. First, WAO of

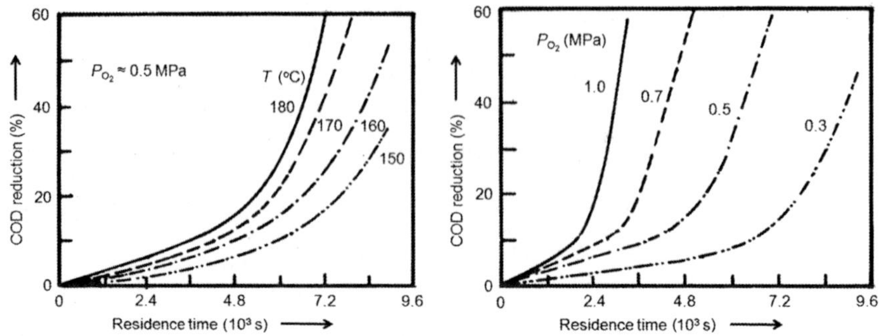

FIGURE 10.5 Oxidation of phenol in high-temperature water. Left: Influence of temperature, $P_{O_2} \approx 0.5$ MPa. Right: Influence of oxygen partial pressure, $T = 180$ °C. *Modified from Ref. [8].*

phenol is considered. Later in this chapter, SCWO of phenol will be discussed.

WAO of phenol in aqueous solutions in the temperature range of $T = 150$–180 °C, at pressures of $P = 0.3$–1.5 MPa with an initial phenol concentration of $x_F = 200$ mg/dm^3 in a stirred autoclave, from which samples were drawn, reveals that the oxidation of phenol in water involves a free-radical mechanism and proceeds in two steps [8]. The reduction of the COD is shown in dependence of temperature and oxygen partial pressure in Figure 10.5. The oxidation reaction is of first order for oxygen and for phenol with the activation energy of $E_A = 50 \times 10^3$ kJ/kmol. Phenol conversion can reach more than 99.9% and COD reduction more than 90% [8].

Reaction time needed for the reduction of the COD by about 60% is in the range of $t = 1.5$–2.5 h. Temperature and oxygen partial pressure are of strong influence on COD reduction, increasing the conversion with increasing values. The two steps of the oxidation process are mass transfer of oxygen from the gas phase to the liquid phase and the reaction between the dissolved oxygen and phenol. Noncatalyzed phenol oxidation shows an induction period followed by a fast reaction phase, as is typical for free-radical reactions. The induction period is inversely proportional to the dissolved oxygen concentration. After the formation of sufficient free radicals, the oxidation of phenol becomes fast. Similar results are found for catalyzed WAO [8].

10.2.2.2 WAO of Refinery Spent Caustic

WAO can be applied for precleaning refinery spent caustic from treatment of refinery products [9]. For final treatment, the effluent from WAO is cleaned in a biological water purification system. WAO removes sulfides, mercaptans, and thiosulfates, and reduces significantly phenols and COD.

During WAO, carboxylic acids and other partially oxidized short-chain organics are formed, in addition to the end products CO_2 and H_2O. About 5–25% of the total organic content (TOC) from the feed remains as acetic acid and other by-products. Carboxylic acids, like acetic acid, are readily degradable in conventional biological water purification systems. The elements nitrogen, phosphorous, sulfur, and chlorine, entering the reactor bound to organic compounds, are reacted to NH_3, PO_4^{3-}, SO_4^{2-}, and Cl^-. The reactions occurring in the reactor can be summarized as [9]:

$$NaHS + 2O_2 \rightarrow NaHSO_4 + H_2O \tag{10.13}$$

$$NaSR + 2O_2 \rightarrow NaHSO_4 + CO_2 + RCOONa\,(\text{unbalanced}) \tag{10.14}$$

$$[\text{Naphthenics}] + O_2 \rightarrow CO_2 + RCOONa\,(\text{unbalanced}) \tag{10.15}$$

$$[\text{Phenols}] + O_2 \rightarrow CO_2 + RCOONa\,(\text{unbalanced}) \tag{10.16}$$

where R is typically CH_3.

In real refinery spent caustic, about 80% of the COD is reduced, and phenols and sulfides are reduced to below detection limits, as listed in Table 10.1.

10.2.2.3 WAO of Wastewater

WAO of wastewater from cotton desizing operations is an example for wastewater-containing products from the processing of a natural product. In Figure 10.6, the reduction of the COD is shown for a temperature range of $T = 150$–$290\ °C$ and for different partial pressures of oxygen from $P_{O_2} = 0.375$–2.25 MPa [10].

A strong influence of temperature and partial oxygen pressure on the reduction of the COD can be concluded from Figure 10.6. Less than 10 wt%

TABLE 10.1 Wet Air Oxidation of Refinery Spent Caustic, $T = 246\ °C$ [9]

	Reactor Inlet	Reactor Effluent
COD (mg/dm^3)	72,000	15,000
COD reduction	–	79.2%
BOD/COD	–	0.515
Phenols (mg/dm^3)	1700	3
Sulfide: S (mg/dm^3)	2700	<1
Mercaptans: CH_3SH (mg/dm^3)	2800	2
Thiosulfate: S_2O_3 (mg/dm^3)	640	<26
pH (-)	13.4	8.2

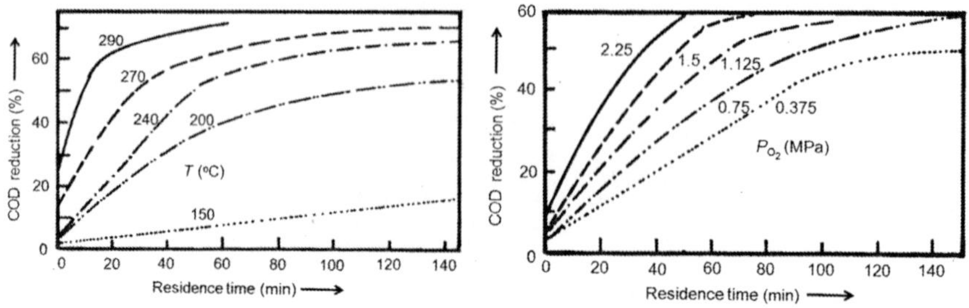

FIGURE 10.6 Wet air oxidation of cotton desizing wastewater. Left: Influence of temperature on COD reduction, $P_{O_2} = 1.5$ MPa (stoichiometric oxygen demand). Right: Influence of oxygen partial pressure on the COD reduction, $T = 240$ °C. *Modified from Ref. [10].*

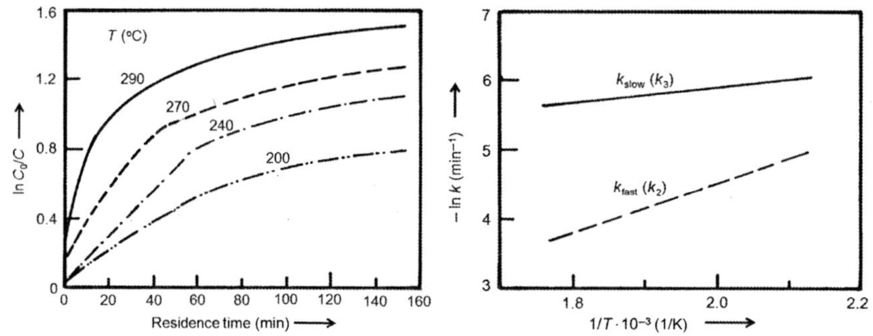

FIGURE 10.7 Wet air oxidation of cotton desizing wastewater. Left: Influence of temperature on COD reduction, $P_{O_2} = 1.5$ MPa (as shown in Figure 10.6). Right: Influence of temperature on the rate constants. *Modified from Ref. [10].*

reduction of COD has occurred after $t = 1$ h treatment, with the stoichiometric amount of oxygen supplied by a partial oxygen pressure of $P_{O_2} = 1.5$ MPa, at a temperature of $T = 150$ °C, while at $T = 290$ °C more than 70 wt% of the COD is reduced. For the influence of the oxygen partial pressure, shown in Figure 10.6 on the right-hand side, about 25 wt% of the COD is reduced at $T = 240$ °C, after $t = 1$ h, at an oxygen partial pressure of $P_{O_2} = 0.375$ MPa, corresponding to 25% of the stoichiometric oxygen demand, while at an oxygen partial pressure of $P_{O_2} = 2.25$ MPa, corresponding to 1.5 times the stoichiometric oxygen demand, about 60 wt% of COD are removed.

WAO of the wastewater proceeds in two steps. A fast initial reaction of large molecules that decompose into intermediate products is followed by a slow reaction that further oxidizes the intermediate products into end products of low-molecular-weight organic acids, carbon dioxide, and water [10]. This is illustrated in Figure 10.7, where the data for the temperature dependence for the WAO of cotton desizing wastewater are plotted and approximated by two distinct reaction regimes of constant slope. The resulting rate constants are shown in Figure 10.7 on the right-hand side, with an activation energy for the fast

TABLE 10.2 Kinetic Parameters of Wet Air Oxidation of Cotton Desizing Wastewater

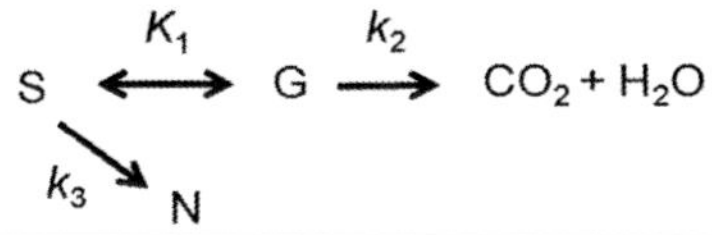

T	°C	150	200	240	270	290
K_1	–	0.0231	0.0667	0.0758	0.0807	0.161
k_2	min^{-1}	0.0428	0.131	0.171	0.329	0.574
k_3	min^{-1}	2.411×10^{-4}	9.85×10^{-3}	8.21×10^{-3}	0.0154	0.0428

S, initial organic components; G, intermediate equilibrium product [10].

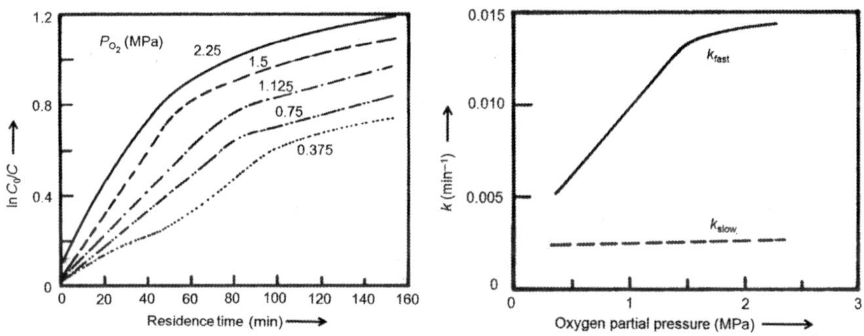

FIGURE 10.8 Wet air oxidation of cotton desizing wastewater. Left: Influence of oxygen partial pressure on COD reduction, $T = 240$ °C (as shown in Figure 10.6). Right: Influence of temperature on the rate constants. *Modified from Ref. [10].*

reaction of $E_{fast} = 30$ kJ/mol, and for the slow reaction of $E_{slow} = 9$ kJ/mol [10] (Table 10.2).

In a similar way, the dependence of COD on oxygen partial pressure is shown in Figure 10.8 [10].

The slow reaction is independent of oxygen partial pressure. The fast reaction strongly depends on the oxygen supply when it is less than the theoretical oxygen requirement of $P = 1.5$ MPa. With excess oxygen, the fast reaction becomes also independent of oxygen partial pressure [10].

WAO can be carried out with catalysts and additional promoters. In Figure 10.9, the effect of various catalysts on the reduction of the COD, and in Figure 10.10, the effect of H_2O_2 as oxidation agent are shown for wastewater from dyeing and printing processes.

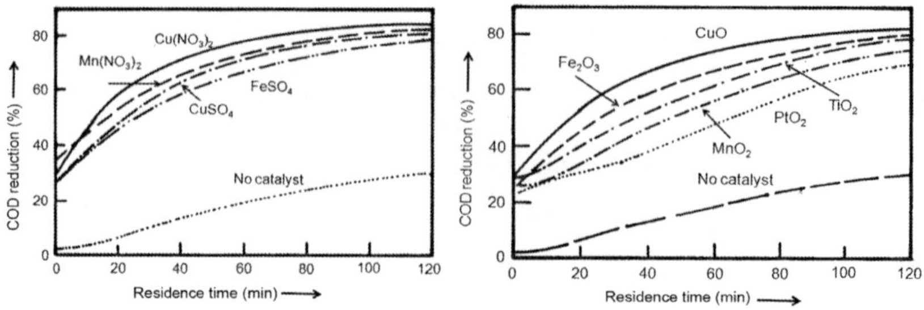

FIGURE 10.9 Wet air oxidation of dyeing and printing wastewater, $T = 200\,°C$, $P_{O_2} = 2.65$ MPa. Influence of various catalysts. Left: Metal salts. Right: metal oxides. *Modified from Ref. [10].*

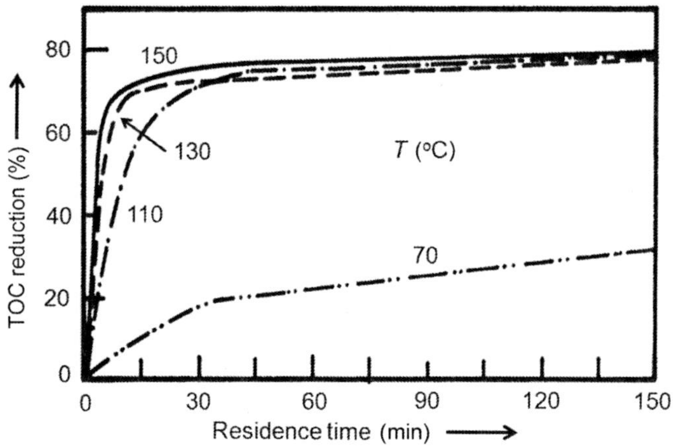

FIGURE 10.10 Wet air oxidation of dyeing and printing wastewater with H_2O_2 as oxidizing agent. *Modified from Ref. [10].*

WAO is promoted substantially by catalysts. For metal salts, the effectiveness is $Cu(NO_3)_2 > CuSO_4 > Mn(NO)_2 > FeSO_4$, and for metal oxides, the effectiveness is $CuO > Fe_2O_3 > TiO_2 > MnO_2 > PtO_2$ [10]. With hydrogen peroxide, the reaction is very fast, but the disadvantage is the higher price for H_2O_2 than that for air or oxygen.

10.2.2.4 WAO of Lignin

WAO is a good method for the destruction of complex mixtures of stable compounds. Degrading lignin with the goal of enhancing biodegradability in a stirred batch reactor brings favorable results. At a temperature of $T = 110\,°C$ about 75% and at $T = 190\,°C$, 100% of the lignin is degraded at a reaction time of $t = 2$ h with oxygen partial pressures of $P = 1$–1.5 MPa and a pH 12, as shown in Figure 10.11. Most of the reactions occur within the first

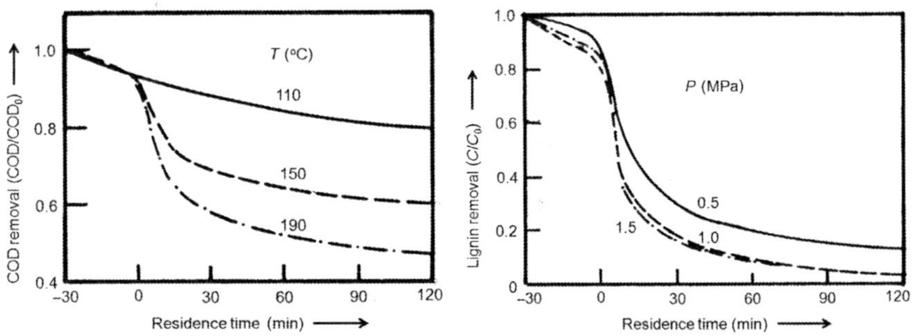

FIGURE 10.11 Wet air oxidation of alkali lignin, reaction time $t=2$ h, $P=1$ MPa, pH 12. Left: Influence of temperature on COD removal, $P_{O_2}=0.5$MPa. Right: Influence of oxygen partial pressure on lignin conversion, $T=150$ °C. *Modified from Ref. [11].*

30–60 min [11]. In general, the effect of temperature is substantial, that of oxygen partial pressure moderate, while increasing the pH value from 9 to 12 considerably improves the conversion efficiency [11].

10.2.2.5 WAO of Animal By-Products

WAO can be an alternative way of disposal of animal by-products by oxidative degradation of the initial total organic carbon (TOC). In a stirred batch reactor, with an adapted titan liner, meat and bone meal (MBM), obtained after fat rendering, crushed raw defatted by-products, and crushed nondefatted raw by-products were treated at $T=230$–280 °C, $P=9.9$–18.6 MPa at an initial TOC concentration of $x_F=5$ g/m^3 with air as oxidant, at a large excess of air, applied by continuous flow, to maintain the oxygen saturation in the aqueous phase [12]. The oxidation of MBM and two different raw animal by-products (before and after rendering) leads to a reduction of the TOC of up to 90 wt% and reaches 99 wt% with platinum as a catalyst for the oxidation of a raw nondefatted sample.

TOC removal increases at higher temperature and higher pressure due to favoring the mass transfer of oxygen and dissolution of organic compounds into the liquid phase. The major organic by-product of the oxidation is acetic acid, representing approximately 50 wt% of the final residual TOC. The major inorganic by-product is ammonia, which is quantitatively removed (98.9 wt%) in the presence of a platinum catalyst on alumina in the basic medium.

WAO leads to a reduction of the total organic matter (TOC) of 85.6% for MBM, 70.7% for raw defatted by-products, and 77.1% for raw nondefatted by-products. A reduction of solid matter of 72.0% is obtained for MBM, of 91.7% for raw defatted by-products, and of 90.5% for raw nondefatted animal by-products [12].

The oxidation is enhanced by pressure, particularly during the initial period of the reaction, as illustrated in Figure 10.12. At the beginning of the reaction, the injection of air induces rapid oxidation of organic compounds

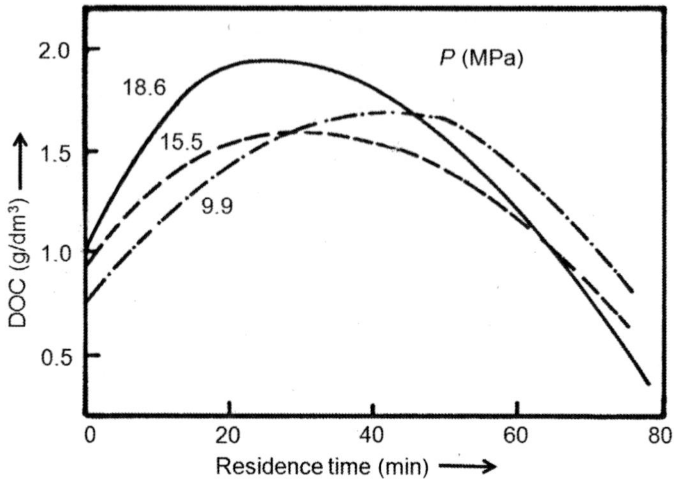

FIGURE 10.12 Wet air oxidation of a raw nondefatted animal by-products, $T=280\,°C$, initial TOC $=5$ g/dm^3. *Modified from Ref. [12].*

into smaller molecules that dissolve in the aqueous phase and increase the amount of dissolved organic compounds (DOC) during the first 30–50 min. After the organic compounds are completely dissolved, oxidation dominates and produces mainly CO_2. Consequently, the DOC in the reaction mixture decreases during the second period [12].

The most stable compounds are ammonia, representing almost 100 wt% of the initial organic nitrogen, and acetic acid, representing about 50 wt% of the residual TOC. In a second step of the WAO, a platinum catalyst on an alumina support leads to higher conversion yields of total organic matter (TOC) of 99 wt% and of ammonia of also 99 wt%, Figure 10.13. In conclusion, raw animal by-products can be treated directly at the slaughterhouse level in such a two-step process [12].

10.2.2.6 WAO of Polyether Solutions

WAO proved to be an effective method to treat nonionic surfactants like polyether [13]. Temperature is of major influence on the COD removal: at a temperature of $T=160\,°C$ 33.5 wt%, at $T=200\,°C$ 57.2 wt%, and at $T=240\,°C$ 94.4 wt% of COD is removed in $t=2$ h. The major intermediate compounds are fatty acids. Oxidation of acetic acid is the rate-determining step during oxidation. Oxidation of fatty acids is enhanced at higher temperatures, and their conversion can be increased by longer reaction times. The concentration of polyether in the range of $x_F=500$–8000 mg/l is of minor influence on the COD removal. Figure 10.14 shows the temperature dependence of the COD removal and Figure 10.15 the concentration of fatty acids and the outlet concentration of the different fractions after $t=2$ h of WAO of polyether solutions [13].

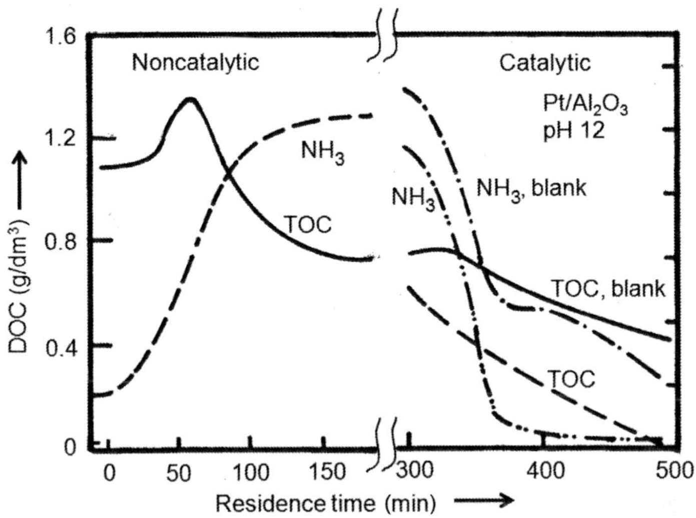

FIGURE 10.13 Two-step wet air oxidation of a raw nondefatted animal by-product, $T=280\ °C$, $P=18$ MPa, initial TOC $=5$ g/dm^3. First step (left) without catalyst and second step (right) with platinum catalyst (5 wt%) on Al_2O_3, pH 12. *Modified from Ref. [12].*

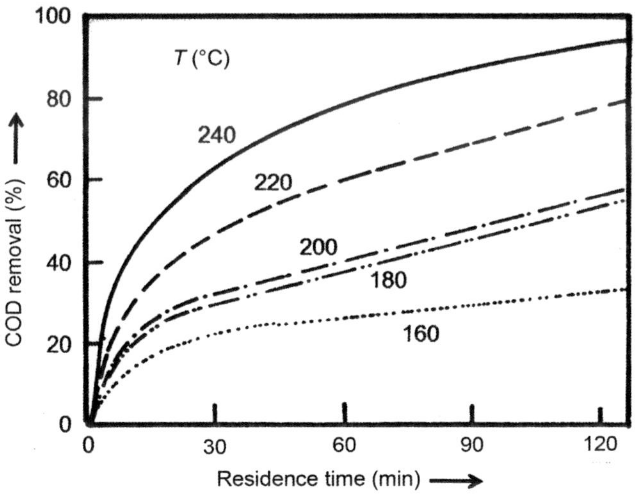

FIGURE 10.14 Wet air oxidation of polyether solutions. Temperature dependence of COD removal. Polyether: $x_F=2$ g/dm^3, $P_{O_2}=0.40$ MPa at $T=25\ °C$. *Modified from Ref. [13].*

10.2.2.7 Chemical Demilitarization

Application of WAO to destroy chemical agents was considered as unsafe because of the combination of high operating pressure and the highly toxic nature of agents [14]. But WAO is considered to be a strong candidate for

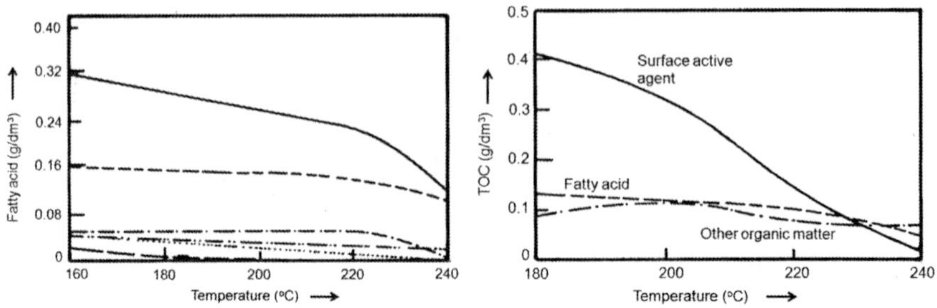

FIGURE 10.15 Wet air oxidation of polyether solutions. Polyether: $x_F = 2000$ mg/l, $P_{O_2} = 0.40$ MPa at $T = 25$ °C, reaction time $t = 2$ h. Left: Fatty acid concentration at the outlet: ———, total; — — —, acetic acid; —•—, butyric acid; —••—, metacetonic acid; ••••, capronic acid; ———, isopropyl acetic acid. Right: Concentration of organic compounds at the outlet. *Modified from Ref. [13].*

treatment of neutralents and hydrolysates since the toxicity of agents is significantly reduced by neutralization. "Neutralent" refers to compounds produced by the destruction of either special chemical agents, including binary chemical agents, or industrial chemicals. According to the test results, WAO can detoxify and demilitarize agents. The remaining organic compounds require further treatment. Pollutants such as NO_x, dioxins, and particulates are not formed at the temperatures used for WAO. Some nitrogen may show up as nitrous oxide or ammonia [14]. Corrosion occurs in WAO reactors, particularly for mixtures containing chlorine, fluorine, sulfur, and phosphorus, see Chapter 12. Results from bench-scale testing of WAO to treat neutralents and binary chemical agents are listed in Table 10.3 [14].

WAO operates at temperatures and pressures below the critical values for water. The process, according to the test results, can be applied to destroy dangerous compounds and to release mainly liquid effluents that can be treated readily with biological methods. WAO is applied to high-diluted aqueous mixtures. It requires reaction times in the range of $t = 1$–3 h or more, for total destruction of components. Consequently, large reaction volumes are needed. WAO is a process that is applied in many plants specifically designed for waste streams that cannot be handled by biological processes or incineration.

10.3 SUPERCRITICAL WATER OXIDATION

10.3.1 Principles and Features of SCWO

Oxidation in supercritical water, known as SCWO, is applied to organic waste streams at temperatures above the critical temperature of water ($T = 374$ °C) and above the critical pressure of water ($P = 22$ MPa). Conditions of temperature and pressure for SCWO are above that of WAO, as is discussed in this chapter, but below that of oxidation in hydrothermal flames, discussed after SCWO in this chapter. The intention of SCWO is to take oxygen and to

TABLE 10.3 Wet air Oxidation of Some Poisonous Agents and Pesticides [14]

Component	Temperature (°C)	Reaction Time (min)	Initial COD	Residual TOC or COD	TOC or COD Reduction (%)	Destruction Efficiency (%)
Mustard agent/monoethanolamine	320	60	Dilute	153	97.7	>99.997
Sarin/monoethanolamine	320	180	Dilute	25	99.6	>98
O-Ethyl-2-diisopropylaminoethyl methyl phosphonate	320	180	Dilute	11	99.8	>98
Methylphosphonyl difluoride	320	360	Dilute	14	99.6	>99.9
Glyphosate	280	60	27,500	6,900	74.9	99.8
Dimethyl methylphosphonate	280	60	13,596	<500	6.9	>97.5
Dinoseb	281	182	–	–	95.3	99.5
Methoxychlor	281	182	–	–	95.3	99.8
Carbaryl	281	182	–	–	95.3	98
Malathion	281	182	–	–	95.3	99.9

destroy in an aqueous stream the organic components and oxidize the organic carbon to CO_2. With SCWO, toxic and high-risk wastes (chemical weapons) can be transformed into compounds that can be safely released to the environment [15].

SCWO has many advantages compared to incineration and WAO. SCWO can be carried out in a totally enclosed treatment plant. The rather high temperature favors reactions that lead to small, thermodynamically stable compounds. The complete miscibility of organic compounds and gases with supercritical water avoids mass transfer limitations and insufficient conversion due to incomplete mixing. No unintended emission of components occurs, unlike with incineration and WAO. SCWO can be carried out in a one-step process to meet standards for gaseous and aqueous effluents [16]. The reaction rates of SCWO are much faster than for WAO, resulting in much smaller reactors and considerably reduced risk potentials. Although costs of the construction materials for the reactor may be higher, the smaller reactor volume reduces the necessary amount of construction material considerably. Economic analyses lead to the conclusion that SCWO has lower costs, based on the amount of destructed organic compounds, than competing technologies, in particular for aqueous waste streams containing $x_F > 10$ wt% organic compounds.

During SCWO, organic compounds are completely oxidized to CO_2 and H_2O. Heteroatoms, such as chlorine, sulfur, or phosphorus, if present in the organic waste stream, are transformed into the mineral acids HCl, H_2SO_4, or H_3PO_3. Nitrogen contained in the feed components mainly forms N_2 and small amounts of N_2O. Dioxins and NO_x are usually not formed [16]. Salts from the neutralization of the acids can be precipitated from the solution, which is useful for purifying the reactor effluent stream [17]. The basic process, and therefore the basic flow sheet of SCWO, is similar to that of WAO, as shown in Figure 10.4. Different are the processing conditions that usually are beyond the critical values for water and the use of oxygen, or for experimental investigations, of hydrogen peroxide as oxidants.

The basic process of SCWO is relatively simple. The waste stream is mixed with water, the reaction mixture is heated to reaction temperature, oxygen is added, and the reaction mixture enters the supercritical water reactor. After a sufficient residence time in the reactor to obtain the necessary destruction efficiency, the reaction mixture leaves the reactor, temperature and pressure are reduced to ambient conditions, the effluent is separated into a liquid and gaseous phase, and the effluent streams are discharged or further processed. For an economic operation, it is necessary to recover heat from the effluent streams and transfer it to the incoming waste stream.

Destruction of waste with supercritical water is carried out over a wide range of temperatures, about $T = 375$–650 °C, at supercritical pressures. The oxidation reactions in water are exothermic reactions ($\Delta h < 0$). Reaction equilibrium basically is shifted to initial reactant compounds with increasing temperature. Yet for the oxidation reactions and temperatures involved in waste destruction, the initial reactants are practically nonexistent at equilibrium. Reactions proceed under an increasing number of moles. High pressures, necessary for maintaining the supercritical environment, therefore favor the initial reactants. But, because reactions proceed in a dilute aqueous system, influence of pressure is not dominating the reaction equilibrium. On the other hand, influence of temperature on reaction rate is important. According to experimental data, the reaction rate can be enhanced and the conversion efficiency raised by a factor of 10 by increasing the temperature by $\Delta T = 100\,\text{K}$. Destruction of chemical compounds in these oxidation reactions can be modeled globally by a first-order reaction and experimental data can be represented sufficiently well [18].

Some examples for global oxidation reactions of waste compounds as they occur under SCWO conditions in the reactor are:

$$\text{Cellulose}\ \ C_6H_{10}O_5 + 6O_2 \rightarrow 6CO_2 + 5H_2O \tag{10.17}$$

$$\text{Dioxin(PCDD)}\ \ Cl_2{-}C_6H_2{-}O_2{-}C_6H_2{-}Cl_2 + 11O_2 \rightarrow 12CO_2 + 4HCl \tag{10.18}$$

$$\text{TNT}\ \ CH_3{-}C_6H_2{-}(NO_2)_3 + 5.25O_2 \rightarrow 7CO_2 + 2.5H_2O + 1.5N_2 \tag{10.19}$$

$$\text{Nerve agent HD}\ \ Cl{-}C_2H_4{-}S{-}C_2H_4{-}Cl + 7O_2 \rightarrow 4CO_2 + 2H_2O + 2HCl + H_2SO_4 \tag{10.20}$$

There has been great interest to investigate the possible applications of SCWO, in particular for toxic and hazardous compounds. Among the compounds and waste streams investigated are pulp and paper mill sludges [19], process wastewaters [20,21], hazardous organics in waste streams [22–24], sludges from water treatment plants [25], waste from manned space missions [26,27], biological wastes from the pharmaceutical industry [28], mixed wastes [29], military wastes [30–32], and others. The examples are cited from Ref. [33], with the exception of Ref. [27].

In the following, some investigations are discussed that stand for many others. At the end of this chapter, reviews are listed that offer an entrance for the interested reader into the vast literature on supercritical water for destruction of compounds.

In SCWO, the processing time for complete conversion is usually seconds or minutes. Only for stable compounds, such as ammonia and acids, longer processing times are needed. At supercritical conditions, ionic reactions are suppressed and radical reactions are favored due to the small values of the dielectric constant and the ion product for water [34].

For global reaction rates in SCWO, in most cases, the following equation, based on the global rate law, has been used

$$\frac{d[S]}{dt} = -k[S]^a[O_2]^b \tag{10.21}$$

with [S] for the concentration of the compound to be oxidized. The rate constant k follows an Arrhenius-type temperature dependence

$$k = A\exp(-E_a/(RT)) \tag{10.22}$$

with A is the preexponential factor, E_a is an activation energy, R is the gas constant, and T is the temperature.

Values for selected reaction orders and kinetic parameters are listed in Table 10.4 [34]. The reaction order for most of the components that are oxidized is close to unity. For oxygen, reaction orders in SCWO vary between 0 and 0.5, different to the value of about 1 for WAO.

10.3.2 Examples: Specific Compounds

10.3.2.1 Carbon Monoxide and Hydrogen

Carbon monoxide and hydrogen have been investigated in SCWO because they are the simplest of the reactive compounds, and their oxidation has already been extensively investigated in the gaseous phase. Carbon monoxide is an important intermediate in the oxidation of carbon compounds to carbon dioxide, which includes two global reaction pathways: (i) the direct oxidation (Equation 10.23) and the water-gas shift reaction (Equation 10.24).

$$CO + \frac{1}{2}O_2 \rightarrow CO_2 \tag{10.23}$$

$$CO + H_2O \rightarrow CO_2 + H_2 \tag{10.24}$$

According to Holgate *et al.* [35], the kinetic pathways can be considered separately. The rate of the direct oxidation and its relative contribution rises

TABLE 10.4 Reaction Orders and Kinetic Parameters of Supercritical Water Oxidation (SCWO) [34]

	Reaction Order			
Compound	*Hydrocarbon*	*Oxygen*	Log *A*	Activation Energy (kJ/mol)
Hydrogen	1.10	0.02	24.4	390
Carbon monoxide	0.96	0.34	8.5	134
Methane	0.99	0.66	11.4	42.8
Methanol	0.89	0.12	28.8	107
Acetic acid	0.72	0.27	9.9	168
Phenol	0.85–1.041	0.38–0.50	2.34–8.69	52–125

more rapidly with temperature than that of the water-gas shift reaction, as shown in Figure 10.16 [35].

The oxidation of hydrogen is of interest since hydrogen is a possible intermediate product and can be used as additional fuel in SCWO processes. For hydrogen oxidation in supercritical water, the reaction rate is proportional to the hydrogen concentration and independent of the oxygen concentration [36]. Experimental results are shown in Figure 10.17.

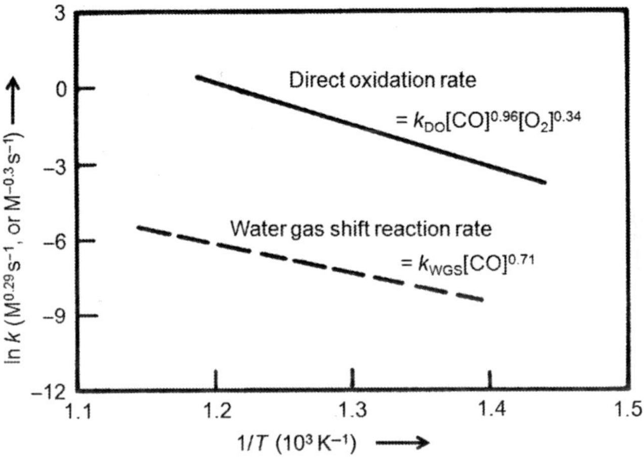

FIGURE 10.16 Arrhenius plot for the direct oxidation and the water-gas shift reaction of carbon monoxide in supercritical water. *Modified from Ref. [35].*

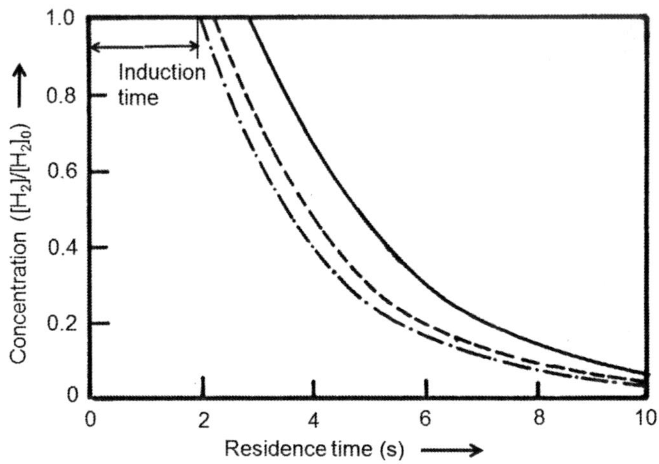

FIGURE 10.17 Oxidation of hydrogen in supercritical water. $T = 550\,°C$, $[H_2O]_0 = 4.25 \times 10^{-3}$ mol/cm^3. ———: $[H_2]_0 = 3.06 \times 10^{-6}$, $[O_2]_0 = 1.55 \times 10^{-6}$ mol/cm^3; – – –: $[H_2]_0 = 2.06 \times 10^{-6}$, $[O_2]_0 = 1.04 \times 10^{-6}$ mol/cm^3; —•—: $[H_2]_0 = 1.065 \times 10^{-6}$, $[O_2]_0 = 0.54 \times 10^{-6}$ mol/cm^3. *Modified from Ref. [36].*

10.3.2.2 Methane

Methane oxidation in supercritical water is first order in methane and of order 0.66 in oxygen, Equation (10.25) [37], see Figure 10.18. The rate constants k' is

$$k' = \frac{-\ln(1-X)}{t(1-X/R)^{0.66}} \tag{10.25}$$

with X as methane conversion and t as the reactor residence time.

The influence of density, determined at a somewhat lower temperature of $T = 400\ °C$ and a ratio of O_2/CH_4 of 0.03, is shown in Figure 10.19 [38]. Methane conversion, as determined in a flow reactor, is much slower, and methane conversion is only in the range of 3% [38].

Catalytic partial oxidation of methane in supercritical water with a Cr_2O_3 catalyst leads to the formation of methanol, as shown in Figures 10.20 and 10.21. Water in high concentration increases the yield of methanol but inhibits

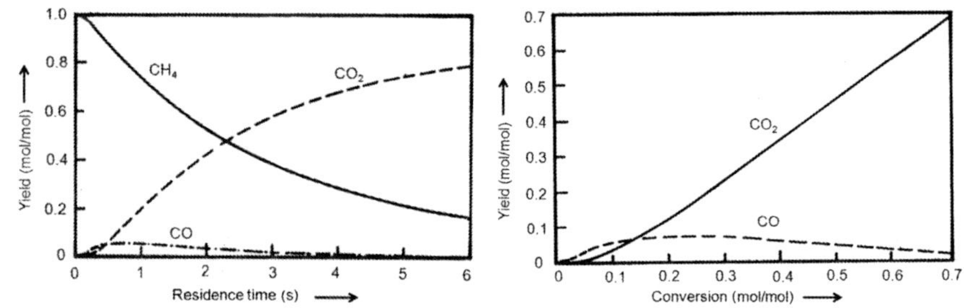

FIGURE 10.18 Oxidation of methane in supercritical water. $T = 586\ °C$, $[CH_4]_0 = 0.7\ mmol/dm^3$, $[O_2]_0 = 5.9\ mmol/dm^3$, $[H_2O]_0 = 4.09\ mol/dm^3$. Left: Product concentration versus time. Right: Product concentration versus conversion. *Modified from Ref. [37].*

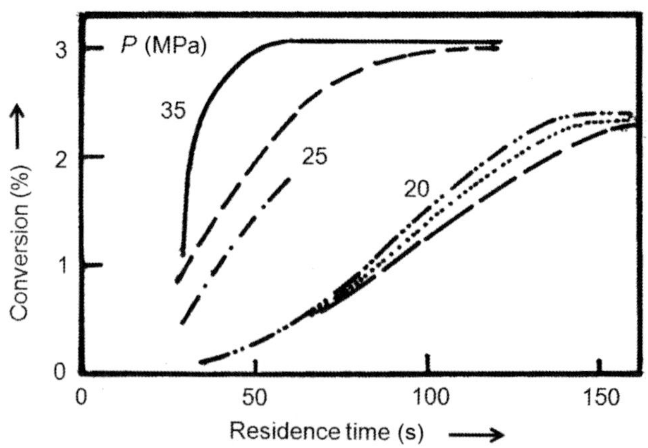

FIGURE 10.19 Oxidation of methane in supercritical water. Influence of density. $T = 400\ °C$, $O_2/CH_4 = 0.03$, ——— 35 MPa, – – – 30 MPa, —•— 25 MPa. *Modified from Ref. [38].*

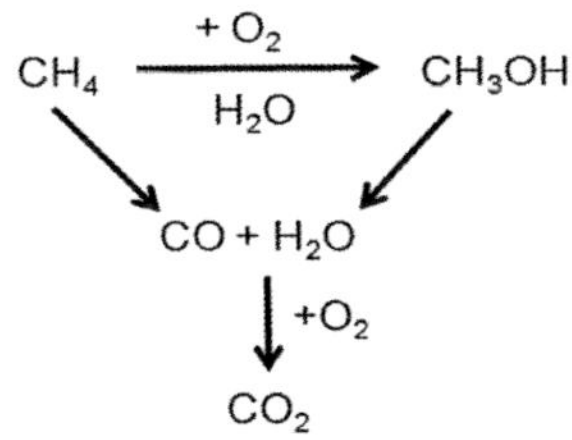

FIGURE 10.20 Reaction pathway for the formation of methanol in partial oxidation of methane [39].

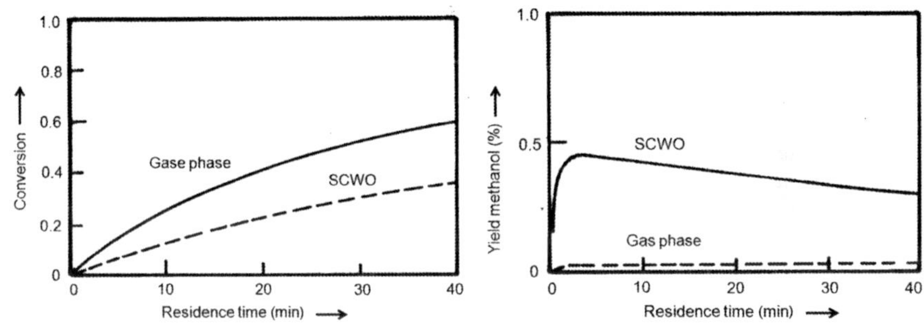

FIGURE 10.21 Catalytic partial oxidation of methane (catalyst Cr_2O_3). $T=450\ °C$, $P=33$ MPa. *Modified from Ref. [39].*

methane conversion. High oxygen concentration reduces the yield of methanol [39]. Compared to reaction in the gaseous phase, the degree of conversion of methane in supercritical water is lower, but the selectivity is substantially higher [39]. Nevertheless, the overall yield of methanol is relatively low.

At high temperatures and high methane (or methanol) concentrations, the oxidation reaction is identical to an incineration in an aqueous medium. In the so-called hydrothermal flames, first demonstrated by Franck and coworker [40], the reaction time is further reduced compared to that for SCWO. The flames ignite at methane or methanol concentrations as low as 6 mol% at temperatures near $T=500\ °C$. The ignition-threshold concentrations rise as temperature is decreased to $T=400\ °C$ [41]. Hydrothermal flames are discussed later in this chapter.

10.3.2.3 Propane

Oxidation of propane in supercritical water leads to the formation of stable products like methanol, acetic acid, or acetone. Total selectivity for these products reaches up to 15% in the conversion range of 30–100% [42]. Transition metal oxides as catalysts are found to be physically stable but undergo severe structural changes and do not influence the overall conversion rates and product distribution, compare Figure 10.22. Transition to the supercritical region is

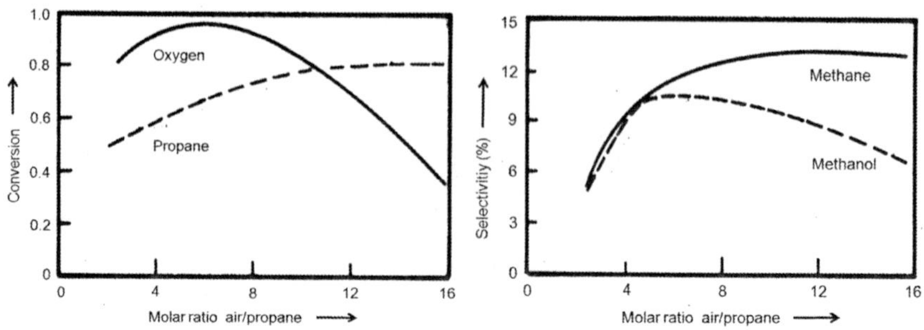

FIGURE 10.22 Oxidation of propane in supercritical water. $T=673$ K, $P=24$ MPa, residence time $t=0.4$ s. Left: Conversion of propane and of oxygen. Right: Selectivity for methanol and for methane. Oxides MnO_2/Al_2O_3, Co_2O_3/Al_2O_3, $MnO_2–Co_2O_3/Al_2O_3$, MoO_3/Al_2O_3 are of negligible influence. *Modified from Ref. [42].*

of significant influence on propane and oxygen conversion rates due to the homogenization of the reaction mixture [42]. Conversion of propane in supercritical water at a temperature of $T=673$ K and a residence time of $t=7.5$ s in a mixture of propane:air:water = 1:2.5:250 reaches 80% at a pressure of $P=20$ MPa. Conversion of propane and selectivity for methane and methanol are shown in Figure 10.22 [42].

10.3.2.4 Nitrogen

The role of nitrogen in SCWO can be derived from an investigation on the conversion of urea [43]. Urea is unstable and readily decomposes to carbon dioxide and ammonia at high temperatures (Equation 10.26).

$$CO(NH_2)_2 + H_2O \rightarrow CO_2 + 2NH_3 \tag{10.26}$$

Regardless of the oxidation state of nitrogen in the starting material, the predominant effluent species of nitrogen is N_2 at a temperature of $T=600$ °C, a pressure of $P=23$ MPa, and a partial pressure of oxygen of $P_{O_2}=2$ MPa [43]. Since high temperatures are required for efficient ammonia oxidation, ethanol was added to the feed to enable ammonia destruction at lower temperatures in the presence of the more easily oxidized ethanol [43]. Figure 10.23 shows the results for the ammonia destruction. Conversion of ammonia together with ethanol leads to high conversion values of up to 100% at high temperatures ($T=690$ K). At those high temperatures, conversion of ammonia to N_2 is favored. Conversion of ammonia together with ethanol at lower temperatures leads to formation of N_2O. Residence time is of insignificant influence [43]. In Figure 10.24, the concentration of NO and NO_x from SCWO and from incineration is compared. Levels of NO and NO_x are much lower for SCWO than that from incineration. It can be assumed that in the typical residence times for SCWO of some 10 s, no detectable quantities of NO_x are formed [43].

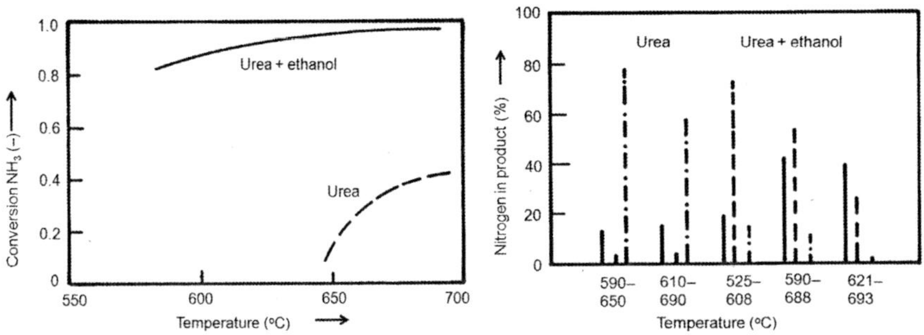

FIGURE 10.23 Oxidation of urea in supercritical water. Left: Conversion, $T = 600\,°C$, $P = 23$ MPa, $P_{O_2} = 2$ MPa. ——— urea + ethanol, – – – urea. Right: distribution of nitrogen products: ——— nitrogen, – – – nitrogen oxide (N_2O), —•— ammonia (NH_3). *Modified from Ref. [43].*

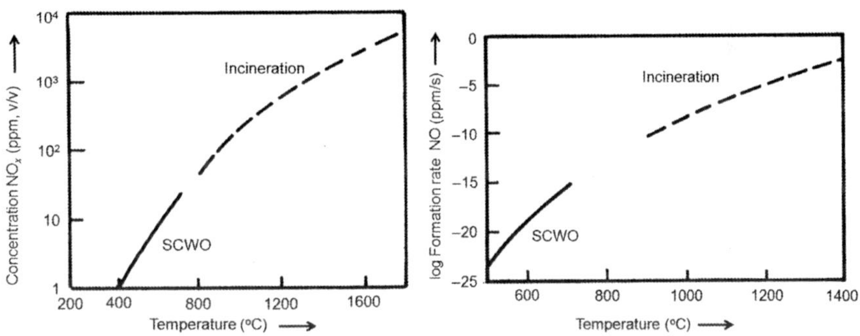

FIGURE 10.24 Oxidation of urea in supercritical water. Comparison of NO_x equilibrium levels (left) and NO formation rates (right) of supercritical water oxidation ($P = 23$ MPa, $P_{O_2} = 2$ MPa), and incineration. ——— Supercritical water oxidation, – – – Incineration. *Modified from Ref. [43].*

10.3.2.5 Methanol

The reaction rate of the oxidation of methanol in supercritical water is proportional to the methanol concentration and independent of the oxygen concentration [44]. Methanol is oxidized consecutively to carbon dioxide via formaldehyde and carbon monoxide [45]. The conversion of methanol can be described with pseudo-first-order reactions for each step, as shown in Figure 10.25. The reaction has an induction period, which decreases with increasing temperature [45].

The apparent time lag in the beginning of the reaction is not due to a physical reason such as mixing time but is due to the chemical kinetics in the radical-chain reaction as can be concluded from *in situ* Raman spectroscopic analysis [46].

A review of kinetic data for the oxidation of methanol in supercritical water [47] showed that variations between the results from the different

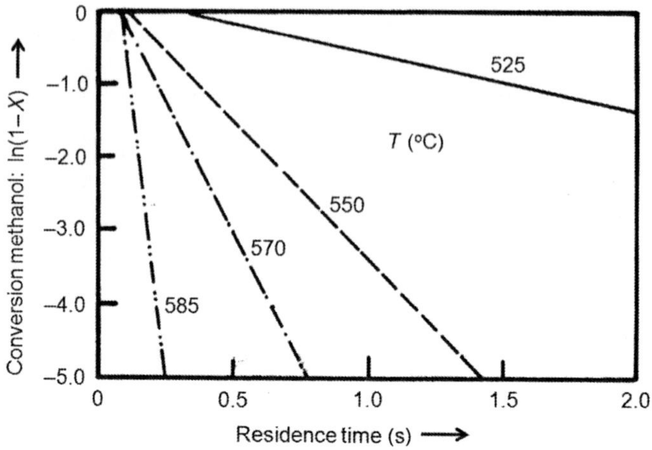

FIGURE 10.25 Oxidation of methanol in supercritical water. Lines represent methanol disappearance $\ln(1-X) = -k(t-t_{ind})$, ——— $T = 525$ °C, – – – $T = 550$ °C, —•— $T = 570$ °C, —••— $T = 585$ °C. *Modified from Ref. [45].*

research groups are due in part to differences in feed preheating and mixing and residual oxygen dissolved in the organic/water feed stream. High initial feed concentrations yield higher apparent first-order rate constants, by producing hot zones in the reactor and also by decreasing the induction time. Experimental data can be affected by reactor wall catalysis. Although consistent apparent first-order kinetics for the postinduction can be derived, recommendation on kinetic data for SCWO of methanol is difficult because of limited information on induction time and insufficient knowledge on the influence of wall catalysis. Simple power law rate expressions are not adequate to describe the complex nature of autocatalytic oxidations over the full range of temperatures, feed concentrations, and residence times studied but can be successfully applied to datasets obtained at similar experimental conditions [18]. First-order Arrhenius parameter values are listed for three groups of kinetic data in Table 10.5.

Nickel metal acts as catalyst in the conversion of methanol in supercritical water [48]. Conversions up to 20% are reached after $t = 2$ h without nickel, while in the presence of a Ni wire, conversions up to 90% are obtained in less than $t = 5$ min, with hydrogen, carbon monoxide, and carbon dioxide as the major products. Deactivation of the catalytic activity occurs with multiple uses of the Ni wire [48].

The presence of methanol during the oxidation of methane/methanol mixtures in supercritical water accelerates the rate of methane disappearance at $T = 540$ °C and $P = 27.3$ MPa and leads to higher methane conversions of up to 50% at a residence time of $t = 1.3$–1.4 s and 13 times the methane concentration in the feed [49]. An example of the conversion of the methane–methanol mixture in supercritical water is shown in Figure 10.26.

TABLE 10.5 First-Order Arrhenius Parameter Values for Different Kinetic Data

Data Source	Initial Concentration $[MeOH]_0$ (mM)	Temperature (°C)	Preexponential Factor A (s^{-1})	Activation Energy E_a (kJ/mol)
High concentration CSTR	33–277	390–475	6.785×10^{11}	167
Low concentration CSTR	0.38–2.4	440–589	1.626×10^{13}	199
Low concentration PFR	1–10	411–475	7.024×10^{15}	234

CSTR, continuous stirred tank reactor; PFR, plug-flow reactor [47].

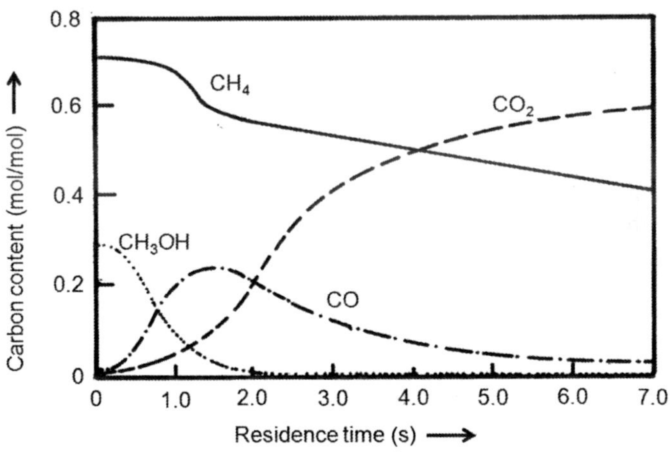

FIGURE 10.26 Oxidation of methane and methanol mixtures in supercritical water. $T = 546$ °C, $P = 24.7$ MPa. $[CH_4]_0 = 1.14$ mmol/dm^3, $[CH_3OH]_0 = 0.49$ mmol/dm^3, $[O_2]_0 = 5.53$ mmol/dm^3. ——— CH_4, •••• CH_3OH, —•— CO, – – – CO_2. *Modified from Ref. [49].*

10.3.2.6 Ethanol

Oxidation of ethanol in subcritical water at relatively low temperatures of $T = 170$–230 °C, $P = 23.5$ MPa shows strong dependence on temperature, see Figure 10.27. Residence time is in the range of $t = 180$–580 s at an initial concentration of ethanol of $x_F = 25$ mmol/dm^3 and a concentration of oxygen

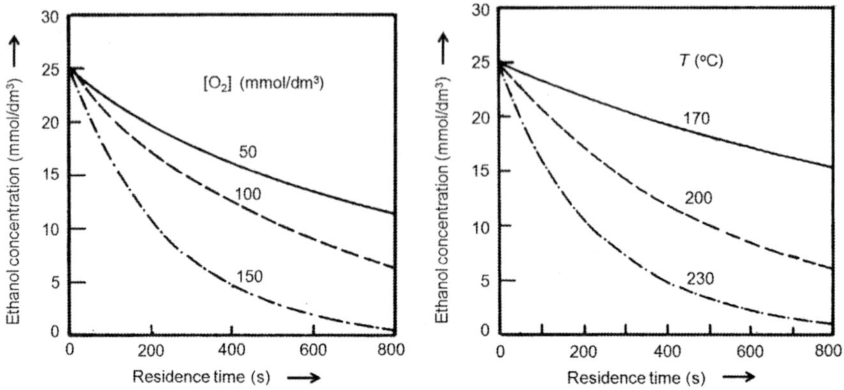

FIGURE 10.27 Ethanol oxidation in subcritical water. $P = 23.5$ MPa, initial ethanol concentration $x_F = 25$ mmol/dm^3. Left: $T = 200$ °C, oxygen concentration: ——— 50 mmol/dm^3, – – – 100 mmol/dm^3, —•— 150 mmol/dm^3. Right: Temperature: ——— 170 °C, – – – 200 °C, —•— 230 °C [50].

from $x_{O_2} = 50–150\,\text{mmol/dm}^3$. The reaction orders are $n = 0.86$ for ethanol and $n = 1.15$ for oxygen with an activation energy of $E_a = 61$ kJ/mol and a preexponential factor of $A = 102$, Equation (10.27) [50].

$$\frac{\mathrm{d[EtOH]}}{\mathrm{d}t} = A\exp\left(\frac{-E_a}{RT}\right)[\mathrm{EtOH}]^a[\mathrm{O_2}]^b \tag{10.27}$$

Residence time is reduced to $t = 2–12$ s for the oxidation of ethanol at supercritical conditions for water, $T = 433–494$ °C, $P = 24.6$ MPa, and an initial concentration of ethanol of $x_F = 1$ mmol/dm^3 [51]. Ethanol conversion is high, >95% with stoichiometric oxygen, compared to 7.4% at $T = 494$ °C in the absence of oxygen. Oxidation products in the liquid phase are acetaldehyde and formaldehyde, and small quantities of acetic acid. In the gaseous phase, carbon monoxide, carbon dioxide, methane, and ethylene are the products [51]. The oxidation reaction can be represented with parameters that are derived from all experimental data, a preexponential factor of $A = 1017$ and an activation energy of $E_A = 214$ kJ/mol [51]. Reaction products are shown in Figure 10.28 and kinetic data in Figure 10.29 [51].

Oxidation kinetics of mixtures of ethanol and methanol show some interference of both components. Methanol conversion is accelerated by ethanol addition, but ethanol oxidation is slightly retarded by methanol [52].

10.3.2.7 Phenol

Phenol is an ubiquitous compound in industrial waste. Several investigations provide information on kinetics of phenol oxidation in supercritical water. Unfortunately, the results from the various investigations differ to a great

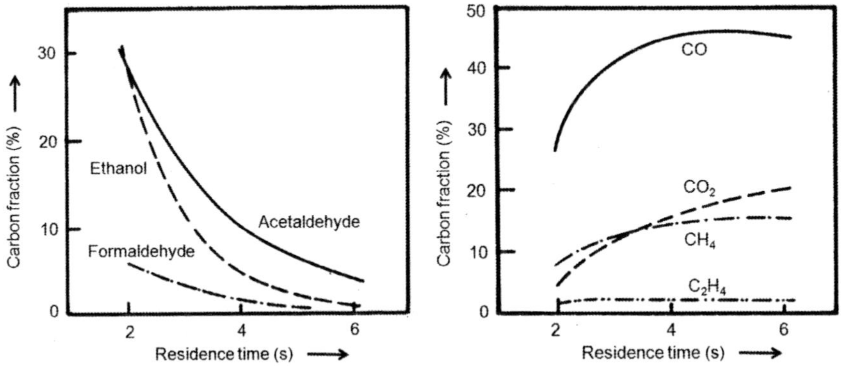

FIGURE 10.28 Products of the oxidation of ethanol in supercritical water. $T=490\,°C$, $P=24.6$ MPa, initial ethanol concentration: $x_F=1$ mmol/dm^3, stoichiometric oxygen. Left: liquid-phase products. Right: gas-phase products. *Modified from Ref. [51].*

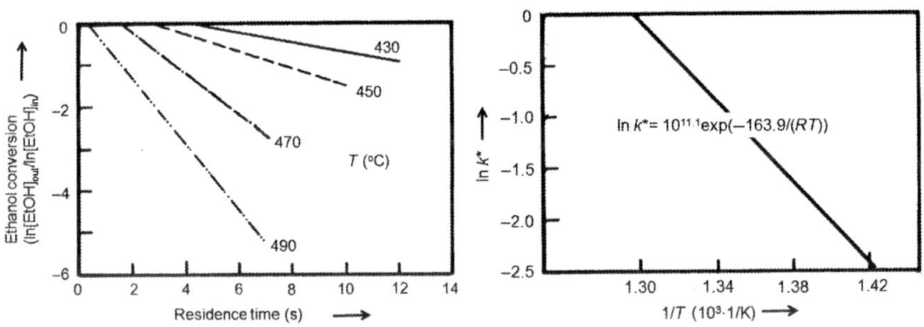

FIGURE 10.29 Kinetics of the oxidation of ethanol in supercritical water. $P=24.6$ MPa, initial ethanol concentration: $x_F=1$ mmol/dm^3, stoichiometric oxygen. Left: Conversion of ethanol. Right: Arrhenius plot for first-order oxidation of ethanol in supercritical water. *Modified from Ref. [51].*

extent. Therefore, first an experimental result illustrates the dependence of the reaction on process parameters, and then kinetics in accordance with experiments for high initial phenol concentration is discussed.

Effect of parameters on phenol oxidation in supercritical water as determined in an isothermal, plug-flow reactor at temperatures from $T=300$–$420\,°C$, pressures from $P=18.8$–27.8 MPa, and residence times from $t=4$–111 s are shown in Figures 10.30 and 10.31. Initial phenol concentrations range from $x_F=2.8\times10^{-4}$ to $x_F=5.3\times10^{-3}$ M, and oxygen concentrations are between $x_{O_2}=6.5\times10^{-5}$ and 6.4×10^{-2} M at reaction conditions [53].

Reaction products include mono- and dicarboxylic acids, dihydroxybenzenes, phenoxyphenols, and dibenzofuran, indicating that the oxidation of phenol in supercritical water involves multiple reactions, compare Figure 10.31 [53].

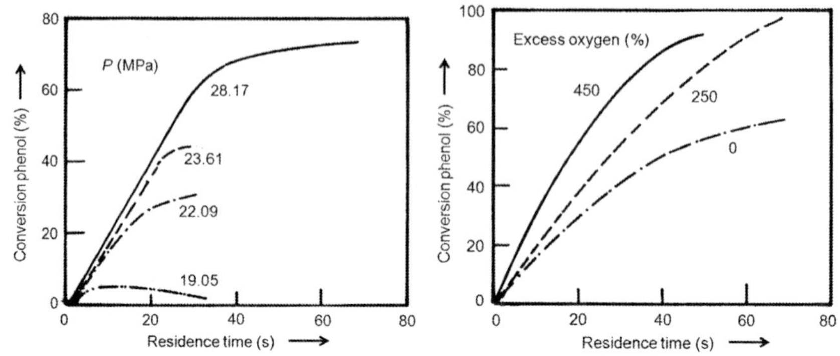

FIGURE 10.30 Oxidation of phenol in supercritical water, $T = 380\ °C$. Left: Influence of pressure, $x_F = 0.1$ wt% phenol, $x_{O_2} = 0\%$ excess oxygen. Right: Influence of oxygen concentration, $P = 28.2$ MPa, $x_F = 0.025$ wt% phenol. *Modified from Ref. [53].*

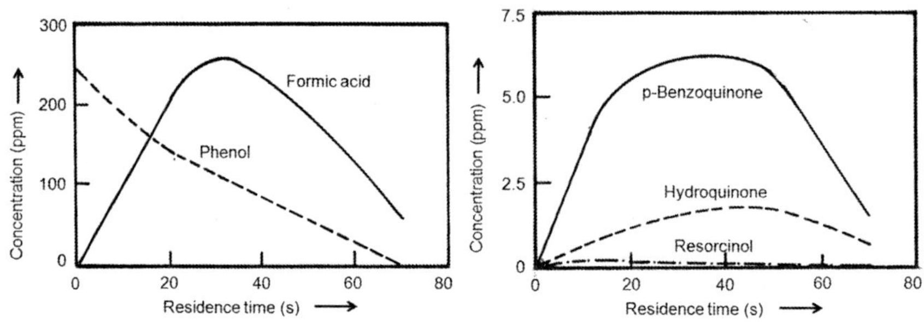

FIGURE 10.31 Oxidation of phenol in supercritical water. Concentration of products. $T = 380\ °C$, $P = 28.2$ MPa, $x_F = 0.025$ wt% phenol, $x_{O_2} = 250\%$ excess oxygen. *Modified from Ref. [53].*

At low initial phenol concentrations of $x_F = (1\text{–}10) \times 10^{-5}$ mol/m^3, conversion is independent of initial phenol concentration but dependent on water density. After a reaction time of $t = 6$ s, conversion of phenol at $T = 390\ °C$ is about 14% at a density of water of $\rho = 200$ kg/m^3, 22% at $\rho = 370$ kg/m^3, and 30% at $\rho = 500$ kg/m^3 [54].

At high concentrations of phenol, the oxidation is complete after $t = 40$ s at $T \geq 390\ °C$, $P = 25$ MPa, and oxygen in excess [55]. Comparison of experimental results with predictions of kinetics of phenol oxidation derived from different investigations show a wide variation of predicted conversions. Experimental results are best represented by two phenol reaction kinetics according to Equation (10.28) and Table 10.6, one derived from experiments at lower phenol concentrations of $x_F = 3\text{–}37$ mg/dm^3 [58], the other from experiments with high phenol concentrations of $x_F = 500\text{–}1000$ mg/dm^3 [56], see Figure 10.32.

TABLE 10.6 Kinetics of Oxidation of Phenol in Supercritical Water [55]

	$[Ph]_0$ (mg/l)	T (K)	P (MPa)	O_2 Excess (%)	A ($mol^{-1-a-b-c}s^{-1}$)	E_a (kJ/mol)	a	b	c
[56]	3–37	693–753	27		219	51.8	0.85	0.5	0.42
[57]	500–1000	653–723	23–26.5	50–1000	4.9×10^8	124.8	1.04	0.38	0
[58]exp.	27,000–40,000	666–778	25	0–35.6	–	–	–	–	–

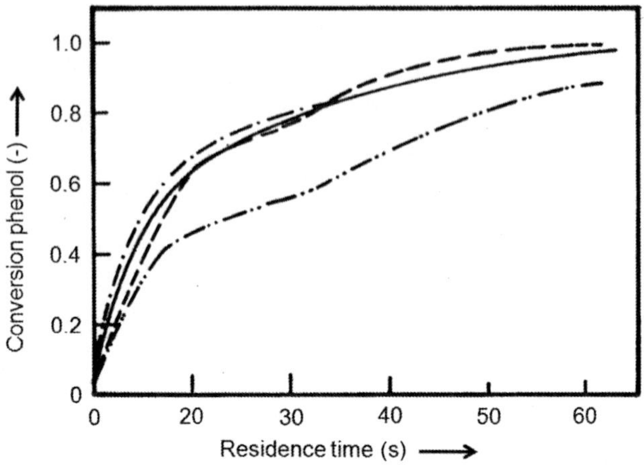

FIGURE 10.32 Oxidation of phenol in supercritical water. $T=681$ K (408 °C), $P=26$ MPa, x_{O_2} : oxygen excess 34%, initial phenol concentration $x_F=2.7$ wt%. Data from: ——— Pérez *et al.* [55], — — — Krajnc and Levec [56], —•— Gopalan and Savage [58], —••— Takahashi *et al.* [26]. *Modified from Ref. [55].*

$$\frac{d[Ph]}{dt}=A\exp\left(\frac{-E_a}{RT}\right)[Ph]^a[O_2]^b[H_2O]^c \tag{10.28}$$

with A as frequency factor, E_a activation energy, and a, b, and c as reaction orders [55].

Catalyzed oxidation of phenol, for example, with activated carbon and NaOH, enhanced the conversion of phenol. Activated carbon at $T=400$ °C and $P=25$ MPa with an equivalent amount of oxygen reduces production of tar and increases phenol conversion [57]. NaOH has little effect on the conversion of phenol in supercritical water but accelerates the conversion under oxidative conditions [59].

The conversion of 2-chlorophenol in supercritical water is accelerated by the addition of NaOH at $T=440$ °C, $P=26$ MPa, an initial concentration of 2-chlorophenolof $x_F=3.89\times10^{-3}$ mol, and residence times of $t=0.13$–0.51 s. Under oxidative conditions, the generation of dimmers is reduced and the dechlorination of 2-chlorophenol is promoted [59]. Hydrogen peroxide is significantly more effective in destruction than oxygen, at least if supplied with air [60], as shown in Figure 10.33. The effect of H_2O_2 can be seen for the conversion of acetic acid, for which destruction is complete with hydrogen peroxide at $T=500$ °C after a residence time of $t=10$ min, but proceeds to only about 65% with oxygen as oxidant (Figure 10.33). Density increase of supercritical water from $\varrho_{H_2O}=0.15$ to $0.35\,g/cm^3$ at $T=400$ °C only slightly improves the conversion of both compounds [60].

It is accepted that oxidation in supercritical water is a radical-chain reaction and consists of many elementary reactions, similar to combustion

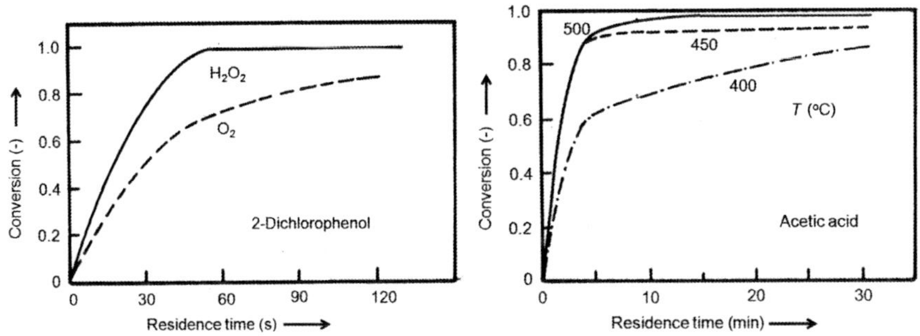

FIGURE 10.33 Conversion of 2-dichlorophenol and acetic acid in supercritical water. $\rho_{H_2O}=0.15\,g/cm^3$, initial 2-dichlorophenol concentration $x_F=2$ g/dm^3, x_{O_2} : 200% stoichiometric oxygen supply. Left: 4-dichlorophenol, $T=500$ °C. Right: acetic acid. *Modified from Ref. [60].*

chemistry [34]. For larger molecules, the application of models based on elementary reactions is difficult because of the complexity of the reaction mechanism. Lumping elementary reactions into one reaction provides good fits to experimental data and can be used as a conventional technique for the modeling of the reaction mechanism [34].

10.3.2.8 Real Waste

The prospect of destructing waste of nearly any kind spurred the extensive research and process development of SCWO. Several processes have been developed or are tested in pilot plants. Examples are discussed in Chapter 13. In the following, experimental results of the conversion of waste streams in supercritical water are presented.

Destruction of unwanted compounds in industrial aqueous waste with supercritical water has been successfully investigated and developed [61]. As a summary, the destruction of compounds in industrial waste, containing salts beside the unwanted compounds, is possible at high pressure conditions of $P=70$ MPa, at moderate temperatures of $T=470$–490 °C. These conditions retain a sufficient salt solubility in supercritical water. Residence times for the destruction of biocides to >99.9% are about $t\approx 7$ min [61].

SCWO is a good candidate for the decomposition of hazardous materials, such as dioxins (see Figure 10.34 for molecular structures) and polychlorinated biphenyls. These chlorinated aromatic compounds are toxic and decomposition resistant. Experimental results for the conversion of several of such compounds are presented in Table 10.7 [62].

Conversion of wastewater from terephthalic acid manufacturing process is only effective with a catalyst, as shown in Figure 10.35.

Oxidative destruction of municipal sludge in supercritical water leads to liquid phase products that are colorless and odorless [64]. The total organic carbon decreases within $t=5$ min at $T=400$ °C, the lowest temperature investigated, see Figure 10.36. With an amount of oxidant (H_2O_2) greater than

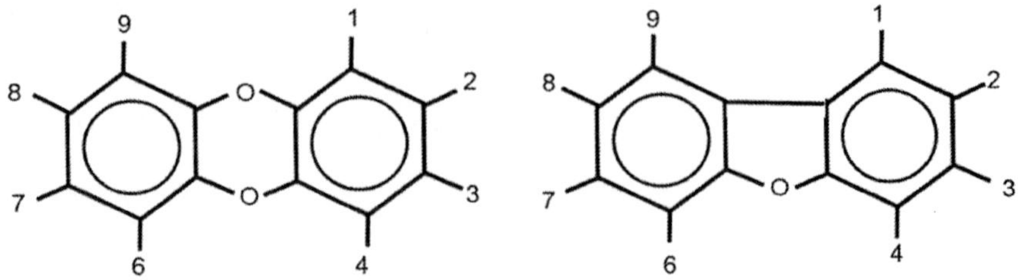

FIGURE 10.34 Structure of polychlorinated dibenzo-*p*-dioxide (PCDD) (left) and polychlorinated dibenzofuran (right) [62].

TABLE 10.7 Oxidation of Dioxins from Fly Ash in Supercritical Water

		Air	Oxygen
		0.1 MPa	*0.5 MPa*
Compounds	Amount in Fly Ash (ng/g)	*Decomposition (%)*	
PCDD	92	98.2	98.5
PCDF	92	96.6	98.5

T = 400 °C, P = 30 MPa, reaction time: t = 30 min, water 7.3–7.4 g, fly ash 2 g [62].

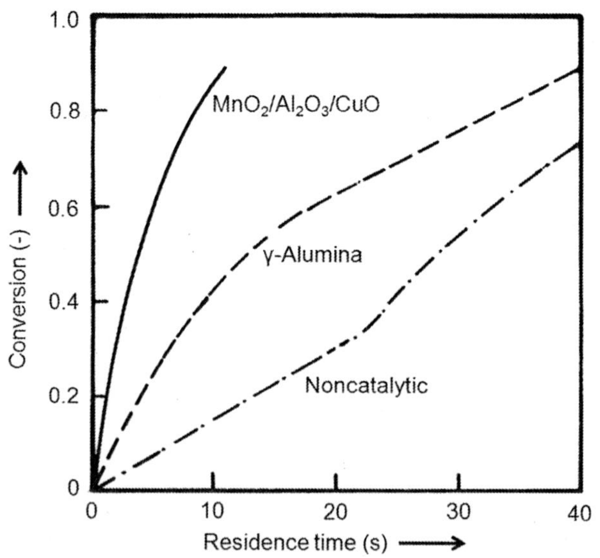

FIGURE 10.35 Oxidation of wastewater from terephthalic acid manufacturing process with different heterogeneous catalysts (T = 420 °C, P = 240 bar, $[COD]_0$ = 7.8 mmol/l, $[H_2O_2]_0$ = 36.6 mmol/l). *Modified from Ref. [63].*

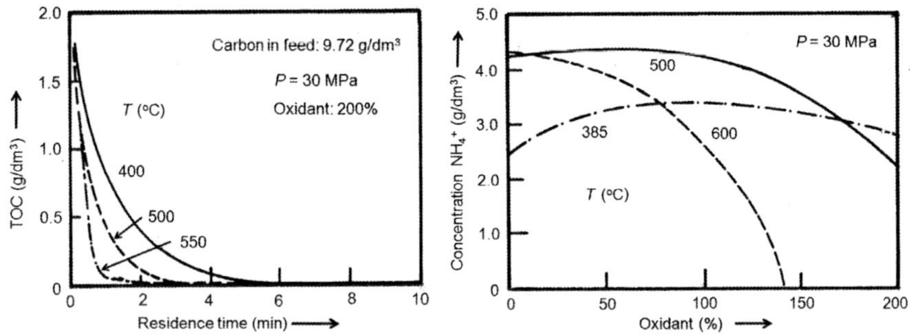

FIGURE 10.36 Oxidation of municipal sludge in supercritical water, $P=30$ MPa. Left: Influence of temperature on TOC reduction. Right: Influence of temperature on ammonium ion concentration in the liquid phase. *Modified from Ref. [64].*

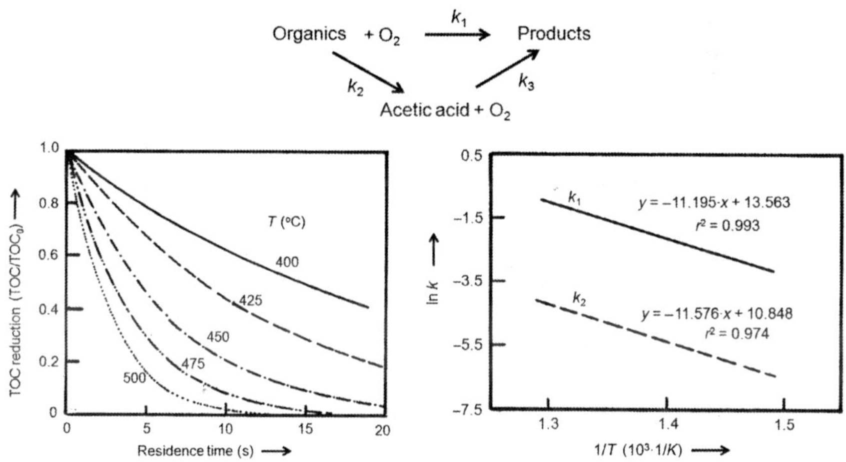

FIGURE 10.37 Reaction scheme for the oxidation of cutting oil wastes in supercritical water, $P \approx 25$ MPa. Cutting oil waste: mineral oil, nonyl phenol, fatty acids (C16–C20), ethoxylated alcohols (C13/C15). Left: Reduction of the total organic carbon (TOC). Right: Arrhenius plot of the reaction rate coefficients. *Modified from Ref. [65].*

stoichiometric demand, organic carbon in the liquid phase was almost completely destroyed [65]. Complete destruction of ammonia needs high oxidant concentration and reaction temperatures of around $T=600$ °C [64].

Total organic carbon of cutting oil wastes is reduced in SCWO after a reaction time of $t=10$ min nearly completely at $T=500$ °C, as shown in Figure 10.37. Global reaction kinetics, based on acetic acid, represent the experimental results with sufficient accuracy (Figure 10.37) [65].

A great number of special compounds has been oxidized using supercritical water, beside the components discussed above, and additional publications are available on waste treatment [18]. Noncatalyzed oxidation in supercritical water is effective for a wide variety of compounds. For some particularly stable compounds, such as ammonia, high temperatures are needed for conversion. Catalysts can increase the selectivity and decrease the temperatures. Much faster reaction rates occur in SCWO that is catalyzed by transition metals. Oxides of Mn, Ti, and Cu reduce reaction times for total destruction and consequently required reactor volumes drastically. An overview on catalysis in SCWO is given by Savage [15].

Several reviews on SCWO address the state-of-the-art in research and commercial verification, including various process and reactor designs, also covering the problems with corrosion and salt precipitation [18,66–68]. Corrosion in high temperature and supercritical water is discussed in Chapter 12, reactor design, process design, and implementation in Chapter 13.

10.4 HYDROTHERMAL FLAMES

10.4.1 Introduction

Oxidation reactions in an aqueous environment, as carried out in WAO and SCWO, can be useful to produce chemicals, but far more for disposing of waste compounds in a controllable way, without further polluting the environment. As has been discussed in this chapter earlier, WAO is mainly applied for waste components in low concentrations from production processes. Incoming components, in particular halogen compounds, must be strictly controlled and limited. Due to the relatively low temperatures below $T = 350\ °C$, in most cases even lower, the reaction time needed for sufficient conversion of the unwanted compounds is in the range of $t = 1–3$ h. Consequently, the reaction volumes are large, and the operating pressure is of major importance for the processing costs. Of advantage for WAO processing is that ionic products are soluble in the liquid reaction phase and do not block the reactor.

Carrying out the oxidation reaction at conditions above the critical pressure and critical temperature of water, in a so-called SCWO process, reduces reaction times to a range of 1 min that is needed for a sufficient conversion of hazardous compounds. Reaction volumes can be much smaller and hazard potential is substantially reduced. On the other hand, interaction of the reaction compounds with the reactor walls, ending in corrosion, may be of some concern, in particular in the near critical temperature regions, discussed in Chapter 12. Furthermore, at supercritical temperatures, the solubility of certain salts drops dramatically, which can cause precipitates that eventually block the reactor.

For high concentrations of organic compounds ($x_F > 5$ wt%), the oxidation reaction leads to higher temperatures in the reactor than the inlet temperature. A combustion-like reaction can be initiated in the aqueous environment, known as hydrothermal flame. With the increase of the reaction temperature and the reduction of the reaction volume, the reaction approaches the typical combustion regime. Differently to that regime, hydrothermal flames do not produce gases in noncontrollable quantity and composition. Nevertheless, the oxidation reaction occurs like combustion reactions at very short reaction times, contrary to the controlled oxidation in supercritical water, which is still a relatively slow chemical reaction.

In the following, the principle of hydrothermal flames is explained together with the main properties of very fast oxidation in an aqueous environment. Consequences for reactor construction will be mentioned but are discussed in detail in Chapter 13.

10.4.2 General Features of Hydrothermal Flames

Hydrothermal flames are very fast oxidation reactions that produce a luminous flame and occur in a supercritical aqueous environment [40,69]. Temperatures in the flame are high ($T > 1000$ °C) and chemical kinetics are very fast. The reactions are completed within $t = 10$–100 ms. Gases and nonpolar organic compounds are completely miscible with water, forming a single homogeneous phase that can be ignited. Because the oxidation process occurs in a single phase, interfacial mass transport limitations are avoided.

Hydrothermal flames are initiated by organic species that enter the reaction zone at significantly lower temperatures, heated to the reaction temperature, mixed with the oxidant, and take part in the combustion process [70]. The amount of compounds that can be oxidized must be high enough (usually $x_F > 5$ wt%) to initiate and sustain the oxidation reaction (combustion). Fuel and oxidant streams are heated to a high enough temperature that autoignition occurs when they are mixed [70]. Hydrothermal flames can occur in SCWO reactions at high reactant and oxidant concentrations, resulting in much higher temperatures and fast oxidation rates than those occurring in SCWO [69].

It was realized that hydrothermal flames avoid two problems: metal corrosion and plugging of the reactor due to precipitating salt solids. These phenomena are of major importance for practical application of SCWO to the treatment of hazardous and toxic compounds in aqueous waste streams. Hydrothermal flames can be used as an internal heat source to destroy organic compounds. The hydrothermal flame reaction zone is confined to the interior of the vessel within a stream of subcritical water that protects the reactor walls from corrosion and salt plugging, see Figure 10.39. Inlet streams enter the hydrothermal flame reactor below the critical point of water, thus circumventing the problems of corrosion and plugging in the preheating section. The hydrothermal flame reaction generates sufficient heat to increase the temperature of the incoming streams to autoignition. Then, the combustion reaction proceeds to complete oxidation of the organic compounds in only $t = 10$–100 ms [69].

Hydrothermal flames can be applied for drilling deep wells in hard rock such as granite, beside for the oxidation reaction to destroy waste. The hydrothermal flame rapidly heats a rock surface and induces thermal stresses that lead to the formation of rock fragments, or spalls (hence "spallation drilling") that are violently ejected from the rock surface and removed from the borehole [69,72].

10.4.3 Type of Flames (Diffusion Flames, Turbulent Flames)

Hydrothermal flames have been investigated in two processing modes: semibatch and continuous. In the semibatch mode, the mixture of fuel and water is maintained at high pressure and temperature in a reactor. The oxidant is continuously delivered to the reaction zone and mixes with the fuel–water solution. After reaching the ignition conditions, the reaction occurs and a flame front is formed. As the flame burns, the concentration of the fuel in the system decreases and the reaction products accumulate, until the fuel concentration is too low to support continued combustion. Since the chemical reaction time scale is much smaller than the time scale for diffusive mass transport, the concentration of reactants at the flame front is nearly zero. Since mass flux of reactants into the flame front occurs by diffusion, this type of flames is named "diffusion flame" [69]. Results of an experiment are shown in Figure 10.38 [73].

Turbulent diffusion flames are produced by turbulent mixing and combustion of separate fuel and oxidant streams. The mixing and combustion process is inherently an unsteady process that does not occur at a well-defined flame front [69]. The fuel and oxidant are both continuously injected as coaxial jets.

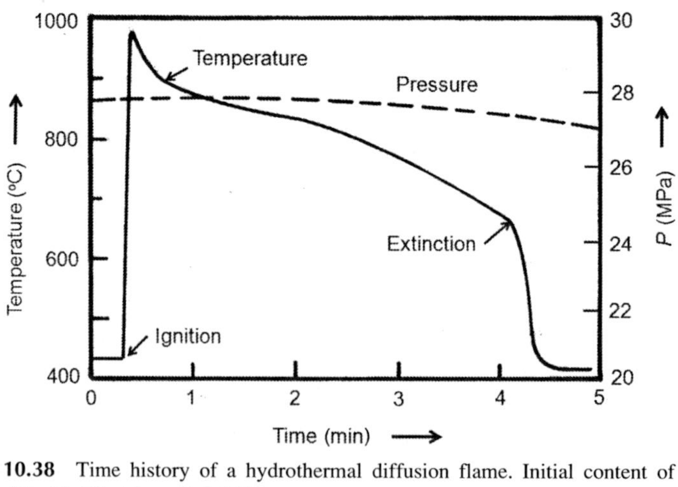

FIGURE 10.38 Time history of a hydrothermal diffusion flame. Initial content of methane: $x_F = 12$ mol%. *Modified from Ref. [73].*

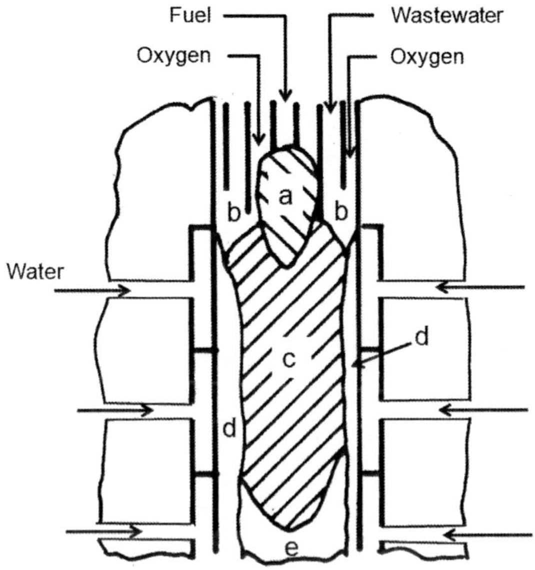

FIGURE 10.39 Principle features of a hydrothermal flame reactor. (a) hydrothermal flame, (b) mixing zone of wastewater and oxygen, (c) reaction zone for the oxidation of organic compounds of the wastewater, (d) protecting film of cold water, and (e) down-cooled reaction mixture. *Modified from Ref. [71].*

The flame front develops where the oxidant is entrained into the moving jet. Reaction products are removed through the reactor exit. Operation of such a hydrothermal flame reactor is not limited by the initial conditions and can be operated for a prolonged period of time and over a range of operating conditions in order to demonstrate the feasibility of a continuous hydrothermal flame process. The principles of design for a hydrothermal flame reactor are shown in Figure 10.39. Figure 10.40 shows temperature profiles for different methanol–water mixtures [71].

10.4.4 Characterization of Hydrothermal Flame Processes

10.4.4.1 Ignition Temperature

The temperature for ignition of a preheated mixture of $x_F = 30$ mol% methane in water is $T = 420$ °C at $P = 20$ MPa. It decreases to $T = 400$ °C at a pressure of $P = 100$ MPa. These ignition temperatures are significantly below the ignition temperature of $T = 550$ °C required for methane under ambient conditions [69,74,75]. The ignition temperature increases when the methane concentration is reduced. Mixtures of ethane and water show a similar ignition behavior at $P = 73$ MPa for a range of $x_F = 5$–20 mol% ethane. A 20-mol% mixture of ethane and water can be ignited at temperatures as low as $T = 350$ °C [69].

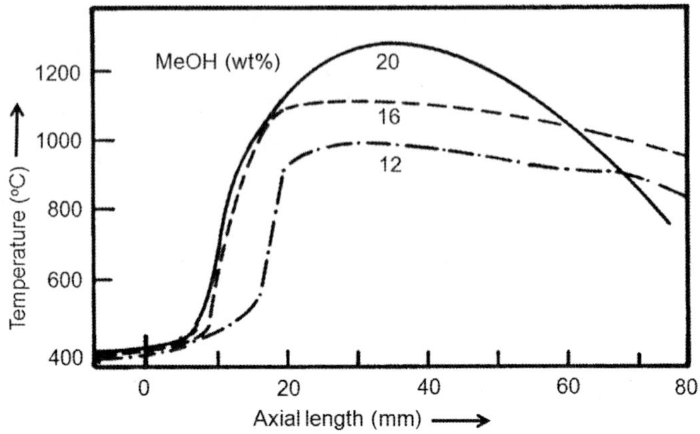

FIGURE 10.40 Hydrothermal flame. Temperature profiles for different methanol–water mixtures. *Modified from Ref. [71].*

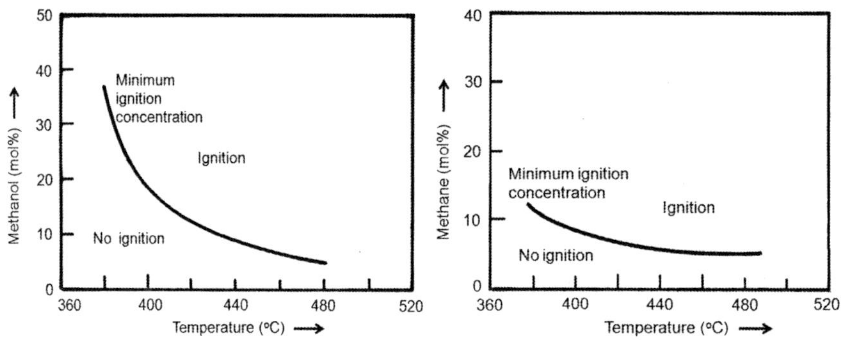

FIGURE 10.41 Limits for spontaneous ignition of methanol (left) and methane (right) in supercritical water and for injection of pure oxygen. $P=27.5$ MPa. *Modified from Ref. [41].*

Methanol–water fuel mixtures can be ignited at temperatures above $T=450\,°C$ at as low as $x_F=6$ mol% methanol [41]. At lower temperatures, higher organic concentrations are needed to induce ignition. Ignition limits for water–methanol and water–methane are shown in Figure 10.41 [41].

10.4.4.2 Combustion Products

For diffusion flames during the oxidation of methane both with and without flames in supercritical water, combustion products CO, CO_2, H_2, and CH_3OH are present [75]. The concentration of formaldehyde is insignificant and higher alkanes are not detected. Similar results are found for the flame oxidation of ethane, with minor traces of alcohols, aldehydes, and carbonic acids [69]. Formation of soot from hydrothermal diffusion flames is suppressed using supercritical aqueous mixtures of *n*-heptane and toluene as fuel, as confirmed with experiments with argon [75]. With air as the oxidant, the formation of NO_x in hydrothermal flames is possible. NO_x was found at concentrations from $x = 18–52$ ppm in supercritical water with methanol as a fuel and significant amounts of CO (800–12,000 ppm) were also detected [76].

Different results are found for turbulent flames. CO concentrations are as low as $x_{CO} = 1$ ppm and no traces of NO_x are found for continuous turbulent diffusion flames, using isopropyl alcohol organic compound and air as oxidant [77,78]. Turbulent flames probably undergo sufficient mixing before combustion to create a fuel–lean mixture that promotes complete combustion to CO_2, at low enough temperatures to avoid the formation of NO_x [69].

10.4.4.3 Destruction Efficiency

Since destruction efficiencies of over 99%, related to the concentration of the initial compounds, are typical for residence times of less than $t = 100$ ms, as they occur in hydrothermal flame processes. For methanol, at subcritical inlet temperatures, a destruction efficiency of 99.8% is obtained in $t = 50–100$ ms [79] or 97.85–99.99% in a different reactor [80]. For reactions between isopropyl alcohol and a water/air mixture at $P = 25$ MPa, destruction efficiencies of over 99.9% are obtained. Dioxins are destructed to more than 99.9% [78].

10.4.4.4 Corrosion and Salt Plugging

For an example, in hydrothermal flame experiments, with 6 wt% methanol, wastewater, and 1–3 wt% added sodium sulfate, carried out in a reactor with sweeping the walls with water (transpiring wall reactor), 65% of the salt introduced into the reactor is in the effluent, regardless of transpiration intensity, type of transpiring wall, or salt content. Salt precipitates on thermocouple tips and in the upper hot zone of the reactor, above the wall protecting transpiration zone. Severe corrosion occurs along the outer diameter of the combustion chamber. The rest of the reactor has been operated without any visible signs of corrosion for $t = 500$ h [81].

10.4.5 Spallation Drilling

The oxidation reaction of hydrothermal flames can be used to rapidly heat a rock surface and induce thermal stresses that lead to the formation of rock fragments, or spalls which are violently ejected from the rock surface and removed from the borehole [69,72].

In boreholes that are more than 2000 m deep and filled with water, the hydrostatic pressure exceeds the critical pressure of water. In such a situation, hydrothermal flames can provide the required heat to spall the rock. A drilling

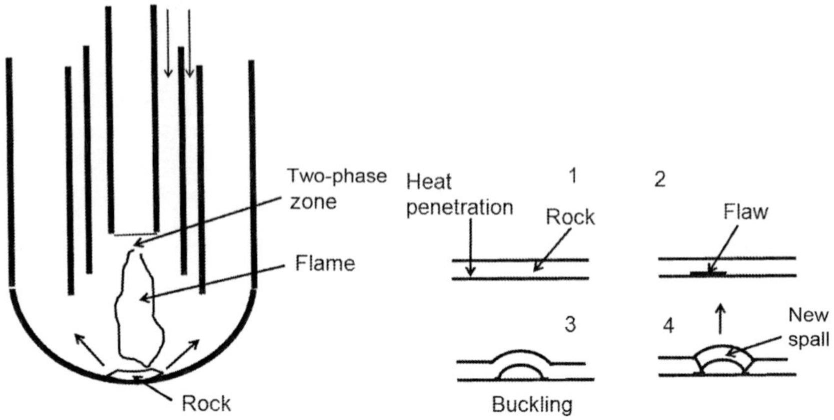

FIGURE 10.42 Characteristic features of a hydrothermal flame for "spallation drilling." *Modified from Ref. [71].*

head may consist of a combustion chamber fed by water, fuel, and oxidant, where the reactants form a hydrothermal flame that exits through a nozzle together with the combustion products, heats the rock surface, and induces fragmentation, as illustrated in Figure 10.42 [82].

The technique of using hydrothermal frames for drilling may be an alternative to rotary drilling for deep boreholes.

REFERENCES

[1] H. Debellefontaine, J.N. Foussard, Wet air oxidation for the treatment of industrial wastes. Chemical aspects, reactor design and industrial applications in Europe, Waste Manag. 20 (2000) 15–25.

[2] F. Luck, A review of industrial catalytic wet air oxidation processes, Catal. Today 27 (1996) 195–202.

[3] D. Duprez, F. Delanoé, J. Barbier, P. Isnard, G. Blanchard, Catalytic oxidation of organic compounds in aqueous media, Catal. Today 29 (1996) 317–322.

[4] A.K. Chowdury, L.W. Ross, Catalytic wet oxidation of strong waste waters, AIChE Symp. Ser. 71 (151) (1975) 46–58.

[5] R.S. Willms, A.M. Balinsky, D.D. Reibel, D.M. Wetzel, D.P. Harrisson, Aqueous phase oxidation. The intrinsic kinetics of single organic compounds, Ind. Eng. Chem. Res. 26 (1987) 148–154.

[6] J.N. Foussard, H. Debellefontaine, J. Besombes-Vailhé, Efficient elimination of organic liquid wastes: wet air oxidation, J. Environ. Eng. ASCE 115 (1989) 367–385.

[7] L. Lixiong, P. Chen, E.F. Gloyna, Generalized kinetic model for wet oxidation of organic compounds, AIChE J. 37 (1991) 1687–1697.

[8] H.S. Joglekar, S.D. Samant, J.B. Joshi, Kinetics of wet air oxidation of phenol and substituted phenols, Water Res. 25 (1991) 135–145.

[9] T.M.S. Carlos, C.B. Maugans, Wet air oxidation of refinery spent caustic: a refinery case study, US Filter, Zimpro Products, Technical Report No. 427, in: NPRA Conference, San Antonio, Texas, September 12, 2000.

[10] X. Hu, Advanced Physio-Chemical Treatment Processes, Course: CENG 5760, Lecture notes, Hong Kong University of Science & Technology, retrieved on March 28, 2013 from http://ihome.ust.hk/kexhu/ceng576/576-03.pdf.
[11] M. Kindsigo, J. Kallas, Degradation of lignins by wet oxidation: model water solutions, Proc. Est. Acad. Sci. Chem. 55 (2006) 132–144.
[12] S. Barbati, V. Fontanier, M. Ambrosio, Wet air oxidation of meat-and-bone meal and raw animal byproducts, Ind. Eng. Chem. Res. 47 (2008) 2849–2854.
[13] W. Tang, X. Zeng, Q. Wang, P. Li, Wet air oxidation of polyether solutions, Chem. Eng. Process. Process Intens. 49 (2010) 815–819. http://dx.doi.org/10.1016/j.cep.2010.07.001.
[14] U.S. Army Chemical Materials Agency (Provisional) Program Manager for Elimination of Chemical Weapons, FY03 Technology Evaluation for Chemical Demilitarization Wet Air Oxidation Technology Assessment, Contract: DAAD13-01-D-0007, Task: T-03-S-002, April 2003, Science Applications International Corporation.
[15] P.E. Savage, A perspective on catalysis in sub- and supercritical water, J. Supercrit. Fluids 47 (2009) 407–414.
[16] E. Dinjus, A. Kruse, Applications of supercritical water, in: R. van Eldik, F.-G. Klärner (Eds.), High Pressure Chemistry, Wiley-VCH, Weinheim, Germany, 2002, pp. 422–446.
[17] T.B. Thomason, M. Modell, Supercritical water destruction of aqueous wastes, Hazard. Waste 1 (1984) 453–467.
[18] G. Brunner, Near and supercritical water. Part II: oxidative processes, J. Supercrit. Fluids 47 (2009) 382–390.
[19] M. Modell, J. Larson, S.F. Sobczynski, Supercritical water oxidation of pulp mill sludges, Tappi J. 75 (1992) 195–202.
[20] L. Li, E.F. Gloyna, J.E. Sawicki, Treatability of DNT process wastewater by supercritical water oxidation, Water Environ. Res. 65 (1993) 250–257.
[21] J.E. Sawicki, B. Casas, Wet oxidation systems—process concept to design, Environ. Prog. 12 (1993) 275–283.
[22] M. Modell, Supercritical water oxidation, in: H.M. Freeman (Ed.), Standard Handbook of Hazardous Waste Treatment and Disposal, McGraw-Hill, New York, 1989, Sec. 8.11.
[23] C.N. Staszak, K.C. Malinowski, W.R. Killilea, The pilot-scale demonstration of the MODAR oxidation process for the destruction of hazardous organic waste materials, Environ. Prog. 6 (1987) 39–43.
[24] K.C. Swallow, W.R. Killilea, Comment on "phenol oxidation in supercritical water: formation of dibenzofuran, dibenzo-p-dioxin, and related compounds" Environ. Sci. Technol. 26 (1992) 1848. T.D. Thornton, D.E. LaDue III, P.E. Savage, Reply to comments on "phenol oxidation in supercritical water: formation of dibenzofuran, dibenzo-p-dioxin, and related compounds" Environ. Sci. Technol. 26 (1992) 1850.
[25] A. Shanableh, E.F. Gloyna, Supercritical water oxidation—wastewater and sludges, Water Sci. Technol. 23 (1991) 389–398.
[26] Y. Takahashi, T. Wydeven, C. Koo, Subcritical and supercritical water oxidation of CELSS model wastes, Adv. Space Res. 9 (1989) 99–110.
[27] G. Lissens, W. Verstraete, T. Albrecht, G. Brunner, C. Creuly, J. Seon, G. Dussap, C. Lasseur, Advanced anaerobic bioconversion of lignocellulosic waste for bioregenerative life support following thermal water treatment and biodegradation by Fibrobacter succinogenes, Biodegradation 15 (2004) 173–183.
[28] J.B. Johnston, R.E. Hannah, V.L. Cunningham, B.P. Daggy, F.J. Sturm, R.M. Kelly, Destruction of pharmaceutical and biopharmaceutical wastes by the Modar supercritical water oxidation process, Nat. Biotechnol. 6 (1988) 1423–1427.

[29] P.C. Dell'orco, B.R. Foy, J.M. Robinson, S.J. Buelow, Hydrothermal treatment of Hanford waste constituents, Hazard. Waste Hazard. Mater. 10 (1993) 221–232.

[30] D.M. Harradine, S.J. Buelow, P.C. Dell'orco, R.B. Dyer, B.R. Foy, J.M. Robinson, J.A. Sanchez, T. Sportarelli, J.D. Wander, Oxidation chemistry of energetic materials in supercritical water, Hazard. Waste Hazard. Mater. 10 (1993) 233–246.

[31] C.K. Rofer, S.J. Buelow, R.B. Dyer, J.D. Wander, Conversion of hazardous materials using supercritical water oxidation, U.S. Patent 5,133,877, July 28, 1992.

[32] S.F. Rice, R.R. Steeper, C.A. LaJeunesse, Destruction of representative Navy wastes using supercritical water oxidation, Sandia National Laboratory Report SAND94-8203, 1993.

[33] P.E. Savage, S. Gopalan, T.I. Mizan, C.J. Martino, E.E. Brock, Reactions at supercritical conditions: applications and fundamentals, AIChE J. 41 (1995) 1723–1778.

[34] T. Adschiri, Y. Oshima, K. Arai, Oxidation and hydrolysis reactions in supercritical water, in: Y. Arai, T. Sako, Y. Takebayashi (Eds.), Supercritical Fluids, Springer, Berlin, Heidelberg, Germany, 2002, pp. 364–374.

[35] H.R. Holgate, P.A. Webley, J.W. Tester, R.K. Helling, Carbon monoxide oxidation in supercritical water: the effects of heat transfer and the water-gas shift reaction on observed kinetics, Energy Fuels 6 (1992) 586–597.

[36] H.R. Holgate, J.W. Tester, Fundamental kinetics and mechanisms of hydrogen oxidation in supercritical water, Combus. Sci. Technol. 88 (1993) 369–397.

[37] P.E. Savage, J. Yu, N. Stylski, E.E. Brock, Kinetics and mechanism of methane oxidation in supercritical water, J. Supercrit. Fluids 12 (1998) 141–153.

[38] T. Sato, M. Watanabe, R.L. Smith Jr., T. Adschiri, K. Arai, Analysis of the density effect on partial oxidation of methane in supercritical water, J. Supercrit. Fluids 28 (2004) 69–77.

[39] C.N. Dixon, M.A. Abraham, Conversion of methane to methanol by catalytic supercritical water oxidation, J. Supercrit. Fluids 5 (1992) 269–273.

[40] W. Schilling, E.U. Franck, Combustion and diffusion flames at high pressures to 2000 bar, Ber. Bunsenges. Phys. Chem. 92 (1988) 631–636.

[41] R.R. Steeper, S.F. Rice, M.S. Brown, S.C. Johnston, Methane and methanol diffusion flames in supercritical water, J. Supercrit. Fluids 5 (1992) 262–268.

[42] U. Armbruster, A. Martin, A. Krepel, Partial oxidation of propane in sub- and supercritical water, J. Supercrit. Fluids 21 (2001) 233–243.

[43] W.R. Killilea, K.C. Swallow, G.T. Hong, The fate of nitrogen in supercritical-water oxidation, J. Supercrit. Fluids 5 (1992) 72–78.

[44] J.W. Tester, P.A. Webley, H.R. Holgate, Revised global kinetic measurements of methanol oxidation in supercritical water, Indus. Eng. Chem. Res. 32 (1993) 236–239.

[45] E.E. Brock, Y. Oshima, P.E. Savage, J.R. Barker, Kinetics and mechanism of methanol oxidation in supercritical water, J. Phys. Chem. 100 (1996) 15834–15842.

[46] S.F. Rice, T.B. Hunter, A.C. Rydén, R.G. Hanush, Raman spectroscopic measurement of oxidation in supercritical water. 1. Conversion of methanol to formaldehyde, Indus. Eng. Chem. Res. 35 (1996) 2161–2171.

[47] F. Vogel, J.L. DiNaro Blanchard, Ph.A. Marrone, S.F. Rice, P.A. Webley, W.A. Peters, K.A. Smith, J.W. Tester, Review: critical review of kinetic data for the oxidation of methanol in supercritical water, J. Supercrit. Fluids 34 (2005) 249–286.

[48] G.J. DiLeo, Ph.E. Savage, Catalysis during methanol gasification in supercritical water, J. Supercrit. Fluids 39 (2006) 228–232.

[49] Ph.E. Savage, J. Rovira, N. Stylski, Ch.J. Martino, Oxidation kinetics for methane/methanol mixtures in supercritical water, J. Supercrit. Fluids 17 (2000) 155–170.

[50] K. Hirosaka, K. Koido, M. Fukayama, K. Ouryoji, T. Hasegawa, Experimental and numerical study of ethanol oxidation in subcritical water, J. Supercrit. Fluids 44 (2008) 347–355.
[51] J. Schanzenbächer, J.D. Taylor, J.W. Tester, Ethanol oxidation and hydrolysis rates in supercritical water, J. Supercrit. Fluids 22 (2002) 139–147.
[52] R. Hayashi, M. Onishi, M. Sugiyama, S. Koda, Y. Oshima, Kinetic analysis on alcohol concentration and mixture effect in supercritical water oxidation of methanol and ethanol by elementary reaction model, J. Supercrit. Fluids 40 (2007) 74–83.
[53] T.D. Thornton, P.E. Savage, Phenol oxidation in supercritical water, J. Supercrit. Fluids 3 (1990) 240–248.
[54] Y. Oshima, K. Hori, M. Toda, T. Chommanad, S. Koda, Phenol oxidation kinetics in supercritical water, J. Supercrit. Fluids 13 (1998) 241–246.
[55] I.V. Pérez, S. Rogaka, R. Branion, Supercritical water oxidation of phenol and 2,4-dinitrophenol, J. Supercrit. Fluids 30 (2004) 71–87.
[56] M. Krajnc, J. Levec, On the kinetics of phenol oxidation in supercritical water, AIChE J. 42 (1996) 1977–1984.
[57] Y. Matsumura, T. Urase, K. Yamamoto, T. Nunoura, Carbon catalyzed supercritical water oxidation of phenol, J. Supercrit. Fluids 22 (2002) 149–156.
[58] S. Gopalan, P.E. Savage, A reaction network model for phenol oxidation in supercritical water, AIChE J. 41 (1995) 1864–1873.
[59] A. Lee, T. Nunoura, Y. Matsumura, K. Yamamoto, Comparison of the effects of the addition of NaOH on the decomposition of 2-chlorophenol and phenol in supercritical water and under supercritical water oxidation conditions, J. Supercrit. Fluids 24 (2002) 239–250.
[60] D.-S. Lee, E.F. Gloyna, L. Li, Efficiency of H_2O_2 and O_2 in supercritical water oxidation of 2,4-dichlorophenol and acetic acid, J. Supercrit. Fluids 3 (1990) 249–255.
[61] S. Baur, H. Schmidt, A. Krämer, J. Gerber, The destruction of industrial aqueous waste containing biocides in supercritical water-development of the SUWOX process for the technical application, J. Supercrit. Fluids 33 (2005) 149–157.
[62] T. Sako, T. Sugeta, K. Otake, M. Sato, M. Tsugumi, T. Hiaki, M. Hongo, Decomposition of dioxins in fly ash with supercritical water oxidation, J. Chem. Eng. Jpn. 30 (1997) 744–747.
[63] T.-J. Park, J.S. Lim, Y.-W. Lee, S.-H. Kim, Catalytic supercritical water oxidation of wastewater from terephthalic acid manufacturing process, J. Supercrit. Fluids 26 (2003) 201–213.
[64] M. Goto, T. Nada, A. Ogata, A. Kodama, T. Hirose, Supercritical water oxidation for the destruction of municipal excess sludge and alcohol distillery wastewater of molasses, J. Supercrit. Fluids 13 (1998) 277–282.
[65] J.R. Portela, E. Nebot, E. Martinez de la Ossa, Generalized kinetic models for supercritical water oxidation of cutting oil wastes, J. Supercrit. Fluids 21 (2001) 135–145.
[66] J.W. Tester, H.R. Holgate, F.J. Armellini, P.A. Webley, W.R. Killilea, G.T. Hong, H.E. Barner, Supercritical water oxidation technology—process-development and fundamental research, in: D.W. Tedder, F.G. Pohland (Eds.), Emerging Technologies in Hazardous Waste Management III, ACS Symposium Series, American Chemical Society, Washington, DC, USA, vol. 518, 1993, pp. 35–76.
[67] J.W. Tester, J.A. Cline, Hydrolysis and oxidation in subcritical and supercritical water: connecting process engineering science to molecular interactions, Corrosion 55 (1999) 1088–1100.
[68] H. Schmieder, J. Abeln, Supercritical water oxidation: state of the art, Chem. Eng. Technol. 22 (1999) 903–908.
[69] C. Augustine, J.W. Tester, Hydrothermal flames: from phenomenological experimental demonstrations to quantitative understanding, J. Supercrit. Fluids 47 (2009) 415–430.

[70] H.L. La Roche, M. Weber, C. Trepp, Rationale for the film cooled coaxial hydrothermal burner (FCHB) for supercritical water oxidation (SCWO), in: Proceedings of the First International Workshop on Supercritical Water Oxidation, Jacksonville, FL, USA, February 6–9, 1995, pp. 125–133.

[71] P.R. von Rohr, K. Príkopský, T. Rothenfluh, Flames in supercritical water and their applications, Strojnícky Časopis 59 (2008) 91–103.

[72] C. Augustine, J. Potter, R. Potter, J.W. Tester, Feasibility of spallation drilling in a high pressure, high density, aqueous environment: characterization of heat transfer from an H_2–O_2 flame jet, Geotherm. Res. Council Trans. 31 (2007) 241–245.

[73] R.R. Steeper, Methane and methanol oxidation in supercritical water: chemical kinetics and hydrothermal flame studiesSandia report No. SAND96-8208, Dissertation. University of California, Davis, 1996.

[74] J.U. Steinle, E.U. Franck, High pressure combustion—ignition temperatures to 1000 bar, Ber. Bunsenges. Phys. Chem. 99 (1995) 66–73.

[75] T. Hirth, E.U. Franck, Oxidation and hydrothermolysis of hydrocarbons in supercritical water at high-pressures, Ber. Bunsenges. Phys. Chem. 97 (1993) 1091–1098.

[76] A. Sobhy, I.S. Butler, J.A. Kozinski, Selected profiles of high-pressure methanol—air flames in supercritical water, Proc. Combus. Inst. 31 (2007) 3369–3376.

[77] H. Sato, S. Hamada, R.M. Serikawa, T. Nishimura, T. Usui, H. Sekino, Continuous flame oxidation in supercritical water, High Pressure Res. 20 (2001) 403–413.

[78] R.M. Serikawa, T. Usui, T. Nishimura, H. Sato, S. Hamada, H. Sekino, Hydrothermal flames in supercritical water oxidation: investigation in a pilot scale continuous reactor, Fuel 81 (2002) 1147–1159.

[79] B. Wellig, Transpiring wall reactor for supercritical water oxidation, Doctor of Technical Sciences Thesis, Swiss Federal Institute of Technology, 2003. Available from, www.e-collection.ethz.ch.

[80] K. Príkopský, Characterization of continuous diffusion flames in supercritical water, Doctor of Technical Sciences ThesisSwiss Federal Institute of Technology, 2007. Available from, www.e-collection.ethz.ch.

[81] K. Príkopský, B. Wellig, P.R. von Rohr, SCWO of salt containing artificial waste water using a transpiring-wall reactor: experimental results, J. Supercrit. Fluids 40 (2007) 246–257.

[82] P. Stathopoulos, T. Rothenfluh, M. Schuler, D. Brkic, P.R. von Rohr, Assisted ignition of hydrothermal flames in a hydrothermal spallation drilling pilot plant, in: Proceedings Thirty-Seventh Workshop on Geothermal Reservoir Engineering, Stanford University, Stanford, California, January 30–February 1, 2012, SGP-TR-194.

Chapter 11

Hydrothermal and Supercritical Water Processing of Inorganic Substances

11.1 INTRODUCTION: WHAT IS HYDROTHERMAL PROCESSING?

Hydrothermal and supercritical water processing of inorganic substances comprises any action of such a substance or a mixture of substances with high-temperature liquid and supercritical water. In terms of temperature and pressure, such hydrothermal processes are in the range from $T = 100$ °C to the critical temperature of water ($T = 374$ °C) and for pressures of $P > 0.1$ to $P > 1500$ MPa. Since the properties of water change considerably in the range of the critical properties, most of the "hydrothermal" processes, as they will be called in this chapter, occur in the range of about $T = 300$–500 °C and pressures of $P = 10$–30 MPa.

Hydrothermal and supercritical water processes are an important part of geochemistry in natural hydrothermal systems. These processes will not be treated in this chapter. There is a vast literature available for this topic that can be approached from Ref. [1].

Hydrothermal processes can be used in the synthesis of fine particles, single crystals, functional metal oxides, and more. For many of these processes, subcritical water conditions are sufficient for the reaction [2]. Processing in near-critical and supercritical water is attractive because the density, viscosity, diffusivity, and dielectric constant of the reaction medium can be controlled through relatively small changes in temperature and pressure. Variation of the properties of the reaction medium makes feasible to control particle size, crystal structure, and morphology. Reaction at supercritical water conditions is a combination of hydrolysis, condensation, and thermal decomposition [2] that can be at least partly also controlled by adjustment of the operating conditions.

These reactions occur in high-temperature and supercritical water due to the specific physical and chemical properties of water at these conditions. In particular, a high solvent power and a high compressibility make possible

Supercritical Fluid Science and Technology, Vol. 5. http://dx.doi.org/10.1016/B978-0-444-59413-6.00011-X

to change and adjust properties to special needs. Special properties of near-critical and supercritical water are a low viscosity, a fast diffusive mass transport, a highly variable dielectric constant and degree of ionization, and the formation of hydrogen-bonded clusters of variable size.

Various processes are possible, such as [3]:

- Synthesis of new phases or stabilization of new complexes,
- crystal growth of inorganic compounds,
- preparation of finely divided materials and microcrystalline materials with well-defined size and morphology,
- leaching of ores in metal extraction,
- decomposition, alteration, corrosion, etching.

Reviews on hydrothermal processes and specific applications [1–5] provide access to more and detailed information that cannot be treated in context of this book. In this chapter, fundamentals for production processes with hydrothermal systems are discussed: First, physical processes for growing crystals, then, chemical reactions of water with inorganic compounds at high temperatures and pressures are addressed, followed by a discussion of the application of hydrothermal processes to produce very different inorganic materials. The examples show the variability of hydrothermal processes.

Hydrothermal reactions date back to the first half of the nineteenth century. An early example is the growth of barium carbonate from high-temperature liquid water [6]. Theoretical background of hydrothermal synthesis, experimental techniques, and all inorganic synthesis in supercritical water up to 1913 are summarized in a review article by Morey and Niggli [7]. Later, the hydrothermal chemistry of known minerals, mainly quartz, was published in 1953 by Morey [8]. Most of this work is related to geochemistry. A wide variety of known minerals, including oxides, silicates, phosphates, and sulfides were grown under hydrothermal conditions [3]. It was also demonstrated that geothermal brines, that is, aqueous mixtures containing cosolvents (mineralizers) such as halides, carbonates, or hydroxides, can solubilize solids like PbS and HgS and enable transport and hydrothermal reactions of the compounds [9–12]. Furthermore, it was realized that naturally occurring zeolites are formed by natural hydrothermal processes [13]. From that time on, new preparative routes to novel solid materials have been investigated and led to a rapid expansion of the field of synthesis of new compounds with hydrothermal methods.

11.2 PRINCIPLE OF HYDROTHERMAL PROCESSING

As mentioned above, hydrothermal processing comprises any action of inorganic substances with high-temperature liquid and supercritical water. Metal compounds are an important example of such compounds.

Metal hydroxides are formed when an aqueous metal salt solution is heated up due to the shift of the reaction equilibrium (Equation 11.1).

At higher temperatures, the equilibrium shifts further toward metal oxide formation (Equation 11.2) [14].

$$M^{x+} + xOH^- \rightarrow M(OH)_x \tag{11.1}$$

$$M(OH)_x \rightarrow MO_{x/2} + (x/2)H_2O \tag{11.2}$$

The method of producing metal oxides from aqueous solutions just by elevating the temperature is hydrothermal synthesis. This method is operated often at around $T = 100–200\ °C$, but operating conditions have been moved to high-temperature and supercritical water conditions due to the advantageous properties of water at these conditions and due to the development and availability of high pressure equipment [14].

Although the dielectric constant of water decreases near the critical point, the fluid remains sufficiently polar that most metal oxy-anions are soluble to some degree (>1%). Therefore, an aqueous base is often the first choice of solvent for most metal oxide-based preparations. Since a large number of mineral crystals were formed in a hydrothermal fluid, great interest was devoted to the chemistry of metal complexes in high-temperature and supercritical water [15].

Reactions of metal salt solutions in an aqueous environment are hydrolytic reactions (compare Chapter 5) that are catalyzed by either acids or bases. Therefore, an increased amount of hydroxyl groups will promote the formation of metal hydroxides. The hydroxyl groups may result from an increase in temperature due to an increase of the ionization of water (compare Chapter 2) or by adding bases. Such added basic compounds will have an influence also on the phase behavior of the aqueous system, in particular, on the solubility of salts, as will be discussed in the following in the context of crystal growth.

11.3 CRYSTAL GROWTH

11.3.1 Metal Oxides

Perhaps the most important hydrothermal process for the development of this topic is the crystallization of α-quartz. It is carried out by dissolution of SiO_2 in H_2O and subsequent precipitation of SiO_2 under adequate conditions for the formation of α-quartz crystals. The problem with conventional quartz crystal growth from melts is that α-quartz is only stable below $T = 580\ °C$ and silicate melts are so viscous that they form glasses long before crystallization occurs [3]. Therefore, methods for controlled crystal growth of α-quartz under hydrothermal conditions were developed. After a first attempt to grow crystals using hydrothermal techniques [16], currently many companies produce α-quartz commercially with a total output of >500 tons per year [3].

The solubility of SiO_2 in H_2O is shown in Figure 3.57. At $T = 400\ °C$ and $P = 30$ MPa, about 0.1 wt% of SiO_2 is soluble in water and 0.2 wt% at $P = 100$ MPa.

For the solubility of inorganic compounds in hydrothermal and supercritical water, compare Sections 3.5–3.7. A discussion of quantitative empirical and semi-empirical approaches to model solubility of inorganic compounds in hydrothermal and supercritical water is given by Leusbrock *et al.* [17]. The approach is based on phase equilibrium between the solid salt and the aqueous phase. The detailed discussion is beyond the scope of this book. Such a semi-empirical correlation is used for modeling the solubility of metal oxides in water.

The solubility of metal oxides in sub- and supercritical water can be described with Equation (11.3) [14,18], taking into account the hydration reaction.

$$\ln K = \ln K_0 + \frac{\Delta H_0}{R}\left(\frac{1}{T} - \frac{1}{T_0}\right) - \frac{\omega}{RT}\left(\frac{1}{\varepsilon} - \frac{1}{\varepsilon_0}\right) - \frac{2Z_A Z_B}{RT}\alpha\sqrt{I} \qquad (11.3)$$

with K, equilibrium constant of the hydration reaction; K_0, equilibrium constant at the reference state; H_0, enthalpy of reaction at the reference state; I, ionic strength; R, gas constant; Z_i, charge of component i; α, ω, constants determined by the system; ε, dielectric constant of the fluid; ε_0, dielectric constant at a reference state.

According to this equation, the reaction equilibrium varies with the change of the dielectric constant occurring at slight variations of temperature and pressure around the critical point [18].

The solubility can be improved by adding compounds. Those compounds act as cosolvents and are named "mineralizer" in context with hydrothermal processes.

Cosolvents (mineralizer) in hydrothermal processes are small soluble molecules at low concentrations that react with the initial inorganic compound and generate soluble intermediate molecules. The solubility of quartz in supercritical water is relatively low (compare Figure 3.57), but the presence of a small amount of hydroxide leads to formation of small silicate anions in concentrations of $x = 1$–3% according to Equation (11.4) [3].

$$3SiO_2 + 2OH^- \leftrightarrow Si_3O_7{}^{2-} + H_2O \qquad (11.4)$$

For crystallization of α-quartz, a nutrient solution is placed in an autoclave. Above the nutrient solution, seed crystals are placed. The nutrient solution is held at a higher temperature than the seed crystals. The anions migrate to a growing crystal site. The transport is facilitated by convection currents in the low viscosity fluid. Super-saturation condition is established at the growing crystal site through careful control of the temperature gradients. The equilibrium of Equation (11.4) is reversed and SiO_2 is deposited on the crystal face (Figure 11.1). Corrosion is of minor importance, since in presence of NaOH, the soluble silica reacts with the iron of the reactor walls to form acmite, $NaFeSiO_4$, and related compounds, which passivate the autoclave walls [15].

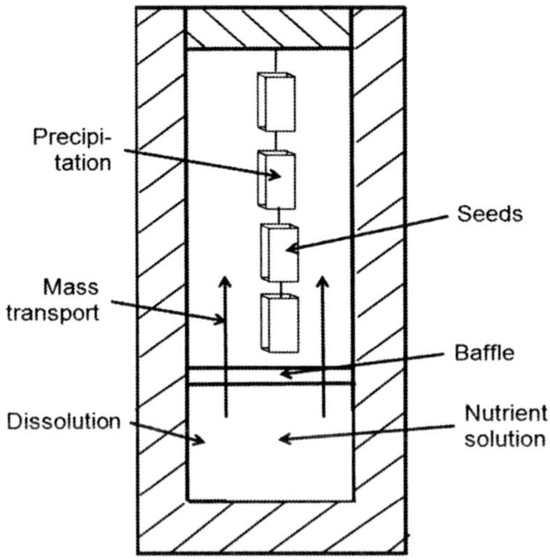

FIGURE 11.1 Principle of quartz crystal growth onto seed crystals.

Cosolvents (mineralizer) are an important factor in inorganic synthesis in hydrothermal and supercritical fluids. The most effective cosolvents are small nucleophiles such as OH^-, halides, sulfides, carbonates, and bicarbonates, added as alkali metal salts or ammonium salts [15]. The solubility of most of the amphoteric metal oxides is a linear function of the cosolvent concentration as shown in Figure 11.2.

Crystallization of quartz is a major commercial process, as the demand for quartz crystals is growing due to the ongoing expansion of mobile phone applications. Earlier reviews on the technology, available in the literature have been reported by the group of Laudise [19–21]. The recent state-of-the-art is reported in the *Handbook on Hydrothermal Technology* [1,3].

Other metal oxides can be prepared as single crystals from hydrothermal solution using methods similar to that for quartz since most of the oxides are amphoteric and behave both as an acid and as a base. Typically, hydroxides are used as cosolvent (mineralizer) at temperatures of $T \approx 400$ °C, a temperature gradient of $\Delta T = 40$–80 °C and pressures of $P \approx 70$–80 MPa. Examples are Al_2O_3 and ZnO [15].

Transition metal oxides are studied to a considerable extent due to their useful magnetic and electronic properties. Especially, iron oxides are suitable for hydrothermal synthesis. Optimum conditions for hematite growth are a crystallization temperature of $T = 385$–445 °C, $\Delta T = 5$ °C, $P > \approx 50$ MPa (71–77% initial filling of the autoclave), reaction solution 5 M NaOH in H_2O [22]. Barium ferrites, as well as ferrites of other iron oxides, such as $AgFeO_2$, synthesized under hydrothermal conditions, can be used in magnetic

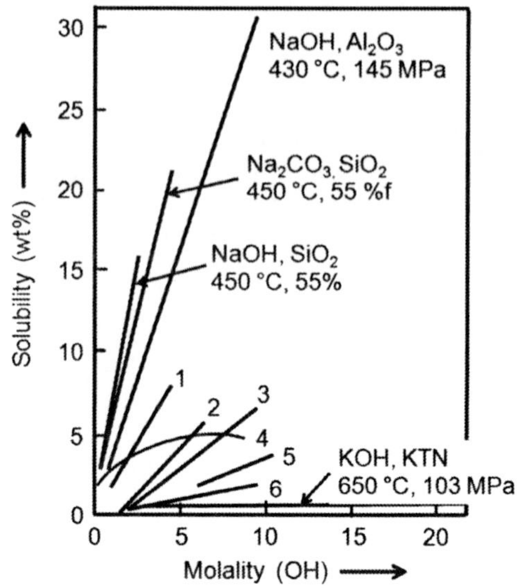

FIGURE 11.2 Solubility of inorganic compounds in aqueous solutions in dependence on cosolvent (mineralizer) concentration. 1: NaOH, $LiGeO_2$, $T=400$ °C, $P=58.6$ MPa; 2: NaOH, ZnO, $T=360$ °C, $P=55$ MPa; 3: KOH, ZnO, $T=360$ °C; 4: Na_2CO_3, Al_2O_3, $T=430$ °C, $P=145$ MPa; 5: KOH, ZnS, $T=450$ °C, $P=55$ MPa; 6: KOH, ZnS, $T=360$ °C, $P=55$ MPa. KTN is $KTa_{1-x}Nb_xO_3$. *Modified from Ref. [15].*

information storage devices. Hydrothermal methods produce finely divided pigments whose particles have a pronounced hexagonal, plate-like habit, a narrow particle size distribution, and advantageous magnetic properties. Ferromagnetic hexagonal ferrites exhibit a far greater coercive force than the conventional magnetic pigments [23]. Below, in this chapter, the generation of fine and nanosize particles is discussed in more detail. Another example is barium ferrite, $BaO{\cdot}2Fe_2O_3$, crystallized under hydrothermal conditions at around $T=260$ °C by heating a suspension of α-Fe_2O_3 in aqueous barium hydroxide solution [24].

Single crystals of oxides of sodium, iron and titanium can be grown hydrothermally at $T=650$ °C, $P=100$ MPa in 10 M NaOH solution, starting with a mixture of TiO_2 (rutile) and metallic Fe powder in the ratio 5:1 (w/w) [25]. After 1 week, single crystals with an orthorhombic cell are observed that undergo transformation to monoclinic freaudenbergite ($Na_2O{\cdot}Fe_2O_3{\cdot}7TiO_2$) [25]. Other examples are manganese oxides [26] and hematite (α-Fe_2O_3) single crystals [27].

Garnets, a group of silicate minerals that have been used as gemstones since ancient times, are nowadays of interest because of their optical and magnetic applications. Garnets can be used as laser hosts and as substrates for thin magnetic garnet films for bubble memory applications, formerly used as computer storage medium, but now outdated by more potent technologies. Although garnets are usually made from fluxes (melts), hydrothermal routes are also feasible from aqueous bases at temperatures of $T > 400\ °C$, for example, for Yttrium aluminum garnet (YAG), chemical composition $Y_3Al_5O_{12}$ (compare Equation 11.5) or $Gd_3Ga_5O_{12}$ [15,28,29].

$$Y_2O_3 + Al_2O_3 \rightarrow Y_3Al_5O_{12} \tag{11.5}$$

Operating conditions for the production of $Y_3Al_5O_{12}$ are $T = 450\ °C$, $P = 80$ MPa, aqueous medium: 6 M Na_2CO_3 in H_2O [15]. The hydrothermal synthesis of these and other related magnetic oxides has been reviewed in detail by Demianets [30] and Emelchenko *et al.* [31].

11.3.2 Gems

Laboratory synthesis of gems is possible under similar conditions as gems were formed from natural geothermal fluids [15]. An early report on hydrothermal synthesis of a gem is on the synthesis of ruby in aqueous base [32]. Sapphire, the corundum modification of aluminum oxide is grown on a seed crystal under hydrothermal conditions, exceeding temperatures of about $T = 395\ °C$, using Na_2CO_3 solutions as solvent. Synthetic ruby can be prepared, if corundum is doped with small quantities of chromium by the addition of sodium dichromate in low concentrations to the solvent [32].

Although high-quality crystals of rubies and sapphires can be prepared this way, melting alumina powders in hot gas flames is most probably more economic (Verneuil process [33]), unless for very specialized applications of these gems [15]. Hydrothermal synthesis for gem synthesis is best used for the quartz-based gems such as topaz, citrine, and amethyst. But low cost and relative abundance of natural materials may still render hydrothermal synthesis uneconomical [15].

Commercial emeralds were prepared hydrothermally by Lechleitner around 1960, and others, using aqueous media with HCl as the mineralizer, Equation (11.6) [33,34].

$$Al(OH)_3 + Be(OH)_2 + SiO_2 + \left[Cr(H_2O)_6\right]Cl_3 \rightarrow Be_3(Al, Cr)_2Si_6O_{18} \tag{11.6}$$

Operating conditions: $T = 450\ °C$, pressure not specified, and aqueous medium: HCl/H_2O [15].

Hydrothermal synthesis of diamond from graphite in supercritical water has been reported for conditions of $T = 800\ °C$ and $P = 170$ MPa [35–37]. Mixtures of SiC powder with diamond seeds are subjected to hydrothermal treatment at $T = 700–750\ °C$ under pressures of $P = 100–500$ MPa. Well-crystallized diamonds are formed in presence of diamond seeds. In a hydrothermal environment, diamond crystals, tens of micrometers in size, can be grown from a mixture of carbon, water, and usually pure nickel [35].

11.4 SYNTHESIS OF NEW PHASES AND COMPOUNDS

11.4.1 Phosphates and Silicates

Phosphates can be dissolved and crystallized under acidic and basic conditions. The polyanionic tetrahedral building block of phosphates in combination with even slight changes of reaction conditions lead to formation of many different products. Most syntheses dealt with metal phosphates at subcritical conditions, usually around $T=200\ °C$ [15]. The vast amount of literature can best be accessed via reviews and handbooks [3,15]. An incentive for the hydrothermal preparation of inorganic metal compounds, in particular, at higher temperatures of about $T>350\ °C$, is the preparation of new microporous solids for catalytic purposes. Microporous aluminum phosphates can be prepared using hydrothermal methods [38] as a new class of crystalline, microporous, alumophosphate phases similar to zeolites. The microporous aluminum phosphates are synthesized hydrothermally at $T=100$–$250\ °C$ from reaction mixtures containing an organic amine or quaternary ammonium salt (R) that is incorporated within the crystalline product of composition $x\mathrm{R}\cdot Al_2O_3(1.0\pm0.2)P_2O_5\cdot yH_2O$, with x and y as the amounts needed to fill the microporous voids within the neutral $AlPO_4$ framework. The species R is essential for the structure obtained in the synthesis, since without R dense $AlPO_4$ hydrates $AlPO_4\cdot yH_2O$ are formed [38].

Aluminum phosphate can be crystallized in near-critical water to well-formed large single crystals of $AlPO_4$, Equation (11.7) [15,21] at operating conditions of $T=275\ °C$ and a 6 M H_3PO_4 solution.

$$AlPO_4(\text{powder}) \rightarrow AlPO_4(\text{crystals}) \tag{11.7}$$

Another metal phosphate, important for electronic applications, is KTP ($KTiOPO_4$), used as a frequency doubler for solid state lasers [15]. Substantial effort was needed to develop the operating conditions, but now KTP can be produced commercially by hydrothermal methods similar to the related compound potassium titanyl arsenate (KTA) that is also used for electronic applications as a frequency doubler for solid state lasers [39]. Seed crystals are grown from a K_2WO_4–Li_2WO_4 flux at a temperature of $T=590\ °C$ and a pressure of $P=170$ MPa. Hydrothermal crystal growth is achieved under a gradient of $\Delta T=30\ °C$ with a 4 M KH_2AsO_4 cosolvent (mineralizer).

Other metal phosphates that can be synthesized hydrothermally in supercritical water are gallium phosphates [40], alkali metal cobalt phosphates, synthesized at relatively low temperatures of $T=250\ °C$, sodium cobalt phosphate, prepared at $T=500\ °C$ [41,42]. If transition metal phosphates are microporous, then catalytically active metals are present in the pore, leading to a new class of highly active catalysts [15]. In such a way, an iron phosphate containing ammonium as the counter-ion [43], and alkaline and alkaline earth iron phosphates with interesting structural types can be prepared [44].

Reactions are all extremely sensitive to reaction conditions. Different reaction conditions lead to different products in the temperature range between $T=200$ and 400 °C, compare Equations (11.8) and (11.9). Changing the cosolvent (mineralizer) or the iron source also leads to different products, as illustrated by Equations (11.10) and (11.11). Thus far, none of these products can be made using classical dry reactions [15].

$$BaHPO_4(\text{or } SrHPO_4) + 1.5FeO \rightarrow BaFeP_2O_7\left(\text{or } SrFe_3(PO_4)_3\right) \quad (11.8)$$

with operating conditions of $T=400$ °C and H_3PO_4 in H_2O [15].

$$BaHPO_4(\text{or } SrHPO_4) + 1.5FeO \rightarrow Ba_3Fe_2(HPO_4)_6(\text{or } SrFe(HPO_4)(PO_4)) \quad (11.9)$$

with operating conditions of $T=200$ °C and H_3PO_4 in H_2O [15].

$$NaOH + BaHPO_4 + 1.5Fe_3O_4 \rightarrow NaBaFe_4(HPO_4)_3 \cdot H_2O \quad (11.10)$$

with operating conditions of $T=400$ °C and H_3PO_4 in H_2O [15].

$$NaOH + BaHPO_4 + 3FeO \rightarrow BaFe_3P_2O_7 \quad (11.11)$$

with operating conditions of $T=400$ °C and H_3PO_4 in H_2O [15].

Metal silicates are of interest for electronic purposes and are prepared mostly in supercritical water at temperatures $T>400$ °C. Laudise and coworkers [45] prepared a series of alkali metal silicates of the rare earths, see Equation (11.12). Syntheses are carried out in aqueous solutions of NaOH, Na_2CO_3, NaF, $NaHF_2$, at temperatures from $T=350$ to 600 °C, pressures of $P=50$–140 MPa, and molarities ranging from 0 to 10 M [45].

$$NaOH + SiO_2 + Y_2O_3 \rightarrow Na_3YSi_6O_{15} \quad (11.12)$$

with operating conditions of $T=500$ °C and $P=80$ MPa in aqueous solution [15,45].

11.4.2 Metal Sulfides

Many metal sulfides are not hydrolyzed in supercritical water and can therefore be used for chemical reactions in hydrothermal fluids [15]. Early work is based on geochemical sulfide-based minerals, such as bornite (Cu_5FeS_5), chalcopyrite ($CuFeS_2$), and tetrahedrite ($Cu_{12}Sb_4S_{13}$) [46]. These sulfo-salts are solubilized under hydrothermal conditions using HS^- or Cl^- for transport. For some components, complex phase equilibrium and solubility are reported in detail in the literature [47]. At $T=350$ °C, under near-neutral to weakly acidic conditions, about 1000 ppm Fe and Cu dissolve in 1.0 M NaCl. At $T=250$ °C, solubility is reduced to about 100 ppm Fe and 10 ppm Cu [47].

Several metal sulfo-salts can be grown in the laboratory as large high-quality single crystals using hydrothermal processing, for example, of metal antimony sulfides and $MnSb_2S_4$ [48]. ZnS can be grown in electronic-grade quality in supercritical water [49]. ZnSe, ZnTe, CdSe, and CdTe form stable solid phases in OH^- solutions under hydrothermal conditions of $T=350$ °C and grow on seeds in small thermal gradients, while under similar conditions, HgS decomposes, forming Hg, and PbS and CdS hydrolyze [50].

11.4.3 Other Compounds

Examples of other hydrothermally crystallized compounds can be accessed via the review by Rabenau [4]. Metal fluorides can be prepared under hydrothermal conditions without hydrolysis, using fluoride as the mineralizer [15]. Thus, compounds such as $K[MF_4]$, $K_2[MF5]$, with M denoting a rare earth, and $[NH_4][FeF_4]$ can be prepared in supercritical water [51]. Crystals of K_2GdF_5, $KGdF_4$, and KGd2F7 can be grown and are stable with excess KF in solution. Phase equilibria of the system $KF\text{-}GdF_3\text{-}H_2O$ at $T=450$ °C and $P\approx 70$ MPa are the basis for determining process conditions [51].

Sillenites, $Bi_{12}MO_{20}$ (M = Si, Ge, or other metal ions), interesting because of their photorefractive properties, are advantageously produced via a hydrothermal route, since growth in an aqueous medium prevents the formation of defects created in the melt processes [15,52].

Formation of such compounds has been rarely considered from a point of reaction kinetics. An example, where the formation of the solid compound is observed *in situ*, is the formation of zinc-silicate in supercritical water [53]. The phase-formation mechanism of $\alpha\text{-}Zn_2SiO_4{:}Mn^{2+}$ crystal in supercritical water was investigated *in situ* with synchrotron radiation X-ray diffraction [53]. Green-emitting $\alpha\text{-}Zn_2SiO_4{:}Mn^{2+}$ crystals are formed at a temperature close to $T=374$ °C, at a pressure of $P=700$ MPa, in a reaction time of $t=400$ s [53].

11.5 HYDROTHERMAL PREPARATION OF MATERIALS, PRODUCTION OF FINE PARTICLES, IMPREGNATION, COATINGS

11.5.1 General Aspects

Synthesis of compounds in hydrothermal and supercritical water leads to stable phases of known compounds and to new compounds, as described above. The morphology of the products is determined by the method of the synthesis. In most cases, synthesis is carried out in autoclaves that are filled with the reactants and products are removed after depressurizing the reactor. If the synthesis is carried out in a continuous way, or the way to remove the products from the reactor is controlled, different morphology of the same product can be obtained, such as fine particles, porous particles, or defined layers on a substrate. This way of controlling not only the reaction conditions but also the removal from the reactor, the precipitation conditions, and the morphology of products is summarized as hydrothermal preparation of materials and will be treated in the following under the aspects of fine particle production, porous particle production, and coating.

Nonpolymeric materials, as discussed above, have many applications as submicron and nanosized particles, catalysts and coatings, for ceramics, electronic devices, and other applications [2]. Processing in hydrothermal and supercritical water is considered to be attractive due to higher diffusivity and lower viscosity of the reaction medium compared to conditions of normal sol–gel processing and because the particle growth mechanism can be controlled to produce submicron size, weakly associated, and uniform size particles. The hydrothermal processing route simplifies the process because particles produced are almost completely crystallized, eliminating additional treatment prior to sintering, such as settling and drying, that is needed with sol–gel processes [2].

The method of producing fine particles from hydrothermal and supercritical water solutions is described in the following. The product solution can be formed in a batch reactor or a continuously operated reactor. Since the particle generation step is the same for both ways, only the continuous way is described.

The method for synthesis at hydrothermal and supercritical conditions of fine particles in a continuously operated apparatus is illustrated in Figure 11.3 and described in the following. The metal salt solution as feed is mixed with supercritical water and fed into the reactor. Separate heating of the water phase and mixing with the mostly smaller amount of metal salt solution ensures a rapid heating of the reaction mixture to reaction temperature and a defined start of the reaction at the entrance of the reactor. The reaction mixture passes the reactor. At the outlet, temperature and pressure of the reaction mixture, containing the product compounds, is reduced in a controlled way. Product compounds precipitate, mostly as fine particles, and are

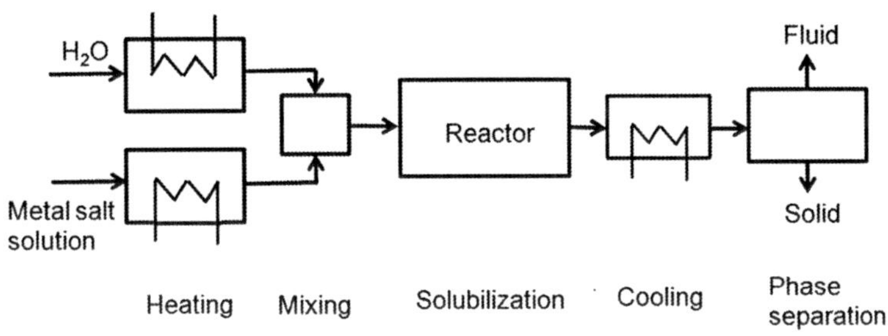

FIGURE 11.3 Scheme of an apparatus for continuous synthesis in hydrothermal and supercritical water.

separated from the aqueous solution. Cooling can be achieved by directly mixing water with the product solution at the outlet or by a cooling water jacket [14]. Thus, the reaction can be limited to the reactor and reactions in the heating-up and cooling-down zone, occurring at different temperatures to the reactor, can be avoided. Reaction conditions in the reactor are adjusted to the values needed for the reaction. Usually, temperature is in the range of $T=300$–550 °C, pressure in the range of $P=20$–70 MPa, and residence time $t=1$–5400 s.

This simple scheme can be varied in several ways by modifying the feed streams, for example, by introducing gases as reducing agents, or by various reactor constructions, and by the method of precipitating the products. These modifications can lead to different product compounds and especially to different product morphology. These aspects will be discussed for several examples.

11.5.2 Hydrolysis

An example for a hydrothermal hydrolysis process is the continuous production of $BaO_6Fe_2O_3$, carried out in a process scheme shown in Figure 11.3. An aqueous solution of barium and iron (III) nitrate is first mixed with an aqueous solution of potassium hydroxide, then, still before the reactor inlet, further mixed with high-temperature water, and at the reactor inlet with supercritical water. Thus, the reaction mixture is rapidly heated to reaction temperature, in particular, since the amount of supercritical water in most cases is substantially greater than the feed solution of the reacting components. A similar example is the production of $YOAl_2O_3$, from a starting solution of aqueous $Y(NO_3)_3$ and $Al(NO_3)_3$ [14].

Barium hexaferrite ($BaO_6Fe_2O_3$) is of interest due to its magnetic properties. It crystallizes as hexagonal plate. Magnetization takes place perpendicular to the plate plane. Barium hexaferrite hexagonal plate particles can be produced, as described above, with a reaction time of approximately $t=1$ min. The $BaO_6Fe_2O_3$ particles obtained at supercritical water conditions have magnetic properties comparable with commercial barium hexaferrite, while those obtained at subcritical conditions have poorer magnetic properties. It is supposed that hydroxyl functional groups are removed at supercritical conditions [14].

The continuous synthesis of fine metal oxide particles using hydrothermal and supercritical water as reaction medium leads from metal salts to metal oxides, according to Equations (11.1) and (11.2), in two overall reactions, first hydrolysis and then dehydration. The first two steps proceed at the same time in supercritical water, while they proceed sequentially in conventional processing. The dehydration step occurs while the particle size is small and the reaction rate is less affected by diffusion in the solid phase. The result is that the overall synthesis rate is very large, which is additionally supported by the

TABLE 11.1 Metal Oxides Synthesized in Hydrothermal and Supercritical Water [14]

Starting Materials	Products	Particle Size (nm)	Morphology
$Al(NO_3)_3$	AlOOH	80–1000	Hexagonal plate, rhombic, needle-like
$Fe(NO_3)_3$	α-Fe_2O_3	<50	Spherical
$Fe_2(SO_4)_3$	α-Fe_2O_3	<50	Spherical
$FeCl_2$	α-Fe_2O_3	<50	Spherical
$Fe(NH_4)2H(C_6H_5O_7)_2$	Fe_3O_4	<50	Spherical
$Co(NO_3)_2$	Co_3O_4	<100	Octahedral
$Ni(NO_3)_2$	NiO	<200	Octahedral
$ZrOCl_2$	ZrO_2 (cubic)	<10	Spherical
$Ti(SO_4)_2$	TiO_2	<20	Spherical
$TiCl_4$	TiO_2 (anatase)	<20	Spherical
$Ce(NO_3)_3$	CeO_2	20–300	Octahedral
$Fe(NO_3)_3$, $Ba(NO_3)_2$	$BaO{\cdot}6Fe_2O_3$	50–1000	Hexagonal plate
$Al(NO_3)_3$, $Y(NO_3)_3$, $TbCl_3$	$Al_5(Y+Tb)_3O_{12}$	20–600	Docahedral

elevated temperature and the high diffusivity in supercritical water [2]. Table 11.1 lists examples for the synthesis of metal oxides in supercritical water [14,54,55].

A great number of compounds have been synthesized for the production of fine particles. In the following, some of the reported compounds obtained are listed:

Lanthanum oxide nanoparticles [56], lithium iron phosphate ($LiFePO_4$) nanoparticles [57], NiO nanoparticles [58], zinc oxide nanoparticles [59], ZnO nanoparticles formation by reactions of bulk Zn with H_2O and CO_2 [60], CFD simulation of ZnO nanoparticle synthesis [61], hafnium oxide nanoparticles [62], effect of cations and anions on properties of zinc oxide particles [63], metallic cobalt nanoparticles [64], Bi_2Te_3 nanoparticles [65], γ-Al_2O_3 nanoparticles [66], Perovskite oxide $Ca_{0.8}Sr_{0.2}Ti_{1-x}Fe_xO_{3-\delta}$ (CTO) nanoparticles [67], anatase TiO_2 nanoparticles [68], nanoparticulate yttrium aluminum garnet [69], $CoFe_2O_4$ nanoparticles [70], YVO_4 and rare earth-doped YVO_4 ultrafine particles [71], lithium iron phosphate ($LiFePO_4$) [72], YAG monodispersed particles [73], luminescent yttrium aluminum garnet ($Y_3Al_5O_{12}$)

nanoparticles, doped with Eu(YAG:Eu^{3+}) [74], copper manganese oxide nanocrystals [75], Zn_2SiO_4:Mn^{2+} fine particles [76], iron nanoparticles [77], iron oxide (α-Fe_2O_3) nanoparticles in activated carbon [78], high-temperature $LiCoO_2$ [78,90], $KNbO_3$ powders [79], $MgFe_2O_4$ nanoparticles [80], Zn_2SnO_4 anode material (synthesized in batch mode) [81], lithium iron phosphate particles [82], $ZnGa_2O_4$:Mn^{2+} nanoparticles [83], magnetite particles [84], and lithium iron phosphate ($LiFePO_4$) nanoparticles [85], boehmite nanoparticles [89].

Formation of fine particles during hydrothermal and supercritical water synthesis of compounds is due to the extremely high hydrolysis reaction rate and the low solubility of produced compounds in supercritical water. This leads to a great number of crystallization germs at a high degree of supersaturation [14] resulting in a great number of fine particles.

Experiments with CeO_2 show that at temperatures above the critical temperature of water, conversion of the precursor cerium nitrate is completed in $t=0.4$–0.6 s, while for completing the reaction at 1ower temperature, several seconds are required. This can be attributed to the drastic reduction of the dielectric constant above the critical temperature, which favors the polarity reducing reactions of hydrotherma1 synthesis [14].

11.5.3 Addition of Oxidizing or Reducing Agents

An oxidizing or reducing compound can be mixed with the reactants before the hydrothermal synthesis reaction starts. Products can thus be oxidized or reduced and form new compounds in one reaction step.

For the production of $LiCoO_2$, a H_2O_2 solution is mixed with the starting solution of $LiNO_3$ and $CO(NO_3)_2$ for the oxidation of Co from Co^{2+} to Co^{3+}, H_2O_2 is decomposed into oxygen and water in the preheating region and forms a homogeneous phase at reaction conditions [14].

When ferrous ammonium citrate is used as a feed, magnetite particles are formed in hydrothermal synthesis. In this case, Fe^{3+} is reduced to Fe^{2+}, due to the reduction of Fe^{3+} by CO that is formed by the thermal decomposition of ammonium citrate. Since CO is miscible with supercritical water, the production of magnetite particles occurs in a single phase. This reducing and oxidizing atmosphere can be controlled by introducing O_2 or H_2 [14].

Oxidation of aluminum with water ($T=665$ K, $P=23.1$ MPa) and H_2O/CO_2 ($T=723$ K, $P=38.0$ MPa) results in the transition of aluminum into AlOOH and Al_2O_3 nanoparticles [86]. If the oxidation occurs at the injection, most α-Al_2O_3 particles are of the size of $d_P=300$–500 nm. If the oxidation is carried out in a uniform temperature increase, smaller particles of γ-$Al_2O_3$of $d_P=20$–70 nm size are formed as well. The addition of CO_2 to water results in the formation of CO, CH_4, CH_3OH, and condensed carbon, increase in size of Al_2O_3 nanoparticles, and significant decrease in time delay. In pure CO_2, an oxide film is formed [86].

Oxidation of solid and liquid zinc by water and CO_2 results under all the following conditions in the formation of ZnO nanoparticles [60]: With pure water at $T=335$–$648\ °C$, $P=13.7$–31.2 MPa, $\rho_w=0.045$–$0.420\ g/cm^3$, with supercritical CO_2 at $T=514$–$600\ °C$, $P=7.1$–24.4 MPa, $\rho_w=0.041$–$0.158\ g/cm^3$, with mixtures of H_2O and CO_2 at $T=140$–$600\ °C$, $P=30$ MPa, $\rho_w=0.026$–$0.569\ g/cm^3$, with an initial ratio of $[H_2O]_0/[CO_2]_0=0.46$–2.75 mol/mol. For the formation of nanoparticles of ZnO, a cluster mechanism is proposed.

Temperature dependence of the rate constant of Zn reacting with H_2O is described by Equation (11.13) [60].

$$k_w(T)=1.09\cdot10^{-4}\exp(-53.23\,[\mathrm{kJ/mol}]/RT) \qquad (11.13)$$

At the same temperature, the rate constant of Zn reacting with CO_2 is 44 times lower than k_W, and the rate constant of Zn reacting with H_2O/CO_2 mixture is 25.4% lower than k_W.[60].

In pure CO_2, besides CO, carbons are formed and in mixtures of H_2O and CO_2 along with H_2 and CO, paraffin and aromatic hydrocarbons, alcohols and formaldehyde are synthesized [60].

11.5.4 Thermal Conversion of Precursors

Precursor compounds can be converted to reactants by high temperature in hydrothermal pyrolysis. Magnetite-dispersed carbon and nickel-zinc ferrite-dispersed carbon particles are formed by hydrothermal pyrolysis of metal-containing divinylbenzene copolymers, or more general, organometallic polymers, in water at pressures of $P=30$–125 MPa and temperatures of $T=300$–$700\ °C$. The magnetic properties of the products processed in supercritical water are improved compared to conventional processed samples [2].

11.5.5 Control of Morphology During Hydrothermal Synthesis

It is of great interest to know the processing conditions for specific forms of the synthesized products, since properties of the product may change considerably with morphology. As an example, the morphology of AlOOH particles changes with temperature and pressure in the neighborhood of the critical point. AlOOH particles are rhombic plates at low concentration or at moderate pH, and hexagonal plates, or sword-like particles at higher pH or at higher concentration. The particle morphology can be related to the chemical species in solution. The distribution of the chemical species is determined using Equation (11.3). For the AlOOH system, the species Al^{3+}, $Al(OH)^{2+}$, $Al(OH)^{2+}$, $Al(OH)_3$, $Al(OH)_4^-$, and NO_3^- are calculated in subcritical ($T=350\ °C$, $P=30$ MPa) and supercritical water ($T=400\ °C$, $P=30$ MPa). Adsorption of positively charged species, $Al(OH)^{2+}$ on the negatively charged surface of the AlOOH crystal seems to determine particle morphology [14], as schematically shown in Figure 11.4.

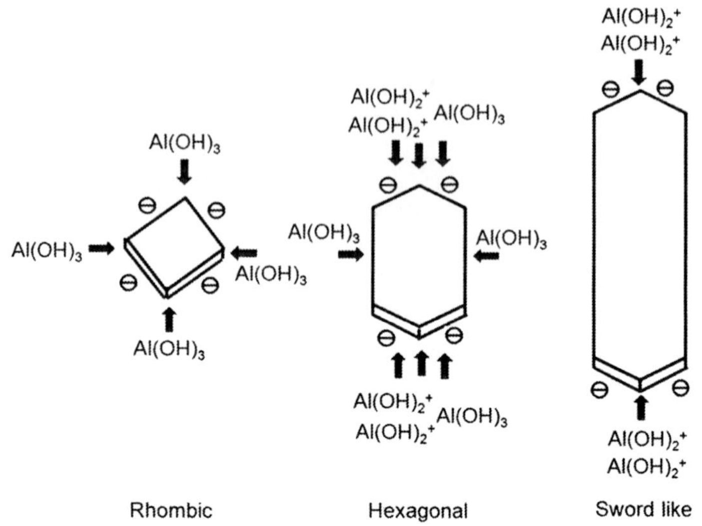

FIGURE 11.4 Morphology of AlOOH particles in hydrothermal syntheses in relation to the availability of different chemical species. *Modified from Ref. [14].*

11.5.6 Impregnation

Fine particles generated by hydrothermal and supercritical water synthesis can be used to impregnate substrates with the product compounds. For example, manganese oxide, silver, and lead oxide can be prepared on alumina supports from metal acetate solutions under supercritical water conditions [87]. Metal or metal oxide particles can be crystallized onto the support surface and in the deeper part of the spherical support. The crystallinity and composition of the deposited products can be controlled by temperature and pressure during the preparation with supercritical water [87]. On the surface and in the pores of activated carbon, metal silver nanoparticles can be deposited by supercritical water [88]. Deposited silver particles have a similar mean particle size as on α-alumina particles, namely, in the range of $d_P = 20$ nm. Deposition inside the support pores occurs without causing particle aggregations [88].

The specific features of hydrothermal synthesis of fine particles at supercritical conditions are summarized as follows:

- Ultrafine particles can be produced due to the rapid hydrolysis reaction and the lower solubility of metal oxides in supercritical conditions compared with subcritical conditions.
- The morphology of the produced particles can be controlled.
- O_2, CO, or H_2 gases can be used for controlling the oxidizing or reducing atmosphere. This method has been applied for the production of magnetic materials, $BaO_6Fe_2O_3$, phosphor, $YOAl_2O_3$, and lithium ion battery cathode materials.

REFERENCES

[1] K. Byrappa, M. Yoshimura, Handbook of Hydrothermal Technology, second ed., William Andrews (Imprint of Elsevier), Oxford, UK, Waltham, MA, USA, 2013.

[2] P.E. Savage, S. Gopalan, T.I. Mizan, C.J. Martino, E.E. Brock, Reactions at supercritical conditions: applications and fundamentals, AIChE J. 41 (1995) 1723–1778.

[3] K. Byrappa, Hydrothermal processing of materials, in: K. Byrappa, M. Yoshimura (Eds.), Handbook of Hydrothermal Technology. A Technology for Crystal Growth and Materials Processing, Noyes Publications/William Andrew Publication, Park Ridge, New Jersey, USA/Norwich, New York, USA, 2001, pp. 754–845.

[4] A. Rabenau, The role of hydrothermal synthesis in preparative chemistry, Angew. Chem. 24 (1985) 1026–1040.

[5] R.A. Laudise, Hydrothermal synthesis of crystals, Chem. Eng. News 65 (1987) 30–43.

[6] R. Bunsen, Bemerkungen zu einigen Einwürfen gegen mehrere Ansichten über die chemisch-geologischen Erscheinungen in Island, Justus Liebigs Ann. Chem. 65 (1848) 70–85 (Currently known as: European Journal of Organic Chemistry).

[7] G.W. Morey, P. Niggli, The hydrothermal formation of silicates, a review, J. Am. Chem. Soc. 35 (1913) 1086–1130.

[8] G.W. Morey, Hydrothermal synthesis, J. Am. Ceram. Soc. 36 (1953) 279–285.

[9] H.L. Barnes, Measuring thermodynamically interpretable solubilities at high pressures and temperatures, Phys. Chem. Earth 13 (14) (1981) 321–343.

[10] T.M. Seward, Metal complex formation in aqueous solutions at elevated temperatures and pressures, Phys. Chem. Earth 13 (14) (1981) 113–132.

[11] A. Rabenau, Methods for the study of hydrothermal crystallization, Phys. Chem. Earth 13 (14) (1981) 361–374.

[12] I.D. Ryabchikov, Mobilization of ore metals by supercritical fluids from crystallizing magmas, Phys. Chem. Earth 13 (14) (1981) 529–536.

[13] R.M. Barrer, Hydrothermal Chemistry of Zeolites, Academic Press, London, 1982.

[14] T. Adschiri, K. Arai, Hydrothermal synthesis of metal oxide fine particles in water, in: Y. Arai, T. Sako, Y. Takebayashi (Eds.), Supercritical Fluids, Springer, Berlin, 2002, pp. 331–340, Chapter 5.4.

[15] J.W. Kolis, M.B. Korzenski, Reactions in SCF. Synthesis of inorganic solids. Hydrothermal chemistry, in: G. Jessop, W. Leitner (Eds.), Chemical Synthesis Using Supercritical Fluids, Wiley-VCH, Weinheim, 1999, pp. 226–234.

[16] R. Nacken, Hydrothermale Mineralsynthese zur Züchtung von Quarzkristallen, Chem. Ztg. 74 (1950) 745.

[17] I. Leusbrock, S.J. Metz, G. Rexwinkel, G.F. Versteeg, Quantitative approaches for the description of solubilities of inorganic compounds in near-critical and supercritical water, J. Supercrit. Fluids 47 (2008) 117–127.

[18] K. Sue, Y. Hakuta, R.L. Smith, T. Adschiri, K. Arai, Solubility of lead(II) oxide and copper(II) oxide in subcritical and supercritical water, J. Chem. Eng. Data 44 (1999) 1422–1426.

[19] A.A. Ballman, R.A. Laudise, in: J.J. Gilman (Ed.), The Art and Science of Growing Crystals, Wiley and Sons, New York, USA, 1963, p. 231.

[20] R.A. Laudise, in: P. Hartman (Ed.), Crystal Growth: An Introduction, North-Holland, Amsterdam, The Netherlands, 1973.

[21] R.A. Laudise, in: E. Kaldis (Ed.), Crystal Growth of Electronic Materials, Elsevier, Amsterdam, The Netherlands, 1985, Ch. 13, cited from [13].

[22] E.D. Kolb, A.J. Caporaso, R.A. Laudise, Hydrothermal growth of hematite and magnetite, J. Cryst. Growth 19 (1973) 242–246.

[23] H. Hibst, Hexagonal ferrites from melts and aqueous solutions, magnetic recording materials, Angew. Chem. 21 (1982) 270–282.

[24] S. Okamoto, H. Sekizawa, S.I. Okamoto, Hydrothermal synthesis, structure and magnetic properties of barium diferrite, J. Phys. Chem. Solid 36 (1973) 591–595.

[25] S. Hirano, M.G.M.U. Ismail, S. Somiya, Crystal growth of sodium iron titanium bronze compounds under hydrothermal conditions, Mater. Res. Bull 11 (1976) 1023–1030.

[26] T. Endo, S. Kume, M. Shimada, M. Koizumi, Synthesis of potassium manganese oxides under hydrothermal conditions, Mineral. Mag. 39 (1974) 559.

[27] L.N. Demianets, S.V. Pouchko, R.V. Gaynutdinov, Fe_2O_3 single crystals: hydrothermal growth, crystal chemistry and growth morphology, J. Cryst. Growth 259 (2003) 165–178.

[28] E.D. Kolb, R.A. Laudise, Phase equilibria of $Y_3Al_5O_{12}$, hydrothermal growth of $Gd_3Ga_5O_{12}$ and hydrothermal epitaxy of magnetic garnets, J. Cryst. Growth 29 (1975) 29–39.

[29] R.A. Laudise, Single crystals for bubble domain memories, J. Cryst. Growth 13 (14) (1972) 27–33.

[30] L.N. Demianets, Organic Crystals, Germanates, Semiconductors, Springer, Berlin, 1980.

[31] G.A. Emelchenko, H. Maier, J. Hesse, Organic Crystals Germanates Semiconductors, Springer, Berlin, ISBN: 978-3-642-67766-3, 1980, 2011.

[32] R.A. Laudise, A.A. Ballman, Hydrothermal synthesis of sapphire, J. Am. Chem. Soc. 80 (1958) 2655–2657.

[33] K. Nassau, Gems Made By Man, Gemological Institute of America, Santa Monica, CA, 1980.

[34] E.M. Flanigen, N.R. Mumbach, Hydrothermal process for growing crystals having the structure of beryl in an acid halide medium, Union Carbide Corp., US Patent 3,567,643, March 2, 1971.

[35] X.-Z. Zhao, R. Roy, K.A. Cherian, A. Badzian, Hydrothermal growth of diamond in metal–C–H_2O systems, Nature 385 (1997) 513–515.

[36] R. Riedel, K.G. Nickel, T. Kraft, Y.G. Gogotsi, Hydrothermal synthesis of diamond, in: Handbook of Ceramic Hard Materials, Wiley-VCH, Weinheim, Germany, 2000, Chapter 3.

[37] Y.G. Gogotsi, K.G. Nickel, P. Kofstad, Hydrothermal synthesis of diamond from diamond-seeded β-SiC powder, J. Mater. Chem. 5 (1995) 2313–2314.

[38] S.T. Wilson, B.M. Lok, C.A. Messina, T.R. Cannan, E.M. Flanigen, Aluminophosphate molecular sieves: a new class of microporous crystalline inorganic solids, J. Am. Chem. Soc. 104 (1982) 1146–1147.

[39] R.F. Belt, J.B. Ings, Hydrothermal growth of potassium titanyl arsenate (KTA) in large autoclaves, J. Cryst. Growth 128 (1993) 956–962.

[40] W.T.A. Harrison, M.L.F. Phillips, G.D. Stucky, Substitution chemistry of gallium for titanium in nonlinear optical KTiOPO4: syntheses and single-crystal structures of KGaF1-δ(OH)δPO4 (δ0.3) and KGa0.5Ge0.5(F, OH)0.5PO4, Chem. Mater. 7 (1995) 1849–1856.

[41] P. Feng, X. Bu, S.H. Tolbert, G.D. Stucky, Syntheses and characterizations of chiral tetrahedral cobalt phosphates with zeolite ABW and related frameworks, J. Am. Chem. Soc. 119 (1997) 2497–2504.

[42] P. Feng, X. Bu, G.D. Stucky, Hydrothermal syntheses and structural characterization of zeolite analogue compounds based on cobalt phosphate, Nature 388 (1997) 735–741.

[43] S. Boudin, K.-H. Lii, Ammonium iron (II, III) phosphate: hydrothermal synthesis and characterization of NH4Fe2(PO4)2, Inorg. Chem. 37 (1998) 799–803.

[44] M.B. Korzenski, G.L. Schimek, J.W. Kolis, G.J. Long, Hydrothermal synthesis, structure, and characterization of a mixed-valent iron (II/III) phosphate, $NaFe_3.67(PO_4)_3$: a new variation of the Alluaudite structure type, J. Solid State Chem. 139 (1998) 152–160.

[45] S.M. Haile, B.J. Wuensch, R.A. Laudise, Hydrothermal synthesis of new alkali silicates II. Sodium neodymium and sodium yttrium phases, J. Cryst. Growth 131 (1993) 373–386.

[46] P.B. Barton, B.J. Skinner, H.L. Barnes, Geochemistry of Hydrothermal Ore Deposits, second ed., Wiley & Sons, New York, 1979, p. 27.

[47] D.A. Crerar, H.L. Barnes, Ore solution chemistry; V. Solubilities of chalcopyrite and chalcocite assemblages in hydrothermal solution at 200 degrees to 350 degrees C, Econ. Geol. 71 (1976) 772–794.

[48] J.W. Kolis, M.B. Korzenski, The synthesis of transition metal sulfides and sulfo-salt crystals in hydrothermal brines, in: P.K. Davies, A.J. Jacobson, C.C. Torardi, T.A. Vanderah (Eds.), Proceedings of the 1996 MRS Fall Meeting, Materials Research Society 453 (1997) 35, 1997.

[49] R.A. Laudise, A.A. Ballman, Hydrothermal synthesis of zinc oxide and zinc sulfide, J. Phys. Chem. 64 (1960) 688–691.

[50] E.D. Kolb, A.J. Caporaso, R.A. Laudise, Hydrothermal crystallization of some II–VI compounds, J. Cryst. Growth 3–4 (1968) 422–425.

[51] P.M. Bridenbaugh, J.O. Eckert, G. Nykolak, G. Thomas, W. Wilson, L.M. Demianets, R. Riman, R.A. Laudise, Hydrothermal growth and properties of $KGdF_4$, J. Cryst. Growth 144 (1994) 243–252.

[52] J. Larkin, M. Harris, E. Cormier, A. Armington, Hydrothermal growth of bismuth silicate (BSO), J. Cryst. Growth 128 (1993) 871–875.

[53] M. Takesue, K. Shimoyama, K. Shibuki, A. Suino, Y. Hakuta, H. Hayashi, Y. Ohishi, R.L. Smith Jr., Formation of zinc silicate in supercritical water followed with in situ synchrotron radiation X-ray diffraction, J. Supercrit. Fluids 49 (2009) 351–355.

[54] T. Adschiri, K. Kanazawa, K. Arai, Rapid and continuous hydrothermal crystallization of metal oxide particles in supercritical water, J. Am. Ceram. Soc. 75 (1992) 1019–1022.

[55] T. Adschiri, K. Kanazawa, K. Arai, Rapid and continuous hydrothermal synthesis of Boehmite particles in subcritical and supercritical water, J. Am. Ceram. Soc. 75 (1992) 2615–2618.

[56] S.J. Nejad, H. Abolghasemi, M.A. Moosavian, A. Golzary, M.G. Maragheh, Fractional factorial design for the optimization of hydrothermal synthesis of lanthanum oxide nanoparticles under supercritical water condition, J. Supercrit. Fluids 52 (2010) 292–297.

[57] S.-A. Hong, S.J. Kim, K.Y. Chung, M.-S. Chun, B.G. Lee, J. Kim, Continuous synthesis of lithium iron phosphate (LiFePO4) nanoparticles in supercritical water: effect of mixing tee, J. Supercrit. Fluids 73 (2013) 70–79.

[58] S.-I. Kawasaki, K. Sue, R. Ookawara, Y. Wakashima, A. Suzuki, Y. Hakuta, K. Arai, Engineering study of continuous supercritical hydrothermal method using a T-shaped mixer: experimental synthesis of NiO nanoparticles and CFD simulation, J. Supercrit. Fluids 54 (2010) 96–102.

[59] R. Viswanathan, R.B. Gupta, Formation of zinc oxide nanoparticles in supercritical water, J. Supercrit. Fluids 27 (2003) 187–193.

[60] A.A. Vostrikov, O.N. Fedyaeva, A.V. Shishkin, M.Ya Sokol, ZnO nanoparticles formation by reactions of bulk Zn with H2O and CO2at sub- and supercritical conditions: I. Mechanism and kinetics of reactions, J. Supercrit. Fluids 48 (2009) 154–160.

[61] A. Leybros, R. Piolet, M. Ariane, H. Muhr, F. Bernard, F. Demoisson, CFD simulation of ZnO nanoparticle precipitation in a supercritical water synthesis reactor, J. Supercrit. Fluids 70 (2012) 17–26.

[62] A. Sahraneshin, S. Takami, D. Hojo, K. Minami, T. Arita, T. Adschiri, Synthesis of shape-controlled and organic-hybridized hafnium oxide nanoparticles under sub- and supercritical hydrothermal conditions, J. Supercrit. Fluids 62 (2012) 190–196.

[63] K. Sue, K. Kimura, K. Murata, K. Arai, Effect of cations and anions on properties of zinc oxide particles synthesized in supercritical water, J. Supercrit. Fluids 30 (2004) 325–331.

[64] G. Seong, S. Takami, T. Arita, K. Minami, D. Hojo, A.R. Yavari, T. Adschiri, Supercritical hydrothermal synthesis of metallic cobalt nanoparticles and its thermodynamic analysis, J. Supercrit. Fluids 60 (2011) 113–120.

[65] J.-L. Mi, T.N. Jensen, P. Hald, J. Overgaard, M. Christensen, B.B. Iversen, Glucose-assisted continuous flow synthesis of Bi2Te3 nanoparticles in supercritical/near-critical water, J. Supercrit. Fluids 67 (2012) 84–88.

[66] T. Noguchi, K. Matsui, N.M. Islam, Y. Hakuta, H. Hayashi, Rapid synthesis of γ-Al_2O_3 nanoparticles in supercritical water by continuous hydrothermal flow reaction system, J. Supercrit. Fluids 46 (2008) 129–136.

[67] J. Lu, Y. Hakuta, H. Hayashi, T. Ohashi, T. Nagase, Y. Hoshi, K. Sato, M. Nishioka, T. Inoue, S. Hamakawa, Preparation of $Ca_{0.8}Sr_{0.2}Ti_{1-x}Fe_xO_{3-\delta}$ ($x=0.1$–0.3) nanoparticles using a flow supercritical reaction system, J. Supercrit. Fluids 46 (2008) 77–82.

[68] S.-I. Kawasaki, Y. Xiuyi, K. Sue, Y. Hakuta, A. Suzuki, K. Arai, Continuous supercritical hydrothermal synthesis of controlled size and highly crystalline anatase TiO_2 nanoparticles, J. Supercrit. Fluids 50 (2009) 276–282.

[69] A. Cabañas, J. Li, P. Blood, T. Chudoba, W. Lojkowski, M. Poliakoff, E. Lester, Synthesis of nanoparticulate yttrium aluminum garnet in supercritical water-ethanol mixtures, J. Supercrit. Fluids 40 (2007) 284–292.

[70] D. Zhao, X. Wu, H. Guan, E. Han, Study on supercritical hydrothermal synthesis of $CoFe_2O_4$ nanoparticles, J. Supercrit. Fluids 42 (2007) 226–233.

[71] Q.X. Zheng, B. Li, M. Xue, H.D. Zhang, Y.J. Zhan, W.S. Pang, X.T. Tao, M.H. Jiang, Synthesis of YVO_4 and rare earth-doped YVO_4 ultra-fine particles in supercritical water, J. Supercrit. Fluids 46 (2008) 123–128.

[72] S.-A. Hong, S.J. Kim, J. Kim, K.Y. Chung, B.-W. Cho, J.W. Kang, Small capacity decay of lithium iron phosphate ($LiFePO_4$) synthesized continuously in supercritical water: comparison with solid-state method, J. Supercrit. Fluids 55 (2011) 1027–1037.

[73] Q.X. Zheng, B. Li, H.D. Zhang, J.J. Zheng, M.H. Jiang, X.T. Tao, Fabrication of YAG mono-dispersed particles with a novel combination method employing supercritical water process, J. Supercrit. Fluids 50 (2009) 77–81.

[74] J.-H. In, H.-C. Lee, M.-J. Yoon, K.-K. Lee, J.-W. Lee, C.-H. Lee, Synthesis of nano-sized YAG:Eu^{3+} phosphor in continuous supercritical water system, J. Supercrit. Fluids 40 (2007) 389–396.

[75] D. Rangappa, S. Ohara, M. Umetsu, T. Naka, T. Adschiri, Synthesis, characterization and organic modification of copper manganese oxide nanocrystals under supercritical water, J. Supercrit. Fluids 44 (2008) 441–445.

[76] K. Shibuki, M. Takesue, T.M. Aida, M. Watanabe, H. Hayashi, R.L. Smith Jr., Continuous synthesis of Zn_2SiO_4:Mn^{2+} fine particles in supercritical water at temperatures of 400–500 °C and pressures of 30–35 MPa, J. Supercrit. Fluids 54 (2010) 266–271.

[77] T. Arita, H. Hitaka, K. Minami, T. Naka, T. Adschiri, Synthesis of iron nanoparticle: challenge to determine the limit of hydrogen reduction in supercritical water, J. Supercrit. Fluids 57 (2011) 183–189.

[78] C. Xu, A.S. Teja, Supercritical water synthesis and deposition of iron oxide (α-Fe_2O_3) nanoparticles in activated carbon, J. Supercrit. Fluids 39 (2006) 135–141.

[79] B. Li, Y. Hakuta, H. Hayashi, Hydrothermal synthesis of $KNbO_3$ powders in supercritical water and its nonlinear optical properties, J. Supercrit. Fluids 35 (2005) 254–259.

[80] T. Sasaki, S. Ohara, T. Naka, J. Vejpravova, V. Sechovsky, M. Umetsu, S. Takami, B. Jeyadevan, T. Adschiri, Continuous synthesis of fine $MgFe_2O_4$ nanoparticles by supercritical hydrothermal reaction, J. Supercrit. Fluids 53 (2010) 92–94.

[81] J.-W. Lee, C.-H. Lee, Synthesis of Zn_2SnO_4 anode material by using supercritical water in a batch reactor, J. Supercrit. Fluids 55 (2010) 252–258.

[82] C. Xu, J. Lee, A.S. Teja, Continuous hydrothermal synthesis of lithium iron phosphate particles in subcritical and supercritical water, J. Supercrit. Fluids 44 (2008) 92–97.

[83] H. Hayashi, A. Suino, K. Shimoyama, M. Takesue, S. Tooyama, R.L. Smith Jr., Continuous hydrothermal synthesis of $ZnGa_2O_4:Mn^{2+}$ nanoparticles at temperatures of 300–500 °C and pressures of 25–35 MPa, J. Supercrit. Fluids 77 (2013) 1–6.

[84] K. Karakama, S.N. Rogak, A. Alfantazi, Characterization of the deposition and transport of magnetite particles in supercritical water, J. Supercrit. Fluids 71 (2012) 11–18.

[85] J. Lee, A.S. Teja, Characteristics of lithium iron phosphate ($LiFePO_4$) particles synthesized in subcritical and supercritical water, J. Supercrit. Fluids 35 (2005) 83–90.

[86] A.A. Vostrikov, O.N. Fedyaeva, Mechanism and kinetics of Al_2O_3 nanoparticles formation by reaction of bulk Al with H_2O and CO_2 at sub- and supercritical conditions, J. Supercrit. Fluids 55 (2010) 307–315.

[87] J. Otsu, Y. Oshima, New approaches to the preparation of metal or metal oxide particles on the surface of porous materials using supercritical water: development of supercritical water impregnation method, J. Supercrit. Fluids 33 (2005) 61–67.

[88] O. Sawai, Y. Oshima, Deposition of silver nano-particles on activated carbon using supercritical water, J. Supercrit. Fluids 47 (2008) 240–246.

[89] T. Mousavand, S. Ohara, M. Umetsu, J. Zhang, S. Takami, T. Naka, T. Adschiri, Hydrothermal synthesis and in situ surface modification of boehmite nanoparticles in supercritical water, J. Supercrit. Fluids 40 (2007) 397–401.

[90] Y.H. Shin, S.-M. Koo, D.S. Kim, Y.-H. Lee, B. Veriansyah, J. Kim, Y.-W. Lee, Continuous hydrothermal synthesis of HT-$LiCoO_2$ in supercritical water, J. Supercrit. Fluids 50 (2009) 250–256.

Chapter 12

Corrosion in Hydrothermal and Supercritical Water

12.1 INTRODUCTION

Corrosion is the deterioration of a material (usually a metal) that results from a chemical or electrochemical reaction with its environment [1]. In the context of hydrothermal and supercritical water processes, corrosion includes all reactions of components of an aqueous reaction mixture with the walls of the equipment of experimental or production size facilities, such as pipes, valves, fittings, and autoclaves. Corrosion comprises general corrosion, intergranular corrosion, pitting, and stress corrosion cracking (SCC) [2], as discussed in the following.

Corrosion reduces the strength of the vessel walls, causes porosity of the construction materials, and contaminates the reaction products with the corrosion products. Initial corrosion, seen as the first reaction of the containment walls with the reaction medium, may produce reaction products that form protecting layers on the walls and prevent further degradation of the autoclave materials. Chapter 5 covers corrosion as a chemical reaction, in addition to the specific aspects discussed in this chapter.

Corrosion in hydrothermal and supercritical water processes depends on the properties of the reaction mixture, on the properties of the construction materials, and on the construction of the processing equipment (reactor). In general, no satisfactory solution for corrosion problems can be achieved by addressing only one aspect of a process. On the contrary, only a comprehensive adjustment of all aspects influencing corrosion can lead to a process configuration that can be operated successfully and for a reasonable, extended time span. The following discussion is based mainly on reviews [2–5] that have addressed the problems in more detail than is possible in this book. Nevertheless, individual investigations will also be discussed in the following and references given that lead to detailed presentation of results.

Influences on corrosion, related to the reaction mixture, are the operating conditions, pressure and temperature, the density of the reaction mixture, the composition of the reaction mixture regarding the type of the components and the range of concentration, and the resulting values of pH, electrochemical

Supercritical Fluid Science and Technology, Vol. 5. http://dx.doi.org/10.1016/B978-0-444-59413-6.00012-1

potential, and reactivity of the involved compounds. Influences on corrosion, related to the construction material, include the composition of the metal, its surface preparation, its heat pretreatment, and the mechanical stress induced by assembling the equipment and by the operating conditions. Construction of the equipment influences contact between the reaction mixture and the surrounding surface with respect to reaction time, flow regime, and the parts of the surface that are mainly contacted.

Density of the aqueous solution is the most important parameter to be considered related to corrosion in hydrothermal and supercritical water, as density directly influences the solubility and the dissociation of the compounds of the reaction mixture and of the corrosion products, and the density is highly variable in this region.

A high solubility and a high ionic product favor ionic reactions and accelerate electrochemical forms of corrosion [2]. At low density, as it occurs under supercritical conditions at moderate pressures, water behaves like a polar solvent, the concentration of H^+ drops down, and solutions become less aggressive. Corrosion products can precipitate on the surface and stop the material loss. For example, the solubility of NaCl decreases at $P=25$ MPa from $x=700$ ppm at $T=400$ °C to about $x=200$ ppm at $T=450$ °C and to $x=100$ ppm at $T=500$ °C.

In the following section, first corrosion phenomena are described as occurring in high-temperature aqueous systems. Then, the parameters influencing corrosion are discussed in some detail together with corrosion mechanisms. The effect of corrosion is illustrated by reporting on a number of detailed investigations on corrosion in aqueous systems. A detailed discussion of measures that can be used against corrosion follows and shows the necessity of combining several approaches and ends in a compilation of means against corrosion.

12.2 TYPES OF CORROSION IN HYDROTHERMAL AND SUPERCRITICAL WATER

Aqueous solutions at hydrothermal and supercritical water conditions are highly oxidizing for wet air oxidation and supercritical water oxidation (SCWO) or highly reducing for supercritical water gasification (SCWG), industrial wastewater treatment, or biomass gasification for hydrogen production at low oxygen concentrations [6]. Different types of corrosion occur in such environments like general corrosion, pitting corrosion, intergranular corrosion, and stress-induced corrosion.

12.2.1 General Corrosion

General corrosion attacks the entire surface. General corrosion occurs if none of the alloy components can form a stabile protective layer as an oxide film.

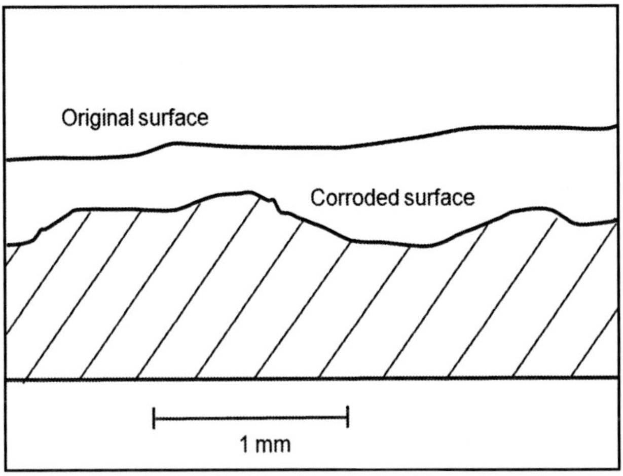

FIGURE 12.1 Typical wave-like form of general corrosion (shallow pitting). *Modified from Ref. [2].*

The typical morphology is a shallow, wave-like pitting as illustrated schematically in Figure 12.1, induced by active and transpassive dissolution of materials [2].

Corrosion rates for general corrosion are linear because they are diffusion controlled and are in the range of 500 μm/100 h (nickel-based alloy 625; conditions: [HCl] = 0.05 mol/kg; $[O_2]$ = 0.48 mol/kg; T = 350 °C; P = 24 MPa; t = 124 h) [7]. Starting points of general corrosion are the weakest points of the surface. Therefore, corrosion may begin as pitting corrosion that turns into general corrosion above a certain temperature, typically in the range of T = 200–250 °C. Due to the diffusion-controlled mechanism, the corrosion rates can be predicted. General corrosion is a common phenomenon in oxidizing hydrothermal and supercritical water [2].

12.2.2 Pitting Corrosion

Pitting corrosion is a local form of corrosion that can occur at passivated surfaces. Inclusions or grain boundaries at the surface can be the starting point of destruction. Aggressive anions penetrate the film at such places and form small pits, as shown in Figure 12.2. The phenomenon is stochastic regarding the corrosion sites, and the corrosion rates are not linear. The surface adjacent to pitting corrosion remains intact. Pitting corrosion rates may be high due to a change in the local chemistry inside the pits. Pitting corrosion is usually in the range of 1000 μm/100 h (nickel-based alloy 625; conditions: [HCl] = 0.05 mol/kg; $[O_2]$ = 0.48 mol/kg; T = 160 °C; P = 24 MPa; t = 124 h) [7].

Reaction of dissolved metal components with water leads to a strong acidification of the reaction mixture inside the pits favoring further corrosion.

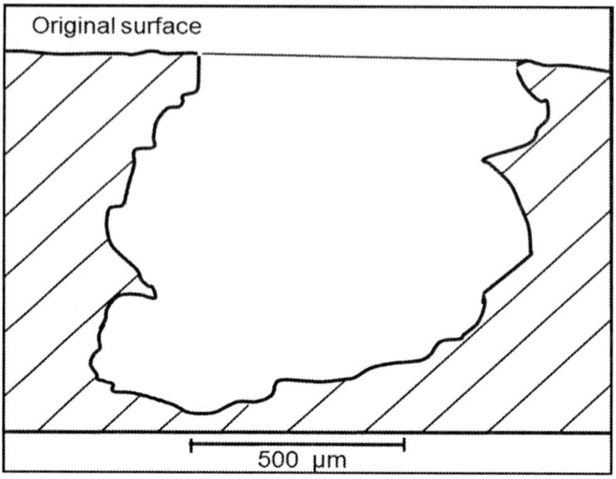

FIGURE 12.2 Pitting corrosion. *Modified from Ref. [2].*

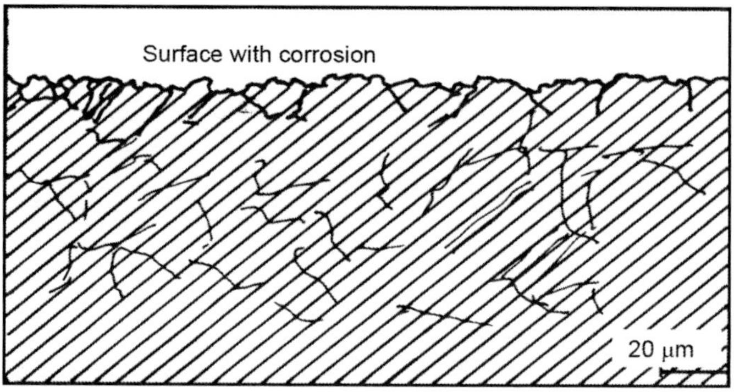

FIGURE 12.3 Intergranular corrosion. *Modified from Ref. [2].*

Pitting corrosion generally occurs in high rates. Increasing temperature additionally weakens the oxide film and further increases pitting corrosion. Due to its stochastic and nonpredictive behavior, pitting corrosion is a dangerous form of corrosion. References for review articles on pitting corrosion are presented by Kritzer [2].

12.2.3 Intergranular Corrosion

Intergranular corrosion occurs preferentially between grain boundaries and grains (Figure 12.3). A significant material loss may not occur. Corrosion rates are in the range of 10 μm/100 h (nickel-based alloy 625; conditions: $[HCl]=0.05$ mol/kg; $[O_2]=0.48$ mol/kg; $T=450$ °C; $P=24$ MPa; $t=150$ h) [7].

Grain boundaries and adjacent grain surfaces may be chemically different from the bulk grains. Furthermore, new compounds can be formed at the grain boundaries, as well as trace compounds may be enriched or removed. Local electrochemical elements might be formed. Corrosion depth and amount of dissolved material in intergranular corrosion usually are low, but extended reaction time may lead to dissolution of whole grains and mechanical stress may lead to the dangerous SCC [2]. References for detailed investigations on intergranular corrosion are given in Ref. [2].

12.2.4 Stress Corrosion Cracking

SCC is a stochastic form of corrosion in which a combination of mechanical stress and chemical reactions leads to fast corrosion that may end in a failure due to leakage. SCC occurs along the grain boundaries or through the grains. It can start at the bottom of a shallow pit and run through the wall of a tubular reactor, as shown in Figure 12.4. SCC is commonly present in the transition ranges between active and passive states as, for example, in high-temperature water in the presence of either hydrogen (active region) or oxygen (transpassive region). Chemical corrosion and mechanical stress both enhance the potential for corrosion. At higher mechanical stress, a reaction solution with low corrosive potential can lead to serious corrosion, while at low mechanical stress, a high corrosive potential of the reaction compounds still can cause the same fatal result due to corrosion. Aggressive compounds of the reaction solution are the anions chloride, bromide, and sulfide [2].

Failure due to SCC occurred in the example of Figure 12.4 in less than $t=50$ h in a solution containing HCl and oxygen (nickel-based alloy 625 at the following conditions: [HCl] $=0.10$ mol/kg; [O_2] $=0.48$ mol/kg; $T=220$ °C; $P=38$ MPa; $t=24.5$ h) [7].

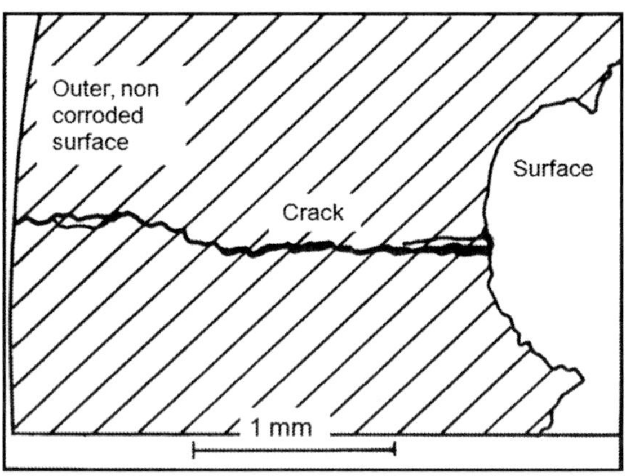

FIGURE 12.4 Stress corrosion cracking. *Modified from Ref. [2].*

TABLE 12.1 Corrosion of Alloy 625 in Oxidizing Aqueous Solutions of Various Acids

Type of Corrosion	Temperature (°C)	Corrosion Rate (μm/100 h)	Reaction Components	Inhibitors
Intergranular	>100	10–20	All	
Pitting	>150–200	500–1000	HCl/O_2 $NaCl/O_2$ HBr/O_2	OH^- CO_3^{2-} PO_4^{3-} F^-
Dissolution	>250–300	500	HCl/O_2 $NaCl/O_2$ HBr/O_2 H_2SO_4/O_2 HNO_3/O_2	OH^- CO_3^{2-} PO_4^{3-} F^-
Stress cracking	250–300	≫1000	HCl/O_2 HBr/O_2	OH^- CO_3^{2-} PO_4^{3-} F^-

[Acid] = 0.05–0.2 mol/kg; $[O_2]$ = 0.5 mol/kg; P = 24 MPa [2].

12.2.5 Example: Corrosion of Nickel-Based Alloys and Stainless Steels in Oxidizing Acidic Solutions

The corrosion of nickel-based alloy 625 in oxidizing high-temperature HCl solution is summarized as a typical example in Table 12.1. Pitting corrosion is observed in the passive state, while general corrosion occurs in the transpassive region (see below). SCC is observed in the transition region between passive and transpassive regions, but it can also start at the bottom of pits in the passive region due to corrosion-induced increased mechanical stress [2].

12.3 SOLUTION PARAMETERS INFLUENCING CORROSION

12.3.1 Ionic Reactions and Oxide Film Stability

Corrosion processes in an aqueous environment are mostly determined by ionic reactions. Rate determining for corrosion is the dissolution of a protective surface compound, in most cases an oxide. At high temperatures, oxides are preferentially formed compared to hydroxides, and the oxide may be different to the surface oxides found under ambient conditions [2].

In hydrothermal and supercritical temperature water, density is the main parameter determining the corrosive potential of the reaction solution. High density of water favors the solubility and the dissociation of ionic species like salts, acids, and bases, and thus favors ionic reactions [2]. Water at low density favors radical reaction pathways. In addition, also nonpolar gases and organic compounds may be completely soluble in hydrothermal and supercritical water and may influence corrosion reactions.

A protective oxide can form a stable protection layer within certain limits of the electrochemical potential and the pH value as the main solution parameters determining corrosion (at constant values of other parameters). At low pH values, acidic dissolution occurs, while at high pH values, alkaline dissolution prevails. In a similar way, at low electrochemical potential, an active dissolution is the corrosive mechanism, while at high electrochemical potential transpassive dissolution occurs [2], in which a passivated metal starts rapid dissolution, if the electrode potential of the metal becomes too positive.

12.3.2 pH Value and Chemical Dissolution

The pH value is an important parameter in discussing corrosion effects as the availability of hydronium ions (H_3O^+) and also hydroxyl ions (OH^-) influences corrosion reactions. Dissociation of water ($H_2O = H^+ + OH^-$) is temperature and pressure dependent (compare Chapter 2). As an endothermic reaction, the equilibrium is shifted to the right hand side with increasing temperature. At about $T = 300$ °C and a pressure of $P = 25$ MPa, the dissociation of water has a maximum that is about three orders of magnitude higher than at ambient conditions. Consequently, the pH value at $T = 300$ °C is three units lower than at ambient conditions, and water acts as a weak acid. At higher temperatures, the structure of water changes (see Chapter 2). The three-dimensional network of hydrogen bonds is weakened and even breaks down. Therefore, self-dissociation of water drastically decreases at temperatures in the neighborhood of the critical point of water.

The dissociation of ionic species, for example, mineral acids, depends on the density of the aqueous solution. Dissociation depends on the density of the solvent, as shown in Figure 12.5 for the dissociation of HCl [8–10]. The dissociation of HCl drops drastically near the critical temperature of water at the moderate pressure of $P = 24$ MPa, but at the higher pressure of $P = 38$ MPa, HCl still acts as acid, with the effect of enhanced corrosion potential. Other strong acids behave similar [2]. High or low values of pH (high or low concentrations of anions and cations) lead to chemical dissolution of oxides due to the amphoteric character of most oxides that can consequently be dissolved either in acidic or in alkaline solutions. Typical examples are the oxides of iron, nickel, and chromium [2]. In that context, corrosion depends strongly on pressure, as density influences solubility and the reaction mechanism. In solutions where no oxidative species are present that could further oxidize after removal of the oxide layer, the metal surface may be stable.

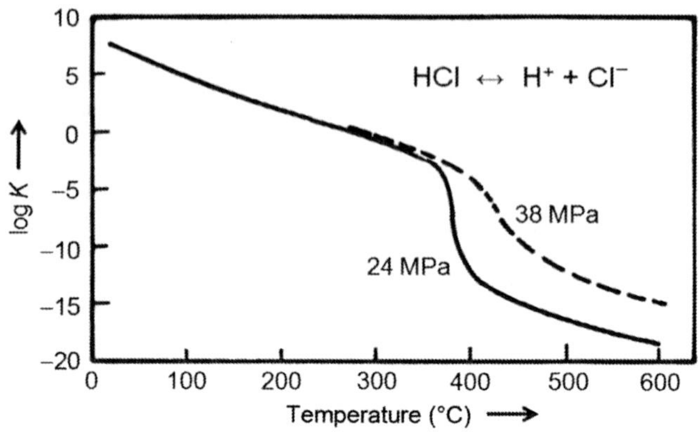

FIGURE 12.5 Dissociation of HCl in water [8,9]. *Modified from Ref. [2].*

TABLE 12.2 Concentrations of Metals Released from a Tube Reactor of Nickel-Based Alloy 625 Corroded by Acidic, Neutral, and Alkaline Chloride Solutions

	Dissolved Compound (ppm)		
Solution	*Ni*	*Cr*	*Mo*
$HCl+O_2$	1560	17	12
$NaCl+O_2$	10	24	6
$NaCl/NaOH+O_2$	0.1 (detection limit)	1.8	0.7

$T_{max}=350$ °C, $P=24$ MPa; $[O_2]=0.5$ mol/kg; $[HCl]=[NaCl]=[NaOH]=0.05$ mol/kg [2,11].

More information on the solubility of metal oxides is presented in Chapter 3, Sections 3.5–3.7 and Chapter 11.

The pH value is of consequence for the solubility of corrosion products. The corrosion products are formed at the corroding surface, where local supersaturation can lead to precipitation of the corrosion products. For example, Cr(III) oxides or hydroxides that have the lowest solubility of compounds forming stainless steels and nickel-based alloys form thick scales on the corroded surface, while nickel remains dissolved [2].

Typical concentrations of metal ions released from a nickel-based alloy 625 reactor (composition: 62 wt% Ni, 22 wt% Cr, 9.0 wt% Mo) under different conditions are shown in Table 12.2 [11].

Ni is dissolved at high concentrations in the acidic solution. At nonacidic conditions, the Ni concentrations in the effluent solution are reduced considerably due to thermodynamic stability of NiO [2]. Chromium and molybdenum are dissolved chemically in acidic solutions and electrochemically in alkaline solutions. Under alkaline conditions, Cr and Mo are more dissolved than Ni, which is protected by NiO [2].

12.3.3 The Electrochemical Potential and Electrochemical Dissolution

Electrochemical corrosion requires the presence of aggressive ionic species as reactants or products which requires an aqueous environment that stabilizes these ionic species [3]. Furthermore, corrosion reactions are sensitive to changes in the electrochemical potential. At low and high values of the electrochemical potential, corrosion can occur by dissolution at essentially constant pH value. For the formation of a protective oxide layer on a metal surface, a minimum electrochemical potential is necessary. Below that potential, the metal can be dissolved (active dissolution) [2].

Chromium is the oxide-forming component in stainless steels and nickel-based alloys. In acidic solutions, the pure metal and its alloys are protected against corrosion by stable, solid Cr(III) compounds: $Cr(OH)_3$, CrOOH, and Cr_2O_3. These oxides cover the underlying metal and protect it from further reaction with the reaction solution. Above $T = 100$ °C, amorphous CrOOH is formed. At higher temperatures above $T = 400$ °C, crystalline CrOOH and Cr_2O_3 predominate [2].

Below the potential needed for stable formation of Cr(III), Cr(II) compounds can be formed that do not form a protective layer, and the metal is dissolved rapidly by active dissolution. On the other hand, high dissolution of the alloy can happen at a state where the electrochemical potential is higher than that of the upper limit of the passive region. At high electrochemical potentials, chromium forms the soluble hexavalent chromate CrO_4^{2-}, hydrogen chromate $HCrO_4^-$, or chromic acid H_2CrO_4 that do not protect the alloy against corrosion. A high dissolution called transpassive dissolution occurs and is observed for many metals. The electrochemical potential for the transpassive transformation of chromium and Cr(III) by forming soluble Cr(VI) compounds decreases with increasing temperature. Oxygen forms chromate at low concentration in the concentration range of parts per million, at high temperatures [2].

Titanium, niobium, and tantalum show no transpassive dissolution, at least at room temperature, which leads to their excellent corrosion performance in highly oxidative environments [2]. But, no experimental results for high temperatures have been published so far. Therefore, care has to be taken to transfer low temperature results to high temperatures.

Nickel-based alloys are corrosion resistant against oxidizing conditions due to formation of protecting NiO. In the presence of ionic chlorides, the oxide film protects no longer against corrosion that can occur with high rates [2].

The corrosion of chromium is the reason for the rapid corrosion of nickel-based alloys and stainless steels in oxidizing acidic high-temperature water of high density, if other compounds are not able to form a protecting layer. Under these conditions, chromium is dissolved oxidatively, while the oxides of the other two main components, iron and nickel, are dissolved chemically. Molybdenum and tungsten behave similar to chromium [2].

Corrosion products from acidic solutions contain iron and chromium but no nickel due to the low solubility of Ni oxides.

12.3.4 Influence of Anions

Anions can destruct an oxide film, dissolve corrosion products, act as oxidizing agents, or support the oxide film by incorporation, and thus increase or inhibit the rates of different corrosion processes.

Oxide film destruction acts locally and is known in high-temperature water for the halides chloride, bromide, and iodide, but not fluoride [2]. At severe conditions, also other anions like sulfide or sulfite can lead to a localized destruction of the oxide. The rate of diffusion-controlled corrosion (general corrosion) reactions is determined by the solubility of the primary corrosion products. For example, corrosion rates of nickel-based alloy 625 differ by an order of magnitude at high subcritical temperatures. But in the same temperature range, in oxidizing solutions of HCl and HNO_3, due to the higher solubility of $Ni(NO_3)_2$ than that of $Ni(Cl)_2$ [2,12], nitrate and sulfate can oxidize and corrode metals in hydrothermal and supercritical water. Thermodynamically favored sulfides, sulfites, or thin sulfur films can be formed and the initial metals undergo a fast active dissolution [2,13,14]. Anions, such as carbonate, phosphate, fluoride, and hydroxide, can be incorporated into the oxide film of nickel-based alloys and lead to increased stability of the film [2,15–17]. Table 12.3 summarizes the influence of different inorganic ions on the corrosion in hydrothermal and supercritical water.

12.3.5 Temperature

Most corrosion processes occur above a minimum temperature. Below that temperature, corrosion is of minor or no importance. Stainless steels are resistant at temperatures below $T = 80$–$100\ ^{\circ}C$, but are attacked by chloride pitting corrosion at higher temperatures [2]. Corrosion rates in water generally increase with temperature as long as the other solution properties remain unchanged. But for hydrothermal and supercritical water, experimental results show that corrosion is lower by orders of magnitude at supercritical temperatures of $T > 500\ ^{\circ}C$ and low water density compared to corrosion at subcritical temperatures at high density [18–25], compare Figure 12.6. This behavior can be related to the change of physical and chemical properties of water, in particular density, as is discussed below.

TABLE 12.3 Influence of Inorganic Anions on the Corrosion of Nickel-Based Alloys and Stainless Steels in Hydrothermal and Supercritical Water

Ion	Action	Possible Result
F^-	Formation of complexes	Homogeneous corrosion
Cl^-, Br^-	Penetration of oxide film	Localized corrosion, pitting, stress corrosion cracking
SO_3^{2-}, SO_4^{2-}, $S_2O_3^{2-}$	Oxidation, forming S^{2-}, S^0	Homogeneous corrosion
S^{2-}	Reduction	Release of H_2, stress corrosion cracking
NO_3^-	Oxidation, corrosion products soluble	General corrosion
CO_3^{2-}, PO_4^{3-}	Formation of slightly soluble salts	Inhibition of corrosion
OH^-	Formation of slightly soluble salts	Passivation, inhibition of corrosion

Modified from Ref. [2].

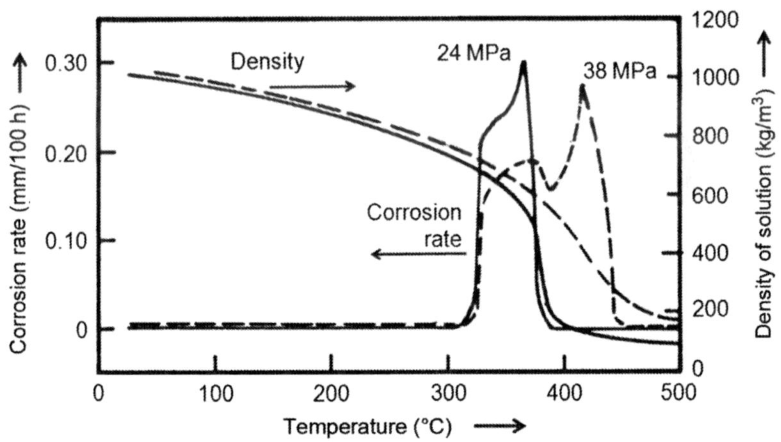

FIGURE 12.6 Corrosion rates of general corrosion of nickel-based alloy 625 (UNS N06625) caused by HCl solutions. $[O_2]=0.5$ mol/kg; $[HCl]=0.05$ mol/kg. *Modified from Ref. [2].*

12.3.6 Solubility of Gases

Above about $T=100$ °C, the solubility of gases in water increases rapidly (Chapter 3). At near critical and supercritical temperatures and supercritical pressures, many gases are completely miscible with water. Consequently, such gases can be available at high concentrations in the homogeneous reaction mixture with the effect, especially for oxygen, of forming a highly corrosive environment.

12.3.7 Density

Density of the reaction mixture is the most important parameter for corrosion. Corrosion rates can be directly correlated with density and ionic product of the solution [26]. In addition, density influences directly the solubility of protecting oxides and corrosion products in the aqueous reaction mixture and thus increases the potential for corrosion.

General corrosion of nickel-based alloy 625 (UNS N06625), considered as representative of other nickel-based alloys and stainless steels, in HCl, is shown in Figure 12.6 [2,27]. General corrosion occurs mainly in the near critical and supercritical temperature range in a region, where the density is high $\left(\rho_{H_2O} > 300\,\text{kg/m}^3\right)$. Corrosion rates below $T = 320\ °C$ are very low. At temperatures $T > 320\ °C$ and pressures higher than the critical pressure, H_2O still has an ionic character and a high solvent power for ionic species. These high temperatures increase the corrosion rates considerably to rates up to 0.3 mm/100 h, while at lower temperatures, corrosion rates are around 0.01 mm/100 h [2]. This temperature range of high corrosion rates is density dependent, as shown in Figure 12.6. At a pressure of $P = 24$ MPa, high corrosion rates drop down to the level observed at lower temperatures just around the critical temperature of water, while at a pressure of $P = 38$ MPa, corrosion rates remain high up to a temperature of about $T = 450\ °C$, since at that temperature density reaches the limit for corrosion due to ionic processes.

The maximum corrosion occurs in the region where temperature is high enough to promote fast kinetics, and concentrations of corrosion-causing species are still sufficiently high for reaction to occur. As Figure 12.6 shows, the end temperature for high corrosion rates shifts to higher temperatures with higher pressures.

In Figure 12.7, a process configuration for a process with supercritical water is shown schematically, together with the connected temperatures and pressures. In this case, the reactor is operated in the region of low corrosion, while the components used to preheat or cool down are operated at least partially in regions of high corrosion [3].

Although the potential for corrosion is greatly reduced under supercritical conditions, still significant corrosion can occur in supercritical water depending on the particular chemical environment. Salts often stick to the walls and a liquid layer can be trapped between the solid or molten salt layer and the wall, creating a microenvironment different in concentration and temperature to that in the bulk fluid phase. Severe corrosion has been observed with rates up to 64 μm/h in a titanium-lined reactor vessel underneath a molten salt layer containing phosphate salts [28].

In addition, even though the supercritical water environment usually does not support ionic-based corrosion mechanisms, there are other pathways by

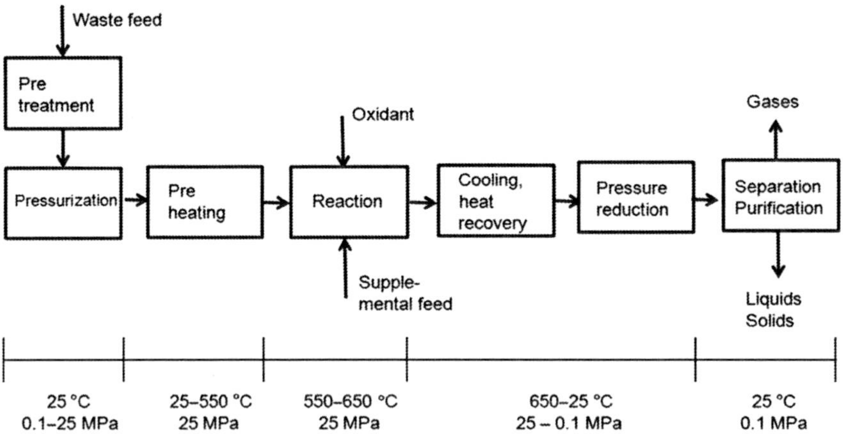

FIGURE 12.7 Temperatures and pressures occurring in a process using supercritical water. *Modified from Ref. [3].*

which corrosion can still occur. For example, at high temperatures ($T > 600$ °C), nondissociated gases attack metal surfaces by direct contact [26]. Furthermore, the high solubility of gases in supercritical water can create strongly oxidizing conditions, as in SCWO, or reducing conditions, as in SCWG [3].

The most severe corrosion in an aqueous system can be expected to occur in high density, high temperature, and high-aggressive ion concentration environments. Results indicate that a material that can withstand all three conditions has not been found. Therefore, the most appropriate material must be found for each set of conditions, which can mean to use different materials in different sections of the system [3]. But, a strategy for corrosion control in a hydrothermal and supercritical water system requires more than choosing a good material of construction. It must include also optimization of operating conditions, appropriate design of the components, and possibly adjustment of the composition of the reaction mixture.

12.4 MEASURES AGAINST CORROSION

Corrosion will be observed in some or all parts of plants for processes with hydrothermal and supercritical water. Therefore, some means of corrosion control are usually essential for operation of an economically viable and versatile process using hydrothermal and supercritical water [3].

12.4.1 General Strategy Against Corrosion

These approaches can be summarized according to their primary objective [3]:

- Prevent corrosive species from reaching a solid surface,
- Form a corrosion-resistant protective layer,
- Minimize corrosion to a controlled and acceptable level,
- Adjust process conditions to avoid or minimize corrosion.

These approaches are discussed below in more detail. They involve different actions with respect to composition of the reaction mixture, operating conditions, construction materials, and reactor constructions. There is no approach that can satisfy all requirements to full satisfaction. On the contrary, a (non-technical) decision on the most important goal of the process will lead to compromises for anticorrosion measures.

12.4.2 Water Treatment in Power Plant Cycles

The most important application of hydrothermal and supercritical water is in power plant cycles. This topic is treated in a highly specialized literature. Only some main features will be discussed here, based mainly on the contribution of Dooley *et al.* [29].

The chemistry of the water cycle is of major influence on the availability and reliability of the key components in any power plant. Boiler tube failures are influenced by hydrogen damage, caustic gouging, acid phosphate corrosion, corrosion fatigue, and pitting [30]. Turbine blade failures are due to corrosion fatigue and disk cracking due to stress corrosion [31]. The owners of plant usually control operation by comparison to guideline and action levels based on the guidance provide by organizations such as VGB, EPRI, and CRIEPI. In power plant cycles, the feed (water) is generally purified and conditioned before it enters the high-temperature zones. This is different to most other applications of hydrothermal and supercritical water in which the feed usually is not changed in composition, besides dilution by the process water, or sometimes by addition of compounds that influence corrosion.

12.4.2.1 *Feedwater Treatment for Conventional Fossil Plants*

Feedwater chemistry is critical to the overall reliability of all fossil plants. Corrosion products are generated in the feedwater train, then transported around the cycle, and finally deposited in various cycle areas acting as initiating centers for most of the major failure mechanisms.

The condition of feedwater for power cycles was determined mainly by the pH value and oxygen content. According to VGB guidelines, the pH at $T = 25$ °C should be 7–8, the oxygen content $x_{O_2} = 0.05–0.25$ mg/l for neutral treatment, pH 8–9 and oxygen content $x_{O_2} = 0.03–0.15$ mg/l for combined treatment, and pH 9–10, with an oxygen content of $x_{O_2} < 0.1$ mg/l for alkaline treatment [32]. This specification was applicable for installations with only ferrous metallurgy. For a mixed metallurgy, a very low oxygen level in the feedwater ($x_{O_2} < 7$ μg/l), backed up by a surplus of a reducing agent, such as hydrazine, was applied [29]. This rule has been modified: The higher the pH of the feedwater, the lower the amount of oxygen that must be added and vice versa. Furthermore, it was recognized that the nature of the feedwater heater tubing metallurgy is decisive when defining the optimum

TABLE 12.4 Feedwater Conditioning According to EPRI Guidelines [29]

	All Volatile Treatment		Oxygenated Treatment
	Mixed metallurgy[a]	*All ferrous*[b]	*All ferrous*[b]
pH at 25 °C	8.8–9.1	9.2–9.6	8.0–8.5[c] 9.0–9.5[d]
Oxygen (μg/l)	<5	1–10	30–150[c] 30–50[d]
Oxidation/reduction potential, ORP (mV)	≪0	0–80	>100

[a]*Some of the tubing is made of admiralty brass, copper–nickel alloys, and Monel.*
[b]*Carbon steel and stainless steel.*
[c]*For once-through units.*
[d]*For drum units.*

feedwater treatment. Table 12.4 shows feedwater treatment limits in dependence of the feedwater heater metallurgy [29].

12.4.2.2 Boiler Water Treatment

Feedwater treatment in once-through and drum plant cycles comprises all-volatile treatment, AVT(O), oxygenated treatment (OT) for cycles with all-ferrous metallurgy, and "reducing" treatment, AVT(R), for cycles with mixed metallurgy. For drum boilers, treatments comprise all-volatile treatment (AVT), equilibrium phosphate treatment (EPT), phosphate treatment (PT), caustic treatment (CT), and OT [33].

AVT is carried out without any additional chemical dosing into the boiler water, the pH of which at $T=25$ °C should be lower than that of the feedwater with a pH value of 9.0–10 for all-ferrous metallurgy and pH 9.1–9.3 for mixed metallurgy.

For EPT, the boiler water should contain $x \leq 2.4$ mg/l PO_4^{3-}, the pH at $T=25$ °C should be 9.0–9.7, with a possible addition of $x \leq 1$ mg/l free NaOH for maintaining pH. This treatment has proved to be very successful in a large number of high-pressure boilers, but risks of hydrogen damage have been reported [34].

PT needs about $x=3$–10 mg/l PO_4^{3-} in the boiler water, a pH at $T=25$ °C of 9.3–9.9 with the Na/PO_4^{3-} molar ratio ≥ 2.8 and $x \leq 1$ mg/l free NaOH. PT can be used even in units with seawater or high total dissolved solids in the cooling tower [29].

CT needs $x=1.0$–1.5 mg/l NaOH or 2.5× the chlorine content in mg/l, at minimum, and the pH value should be 9.5–10. CT is applicable in units with seawater or at high values of total dissolved salts in the cooling tower [29].

OT represents the optimum treatment for drum boiler units with all-ferrous feedwater metallurgy [29]. For a pH value of the feedwater in the range 9.0–10.0, the pH of the boiler water is kept lower than this value, with an oxygen content of $x_{O_2} < 100\,\mu g/l$ [32]. Another source reports pH 8.5–9.2 for the boiler water and a pH of 9.0–9.6 for the feedwater with an oxygen content of $x_{O_2} = 30–50\,\mu g/l$ [35], cited from Ref. [29].

Organic additives for boiler water treatment, such as aliphatic polyamines, that are relatively stable under drum unit conditions provide an increased tolerance against sudden cycle contamination.

For a more detailed discussion of process water treatment in power plants, the reader is referred to the review article of Dooley *et al.* [29].

12.4.2.3 Supercritical Water in Nuclear Reactors of Generation IV

In the application of supercritical water for the moderation and cooling of nuclear reactors, corrosive solutes are avoided. But the very pure water itself has corrosive properties under those conditions [36]. Additional aspects in this application are the very thin coatings, the radiolysis products of water such as O_2 and OH^- [37], and irradiation-induced changes in microstructure and microchemistry [38]. Materials that have been proposed include ferritic–martensitic and austenitic stainless steel, nickel-based alloy 625, alumina as thin coating on an iron-based alloy (P91), and zirconia deposited by plasma electrolytic oxidation on a zircaloy cladding [36]. Ferritic–martensitic steels generally have the best resistance to SCC, but suffer from the worst oxidation. Austenitic stainless steels and Ni-based alloys have better oxidation resistance but are more susceptible to SCC [37]. Cabet *et al.* [39] and Sun *et al.* [38] review these aspects.

12.4.3 Measures to Influence the Composition of the Feed

The feed for hydrothermal and supercritical water processes is different for the various process applications. For water in power generation cycles, there is a very different situation than for waste destruction with SCWO. Nevertheless, some basic features are common to all fluids for the management of corrosion. These comprise:

- restrict concentration of certain components,
- remove components that can cause corrosion,
- dilute corrosion products,
- add compounds that inhibit the corrosive action of other components in the reaction mixture.

12.4.3.1 Neutralization

For a feed that is known to contain acids or species that will generate acids during the oxidation reaction (e.g., materials containing halogens, sulfur, or phosphorus), neutralization is an easy and common means. In this approach, a base (e.g., sodium hydroxide) is added to the feed before the reaction, usually inline before reaching the reactor or directly to the reactor through a separate feed line. The amount of base to be added is predetermined to balance the expected acid concentration. A surplus of base is to be avoided since it reacts with carbon dioxide to carbonates that can enhance corrosion or cause plugging [3]. The base neutralizes acid species within the reactor and removes corrosive ions as salts that are nondissociated due to their low solubility in the low-density aqueous phase. Thus, the problem is shifted from corrosion to salt precipitation. Such means are discussed in Ref. [40]. Neutralization significantly reduces corrosion. But corrosion can still occur, for example, beneath a salt layer that has been formed [3].

12.4.3.2 Cold Feed Injection

Cold feed injection protects the process feed section from severe corrosion in the temperature region where the corrosion is at its highest. Although a continuously heated feed stream provides the wanted feed temperature at the inlet of the reactor, corrosion in the preheater may be severe, in particular when the feed contains or sets free aggressive species during the preheating, which has to be compensated by using expensive construction materials. Furthermore, feed compounds can decompose when heated, which leads to char formation and fouling in the preheater.

Cold feed injection can be operated such that only the feed is injected cold and mixed with a preheated process stream, or feed and process fluid are both injected cold. In both cases, sufficient mixing within the reactor must be achieved to utilize the heat of reaction in order to reach the required temperature fast. Appropriate design of the feed nozzle or feed injection ports have proved that this can be achieved in minimal time and without any loss of organic destruction efficiency [3]. Commercial plants for SCWO often apply cold feed injection because of the reduction in corrosion and material cost [28,41–43].

12.4.3.3 Feed Dilution with Noncorrosive Wastes

Waste streams of different compositions may be mixed with water or a less corrosive feed stream and thus dilute the more corrosive feed with a less corrosive one. This combination of feeds effectively reduces the concentration of corrosive species without increasing the total amount of feed or the total process time [3].

Success of this approach depends mainly on how much less corrosive feed is available. An example is the processing of highly corrosive chemical agent hydrolysates together with a noncorrosive energetic hydrolysate [3,44].

12.4.3.4 Effluent Dilution or Cooling

Diluting and cooling the effluent of a reactor operated at supercritical temperatures is mainly intended as a means of quickly redissolving salts to prevent plugging of the reactor exit and the smaller diameter downstream piping [3]. In addition, corrosion is reduced substantially as well as the length of piping exposed to subcritical temperatures with high potential for corrosion. A base, such as sodium hydroxide, can be added to the quench water to further reduce corrosion with acidic reaction products. The more water is added, the greater is the temperature drop and the reduction of the corrosion endangered zone. However, the more water is added, the greater is the volume of effluent that must be processed downstream. When water constraints are not significant, quench water is an effective option that is used in commercial applications [3,28,41,43–46].

12.4.4 Materials Withstanding Corrosion

For carrying out processes in hydrothermal and supercritical water, it is essential to choose a material that can withstand the operating conditions and the interaction with the reaction mixture. This is tried in all cases of such processes, but may not be sufficient. In addition, other measures must be taken to control corrosion. But the choice of an appropriate construction material still remains the first requirement for construction of such a plant. In general, stainless steel alloys and nickel-based alloys are used. Other materials, such as titanium, noble metals, ceramics, and others, are selected for special purposes.

12.4.4.1 Nickel-Based Alloys

Nickel-based alloys (e.g., 600, 625, 718, C-276) are the most common base materials of construction for oxidative hydrothermal and supercritical water system components at high temperature and pressure. These alloys show a good overall strength and corrosion resistance to a wide range of substances and conditions [3]. Under certain conditions, they are not sufficient for an acceptable reduction of corrosion. Their applicability depends on the individual conditions. In general, alloy 625 is more corrosion resistant at supercritical temperatures, while alloy C-276 is more corrosion resistant at subcritical temperatures [3]. Experimental data show that these materials may not be able to handle very aggressive oxidative supercritical water feed streams due to significant weight loss and localized effects including pitting, SCC, and dealloying. Nevertheless, nickel-based alloys are often used with liners or coatings as the pressure-bearing shell material [3].

In deionized water at elevated temperatures ($T=450$–$500\,°C$), a potentially protective film forms for both alloys 625 and C-276 [4]. But even under such conditions, for alloy 625, minor pit development and grain boundary carbide formation have been observed [4]. Dealloying of Cr and Mo (I-625) or

Cr, Mo, and W (C-276) is recognized as a potential contributor to degradation within oxidative supercritical systems. Loss of chromium occurs with non-chlorinated feeds, while selective dissolution of Ni is recognized for chlorinated feeds, especially for acidic chlorinated ones.

At supercritical temperatures, for an acidic chlorinated feed, corrosion rate of high-nickel alloys I-625, C-22, and I-686 between $T=400$ and 600 °C increases significantly above $T=600$ °C. In the same temperature range of $T=400$–600 °C, the corrosion rate of C-276 increases with increasing temperature. Corrosion phenomena depend essentially on the composition of the reaction mixture and the density of the aqueous solution as discussed above.

12.4.4.2 Stainless Steel Alloys

Stainless steel alloys are much less expensive than nickel-based alloys and easily workable. They can be an adequate material if the feed is not aggressive and they can be used safely for components that are not exposed to high temperatures. Resistance of stainless steels (and nickel-based alloys) can be improved against stress-related corrosion by annealing heat treatment for stress relief [4]. In oxidative supercritical water systems, stainless steel alloys are riskier and more restricted in their use than nickel-based alloys.

The components of an alloy are of major influence on corrosion resistance. Chromium improves the resistance against acidic and oxidizing media and reduces pitting corrosion, but at higher temperatures, it tends to form chromate and loses its protective effect. Nickel improves the corrosion behavior in neutral or alkaline environments. It forms the protective layer under these conditions and is stable in low-density supercritical solutions, while chromium is unstable. Molybdenum protects alloys at low and reducing potentials at which other alloys show active dissolution, for example, during the reductive conversion of biomass [2].

Nonmetallic inclusions in the metal, such as manganese sulfide and titanium nitride, lead to weak points of the oxide layer and are starting points of pitting and general corrosion. Due to heat treatment of the alloy, the grain boundaries may be depleted of one or more of the alloy components. Chromium depletion can occur by the formation of chromium carbides that decrease chromium concentrations of adjacent areas up to 50% and increase the concentration of impurities like phosphorus and sulfur. This is a problem at high temperatures of $T>600$ °C, while at lower temperatures the formation of these phases is too slow [2].

12.4.4.3 Ceramics and Ceramic/Alloy Combinations

Ceramics have exhibited poor resistance to chlorinated waste over a wide range of pH (2–12) and temperature $T=350$–500 °C, with the possible exception of monolithic alumina and stabilized zirconia. In chlorinated and nonchlorinated solutions, ceramic materials like Al_2O_3, AlN, Sapphire, Si_3N_4, SiC, and ZrO_2

show very poor corrosion resistance. Zirconia ceramics exhibit poor resistance in aqueous sulfuric acid feeds [4]. These materials typically have inadequate physical strength for the primary, pressure-bearing material of construction [3]. They can be applied inside an outer pressure-bearing reactor.

12.4.4.4 Noble Metals

Materials like gold or platinum increase the initial investment costs. In addition, lack of mechanical strength often limits their use as a primary construction material in high-pressure systems at hydrothermal and supercritical temperatures. In small quantities, they can be used as liners (see below) and for small setups for very aggressive waste streams.

Platinum and two platinum alloys (Pt-10Ir, Pt-30Rh) reveal excellent corrosion resistance with rates of 0.03–0.08 mm/a in a chlorinated feed stream with small additions of Zn, Pb, and Ce at a temperature of $T=610\,°C$. At the lower temperature of $T=400\,°C$, corrosion rates were increased to a level of 1.14–4.83 mm/a, resulting possibly in high replacement cost of hundreds of thousands of dollars a year, even at the lowest corrosion rate [4].

Gold was found to be unsuitable for use in a chlorinated acidified feed. An Inconel-625 tube coated with a 30-μm gold layer showed intergranular SCC and failed within $t=34$ h, while the uncoated tube did not fail even after $t=150$ h [4].

12.4.4.5 Titanium-Based Alloys

Titanium provides outstanding performance at subcritical temperatures and is as resistant as the Ni alloys at supercritical temperatures. Corrosion tests of titanium revealed poor resistance against nonchlorinated acidic chemical feeds and an acceptable corrosion rate of less than 3.5 mm/a [4]. Problems have been experienced with titanium, in particular pitting corrosion [4]. Although titanium is not prohibitively expensive, its use for the reactor shell is not recommended because of the risk of fire in contact with high concentrations of oxidant [3].

A more detailed discussion of materials for hydrothermal and supercritical water systems can be found in the review articles [2–5], which provide also access to many of the original investigations on corrosion.

12.4.5 Constructive Means

Constructive means against corrosion are among the important issues to achieve a reasonable performance of processing plants with hydrothermal and supercritical water. Among these means are the application of liners, coatings, special construction of reactors such as transpiring wall reactors, and constructive means to influence the flow pattern in the reactor.

12.4.5.1 Liners

Liners overlay the main structural material of a plant component and protect it in cases when the structural material is not corrosion resistant. The liner therefore has not to withstand the system pressure and can be thin. That enables the use of a greater number of materials with high corrosion resistance but poor strength properties or high cost [3]. Liners are mostly used in reactors and heat exchangers. Liners are solid coatings of the surface of a structural material and are used in a variety of types, metal or ceramic, fixed or removable, corrosion resistant or sacrificial, and solid or porous [3].

Fixed liners consist of corrosion-resistant metals and are permanently combined with the construction material. They are often bonded or welded to the structural material. Examples have been reported by Modar Inc. [3] and General Atomics [44,47,48] for destruction of chemical agent and energetic hydrolysates. Researchers at Los Alamos National Laboratory utilized a nonremovable, titanium (grade 2)-lined, stainless steel tube reactor for destruction of chlorinated hydrocarbons [49].

Removable liners are placed concentric inside a tubular component that they protect. An annular region is created between the liner and the structural shell, where a fluid or an insulating material supports the liner. Thus, the pressure-bearing structure is isolated from the corrosive reaction medium inside the liner, and the liner only has to withstand the corrosion. Various such dual shell reactors have been described in the patent literature [3,50–53].

Liners designed to be corrosion resistant can be manufactured from expensive noble metals like platinum, gold, niobium, or tantalum. These liners can be thin ($<$3 mm) and may need some mechanical support, as is the case with platinum because its high cost restricts its use to a thin lining.

Platinum has been shown to be highly corrosion resistant to most oxidative hydrothermal and supercritical media, with the exception of hydrochloric, sulfuric acids, and neutral chloride salts [3]. A platinum removable liner has been used by General Atomics for bench- and pilot-scale reactors for investigation of the destruction of aggressive chemical agents [54].

A gold liner was used successfully for destruction of explosive materials with good corrosion resistance, except at low pH values and with chlorides [3,54,55].

Niobium and tantalum show good corrosion resistance at subcritical but not supercritical conditions, due to a change in oxide structure at higher temperatures, and are therefore good materials for heat exchangers before and after the reactor [56,57,85].

Sacrificial metal liners use inexpensive materials that can be periodically replaced [3]. Nickel-based alloys can be used as good sacrificial liners and the liner can be of the same material as the pressure shell. Corrosion products must cause no additional problems. In such a way, titanium can be used as a sacrificial liner when exposed to phosphate feeds, while it is corrosion resistant in chloride feeds. An example for the possible double role of a liner is a removable titanium liner for nuclear-contaminated organic wastes that is corrosion resistant with sulfur-containing feed and of sacrificial behavior with phosphorus-containing feed [57]. For high concentrations of salts, General Atomic used alloy C-276 and titanium as sacrificial removable liners [40].

Ceramic materials are brittle but can be used as liners because of their good inertness to many forms of chemical attack and good thermal stability [3]. Oxide and nonoxide ceramics, alumina (Al_2O_3), and partially stabilized zirconia (ZrO_2) are good materials for liners in oxidative supercritical water containing HCl [60]. A removable alumina liner was developed for corrosion control in SCWO at high pressures [58,59,86]. With halogenated feeds, no corrosion was found for the alumina liner in long-term tests with 2,4-dichlorophenol [60,87].

Zirconia was used successfully without corrosion for the SCWO product stream by Modar SCWO effluent [3,60]. The best stability was experienced with zirconia stabilized by a small amount of magnesia or magnesia and yttria [61].

12.4.5.2 Coatings

Coatings are thin layers (in the range of $\leq$0.1 mm) for protection of less corrosion-resistant construction materials. Coatings are versatile and can be applied to small areas and complex shapes of plant components. Usually coatings are made from the same materials as liners, but they can also be manufactured from the structural material itself by formation of an oxide layer. The formation of an oxide layer is the simplest way to form a protective coating and occurs during the first operation of a plant or is produced by oxidative treatment prior to the application in a process [3]. Nickel-based alloys form strong oxide coatings on their surface, provided that the pH value is not too low, which is the reason for their corrosion resistance [3].

Coating materials that are different from the base metal can be produced by chemical vapor deposition, thermal spraying, and electroplating. Examples of metal coatings that showed excellent corrosion resistance are iridium and iridium alloys with platinum and rhodium in an SCWO reactor exposed to chloride and sulfate acids and salts at temperatures $T>600$ °C [62].

Ceramic coatings must be free of surface defects and must be flexible enough to expand and contract with the base material without forming cracks. An intermediate layer can be applied to enable the transition between base metal and ceramic coating.

Tests with yttria-stabilized zirconia, magnesia-stabilized zirconia, and titania coatings on alloy 625 and titanium [62] showed that a protection of the alloy 625 was not achieved, but the combination of titania on titanium showed no degradation and good resistance in a temperature range of $T=300$–650 °C and a mixture of machine cutting fluid and radioactive mixed wastes [46].

12.4.5.3 Reactors and Protective Process Modes

Constructive measures and protective process modes mainly are applied for keeping the corrosive process medium away from the surface of the reactor. Generally, a noncorrosive fluid is introduced that forms a film on the surface of the reactor and flows in direction of the exit, keeping the reactive mixture away from the construction material of the reactor and keeping in solution salts formed during the reaction. This protective layer can be produced by injection of the noncorrosive fluid and distributing it without further constructive means on the walls of the reactor, or by introducing the nonreactive fluid via a porous liner. In addition, the noncorrosive fluid can be cooler than the inner reaction mixture, thus allowing to use temperatures that would not be tolerated by the construction material. Furthermore, a diluting effect occurs along the mixing zone, contributing to keep salts in solution.

12.4.5.3.1 Transpiring Wall Reactor

In the so-called transpiring wall reactors, the water flow is introduced through a liner that extends the length of the reactor. Examples of such reactors have been used, for example, by Foster Wheeler [63,64] using liner formed from several thin metal sheets of alloy 600. The protective water flow was about two to three times greater than the feed flow. Tests with Navy shipboard wastes, chemical agent, and energetic hydrolysate mixtures and longer-term testing ($\sim$500 h) showed significantly less corrosion than experienced by a solid wall reactor [28,48,65].

Another design uses sintered metal or sinter ceramic porous liners [66]. Examples of this type have been used [3] by Forschungszentrum Karlsruhe, Germany (stainless steel liner, 0.06 m ID $\times$ 0.95 m length; $m = 20$ kg/h waste feed rate) [67–69], the University of Valladolid, Spain (Alloy 600 liner, 0.074 m ID $\times$ 1.5 m length; up to $m = 40$ kg/h waste feed rate) [70–72], ETH, Zürich, Switzerland (Alloy 625 liner, 0.022 m ID $\times$ 0.375 m length, $m = 10$ kg/h waste feed rate) [73,74], and CEA in France (alumina liner, 0.015 m ID $\times$ 0.50 m length, waste feed rate of $m = 3$ kg/h) [75,76].

A tubular reactor with water film cooling, where the flow of clean water introduced at the top of the reactor forms a protective film, was developed at ETH [77,78]. Significant corrosion was observed at the center tube end and the inner surface of the coaxial tubes [79]. Therefore, research was focused on porous liners according to the transpiring wall design. Figure 12.8 shows the principal features of a transpiring wall reactor and a reactor with a protecting film.

Another option is to use the fluid motion itself to keep the aggressive and high-temperature reactive fluid away from the shell by introducing the feed tangentially to the reactor wall, causing the fluid to rotate or spiral toward the center of the reactor in a vortex-like flow. The reaction generates heat and the less dense reacting fluid containing the aggressive species formed

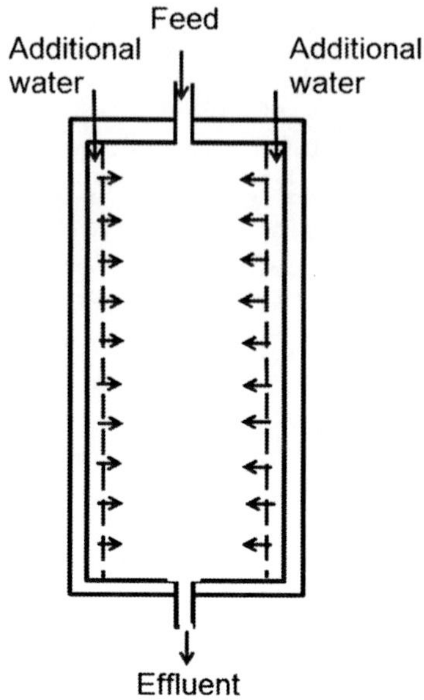

FIGURE 12.8 Scheme of a transpiring wall reactor (left) and a reactor with a protective film. *Modified from Ref. [4].*

during reaction, concentrates in the center of the reactor vessel by centrifugal force, leaving cooler fluid in contact with the reactor wall [3,80]. Such a design is not applicable to reaction mixtures containing undissolved salts, because these would be transported to the reactor wall.

12.4.5.3.2 Fluidized-Bed Hydrothermal Oxidation

A fluidized bed of solids, for example, sodium carbonate, provides the reaction zone and the adsorptive surface for removing corrosive reaction products from the reaction mixture. This design was developed by SRI's Assisted Hydrothermal Oxidation process [81]. Oxidation of chlorinated and aromatic compounds is significantly accelerated in the presence of sodium carbonate particles [81–83] that probably act as a catalyst and neutralizing agent.

Mitsubishi Heavy Industries proposed a version of this process where the sodium carbonate is generated and replaced *in situ* during operation by addition of sodium hydroxide [84], and constructed a plant for processing PCB with a capacity of about 1800 kg PCB/day.

12.5 SUMMARY

There is no general solution for handling corrosion. As discussed above, a strategy consisting of different measures must be combined to achieve the optimal solution. This solution must be based on the composition of the feed that must be known. Then, the appropriate material and conditions can be chosen. For varying feed composition or previously unknown type of feeds, corrosion cannot be handled in a reasonable way.

REFERENCES

[1] P.R. Roberge (Ed.), Corrosion Basics: An Introduction, second ed., NACE Press Book, Houston, 2006, ISBN: 1-57590-198-0.

[2] P. Kritzer, Corrosion in high-temperature and supercritical water and aqueous solutions: a review, J. Supercrit. Fluids 29 (2004) 1–29.

[3] P.A. Marrone, G.T. Hong, Corrosion control methods in supercritical water oxidation and gasification processes, J. Supercrit. Fluids 51 (2009) 83–103.

[4] D.B. Mitton, N. Eliaz, J.A. Cline, R.M. Latanision, An overview of the current understanding of corrosion in SCWO systems for the destruction of hazardous waste products, Mater. Technol. Adv. Perform. Mater. 16 (2001) 30–53.

[5] G. Brunner, Near and supercritical water. Part II: oxidative processes, J. Supercrit. Fluids 47 (2009) 382–390.

[6] C. Sun, R. Hui, W. Qu, S. Yick, Progress in corrosion resistant materials for supercritical water reactors, Corros. Sci. 51 (2009) 2508–2523.

[7] P. Kritzer, Die Korrosion der Nickel-Basis-Legierung 625 unter hydrothermalen Bedingungen: Report FZKA 6168, Forschungszentrum Karlsruhe GmbH, Karlsruhe, Germany, 1998, p. 180.

[8] P. Kritzer, N. Boukis, E. Dinjus, Factors controlling corrosion in high-temperature aqueous solutions. A contribution to the dissociation and solubility data influencing corrosion processes, J. Supercrit. Fluids 15 (1999) 205–227.

[9] J.D. Frantz, W.L. Marshall, Electrical conductances and ionization constants of salts, acids, and bases in supercritical aqueous fluids: I. Hydrochloric acid from 100 to 700 °C and at pressures to 4000 bars, Am. J. Sci. 284 (1984) 651–667.

[10] X. Chen, R.M. Izatt, J.L. Oscarson, Thermodynamic data for ligand interaction with protons and metal ions in aqueous solutions at high temperatures, Chem. Rev. 94 (1994) 467–517.

[11] P. Kritzer, N. Boukis, E. Dinjus, Corrosion of alloy 625 in aqueous chloride and oxygen containing solutions, Corrosion 54 (1998) 824–834.

[12] P. Kritzer, N. Boukis, E. Dinjus, The corrosion of nickel-base alloy 625 in oxidizing sub- and supercritical aqueous solutions of HNO_3, J. Mater. Sci. Lett. 17 (1999) 771–773.

[13] P. Marcus, E. Protopopoff, Potential pH diagrams for sulfur and oxygen adsorbed on nickel in water at 25 and 300 °C, J. Electrochem. Soc. 140 (1993) 1571–1575.

[14] P. Marcus, E. Protopopoff, Potential–pH diagrams for sulfur and oxygen adsorbed on chromium in water, J. Electrochem. Soc. 144 (1997) 1586–1590.

[15] W.F. Bogaerts, A.A. Van Haute, M.J. Brabers, Relative critical potentials for pitting corrosion of 304 stainless steel, Incoloy 800 and Inconel 600 in alkaline high-temperature aqueous solutions, J. Nucl. Mater. 115 (1983) 339–342.

[16] P. Kritzer, N. Boukis, E. Dinjus, The corrosion of alloy 625 ($NiCr_{22}Mo_9Nb$, 2.4856) in high-temperature, high-pressure aqueous solutions of phosphoric acid and oxygen. Corrosion at sub- and supercritical temperatures, Mater. Corros. 49 (1998) 831–839.

[17] P. Kritzer, N. Boukis, E. Dinjus, The corrosion behavior of nickel-base alloy 625 ($NiCr_{22}Mo_9Nb$, 2.4856) and ceria stabilized tetragonal zirconia polycrystal (Ce–TZP) against oxidizing aqueous solutions of hydrofluoric acid (HF), hydrobromic acid (HBr), and hydriodic acid (HI) at sub- and supercritical temperatures, Mater. Corros. 50 (1999) 505–516.

[18] S. Huang, K. Daehling, T.E. Carleson, P. Taylor, C. Wai, A. Propp, Thermodynamic analysis of corrosion of iron alloys in supercritical water, in: Supercritical Science and Technology, ACS Symposium Series, vol. 406, American Chemical Society, Washington, DC, 1989, p. 276.

[19] L.B. Kriksunov, D.D. Macdonald, Corrosion in supercritical water oxidation systems: a phenomenological analysis, J. Electrochem. Soc. 142 (1995) 4069–4073.

[20] L.B. Kriksunov, D.D. Macdonald, Corrosion testing and prediction in SCWO environments, in: Proceedings of the ASME Heat Transfer Division, The American Society of Mechanical Engineers, New York, 1995, p. 281.

[21] F. Matthews, E.F. Gloyna, Corrosion behavior of three high-grade alloys in supercritical water oxidation environments: Report CRWR 234, Center for Research in Water Resources, Bureau of Engineering Research, University of Texas in Austin, Austin, TX, 1992, p. 73.

[22] K.W. Downey, R.H. Snow, D.A. Hazlebeck, A.J. Roberts, Corrosion and chemical agent destruction, in: K.W. Hutchenson, N.R. Foster (Eds.), Research on Supercritical Water Oxidation, Oxidation of Hazardous Military Wastes, in: ACS Symposium Series, vol. 608, American Chemical Society, Washington, DC, 1995, p. 313.

[23] D.M. Harradine, S.J. Buelow, P.C. Dell'Orco, R.B. Dyer, B.R. Foy, J.M. Robinson, J.A. Sanchez, T. Spontarelli, J.D. Wander, Oxidation chemistry of energetic materials in supercritical water, Hazard. Waste Hazard. Mater. 10 (1993) 233–246.

[24] D.B. Mitton, J.C. Orzalli, R.M. Latanision, Corrosion phenomena associated with SCWO systems, in: G. Brunner, M. Perrut (Eds.), Proceedings of the 3rd International Symposium on Supercritical Fluids, Strasbourg, France, 1994, p. 43.

[25] S.F. Rice, R.R. Steeper, C.A. LaJeunesse, Destruction of representative Navy wastes using supercritical water oxidation, Final report, Sandia National Laboratories, Livermore, CA, 1994, 35 pp.

[26] L.B. Kriksunov, D.D. Macdonald, Corrosion in supercritical water oxidation systems: a phenomenological analysis, J. Electrochem. Soc. 142 (1995) 4069–4073.

[27] P. Kritzer, N. Boukis, E. Dinjus, Review of the corrosion of nickel-based alloys and stainless steels in strongly oxidizing pressurized high-temperature solutions at subcritical and supercritical temperatures, Corrosion 56 (2000) 1093–1104.

[28] P.A. Marrone, S.D. Cantwell, D.W. Dalton, SCWO system designs for waste treatment: application to chemical weapons destruction, Ind. Eng. Chem. Res. 44 (24) (2005) 9030–9039.

[29] R.B. Dooley, M. Bull, A. Bursik, M. Rziha, R. Svoboda, Water chemistry in commercial water-steam cycles, in: D.A. Palmer, R. Fernández-Prini, A.H. Harvey (Eds.), Aqueous Systems at Elevated Temperatures and Pressures, Elsevier, Amsterdam, 2004, pp. 676–716.

[30] R.B. Dooley, W.B. McNaughton, Boiler tube failures: theory and practice, EPRI Technical Report TR-105261, 1996.

[31] T. McCloskey, R.B. Dooley, W.B. McNaughton, Turbine steam path damage: theory and practice, EPRI Technical Report TR-108943, 1999.

[32] Guideline for boiler feedwater, boiler water and steam of steam generators with a permissible operation pressure >68 bar. VGB Kraftwerkstechnik GmbH, Essen, Germany, VGB R 450 Le, 1988.

[33] R.B. Dooley, A. Bursik, State of the art in fossil plant cycle chemistry, in: H.J. White Jr., J.V. Sengers, D.B. Neumann, J.C. Bellows (Eds.), Physical Chemistry of Aqueous Systems: Meeting the Needs of Industry, Proceedings of the 12th International Conference on the Properties of Water and Steam, Begell House, New York, 1995, pp. 33–47.

[34] R.B. Dooley, W.B. McNaughton, Appropriate controls for phosphate boiler water treatments to avoid acid phosphate corrosion, Power Plant Chem. 3 (2001) 127–134.

[35] A. Bursik, R.B. Dooley, D. Gunn, B.A. Larkin, I. Oliker, F.J. Pocock, D. Ryan, L.C. Webb, Cycle chemistry guidelines for fossil plants: oxygenated treatment, EPRI Technical Report TR-102285, 1994.

[36] Y. Marcus, Supercritical Water. A Green Solvent: Properties and Uses, John Wiley & Sons, Hoboken, NJ, 2012.

[37] G.S. Was, P. Ampornrat, G. Gupta, S. Teysseyre, E.A. West, T.R. Allen, K. Sridharan, L. Tan, Y. Chen, X. Ren, C. Pister, Corrosion and stress corrosion cracking in supercritical water, J. Nucl. Mater. 371 (2007) 176–201.

[38] C. Sun, R. Hui, W. Qu, S. Yick, Progress in corrosion resistant materials for supercritical water reactors, Corros. Sci. 51 (2009) 2508–2523.

[39] C. Cabet, J. Jang, J. Konys, P.F. Tortorelli, Environmental degradation of materials in advanced reactors, MRS Bull. 34 (2009) 35–39.

[40] P.A. Marrone, M. Hodes, K.A. Smith, J.W. Tester, Salt precipitation and scale control in supercritical water oxidation—part B: commercial/full-scale applications, J. Supercrit. Fluids 29 (2004) 289–312.

[41] A. Gidner, L. Stenmark, High pressure and high temperature reaction system, U.S. Patent No. 6,958,122, 2005.

[42] G.T. Hong, Process for oxidation of materials in water at supercritical temperatures and subcritical pressures, U.S. Patent No. 5,106,513, 1992.

[43] D.A. Hazlebeck, K.W. Downey, M.H. Spritzer, Downflow hydrothermal treatment, U.S. Patent No. 6,054,057, 2000.

[44] National Research Council (NRC), Interim Design Assessment for the Blue Grass Chemical Agent Destruction Pilot Plant, National Academies Press, Washington, DC, 2005.

[45] Stone & Webster Engineering Corp., Supercritical water oxidation data acquisition testing, Contract No. DE-FC07–94ID13303, Final report, vol. 1, 1996, www.osti.gov/bridge/servlets/purl/477764-hJAklF/webviewable/477764.PDF.

[46] P.J. Crooker, K.S. Ahluwalia, Z. Fan, J. Prince, Operating results from supercritical water oxidation plants, Ind. Eng. Chem. Res. 39 (12) (2000) 4865–4870.

[47] D.A. Hazlebeck, K.W. Downey, D.D. Jensen, M.H. Spritzer, Supercritical water oxidation of chemical agents, propellants, and other DOD hazardous wastes, in: H.G. White, J.V. Sengers, D.B. Neuman, J.C. Bellows (Eds.), Physical Chemistry of Aqueous Systems: Meeting the Needs of Industry, Begell House, New York, 1995, pp. 632–637.

[48] National Research Council (NRC), Analysis of Engineering Design Studies for Demilitarization of Assembled Chemical Weapons at Blue Grass Army Depot, National Academies Press, Washington, DC, 2002.

[49] B.R. Foy, K. Waldthausen, M.A. Sedillo, S.J. Buelow, Hydrothermal processing of chlorinated hydrocarbons in a titanium reactor, Environ. Sci. Technol. 30 (1996) 2790–2799.

[50] A.G. Fassbender, Dual shell pressure balanced vessel, U.S. Patent No. 5,167,930, 1992.

[51] D.A. Hazlebeck, System and method for hydrothermal reactions—three layer liner, U.S. Patent No. 6,576,185, 2003.

[52] R.N. McBrayer, J.E. Deaton, J.M. Eller, Turbulent flow cold-wall reactor, U.S. Patent No. 5,552,039, 1996.

[53] C. Joussot-Dubien, G. Didier, H.A. Turc, Method for oxidizing materials in supercritical water, U.S. Patent No. 6,878,290, 2005.

[54] S.J. Buelow, D. Allen, G.K. Anderson, F.L. Archuleta, J.H. Atencio, G.T. Baca, W.D. Breshears, T.J. Butenhoff, P.C. Dell'Orco, R.B. Dyer, B.R. Foy, K.A. Funk, D.M. Harradine, K.C. Knutsen, J.L. Lyman, D.A. Master, T.G. McGuinness, R.E. McInroy, C.J. Monahan, R.C. Oldenborg, J.M. Robinson, M.A. Sedillo, D.A. Counce, C.K. Rofer, P.E. Trujillo, R.L.

Brewer, G.A. Buntain, R.L. Flesner, J.A. Sanchez, T. Spontarelli, L.L. Sprouse, C.A. Vecere, G.R. Brewer, R.D. McFarland, W.J. Parkinson, R.P. Courier, S.M. Chitanvis, C.W. Patterson, L.R. Pratt, J.C. Oxley, J.D. Wander, Final report on the oxidation of energetic materials in supercritical water, Los Alamos National Laboratory Report No. LA-UR-95-1164, April 1995, www.fas.org/sgp/othergov/doe/lanl/lib-www/la-pubs/00413602.pdf.

[55] D.M. Harradine, S.J. Buelow, P.C. Dell'Orco, R.B. Dyer, B.R. Foy, J.M. Robinson, J.A. Sanchez, T. Spontarelli, J.D. Wander, Oxidation chemistry of energetic materials in supercritical water, Hazard. Waste Hazard. Mater. 10 (1993) 233.

[56] C. Friedrich, P. Kritzer, N. Boukis, G. Franz, E. Dinjus, The corrosion of tantalum in oxidizing sub- and supercritical aqueous solutions of HCl, H_2SO_4, and H_3PO_4, J. Mater. Sci. 34 (1999) 3137–3141.

[57] Y. Calzavara, C. Joussot-Dubien, H.A. Turc, E. Fauvel, S. Sarrade, A new reactor concept for hydrothermal oxidation, J. Supercrit. Fluids 31 (2004) 195–206.

[58] V. Casal, H. Schmidt, SUWOX—a facility for the destruction of chlorinated hydrocarbons, J. Supercrit. Fluids 13 (1998) 269–276.

[59] S. Baur, H. Schmidt, A. Krämer, J. Gerber, The destruction of industrial aqueous waste containing biocides in supercritical water—development of the SUWOX process for the technical application, J. Supercrit. Fluids 33 (2) (2005) 149–157.

[60] G.T. Hong, Hydrothermal oxidation: pilot scale operating experiences, in: Proceedings of 56th International Water Conference, Paper No. IWC-95-51, Pittsburgh, PA, 1995.

[61] N. Boukis, N. Claussen, K. Ebert, R. Janssen, M. Schacht, Corrosion screening tests of high-performance ceramics in supercritical water containing oxygen and hydrochloric acid, J. Eur. Ceram. Soc. 17 (1997) 71–76.

[62] K.M. Garcia, R. Mizia, Corrosion investigation of multilayered ceramics and experimental nickel alloys in SCWO process environments, Idaho National Engineering Laboratory Report No. INEL-94/0017, February 1995, www.osti.gov/bridge/servlets/purl/31690-E7zfIS/webviewable/31690.PDF.

[63] E.L. Daman, Process and apparatus for supercritical water oxidation, U.S. Patent No. 5,571,423, 1996.

[64] E.L. Daman, Process and apparatus for supercritical water oxidation, U.S. Patent No. 5,723,045, 1998.

[65] P.J. Crooker, K.S. Ahluwalia, Z. Fan, J. Prince, Operating results from supercritical water oxidation plants, Ind. Eng. Chem. Res. 39 (12) (2000) 4865–4870.

[66] T.G. McGuinness, Supercritical oxidation reactor, U.S. Patent No. 5,384,051, 1995.

[67] J. Abeln, M. Kluth, G. Petrich, H. Schmieder, Supercritical water oxidation (SCWO): a process for the treatment of industrial waste effluents, High Press. Res. 20 (2001) 537–547.

[68] J. Abeln, M. Kluth, M. Böttcher, W. Sengpiel, Supercritical water oxidation (SCWO) using a transpiring wall reactor: CFD simulations and experimental results of ethanol oxidation, Environ. Eng. Sci. 21 (2004) 93.

[69] J. Abeln, M. Kluth, M. Pagel, Results and rough cost estimation for SCWO of painting effluents using a transpiring wall and a pipe reactor, J. Adv. Oxid. Technol. 10 (2007) 169–176.

[70] M.D. Bermejo, F. Fdez-Polanco, M.J. Cocero, Experimental study of the operational parameters of a transpiring wall reactor for supercritical water oxidation, J. Supercrit. Fluids 39 (2006) 70–79.

[71] M.D. Bermejo, F. Fdez-Polanco, M.J. Cocero, Effect of the transpiring wall on the behavior of a supercritical water oxidation reactor: modeling and experimental results, Ind. Eng. Chem. Res. 45 (10) (2006) 3438–3446.

[72] M.D. Bermejo, M.J. Cocero, Destruction of an industrial wastewater by supercritical water oxidation in a transpiring wall reactor, J. Hazard. Mater. 137 (2006) 965–971.
[73] K. Príkopsky´, B. Wellig, P.R. von Rohr, SCWO of a salt containing artificial wastewater using a transpiring-wall reactor: experimental results, J. Supercrit. Fluids 40 (2007) 246–257.
[74] B. Wellig, K. Lieball, P.R. von Rohr, Operating characteristics of a transpiring-wall SCWO reactor with a hydrothermal flame as internal heat source, J. Supercrit. Fluids 34 (2005) 35–50.
[75] E. Fauvel, C. Joussot-Dubien, E. Pomier, P. Guichardon, C. Charbit, F. Charbit, S. Sarrade, Modeling of a porous reactor for supercritical water oxidation by a residence time distribution study, Ind. Eng. Chem. Res. 42 (10) (2003) 2122–2130.
[76] E. Fauvel, C. Joussot-Dubien, P. Guichardon, G. Charbit, F. Charbit, S. Sarrade, A double-wall reactor for hydrothermal oxidation with supercritical water flow across the inner porous tube, J. Supercrit. Fluids 28 (2004) 47–56.
[77] H.L. LaRoche, M. Weber, B. Zehnder, Purification of salt charges waste water by wet oxidation under super-critical conditions, U.S. Patent No. 5,437,798, 1995.
[78] H.L. LaRoche, M. Weber, C. Trepp, Design rules for the wall cooled hydrothermal burner (WHB), Chem. Eng. Technol. 20 (1997) 208–211.
[79] M. Weber, B. Wellig, P.R. von Rohr, SCWO apparatus design—towards industrial availability, CORROSION 99, Paper No. 258, NACE, Houston, TX, 1999.
[80] J.S. Barber, Methods of controlling flow of fluids reacting at supercritical conditions, U.S. Patent No. 5,427,764, 1995.
[81] P. Muthukumaran, R.B. Gupta, Sodium-carbonate-assisted supercritical water oxidation of chlorinated waste, Ind. Eng. Chem. Res. 39 (12) (2000) 4555–4563.
[82] D.S. Ross, I. Jayaweera, D. Bomberger, On-site disposal of hazardous waste via assisted hydrothermal oxidation, Rev. High Press. Sci. Technol. 7 (1998) 1386.
[83] M. Tateishi, Y. Tsuchiyama, Y. Yamauchi, T. Fukuzumi, T. Hatano, PCB decomposition process, U.S. Patent No. 6,162,958, 2000.
[84] K.W. Downey, R.H. Snow, D.A. Hazlebeck, A.J. Roberts, Corrosion, chemical agent destruction, in: K.W. Hutchenson, N.R. Foster (Eds.), Innovations in Supercritical Fluids, in: ACS Symposium Series, vol. 608, American Chemical Society, Washington, DC, 1995, pp. 313–326.
[85] P. Kritzer, N. Boukis, G. Franz, E. Dinjus, The corrosion of niobium in oxidizing sub- and supercritical aqueous solutions of HCl and H_2SO_4, J. Mater. Sci. Lett. 18 (1999) 25–27.
[86] H.C. Lee, J.H. In, S.Y. Lee, J.H. Kim, C.H. Lee, An anti-corrosive reactor for the decomposition of halogenated hydrocarbons with supercritical water oxidation, J. Supercrit. Fluids 36 (2005) 59–69.
[87] G.T. Hong, V.A. Zilberstein, Iridium material for hydrothermal oxidation environments, U.S. Patent No. 5,527,471, 1996.

Chapter 13

Process Components and Processes

13.1 INTRODUCTION

Hydrothermal and supercritical water processes are carried out in a high-temperature and high-pressure environment that needs to be enclosed in a vessel. The vessel walls have to withstand the effects of processing, including the interactions of the reaction mixture with the material of the vessel walls. In the context of this chapter, the term "reaction" stands for all physical and chemical processes in the pressure vessel, including equilibrium states. Depending on the scale of the process, different constructions for the processing equipment will be needed, but the function of all these is comparable and essentially independent on size.

In this chapter, similarities of design are used to facilitate an easier approach to the first design of experimental and larger scale equipment. The basic features are illustrated with examples. No systematic overview or a comprehensive discussion of the subject of reactors, processing equipment, and commercial processes is intended.

Processing steps, common to hydrothermal and supercritical water systems are: The feed has to be introduced into the high-pressure vessel, operating conditions have to be established and maintained, the process has to be carried out that leads to the products, the reaction products have to be retrieved and separated, water as the working fluid has to be recycled or cleaned for discharge.

Obviously, differently designed components are needed for the laboratory scale and for demonstration or the production scale, although the basic process remains the same. At all scales, the function of the components must be fulfilled. While at the laboratory scale the main effort is concentrated on the reaction, at the larger scale new problems arise and need solutions, such as handling of feed and of products in large quantities, heat transfer, fluid dynamics, and treatment of solids.

In this chapter, first the basic features of different reactors and of major components, like mixing tees, are discussed. The general layout is illustrated by examples. Features of typical experimental setups are then discussed and

Supercritical Fluid Science and Technology, Vol. 5. http://dx.doi.org/10.1016/B978-0-444-59413-6.00013-3

illustrated by several examples. Some of them come from the experience of the author over the years.

A process scheme for high-pressure and high-temperature processes is proposed that can be applied in principle for all the processes carried out with hydrothermal and supercritical water, due to the fact that many of the process steps are similar. Some process designs are outlined and are illustrated also by examples.

Design and construction of pressure vessels is beyond the scope of this book. The reader is referred to the appropriate literature, for example 1 as a first step. It should be borne in mind that the construction and operation of high-pressure vessels beyond small experimental devices, is regulated by safety codes. The main ones are [1]:

- A.S.M.E. Code, section VIII, Div. 2 and 3, American Society of Mechanical Engineers, United States of America,
- AD-Merkblatt, Technischer Überwachungsverein, TÜV, Germany,
- CODAP, Service de Mines, France,
- Unified Pressure Vessel Code, Japanese Industrial Standards Committee, Japan.

13.2 REACTORS

13.2.1 General Types of Reactors and Their Properties

Reactors enclose the pressurized feed that is to be processed. For experimental purposes, reactors of very different size are used. Their volumes range from fractions of 1 cm^3 to more than 10,000 cm^3. The smaller volumes are used mainly for experimental screening purposes, model mixtures, and the determination of reaction kinetics, while the larger ones are used for experiments with real mixtures and for acquiring sample products in larger quantities. Obviously, reactors for production are much bigger. Volumes of production-scale pressure vessels range from several dm^3, for very high pressures or low quantity high value compounds, to many m^3, for example for pressure vessels in power plants.

Reactors are the central component in an experimental or production plant. In the pressure vessels that serve as reactors, properties of mixtures and physical and chemical processes are determined, such as phase equilibrium, thermodynamic properties, transport properties, as discussed in Chapter 2 and 3, heat and mass transfer, as discussed in Chapters 4, 6, and 11, and chemical reactions, as discussed in Chapters 5, 7–10, and 12.

Several basic types of reactor vessels can be discerned. Reactors can be simple pressure vessels, operated in batch mode and often named "Autoclave" or "Bomb", see Figure 13.1a, or pressure vessels operated in batch mode also, but with a stirrer, see Figure 13.1b, continuously operated pressure vessels with a stirrer, see Figure 13.1c, or continuously operated tubular reactors,

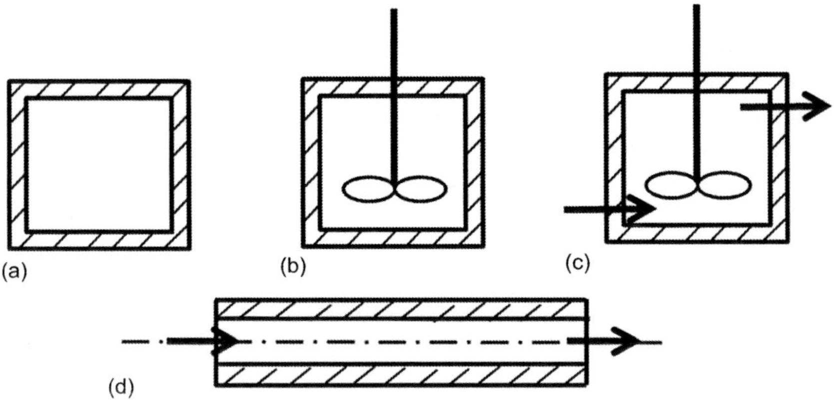

FIGURE 13.1 General types of pressure vessels. (a) Simple pressure vessel, operated in batch mode, (b) pressure vessel, operated in batch mode, stirred, (c) continuously operated pressure vessel, stirred, (d) continuously operated tubular reactor.

see Figure 13.1d. The internal volume can be constructed to the needs of the special process, as will be mentioned for certain applications, such as oxidation with supercritical water (SCWO), or hydrothermal processing.

In the following discussion, the application of the various types of pressure vessels and their ability to determine essential processing parameters is addressed.

13.2.1.1 Simple Pressure Vessel, Operated in Batch Mode

Simple pressure vessel, operated in batch mode, consist mainly of the pressure-bearing hull and the opening and closure arrangements (Figure 13.1a). Optical windows can be included in the construction to enable observation.

Pressure vessels of this type have often been used for carrying out chemical reactions. Construction and operation of such pressure vessels is simple. They are filled with the reactants, then closed and brought to operating temperature. Pressure is established according to the vapor pressure of the feed components and the filling rate. Often, no measurement of pressure is carried out, and pressure has to be estimated from the component properties and the filling rate. After a predetermined residence time, the reactor is cooled and ambient pressure is reestablished. Then, the pressure vessel is opened and the content is analyzed.

Procedure of the experiment and construction of the pressure vessels are simple. Reaction volumes can be kept small, and reaction conditions can be achieved easily that are difficult to reach with more sophisticated equipment. If needed, due to the simple procedure and equipment, several experiments can be operated in parallel without great effort.

The disadvantage is that no information on the processes during operation can be obtained, in particular no concentrations can be related to a certain reaction time, if not indirect information can be retrieved, for example through an optical window in the autoclave, or through signal-transmitting lines that lead to the inside of the reactor. Furthermore, due to the relatively high mass of the pressure vessel, the time of heating and cooling the pressure vessel and the content may be in the same order of magnitude, or even much longer, than the time needed for the reaction. Mixing of components and phases occurs only through diffusion and convection processes induced by the reaction conditions. Better mixing conditions can be obtained, if the content of the pressure vessel is mixed by periodically moving the autoclave in total (rocking autoclave). Nevertheless, without indirect measurement of properties of the ongoing processes within the autoclave, reaction paths and reaction time cannot be determined with reasonable accuracy. Therefore, such procedure and such equipment usually are of limited use for the investigation of reactions.

13.2.1.2 Pressure Vessels, Operated in Batch Mode, Stirred

Stirred pressure vessels (Figure 13.1b) that are operated in batch mode reduce the limitations of mass transfer during the reactions considerably. Still, the information on the ongoing reactions can only be retrieved by indirect methods, usually through an optical window and using spectroscopic methods. Furthermore, the amount of components in the reaction is fixed by the composition of the feed. Process parameters that are essential, such as the relative concentration of the components, can be varied only in different experiments.

Nevertheless, such stirred pressure vessels are of great importance for acquiring properties of equilibrium at different operating conditions. In this case, stirring is applied, and necessary, to establish equilibrium in a reasonable time. After the mixing time, the stirrer is shut off, the content of the pressure vessel is given time to adjust to the equilibrium situation, for example to separate in discrete phases. Then, the properties of the content of the pressure vessel are determined, either directly by taking samples from the different phases, or indirectly via an optical window using for example spectroscopic methods. An important feature of such an equipment can be, if a variable volume is part of the installation, to retrieve directly information on the volume of the content of the pressure vessel, which allows to determine the P, V, T data for the content of the pressure vessel. An example of such a device is shown below.

13.2.1.3 Continuously Operated Pressure Vessel, Stirred

A stirred pressure vessel can be used as a continuously operated reactor (Figure 13.1c). Then, feed is introduced continuously and effluent is removed continuously from the pressure vessel. The characteristics of such a reactor are those of an ideally mixed reaction vessel with a reaction time for the individual molecules that is distributed around an average residence time. Such reactors are very flexible and have been treated extensively in the literature on reaction kinetics. They can be applied for high-pressure processes with the aid of a pressure pump for the delivery of the feed and a pressure reduction valve for the outlet.

Average reaction times can be varied in wide ranges depending on the size of the vessel and the flow rate of the feed. The higher the ratio of flow rate to size of the vessel, the narrower is the distribution of the residence time around the average value. The residence time can be adjusted from the range of seconds to many hours. Due to the retrieval of the effluent flow, information can be directly derived on the reaction and closely related to the reaction time in the vessel.

13.2.1.4 Continuously Operated Tubular Reactor

Basically, a continuously operated tubular reactor consists of a pipe of certain length to which the reactants are fed continuously and the product components are removed in the same quantity (Figure 13.1d). Construction and operation of such a reactor is relatively simple.

In a continuously operated tubular reactor, ideally the feed is transported in plug-flow through the reactor. The effluent components all have the same process history and samples can be analyzed. Results are directly connected to the residence time, which is equal to the reaction time. Furthermore, during the reaction, the reaction parameters, for example temperature, can be changed. In such a way, nonisothermal processes can be simulated. The residence time typically is on the order of fractions of a second to a few minutes. Extended residence time is difficult to achieve because of practical problems arising from the necessary length of the tubular reactor, mainly plugging problems. Fast reactions can be investigated with such reactors.

13.2.2 Examples for Reactors

Pressure vessels, or reactors, can be used for different purposes, such as equilibrium property measurements, for measurements of process parameters, such as chemical reactions, dissolution of compounds for further processing, extraction of compounds, and heat transfer, and for hydrothermal processing. Examples of reactors are shown in the following. Many more examples can be accessed in the original literature and in reviews. For example, an introduction into high-pressure technology, and design and construction of high-pressure equipment for research and production, can be found in [1,2].

An example of an autoclave that can be used for phase equilibrium measurements as well as for the investigation of reactions is shown in

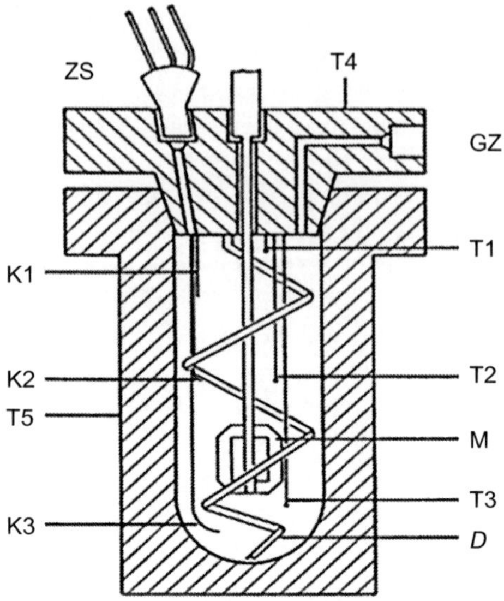

FIGURE 13.2 Pressure vessel for determining phase equilibrium or investigating reactions. M, magnetically driven rotating stirrer; K1–K3, sampling lines; T1–T5, thermocouples; D, flushing line for liquid phase, GZ, flushing line for gaseous phase; ZS, plug for sampling lines [3,4].

Figure 13.2 [3,4]. The vessel size is $V = 1000\ \text{cm}^3$, the maximum operating pressure is $P = 100$ MPa, the maximum temperature $T = 400$ °C.

The vessel content is stirred with a magnetically driven blade-stirrer. Temperature is measured by three thermocouples, placed at different locations in the vessel. Samples are taken by three sample lines of $d_\text{I} = 0.1$ mm, placed at locations for sampling from the upper, middle, and lower part of the vessel. During the experiment, the vessel can be flushed with liquid or gas, Figure 12.2 [3,4]. A similar device has been used to determine the phase equilibrium of water–carbon dioxide, see Chapter 3, Figure 3.44 [5].

Pressure vessels constructed with optical windows make possible the visual and spectrometric observation of the content during the experiment. Such optical cells are used to investigate phase equilibria at high pressures, and reactions, and are used for analytical purposes. Such an optical cell is shown in Figure 13.3 [6]. This cell has been used to investigate phase equilibria of water–hydrocarbon mixtures, shown in Chapter 3, Figures 3.28–3.31. The volume of this view cell is $V = 110\ \text{cm}^3$, big enough to work with representative quantities of the mixture components. It has an excellent viewing possibility to detect coexisting phases and their appearance and disappearance [6].

It may be of use in many cases to adjust the volume of the experimental cell during the experiment or to register volume changes. This is possible with variable volume bellows as shown in Figure 13.4 in a view cell design that has

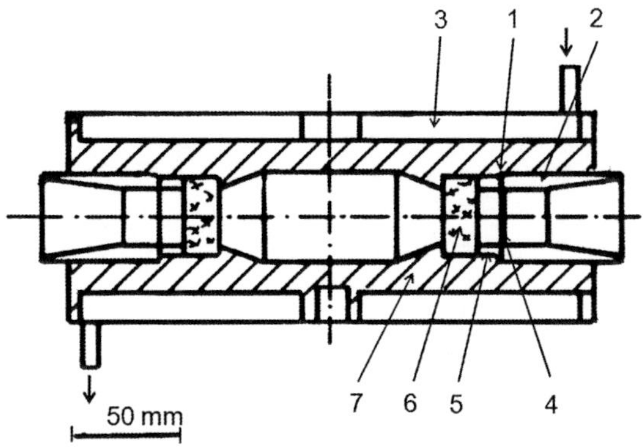

FIGURE 13.3 Optical cell for high-pressure and high-temperature experiments. Maximum operating pressure $P = 200$ MPa, maximum operating temperature $T = 350\,°C$, internal volume $V = 110\,cm^3$. 1, pressure distribution ring; 2, pressure screw; 3, heating jacket; 4, seal ring; 5, pressure ring; 6, window; 7, pressure vessel. *Modified from [6].*

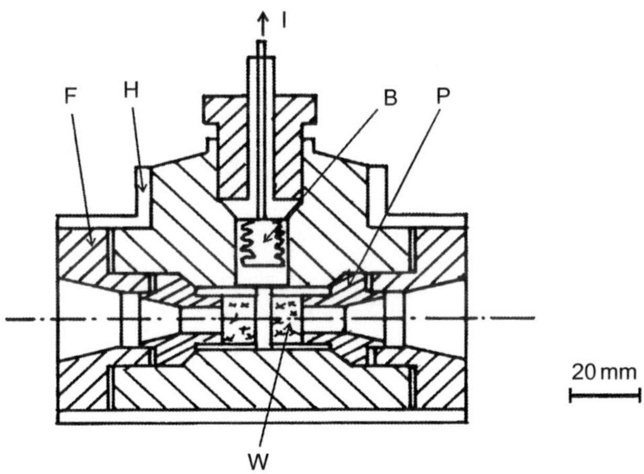

FIGURE 13.4 Optical high-pressure cell with internal bellows. W, window; B, bellows; F, flange; H, heating; P, plug; I, inductive detector. *Modified from [7].*

been extensively used for investigating reaction kinetics by IR-spectroscopy [7]. Further examples of reactors are, for example, presented in [8,9].

A reactor of very small size that can be used to verify extreme conditions up to about 100,000 MPa, is the Diamond Anvil Cell. An example is shown in Figure 13.5 [10]. The essential features of such a cell are: A miniature mechanical press applies a balanced force onto two opposing anvils (1 mm diameter), typically made of diamond. A metal sheet with a hole is placed between the anvils. The diamonds and the metal sheet confine liquids and solutions. Modern cells are compact (2–10 cm diameter, 3–7 cm high) and

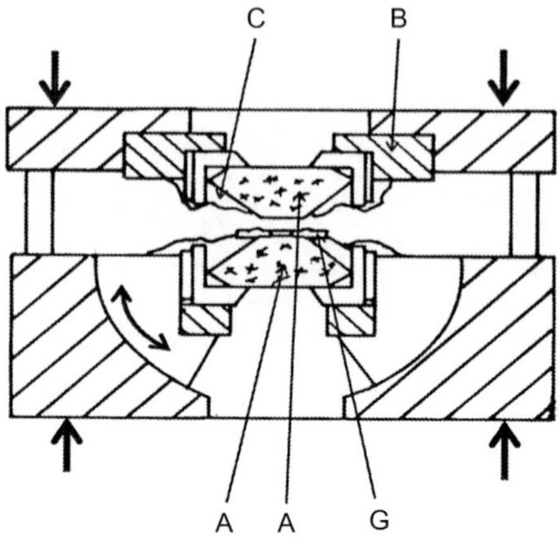

FIGURE 13.5 Diamond anvil cell for the investigation of mixtures at very high pressures. A, Anvil; B, Ceramic barrier; C, Cement; G, Gasket. *Modified from [10].*

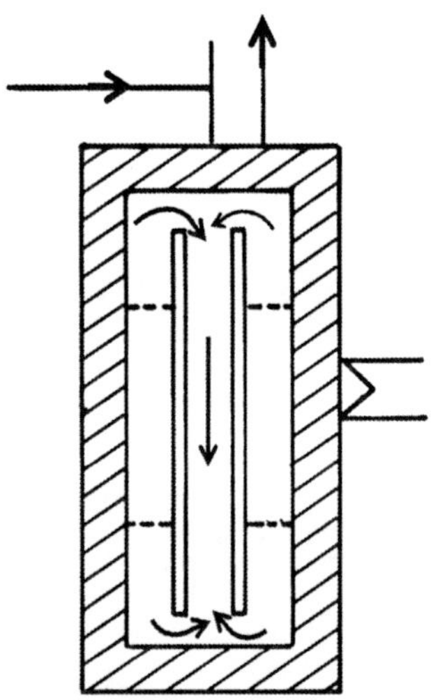

FIGURE 13.6 Jet loop reactor. Reactor volume $V=16\ cm^3$. *Modified from [11].*

are commercially available. They can be used to safely study aqueous systems up to $T=800\ °C$ and pressures of >3000 MPa [10].

An example for a special reactor for investigating chemical reactions is the jet loop reactor as shown in Figure 13.6 [11]. It realizes an internal cycle of the reacting components. The feed stream enters the reactor at the top and flows downwards. It turns around at the bottom and passes upwards through the annular gap. The entering flow entrains the reaction mixture and forces it downwards, generating thorough mixing by a circulating flow.

Other reactors shown in this book are: Reactor for hydrothermal processing, Chapter 11, Figure 11.1, transpiring wall reactor for minimizing corrosion, Chapter 12, Figure 12.8. More on reactors can be found, for example, in [12].

13.3 EXPERIMENTAL PROCESSES

13.3.1 General Features of Experimental Processes

An experimental process typically is concerned mainly with the process step itself. Handling of reactants and product components is of secondary importance, as long as conditions are maintained for operating the process appropriately.

An experimental process can be divided into several steps that support the process that is in the center of interest, as shown in Figure 13.7. Components, that are processed, are taken from a reservoir (not shown), pressure and temperature are adjusted to processing conditions, components and water, as the processing medium, are mixed in such a way that the process proceeds as intended. For example, mixing is carried out such that the reaction starts only at the entrance to the reactor. In case of investigating heat transfer, the transport process has to begin at a known position and at known conditions. After the process step is carried out, the product components are removed, temperature, and pressure are reduced to usually ambient conditions, and the product flow is separated into gaseous, liquid, and solid products (Figure 13.7).

The process can be carried out in sequence in a single reactor. One process step follows the other at the same place. Otherwise, it can be carried out in sequence, in such a way that process steps take place in different parts of the experimental setup. Even in experimental scale, process steps can be carried out in the same way as in larger plants. Such a technique is called "Mini-Plant Technique".

The general procedure for experimental processes can be applied to all types of experiments, not only for hydrothermal and supercritical water process experiments. These are characterized by the fact that water is the main processing medium which is added to the reactants, takes part in the process, and is removed from the product components. Most of such experimental

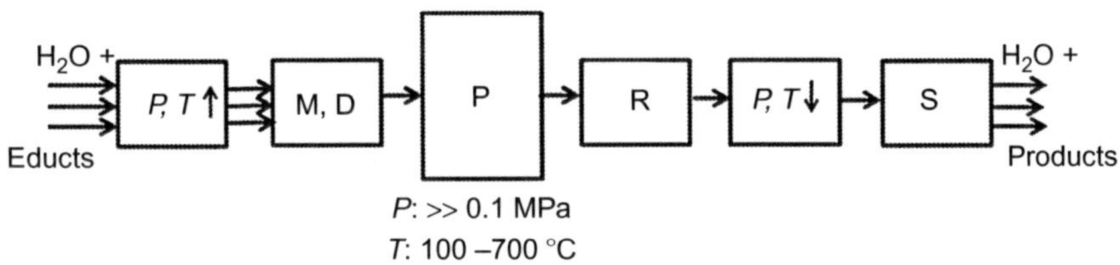

FIGURE 13.7 Basic flow sheet of experimental processes carried out with hydrothermal and supercritical water. *P*, *T* ↑, Pressure and temperature increase; M, D, mixing and delivery; *P*, process, R, product recovery, *P*, *T* ↓, pressure and temperature reduction; S, separation.

processes are run as once-through processes without any recycling of products or processing medium.

In the following, some examples of experimental setups for processes with hydrothermal and supercritical water are presented. Literature will be cited for further access to experimental equipment and methods.

13.3.2 Examples of Experimental Processes

13.3.2.1 Experimental Setup for Phase Equilibrium Measurements

An example for an experimental apparatus for determining phase equilibrium in hydrothermal and supercritical water systems is shown in Figure 13.8 [4]. It consists of a pressure vessel *A* of $V = 1200\ \text{cm}^3$ volume, used as equilibrium cell, a second pressure vessel *B* of $V = 300\ \text{cm}^3$, used as buffer autoclave, both with maximum operating conditions of $T = 450\ °\text{C}$, and $P = 30\ \text{MPa}$. The autoclaves are kept at constant temperature. The setup consists further of a delivery system (left side of Figure 13.8) for an inert gas that is pumped with a long stroke pump *S* and delivers via a reservoir autoclave *C* gas to the buffer autoclave *B*, from which liquid phase is such transported to the equilibrium autoclave *A*. In such a way, pressure can be kept constant if samples are

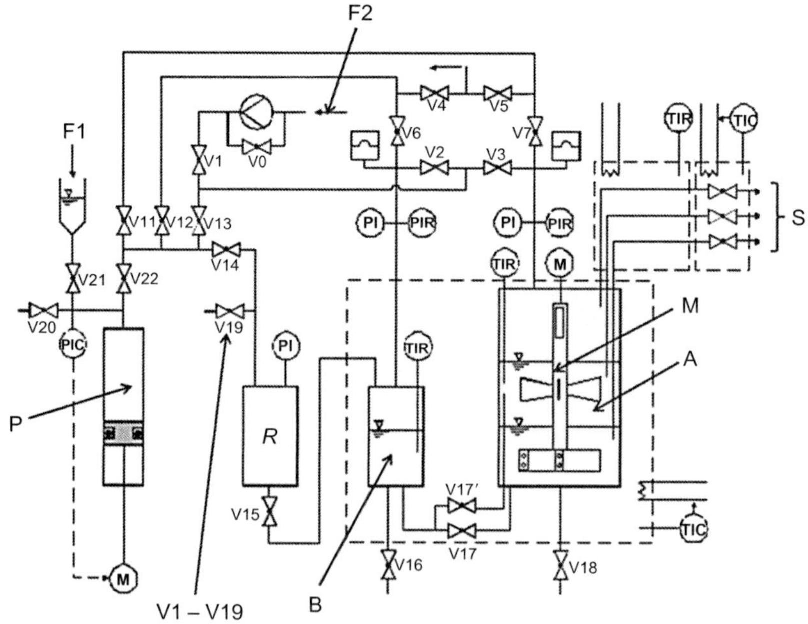

FIGURE 13.8 Experimental setup for determining phase equilibria in hydrothermal and supercritical water systems. A, equilibrium autoclave; M, magnetically driven rotating stirrer; *B*, buffer autoclave; V1–V19, valves; P, long-stroke pump; R, additional buffer autoclave; F1, F2, feed lines; S, sampling lines. *Modified from [4].*

withdrawn. Samples are withdrawn from the equilibrium autoclave with three sampling lines *D* from three positions in the equilibrium cell.

With such an experimental apparatus, multicomponent phase equilibria can be determined with high accuracy, if the experimental procedure and analyses are carried out with care. Phase equilibria, shown in Chapter 3 for aqueous systems with hydrocarbons and gases like hydrogen and carbon dioxide, have been determined with this apparatus.

Although this method is very flexible, it is also time consuming. There are other methods for more rapid determination of phase equilibrium, in particular such ones that apply indirect measurement of the composition of the phases. A discussion of other equipment and methods is beyond the scope of this book. An extensive literature exists on this topic. The author has published a discussion [3] that may be considered as an introduction to this topic. Thorough information on methods and on available experimental data has been published by Dohrn and coworkers [13–16].

13.3.2.2 Experimental Setup for Heat Transfer

An experimental setup for investigation of heat transfer normally needs two reactors. In one reactor, the heat energy is transferred to the heat transporting medium, and in the other, heat is removed from this medium in a controlled process. Such equipment is schematically shown in Figure 13.9. In addition to the parts common also to other setups, such as a delivery sand recovery system for the fluid, this apparatus consists of a reactor that is used to

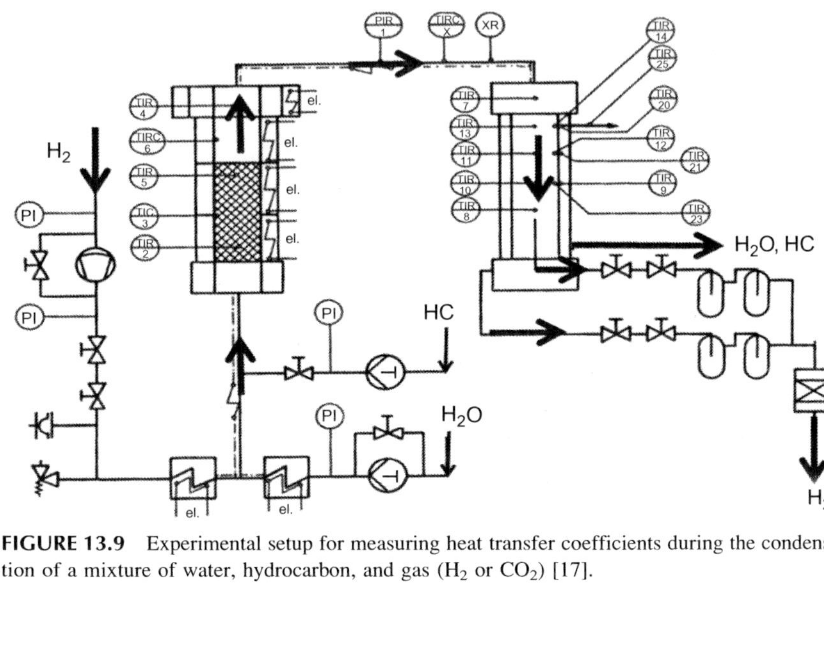

FIGURE 13.9 Experimental setup for measuring heat transfer coefficients during the condensation of a mixture of water, hydrocarbon, and gas (H_2 or CO_2) [17].

produce the gaseous mixture (at high temperature) that is transferred to the second reactor (shown on the right hand side in Figure 13.9), where it is condensed by reducing the temperature, while the mixture continuously flows from upside down. Heat transfer coefficients are derived from the measurement of temperatures and compositions at various places within the reactor [17]. The procedure is explained in Chapter 4.4.2. Composition of the phases during condensation can be derived from phase equilibrium, discussed in Chapter 3.4.3.4. Other examples are presented in the literature, for example [18] and from the investigations cited in Chapter 4.

13.3.2.3 Experimental Setup for Investigations of Chemical Reactions

As an example, the continuous decomposition of glucose in hydrothermal and supercritical water is cited [11] (Figure 13.10). The special feature of this experiment is the jet loop reactor. Such a reactor enables to maintain a relatively extended reaction time, if needed, and provides thorough mixing of the components. The entering jet creates a pressure drop at the top and entrains the reaction mixture. Thereby the reaction mixture is forced downwards again and generates thorough mixing by a circulating flow [11]. The reactor can be of relatively small size, in this case about $V = 16\ cm^3$.

More on experimental setups for the investigation of chemical reactions are given, for example, in the book of Jessop and Leitner [12], and in the literature cited in Chapter 5.

13.3.2.4 Experimental Setup for Extraction with Hydrothermal and Supercritical Water

Experimental setup for extraction with hydrothermal and supercritical water can be relatively simple, as shown in Figure 13.11. With such equipment, the extraction of various substrates with water was investigated. The experimental results have been discussed in Chapters 6 and 8. The setup consists

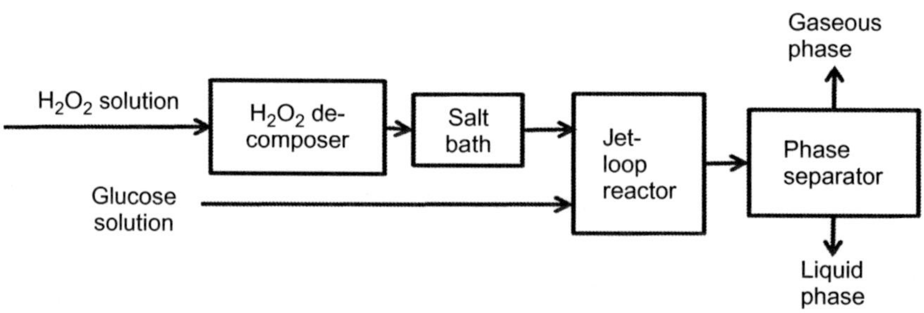

FIGURE 13.10 Experimental setup for continuous decomposition of glucose in hydrothermal and supercritical water. *Modified from [11].*

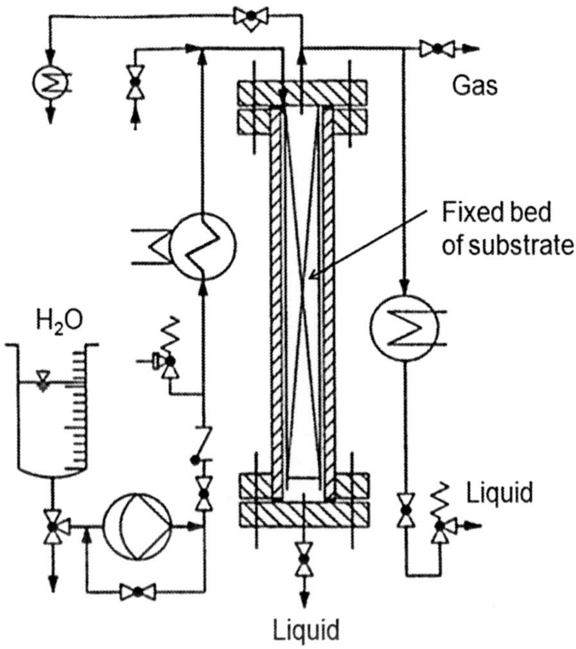

FIGURE 13.11 Experimental setup for extraction with hydrothermal and supercritical water in a semibatch reactor. *Modified from [19].*

of the elements shown basically at the beginning of this chapter. The reactor is an autoclave in which the solid substrate is placed and the extraction is carried out as a semibatch process [19,20].

The setup consists of a reservoir, a membrane metering pump, a preheater, the extraction vessel, containing the fixed bed of substrate ($T_{max}=400$ °C, $P_{max}=27$ MPa, $V=460$ cm^3, length/diameter $L/d=10$), a heat exchanger to cool the effluent, a back-pressure valve, a sampling line, the gas supply, the release valve, and a high-temperature valve, used as connection to an optional tubular reactor [19].

A combination of a reactor, operated semicontinuously as fixed bed reactor, or continuously as stirred reactor with a tubular reactor, allows the very flexible combination of the characteristics of the two reactor-types. In the reactor vessel, residence time can be varied in a wide range up to hours, but not reduced to a very short residence time in the range of seconds or less. That is possible with the attached tubular reactor that can be operated continuously and cocurrently at conditions different to that in the vessel reactor. Such an experimental setup for extraction and reaction (combined autoclave and tubular reactor) [21,22] is shown in Figure 13.12. This setup has been used extensively for the investigation of extraction and reaction processes as discussed in Chapters 6–8 and 10 for the oxidation of waste.

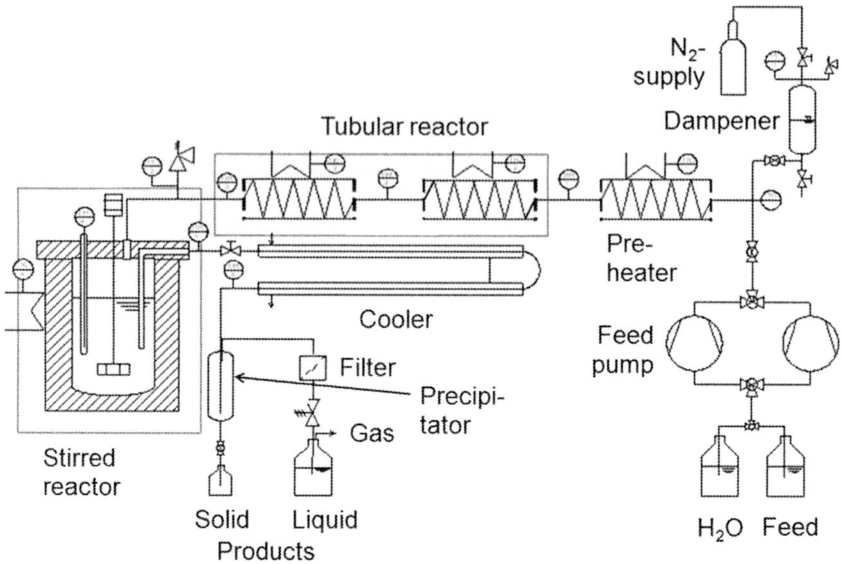

FIGURE 13.12 Experimental setup for extraction and reaction (combined autoclave and tubular reactor) [21,22]. *Modified from [22].* (For the color version of this figure, the reader is referred to the online version of this chapter.)

For supercritical water oxidation SCWO, a setup with a tubular reactor has been used successfully [23–26], as shown in Figure 13.13. The main specifications of this plant are: Maximum pressure: $P_{max} = 30$ MPa, maximum temperature: $T_{max} = 400\ °C$, maximum throughput: $\dot{m} = 5\,kg/h$, initial solid concentration: $x_F = 1–2$ wt%, reaction volume: $V = 100$ ml [24].

An area of increasing importance is the production of nanoparticles with hydrothermal and supercritical water processes. Such processes are discussed in Chapter 11. In such a process, the feed components are mixed, the reaction takes place in the reactor, the product component is kept in solution in the reactor, and then the end-product is produced by precipitation, for example by rapid expansion in a so-called RESS process. Such an experimental setup of a hydrothermal process for the production of nanoparticles is shown as basic flow-sheet in Figure 13.14 [27].

In hydrothermal and supercritical water processes, a main issue is the heating of the feed solution to operating temperature. The reasons are: Increased corrosion in the subcritical region, uncontrolled reactions at nonoptimal conditions, insufficient mixing of the reaction compounds. Mixing obviously is important, but may cause problems due to a very different amount of the streams to be mixed, or a different density, or concentration-dependent reactions. For a sufficient mixing behavior, a tee-mixing device is applied in most cases. A number of possible arrangements are shown in Figure 13.15 [28].

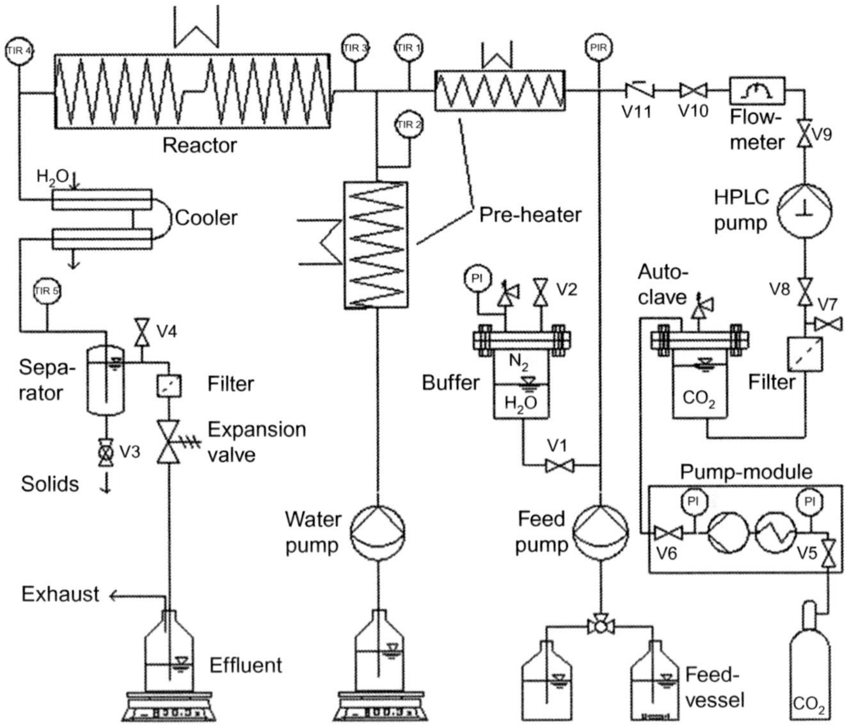

FIGURE 13.13 Experimental setup with a tubular reactor for supercritical water oxidation SCWO in a tubular reactor [24].

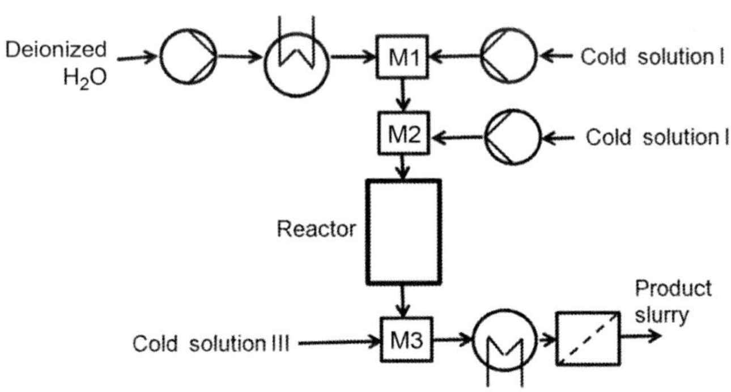

FIGURE 13.14 Flow sheet of an experimental setup of a hydrothermal process for the production of nanoparticles. M1–M3, mixing units. *Modified from [27].*

The geometry of a mixing tee and flow rates affect the properties of the synthesized product ($LiFePO_4$ [28]), such as particle size, surface area, crystalline structure, morphology, and electrochemical performance. For example, for an increasing flow rate, the particle size decreases; but amount of particles

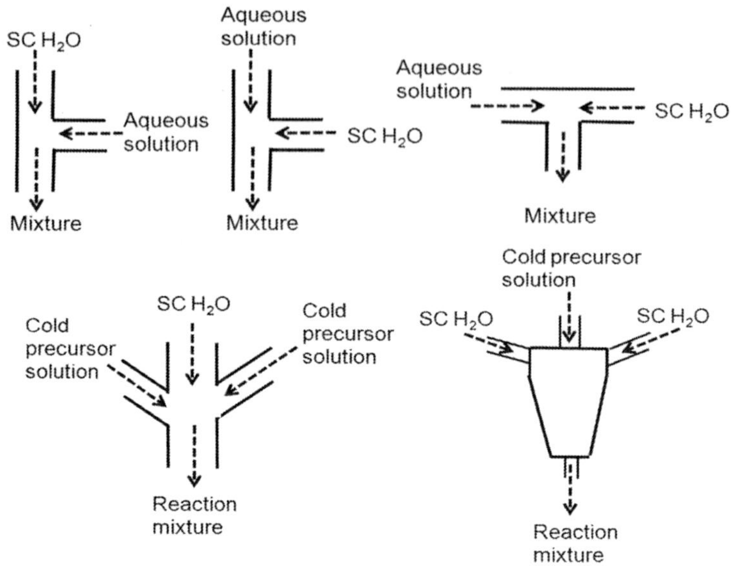

FIGURE 13.15 Tee-mixing arrangements. A 90° tee, 50° tee, and vortex-mixing tee. *Modified from [28].*

synthesized decreases due to enhanced formation of Fe^{3+} impurities [28]. A vortex-mixing (swirling) tee that leads to smaller particles with fewer impurities is superior to the 90° and 50° tees [28].

Although the basic setup for experimental processes with hydrothermal and supercritical water is relatively simple, there is no standard design. Experimental plants differ by the type of the reactors, the type of additional plant components, the complexity of the design, and according to size of the equipment. It is beyond the scope of this book to discuss the experimental plants in more detail. Information may be accessed via the original investigations and via reviews. Some references, in which information on experimental setups for processing with hydrothermal and supercritical water are discussed, are given in the following: [1,2] for high-pressure equipment and chemical reactions, [12] for most types of processes with supercritical fluids, [29] for extraction of natural substances with hydrothermal water, [30] for reaction of cellulose, [31] on conversion of biomass (corn stover), [32] for polymer recycling, [33] for fast hydrolysis and oxidation reactions, [34] for a scheme of a continuous anticorrosive SCWO reactor system, [35] for a plug flow and batch reactor system for SCWO, [36] for a hydrothermal reaction system for the production of nanoparticles, [37] for hydrothermal crystallization, [38] for the design of an experimental plant for antisolvent precipitation.

13.4 PRODUCTION PROCESSES

13.4.1 General Features of Production Processes

Production processes are obviously operated to deliver products, contrary to experimental processes, for which the determination of optimal process parameters is the main objective. Produced sample products are used for analysis and for presentation. In the context of this book, production processes are only treated to characterize the differences to experimental processes. No systematic compilation of production processes with hydrothermal and supercritical water is presented due to limitation of space.

It seems, at a first glance, that production processes differ from experimental processes mainly by size. While in experimental processes samples of the size of less than grams to maximal kilograms are produced, production processes may begin at a production rate of hundreds of kilograms for very high-priced materials and may reach a yearly production rate of hundreds of thousands of tons. This difference in size brings along a number of problems, mostly engineering problems, but not only. If the product is attractive and can be sold, other problems to be solved are the supply of the feed and the disposal of wastes, as the major logistic problems, financing as an economical problem, and acceptance as a sociological problem.

From the point of view of technology, the main difference between an experimental process and a production process is complexity. This means that solving a major processing problem with hydrothermal and supercritical water, does not mean that a production process is established. Many more steps, some of them of total other aspects than hydrothermal processing, become important, and in many cases are decisive. One of the aspect that make production processes much more complex than experimental processes is the necessity to treat the process fluid, water, in order to recycle it into the process or to dispose it into the environment. Some examples for processes with hydrothermal and supercritical water are presented in the following.

13.4.2 Examples of Processes

13.4.2.1 Heat Transfer

The main application of hydrothermal and supercritical water is in heat transfer processes. Those processes occur mostly in power plants. In conventional power plants, supercritical conditions are realized for some time already. For nuclear power plants, supercritical water conditions are planned for the future generation. The reason is obvious: enhancement of the process efficiency, which is determined by the upper temperature of the process. In Figure 13.16, the process conditions for nuclear power plants are shown.

In a typical fossil fuel power plant the chemical energy stored in fossil fuels and oxygen in the air is converted into thermal energy. This is used to

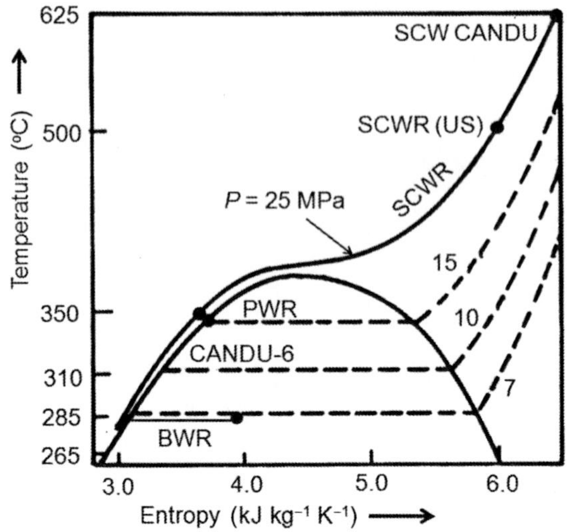

FIGURE 13.16 *T,s*-diagram of water with typical operating conditions of power plants. BWR, boiling water reactor; CANDU-5, CANada Deuterium Uranium nuclear reactor; PWR, pressurized water reactor; SCWR, supercritical water reactor. *Modified from [18].*

heat water in a boiler, producing steam that drives a turbine connected to the electrical generator, which in turn produces electricity. Each fossil fuel power plant is a highly complex, custom-designed system, designed on a large scale for continuous operation. Such plants can operate beyond the critical pressure of water and at temperatures up to $T=700$ °C. Supercritical power plantsare once-through plants since boiler water does not circulate multiple times, see Figure 13.17 [40].

Hydrothermal and supercritical water processes are carried out at a relatively high-temperature level. Therefore, heat transfer is of major importance in hydrothermal and supercritical water processes. Fundamentals of heat transfer have been discussed in Chapter 4. The heat needed to reach the operating temperature of the high-temperature processes with water must be recovered as far as possible from the effluent stream. During heat recovery, some problems can occur that are due to the corrosive character of the effluent (see Chapter 12), the unfavorable temperature region regarding corrosion (see Chapter 12), the formation of more than one phase during temperature reduction (see Chapter 3), and in particular the formation of solid phases that may stick to the heat transferring surfaces, or even block the heat exchanger. Design and operation of such heat exchangers has been carried out successfully, for example for biomass processing [41,42]. It is claimed that an efficiency of 80% is reached for the main heat exchanger in recovering process heat from the effluent flow [42].

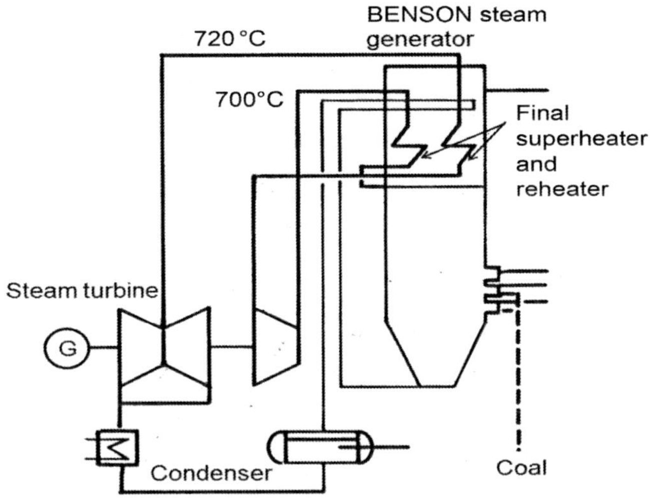

FIGURE 13.17 Power plant with supercritical water steam generator (BENSON-type). *Modified from [39].*

13.4.2.2 Wet Air Oxidation

Fundamentals of wet air oxidation are treated in Chapter 10. Wet air oxidation is a process for treating waste in aqueous solution by oxidation with air at conditions below the critical conditions of water. Typical operating conditions are for temperature 150 °C $< T <$ 300 °C, and for pressure 1.5 MPa $< P <$ 10 MPa. Residence time is in the range of 1 h or more.

Wet air oxidation is a large scale process with flow rates up to 50 m^3/h, mainly applied to treat wastewater by reducing chemical oxygen demand (COD) of about x_F = 10,000–150,000 ppm to acceptable values, in particular to enable conventional water cleaning processes. The principal flow sheet is shown in Chapter 10, Figure 10.4. Process examples for wet air oxidation are: Spent caustic treatment, treatment of wastewater from special plants such as the methyl methacrylate production, treatment of effluent from the polystyrene production, or treatment of effluent generated from the naphtha cracker for ethylene production [43].

13.4.2.3 Supercritical Water Oxidation

Fundamentals of supercritical water oxidation are treated in Chapter 10. Supercritical water oxidation is a process mainly for treating waste in aqueous solution by oxidation with oxygen. Operating conditions are higher than the critical values for water, typically for temperature 400 °C $< T <$ 600 °C, and for pressure 22 MPa $< P <$ 40 MPa. Residence time is in the range of seconds up to 1 min. A basic flow sheet is shown in Figure 13.18.

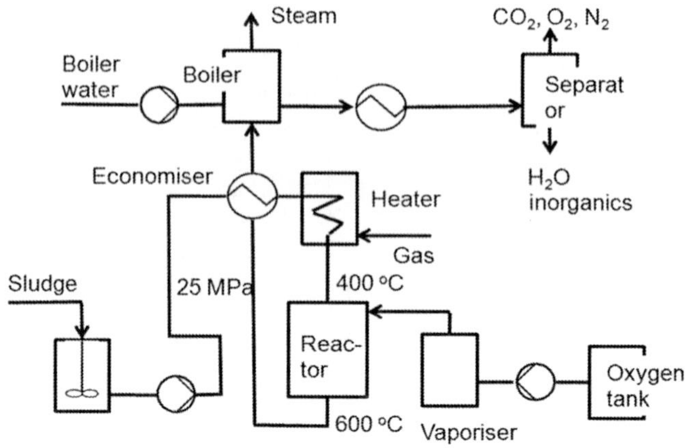

FIGURE 13.18 Basic flow diagram for a supercritical water oxidation process (SCWO). *Modified from [44].*

For supercritical water oxidation of sewage sludge [44], as an example, the sludge is pumped from the feed tank, after eliminating big particles, by the high-pressure pump through the supercritical water oxidation system. The feed is preheated in a heat exchanger by the reactor effluent and further in the heater to initial operating temperature of about $T = 400$ °C, and then enters the reactor. In the reactor, oxygen is injected and the oxidation reaction proceeds. The reactor temperature increases and must be kept within limits for highly concentrated feed streams. The waste is oxidized in the reactor and the reaction product flows through the heat exchanger and preheats the incoming feed. The effluent further passes through a steam generator and is further cooledto its final exit temperature, before pressure is reduced. Then, the effluent is separated into gas and liquid [44].

Supercritical water oxidation has some history. Early plants have been used mainly for the destruction of hazardous waste components, such as dioxins. Problems arose with corrosion and with plugging, caused by precipitation of salts. Corrosion has since then been intensively investigated and solutions are available, as discussed in Chapter 12. Precipitation of salts and plugging is fought by special construction of the reactors and by sweeping the reactor walls with lower temperature water (Chapter 12).

More recently, beside treatment of sewage slugde and similar wastes, the processing of energetic materials and military wastes has been a major objective for applying supercritical water oxidation. Examples are the treatment of pink and red water from TNT operations, destruction of hydrolysate from chemical agent and energetics neutralization, destruction of hydrolysate from the hydrolysis of excess explosives and propellants, destruction of solid rocket propellant and chemical warfare agents [44,45].

13.4.2.4 Biomass Processing

Fundamentals of biomass processing are discussed in Chapter 8. Beside separation of biomass into useful compounds and liquefaction of biomass, mainly biomass gasification has found interest. There are some good reasons for that. Gasification technology is known for some time, and large scale installations have been operated. Furthermore, breaking down the biomass feed into a few simple molecules makes the separation of the product flows much easier. On the other hand, valuable chemical components are destroyed that could be used as feed-stock. Biomass gasification is shortly discussed in the following as an example for large scale biomass treatment. Other processes may be approached from the literature cited in Chapter 8.

Biomass gasification with supercritical water is a process that has been investigated at pilot-plant scale [42], see Figure 13.19. Wet biomass and organic wastes can be efficiently gasified under hydrothermal conditions to produce a hydrogen rich fuel gas at conditions of $T=650\,°C$, $P=28$ MPa, residence time $t=60$ s [41].

Biomass is pretreated and can then be mixed with water and pumped into the reactor, passing the main heat exchanger to approach the operating temperature, as shown in Figure 13.19. In the reactor, a gaseous product is formed consisting of mainly H_2 and CO_2, with CO and CH_4 as byproducts. The reaction product flow heats the incoming biomass flow in the main heat exchanger. The gaseous product is then separated from the aqueous liquid. CO_2 is removed in a high-pressure scrubber from hydrogen. Thermal efficiency is in the range of 60% for an initial concentration of $x_F=5$ wt% in the feed and up to 80% for an initial concentration of $x_F=20$ wt% [42].

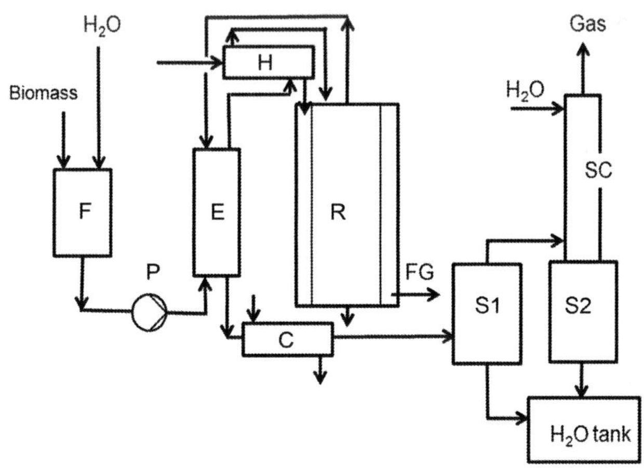

FIGURE 13.19 Simplified flow scheme for gasification of biomass in a pilot-plant (VERENA). $T_{max.}=650\,°C$, $P=28$ MPa, residence time $t=60$ s, $\dot{m}=100$ kg/h. F, feed tank; P, high-pressure pump; E, heat recovery; H, heater; C, cooler; R, reactor; FG, flue gas; S1, S2, separator; SC, CO_2-scrubber. *Modified from [42].*

13.4.2.5 *Fuel Processing*

An interesting case is the application of hydrothermal and supercritical water to fuel processing. Fossil fuel sources are mostly accompanied by water. Therefore, the question arises, whether water can be used economically to process such fuel materials. As examples, the extraction of bitumen from oil sands and the conversion of kerogen from oil shale are mentioned in the following.

Extraction of bitumen from oil sands is a process that is carried out with hot water. Temperatures are much lower than necessary for a hydrothermal process, but a simplified process scheme, as shown in Figure 13.20 shows that water is intensively involved in such a process and the complexity is relatively high. A hydrothermal process could probably solve some problems, in particular with froth treatment and wastewater processing, and additionally enhance yield, as discussed in Chapter 7.

For the conversion of kerogen in oil shale, an extensive investigation explored the use of water as processing medium [47,48]. The advantage of processing oil shale with supercritical water is the conversion of asphaltenes and preasphaltenes. For oil shales with kerogen, rich in these components, such a process may be useful. For paraffinic kerogen, as in Green River oil shale, the process is probably not advantageous [47,48].

As a result of the experimental investigations in the laboratory, it was found that water at sub- and supercritical conditions can be used for the processing of solid carbonaceous material with the goal of producing liquid fuels. A process has been proposed, as shown in Figure 13.21. Conversion of the organic material is high and liquid yield is also high (compare results in Chapter 7). The proposed process is shown here to illustrate the necessary

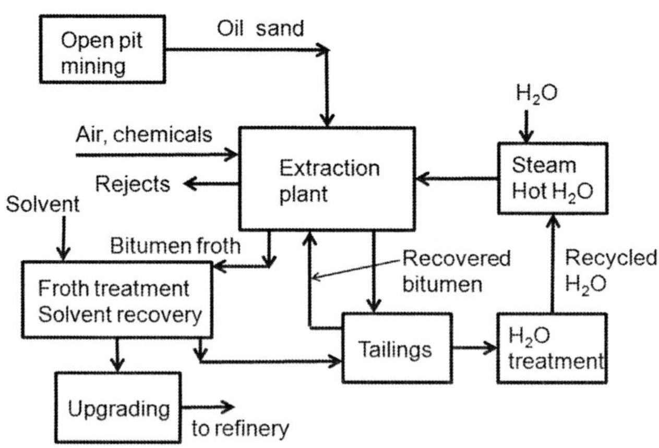

FIGURE 13.20 Generalized scheme for oil sands processing using water-based extraction processes. *Modified from [46].*

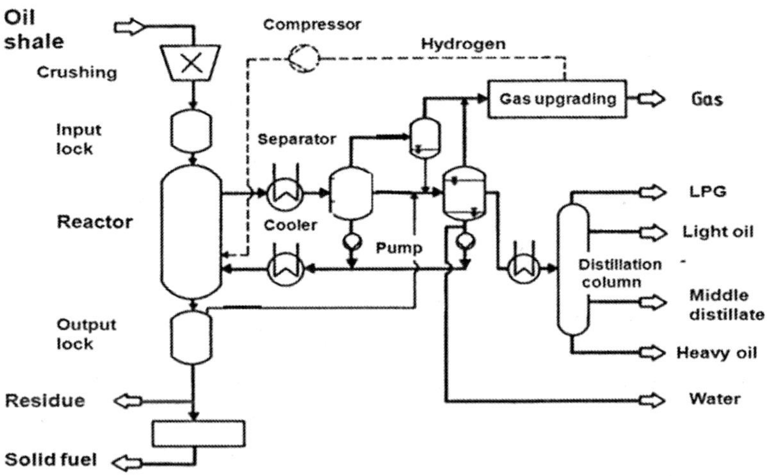

FIGURE 13.21 Flow scheme of kerogen conversion with supercritical water and downstream processing to fuels. *Modified from [47,48].*

downstream processing after treatment with supercritical water. Similar treatment is necessary for all other types of such processes.

13.4.2.6 Synthetic Polymer Processing

Hydrothermal water has a significant capability to hydrolyze polymers, as is discussed in Chapter 9. To illustrate the possibility of using water at high temperatures to recycle polymers, a composite material of thermo-setting polyester resin, inorganic filler, such as calcium carbonate, and glass fiber, so-called fiber-reinforced plastics (FRPs), is considered.

The fundamental experimental results are discussed in Chapter 9 and are shortly repeated: At a temperature of $T = 230\,°C$, thermo-setting polyester resin of FRPs is completely dissolved in subcritical water and an alkali catalyst after a reaction time of $t = 4$ h at saturation pressure ($P = 2.8$ MPa). Inorganic filler and milled glass fiber are separated from the reaction liquid. Solidstyrene-fumaricacid copolymer (SFC), amounting to 75 wt% of the initialresin, glycol and fumaric acid, to 71 and 22 wt%, respectively, are recovered adding up to 96 wt% of the initial resin [45,49].

Based on these results, a recycling process was designed and operated [45,49], see Figure 13.22. FRPs feed material is hydrolyzed. After the reaction, inorganic filler and milled glass fiber are separated with a filter-press from the effluent stream, styrene-fumaric acid copolymer is extracted by adjusting the pH of the reaction liquid, and glycol is recovered by evaporation of water. Recovered glycol is mixed with new glycol and new organicacid and polymerized to unsaturated polyester resin. To adjust the polymer properties to the properties of the FRP feed material, it is necessary to modify the

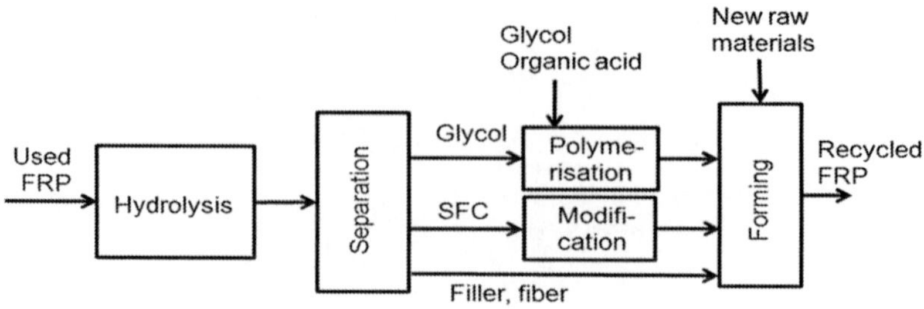

FIGURE 13.22 Recycling of fiber-reinforced resin. SFC, solidstyrene-fumaricacid copolymer. *Modified from [45].*

carboxylic acid group of the styrene-fumaric acid copolymer. Using benzyl chloride for the modification reaction, the necessary shrinkage control effect can be introduced [45,49]. Thus, nearly all the material can be recycled to FRPs.

13.4.2.7 Hydrothermal Processing

Hydrothermal manufacturing of single crystals is carried out on a relatively large scale. Fundamentals are discussed in Chapter 11. A basic flow diagram of hydrothermal processing is shown in Figure 11.3. Information on large scale application for the production of crystals is available from [50], published recently in a second edition, and is not repeated here.

For the future, large scale hydrothermal processing of functional nanoparticles is an interesting topic. Still, there are problems to overcome, mainly due to the recovery of the nanoparticles produced. Particles can be recovered, for example as a slurry solution, then separated with a membrane filter, and washed with pure water [51]. Clearly, particles aggregate and lose some of the inherent properties of nanoparticles. Precipitation of the particles may cause blocking. A design, consisting of a ring-shaped reactor, seems to avoid these problems [36]. These problems will be solved and the technique can be considered as one of the major new production technologies.

REFERENCES

[1] A. Bertucco, G. Vetter, In: High Pressure Process Technology: Fundamentals and Applications, Elsevier, Amsterdam, The Netherlands, 2001, pp. 141–200, Chapter 4.

[2] G. Vetter, G. Luft, S. Maier, Design and construction of high-pressure equipment for research and production, in: A. Bertucco, G. Vetter (Eds.), High Pressure Process Technology: Fundamentals and Applications, Elsevier, Amsterdam, The Netherlands, 2001, pp. 141–200, Chapter 4.

[3] G. Brunner, Gas Extraction. An Introduction to Fundamentals of Supercritical Fluids and the Application to Separation Processes, Steinkopff, Darmstadt, 1994, Chapter 3.

[4] O. Pfohl, Messung und Berechnung von Phasengleichgewichten mit nahe- und überkritischem Kohlendioxid sowie assoziierenden Komponenten im Hochdruckbereich, Hamburg University of Technology, Hamburg, Germany, 1998, Ph. D. Dissertation.

[5] K. Tödheide, E.U. Franck, Das Zweiphasengebiet und die kritische Kurve im System Kohlendioxid - Wasser bis zu Drucken von 3500 bar, Zeitschr. Phys. Chem. Neue Folge 37 (1963) 387–401.

[6] E. Brunner, S. Maier, K. Windhaber, Three high-pressure optical cells for phase equilibrium investigations, J. Phys. E Sci. Instrum. 17 (1984) 44–48.

[7] M. Buback, Absorption spectroscopy in fluid phases, in: E. Kiran, J.M.H. LeveltSengers (Eds.), Supercritical Fluids, Kluwer Academic Publishers, Dordrecht, The Netherlands, 1994, pp. 499–526.

[8] H.H. Buchter, Apparate und Armaturen der Chemischen Hochdrucktechnik, Springer, Berlin, Heidelberg, New York, 1967.

[9] S. Koda, Oxidation reactions of solid carbonaceous and resinous substances insupercritical water, J. Supercrit. Fluids 47 (2009) 400–406.

[10] R.L. Smith Jr., Z. Fang, Techniques, applications and future prospects of diamond anvil cells for studying supercritical water systems, J. Supercrit. Fluids 47 (2009) 431–446.

[11] D. Klingler, H. Vogel, Influence of process parameters on the hydrothermal decomposition and oxidation of glucose in sub- and supercritical water, J. Supercrit. Fluids 55 (2010) 259–270.

[12] Ph.G. Jessop, W. Leitner (Eds.), Chemical Synthesis Using Supercritical Fluids, Wiley-VCH, Weinheim, Germany, 1999.

[13] R. Dohrn, G. Brunner, High-pressure fluid-phase equilibria: experimental methods and systems investigated (1988–1993), Fluid Phase Equilib. 106 (1995) 213–282.

[14] M. Christov, R. Dohrn, High-pressure fluid phase equilibria: experimental methods and systems investigated (1994–1999), Fluid Phase Equilib. 202 (2002) 153–218.

[15] R. Dohrn, S. Peper, J.M.S. Fonseca, High-pressure fluid-phase equilibria: experimental methods and systems investigated (2000–2004), Fluid Phase Equilib. 288 (2010) 1–54.

[16] J.M.S. Fonseca, R. Dohrn, S. Peper, High-pressure fluid-phase equilibria: experimental methods and systems investigated (2005–2008), Fluid Phase Equilib. 300 (2011) 1–69.

[17] E. Liebold, Zur Abscheidung von Kohlenwasserstoffen aus wasserdampfhaltigen Gasgemischen bei höheren Temperaturen und Drücken, Hamburg University of Technology, Hamburg, Germany, 1987, Ph. D. Dissertation.

[18] I.L. Pioro, R.B. Duffey, Heat Transfer and Hydraulic Resistance at Supercritical Pressures in Power-Engineering Applications, ASME Press, New York, NY, 2007, p. 128.

[19] K. Nowak, Reinigung kontaminierter Bodenmaterialien mit überkritischem Wasser, Hamburg University of Technology, Hamburg, Germany, 1995, Ph. D. Dissertation.

[20] G. Brunner, B. Misch, A. Firus, K. Nowak, Cleaning of soil with supercritical water and supercritical carbon dioxide, in: R. Stegmann, G. Brunner, W. Calmano, G. Matz (Eds.), Treatment of Soil—Fundamentals, Analysis, Applications, Springer, Berlin, Germany, 2001, pp. 491–517.

[21] D. Doncheva, G. Brunner, Cleaning of animal-derived bone material for implantation by combined extraction/reaction process of organic matrix with subcritical water and characterisation of hydrolysates, in: Proceedings of the 5th International Symposium on High Pressure Process Technology and Chemical Engineering, Segovia, Spain, June 24–27, 2007.

[22] D.A. Doncheva-Albrecht, Untersuchung der vollständigen Extraktion organischer Stoffe aus Knochenmaterial mit Wasser und überkritischen Fluiden, Hamburg University of Technology, Hamburg, Germany, 2009, Ph. D. Dissertation.

[23] G. Lissens, W. Verstraete, T. Albrecht, G. Brunner, C. Creuly, J. Seon, G. Dussap, C. Lasseur, Advanced anaerobic bioconversion of lignocellulosic waste for bioregenerative life support following thermal water treatment and biodegradation by Fibrobactersuccinogenes, Biodegradation 15 (2004) 173–183.
[24] T. Albrecht, Conversion of Lignocellulosic Materials and Model Compounds in Sub- and Supercritical Water, Hamburg University of Technology, Hamburg, Germany, 2006, Ph. D. Dissertation.
[25] T. Albrecht, G. Brunner, Production of valuable organic compounds by hydrolyzing biopolymers in water at elevated temperatures and pressures, in: G. Brunner, I. Kikic, M. Perrut (Eds.), Proceedings 6th International Symposium on Supercritical Fluids, Versailles, France, vol. II, 2003, pp. 1411–1416, ISBN 2-905-267-37-02.
[26] T. Rogalinski, K. Liu, T. Albrecht, G. Brunner, Hydrolysis kinetics of biopolymers in subcritical water, J. Supercrit. Fluids 46 (2008) 335–341.
[27] M. Daschner de Tercero, I.G. Martínez, M. Herrmann, M. Bruns, Ch. Kübel, S. Jenneweind, U. Fehrenbacher, L. Barner, M. Türk, Synthesis of in situ functionalized iron oxide nanoparticles presenting alkyne groups via a continuous process using near-critical and supercritical water, J. Supercrit. Fluids 82 (2013) 83–95.
[28] S.-A. Hong, S.J. Kim, K.Y. Chung, M.-S. Chun, B.G. Lee, J. Kim, Continuous synthesis of lithium iron phosphate ($LiFePO_4$) nanoparticles in supercritical water: effect of mixing tee, J. Supercrit. Fluids 73 (2013) 70–79.
[29] A.A. Clifford, Separations using Superheated Water, in: M. Goto, T. Adschiri, B. Han, Y.-W. Lee (Eds.), Proceedings 8th International Symposium on Supercritical Fluids, Kyoto, Japan, 2006, Presentation OC 2 10.
[30] D.A. Cantero, M.D. Bermejo, M.J. Cocero, Kinetic analysis of cellulose depolymerization reactions in near critical water, J. Supercrit. Fluids 75 (2013) 48–57.
[31] Ch. Liu, Ch.E. Wyman, The effect of flow rate of compressed hot water on xylan, lignin and total mass removal from corn stover, Ind. Eng. Chem. Res. 42 (2003) 5409–5416.
[32] T. Nakagawa, T. Urabe, T. Miyazaki, M. Hidaka, K. Oka, T. Maekawa, T. Yoshimura, H. Yoshida, FRP recycling technology using sub-critical water hydrolysis, in: M. Goto, T. Adschiri, B. Han, Y.-W. Lee (Eds.), Proceedings 8th International Symposium on Supercritical Fluids, Kyoto, Japan, 2006.
[33] T. Adschiri, K. Arai, Y. Oshima, Oxidation and hydrolysis reactions in supercritical water, in: Y. Arai, T. Sako, Y. Takebayashi (Eds.), Supercritical Fluids, Springer, Berlin, 2002, pp. 364–381, Chapter 6.3.
[34] J.-H. Lee, S.-H. Son, T.T. Viet, Ch.-H. Lee, Decomposition of 2-chlorophenol by supercritical water oxidationwith zirconium corrosion, Korean J. Chem. Eng. 26 (2009) 398–402.
[35] Ch.K. Rofer, G.E. Streit, Oxidation of hydrocarbons and oxygenates in supercritical water-Phase II Final Report, LA-11700-MS, (1989), DOE/HWP-90, UC-721.
[36] E. Lester, P. Blood, J. Denyer, D. Giddings, B. Azzopardi, M. Poliakoff, Reaction engineering: the supercritical water hydrothermal synthesis of nano-particles, J. Supercrit. Fluids 37 (2006) 209–214.
[37] J. Hulliger, R. Gutmann, Exploratory technique in liquid phase epitaxy of potassium tantalateniobate, J. Cryst. Growth 99 (1990) 634–637.
[38] G.H. Chong, R. Yunus, T.S.Y. Choong, N. Abdullah, S.Y. Spotar, Simple guidelines for a self-built laboratory-scale supercritical anti-solvent system, J. Supercrit. Fluids 60 (2011) 69–74.
[39] L. Balling, CO_2-Minderungstechnologien, DPG, Bad Honnef, Germany, 2005.

[40] http://www.energy.siemens.com/hq/en/fossil-power-generation/power-plants/steam-power-plants/. http://www.energy.siemens.com/hq/de/fossile-stromerzeugung/kraftwerke/dampfkraftwerke/benson-dampferzeuger.htm.

[41] H. Schmieder, J. Abeln, N. Boukis, E. Dinjus, A. Kruse, M. Kluth, G. Petrich, E. Sadri, M. Schacht, Hydrothermal gasification of biomass and organic wastes, J. Supercrit. Fluids 17 (2000) 145–153.

[42] N. Boukis, A. Kruse, U. Galla, V. Diem, E. Dinjus, Biomassevergasung in überkritischem Wasser, Nachrichten-Forschungszentrum Karlsruhe 35 (2003) 99–104.

[43] http://www.water.siemens.com/SiteCollectionDocuments/Product_Lines/Zimpro/Brochures/Zimpro-Wet-Air-Oxidation-System-The-Cleanest-Way.pdf.

[44] http://www.scfi.eu/products/.

[45] (a) D.A. Hazlebeck, K.W. Downey, D.D. Jensen, M.H. Spritzer, Supercritical water oxidation of chemical agents, propellants, and other DOD hazardous wastes, in: H.J. White Jr., J.V. Sengers, O.B. Neumann, J.C. Bellows (Eds.), Physical Chemistry of Aqueous Systems: Meeting the Needs of Industry, Proceedings ofthe 12th International Conference onthe Properties of Water and Steam, Begell House, New York, 1994, pp. 632–637. (b) T. Nakagawa, T. Urabe, T. Miyazaki, M. Hidaka, K. Oka, T. Maekawa, T. Yoshimura, H. Yoshida, FRP recycling technology using sub-critical water hydrolysis, in: 10th International Symposium on Supercritical Fluids, Kyoto, Japan, 2006.

[46] J. Masliyah, Z. Zhou, Z. Xu, J. Czarnecki, H. Hamza, Understanding water-based bitumen extraction from Athabasca oil sands, Can. J. Chem. Eng. 82 (2004) 629–654.

[47] P. Missal, Extraktion eines Colorado-Ölschiefers mit Wasser in unter- und überkritischer Phase, VDI-Verlag, Düsseldorf, Germany, 1988, Ph.D. Dissertation, University of Karlsruhe, 1988. Fortschritts-Berichte VDI Series 3, No. 164.

[48] P. Missal, K. Hedden, Extraction of a Colorado oil shale by water in the sub-and supercritical phases, Erdoel & Kohle, Erdgas, Petrochemie 42 (1989) 346–352.

[49] T. Nakagawa, FRP recycling technology using sub-critical water hydrolysis, JEC Compos. Magaz. (March–April) (2008) 40–43.

[50] K. Byrappa, M. Yoshimura, Handbook of Hydrothermal Technology, second ed., William Andrews (Imprint of Elsevier), Oxford, UK, Waltham, MA, USA, 2013.

[51] H. Hayashi, Y. Hakuta, Hydrothermal synthesis of metal oxide nanoparticles in supercritical water, Materials 3 (2010) 3794–3817.

Index

Note: Page numbers followed by "*f*" indicate figures, and "*t*" indicate tables.

C

D

H

N

O

P

R

S

T

U

V

W

X

Y